Biology of Disease

Biology of Disease

Nessar Ahmed
Manchester Metropolitan University, UK

Maureen Dawson
Manchester Metropolitan University, UK

Chris Smith
Manchester Metropolitan University, UK

Ed Wood
University of Leeds, UK

Published by:

Taylor & Francis Group

In US: 270 Madison Avenue
New York, NY 10016
In UK: 2 Park Square, Milton Park
Abingdon, OX14 4RN

ISBN: 0-7487-7210-3

A catalog record for this book is available from the British Library.

Library of Congress Cataloging-in-Publication Data

Biology of disease / Nessar Ahmed ... [et al.].
p. ; cm.
Includes bibliographical references and index.
ISBN 0-7487-7210-3 (alk. paper)
1. Pathology--Textbooks. 2. Diseases--Textbooks. 3. Medical sciences--Textbooks.
I. Ahmed, Nessar.
[DNLM: 1. Pathology--methods. 2. Disease. QZ 4 B615 2007]

RB111.B482 2007
616--dc22

2006026001

Senior Publisher: Jackie Harbor
Editor: Elizabeth Owen
Editorial Assistant: Kirsty Lyons
Production Editor: Karin Henderson
Illustrated by: Mick Hoult
Typeset by: Phoenix Photosetting, Chatham, Kent, UK
Printed by: Cromwell Press, Trowbridge, Wilts, UK

Printed on acid-free paper

10 9 8 7 6 5 4 3 2 1

Taylor & Francis Group, an informa business

Visit our web site at http://www.garlandscience.com

To our families

CONTENTS

PREFACE

Many students decide to follow a career path that is related to medicine, for example Biomedical Sciences (Medical Technology in the USA), Clinical Sciences, Clinical Physiology and Nursing. While there are a number of textbooks for medical students and nurses as well as a number covering the clinical biochemistry area, there did not seem to us to be a book that dealt with disease from a more scientific standpoint. Such a book would cover a range of disease conditions, their causes and diagnoses, and outline treatment but set at an appropriate level. The idea of writing *Biology of Disease* arose from discussions between ourselves and various colleagues and students over a number of years regarding the absence of a single book, which summarized most of the clinical material studied by first- and second-year Biomedical Sciences and related-degree students. It was felt that such a textbook would assist students during the first and second years of their degree programmes, particularly if it covered aspects of the Biomedical Sciences appropriate to courses accredited by the Institute of Biomedical Science of the UK and similar courses elsewhere.

When writing *Biology of Disease* it was decided to include appropriate background material, describing it in relatively simple terms. However, in practice it was not possible to cover all these aspects; hence some preknowledge of biochemistry and cell and molecular biology has been assumed. Although *Biology of Disease* is not primarily a *medical* textbook, we do think it could also be of value to students participating in foundation courses for medical schools. The areas and topics covered by the term *biomedical sciences* are vast and we have had to curtail the number of topics included – modern medicine and its ancillary subjects represent a huge body of knowledge.

The structure of the book aims to help students plan their learning and navigate their way through complex topics. For example, all chapters begin with clearly stated *Objectives*, followed by a short *Introduction* to set the scene for the ensuing contents. The main body of text of all the chapters includes '*Boxes*' of one or two pages length to highlight a medical or scientific aspect we felt to be of special interest. *Margin Notes* emphasize material worth highlighting and provide a little extra explanation to the text. Each chapter has a concluding short *Summary*. Students (and staff) can test themselves against the chapter material using the simple *Case Studies* and relevant *end-of-chapter questions* provided. Each chapter is illustrated by simple line diagrams, light and electron micrographs and tables, as well as molecular models of compounds of interest.

Units and drug nomenclature proved somewhat troublesome! For the sake of consistency, we have generally tried to apply SI units whenever possible. Within the UK, differing hospitals seem to apply units on a rather arbitrary basis and, indeed, other countries, such as the USA, use differing units and systems of nomenclature. Thus, for example, in the UK a serum cholesterol concentration greater than 5 mmol dm^{-3} is regarded as clinically dangerous whereas the corresponding unit used in the US is 190 mg dL^{-1}. In such cases, we have generally made somewhat arbitrary decisions and used whatever units seemed best suited to the purpose. We have applied a similar *ad hoc* system for naming some biological materials and drugs. Thus, for example in the USA adrenaline and paracetamol are called epinephrine and acetaminophen respectively. We have used UK nomenclature.

We are grateful to the many people who helped in producing *Biology of Disease*, especially to our graphic artist, Mick Hoult of Manchester Metropolitan University, who not only drew and re-drew all of the many figures in the text and frequently improved upon our initial hazy notions and crude sketches. We would like to thank the numerous colleagues who offered helpful advice and suggestions, and to the many individuals who supplied illustrations. We are also grateful to our reviewers; in particular, Dr David Holmes of the University of Northumbria at Newcastle-upon-Tyne who commented on nearly every chapter but also to the individual reviewers of each respective chapter. Thanks must also go to the team at Taylor & Francis, especially editorial assistant Chris Dixon and editor Liz Owen, for their constant encouragement, and to our production editor Karin Henderson. However, the authors themselves take responsibility for any errors and omissions as well as for those parts of the text that are still not as clear or comprehensive as they might be. We hope that you will write and tell us how we can still further improve the book for the benefit of future editions.

Nessar Ahmed
Maureen Dawson
Chris Smith
School of Biology, Chemistry and Health Science,
Manchester Metropolitan University, UK
and
Ed Wood
School of Biochemistry and Microbiology,
University of Leeds, UK

List of reviewers

David Holmes, University of Northumbria
Joyce Overfield, Manchester Metropolitan University
Richard Luxton, University of the West of England
Len Seal, Manchester Metropolitan University
Eileen Ingham, University of Leeds
David Bender, University College London
Khalid Rahman, Liverpool John Moores University
Howard Foster, University of Salford
John Gaffney, Manchester Metropolitan University
Brenda Price, Macclesfield District General Hospital

GLOSSARY

Accuracy The ability of a test for an **analyte** to give results comparable to the true value. Compare with **precision**.

Achlorhydria Low or absent secretion of acid in the stomach.

Acquired immunity The development of true immunity arising from a **specific immune response** to an **immunogen**.

Actinomycosis A fungal infection that may occur following injuries to oral tissues and causes inflammation.

Acute Diseases or clinical events that have a rapid, often severe onset, but of short duration (compare with **chronic**).

Acute intravascular hemolysis **Hemolysis** that results from an incompatible blood transfusion, particularly involving the ABO blood group system.

Acute phase proteins A set of plasma proteins associated with an **acute phase response**, for example C-reactive protein and fibrinogen.

Acute phase response A systemic response involving changes in the composition of the blood, including an increased neutrophil count and increase in amount of the **acute phase proteins**, which occur in certain clinical conditions such as trauma, infections, necrosis, tumors and other inflammatory events.

Acute rejection of an allograft is one that occurs usually within a few weeks following the transplant.

Agammaglobulinemia (Or hypogammaglobulinemia) clinical conditions in which the γ-class of **immunoglobulins** is deficient.

Agglutination The clumping together of blood cells.

Aging Most widely understood to be the decrease in the ability to survive on growing older.

Alleles Different forms of the same gene. Any one individual will normally have two alleles for each gene, one from each parent, since, in general, chromosomes occur in matching or homologous pairs where each member of a pair contains an allele.

Allergens These are immunogens that cause an allergy.

Allergy A term often used to describe a Type I hypersensitivity, which results in inflammation on exposure to an allergen.

Allogeneic Transplants that involve donors and recipients who are genetically nonidentical (*see also* **autologous** and **syngeneic**).

Allograft A transplant between two genetically different people (*see also* **isograft** and **xenogeneic transplants**).

Allotopic expression A technique whereby a mitochondrial gene is inserted into the nucleus where it is transcribed to produce a product that is imported into the mitochondria where it alleviates the effects of a mtDNA mutation.

Alternative pathway Activation of complement proteins in the absence of antibody by cell wall components of bacteria and yeasts.

Amenorrhea An absence or stopping of the menstrual cycle. Compare with **galactorrhea** and **oligomenorrhea**.

Analytes Substances that are measured to assist diagnosis and monitor the treatment of disease.

Analytical sensitivity The smallest quantity or concentration of an **analyte** that can be detected by an analytical method. Compare with **analytical specificity**.

Analytical specificity The ability of an analytical method to detect only the **analyte** in question. Compare with **analytical sensitivity**.

Anemia A reduced amount of hemoglobin in the blood because of insufficient red cells or insufficient hemoglobin in the cells.

Aneuploidy A situation in which the number of chromosomes is fewer or greater than an exact multiple of the **haploid** number (*see* **euploidy** and **polyploidy**).

Aneurysm A bulge that usually develops in weak areas of an arterial wall.

Anosmia Loss of the sense of smell.

Anoxia The absence of oxygen.

Antibiotics Drugs used to treat bacterial and other infectious diseases, but which do not affect viruses. Most antibiotics are derived from microorganisms.

Antibodies Glycoproteins produced by the immune system that bind to **antigens** (*see* **immunoglobulins**).

Antigen A macromolecule (protein, lipopolysaccharide) that provokes an immune response and binds specifically to an **antibody**.

Antigen presenting cells (APC) White blood cells that take up an **immunogen** and process it to form a peptide that can be recognized by T_H cells resulting in its activation.

Antinuclear factors **Antibodies** against nuclear components that occur in some **autoimmune** conditions such as rheumatoid arthritis and systemic lupus erythematosus (*see* **extractable nuclear antigens**).

Anuria Failure to pass urine (compare with **oliguria**).

Apheresis Isolation and concentration of components of the blood for clinical use (*see* **plasmapheresis**).

Apoptosis (programmed cell death) This is also called cell suicide. A genetically programmed release of enzymes and other proteins that eventually bring about the death of a cell.

Arrhythmias Disturbances to the heart's rhythmic contractions.

Arteriosclerosis *See* **atherosclerosis**.

Ascites An abnormal accumulation of fluid ("ascites fluid") in the abdomen.

Aspergilloma An infective mass of the fungus *Aspergillus fumigatus* growing within an old wound within the lungs.

Ataxia Inability to control muscular movements.

Atheroma A hard yellow plaque consisting of a necrotic (dead) core, rich in cholesterol and surrounded by fibrous tissue that gradually builds up on the inside of medium-sized arteries (*see* **atherosclerosis**).

Atherosclerosis or **arteriosclerosis** The simultaneous development of an **atheroma** in an artery wall and its sclerosis (abnormal hardening or fibrosis).

Atrioventricular or **AV node** The node situated between the **atria** and the **ventricles** of the heart which delays the transmission of impulses from the **sinoatrial node** to allow the atria to contract completely and the ventricles to fill with as much blood as possible before **systole** occurs.

Atrium (plural atria) The upper chamber(s) of the heart (*see* **ventricle**).

Autoimmune disease A situation in which the immune system mounts an immune response against **"self"** tissues.

Autologous stem cell transplants Use of the patient's own stem cells, which were harvested and stored prior to a treatment, for example by radiation or high-dose chemotherapy that destroys the patient's own stem cells (*see also* **allogeneic** and **syngeneic**).

Autosomes The non-sex chromosomes that comprise 22 homologous pairs of the 46 chromosomes found in human **diploid** cells.

B lymphocytes **Small lymphocytes** that develop fully in the bone marrow. On stimulation by an immunogen they develop into antibody producing cells.

Bactericidal An antibiotic or other drug that kills its target bacteria.

Bacteriostatic An antibiotic or other drug that prevents its target bacteria replicating.

Balanced diet A diet that supplies adequate energy distributed appropriately between carbohydrates, lipids, and proteins along with the necessary amounts of vitamins, minerals, trace elements, water and non-digestable fiber.

Basophil One of the three types of **polymorphonuclear leukocyte** (*see* **eosinophil** and **neutrophil**).

Biopsy The removal of a piece of tissue from a patient for clinical analysis.

Black Death The **pandemic** of bubonic plague that spread across Asia and Europe in medieval times. So called because of the black spots (buboes) that occurred on the skin of patients.

Bleeding time The period of time following wounding before the flow of blood stops (compare with **clotting time**).

Body mass index (BMI) The generally accepted way to assess the weight of a patient, where BMI = weight (kg)/height (m)^2 (*see also* **ideal body weight**).

Bone resorption The release of Ca^{2+} from bone.

Bradycardia An abnormally low heart rate.

Broad-spectrum antibiotics Antibiotics that are effective against Gram-positive and Gram-negative bacteria.

Bronchiectasis A condition in which the bronchi and bronchioles are abnormally dilated.

Brush border The microscopic extensions called microvilli on the apical surfaces (i.e. luminal sides) of the plasma membranes of enterocytes lining the intestine.

Bundle of His A group of modified cardiac muscle fibers, called **Purkinje fibers** that carry electrical impulses in the heart.

C3 convertase A proteolytic enzyme consisting of activated complement proteins C4b2a that cleaves complement protein C3 into two fragments: C3 and the larger C3b molecule that is an opsonin. *See* **opsonization**.

C5 convertase A proteolytic enzyme consisting of complement proteins C3b and C3 convertase involved in the formation of the **Membrane Attack Complex**.

Cachexia Severe and prolonged muscle wastage (weight loss) associated with a number of clinical conditions including cancer.

Cancer A general term covering a number of diseases in which the growth of certain body cells becomes uncontrolled forming a tumor, which may be benign or **malignant**.

Capsid The complex of proteins that forms a shell enclosing the nucleic acid of a virus.

Carbuncle An amalgam of several abcesses or boils.

Carcinogens Environmental agents (compounds, radiation) that cause cancers.

Carditis General inflammation of the heart.

Carriers (asymptomatic) Individuals who harbor a pathogen but are symptom free.

Cell cycle This is an orderly sequence of biochemical and morphological events that constitute the growth phases of a reproducing cell. *See* **interphase**, **mitosis** and **meiosis**.

Cell-mediated immunity (CMI) Production of cells that kill or recruit other cells to kill host cells that are infected by viruses or other intracellular parasitic microorganisms (compare with **humoral immunity**).

Cerebrovascular accident *See* **stroke**.

Chemotactic The ability of substances secreted by cells to attract other cells to the area.

Chloride shift The exchange of HCO_3^- for Cl^- across the membrane of the erythrocyte.

Cholelithiasis The formation of gallstones.

Cholestasis The failure of bile to reach the small intestine.

Chromatid One of the pair constituting a condensed replicated chromosome present at **mitosis** or **meiosis**.

Chronic Diseases or clinical events that develop gradually over a relatively long time and persist (compare with **acute**).

Chronic rejection of a transplantation Immunological rejection that occurs over months or years by a combination of **cell-mediated** and **humoral immunity**.

Chylomicron Large lipoprotein complexes that transport triacylglycerols, phospholipids and lipid-soluble vitamins from the gastrointestinal tract to the lacteals of the lymphatic system.

Chyme The watery mixture of gastric juice and partially digested food released from the stomach into the small intestine.

Classical pathway of complement activation The pathway initiated when certain classes of **antibody** binds to an **immunogen** (compare with **alternative pathway**).

Clinical audit A process whereby practices and procedures involved in patient care are monitored and, if necessary, revised to provide a more effective, efficient and cost-effective service.

Clinical biochemistry The science concerned with investigating the biochemical changes associated with diseases.

Clinical genetics The science concerned with the identification of genetic abnormalities.

Clinical manifestations of disease These include the **signs** and **symptoms** associated with the disease.

Clotting time The period required to seal a wound with a clot that prevents further blood loss (usually 6 to 12 minutes). Compare with **bleeding time**.

Codons These are the three bases in a nucleic acid each of which code for a single amino acid residue in a polypeptide.

Cofactor proteins Those that bind to **zymogens** and their protease products and increase the specificity and speed of the activation of the zymogen.

Colitis An **inflammation** of the colon.

Compensation In acid–base disorders, the physiological mechanisms that attempt to return the pH of the blood to values within the reference range.

Conjugate vaccines These are those formed when a relatively poorly immunogenic material is attached to a protein that is much more immunogenic.

Contact allergies Reactions produced by a number of chemicals that directly affect the skin leading to a delayed type **hypersensitivity**.

Control samples Samples that are identical in composition to test samples except that they contain known concentrations of the test **analyte**.

Corneal arcus Opaque fatty deposits around the periphery of the cornea that occur in **familial hypercholesterolemia**.

Cretinism A congenital condition in which a child suffers from mental retardation, muscle weakness, short stature, neurological signs and is often dumb and mute because of the failure of the thyroid gland to develop normally during embryonic growth.

Cytogenetics The microscopic study of chromosomes.

Cytokines A general name given to a large family of proteins which resemble hormones in being secreted by cells and stimulating activities in other cells after binding to receptors on their surfaces.

Cytotoxic cells White blood cells that directly destroy infected cells in **cell-mediated immunity**.

Cytotoxic T lymphocytes They develop from T_C cells, which have been stimulated by an **immunogen**.

Debridement The surgical removal of infected or necrotic tissue.

Degenerative diseases Diseases involving the progressive loss of body tissues and impairment of their functions: usually associated with aging.

Dehydration This occurs when water loss from the body exceeds intake (compare with **overhydrated**).

Diarrhea The frequent passage of feces that are larger in volume and more fluid than normal. It is not a disease but a symptom of an underlying condition that increases intestinal movements, accelerating the passage of the contents through the gastrointestinal tract leaving insufficient time for fluid reabsorption.

Diastole Relaxation of the heart's ventricles.

Dietary reference values Guidance values as to the amounts of nutrients that should be ingested, rather than exact recommendations.

Diploid The presence of pairs of chromosomes in a cell. Normal human somatic (body) cells contain 23 pairs of chromosomes (compare with **haploid**).

Dissection (with respect to arteries) The separation of the layers of the wall of an artery.

Dominant Genetically controlled factors that are expressed phenotypically in both the **homozygous** and **heterozygous** conditions (*see* **recessive**).

Dysmorphic Features that differ from the ones generally accepted in appearance.

Dystonia A neurological disorder associated with muscular spasms.

Edema An excessive accumulation of water in the intercellular spaces.

Electrocardiogram (ECG) The record of the electrical potentials associated with the spread of depolarization and repolarization through the muscle mass of the heart during its beat cycle.

Emesis *See* **vomiting**.

Endocardium The layer of smooth tissue lining the inside of the chambers of the heart.

Endocrine diseases Diseases that arise from the over- or under-production of hormones, or from resistance of a target tissue to a particular hormone.

Endogenous A general term meaning originating from within the body (compare with **exogenous**).

Envelope The lipid membrane that some viral particles acquire as the viral particle leaves its host cell.

Enzyme immunohistochemistry The use of antibodies labeled with an enzyme, such as horseradish peroxidase or alkaline phosphatase. The binding of the **antibody** to its **antigen** can be localized in cells as the enzyme converts a colorless substrate into an insoluble, colored compound that can be seen under the microscope.

Eosinophil One of the three types of **polymorphonuclear leukocyte** (*see* **basophil** and **neutrophil**).

Epitopes Small regions of immunogenic molecules that are specifically recognized by components of the immune system.

Erythrocytes Red blood cells; anucleate cells rich in hemoglobin that transport oxygen around the body.

Essential Nutrients that the body cannot make from other compounds and must therefore be supplied in in the diet (*see* **essential fatty acids**, compare with **non-essential**).

Essential fatty acids Fatty acids that must be supplied in the diet since humans lack the enzymes necessary to produce them.

Estimated average requirement (EAR) The mean daily amount of nutrient (or energy) thought to be needed by a population (compare with **lower reference nutrient intake**, **reference nutrient intake** and **safe intake**).

Euploidy The situation in which the number of chromosomes is an exact multiple of the **haploid** number (*see* **aneuploidy**, **polyploidy**).

Exogenous A general term meaning originating outside the body (compare with **endogenous**).

Exons Coding sequences of bases found within a gene that are separated by intervening noncoding sequences called introns.

Exophthalmos Protruding eyeballs.

Extended spectrum β-lactamases (ESBLs) Enzymes produced by certain bacteria, for example *Klebsiella* spp, *E. coli*, that can hydrolyze extended spectrum cephalosporins (antibiotics).

Extractable nuclear antigens (ENA) Antigens that lead to the formation of autoantibodies in certain diseases such as rheumatoid arthritis and systemic lupus erythematosus (*see* **antinuclear factors**).

Extravascular hemolysis The uptake and subsequent destruction of antibody-coated erythrocytes in the spleen and liver (*see also* **immune adherence**).

Fainting Common term for **syncope**.

Familial hypercholesterolemia An autosomal **dominant** condition with a frequency of 1 in 500 that affects heterozygotes; associated with defective receptors for LDL on the liver cells.

Favism The hemolytic crisis brought on by the consumption of fava or broad beans (*Vicia fava*) in persons with G6PDH deficiency.

Fistula An abnormal passage from a cavity or tube to another cavity or free surface.

Fluorosis The mottling of the teeth as they form in the jaws caused by fluoride in drinking water at concentrations greater than 12 ppm.

Fragile sites Particular regions of chromosomes that are susceptible to breakage when the cells are cultured in the absence of certain chemicals such as folic acid.

Fulminant cholera The massive leakage of intracellular water into the gastrointestinal tract and the subsequent **diarrhea** following infection with *Vibrio cholerae*.

Galactorrhea An inappropriate secretion of milk. *See also* **amenorrhea** and **oligomenorrhea**.

Gene mutation theory of aging This suggests that an accumulation of mutations during the course of life leads ultimately to tissue and organ malfunctions and eventually death.

Genetic diseases Diseases that arise due to defects in the genes or chromosomes.

Genetic imprinting A condition that depends upon whether paternal or maternal inheritance of a trait has occurred.

Genotype The genetic or hereditary constitution of an individual. The term can also be applied to any particular pair of alleles that an individual possesses at a specific locus on a chromosome (contrast with **phenotype**).

Glossitis **Inflammation** of the tongue.

Graft versus host disease (GVHD) The condition associated with the transfusion of whole blood, blood products or bone marrow particularly to an immunodeficient patient, which contain residual lymphocytes. The **small lymphocytes** present in the donated blood recognize the **immunogens** of the recipient as foreign and mount an immune response. The donor lymphocytes proliferate in the patient and attack tissues.

Gummas Tumor-like growths of a rubbery consistency that some patients develop in benign late syphilis.

Gynecomastia The development of enlarged breasts in males.

Haploid (N) The presence of half the **diploid** number of chromosome (compare with **diploid**).

Health The state of physical, mental and social well-being; not merely the absence of disease.

Heart attack *See* **myocardial infarction**.

Helminths Parasitic worms such as nematodes, cestodes and trematodes.

Helper T lymphocytes *See* **T_H cells**.

Hematemesis The presence of blood in vomit.

Hematology The study of blood and its disorders.

Hemoglobinopathies Clinical conditions that result from mutations in the genes for globin molecules.

Hemolysis The rupture or lysis of **erythrocytes** and release of their hemoglobin.

Hemoptysis Coughing up of blood-containing fluid from the lungs.

Hernia The protrusion of an organ or tissue out of the body cavity in which it is normally found.

Heteroplasmy The condition in which a cell has a mixed population of mitochondria containing normal and mutant mitochondrial DNA (compare with **homoplasmy**).

Heterozygous The condition in which the two alleles of a gene are different (compare with **homozygous**).

Hirsutism The development of inappropriate increased body hair on the face, chest, upper back and abdomen in females.

Histopathology laboratories Laboratories concerned with the investigation of disease by examining specimens of cells and tissues.

Homeostatic diseases Diseases that arise when mechanisms for controlling homeostasis are disrupted or defective.

Homoplasmy The condition in which every mitochondrial DNA molecule in a cell has the same mutation (compare with **heteroplasmy**).

Homozygous The condition in which the two alleles are identical (compare with **heterozygous**).

Hordeolums "Styes" that are caused by infections of the eyelid generally involving the lid margins, eyelid glands or follicles.

Human Herpes Viruses (HHV1-8) There are at least eight herpes viruses that can infect humans and cause clinical disease.

Humoral immunity The production of antibodies to protect against pathogens and parasites (compare with **cell-mediated immunity**).

Huntingtin A protein from a single mutated gene that accumulates in the nucleus rather than in the cytoplasm of cells in patients with Huntington's disease.

Hybridomas or **hybrid myelomas** Antibody-producing cells made by fusing cultured **plasma cells** with cells derived from a myeloma (a **plasma cell** tumor) *in vivo*. The resulting cell is immortal and produces a **monoclonal antibody**.

Hyperacute rejection Failure of a graft within hours or minutes of transplantation once the tissue has become revascularized due to the presence of preformed antibodies against graft antigens present in the plasma of the recipient.

Hypercapnia A high partial pressure of CO_2 (PCO_2) (compare with **hypocapnia**).

Hypersensitivity *See* **immunological hypersensitivity**.

Hypertension Blood pressure higher than that regarded as normal.

Hypervitaminoses Clinical conditions caused by a toxic excess of a given vitamin (compare with **hypovitaminoses**).

Hypocapnia A low partial pressure of CO_2 (PCO_2) (compare with **hypercapnia**).

Hypogammaglobulinemia *See* **agammaglobulinemia**.

Hypogonadism (In males) testes small and soft, deficient production of sperm and decreased testosterone; (in females) impaired ovarian function; estrogen deficiency and abnormalities of the menstrual cycle.

Hypovitaminoses Clinical conditions caused by an inadequate dietary intake of a vitamin, its impaired absorption, or insufficient utilization of an adequate intake, increased dietary requirements, for example in pregnancy, without a corresponding increased intake or increased excretion of a vitamin (compare with **hypervitaminoses**).

Hypovolemia A reduction in the volume of blood in the body.

Hypoxemia A low partial pressure of O_2 (PO_2).

Iatrogenic diseases Clinical conditions that arise as a consequence of treatment.

Ideal body weight (IBW) A way to assess the weight of patients. For males of height 5 feet it is 106 pounds and this increases by an additional 6 pounds for each inch over the height of 5 feet. For females, the IBW at 5 feet tall is 100 pounds and this increases by 5 extra pounds for each additional inch. *See* **body mass index**.

Idiogram A diagrammatic representation or interpretive drawing of the chromosomes based on the physical features as seen in a **karyogram** (*see also* **karyotype**).

Idiopathic diseases Diseases of unknown cause.

Immune adherence The binding of antibody-coated erythrocytes to receptors on **monocytes** and macrophages as the blood passes through the spleen and liver (*see also* **extravascular hemolysis**).

Immune system The set of organs, tissues, cells and molecules that protect the body from diseases caused by **pathogens** and **parasites**.

Immunoassay The use of antibodies to quantify the amount of an **analyte** in a clinical sample.

Immunodeficiency diseases Conditions that occur due to an inadequate **immune system**.

Immunofluorescence A type of **immunohistochemistry** in which the antibody is labeled with a fluorescent dye.

Immunogen Any molecule or organism that stimulates a specific immune response.

Immunoglobulins (Igs) A collective name for the five classes of **antibodies**.

Immunohistochemical techniques This refers to the use of labeled **antibodies** to detect **antigens** on or in cells.

Immunological diseases Clinical conditions that occur when the immune system causes damage to the body's own tissues or when it is deficient and unable to respond appropriately to **pathogens** and **parasites**.

Immunological hypersensitivity A state in which an immune response, often to a seemingly innocuous immunogen, results in tissue damage.

Immunological memory A molecular memory that allows a rapid protective response on second or subsequent contact with an immunogen.

Immunology The study of the immune system and how it works.

Immunology laboratories Those involved in the study of the **immune system** and its responses in both healthy and diseased states.

Immunotherapy Treatments that manipulate the immune system in order to improve its response against for example a **cancer**.

Incidence rate The number of new cases of a disease in a population occurring within a specified period of time.

Incubation period The time that must elapse before a disease becomes apparent following exposure to the etiological agent.

Infarct A segment of tissue damaged by a disruption of its blood supply (*see* **infarction**).

Infarction The death (necrosis) of a section of tissue following a disruption to its blood supply (*see also* **infarct**).

Infection The successful persistence and/or multiplication of a pathogen on or within the host.

Infectious diseases Diseases caused by **pathogens**, such as viruses and microorganisms (bacteria, fungi, protozoa and helminths).

Infective dose A measure of virulence, being the minimum number of organisms or virus particles that can cause an **infectious disease**.

Inflammation The array of responses to infection and tissue damage, such as localized pain, redness, swelling and heat.

Interphase The period of the cell cycle between **mitosis** or **meiosis**.

Intervention studies Studies aimed at changing diet and lifestyle in order to attempt to reduce the **incidence** of a disease associated with current activities.

Intrinsic factor A glycoprotein secreted by the gastric parietal cells that binds vitamin B_{12} in the stomach and carries it to specific receptors on the mucosal surface of the ileum.

Intron *See* **exon**.

Isograft A transplant of a piece of tissue from one site to another on the same patient such as skin grafting in burns (*see also* **allograft** and **xenogeneic transplants**).

Isohemagglutinins Antibodies against blood group antigens that cause the **agglutination** of **erythrocytes**.

Karyogram The photographic representation of an individual's chromosome complement stained and arranged in order (*see also* **idiogram** and **karyotype**).

Karyotype The characteristic number, size and shape of the chromosome complement of an individual or species (*see also* **idiogram** and **karyogram**).

Kernicterus Brain damage caused by the accumulation of bilirubin in the brain to concentrations of 200 μmol dm^{-3} and above.

Ketone bodies Acetoacetate, acetone, and β-hydroxybutyrate.

Ketonemia Accumulation of ketone bodies (e.g. acetoacetate, β-hydroxybutyrate) in the blood.

Ketonuria Excretion of ketone bodies in the urine.

Korotkoff sounds The clear characteristic sounds that can be heard with a stethoscope due to the turbulence generated as the blood flows through a partially occluded artery, for example when measuring blood pressure.

Kussmaul respiration Deep sighing breathing associated with the hyperventilation of metabolic acidosis.

Kwashiorkor A protein-energy deficient nutritional condition that develops in children typically 3–5 years old who have been abruptly weaned when a new sibling is born. Derived from one the Kwa languages of Ghana meaning *the one who is displaced. See also* **marasmus**.

Large granular lymphocytes **Mononuclear leukocytes** involved in nonspecific immunity.

Latent period The time when an illness is developing but overt signs and symptoms are not apparent.

Lectins Glycoproteins that bind to carbohydrate residues on macromolecules or cell membranes.

Lesions Structural or functional abnormalities.

Leukocyte One of several types of white blood cell each with its own function.

Leukocytosis **Leukocyte** count in excess of 20×10^9 dm^{-3} compared with reference values of $4–11 \times 10^9$ dm^{-3}.

Leukodepletion Removal of **leukocytes** from blood by filtering it through leukocyte-specific filters which trap the leukocytes but not the smaller **erythrocytes** or platelets, and reduces the leukocyte count to less than 5×10^6 dm^{-3}.

Leukopheresis The removal of leukocytes from blood with the resulting plasma and erythrocytes being returned to the blood donor (*see* **apheresis**, **plasmapheresis** and **plateletpheresis**).

Life expectancy The average length of life of individuals in a population.

Lipofuscins Yellow-brown pigments produced by degeneration of cell membranes and organelles, probably by the free radical peroxidation of membrane lipids.

Lipoprotein particles Complexes of cholesterol, triacylglycerols and proteins involved in transport of these lipids in the blood. They occur as high density lipoprotein (HDL), low density lipoprotein (LDL) and very low density lipoprotein particles (VLDL). Chylomicrons formed from fatty materials absorbed by the gastrointestinal tract are also lipoprotein particles.

Lower reference nutrient intake (LRNI) The amount of a nutrient sufficient for the requirements of 2.5% of a population (compare with **estimated average requirement**, **reference nutrient intake** and **safe intake**).

Lymphocytes **Mononuclear leukocytes** that occur in the blood as large granular lymphocytes and small lymphocytes.

Lymphoid stem cells Hemopoietic stem cells found in bone marrow which can divide to produce the precursors of small lymphocytes.

Lymphoma A tumor that arises from lymphoid tissue, such as a lymph node.

Lymphopenia A reduced **lymphocyte** count.

Macrophages Immune cells that differentiate in solid tissues from immature **monocytes**.

Major Histocompatibility Complex (MHC) A region of the chromosome that encodes membrane proteins which present antigenic peptides derived from proteins on the surfaces of antigen-presenting cells to T lymphocytes.

Malignant tumor *See* **cancer**.

Mammography An X-ray examination of the breast.

Marasmus A chronic disorder that develops over a period of months to years caused by an inadequate protein-energy intake (*see* **kwashiorkor**).

Mast cell A type of granule-containing white blood cell that resembles a **basophil** but which is found in solid tissues.

Medical microbiology The science concerned with the detection and identification of pathogenic microorganisms.

Megakaryocytes Large precursor cells from the bone marrow that eventually fragment to form **platelets** in the blood.

Meiosis Reduction division. The type of cell division in which the daughter cells receive only one of each type of chromosome and which results in the formation of **haploid** gametes (*see* **mitosis**).

Membrane Attack Complex (MAC) This is a large, cylindrical, hydrophobic structure constructed from single complement protein molecules (C5b, C6, C7, C8) along with several molecules of C9. It can insert itself into a cell membrane leading to cell lysis.

Mesentery The double-layered, fan-shaped portion of **peritoneum** that extends from the posterior abdominal wall and wraps around the small intestine and returns to its origin and contains the blood and lymphatic vessels and nerves that supply the small intestine (*see also* **mesocolon**).

Mesocolon The structure with similar functions to the **mesentery** but which holds the large intestine.

Metabolic acid–base disorder This is the occurrence of an abnormal blood pH because of a metabolic or renal dysfunction (*see* **respiratory acid–base disorder**).

Metastasis The detachment of cells from a **primary tumor** allowing it to spread and form **secondary tumors**.

Methemoglobin Oxidation of ferrous (Fe(II)) iron in the heme group of hemoglobin to Fe(III) iron produces methemoglobin which is incapable of carrying oxygen.

Mitogens Substances that stimulate cells to divide. *See* **mitosis**.

Mitosis The type of cell division in which each of the daughter cells receives a full complement of chromosomes (*see* **meiosis**).

Mixed acid–base disorder The situation in which a patient presents with more than one acid–base disorder.

Monoclonal antibody An antibody that recognizes only a single specific **epitope** on an antigen and which is produced by a clone of identical plasma cells. Compare with **polyclonal antibody**.

Monocyte A type of **mononuclear leukocyte** that circulates in the blood for about 72 hours before entering solid tissues and developing into a **macrophage**.

Mononuclear leukocytes (MNs) One of the two main groups of **leukocytes**, which can be distinguished from the other **polymorphonuclear leukocytes** in having a more rounded nucleus. They are subdivided into **monocytes**, **large granular lymphocytes** and **small lymphocytes**.

Morbidity The effects of a disease on the patient.

Mortality This describes the possibility of a disease causing death: usually expressed as a percentage.

Murmurs The characteristic sounds formed when abnormalities of the valves and heart structures cause a turbulent blood flow.

Mutation A change in the **genotype**, which occurs as a result of incorrect replication of DNA or from a random change to it caused by physical or chemical agents.

Myocardial infarction ("heart attack") A medical emergency that occurs when some or all of the blood supply to the heart muscle through the coronary arteries is cut off.

Myocardium The cardiac muscle-rich wall of the heart.

Natural killer (NK) cells A type of **large granular lymphocyte** that kills virus-infected cells and some tumor cells nonspecifically.

Neoplasm or **neoplastic diseases** Often used synonymously with **cancer**, means, literally, "new tumor" or "new mass".

Nephrons The functional units of the kidney: each is composed of a tuft of capillaries called a glomerulus, and a tubule.

Neutrophil One of the three types of **polymorphonuclear leukocyte** (*see* **basophil** and **eosinophil**).

Nocosomial infections Those acquired in hospital.

Nonessential A nutrient that the body can make from other nutrients and which does not need to be supplied in the diet (compare with **essential**).

Nonhemolytic febrile transfusion reaction Transfusion of leukocytes may lead to adverse reactions, such as **nonhemolytic febrile transfusion reactions**. Such patients exhibit flushing, fever, rigors and hypotension. These may be caused by the reaction between antibodies and leukocyte antigens in the recipient, resulting in the lysis of donated leukocytes and release of cytokines from them.

Nonself Used in **immunology** to signify macromolecules or cells that are foreign to the body (compare with **self**).

Nonspecific defences The first line of immunological defence including responses such as **inflammation** and the **acute phase response**.

Normal range *See* **reference ranges**.

Notifiable diseases Diseases for which clinicians are legally required to supply details of all new cases to a central register.

Nuclear envelope The outer and inner nuclear membranes that separate the nucleus from the cytoplasm.

Nucleocytoplasmic transport The movement of materials between the nucleus and cytoplasm of a cell.

Nutrition The study of food and nutrients and how they are used by the body.

Nutritional diseases Diseases that result from an inappropriate intake of nutrients.

Oligomenorrhea Infrequent or irregular menstruation associated with a number of clinical disorders. Compare **amenorrhea** and **galactorrhea**.

Oliguria Passing less than 400 cm^3 of urine per day (compare with **anuria**).

Oncogenes One of the two major groups of genes linked to **cancer**: they are mutated forms of normal genes called proto-oncogenes (*see* **tumor suppressor genes**).

Oncology The branch of medicine involved with the study of **malignant tumors**, their epidemiology, diagnosis and treatment.

Opisthotonos An involuntary arching of the head, neck and spine backwards.

Opportunistic infections Infections that occur when a decline in the immunological functions of an individual makes them susceptible to infections by microorganisms that are normally nonpathogenic.

Opportunistic pathogen A microorganism that causes **opportunistic infections**.

Opsonization The process by which phagocytic cells bind readily to bacterial cells coated with activated complement protein, C3b, promoting phagocytosis of the cells.

Ototoxicity The loss of hearing or balance, tinnitus or dizziness caused by drug or chemical damage to the inner ear.

Overhydrated The accumulation of water in body compartments (compare with **dehydrated**).

Pacemaker *See* **Sinoatrial** or **SA node.**

Palpitations Abnormal or irregular heartbeats of such severity that the patient is conscious of them.

Pandemic A disease epidemic that affects patients over a wide geographical area.

Papules These are solid, limited, raised areas of skin about 5 mm in diameter whose shape and color may vary.

Paraparesis A progressive weakness of the legs.

Parasite An organism that lives at the expense of another: often applied to viruses, protozoa and helminths. The terms **pathogen** and parasite are virtually interchangeable.

Paresthesia Abnormal sensations such as itching, prickling, pins and needles anywhere in the body.

Pathogens These are organisms that cause disease. *See* **parasite.**

Penicillin An antibiotic first discovered (in the mold *Penicillium notatum*) in 1928 but only used clinically since 1940.

Pericarditis An inflammation of the pericardium; can be **acute** or **chronic.**

Peristalsis The rhythmic waves of smooth muscle contractions that propel food along the gastrointestinal tract.

Peritoneum The strong, transparent membrane with a smooth surface that lines the abdomen and binds the gastrointestinal tract and accessory organs to each other and to the inner wall of the abdomen.

Petechial hemorrhages These are small, round, dark red spots caused by bleedings under the skin.

Pharmacology The study of the effects of drugs in the prevention, diagnosis, and treatment or cure of disease.

Phenotype The visible or measurable characteristics of an individual, or indeed, any observable biological trait (contrast with **genotype**).

Photoaging The characteristic fine and coarse wrinkling, blotchy pigmentation and rough skin texture induced by the ultraviolet light from the sun that is a major factor contributing to the premature aging of skin.

Phototherapy The use of light in treating clinical conditions, for example the repeated exposure for 36 h of full spectrum bright light to treat **seasonally affective disorder.**

Plague A term originally applied to any widespread disease causing great mortality but now confined to bubonic plague, an infectious disease of animals and humans caused by the bacterium *Yersinia pestis.*

Plasma The protein-rich fluid remaining when **erythrocytes**, **leukocytes** and **platelets** have been removed from blood, but with an intact clotting system (*see* **serum**).

Plasma cells Antibody-secreting cells that develop from **B lymphocytes** under the influence of cytokines released from **T_H cells.**

Plasmapheresis The collection of plasma, with the return of the **erythrocytes** and **leukocytes** to the blood donor (*see* **apheresis**, **leukopheresis** and **plateletpheresis**).

Platelet aggregation This occurs upon wounding when **platelets** extend pseudopodia and become sticky and clump together when stimulated by ADP to form a plug that prevents further blood loss.

Plateletpheresis The collection of **platelets** from blood with all other components being returned to the blood donor (*see* **apheresis**, **leukopheresis** and **plasmapheresis**).

Platelets Vesicle-like, subcellular fragments about 3 μm in diameter with a volume of 7 fdm^3 (femtodecimeters^3 or femtoliters) formed by the fragmentation of **megakaryocytes.**

Polyallelic The occurrence of many different **alleles** of a gene, although any one individual can only posses a maximum of two alleles at each locus, given that chromosomes occur in pairs.

Polyclonal antibodies Each mature, differentiated B lymphocyte or **plasma cell** produces a single type of antibody. However, the response of the immune system to a single **immunogen** yields hundreds of different clones of B cells so that the response of the system is polyclonal, resulting in the appearance of a heterogeneous array of **antibodies** in the blood. Compare with **monoclonal antibody**.

Polydipsia Intense thirst.

Polymerase chain reaction An experimental, *in vitro* method for duplicating in an exponential way short strands of specific DNA fragments.

Polymorphonuclear leukocyte One of the two main types of **leukocytes** that can be distinguished from the other type (**mononuclear leukocytes**) in having lobed nuclei and granular cytoplasm.

Polymorphonucleocyte *See* **polymorphonuclear leukocyte**.

Polyploidy The situation in which the chromosome number is an exact multiple of the haploid number but exceeds the diploid number (*see* **aneuploidy** and **euploidy**).

Polyps Benign tumors that arise in the nose or from the endothelium of the gastrointestinal tract.

Polyuria The production of abnormally large volumes of urine. Compare with **anuria** and **oliguria**.

Portals of entry Sites at which **pathogens** can enter the body.

Portals of exit Sites at which **pathogens** can leave the body.

Precision The ability of a test for an **analyte** to give the same result every time the same sample is analyzed. Compare with **accuracy**.

Predisposing factors Conditions or situations that make an individual more susceptible to disease: they include age, sex, heredity and environmental factors.

Premalignant This refers to clinical conditions that increase the risk of the patient developing a malignant tumor.

Prevalence The prevalence of a disease refers to the proportion of people in a population affected at a specific time.

Primary immunodeficiency diseases (PID) Clinical conditions that result from a genetic defect leading to a failure of one or more components of the immune system (*see also* **secondary immunodeficiencies**).

Primary tumor Malignant tissue growing in the site where it initially started (compare with **metastasis**, **secondary tumor**).

Prognosis The likely outcome of a disease.

Programmed theory of aging Suggests that each species has an inbuilt biological clock and that aging is a genetically programmed series of events.

Proinflammatory Description of a number of cytokines, which when secreted promote **inflammation**.

Prophylactic bilateral mastectomy The removal of apparently healthy breasts to prevent the development of breast cancer in genetically susceptible women.

Psychogenic diseases Clinical conditions that originate in the mind.

Ptosis An inability to raise the eyelids.

Pulmonary circulation The circuit which the blood follows between the right side of the heart, the lungs and the left **atrium** (compare with **systemic circulation**).

Pulse The bulging of the arteries that can be felt at a number of sites in the body during systole.

Purified protein derivative (PPD) A mycobacterial extract used in the Mantoux reaction to test if an individual is already sensitized by infection or previous vaccination to TB.

Purkinje fibers *See* **bundle of His**.

Pyogenic infections Infections with bacteria that cause the formation of pus.

Recessive mutations Mutations that are only expressed in an individual who is **homozygous** for the condition.

Recommended daily amount (RDA) This was originally used by the UK Department of Health in 1979 to define the amounts of certain nutrients needed by different groups within the population.

Reference nutrient intake (RNI) An amount of a nutrient that is greater than the dietary needs of 97.5% of a population (compare with **estimated average requirement**, **lower reference nutrient intake** and **safe intake**).

Reference ranges Numerical reference limits based on the mean value of an **analyte** plus or minus two standard deviations against which test results can be compared. The term **normal range** is still used synonymously with reference range.

Reflux esophagitis A back flow of gastric juice from the stomach into the lower end of the esophagus.

Regurgitation An imperfect closing of the heart valves leading to leakage and backflow (compare with **stenosis**).

Relapse A relapse occurs when, following a period of apparent recovery, the symptoms of the disease return.

Remission A period of good health with a reduction or disappearance of symptoms following a disease; may be permanent or temporary.

Respiratory acid–base disorder This is the occurrence of an abnormal blood pH because of lung dysfunction. Compare with **metabolic acid–base disorder**.

Reticulocytes The immediate precursors of erythrocytes, which have nuclei and are still able to synthesize hemoglobin.

Retinoids Name for a group of naturally occurring as well as many synthetic forms of vitamin A (retinol).

Reversion This occurs in some patients who do not develop an effective immune response to an attenuated vaccine and the weakened microorganisms in it become virulent.

Rheumatoid factors (RFs) Autoantibodies present in the plasma of patients with rheumatoid arthritis which seem to be directed against the patient's own IgG molecules.

Risk factors Certain dietary, occupational or lifestyle conditions that increase the chances of developing a disease.

Safe intake The amount of a nutrient judged to be sufficient for the needs of most people (compare with **estimated average requirement**, **lower reference nutrient intake** and **reference nutrient intake**).

Seasonally affective disorder (SAD) A condition characterized by changes of mood and eating and sleeping patterns in winter associated with an inappropriate secretion of melatonin.

Secondary A term often used to refer to a condition that arises from an existing disease. It is also used to describe a stage of a disease.

Secondary immunodeficiency diseases (SIDs) These arise as a consequence of other conditions (compare with **primary immunodeficiency diseases**).

Secondary tumors These are formed when a primary tumor metastasizes and spreads to other tissues and organs. Often loosely referred to as "secondaries".

Secretors Individuals who express the A, B and H blood antigens on soluble proteins and secrete them into saliva and other body fluids.

Self Used in **immunology** to refer to the components that make up the body itself (compare with **nonself**).

Senescence The decline in the functions of almost all parts of the body and at all levels of organization, from cells to organ systems, seen on aging.

Sepsis The presence of the **pathogens** in the blood of a patient.

Serum The clear straw-colored liquid remaining after the blood has been allowed to clot and the clot and all the cells have been removed.

Severe combined immunodeficiency (SCID) A life threatening clinical condition involving deficient numbers of both B and T lymphocytes.

Shock A patient goes into shock if the blood pressure falls too low.

Sickle cell anemia A hemoglobinopathy characterized by the presence of crescent-shaped erythrocytes.

Sign Something the clinician specifically looks or feels for, such as redness or swelling of the skin, when examining the patient.

Sinoatrial or **SA node** The "pacemaker" that generates electrical impulses that regulate the rate of contraction of the heart.

Small lymphocytes **Mononuclear leukocytes** involved in the **specific immune response**.

Specific immune response The response that allows the development of true specific immunity against an infectious agent.

Stenosis Failure of the heart valves to open appropriately (compare with **regurgitation**, **murmur**).

Strabismus A squint or abnormal alignment of the eyes.

Stroke or **cerebrovascular accident** This occurs if the blood supply to the brain is disrupted for sufficient time that brain cells are permanently damaged or die due to lack of oxygen. Strokes can be hemorrhagic when a blood vessel bursts disrupting the normal flow of blood and allowing it to leak into an area of the brain and destroy it, or ischemic, which occurs when the blood supply to part of the brain is cut off either because of **atherosclerosis** or because a clot is blocking a blood vessel.

Subclinical A stage at which a disease is established even though any overt **signs** and **symptoms** are not apparent.

Superantigens Toxins secreted by bacteria such as *Streptococcus pyogenes* or *Staphylococcus aureus* that stimulate the release of cytokines from immune cells and produce an excessive inflammatory response called systemic inflammatory response syndrome (SIRS).

Symptom An indication that a disease is present and something of which the patient complains, for example, nausea, malaise or pain.

Syncope Fainting with a temporary loss of consciousness resulting from a temporarily inadequate supply of oxygen and nutrients to the brain.

Syndrome A term applied to describe certain diseases that are characterized by multiple abnormalities and features that form a distinct clinical picture.

Syngeneic stem cell transplants These involve donors and recipients who are genetically identical, for example between identical twins (*see also* **allogeneic** and **autologous**).

Systemic circulation The flow of blood from the left ventricle to all the tissues of the body with the exception of the lungs and back to the right atrium (compare with **pulmonary circulation**).

Systemic disease One that affects the whole body.

Systemic inflammatory response syndrome (SIRS) (*see* **superantigens**).

Systole Contraction of the heart ventricles.

T lymphocytes These are derived from **small lymphocytes** that leave the bone marrow in an immature state and mature into $\mathbf{T_C}$ or $\mathbf{T_H}$ **cells** in the fetal thymus.

Tachycardia An inappropriately fast heartbeat.

$\mathbf{T_C}$ **cells** Blood cells that develop into **cytotoxic** T lymphocytes when appropriately stimulated, and which are capable of killing virus-infected cells.

$\mathbf{T_H}$ **cells** or **helper T lymphocytes** Blood cells that develop into cytokine-secreting T_H cells when stimulated by an **immunogen**, and which produce an array of cytokines that control the activities of cells of the immune system.

Thalassemias A large group of related hemoglobinopathies of varying severity that originate from point mutations or deletions in globin genes.

Toxic diseases Diseases caused by the ingestion of a variety of poisons that may be encountered in the environment.

Toxicology The study of the adverse effects of toxic chemicals on organisms.

Transgenic organisms Organisms that express genes from a different species that have been inserted into their cells.

Traumatic diseases Diseases caused by physical injury including mechanical trauma.

Tumor suppressor genes One of two major groups of genes whose mutated forms are linked to the formation of cancers (compare with **oncogenes**).

Uricolysis The degradation of urate (uric acid) to CO_2 and NH_3.

Vaccination The process of protecting people from infection by deliberately exposing them to parasite components that initiate a protective immune response.

Vegetation An accumulation of bacteria and blood clots on the valves of the heart.

Ventricle Lower chamber of the heart.

Virilism The development in females of inappropriate male characteristics, for example hirsutism, increased muscle mass, deepening of the voice and male psychological characteristics.

Virulence factors Features of **pathogens** that allow them to cause **infectious diseases.**

Vitamers A term applied to a group of closely related compounds that all posses the activity of the named vitamin.

Vomiting or **emesis** The forced expulsion of food from the stomach (and sometimes duodenum) through the esophagus and mouth.

Winter depression *See* **seasonally affective disorder**.

Xanthomas Yellowish deposits on the tendons that occur in **familial hypercholesterolemia**.

Xenobiotic A substance that does not originate in the body but is pharmacologically, endocrinologically or toxicologically active.

Xenogeneic transplant The use of a graft from another species (*see also* **allograft**).

Zoonoses Diseases that can be transmitted from animals to humans.

Zymogens Precursor molecules that turn into active proteases when subjected to a specific proteolytic cleavage.

chapter 1:

THE NATURE AND INVESTIGATION OF DISEASES

OBJECTIVES

After studying this chapter you should be able to:

- define appropriate terms associated with health and disease;
- give a simple classification of diseases;
- list the ways pathological investigations help in the diagnosis, treatment and management of diseases;
- discuss the clinical and analytical evaluations of laboratory tests.

1.1 INTRODUCTION

Disease can be defined as any abnormality or failure of the body to function properly and this may require medical treatment (*Figure 1.1*). The scientific study of diseases is called **pathology**.

Every disease has a distinct set of features that include a cause, associated clinical symptoms and a characteristic progression, with associated morphological and functional changes in the patient. The presence of an abnormality on its own, however, does not necessarily indicate disease since the affected individual must also suffer from ill health. **Health** can be defined as an absence of signs and symptoms associated with any disease. This definition has limitations in that there are circumstances where individuals believe they are ill even though detectable indications of disease are not present. Conversely, there are individuals who believe they are healthy but on detailed examination are found to suffer from a serious disease. For this reason, the World Health Organization (WHO) devised the more appropriate definition of health as *a state of physical, mental and social well-being and not merely the absence of disease*. Currently there is considerable interest, particularly in developed countries, in promoting health by improving lifestyle and reducing mental and social factors associated with ill health.

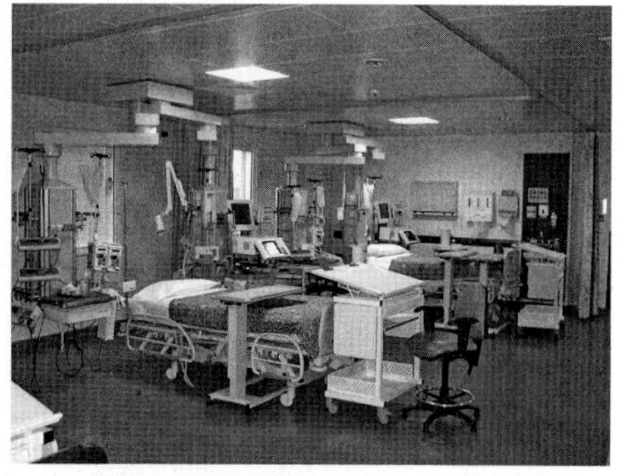

Figure 1.1 A modern intensive care unit of a hospital.

Individuals differ in their physical appearance and also internally in the composition of the biological materials of which they are made. Differences between individuals are due to biological variation. Variation in values for measurable body features or biological substances (**analytes**) occurs for a number of reasons. Differences in the genotypes of individuals ensure that no two individuals are the same (other than identical twins). Biological variation arises due to differences in lifestyle, experiences, dietary and other factors as well as genes. There are also differences in physiological processes arising from bodily control mechanisms. For example, the concentration of blood glucose varies between individuals with diet, time of day and physical activity.

1.2 CHARACTERISTIC FEATURES OF DISEASES

Every disease has a number of characteristic features. These features allow diseases to be categorized and allow a better understanding of the disease, its diagnosis and management. A correct diagnosis should mean that appropriate treatment is given.

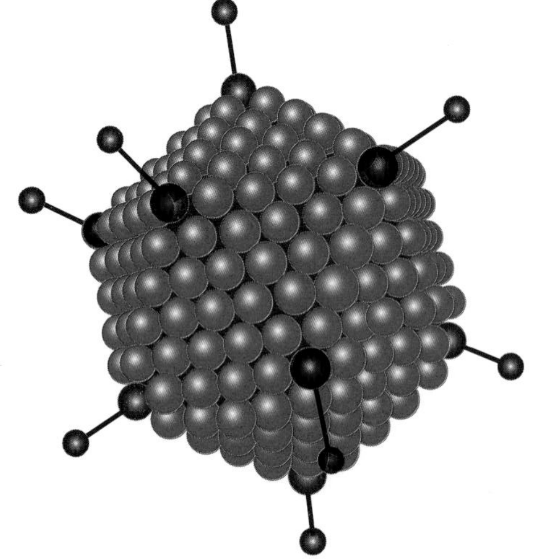

Figure 1.2 Schematic of an adenovirus (diameter about 70nm), which can cause respiratory infections.

ETIOLOGY

Etiology refers to the cause of a disease. Etiological agents can be **endogenous**, in other words originating from within the body, or **exogenous**, coming from outside the body. Endogenous agents include genetic defects and endocrine disorders, while exogenous agents include microorganisms such as viruses (*Figure 1.2*), bacteria (*Figure 1.3*) and fungi that cause infections, chemicals, physical trauma and radiation. Many diseases are said to be predictable and arise as a direct consequence of exposure to the causative agent. Other diseases are considered probable in that they may be a consequence of the causative agent but the development of illness is not inevitable. An individual can be infected with a pathogenic microorganism but the outcome of the disease may depend on other factors such as the nutritional and immune status of the affected person.

Some diseases have more than one etiological agent and may, indeed, be caused by a range of factors. Such diseases are said to be multifactorial in origin. Diabetes mellitus type 2, a disorder of carbohydrate, fat and protein metabolism, is believed to have a multifactorial origin involving several genetic, dietary and environmental factors (*Chapter 7*). Many diseases are of unknown cause and are said to be **idiopathic**. An example of this is hypertension, where more than 90% of cases are of unknown cause. The treatment of idiopathic diseases is restricted to alleviating the symptoms. Some conditions are caused by the effects of treatment and are called **iatrogenic** diseases (from the Greek word *iatros*, doctor). The treatment of some cancers with cytotoxic drugs, for example, can cause a severe iatrogenic anemia although they may be curing the cancer.

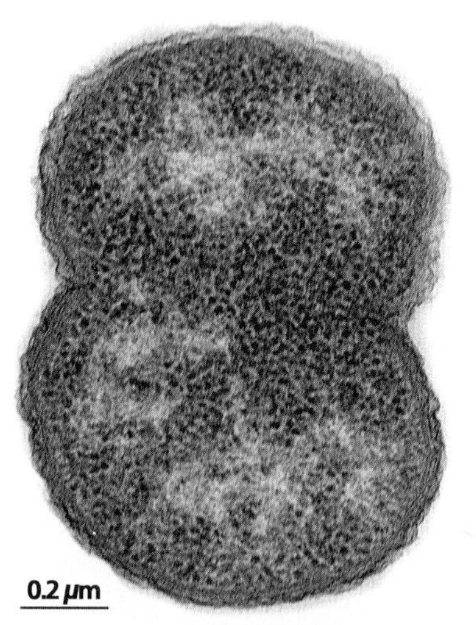

Figure 1.3 Electron micrograph of the bacterial pathogen, *Neisseria meningitidis*, a causative organism of meningitis (*Chapters 2* and *3*). Courtesy of Dr A. Curry, Manchester Royal Infirmary, UK.

Occasionally a disease of unknown etiology is more commonly found in populations with certain dietary, occupational or lifestyle conditions called **risk factors.** Smoking is a significant risk factor in the development of heart disease and lung cancer (*Chapters 14* and *17*). Some risk factors may be important in the development of the disease whereas others may make the individual more susceptible to disease. **Predisposing factors** are conditions or situations that make an individual more susceptible to disease. They include age, sex, heredity and environmental factors. For example, the immune system in a newborn is not fully developed and, as a consequence, babies are more susceptible to infections. However, during aging the immune system undergoes a progressive decline in function making the elderly also more susceptible to infections. Sex may also be a predisposing factor: men are more likely to suffer from gout than women whereas osteoporosis is more common in the latter.

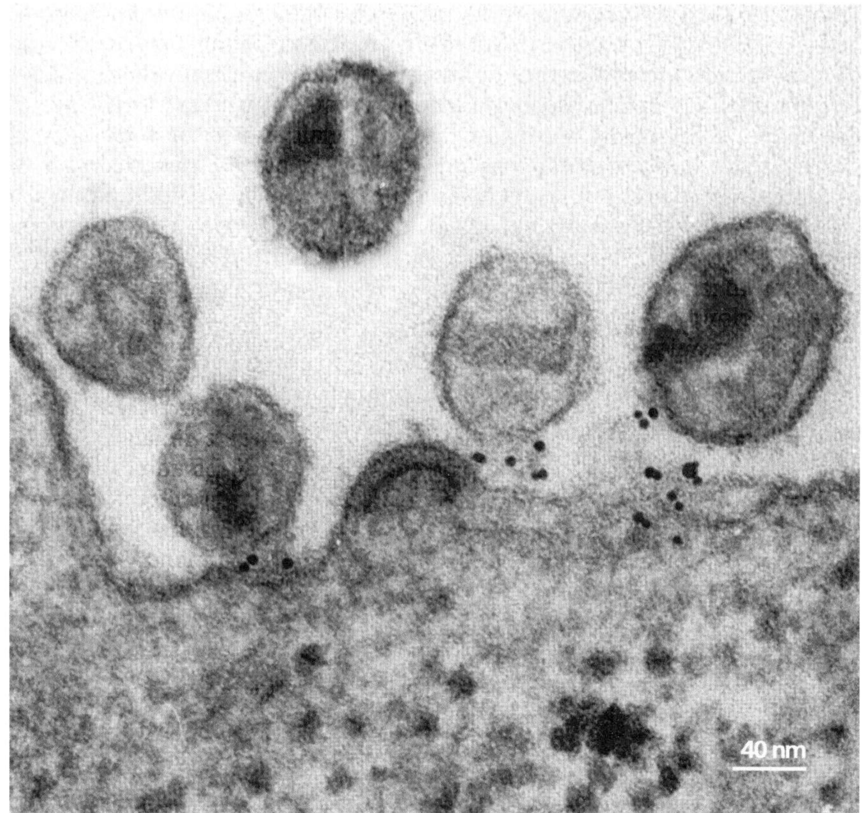

Figure 1.4 Electron micrograph of HIV particles escaping from a cultured human cell. Courtesy of Dr. D. Robertson and Professor R.A. Weiss, Institute for Cancer Research, Royal Cancer Hospital, UK.

Some diseases increase the risk of developing others. Thus some conditions increase the risk of someone developing cancer and are said to be **premalignant**. This is seen in ulcerative colitis, an inflammatory condition affecting the large intestine that increases the probability of developing bowel cancer in sufferers. Some diseases predispose the patient to other conditions by allowing infectious agents, not normally pathogenic, to cause disease. This is seen in **opportunistic infections** where a decline in the immunological functions of an individual makes them susceptible to infections by microorganisms that are normally nonpathogenic. In acquired immunodeficiency syndrome (AIDS), the individuals infected with the human immunodeficiency virus (HIV), (*Figure 1.4*) have little resistance to infections by microorganisms, such as those responsible for causing pneumonia, as well as microorganisms such as the yeast *Candida*, which are part of the normal flora of the body (*Chapter 2*).

Diseases are often described as being **primary** or **secondary**. Primary may refer to a disease of unknown cause or **idiopathic**, whereas secondary is used to refer to a condition that arises from an existing disease. However, these terms are also often used to describe the stages of a disease. For example, in cancer the primary tumor is the initial tumor whereas secondary tumors arise following metastasis of the primary tumor to other tissues (*Chapter 17*).

PATHOGENESIS

Every disease has a pathogenesis that describes the development of the disease or, more specifically, how the etiological agent(s) acts to produce the clinical and pathological changes characteristic of that disease. Some examples of how diseases undergo pathogenesis include inflammatory reactions in response to harmful agents and carcinogenesis where the formation of tumors occurs as a result of exposure to carcinogens (cancer-inducing substances).

Diseases have 'natural histories' that describe the typical patterns of how each disease usually progresses, its effects and its duration. The effects of the disease on the patient are referred to as **morbidity**. Occasionally the morbidity of a disease may cause disability that, in turn, may restrict the activities of the patient. The **mortality** of a disease describes its possibility of causing death. This is usually expressed as a percentage. Some diseases have a rapid, often severe onset that is described as **acute**. However, other diseases have a **chronic** onset and develop gradually over a relatively longer time. Thus acute renal failure is characterized by rapid deterioration of kidney function over a matter of days, while chronic renal failure develops over months or even years (*Chapter 8*).

Diseases rarely occur immediately following exposure to an etiological agent. In most cases, a period of time, the **incubation period**, must elapse before the disease becomes apparent. In carcinogenesis, this period may last several decades and is referred to as the **latent period**. With infectious diseases, the time between exposure and development of the disease is often characteristic of the infectious agent involved.

MANIFESTATIONS OF DISEASES

The etiology of a disease and its pathogenesis produce **clinical manifestations** that include signs and symptoms of the disease. A **symptom** is an indication that a disease is present and something of which the patient complains, for example, nausea, malaise or pain (*Table 1.1*). A **sign** is something that the clinician specifically looks or feels for, such as redness or swelling of the skin, when examining the patient. Some diseases present with a **subclinical** stage where these signs and symptoms are not apparent, even though the disease is established and characteristic biochemical and cellular changes that are detectable by laboratory analysis of, for example blood or urine have taken place.

Clinical signs and symptoms are often accompanied by structural or functional abnormalities, called **lesions**, in affected tissues that are responsible for ill health and usually cause the signs and symptoms of disease. Lesions may be biochemical in nature, such as defective hemoglobin in patients with hemoglobinopathies (*Chapter 13*). Alternatively, a lesion may include deposition of abnormal substances in cells, tissues and organs such as deposition of amyloid in the brain in patients with Alzheimer's disease (*Chapter 18*). Loss of healthy surface tissue, for example, in gastric ulceration, may also

Symptom	Cause
Pain	stimulation of nerve endings by trauma, chemicals and heat
Swelling	increase in number or size of cells or an accumulation of tissue fluid
Fever	actions of interleukin-1 and prostaglandins stimulate thermoregulatory center in the brain
Weight loss	decreased intake of food or a catabolic state stimulated by release of factors from tumors
Diarrhea	inadequate absorption of food by the GIT leads to an osmotic retention of water and production of watery stools (*Chapter 11*)
Cough	release of neuropeptides following irritation of the respiratory mucosa
Cyanosis	reduced supply of oxygenated hemoglobin to the skin

Table 1.1 Disease symptoms and their causes

be a feature of some diseases. Diseases may involve functional abnormalities such as an inappropriate secretion of hormones. Examples of these include the excessive production of thyroid hormones in hyperthyroidism or the inadequate secretion of insulin in type 1 diabetes mellitus. Other functional defects might include impaired nerve conduction and muscular contraction.

The term **syndrome** is often applied to describe certain diseases that are characterized by multiple abnormalities that form a distinct clinical picture. For example, Cushing's syndrome (*Chapter 7*) occurs when an excess of the hormone cortisol produces a combination of clinical features that include **hirsutism** (excessive growth of facial hair), obesity, hypertension and characteristic facial and body features.

Some diseases frequently present with complications, that is, new or separate processes secondary to, and a consequence of, the initial disease. Diabetic cataracts, retinopathy and nephropathy are all chronic or long-term complications of diabetes mellitus.

The manifestations of a disease in a given person are not static and are affected by compensatory mechanisms in the body as well as by environmental influences and responses to treatment. Diseases often have a range of manifestations and their presence and severity may vary from patient to patient. In addition to differences between individuals, differences occur within an individual at different stages of development, from infancy to old age. *Figure 1.5* outlines the key features of some diseases in terms of their etiology, pathogenesis and manifestations.

OUTCOME OF A DISEASE

The **prognosis** of a disease is its likely outcome. Prognoses can vary considerably between different diseases and, of course, can be influenced by treatment. Hence when giving a prognosis, it is necessary to clarify if the disease is following its natural course or whether there is, or needs to be, medical or surgical intervention. A viral disease such as German measles (rubella) will normally resolve of its own accord, whereas a broken leg or a heart attack needs treatment. With some diseases, especially some cancers, patients may go through a period of good health with a reduction or disappearance of the symptoms and the disease is said to be in **remission**. However, a **relapse** may occur with a return of the disease symptoms following this period of apparent recovery. Diseases with a tendency towards remission and relapse include acute lymphoblastic leukemia and ulcerative colitis (*Chapters 17* and *11*).

	Etiology	Pathogenesis	Morphological and functional features	Complications and sequelae
Local skin infection *(Chapters 2 & 3)*	*Staphylococcus aureus*	acute inflammation	local skin infection visible	septicemia
Neoplasms, for example lung cancer *(Chapter 17)*	smoking	mutation *(Chapter 15)*	tumor of the lung	metastases leading to secondary tumors
Cirrhosis of the liver *(Chapters 2 & 12)*	hepatitis B virus	immune reaction to virus-infected cells *(Chapter 4)*	cirrhosis of the liver	liver failure

Figure 1.5 Characteristics of some diseases showing relationship between etiology, pathogenesis, morphological and functional features and complications and sequelae.

Class of disease
Infectious
Immunological
Endocrine
Homeostatic
Nutritional
Toxic
Genetic
Congenital
Neoplastic
Traumatic
Degenerative
Psychogenic
Iatrogenic
Idiopathic

Table 1.2 A classification of diseases

1.3 CLASSIFICATION OF DISEASES

Some diseases share common features and can be grouped together in a classification system. One way of classifying diseases is on the basis of their cause (*Table 1.2*). This is by no means perfect as some diseases have multiple causes and there is likely to be an overlap between the different categories.

Infectious diseases are caused when microorganisms such as viruses, bacteria, fungi, protozoa and helminths enter and spread within the body (*Chapters 2* and *3*).

Immunological diseases (*Chapters 4* and *5*) occur in circumstances in which the immune system can cause damage to the body's own tissues. In autoimmune conditions, for example autoimmune thyroiditis, antibodies are produced that attack the body's own tissues. Alternatively, there are diseases associated with immunodeficiency that increase the susceptibility of the patient to infectious agents. This occurs in severe combined immunodeficiency (SCID) and in AIDS.

Endocrine diseases arise from the over- or underproduction of hormones or from resistance to a particular hormone perhaps because the cellular receptor is absent as the result of a mutation (*Chapter 7*). Thus, for example, acromegaly is caused by the overproduction of growth hormone in adults, whereas type 2 diabetes mellitus is a consequence of insulin resistance, when the appropriate target cells fail to respond to the hormone (*Figure 1.6*).

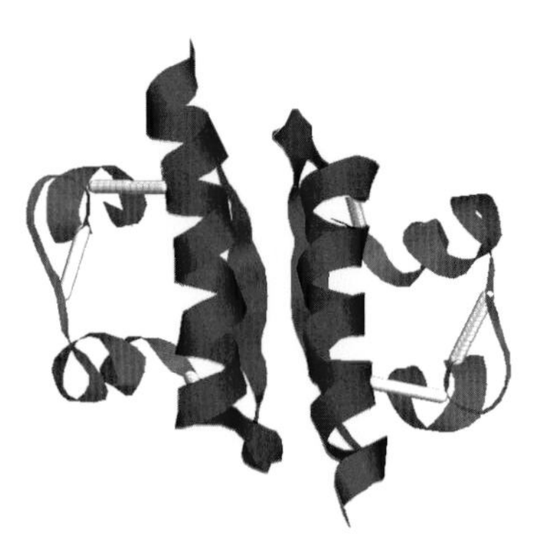

Figure 1.6 Model of a dimer (two molecules) of insulin. The represent disulfide bonds. PDB file 1ZEH.

Homeostatic diseases arise when mechanisms for controlling homeostasis are disrupted. For example, in the syndrome of 'inappropriate ADH secretion' (*Chapter 8*) diminished urine production leads to an increase in body fluids.

Nutritional diseases result from an inadequate intake of nutrients, such as proteins (which supply essential amino acids), carbohydrates, essential fatty acids, vitamins or trace elements. Inadequate nutrition is a major cause of disease, particularly in developing countries. Such deficiencies may be generalized, as in protein-energy malnutrition (*Chapter 10*) where there is simply not enough food, or there may be a lack of a specific nutrient, for example, vitamin A (*Figure 1.7*) leading to several disorders including night blindness. In contrast, in many developed countries an excessive intake of energy combined with a lack of exercise is responsible for a worrying increase in obesity.

Toxic diseases (*Chapter 12*) are caused by the ingestion of a variety of poisons that may be encountered in the environment. Ingestion may be accidental or deliberate. Carbon monoxide can be inhaled from car exhausts or faulty gas fires or water heaters, causing tissue hypoxia and death.

Genetic diseases arise due to defects in the genes or chromosomes of individuals (*Chapter 15*). A defective gene may result in the inadequate production of a key enzyme, such as phenylalanine hydroxylase in phenylketonuria. Down syndrome is an example of a disorder which arises due to an abnormal chromosome complement. Some genetic disorders are not inherited from parents and may arise from a new genetic mutation in the offspring, as in the disease progeria. Congenital diseases are present at birth and may or may not have been inherited. They may arise due to a developmental defect of known or unknown cause. Thus, a newborn may suffer from fetal alcohol syndrome, a congenital condition arising as a consequence of excessive alcohol intake by the mother during pregnancy.

Neoplastic diseases are characterized by the uncontrolled and abnormal growth of cells. These cells form benign or malignant tumors (*Chapter 17*). Malignant neoplasms are a major cause of death in many developed countries. Moreover, their incidence is increasing as people live longer.

Traumatic diseases are caused by physical injury and include mechanical trauma, extremes of heat or cold, electrical shock and radiation. Apart from the obvious problems caused by extensive damage to tissues, traumatic diseases may render an individual more prone to infection by compromizing the immune system (*Chapter 4*).

Degenerative diseases involve the progressive loss of body tissues and impairment of their functions usually associated with aging (*Chapter 18*). Examples include neurodegenerative diseases, such as the relatively common Alzheimer's disease and muscular dystrophy.

Psychogenic diseases originate in the mind. They may have a significant psychological or emotional component as seen, for example, in schizophrenia.

Iatrogenic diseases arise as a consequence of treatment. For example, patients who are receiving drugs, such as thiazide diuretics to control their blood pressure, may suffer from low serum K^+ (hypokalemia, see *Chapter 8*) caused by an excessive renal loss. If untreated, hypokalemia may, in turn, cause cardiac arrhythmias (*Chapter 14*). Finally, **idiopathic diseases** are those of unknown cause.

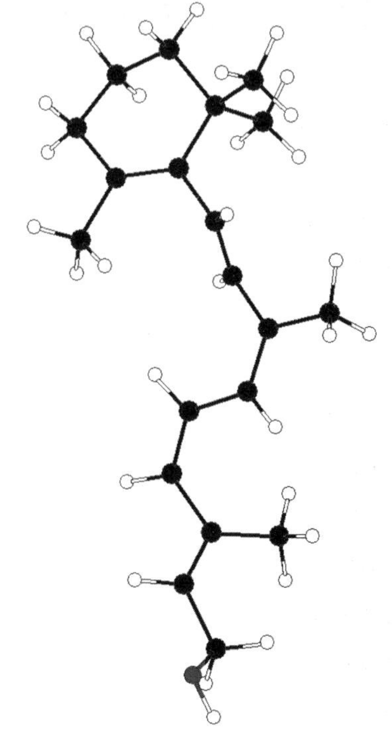

Figure 1.7 Computer generated model of vitamin A (retinol).

1.4 EPIDEMIOLOGY OF DISEASE

Epidemiology is the study of how diseases spread in populations in relation to their causal factors. Consequently, epidemiology is largely concerned with the collection and interpretation of data about diseases in groups of people rather than in individuals. The types of data collected in epidemiological studies provide information about the etiology of the diseases, whether there is a need for screening or the introduction of other preventative measures and whether health care facilities are appropriate.

The **prevalence** of a disease refers to the proportion of people in a population affected at a specific time. The **incidence rate** is the number of new cases of a disease in a population occurring within a specified period of time.

Epidemiological studies can often provide information about the cause(s) of diseases. Thus if a disease has a high incidence in a particular region or population, then the disease may have a genetic origin or it may be caused by environmental factors peculiar to that area. Epidemiological studies of migrant populations are especially useful since they can provide valuable information on the etiology of a disease. A case in point might be where a migrant population has a high incidence of a particular disease and then moves to another geographical area where the incidence of the same disease is low. If the incidence of disease in the migrant population remains high, then it is likely that the disease has a genetic basis. If, however, the incidence in the migrant population decreases to the level of the new geographical region, then environmental factors probably play a role in its etiology.

The data on the incidence of some diseases are very reliable. This is especially so for some infectious diseases and cancers that are **notifiable**. Clinicians are legally required to supply details of all new cases of diseases on the notifiable list to a central register. However, obtaining data on the incidence of other diseases can be difficult. For most diseases, the data obtained refer to mortality rates for that disease based on the causes listed on death certificates. This method of obtaining data has the major limitation of underestimating the incidence if the disease does not have a fatal outcome.

The incidence of certain diseases changes with time and also can vary considerably from one country to another and even within different regions of the same country. These differences are particularly marked between developing and developed countries. Infectious diseases and malnutrition are still more prevalent in developing countries, while in the developed world, the incidence of many infectious diseases has been reduced dramatically in the last 100 years. The infant mortality rate is often used as a measure of health related to socioeconomic status. In general, the infant mortality rate is higher in developing compared with developed countries.

The decreased incidence of many diseases in developed countries may reflect changes in exposure to causative agents as well as the effects of preventative measures. For example, the reduction of diseases such as cholera is associated, in a large part, with improved public health measures. Improvements in sanitation, sewage and hygiene have had a considerable impact in reducing the incidence of many infectious diseases. Mass immunization against infectious diseases, such as polio, has had enormous beneficial effects in reducing disease in the population as a whole. Unfortunately the reduction in infectious diseases has been accompanied by an increasing incidence of other diseases, such as cardiovascular diseases, diabetes, several types of cancers and psychiatric diseases. All are associated with aging and, to a certain extent, this may reflect the increased life expectancy in the developed countries: people are not killed by infectious diseases and live longer. Some evidence does suggest that the increased incidence of these diseases is also due to changes in diet such as increased consumption of saturated fats and other lifestyle factors, for instance a lack of exercise. **Intervention studies** aimed at changing diet and lifestyle factors in an attempt to reduce the incidence of these diseases are already proving beneficial.

Socioeconomic factors can also influence the incidence of many diseases. Poverty tends to be associated with an increased incidence of malnutrition and malnourished individuals are more susceptible to infectious diseases. Overcrowding is known to promote the spread of infectious diseases resulting in epidemics.

Some diseases have a high incidence in populations associated with certain occupations. For example, coal workers have a high incidence of pneumoconiosis caused by inhalation of coal dust and, in the past, workers with asbestos faced a high risk of asbestosis, and of developing mesothelioma of the lung. Occupational hazards need to be identified and minimized to reduce the incidence of these diseases.

1.5 INVESTIGATING DISEASES

For the majority of diseases, the clinical outcome is likely to be improved if treatment is started at an early stage. Consequently the proper investigation of disease is necessary to ensure a rapid and accurate diagnosis and to allow appropriate treatment to be initiated as soon as possible. The procedure for investigating a disease is outlined in *Figure 1.8*. It starts with the affected person presenting symptoms and visiting his or her physician when feeling unwell or after a period of ill health. The examination usually begins with the clinician asking the patient about his or her current and past medical histories, current and previous medications, use of alcohol and tobacco, any family history of disease and possibly occupational history. This is usually followed by a clinical examination to look for signs of any abnormality. This may involve visual examinations of the skin, eyes, tongue, throat, nails and hair to detect abnormalities together with tests to assess cardiovascular, respiratory, gastrointestinal, genitourinary, nervous and musculoskeletal functions. Since diseases typically present with recognizable signs and symptoms, the clinician may make a diagnosis of the disease based on the clinical history and the examination and then initiate treatment. Sometimes this may not be possible, given that many clinical symptoms and signs are not specific to any

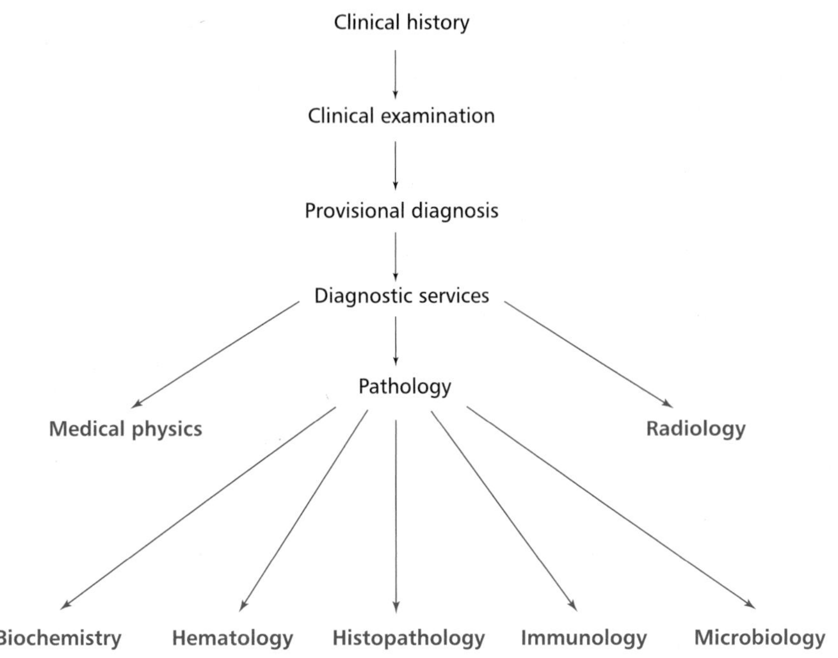

Figure 1.8 An outline of the steps involved in investigating diseases.

Types of pathology laboratories
Medical microbiology
Immunology
Clinical biochemistry
Hematology
Histopathology
Clinical genetics

Table 1.3 Types of pathology laboratories

one disease. However, a range of diagnostic services is also available to the clinician in modern health care systems. These include imaging techniques, physiological function tests, radiographic examinations (X-rays) and pathology laboratory investigations that can be applied to confirm, reject or distinguish between the various provisional diagnoses. The clinician may only be able to make a provisional diagnosis or a shortlist of possible diagnoses and then request additional investigations that rely on the diagnostic services available at the surgery, clinic or hospital.

1.6 TYPES OF PATHOLOGY LABORATORIES

The function of hospital pathology laboratories (*Table 1.3*) is to make scientific investigations of disease. The typical pathology service offered by hospitals has six main branches: medical microbiology, immunology, clinical biochemistry, hematology, histopathology and clinical genetics.

Medical microbiology is concerned primarily with the detection and identification of pathogenic microorganisms. For clinical purposes, these consist of viruses, bacteria, protozoa, fungi and helminths (worms). Microorganisms are detected directly in specimens obtained from the patient or on swabs (for example throat, nasal) that are cultured in growth medium to increase the number of microorganisms and allow their easier detection. The presence of microorganisms may also be determined indirectly by detecting antibodies produced by the patient in response to the infection. Medical microbiology laboratories also investigate the responses of pathogenic microorganisms to antibiotics.

Immunology laboratories are concerned with studying the body's immune response in both healthy and diseased states. Immune responses are 'cell-mediated' or 'humoral'. The former involves T lymphocytes, the latter the production of antibodies by specialized B lymphocytes. The presence or absence of antibodies in plasma can be determined, for example, by serum electrophoresis (*Figure 1.9*) to assess generalized immunodeficiencies and other diseases. However, of more diagnostic value during the investigation

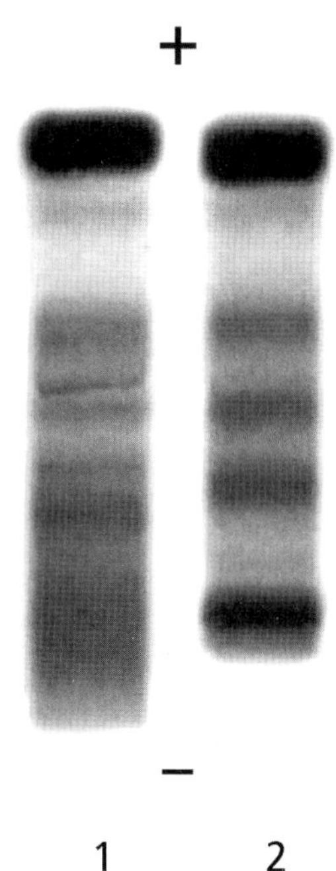

Figure 1.9 Excessive immunoglobulin (*Chapters 4* and *5*) is produced in multiple myeloma, a tumor of the B lymphocytes; the white blood cells that produce immunoglobulins. The proteins in samples of serum can be separated by electrophoresis and stained with dye. Lane 1 shows the separated proteins from normal serum. The most abundant protein is serum albumin which shows as the strongly staining band near the positive end. Lane 2 shows a myeloma serum sample with a second dense band at the negative end. This band shows the enormous amount of a single type of immunoglobulin, produced by the tumor. In multiple myeloma, the serum protein concentration increases making the blood thick and difficult to pump around, putting a strain on the heart and kidneys. In addition, so much effort is put into synthesizing one type of useless antibody that the concentration of other antibody molecules decreases and patients become prone to infection.

Figure 1.10 A typical automated analyzer in a hospital clinical chemistry laboratory capable of performing most of the major investigations. Courtesy of the Department of Clinical Biochemistry, Manchester Royal Infirmary, UK.

of immune diseases may be the measurement of specific antibodies that are produced in response to a particular antigen (which may be an infectious agent or an autoantigen). The number of cells involved in immunity, such as T-cells, B-cells, T-helper and T-suppressor cells, are often determined as this can provide valuable information about the immune status of an individual.

Clinical biochemistry is concerned with investigating the biochemical changes associated with diseases. A wide range of substances or **analytes** are measured in clinical biochemistry laboratories. Some of these analyses are carried out routinely on all samples (blood, urine) coming into the laboratory using automated methods (*Figure 1.10*); others need to be requested specially. Analyses include those for proteins, enzymes, hormones, lipids, tumor markers, blood gases, sugars and inorganic ions to investigate a variety of disorders, including those associated with abnormal renal, respiratory, metabolic, bone and endocrine function. In addition, analytes are measured during investigations of genetic disorders both to diagnose and to monitor the effectiveness of therapies.

Hematology is concerned with the study of disorders of blood cells, including blood clotting (coagulation) defects. Hematological investigations can involve determining the concentrations of blood proteins, such as hemoglobin, to aid in the diagnosis of diseases. The microscopic examination of blood films, thin layers of blood spread out on a microscope slide and stained (*Figure 1.11*) and marrow removed from bone cavities by aspiration (*Figure 1.12*) may also be helpful. Some hematology laboratories may also be involved in the provision of blood and blood products for transfusion services, but these are often run as separate services.

Histopathology is concerned with the investigation of disease by examining cells and tissues. This involves the macro- and microscopic investigation of body tissues for the identification of disease. Amongst other things, histopathology laboratories are usually involved in the diagnosis of malignancies and can also provide information on how far a tumor has progressed ('staging') and therefore can suggest a likely prognosis. In addition, histopathology laboratories may also assist with investigation of a range of infectious and inflammatory conditions affecting body tissues.

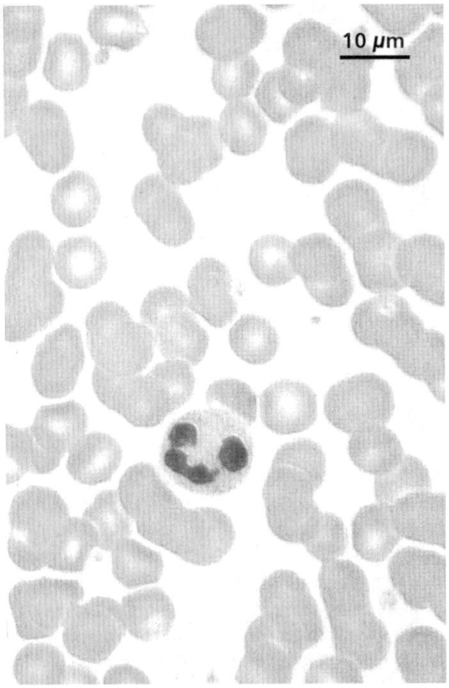

Figure 1.11 Blood film showing a single white blood cell surrounded by erythrocytes (*Chapter 13*). Courtesy of Dr L. Seal, School of Biology, Chemistry and Health Science, Manchester Metropolitan University, UK.

Clinical genetics is a growing area in the investigation of diseases. A major focus of clinical genetics laboratories is the identification of genetic abnormalities (*Chapter 15*). This could include, for example, identifying the number and form of chromosomes (*Figure 1.13*) in blood films to identify any numerical and structural abnormalities.

1.7 ROLE OF HOSPITAL LABORATORY TESTS

Tests performed by the pathology laboratory can assist clinicians in investigating disease. The tests may only give a subjective assessment, such as when a pathologist assesses the types of cells obtained from a fine needle aspirate of a suspected breast tumor when investigating breast cancer. However, tests may provide quantitative information, such as the concentration of thyroid hormones in the serum, that can then be compared with a **normal** value. Unfortunately, the term normal is often difficult to define in clinical terms. To alleviate this problem, **reference ranges** have been widely adopted. Numerical reference limits are based on the mean value plus or minus two standard deviations against which test results can be compared. The uses of reference ranges are explored more thoroughly later in the chapter. The term **normal range** is still used synonymously with reference range.

In general, the roles of laboratory tests include:

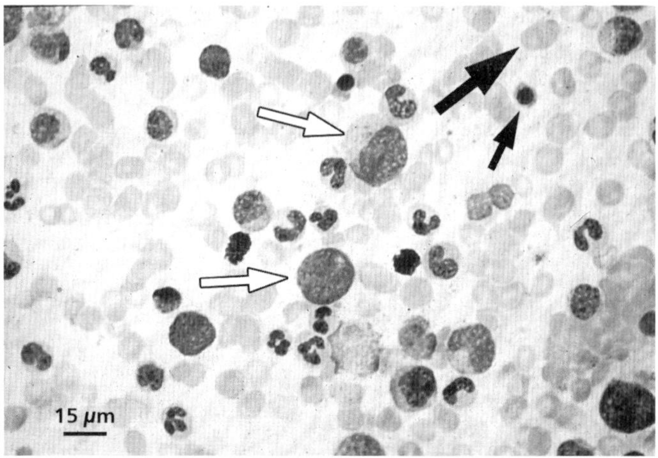

Figure 1.12 The bone marrow site is the site of blood cell formation. This light micrograph shows its normal cells with a range of immature erythrocytes (smaller solid arrow) and leukocytes (open arrow). The large proportion of mature erythrocytes (larger, solid arrow) in the background is due to unavoidable contamination with peripheral blood during sample collection. Courtesy of J. Overfield, School of Biology, Chemistry and Health Science, Manchester Metropolitan University, UK.

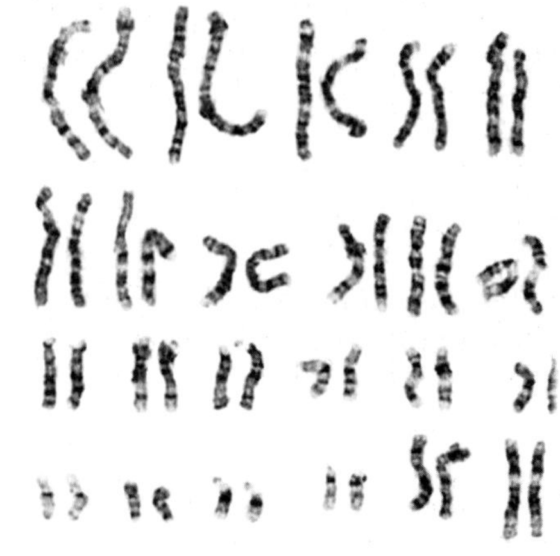

Figure 1.13 A spread of human chromosomes from a female (*Chapter 15*).

- the diagnosis, to identify the disease;
- monitoring of treatment;
- screening and assessment of risk;
- the prognosis, to inform the physician and the patient of the likely outcome;
- detection of complications.

The results of laboratory investigations are used in conjunction with the patient's clinical history and examination to determine the nature of the disease affecting the patient. Thus a low value for the concentration of glucose in the blood of a patient can confirm hypoglycemia and the clinician can start palliative treatment, even though the cause of the hypoglycemia may be unknown at this stage.

Laboratory tests may be used to monitor the course of an illness or the effects of its treatment. For example, the concentration of glycated hemoglobin in erythrocytes is measured in diabetics on a regular basis. The higher the concentration of glucose in the serum, the more readily the sugar reacts with proteins in a nonenzymic reaction to form glycated hemoglobin, in which sugar molecules are covalently attached to the protein. Thus the amount of glycated hemoglobin is an indicator of average glycemia in such patients over a period of days or months. Diabetics who are not complying with their treatment by not taking their insulin regularly or giving themselves the wrong dose, or whose treatment is ineffective, can be identified because poor control of blood glucose gives rise to higher concentrations of glycated hemoglobin than is normal or even found in well-controlled diabetics (*Figure 1.14*).

Laboratory tests may be used to detect a disease *before* it presents clinically. This is referred to as screening. For example, the concentration of phenylalanine in the serum of all newborn babies in the UK is measured to detect phenylketonuria (*Chapter 15*). Affected children have a high concentration of serum phenylalanine (hyperphenylalaninemia) and metabolites of phenylalanine, such as phenylpyruvic acid are usually present. If untreated, this condition leads to irreversible brain damage but if caught early and treated by diet, individuals develop normally. Other examples of screening tests include smears taken from the lining of the uterine cervix (*Margin Note 1.1* and *Figure 1.15*). Screening tests may also be

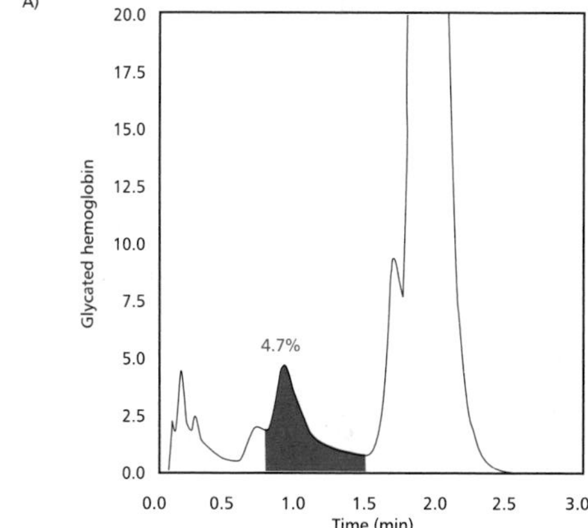

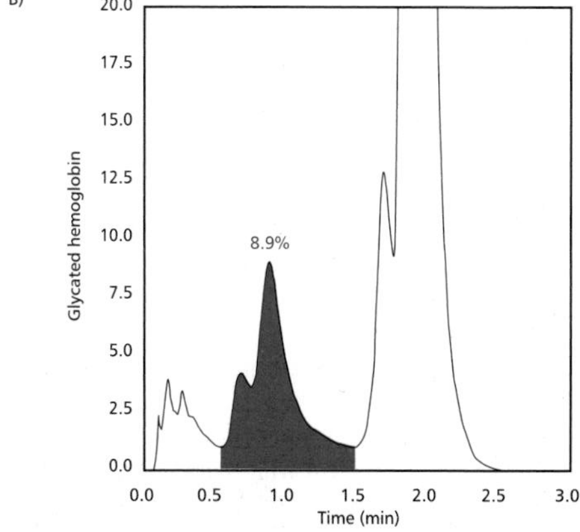

Figure 1.14 A chromatogram glycated hemoglobin determined by HPLC. The results are shown for (A) a normal person and (B) a patient with diabetes mellitus. Courtesy of Department of Clinical Biochemistry, Manchester Royal Infirmary, UK.

BOX 1.1 Clinical specificity and sensitivity

If a test for a particular disease gives a positive result in affected patients, the result is referred to as a true positive (TP). However, if a positive result is obtained in an individual who does not have the disease, this is referred to as a false positive (FP). In individuals without the disease, the test results should be a true negative (TN) but occasionally a negative result is obtained in a patient who has the disease and this is referred to as a false negative (FN). The ability of a test to discriminate between diseased and healthy states is described by its *clinical* specificity and sensitivity.

The specificity of a test is the measure of the incidence of negative results in individuals free of the disease, and defined as:

$$\textit{Specificity} = TN \times 100 / TN + FP$$

A test with a specificity of 90% means that, on average, 90 out of 100 individuals without the disease would give a negative test. Conversely, 10 of these individuals would give a positive result even though they do not have any disease.

The sensitivity of a clinical test is a measure of the incidence of positive results in individuals affected by the disease. Sensitivity can be expressed as:

$$\textit{Sensitivity} = TP \times 100 / TP + FN$$

A test of 90% sensitivity means that, on average, 90% of individuals with the disease will give a positive test while the remaining 10% of individuals with the disease would give a negative result.

Ideally, a test should have 100% specificity and sensitivity, that is, it should give a negative result in all individuals without disease and a positive result in all patients affected by the disease. Such a test would discriminate completely between the diseased and healthy states. Unfortunately, such perfection rarely occurs and tests almost always have some degree of overlap (*Figure 1.16(A)*). Indeed, factors that increase the specificity of a test often decrease its sensitivity and *vice versa*.

When using a clinical test, it should also be appreciated that its ability to detect a disease is influenced by the prevalence of that disease in the population being studied. This ability is described by the predictive values of the test. The predictive value of a positive test is defined as:

$$\textit{Predictive value of positive test} = TP \times 100 / TP + FP$$

and of a negative test as:

$$\textit{Predictive value of negative test} = TN \times 100 / TN + FN$$

The range of values obtained for any test in healthy individuals usually overlaps with those obtained from patients with the disease. Hence, some patients who are genuinely ill will give test results that imply they are healthy (FN), whereas others who are not ill appear to have the disease (FP). However, if extreme values are used in the test for comparison, then the number of FN results will be reduced or eliminated. The method will, however, detect more FP results (*Figure 1.16(B)*). Thus the test will have a high specificity but low sensitivity. If the cut-off value is reduced, then the number of FP results would be reduced but at the expense of increasing the number of FN results. Thus the test has a high sensitivity but only by decreasing its specificity (*Figure 1.16(C)*).

Whether the sensitivity or the specificity of a test should be increased depends on the disease under investigation and the

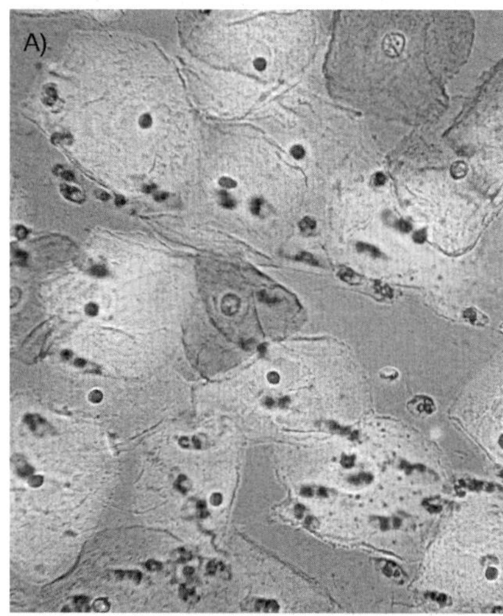

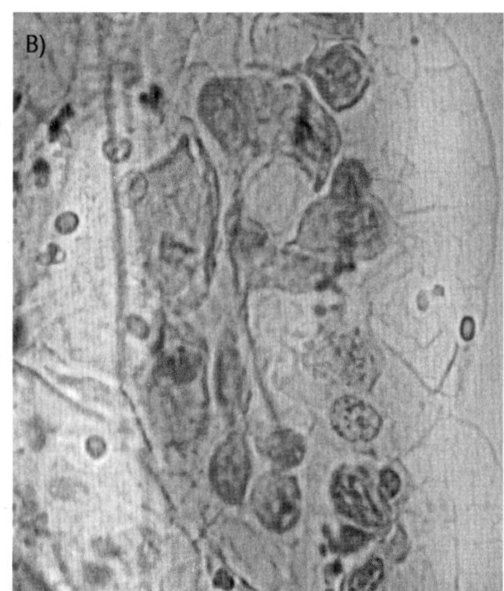

Figure 1.15 (A) Light micrograph showing normal squamous cells in a cervical smear from the superficial layer of the cervix and are from the layer of the cervical wall immediately below that of the squamous cells. The cells are healthy and have comparatively small nuclei. (B) Light micrograph showing abnormal cells from a different patient. Note the comparatively large nuclei compared with the healthy cells. Courtesy of H. Glencross, Manchester Cytology Centre, Manchester Royal Infirmary, UK.

consequences of making an incorrect diagnosis. Thus when screening for a disease with severe or fatal consequences, the test must have a high sensitivity. This ensures it will detect all results that are TP although some FP results will also be detected. Patients with positive results can then be investigated further to identify those with FP results. This test should also have a high predictive value for a negative test so that affected individuals are not missed when screening for the condition. Conversely, in some circumstances, it might be more important to have a test with high specificity. For example, if the purpose of the test is to identify and select patients for treatment with a new drug then it is necessary that the test has a high specificity. This will ensure that individuals without the disease are not selected and treated. This type of test should have a high predictive value for a positive result so that number of individuals with FP results are minimized and not subjected to any unnecessary treatment.

Physicians have to be very careful when interpreting the results of tests. It is obviously very unsatisfactory to tell a patient that he or she is suffering from cancer when this is not the case and *vice versa*. The wrong diagnosis may well lead to the wrong treatment being given.

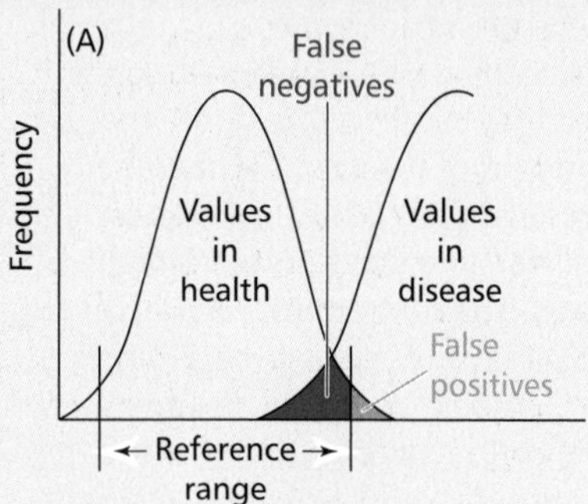

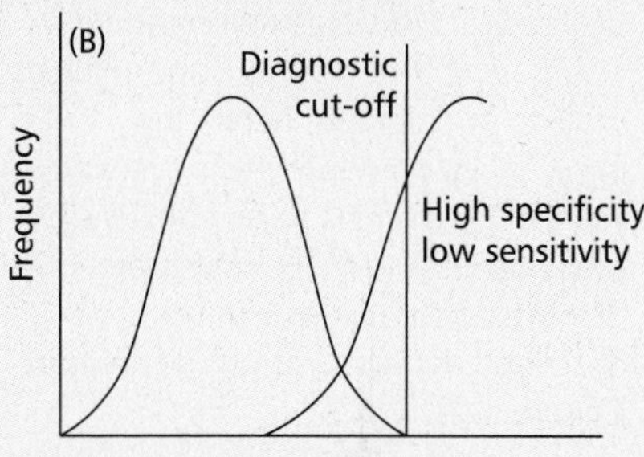

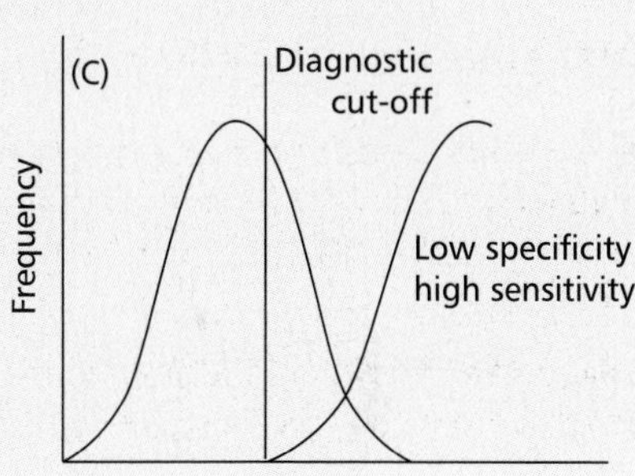

Figure 1.16 (A) The range of results for tests in a healthy and diseased population overlap and so some patients with disease will have results within the reference range (false negatives) whereas others without disease will fall outside the reference range (false positives). (B) If the diagnostic cut-off value is set too high, then this will reduce false positives but increase the number of false negatives, that is, the test will have high specificity but low sensitivity. (C) If the diagnostic cut-off value is set too low then the number of false positives increase whereas number of false negatives decrease, that is, the test has low specificity and high sensitivity.

used in certain groups of people to assess occupational exposure to harmful substances such as lead and radiation.

Laboratory tests can be used to indicate the risk of developing a disease. The risk of developing coronary artery disease increases with increasing concentration of blood cholesterol, more so if other risk factors, such as smoking, obesity or diabetes, are present.

Tests in pathology laboratories can indicate the likely outcome of a disease. Renal failure (*Chapter 8*) is a progressive disease that leads to a gradual build up of creatinine in the serum (*Figure 1.17*). Measurements of serum creatinine may therefore indicate end-stage renal disease when the patient may need dialysis to survive.

Vital information regarding the development and complications in a particular disease may be provided by laboratory tests. Urine is normally essentially free of protein; hence the presence of 30 to 200 mg dm^{-3} of serum albumin in the urine (microalbuminuria) of diabetic patients may indicate the development of nephropathy, a common secondary complication of diabetes.

Margin Note 1.1 Cervical smear screening 

Cervical smear (*Figure 1.15*) testing is a screening that looks for abnormal changes in cells of the cervix, that is the neck of the uterus (*Chapter 17*). Some of these abnormal cells can develop into cancer over 10 or more years. The commonest cervical cancer, squamous cell carcinoma, is largely preventable given that treatment of the abnormal cells will remove them in more than 90% of cases although occasionally further treatment may be needed. However, no screening is 100% reliable and some abnormalities may go undetected, hence the importance of regular tests every three years if screening is to be effective in preventing cancer. The development of a precancerous state in the cervix is described and illustrated more fully in *Chapter 17*.

(A)

(B)

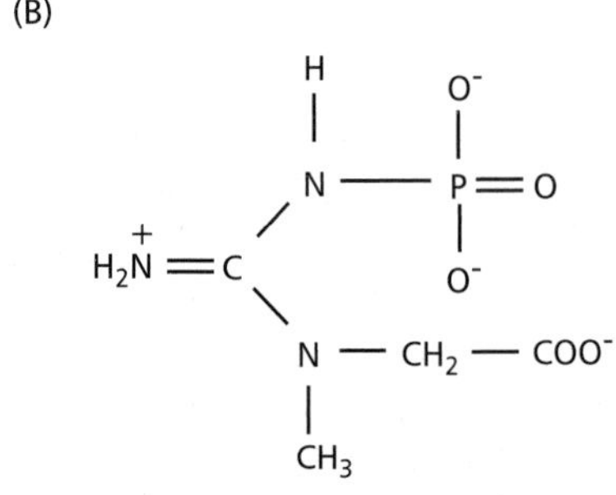

Figure 1.17 (A) Creatinine is formed from (B) creatine phosphate by the body at a relatively constant rate and excreted in urine. It is produced in amounts that are essentially proportional to muscle mass and so its concentration in blood is commonly used as an indicator of kidney function (*Chapter 8*).

1.8 HOSPITAL LABORATORY TESTS

Hospital laboratories routinely offer a wide range of clinical tests all of which must undergo a thorough evaluation for both analytical and clinical performance. The clinical demand for the test has to be established and its clinical relevance is subject to review. In addition to thorough evaluation of analytical methods, other aspects, such as the stability of samples, needs to be considered. For example, some samples must be assayed immediately while others can be stored at an appropriate temperature. All laboratory staff must be appropriately trained to ensure high analytical standards and produce valid data. This usually means that control urine and sera are run through the analyzers at regular intervals to check that the methods are working properly and reproducibly. Many of the companies that supply apparatus and reagents also supply standard sera for example, which are tested in laboratories all over the country, and the results are recorded in a nationwide database so that comparisons can be made and laboratories can check that their methods are all giving the same results.

Figure 1.18 illustrates the overall procedure routinely followed when a clinician requires a hospital test. Note that this procedure can be divided into a number of distinct steps:

- request for the test (a form; patient's details recorded);
- specimen collection, labeling, transport and storage (instructions);
- analysis (obtaining the results);
- interpretation (the results are often printed out with the range of values to help the doctor make an interpretation).

Prior to any test being requested, careful thought should be given as to whether the test is necessary and how its results will affect the management of the disease for the benefit of the patient. If this is not the case, then one has to consider the value in requesting and performing the test. Unfortunately clinicians sometimes request clinical tests that are unnecessary and will not be of benefit in treating the patient. This problem often occurs when using forms on which tests can be requested simply by ticking a box. Requesting unnecessary tests poses a number of problems for the patients, clinicians, the biomedical scientists (medical technologists in the USA) and the hospital. The test means that the patient is put to an inconvenience, as extra specimens are required. Unnecessary tests can be misleading and result in poor patient management while imposing a financial burden on the hospital. The increased workload for laboratory scientific staff may make the clinician in question rather unpopular! However, set against this is the fact that many of the machines used in the hospital for clinical analysis routinely test for a number of analytes, whether they are asked to or not, since it is easier to set up the machine in this way rather than to adjust them for individual patients.

Specimens are collected in a variety of ways (*Figure 1.19*) from the collection of blood using a simple thumbprick or, more usually a syringe, to surgery to obtain a **biopsy,** where a small piece of tissue is taken from the patient.

Blood needs to be collected with care (*Chapter 13*). If too many erythrocytes burst, this is known as **hemolysis**; the specimen will be unsuitable for the determination of some analytes: for example, the value obtained for 'serum K^+' will not be a true value since potassium is released from hemolyzed blood cells. Blood should not be collected from an arm that is receiving an infusion as a drip, since this will dilute the blood. Often, in such cases, the measured concentrations of electrolytes and glucose in the blood samples resemble those of the infusion fluid.

For some analytes, the blood must be collected into a tube containing an anticoagulant or preservative. Specimens of blood for glucose determination must be collected into a tube containing fluoride ions (F^-) since this inhibits glycolysis and prevents the utilization of serum glucose by blood cells. Occasionally blood is collected into the wrong tube and then decanted into the correct tube. This can cause a number of problems. For example, blood collected into a tube containing ethylene diamine tetraacetate (EDTA) will be unsuitable for the determination of serum Ca^{2+}, since EDTA is an anticoagulant, works by chelating and removing available Ca^{2+}.

The transport of specimens and their storage must be considered carefully since an inappropriate environment can influence the values of clinical test results. Swabs, for example, obtained during a microbiological investigation of an affected site, contain only a small volume of specimen and dry easily. They therefore need to be transported to the laboratory as quickly as possible

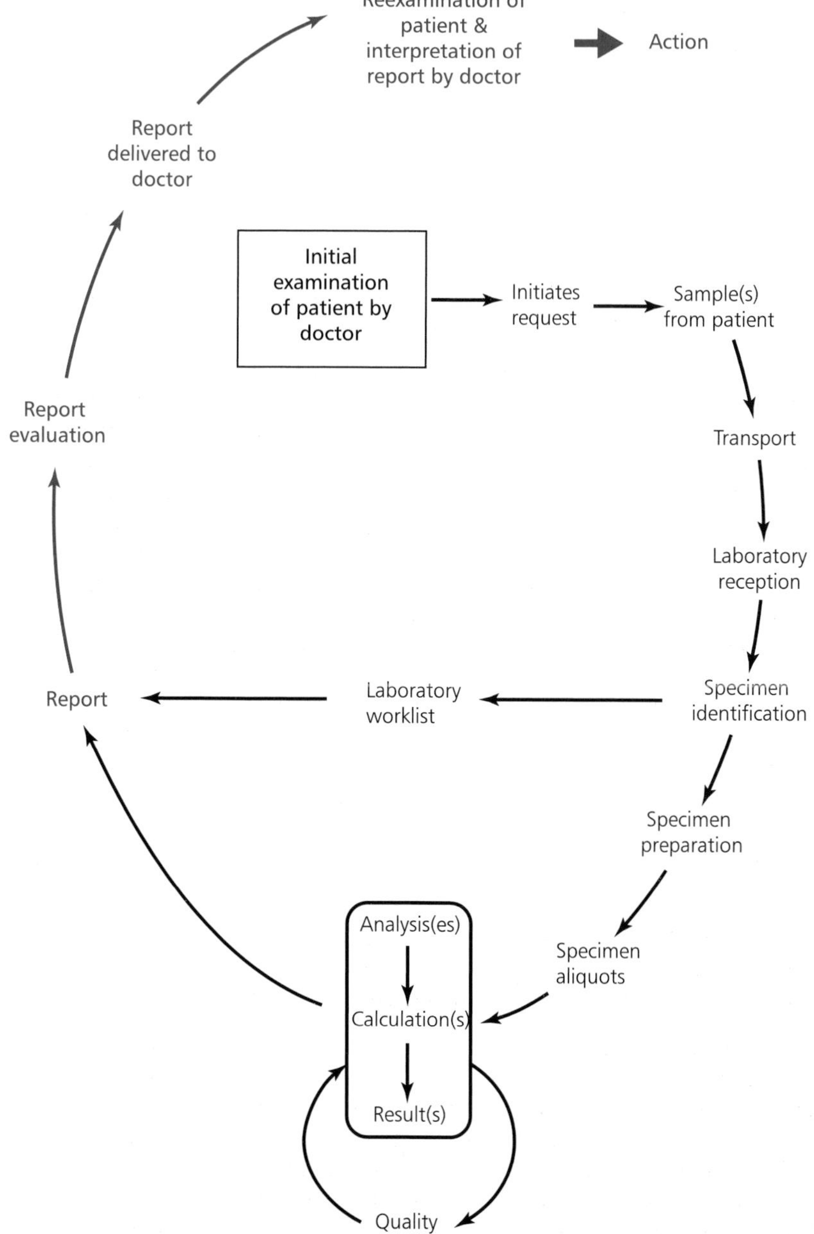

Figure 1.18 Sequence of events involved in obtaining a clinical test result.

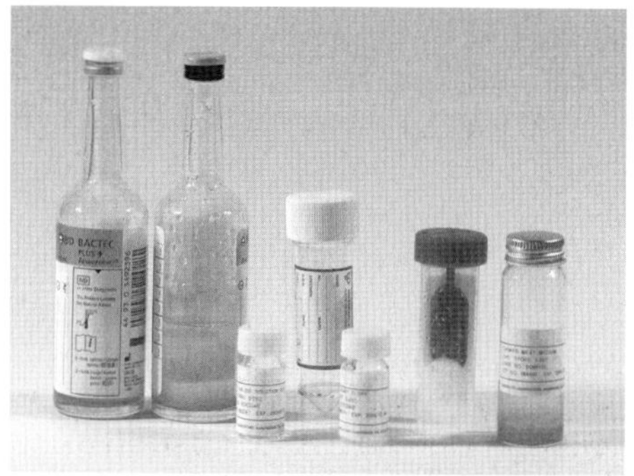

Figure 1.19 Photograph of some tubes used in the collection of clinical specimens. Courtesy of BUPA Hospital, Manchester, UK.

Margin Note 1.2 Mean, standard deviation and coefficient of variance

The mean ($\bar{x}$) is the arithmetic average value of a particular group of measurements. It can be calculated from:

$$\bar{x} = \Sigma x / n$$

where n is the number of individual measurements and Σ is the total of individual values (x).

The standard deviation (SD) is a measure of the dispersion of the data. This is defined as the square root of $\Sigma (x-\bar{x})^2 / n-1$. Hence the smaller the value of the SD relative to that of the mean, the less dispersed the data. It can be seen from *Figure 1.20* that the $\bar{x} \pm$ SD include about 68% of the samples, the $\bar{x} \pm$ 2SD will include approximately 95% of the samples.

The coefficient of variance (cv) is also a measure of dispersion. It is related to both the $\bar{x}$ and SD:

$$cv = 100.SD / \bar{x}$$

Since both the mean and standard deviation have the same units, the coefficient of variance is a percentage; the lower its value the lower the dispersion.

to preserve the pathogens. Blood specimens that have been stored overnight may show erroneously high concentrations of serum K^+, phosphate and activities of erythrocyte enzymes because these all leak from the cells during storage. To prevent this happening, the serum should be separated from blood cells immediately following collection and stored separately if it is to be analyzed the following day.

Certain samples require a timed collection, for example, the collection of urine specimens over a 24-hour period for determination of creatinine clearance values (*Chapter 8*) or the collection of stools over a three-day period for fecal fat determination to assess malabsorption (*Chapter 11*). The results obtained for such tests often lack accuracy because of the practical difficulties in obtaining accurately timed specimens from the patient.

1.9 EVALUATION OF LABORATORY TESTS

The clinical tests used in hospital laboratories undergo thorough evaluation prior to usage in a laboratory. A number of factors are assessed including accuracy, precision, reliability, practicality, safety, ease of use, duration and cost.

Accuracy refers to the ability of a method to give results that are close to the true value of the substance (analyte) being measured. The results obtained from laboratory tests may be based on subjective assessment as, for example, following the microscopic examination of a tissue section obtained after biopsy during the investigation of a possible malignancy. These types of assessments rely heavily on the experience of the practitioner in recognizing and identifying key changes. Many tests, however, provide quantitative data such as blood glucose concentrations in diabetics, or the concentration of thyroid stimulating hormone (TSH) in the serum of a patient with suspected hypothyroidism. The interpretation of many clinical tests for analytes requires referral to its reference range (*see Section 1.9*).

The **precision** of a method refers to its ability to provide the same result every time it is used. Precision is assessed by repeatedly measuring samples taken from a single specimen and from batches of samples. The variation in the results may be assessed by calculating statistical parameters, such as standard deviation (SD) or coefficient of variance (cv).

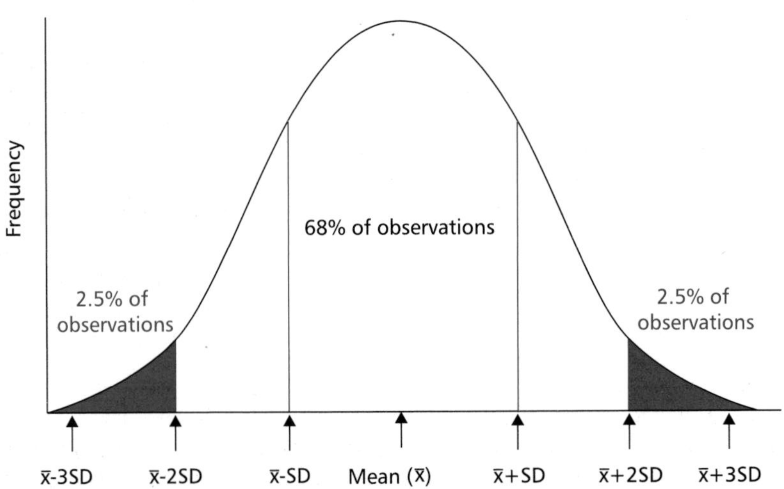

Figure 1.20 Gaussian distribution for results of a test in a healthy population. The reference range encompasses 95% of these results within –2 and +2 SD of the mean.

Often substances in biological materials are present in extremely low concentrations and it may be necessary for the clinical test to detect such low concentrations of analyte and, indeed, monitor changes in its concentration. The **analytical sensitivity** of a method is its ability to detect small amounts of the analyte under investigation. A related term is the limit of detection, which is the smallest amount of a substance that can be distinguished from the zero value. Biological material contains many components and some of these may interfere with the test being used, giving rise to unreliable results. For this reason, the **analytical specificity** of a test, that is, the ability of the method to detect only the test substance, may be determined.

In addition to analytical variation, test results are subject to biological variation. The discrimination between normal and abnormal results can be influenced by a number of biological factors. These include the sex and age of the patient, his or her diet, the time of collecting the sample, the posture adopted, whether the patient is stressed or had been exercising, the menstrual state of a female patient and whether she is pregnant, whether the person is taking drugs (legal or illegal). All these influence the results of the test. Thus, for example, plasma iron and urate values are higher in male than female patients; the activity of serum alkaline phosphatase is greater in growing children than in adults. Variations in diet may affect concentration of certain analytes, such as cholesterol (*Figure 1.21*). The values of some analytes, such as the concentration of cortisol in the plasma, show a diurnal variation. Blood should be collected from a seated patient since differences in posture at the time of blood collection can influence the concentrations of a number of analytes including the concentration of plasma proteins. Stress influences the release of a number of hormones, such as adrenaline and cortisol, while the concentrations of serum analytes, like creatine kinase and lactate, increase following exercise. The concentrations of ovarian hormones are strongly influenced by the menstrual cycle and corticosterone is known to vary by as much as 50% during different stages of the cycle. The nature and concentrations of many hormones change during pregnancy. Lastly, some drugs can influence results. For example, patients on estrogen-containing oral contraceptives often have an increased concentration of total plasma protein.

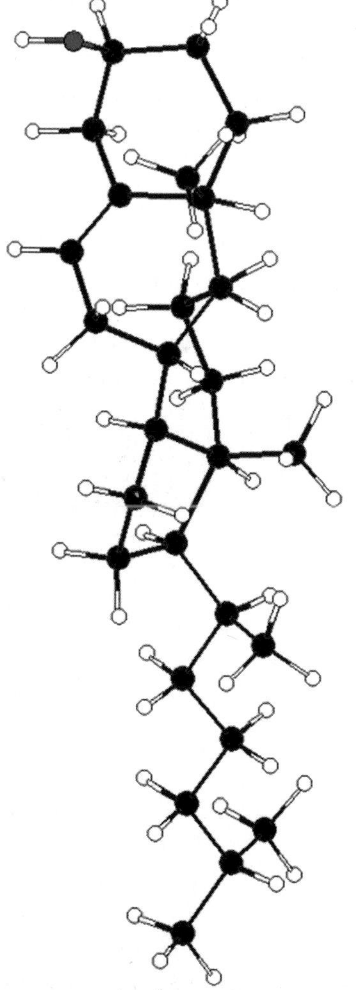

Figure 1.21 Computer generated model of cholesterol.

Analytical methods need to satisfy certain criteria to ensure they are practical and suitable for use in the laboratory. New methods are introduced into the laboratory only if they offer significant advantages over existing methods. New methods are assessed for their speed, that is, how many specimens can be processed in a given time and how long it takes to produce a result. The time a test takes may, of course, be vitally important in the care and treatment of a patient when urgent intervention is necessary.

Hospital laboratories have witnessed an increasing workload in recent years, and although they are helped by automation (*Figure 1.10*), they process very large numbers of specimens on a daily basis. Like all organizations, they have a finite budget and it is vital that the cost of tests is kept to a minimum. As well as direct costs, such as those for reagents, equipment and labor, there are indirect costs, such as the heating and lighting of laboratories. Safety is also of importance and biological, chemical, mechanical and electrical hazards associated with the method need to be assessed to ensure the safety of hospital laboratory staff.

1.10 REFERENCE RANGES

Reference ranges were mentioned earlier (*Section 1.7*). The reference range for any particular analyte can be obtained by measuring it in healthy individuals from a representative sample of the local population. Most laboratories use healthy blood donors. It may be important to know the normal range in adults, children, males, females, particular ethnic groups, or pregnant or postmenopausal women. However, since no analytical method is 100% accurate and since individuals vary, care must be taken when establishing a reference range. To determine the reference range, the values for the measured analyte are plotted against their frequency in the selected population. In most cases, the resultant graph shows a normal or Gaussian distribution with most values clustered around the center as seen in *Figure 1.20*. The mean ($\bar{x}$) and standard deviation (SD) can be determined from these data.

In general, the reference range is taken to be between two standard deviations either side of the mean. This will cover 95% of the values obtained for the selected sample (provided the curve is Gaussian). The 95% reference range was selected as this minimizes any overlap between the results for a healthy population and those for a population with the disease. However, choosing a 95% range is one of the major limitations of reference ranges, since 5% of healthy individuals will, by definition, give results that are outside these values. Thus a test result outside the reference range does not necessarily imply that the individual is ill although it does indicate that there is a greater likelihood of the presence of disease.

The profile of test results for some substances, for example serum bilirubin, does not give a Gaussian curve but shows a skewed distribution. This skewed distribution can be transformed mathematically to a Gaussian distribution and a normalized reference range calculated.

The values of a number of analytes, such as serum iron and alkaline phosphatase vary with the age or sex of the patient. In such cases, age- and sex-matched reference ranges are required. When interpreting results for a particular patient, the ideal reference value would be obtained from the same patient before their illness and this is sometimes possible. For example, the concentrations of electrolytes in serum can be measured in a patient before an operation for comparison with those obtained postoperatively. However, in most cases the results for the patient before they became ill are not available.

1.11 QUALITY OF TEST RESULTS AND CLINICAL AUDITING

The results of tests performed in pathology laboratories assist with diagnosis of disease or the monitoring of treatment. Thus they can greatly influence the management of patients, and it is essential to assure the quality of laboratory results. Erroneous results have the potential to cause considerable harm (both physical and psychological) to patients and must be avoided. All laboratories have practices and procedures to ensure erroneous results are minimized and that good quality results are provided. Errors can arise at the three different stages of analysis, that is, preanalytical, analytical and postanalytical. Preanalytical errors occur before the sample has been analyzed. Analytical errors arise during the laboratory testing procedure. Postanalytical errors arise after the specimen has been analyzed.

In general, preanalytical mistakes result from inappropriate methods of collection or incorrect labeling, handling, transport or storage of the specimen. Experimental mistakes that can give rise to analytical errors are detected by introducing systems for each clinical test to warn when errors occur. This is normally achieved by analyzing a **control** sample within each batch of tests. A control sample is one that is identical in composition to the test samples except that it contains a known concentration of the test analyte. All samples, including the control sample, must be treated identically. For example, if the concentration of glucose in serum is being determined, then the control should be serum, not water, containing a known concentration of glucose.

Margin Note 1.3 False negatives and false positives 

It is clear that false negative or false positive results from a test can have serious consequences (and no test is 100% reliable). This applies even more to tests that patients can carry out at home. One can imagine the anguish or joy that a false positive in a pregnancy test might cause. The recent development of a quick home test for HIV based on a saliva test is perhaps a more severe example. This HIV test, approved in the US, has raised fears that people who find that they are infected (or obtain a positive result) may kill themselves.

The control sample is usually an aliquot from a larger sample for which the mean and standard deviation have already been determined. The results for control samples are usually recorded graphically so that changes in the quality of the method are detected as soon as they arise. A common chart used for quality control purposes is the Levey-Jennings chart (*Figure 1.22*) in which the control limits are set at the mean ± 2 SD and ± 3 SD. If the control values fall outside the ± 2 SD limit, that is, there is drift away from the accepted limits, there is only a 5% probability that the result lies in a normal distribution around the mean and is still valid. Any results that lie outside the 3 SD warning limit suggests that a problem is occurring with the method. Problems could include unstable reagents in the analyzer, problems with temperature control or contamination, all of which require investigation. Occasionally there are gross (and usually very obvious) inaccuracies in the value of a test result such that it bears no resemblance to values seen in health or disease. These are referred to as 'blunders'. Blunders usually arise because of transcriptional errors in reporting the result. To reduce the number of blunders, results should be checked thoroughly by senior staff before being sent to clinicians.

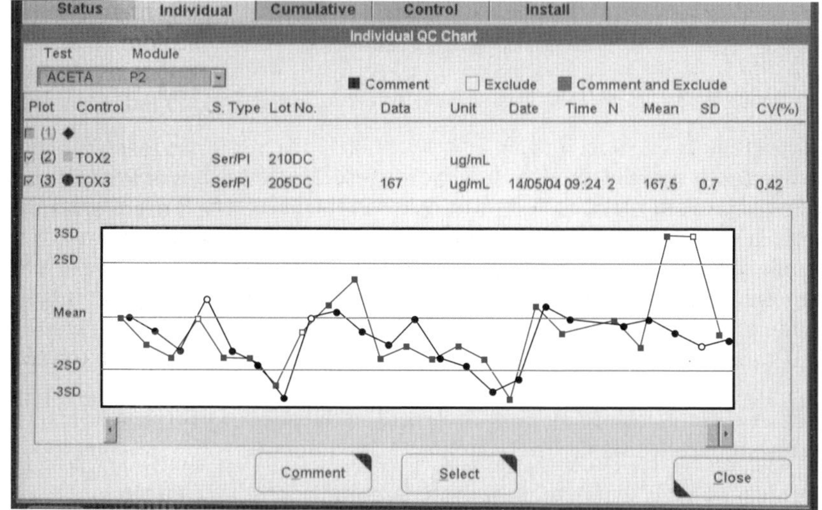

Figure 1.22 Levey-Jennings Chart used to assess quality control of quantitative results of tests. The values of controls are plotted as means ± 2 and 3 SD. When results are plotted out like this, any trend affecting quality of test can be detected, for example the values for one control (red) lie outside ± 3 SD. Courtesy of Department of Clinical Biochemistry, Manchester Royal Infirmary, UK.

Most laboratories have their own quality control samples and these are used for internal quality assurance purposes. Many countries now participate in external quality assurance, whereby quality control samples are sent to participating laboratories from a central source to assess the analytical performance of their methods for particular analytes. Furthermore, to ensure quality of service provision, many laboratories follow a set of procedures required for accreditation by external agencies. These procedures ensure good laboratory practice (GLP) and cover all aspects of the laboratory that are involved in the production of test results. These procedures ensure that all laboratory staff are adequately trained and have clearly defined responsibilities. The equipment used should be of adequate standard with a logbook showing a full record of maintenance and faults. All methods used in the laboratory are standardized, fully documented and appropriate for the analysis. Full details of each method are provided as a standard operating procedure (SOP) that includes details of specimen handling, the analytical method, equipment used and quality control procedures.

To improve the quality of the services they provide, many laboratories participate in some form of audit. **Clinical audit** (*Figure 1.23*) is a process whereby practices and procedures involved in patient care are monitored and, if necessary, revised to provide a more efficient and cost-effective service that should ultimately benefit the patient. Audit is part of the process of ensuring

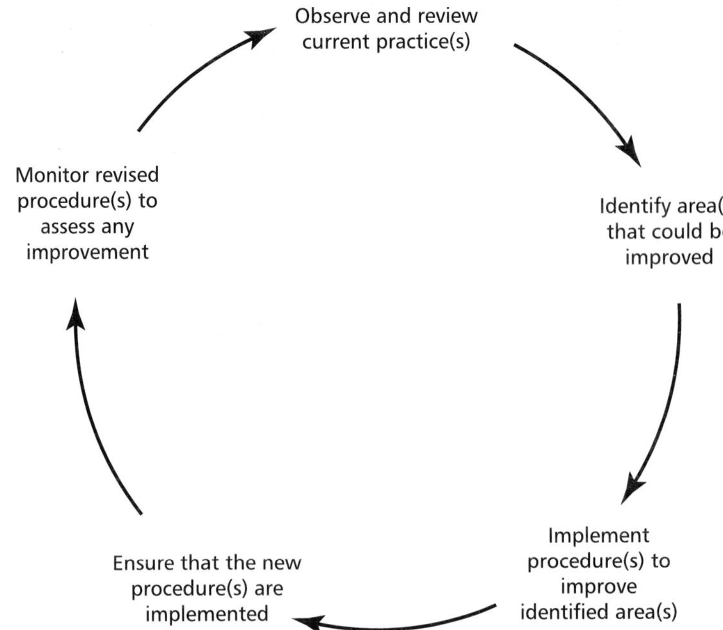

Figure 1.23 Outline of the five stages of a clinical audit.

quality and is usually divided into five stages. The first is the observation and review of current practice and procedures in the laboratory. The second stage involves the identification of areas of concern that could be improved and questions are asked as to whether the current service can be provided more economically. Third, a series of changes are devised to rectify and improve the identified area(s) of concern. Fourth, these changes are implemented and steps taken to ensure compliance and, finally, at the fifth stage, the changes are monitored and compared with previous procedures in order to assess whether there is indeed any improvement in the service provided or whether the revised procedures are actually more cost effective.

A clinical audit is usually followed by a re-audit after an appropriate period of time. The audit may include several processes such as the initial stages of test requesting, specimen collection and transport. The audit may wish to investigate whether appropriate advice is available to clinicians requesting tests, whether the test request forms are easy to use or whether appropriate containers are provided for specimen collection. Other types of audit processes may relate to the analytical service provided by the laboratory, such as whether the repertoire of tests offered is appropriate to the needs of the clinical service. A clinical audit may wish to investigate whether provision of the laboratory service out of hours is efficient and cost-effective and whether test results are being returned to the clinicians at the right place and within an appropriate time.

CASE STUDY 1.1

The serum K^+ concentration of Emma, a 22-year-old patient, about to undergo surgery was determined. The laboratory sent a value of 35 mmol dm^{-3} to the ward. The reference range for serum K^+ is 3.5 to 5.0 mmol dm^{-3}.

Question

What is the most likely explanation of this result? Discuss its implications.

CASE STUDY 1.2

A new test for the detection of prostate cancer has been developed and is undergoing clinical evaluation to determine how effective it is in the diagnosis of this condition. The study included 100 healthy men and 100 men with known prostate cancer. All were screened using the new test, which involved the measurement of a serum tumor marker. The patients were rated either positive or negative depending on whether the determined concentration of the tumor marker was above or below a certain cut-off value. The results obtained were as follows:

	Positive	Negative
Patients with prostate cancer	95	5
Healthy individuals	5	95

Question

What are the sensitivity and specificity of the test? Discuss the values.

1.12 SUMMARY

Health is a 'state of physical, mental and social well-being, not merely the absence of disease'. Disease refers to any abnormality or failure of the body to function properly. For medical treatment to commence the disease needs to be diagnosed and its etiology established. This also helps to define the prognosis, that is, the likely course of the disease and its outcome. Diseases have a number of possible causes. They may be exogenous, such as infections or trauma (accidents) or they may be endogenous such as diabetes or cancer. Some diseases may be caused by a range of factors and are called multifactorial and, for some, the cause may be unknown and these are referred to as idiopathic.

The pathogenesis describes how the etiological agent produces the clinical signs and symptoms and the pathological changes characteristic of that disease. This enables the physician to make a diagnosis and prescribe treatment. The prognosis should also emerge at this stage.

Diseases may be classified into a number of types: genetic, infectious, endocrine, traumatic, degenerative, immunological, nutritional, homeostatic, neoplastic (cancer), toxic, psychogenic and iatrogenic (caused by the treatment itself). Epidemiology is the study of how diseases spread in populations. This is of importance in the control of diseases.

Various types of laboratories specialize in investigating the pathology of diseases. A laboratory may measure the concentrations of analytes in blood and urine, identify infectious agents such as bacteria and viruses, characterize genetic diseases by looking at the patient's chromosomes, or identify problems with the blood, for example those involving defective hemoglobin or clotting factors. The laboratories report their findings to the clinician in charge of the patient to help in the process of diagnosis and in making decisions on treatment and how well any treatment is working. Pathology laboratories must work to high standards of accuracy, otherwise wrong treatments may be given and patients may be misinformed about their disease and its prognosis. Therefore pathology laboratories take steps to standardize and check their procedures on a daily basis and investigate new tests exhaustively before they are introduced.

QUESTIONS

1. Which of the following terms best describes a test used to detect disease before it presents clinically?
   a) diagnostic test;
   b) sensitive test;
   c) screening test;
   d) prognostic test;
   e) specific test.

2. The term used to describe the cause of a disease is:
   a) manifestation;
   b) etiology;
   c) pathogenesis;
   d) mortality;
   e) epidemiology.

3. A Na^+ solution of 10 mmol dm^{-3} was measured by two methods X and Y. The value obtained with method X was 11.8 mmol dm^{-3} and for method Y it was 9.7 mmol dm^{-3}. Which one of the following statements is correct?
   a) Method X is more accurate than method Y.
   b) Method X is more precise than method Y.
   c) Method Y is more accurate than method X.
   d) Method Y is more precise than method X.
   e) Method X is as accurate as method Y.

4. The concentration of Mg^{2+} in the serum of an individual is 0.5 mmol dm^{-3}. The reference range is 0.7 to 1.0 mmol dm^{-3}. Is it possible to be 100% certain that this individual has a deficiency of Mg^{2+}? Explain your answer.

FURTHER READING

Henny, J and Hyltoft Petersen, P (2004) Reference values: from philosophy to a tool for laboratory medicine. *Clin. Chem. Lab. Med.* **42:** 686–691.

Hooper, J, McCreanor, G, Marshall, W and Myers, P (1996) *Primary Care and Laboratory Medicine.* ACB Venture Publications, Cambridge.

Irjala, KM and Gronroos, PE (1998) Preanalytical and analytical factors affecting laboratory results. *Ann. Med.* **30:** 267–272.

Khan, KS, Dinnes, J and Klijnen, J (2001) Systematic reviews to evaluate diagnostic tests. *Eur. J. Obstet. Gynecol. Reprod. Biol.* **95:** 6–11.

Libeer, J-C (2001) Role of external quality assurance schemes in assessing and improving quality in medical laboratories. *Clin. Chim. Acta* **309:** 173–177.

Moyer, VA and Kennedy, KA (2003) Understanding and using diagnostic tests. *Clin. Perinatol.* **30:** 189–204.

Stewart, A (2002) *Basic Statistics and Epidemiology*. Radcliffe Medical Press, UK.

Stolley, PD and Lasky, T (1998) *Investigating Disease Patterns*. Scientific American Library, NY.

Thagard, P (1996) The concept of disease: structure and change. *Commun. Cogn.* **29:** 445–478.

Wachel, M, Paulson, R and Plese, C (1996) Creation and verification of reference intervals. *Lab. Med.* **26:** 593–597.

Weinstein, S, Obuchowski, NA and Lieber, ML (2005) Clinical evaluation of diagnostic tests. *Am. J. Roentgenol.* **185:** 14–19.

chapter 2:

PATHOGENS AND VIRULENCE

OBJECTIVES

After studying this chapter you should be able to:

- define the terms pathogen, parasite and infection;
- list the types of parasites and pathogens;
- outline the types and effects of microbial virulence factors.

2.1 INTRODUCTION

The body is exposed to many pathogenic microorganisms and multicellular parasites. Most microorganisms found associated with the body are harmless and live as commensals or symbionts but others cause disease and are known as **pathogens**. **Infection**, from the Latin 'inficere' (to put in), is the successful persistence and/or multiplication of the pathogen on or within the host. In this respect, **pathogenesis** may be defined as the molecular and biochemical mechanisms that allow pathogens to cause diseases. Mummies preserved in ancient Egypt and elsewhere display evidence that infectious diseases have always been a threat. The emergence of new pathogens and the development of resistance to current treatments for existing pathogens means that infectious diseases will probably always be with us.

Many antibiotic drugs (*Chapter 3*) are available to treat infectious diseases and, in many cases, will effect a cure. It is, however, possible for them to exacerbate the problem since the drug can remove commensals allowing antibiotic resistant pathogens to flourish. Also, an adequate immune response is often necessary, since drugs alone may fail to eliminate the infection. Thus all pathogens must overcome the defense systems present in their hosts (*Chapter 4*).

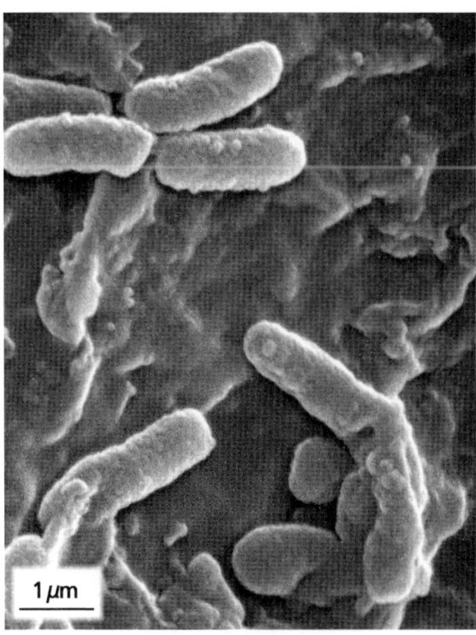

Figure 2.1 Electron micrograph of *Pseudomonas aeruginosa*. Courtesy of Dr J. Carr, Public Health Image Library, Centers for Disease Control and Prevention, USA.

Some pathogens regularly cause diseases while others do not. For example, *Pseudomonas aeruginosa* (*Figure 2.1*) can cause overwhelming disease in patients whose defense systems are compromized but not in those with intact defenses. It is likely that any microorganism with the ability to live in or on humans will sometimes become an **opportunistic pathogen** especially if the balance between the usual microorganisms present and the immune

Figure 2.2 Electron micrograph of dividing *Escherichia coli* cells. Courtesy of Dr A. Curry, Manchester Royal Infirmary, UK.

system is disturbed. Thus bacteria which are normally harmless, but which are opportunistic pathogens, can cause infections under certain conditions. For example, wounds can become badly infected with bacteria that normally exist on the skin, and bacteria that normally live in the gut can cause serious infections if peritonitis (*Chapter 11*) allows the gut contents to enter the peritoneum. In general, these infections are not transferable to other healthy humans.

The success of pathogenic microorganisms depends on their ability to colonize host tissues and to counter the host's defense mechanisms. Virulence is measured by the **infective dose** and the severity of the disease caused. For example, as few as 10 to 100 *Shigella dysenteriae* cells can cause shigellosis but more than 10 000 cells are needed of the less virulent salmonella or cholera bacteria. True pathogens are equipped with a range of **virulence factors**. The strains of some bacterial species, such as pathogenic forms of *Escherichia coli* (*Figure 2.2*) can produce different virulence factors that cause, for example, diarrhea, urinary tract infections or sepsis. Other strains, however, do not produce virulence factors or do so to a lesser extent and are therefore not pathogenic, except when they infect an immunocompromized host.

A pathogen must be transmitted from a source to the patient. Direct contact between hosts is the most obvious form of transmission but coughs and sneezes (aerosols), food, water and arthropod vectors are all used by various pathogens. The long-term survival of pathogenic microorganisms also depends on maintaining their infectivity during transmission from host to host. Diseases that are transmitted from animals to humans are called **zoonoses**, while humans who harbor a pathogen but are asymptomatic are called **carriers.**

> **Margin Note 2.1**
>
> **Communicable pathogens** are transmitted from host to host and, in general, cannot live for extended periods outside a host. **Noncommunicable pathogens** normally live in the abiotic environment but may infect a host if they are transmitted to him or her. **Contagious pathogens** are those that are easily transmitted. The sources of pathogens can be abiotic: soil, water, or animals or other humans.

2.2 TYPES OF PATHOGENS

Infectious diseases are caused by pathogens that have the ability to infect humans. They may be subcellular, such as prions and viruses, single-celled prokaryotic bacteria, single-celled eukaryotic protozoa and yeasts, or multicellular organisms such as, fungi, certain worms, such as nematodes and flukes (generally referred to as **helminths**) and arthropods, such as mites. The term **parasite**, an organism that lives at the expense of another, is often applied to viruses, protozoa and helminths, although the terms pathogen and parasite are virtually interchangeable.

PRIONS

Spongiform encephalopathies or prion diseases (*Chapter 15*) are all fatal diseases for which there is no cure. They include Creutzfeldt-Jacob disease (CJD), Gerstmann-Sträussler-Scheinker syndrome (GSS) and fatal familial insomnia (FFI). These diseases generally develop slowly over 10 to 20 years in older individuals. Prion diseases occur sporadically, or they can be familial, that is genetic, or they can be acquired, that is, infectious.

Prions are degenerate host proteins. The normal form of the protein adopts a largely α helical conformation that is harmless, but can refold to a β sheet-rich form that is a pathological conformation. Such misfolded proteins aggregate to form deposits in the brain leading to a lethal spongiform condition where holes develop in the brain. A misfolded prion protein, in some poorly understood way, induces a conformational change in a native α prion protein to produce a β type conformation. This new misfolded protein, in turn, can catalyze conformational changes in other native proteins, eventually forming a chain reaction and produces deposits of prions in the brain. The sporadic form of the disease occurs in individuals with mutations in the prion gene that predisposes them to produce the misfolded form of the protein. Since the diseases usually only occur after reproductive life is over, they can run in families giving the familial form. If misfolded prions enter the body in the diet, they resist digestion. They may also enter by iatrogenic means, through surgery or blood transfusions for instance, and can initiate the infectious form of the disease.

VIRUSES

Viruses are obligate intracellular parasites. They are complexes of proteins, which form a **capsid,** and nucleic acid (RNA or DNA), comprising their genome, that together form a viral particle or virion (*Figure 2.3*). Some viral particles also have a lipid membrane or **envelope** acquired when the viral particle leaves its host cell (*Figures 1.4* and *2.7*). Viruses must enter a target or host cell to replicate. They bind to the target cell by attaching to specific proteins or carbohydrates on the cell's surface (*Figure 2.4*). For example, the human immunodeficiency virus (HIV) attaches to a protein called CD4 found on the surfaces of certain T lymphocytes and macrophages of the immune system (*Chapter 4*). Other examples of viruses and the cellular receptors to which they bind are shown in *Table 2.1*.

Once infected, the host cell then manufactures new viral particles. In some cases, replication may include an inactive latent state. For example, the virus *Varicella zoster* (*Figure 2.5*), which causes chicken pox, enters nerve cells and, after the initial infection, remains dormant. If, however, the host immune system becomes weakened, *Varicella* can reactivate and cause painful attacks of shingles in the area served by that nerve.

In some cases the viral nucleic acid can be integrated into that of the host and eventually lead to cell transformation and the formation of cancers (*Chapter 17*). Thus, for example, hepatitis B virus can contribute to primary hepatocellular carcinoma, while certain strains of the human papillomavirus that cause genital warts may contribute to the development of cervical carcinoma.

Virus	Cell membrane protein used as virus receptor
HIV	CD4 (*Chapter 4*)
Human rhinovirus 91	ICAM (Intracellular adhesion molecule) I
Sindbis virus	High density laminin receptor
Coxsackie A	ICAM I
Human coronavirus 229E	aminopeptidase N
Hepatitis B virus	HBV Binding factor (a metalloprotease)

Table 2.1 Some cell membrane proteins used for viral attachment

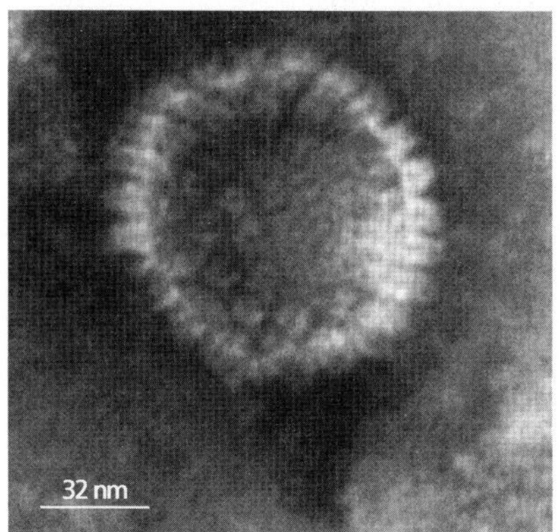

Figure 2.5 Electron micrograph of *Varicella zoster*. Courtesy of H. Cotterill, Manchester Royal Infirmary, UK.

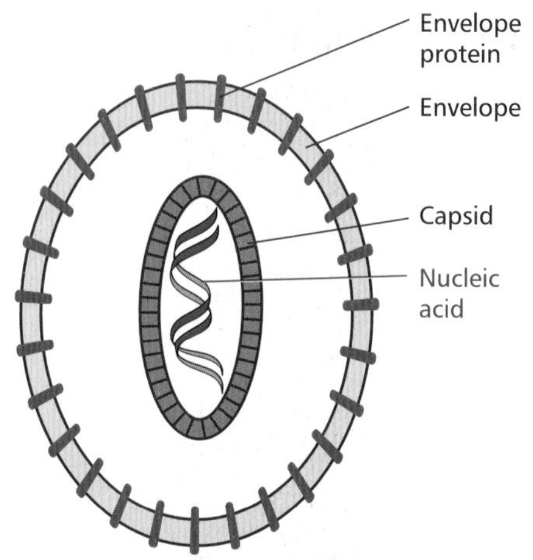

Figure 2.3 Schematic showing the structure of a typical viral particle.

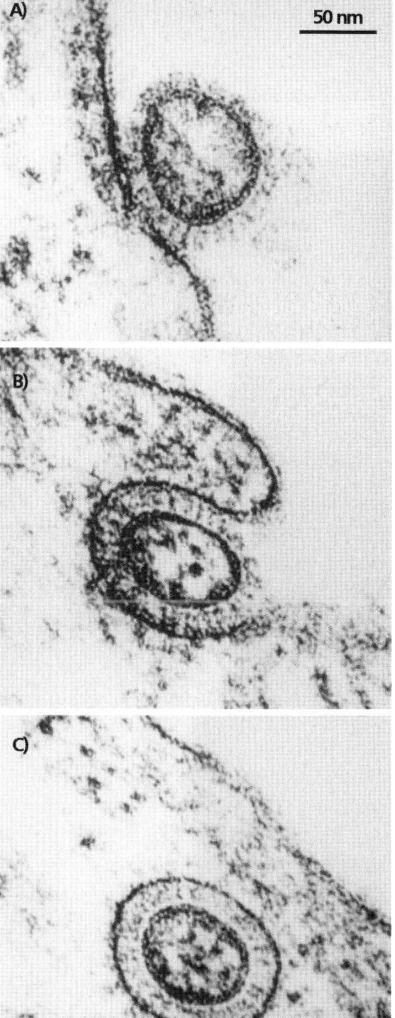

Figure 2.4 Electron micrographs showing (A) the binding, (B) enveloping and (C) internalization of an influenza virus. Courtesy of Dr J.J. Shekel, National Institute for Medical Research, London.

BOX 2.1 Infectious diseases in history

The earliest human skeletons show evidence of a variety of infectious diseases. Evidence from mummies preserved in ancient Egypt 4000 years ago shows that the population suffered from diseases such as tuberculosis, trachoma and dental caries. The skeletons of monks buried in medieval Northern monasteries show the characteristic signs of syphilis!

Perhaps the best known example of an infectious disease is **plague,** a term originally applied to any widespread disease causing great mortality. The name is now confined to bubonic plague, an infectious disease of animals and humans caused by the bacterium *Yersinia pestis* (*Figure 2.6*). This bacterium mainly affects rodents but their fleas can transmit the bubonic form of the disease to humans when they bite to feed on blood. Transmission occurs readily in crowded urban areas with poor hygiene. Infection usually results from a bite from a rodent flea carrying the plague bacterium or by handling an infected animal. Once humans are infected, they infect others very rapidly. Plague causes fever and painful swellings of the lymph glands called buboes, which is how the disease derives its name. It also causes spots on the skin that are initially red but then turn black. The coughing and sneezing of infected individuals spreads pneumonic plague; a much more lethal form of the disease. Bubonic plague is fatal in about 30% of cases but is readily treatable with antibiotics. In contrast, pneumonic plague is often fatal even with antibiotic therapy.

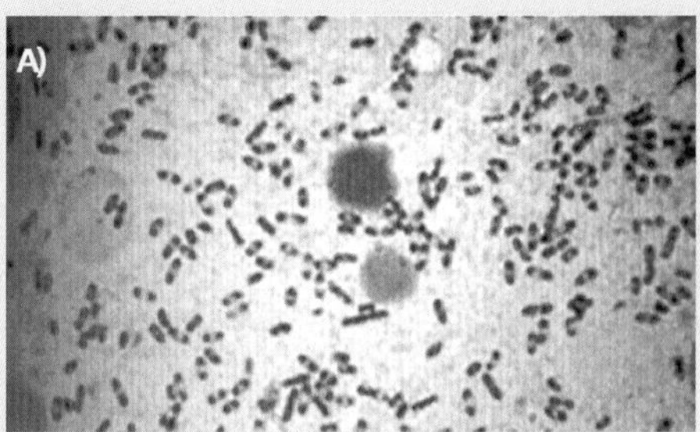

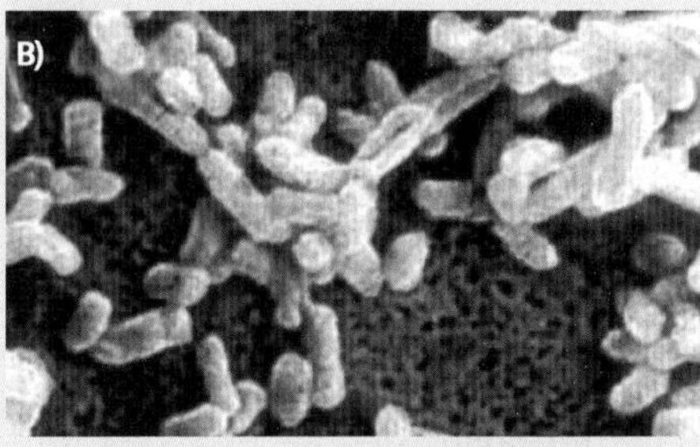

Figure 2.6 (A) Light and (B) electron micrographs of *Yersinia pestis*. The cells are 1 to 3 μm in length. Courtesy of Dr M.K. Khalid, Editor, *Middle East Journal of Emergency Medicine* and Dr M. Schneider, *Kryptozoologie* respectively.

Bubonic plague has had a profound impact on humans throughout recorded history. In AD 541, the first great plague **pandemic**, that is a disease affecting patients over a wide geographical area, spread throughout the world from its origins in Egypt. It is thought to have killed between 50% and 60% of the population over four years, being spread by the flea-infested rats that inhabited human homes and workplaces and by human sufferers of the disease. In the early 1330s, a second pandemic originated in China. At the time China was one of the world's busiest trading nations and the disease rapidly spread to western Asia, the Middle East and Europe. It entered Europe in October 1347 when several Italian merchant ships, returning from a trip to the Black Sea (a key link in trade with China) arrived in Sicily with many people on board already dying of plague. Within days the disease spread to the city and the surrounding countryside. In August the following year, the plague had spread as far north as England, where people called it the **Black Death** because of the black spots on the skins of patients. Medieval medicine was ineffective in treating the disease. However, in winter the disease declined simply because the fleas that carried the bacteria were dormant. Each spring, the plague returned. After five years more than 13 million people in China and 20–30 million in Europe, one third of the European population, were dead.

In addition to the terrible fatalities, the Black Death produced enormous social changes from which medieval society never recovered. The resulting serious labor shortages all over Europe led to a demand for higher wages by workers and peasant revolts broke out in Belgium, England, France and Italy. The end of the 1300s saw the eventual collapse of the prevailing social system of tied serfs.

Even when the Black Death pandemic ended, smaller outbreaks of bubonic plague continued for centuries. The Great Plague of London in 1664 to 1666 is estimated to have killed 70 000 people out of a total population of 460 000. In 1894, outbreaks occurred in Canton and Hong Kong in which 80 000 to 100 000 died. Within 20 years the plague had spread and killed 10 million people worldwide.

Advances in living conditions, public health and antibiotic therapy make future pandemics of *Yersinia pestis* unlikely. However, if an infected person is not treated promptly, the disease is likely to cause severe illness or death. Outbreaks of plague still occur in some rural communities or in some cities that are still associated with infected rats and their fleas. The World Health Organization reports 1000 to 3000 cases of plague annually. It is unlikely that the disease will ever be completely eradicated because wild animals hold a huge reservoir of the bacterium.

In more recent years, pandemics of a number of other diseases have occurred. The most recent is associated with the human immunodeficiency virus (HIV) virus (*Figures 1.4* and *2.7* and *Chapter 4*), although the most numerous are associated with influenza (*Figure 2.8*). The influenza pandemic of 1918–1919, named Spanish Flu though the name has little relevance, killed up to 40 million people, many more than the nine million fatalities of World War 1 (WW1). It was one of the most devastating epidemics in history and is comparable with the Black Death in sheer numbers of people killed. In two years, a fifth of the world's population was infected. The virus had a mortality rate of 2.5% compared with the previous influenza epidemics of less than 0.1%. Further, most deaths were of people aged 20 to 40 years, which is unusual given that influenza normally kills the elderly and the very young. Indeed, the death rate for 15- to 34-year-olds from the influenza and associated pneumonia were 20 times higher in 1918 than in previous years.

The origin of this influenza variant is uncertain but is thought to have originated in China where mutations led to an influenza virus with novel surface proteins making it relatively unrecognizable to immune defense. The influenza pandemic swiftly followed trade routes and shipping lines. The mass movement of people as a result of WW1 also enabled the virus to spread rapidly. Outbreaks swept through North America, Europe, Asia, Africa, Brazil and the South Pacific. The shortage of medical facilities created by WW1 accentuated problems. As with the Black Death, social problems occurred with shortages of coffins, morticians and gravediggers so that bodies had to be stored in piles until they could be buried.

The genes of all influenza viruses are maintained in wild aquatic birds. Periodically these viruses are transmitted to other species. Thus the potential for outbreaks of influenza is still present. Influenza viruses mutate constantly but usually in such a way that one year's vaccine offers some protection against the next year's strain. However, every 10 to 20 years, a major mutation produces a particularly new virulent strain against which current vaccines offer little protection. Such viruses are associated with epidemics and pandemics like that of 1918 to 1919. Indeed, in 1997 epidemiologists and public health officials recognized a new variety of influenza virus, known as subtype H5N1 from its surface proteins. Given its lethal effects on poultry it was called Chicken Ebola, but is now more commonly called 'bird flu'. When it infected the human population of Hong Kong it killed six of the first 18 confirmed cases. Fortunately the H5N1 subtype cannot be transmitted through the air from one human host to another. In Hong Kong, bird-to-human contact is relatively easy given the often close proximity of the two and is believed to have been the route of transmission and Hong Kong officials ordered the slaughter of Hong Kong's entire poultry population in 1997. However, by 2005/6, cases of bird flu were being reported in diverse parts of the world.

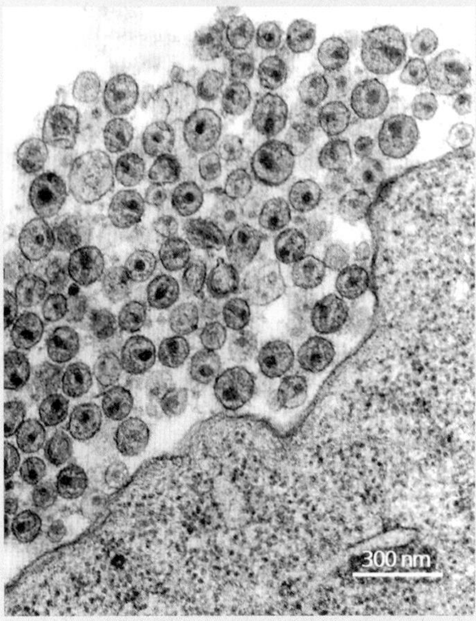

Figure 2.7 HIV viral particles being released from the surface of a human cell. Courtesy of H. Cotterill, Manchester Royal Infirmary, UK.

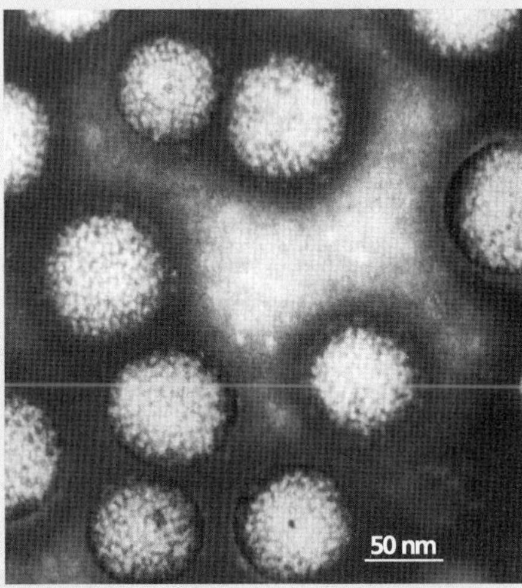

Figure 2.8 Electron micrograph of influenza viral particles. Courtesy of H. Cotterill, Manchester Royal Infirmary, UK.

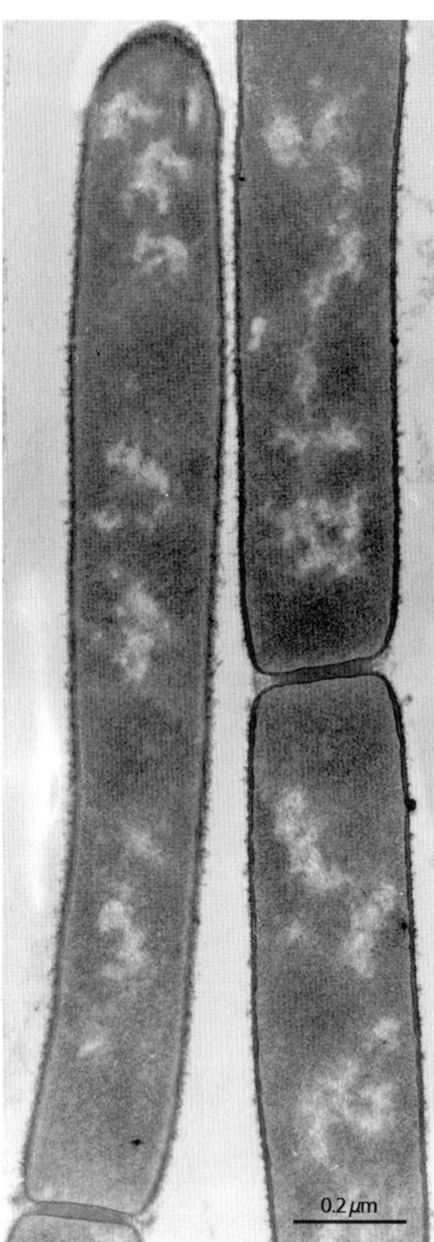

Figure 2.9 Electron micrograph of bacterial cells. Note the lighter staining nuclear material free in the cytoplasm and not contained in a nucleus. Courtesy of Dr I.D.J. Burdett, National Institute for Medical Research, London.

BACTERIA

Bacteria are cellular microorganisms that do not have a discrete nucleus and are described as prokaryotic (*Figure 2.9*). They are responsible for a large proportion of all infectious diseases and although prokaryotes share considerable biochemistry with eukaryotes, their metabolism differs in numerous ways, making them susceptible to chemical agents that do not significantly damage human cells. This is the basis of antibiotic therapy (*Chapter 3*). While most species of pathogenic bacteria do not enter host cells and are described as extracellular pathogens, some significant groups of bacteria are intracellular parasites. Examples of the latter include Mycobacteria and Listeria.

PROTOZOA, FUNGI AND HELMINTHS

Protozoa, fungi and helminths are eukaryotic organisms. They therefore share many biochemical features with humans. Also, they have many seemingly sophisticated ways of countering the host immune system. This often makes them difficult to eradicate when they cause disease. Fungi, protozoa and helminth parasites are responsible for a large proportion of infective diseases, particularly in the developing world.

Protozoa consist of single eukaryotic cells although in some species these may group together as loose aggregates. Malaria, caused by species of *Plasmodium*, is easily the most clinically important protozoal infection worldwide and is responsible for many millions of deaths annually (*Box 2.2*). Protozoa are, however, also the causative agents for a number of other infections including amebiasis, giardiasis, leishmaniasis, toxoplasmosis, trichomoniasis and trypanosomiasis.

Fungi are a heterogeneous group of organisms, ranging from unicellular yeasts to elongated chains of cells, known as hyphae (*Figure 2.10*). Fungal infections, or mycoses, are relatively common and may be superficial or systemic. Fungi cause disease in humans by invading tissues, by being toxic or by initiating an allergic response. Clinically important fungal infections include those of *Epidermophyton*, *Microsporum* and *Trichophyton* species that cause ringworm, athlete's foot and nail infections (*Figure 2.11*). Infections with yeasts and other fungi can result in, for example, candidiasis,

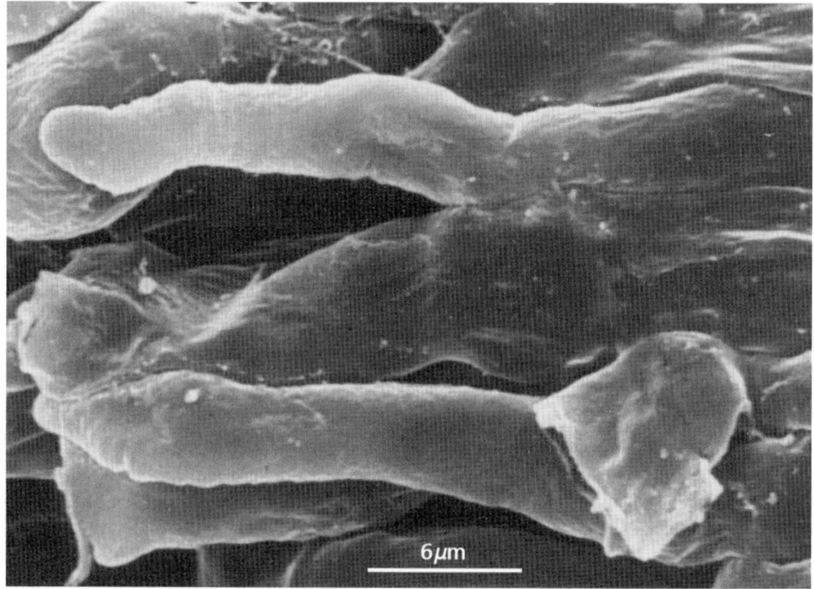

Figure 2.10 Electron micrograph of fungal hyphae.

aspergillosis and cryptococcosis. Systemic fungal infections are, in general, much more common in immunocompromized individuals, for example those with AIDS (*Chapter 3*), those undergoing cancer chemotherapy (*Chapter 17*) or those being treated with immunosuppressive drugs to prevent rejection of a transplant (*Chapter 6*).

The clinically important helminths (worms) can be divided into three main groups: nematodes (roundworms), cestodes (flatworms) and trematodes (flukes). Helminths often have complex life cycles that may involve other hosts in addition to humans. In humans, they may infect the alimentary canal, blood vessels, lymphatics or other tissues such as skeletal muscle. Helminths are significant parasites in tropical climates. Examples of disease-causing helminths are the pork tapeworm *Taenia solium* (*Figure 2.12*) that can live in the gut and *Schistosoma* the cause of bilharzia.

ARTHROPODS AND VERTEBRATES

Arthropods may be directly parasitic, but many are also relevant to infectious diseases as vectors of pathogens. The mite *Sarcoptes scabiei* lives in the outer layers of the skin and can cause scabies while fleas (*Pulex*) and head and pubic lice, *Phthirus capitans* and *pubis* respectively, are blood sucking parasites (*Figure 2.13*). Houseflies and cockroaches are noted carriers of food poisoning organisms. More specific vectors include ticks that transmit *Borrelia burgdorferi*, the cause of Lyme disease. *Yersinia pestis*, the bubonic plague organism is spread by the fleas on black rats (*Box 2.1*). Malarial parasites are spread by female *Anopheles* mosquitos (*Box 2.2*) and tsetse flies are vectors for *Trypanosoma brucei* which causes sleeping sickness. Several disease-causing organisms use mammals as vectors with perhaps the best known being the rabies virus.

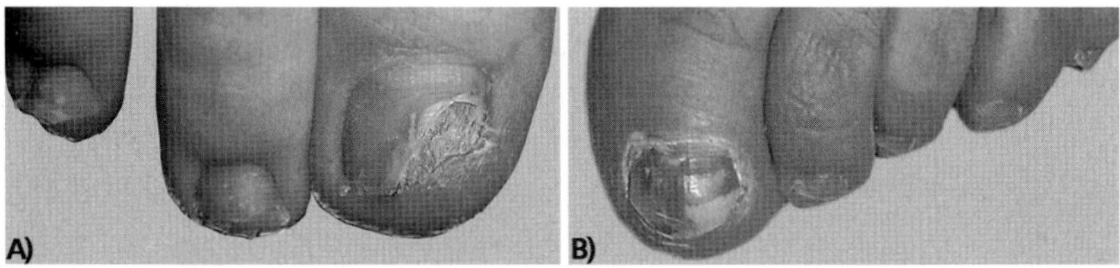

Figure 2.11 Infections of the toenails (onychomycosis) with (A) *Trichophyton tonsurans* and (B) *Trichophyton rubrum* with secondary infections with the fungus, *Scopulariopsis brevicaulis* and the yeast, *Candida guilliermondii*. Courtesy of Dr Pavel Dubin, Israeli Board Certified Dermatologist.

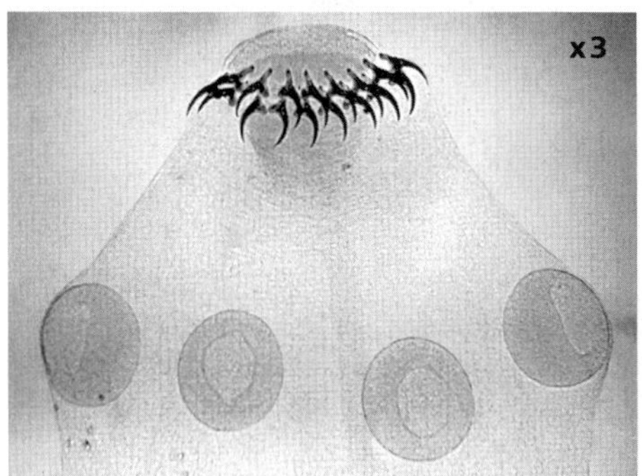

Figure 2.12 The 'head' of the pork tapeworm, *Taenia solium*. Note the hooks and suckers that allow it to remain attached to the intestine wall. Courtesy of Public Health Image Library, Centers for Disease Control and Prevention, USA.

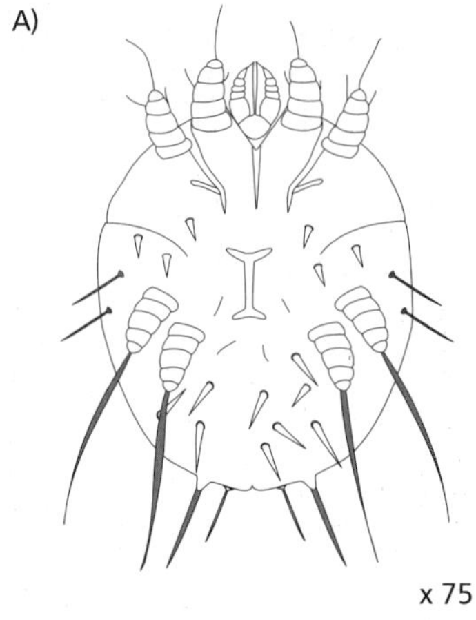

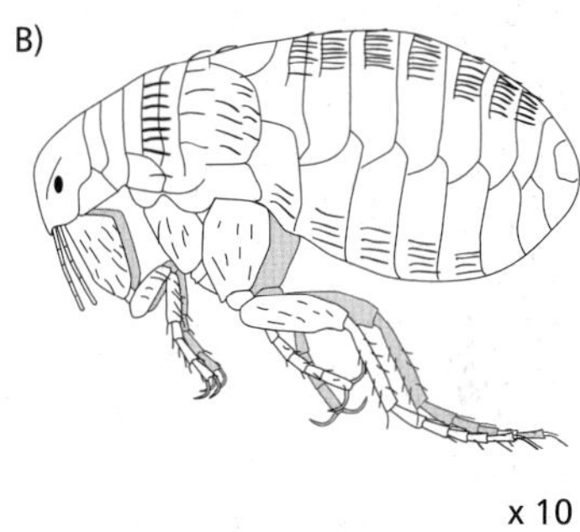

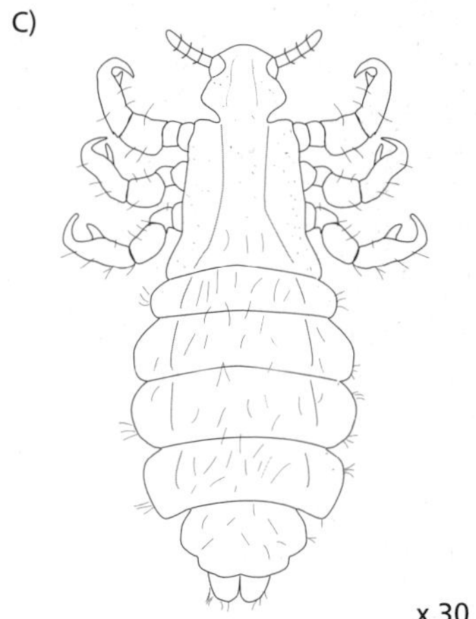

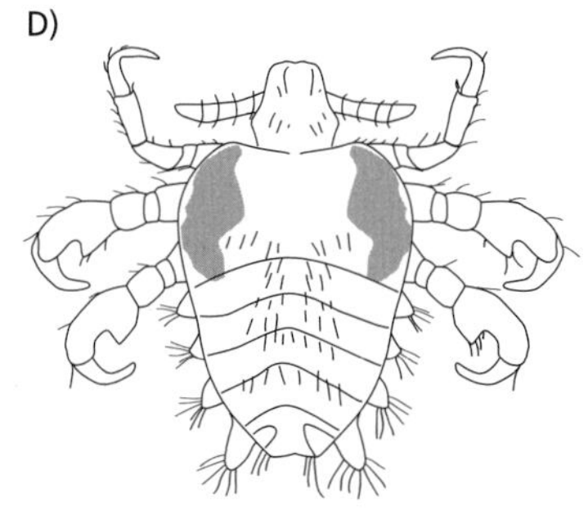

Figure 2.13 Drawings of (A) the scabies mite, (B) flea, (C) head and (D) pubic lice.

BOX 2.2 Malaria – the bad air disease

Malaria infects hepatocytes and blood. It is named malaria from the eighteenth-century Italian 'mala aria', meaning bad air, from the belief the disease was caused by the unwholesome air of swampy districts. Malaria is caused by four main *Plasmodium* species, *falciparum, malariae, ovale* and *vivax*, and is responsible for a significant proportion of mortality and morbidity worldwide, particularly in tropical climates.

The life cycle of *Plasmodium* (*Figure 2.14*) is complicated by any standards! A feeding female *Anopheles* mosquito infects a human with a haploid stage of the parasite called a sporozoite. These infect liver cells, replicate and develop, and when released as merozoites can infect other liver cells or erythrocytes. Here the parasite, usually called a trophozoite, also grows and divides leading to the eventual lysis of the blood cell. The released trophozoites can invade other erythrocytes in 48 h cycles of invasions and lyses that cause the characteristic fevers and chills associated with malaria. The symptoms of the disease are usually most severe during the lysis of erythrocytes. However, eventually some trophozoites develop into male and female gametocytes that can be ingested by a feeding mosquito. Within the insect gut, the gametocytes fuse forming a diploid zygote. The zygote enters the gut wall forming a cyst. Within the cyst, the zygote develops and forms sporozoites that migrate to the salivary gland. Thus the cycle of events can be repeated when the mosquito next feeds on a human.

The symptoms of malaria may present after a variable incubation period and include headaches, general malaise, sweating, muscular pains and rigors and anorexia. The clinical course and symptoms of different forms of malaria are variable although infections are characterized by the recurring attacks of chills and fevers described above.

A variety of antimalarial drugs is available (*Chapter 3*). Some prevent the hepatic forms of *Plasmodium* from invading erythrocytes; others destroy the erythrocytic or gametocyte forms of the parasite in the patient's blood preventing transmission of the parasite by the mosquito. Chloroquine is the usual choice for treating malaria because it is cheap, safe and normally effective. However, chloroquine-resistant strains of *Plasmodium falciparum* are now endemic in sub-Saharan Africa and elsewhere. Drugs available to treat these resistant parasites include halofantrine, mefloquine, quinine and quinidine and others. However, in many cases the molecular mechanisms of action of antimalarial drugs are not well understood.

The remarkably complicated life cycle of the malarial parasites with their prehepatic, hepatic, pre-erythrocyte and erythrocyte stages, means there is a large choice of antigenic targets to use for vaccine development (*Chapter 3*). Despite a number of efforts and trials to develop an effective malarial vaccine, to date none has been successful, although the complete sequencing of its genome in 2005 should assist in this.

A)

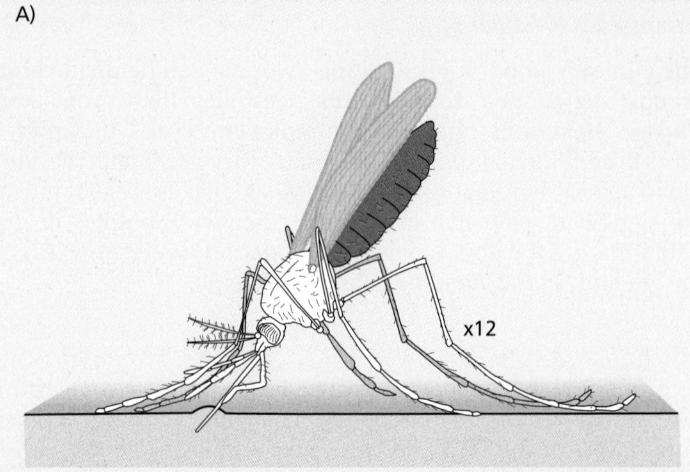

B)

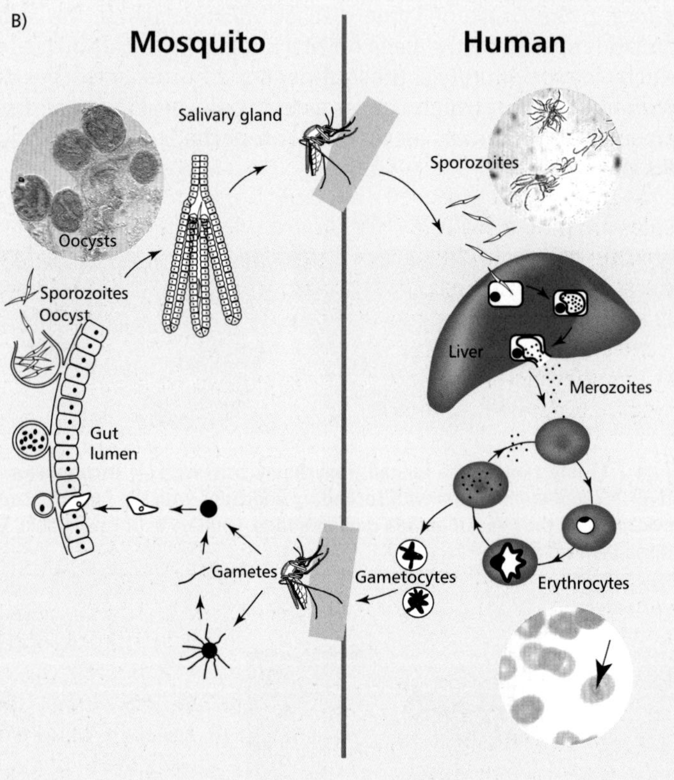

Figure 2.14 (A) Drawing of an *Anopheles* mosquito. Courtesy of Public Health Image Library, Centers for Disease Control and Prevention, USA. (B) Diagram showing the life cycle of *Plasmodium*. See text for details. **The three inserts are light micrographs showing from top left in a clockwise direction: cysts in the *Anopheles* gut, free trophozoites in the blood of a human patient and a young *P. falciparum* trophozoite in an erythrocyte. Note its characteristic signet ring shape.** Courtesy of J.R. O'Kecha, Homerton University Hospital, London.

2.3 VIRULENCE FACTORS

Evolution has provided pathogens and parasites with a wide range of factors that allow them to invade and colonize their host while at the same time avoiding and/or neutralizing host defense mechanisms. Many virulence factors of pathogens have been identified; some are relatively nonspecific in action: some microorganisms, for example, possess specialized iron uptake systems. Microorganisms require iron for oxygen transport, mitochondrial energy metabolism, electron transport, the synthesis of nucleic acids and gene expression. However, although an essential element, iron is often only available in limited quantities and microorganisms that possess a variety of iron uptake systems are able to grow in regions of the host that would otherwise be expected to be sterile since little iron is available. Other virulence factors have rather more defined defensive or offensive actions.

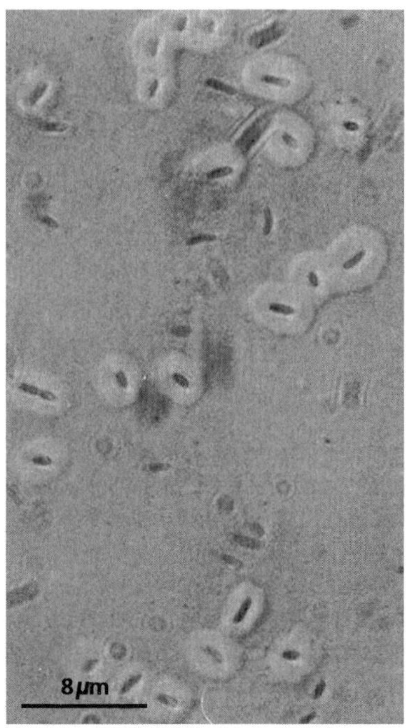

Figure 2.15 Light micrographs of *Bacteroides fragilis* an obligately anaerobic bacterium that is normally found in the gastrointestinal tract. It is most frequently isolated from clinical infections such as peritonitis (*Chapter 11*). Note the prominent capsules surrounding each cell. Courtesy of Dr S. Patrick, Queen's University Belfast, UK.

DEFENSIVE VIRULENCE FACTORS

Numerous pathogens evade the host's defenses by producing slime layers or possessing polysaccharide capsules (*Figure 2.15*). Slime layers consist of exopolysaccharides (EPSs) that bind large quantities of water. Slime production is particularly important in bacteria that form biofilms since it forms a protective coat around the bacterial population. For example, the biofilm formed by the opportunistic *Pseudomonas aeruginosa* in the respiratory tract of cystic fibrosis patients protects it from the immune system and antibiotics.

Capsules generally consist of a single polysaccharide structure that also binds considerable quantities of water and forms a protective layer around the bacterial cell. The polysaccharide is often negatively charged which renders it resistant to uptake by phagocytic cells. Capsules can also protect the bacterium from attack by the immune system (*Chapter 4*). Some polysaccharide capsules are molecular mimics of host cell surface structures. The capsule of *Escherichia coli* K1 and the type B capsule of *Neisseria meningitidis* consists of α-2, 8-*N*-acetylneuraminic acid residues. This is identical to neuraminic acid residues on the neuronal adhesion molecule N-CAM and other sialylated molecules of the nervous system. Consequently host immune systems do not recognize the bacteria as foreign and both pathogens can invade the CNS causing meningitis.

Other examples of molecular mimicry are the many proteins of pathogenic bacteria that are homologous to specific regions of host proteins. *Yersinia* induces the production of antibodies that cross-react with part of a particular variant of a host protein called HLA-B27 (*Chapters 4 and 5*). Cross-reactivity between other bacterial species and HLA-B27 is thought to be involved in the development of types of arthritis known as Reiter's syndrome and ankylosing spondylitis.

Some bacteria directly or indirectly activate or suppress actions of the immune system by producing pathogenicity factors called modulins or microkines. The P fimbriae of the uropathogenic *Escherichia coli*, for example, induces an increase in the release of interleukin 4 (IL-4) by uroepithelial cells. Modulation of cytokine production may lead to increased pathogenicity (*Chapter 4*).

A number of microorganisms prevent their hosts mounting an effective immune response by changing their surface antigens. This can occur in a number of ways. For example, several viruses, including influenza and HIV, have genes coding for surface proteins that mutate at relatively fast rates. This is referred to as hypermutability. Thus, the antigenic structure of these surface proteins is prone to change at intervals, leaving a population that is no longer immune to that virus. Other microorganisms that undergo antigenic variation include the trypanosome that causes sleeping sickness. These regularly change the structure of their surface glycoproteins during the course of an infection. The malarial parasites *Plasmodium* also express different surface antigens during the infection process. Some parasitic worms, for example, schistosomes, become coated with patient antigens, such as MHC molecules and common blood group antigens, and so avoid recognition by the host.

OFFENSIVE VIRULENCE FACTORS

Bacterial offensive virulence factors include adhesins, invasins and toxins. Adhesins are proteins found on the surfaces of microbial cells that bind to specific sites on the cells of the host. The best studied are those in the pili of, for example, certain strains of *Escherichia coli* and *Vibrio cholerae*. Pili are fibers about 2 μm long and 2 to 8 nm in diameter that extend from some bacterial cells (*Figure 2.16*). They consist of about 1000 protein molecules and include a type of adhesin that belongs to a group of biomolecules called lectins. These are glycoproteins that bind specifically to certain sugars or the glycosidic bonds found in some carbohydrates. In the case of *Escherichia coli* and *Vibrio cholerae*, these are mannose and fucose sugar residues respectively which may be found on the surfaces of host cells.

Invasins are also proteins. They allow pathogens that have bound to the host to be internalized, that is, enter the host cell preventing it from being removed by ciliary action or washing and ensuring that the pathogen is protected from direct immune attack. Once internalized, the microorganisms may remain in membrane-bound vesicles. Others escape into the cytosol and so avoid the killing mechanisms associated with phagocytosis. Some microorganisms are so adapted to intracellular life that they are unable to reproduce outside the host cell. These include species of *Chlamydia*, *Rickettsia* and some mycobacterial pathogens. These organisms are therefore obligate intracellular parasites and may cause infectious diseases, for example, *Chlamydia pneumoniae*, *Rickettsia typhi* and *Mycobacterium leprae*. However, others, such as strains of Listeriae, Salmonellae, Shigellae and Yersiniae, are facultative and can live outside their host cells.

Pathogenic microorganisms produce different types of toxin. They can be classified into two types: cell-associated toxins, for example endotoxins, and those secreted by the bacterium called exotoxins.

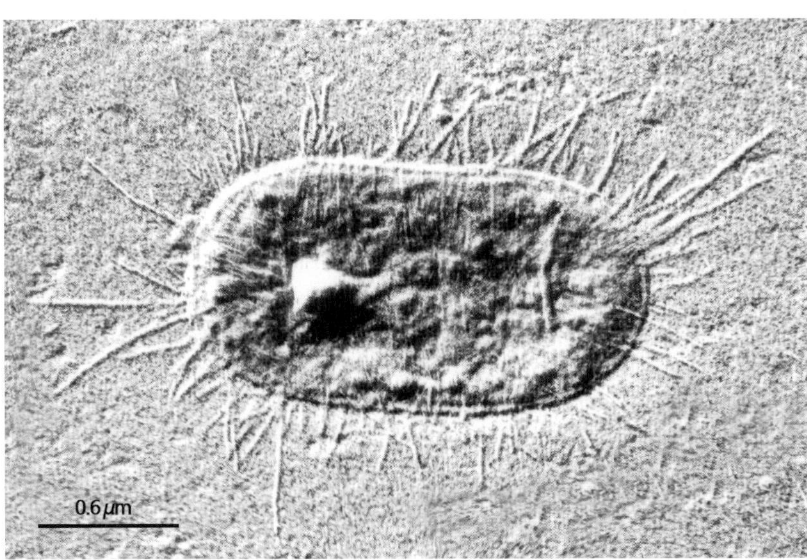

Figure 2.16 Electron micrograph of an *Escherichia coli* cell with numerous pili.

Endotoxins are produced by bacteria with Gram-negative cell walls and are lipopolysaccharides embedded in the outer membrane of the cell walls of Gram-negative bacteria. The toxic portion is called lipid A and the damage it causes varies with the susceptibility of the host. Fever is common because endotoxins stimulate host cells to release cytokines that affect the thermoregulatory center of the hypothalamus. In serious cases, endotoxic shock can result.

Exotoxins are proteins secreted by both Gram-positive and negative bacteria. They may be subdivided into three groups: those that damage membranes, those with specific host targets and superantigens.

Exotoxins that damage membranes cause the cell to lose water and ions, disrupting ion gradients across the membrane. In high doses cell lysis occurs, hence they are sometimes called hemolysins or cytolysins. A clinically significant feature of such toxins in an infection is their antiphagocytic activity. Some of the hemolysins and cytolysins of bacteria, such as *Staphylococcus aureus, Streptococcus pyogenes* and *Bordetella pertussis* are polypeptides that aggregate in the membranes of host cells forming pores. Thiol-activated lysins are predominantly produced by Gram-positive bacteria. These toxins are proteins that contain a large number of cysteine residues. They bind to cholesterol molecules in the membranes of target cells in oligomers of 25 to 100 toxin molecules. These form large toxin-lined aqueous pores in the membrane that constitute the lesions of membrane damage.

Figure 2.17 Molecular model of the α toxin of *Clostridium perfringens*. PDB file 1CA1.

Phospholipases catalyze the hydrolysis of phospholipids in the membranes of host cells. For example, the α toxin of *Clostridium perfringens* (*Figure 2.17*) is a phospholipase C which catalyzes the following reaction:

$$\text{Phosphotidylcholine} + H_2O \xrightarrow{\text{Phospholipase C}} \text{phosphocholine} + \text{diacylglycerol}$$

Similarly, the β hemolysin of *Staphylococcus aureus* is a sphingomyelinase C that catalyzes the following reaction:

$$\text{Sphingomyelin} + H_2O \xrightarrow{\text{Sphingomyelinase C}} \text{phosphocholine} + \text{ceramide}$$

The degradation of membrane lipids, naturally, results in a loss of membrane integrity and function. Exotoxins that target specific sites in the host do so in a wide variety of ways, for example, they may act on cells to deregulate or kill them or they may have an extracellular target. Gram-negative bacteria, such as Enterobacteriaceae (*Escherichia coli, Citrobacter freundii, Yersinia enterocolitica*), secrete heat-stable enterotoxins as small as M_r 2000 (*Figure 2.18*). These toxins bind to specific receptors that are part of a cyclic GMP-dependent signal transduction system of enterocytes in the upper intestinal epithelium. This system regulates the concentration of intracellular cyclic GMP that, in turn, is involved in the activation of intracellular enzymes, for example, protein kinase G (*Chapter 7*). The binding of the toxins interrupts the secretion of Na^+ and Cl^- and this results in a watery diarrhea.

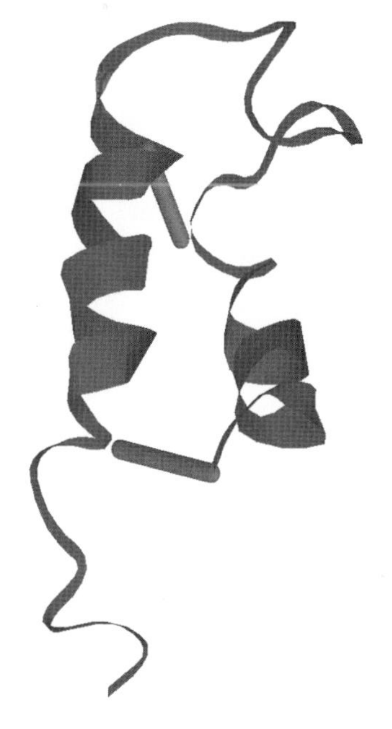

Figure 2.18 Molecular model of the heat stable enterotoxin of *Escherichia coli*. The bars represent disulfide bonds. PDB file 1EHS.

A number of exotoxins consist of two dissimilar polypeptides usually referred to as the A and B subunits. The B subunit recognizes and binds to specific target cells and facilitates entry of the A subunit which has an intracellular toxicity. The tetanus and botulinum toxins are Zn-dependent proteases (*Figure 2.19*) that act as neurotoxins. They catalyze the hydrolysis of synaptobrevin 2, a protein involved in docking and fusion of vesicles containing neurotransmitters. Thus their actions inhibit the release of neurotransmitters. Following its internalization into neurons of the CNS, the tetanus A subunit, tetanospasmin, migrates to peripheral nerve endings by

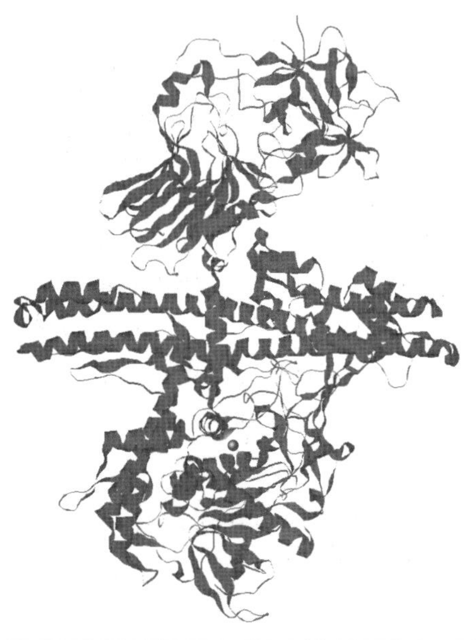

Figure 2.19 Molecular model of the botulinum toxin. The red sphere represents a bound Zn. PDB file 1S0F.

retrograde axonal transport. Here it is released by postsynaptic dendrites and diffuses to the presynaptic neurons where its action prevents the release of the inhibitory neurotransmitters, γ-aminobutyric acid and glycine. This leads to unchecked excitatory impulses with a continuous stimulation of muscles and spastic paralysis. In contrast, *botulinum toxin* is absorbed from the gastrointestinal tract and is transported to susceptible neuromuscular and peripheral autonomic synapses where it inhibits the release of acetylcholine, causing flaccid paralysis.

Some exotoxins are ADP ribosyl transferases. The cholera, pertussis and diphtheria toxins (*Figure 2.20*) use NAD^+ as a donor substrate so that the ADP ribosyl portion of NAD^+ is transferred to the target protein releasing nicotinamide. Cholera and pertussis toxins attack G proteins and interfere with signal transduction so that receptor-mediated signal transduction pathways are activated or inhibited (*Chapter 7*).

ADP ribosylation by cholera toxin fixes the Gα protein in its active form. This leads, in turn, to a long-lasting activation of adenylate cyclase and synthesis of cyclic AMP and activation of protein kinase A. The net result is a long-lived opening of the chloride channel of the cystic fibrosis transmembrane conductance regulator, (CFTR, *Chapter 16*), that increases secretion of hydrogen carbonate (HCO_3^-) and Cl^- into the intestinal lumen but which inhibits the absorption of Na^+ and Cl^-. The resulting osmotic effect causes a massive leakage of intracellular water into the intestinal lumen and subsequent diarrhea. This is called **fulminant cholera**.

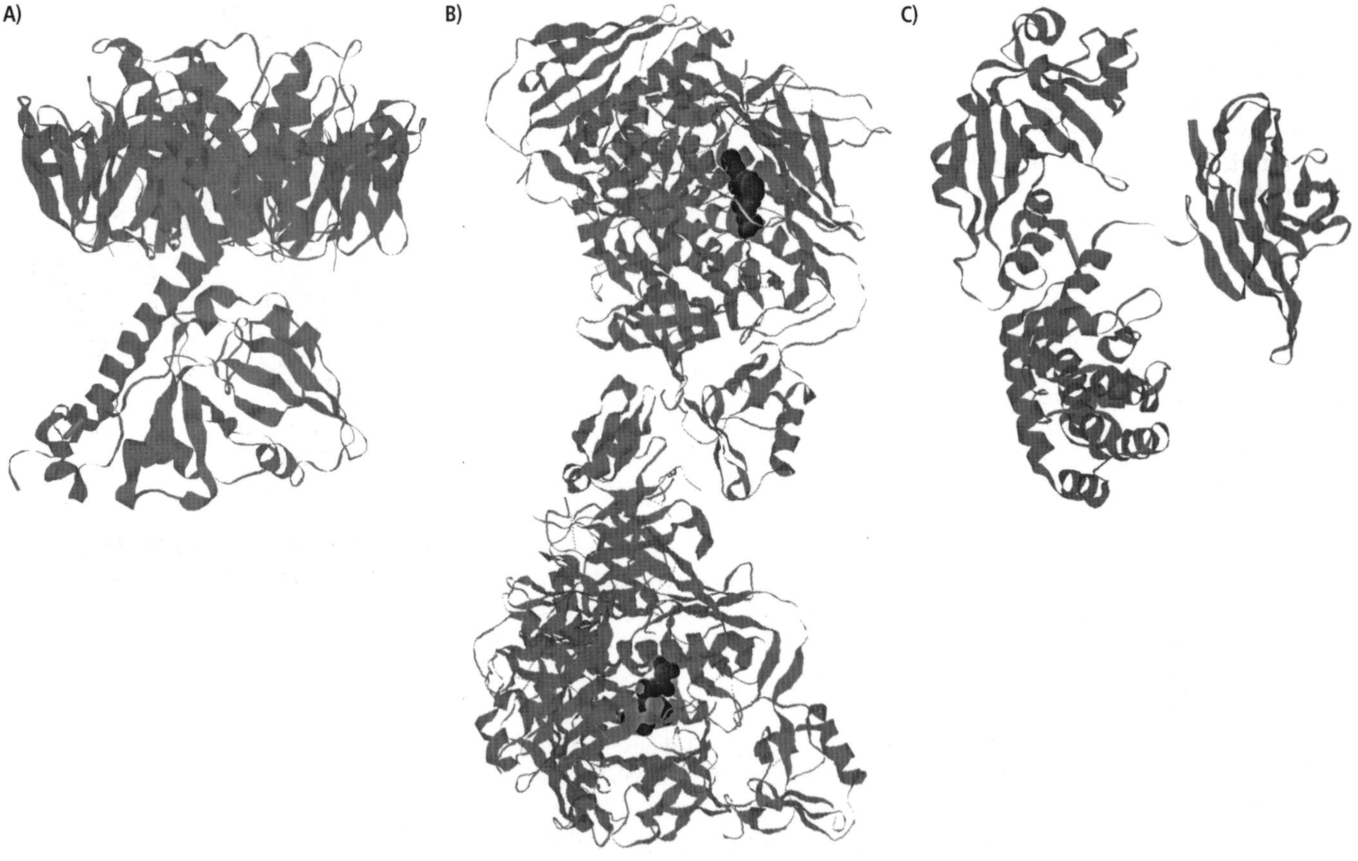

Figure 2.20 Molecular models of (A) cholera, (B) pertussis, with the bound ATP shown in red, and (C) diphtheria toxins. The structure of many bacterial toxins are known, which helps in understanding how they function and in the design of antidotes. PDB files 1XTC, 1BCP and 1DDT respectively.

Diphtheria toxin kills cells by blocking protein synthesis. The target of the toxin is a single protein, elongation factor EF2, involved in protein synthesis. Its activities are inhibited by ADP ribosylation. The Shiga family of toxins produced by *Shigella dysenteriae* and *Escherichia coli* (*Figure 2.21*) also inhibit protein synthesis but by a different mechanism. They are *N* glycosidases that hydrolyze an *N*-glycosidic bond between specific adenine and ribose residues of the 28S ribosomal RNA of the 60S ribosomal subunit.

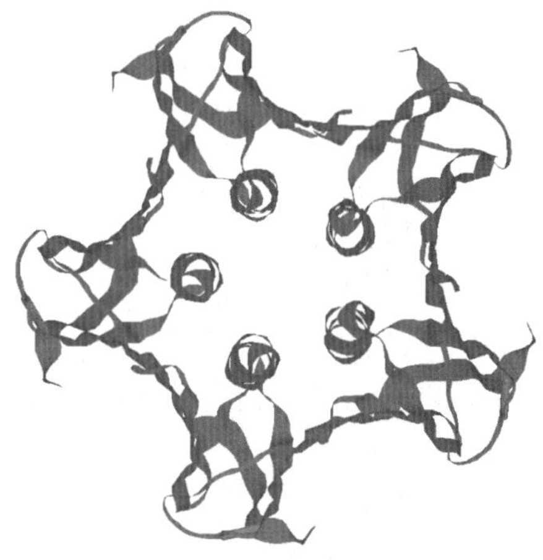

Figure 2.21 Molecular model of a Shiga-like toxin subunit. PDB file 1CZG.

Microorganisms secrete a number of toxins that are enzymes which degrade components of the connective tissues. Staphylococci and Streptococci secrete enzymes that degrade the extracellular matrix allowing them to invade and colonize tissues. *Clostridium perfringens* secretes a variety of carbohydrases and proteases that degrade connective tissues, allowing it to colonize and spread through subcutaneous tissues. If the bacteria invade the deeper muscles tissue they may cause necrosis and gas production from anaerobic fermentations (gas gangrene). This type of infection is extremely painful and can spread rapidly. Much of the direct damage is caused by the α toxin described above. Other pathogens secrete proteases that hydrolyze specific components of the immune defense, such as IgA or enzymes that detoxify catalases and superoxidase dismutases (SODs) that are used by some immune cells to kill microbial pathogens (*Chapter 4*).

Superantigens are polypeptides synthesized by Gram-positive pathogens, such as some strains of *Staphylococcus aureus* and *Streptococcus pyogenes*. They are extremely effective and potent stimulators of the immune system because of their unique ability to stimulate large numbers of its cells simultaneously. This leads to a massive release of molecules called cytokines that activate numerous physiological systems, such as the temperature regulatory system. The superantigen of *Streptococcus pyogenes* is responsible for streptococcal toxic shock syndrome (STSS). The fever, shock and tissue damage associated with STSS is thought to be the result of an overproduction of the cytokines Tumor Necrosis Factor α, Interleukin 1β and Interleukin 6. The superantigen of *Staphylococcus aureus*, toxic shock syndrome toxin 1 (*Figure 2.22*) causes symptoms that can lead to a rapid failure of many body organs.

Figure 2.22 Molecular model of TSST-1 from *Staphylococcus aureus*. PDB file 2QIL.

2.4 COURSE OF INFECTION

The course of an infection can be considered to follow up to four major stages namely: adhesion, entry, localized infection and generalized infection.

A virulent pathogen is one that is well adapted to establish an infection. Most pathogens are adapted to adhere to cells, usually epithelial, that line the site of entry. Adherence is the first stage of infection, given that it prevents the pathogen being swept away and eliminated from the body.

Pathogens enter the body through one of a number of so-called **portals of entry**. These include the skin and conjunctiva, respiratory tract, gastrointestinal tract (GIT), urogenital system and, in the case of fetuses, the placenta. Once entry has been gained, conditions for growth, such as temperature, nutrients, must be favorable, but the pathogen must also be able to overcome the local defenses. Pathogens are often adapted to enter their hosts through a single portal of entry and do not cause infectious diseases if they enter through a different portal.

A localized infection acts as a focus of infection and, indeed, many remain local in nature or are prevented from spreading by host defenses. If the pathogenic organisms penetrate tissues and reach the blood or lymphatic systems or enter cells such as phagocytes, they can be distributed throughout the body and infect other tissues and organs causing a generalized infection. Again, certain pathogens are particularly well adapted to infect specific target tissues and organs. During infections, the pathogen multiplies and can be shed from the host into the environment through **portals of exit**. In localized infections these are the same as that of entry; in a generalized infection, other portals are used. These may include loss from the skin in pus, scales and blood, from the respiratory tract as droplets and aerosols of mucus and saliva, from the GIT in feces and bile, from the urogenital system in urine, mucus and genital secretions and, in pregnancy, from the placenta with direct transfer to the fetus.

A pathogen may be eliminated at any stage of the infective process. In some cases, the growth may be held in check but not eliminated. Such latent infections can be activated later giving recurring infections. In some cases, individuals can recover from a disease but the pathogen may remain in the body for considerable periods. Such people are symptomless carriers and are reservoirs of the disease.

Some viruses which exhibit latency, with or without sporadic reactivation are shown in *Table 2.2*.

Microorganism	Disease
Herpes simplex virus types 1 and 2 (HSV-1 and HSV-2)	Oral and genital herpes
Epstein-Barr virus	Glandular fever
Hepatitis B virus	Hepatitis Hepatocellular carcinoma
Varicella zoster	Chicken pox, shingles
Measles virus	Measles, subacute sclerosing panencephalitis (SSPE)

Table 2.2 Some viruses that exhibit latency

CASE STUDY 2.1

A 38-year-old male, Brian was admitted into hospital two weeks after spending a year working in several African countries. Seven days before admission Brian had developed coughing, muscular pains and recurrent chills and fevers approximately three times daily. He was also slightly jaundiced.

Questions

(a) What is the most likely disease affecting Brian?

(b) Which organism(s) causes this disease? How could this be confirmed?

(c) How could this disease have been prevented?

2.5 SUMMARY

Organisms that cause diseases in humans are found in all microbial groups, including viruses, bacteria as well as fungi and helminths. In addition, prion proteins can cause infectious disease since these aberrant proteins can be passed on, for example, in food. A number of microorganisms are not normally pathogenic, but can become so when a new 'niche' becomes available to them, for example a burn wound, or when the host is immunocompromized in some way. Pathogenic microorganisms display a variety of virulence factors that aid entry into the host and which help them to overcome host defenses. These virulence factors can have a generalized action or be more specifically defensive or offensive in nature. An example of the former is the possession of hypermutable genome that leads to changes in the nature of the surface antigens of the microorganism and enables them to escape the host's immune system. Examples of offensive virulence factors include the production of exotoxins.

QUESTIONS

1. To which class of pathogen or parasite do the following belong:

   Trypanosoma brucei, HIV, Microsporum species *Treponema pallidum, Candida albicans*?

2. Arrange the two following lists into their most appropriate pairings.

Clostridium perfringens	shigellosis
Epstein-Barr virus	sleeping sickness
HIV	gas gangrene
Human rhinovirus 91	chicken pox
Microsporum spp, *Trichophyton* species	bubonic plague
Plasmodium	CD4
Sarcoptes scabiei	malaria
Shigellae dysenteriae	glandular fever
Trypanosoma brucei	scabies
Varicella zoster	intracellular adhesion molecule
Yersinia pestis	athlete's foot

3. Compare and contrast the actions of endo- and exotoxins.

FURTHER READING

Alouf, JE and Freer, JH (eds) (1999) *The Comprehensive Sourcebook of Bacterial Protein Toxins*, 2nd edn, Academic Press, London.

Braun, V and Killman, H (1999) Bacterial solutions to the iron supply problem. *Trends Biochem. Sci.* **24:** 104–109.

Caughey, B (2001) Interactions between prion protein isoforms: the kiss of death? *Trends Biochem. Sci.* **26:** 235–242.

DeArmond, SJ and Bouzamondo, E (2002) Fundamentals of prion biology and diseases. *Toxicology* **181–182:** 9–16.

Deitsch, KW, Moxon, ER and Wellems, TE (1997) Shared themes of antigenic variation and virulence in bacterial, protozoal and fungal infections. *Microbiol. Mol. Biol. Rev.* **61:** 281–293.

Finlay, BB and Falkow, S (1989) Common themes in microbial pathogenicity. *Microbiol. Rev.* **53:** 210–230.

Glatzel, M, Stoeck, K, Seeger, H, Luhrs, T and Aguzzi, A (2005) Human prion diseases: molecular and clinical aspects. *Arch. Neurol.* **62:** 545–552.

Greenwood, BM, Bojang, K, Whitty, CJ and Targett, CA (2005) Malaria. *Lancet* **365:** 1487–1498.

Irving, W, Boswell, T and Ala' Aldeen, D (2005) *Instant Notes on Medical Microbiology*, BIOS/Taylor and Francis, London, UK.

Jedrzejas, MJ (2001) Pneumococcal virulence factors: structure and function. *Microbiol. Mol. Biol. Rev.* **65:** 187–207.

Jerse, AE and Rest, RF (1997) Adhesion and invasion by the pathogenic neisseria. *Trends Microbiol.* **5:** 217–221.

Maitland, K, Bejon, P and Newton, CRJC (2003) Malaria. *Curr. Opin. Infect. Dis.* **16:** 389–395.

Perry, RD and Fetherston, JD (1997) *Yersinia pestis* – etiologic agent of plague. *Clin. Microbiol. Rev.* **10:** 35–66.

Sears, CK and Kaper, JB (1996) Bacterial toxins: mechanisms of action and linkage to intestinal secretions. *Microbiol. Rev.* **60:** 167–215.

Smith, AE and Helenius, A (2004) How viruses enter animal cells. *Science* **304:** 237–242.

Spicer, JW (2000) *Clinical Bacteriology, Mycology and Parasitology*, Churchill Livingstone, London, UK.

Turton, K, Chaddock, JA and Acharya, KR (2002) Botulinum and tetanus neurotoxins: structure, function and therapeutic utility. *Trends Biochem. Sci.* **27:** 552–558.

Useful web site:

www.ph.ucla.edu/epi/snow.html

chapter 3:

INFECTIOUS DISEASES AND TREATMENTS

OBJECTIVES

After studying this chapter you should be able to:

- describe the types and symptoms of some of the more common infectious diseases;
- outline preventative measures for infectious diseases;
- list the general effects of antibiotics on infectious organisms;
- discuss some general aspects of the management and treatment of specific infectious diseases.

3.1 INTRODUCTION

The presence of virulence factors, described in *Chapter 2*, allows pathogenic microorganisms to infect specific body systems and cause a vast range of diseases. A small number of such organisms are also able to cause **systemic** disease, that is one affecting the whole body. A pathogen must be transmitted from a source to the patient. Direct contact between hosts is the most obvious form of transmission but coughs and sneezes (aerosols), food, water and arthropod vectors are all used by various pathogens. The long-term survival of pathogenic microorganisms also depends on them maintaining their infectivity during transmission from host to host. The sources of pathogens can be abiotic, soil, water for example, or animals or other humans. Diseases that infect animals and humans are called **zoonoses**, while humans who harbor a pathogen but are symptomless are called **carriers**.

In a text of this length, it is simply not possible to describe all diseases caused by microorganisms. This chapter will concentrate on selected, representative examples of infections of the major body systems and those microorganisms that can produce a generalized infection. In addition, the ways in which infectious diseases are prevented, investigated and treated will be outlined.

3.2 INFECTIONS OF THE SKIN

The skin is a major element of the innate immune defense (*Chapter 4*). It is normally colonized by a variety of microorganisms although the numbers and types vary between different areas of the body. In normal circumstances it forms an effective barrier to invading microorganisms.

Papillomaviruses can infect epidermal cells and stimulate their proliferation to form warts. Numerous herpes viruses have been described and at least eight of them, the **Human Herpes Viruses** (**HHV1-8**), infect humans and can cause clinical disease. Largely due to historical reasons, they are also known by other names (*Table 3.1*). For example, *Herpes simplex* viruses 1 and 2 may infect skin of the genitalia causing Herpes labialis and Herpes genitalis respectively and *Varicella zoster (Figure 2.5)* causes chickenpox and shingles.

Human herpes viruses	Historical name
HHV1	*Herpes simplex* virus 1 (HSV1)
HHV2	*Herpes simplex* virus 2 (HSV2)
HHV3	*Varicella zoster* virus (VZV)
HHV4	Epstein Barr virus (EBV)
HHV5	Cytomegalovirus (CMV)
HHV6	B-lymphotropic virus (B-LV)
HHV7	T-lymphotropic virus (T-LV)
HHV8	Kaposi's sarcoma virus (KSV)

Table 3.1 Nomenclatures of the Herpes viruses 1–8

Figure 3.1 ***Propionibacterium acnes*, a bacterium associated with skin infections such as acne.** Courtesy of Dr S. Patrick, Queen's University Belfast, UK.

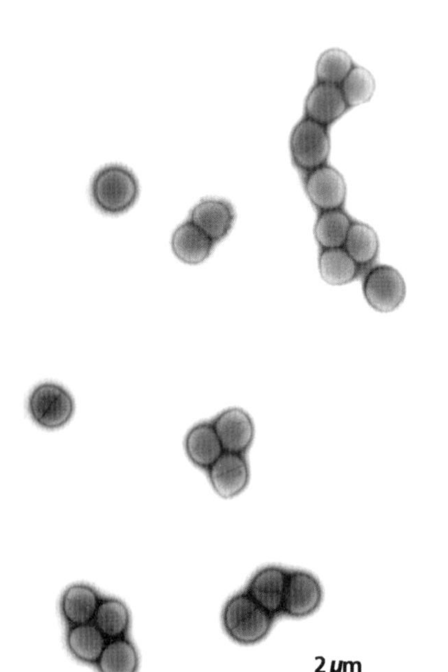

Figure 3.2 Electron micrograph of *Staphylococcus aureus*. Courtesy of Dr A. Curry, Manchester Royal Infirmary, UK.

Bacterial skin infections normally occur only if the normal balance between the skin environment and these organisms is disturbed. The outbreaks of acne caused by *Propionibacterium* species (*Figure 3.1*) during the hormonal changes associated with puberty are typical. Breaks in the skin from wounds or surgery or lesions from insect bites or chickenpox may also lead to infections of the skin by *Staphylococcus aureus* (*Figure 3.2*) and *Streptococcus pyogenes*. In children especially, they may cause impetigo contagiosa, an extremely contagious skin infection. *Staphylococcus aureus* can colonize hair follicles leading to inflammations that can develop into abscesses (boils) or even, in extreme cases, **carbuncles**: an amalgam of several abscesses. Enterobacteria, *Pseudomonas aeruginosa*, *Staphylococcus aureus* and *Streptococcus pyogenes*, are all associated with a variety of skin problems following weakening of the patient, for example, by diabetes (*Chapter 7*), by a deficiency of the immune system (*Chapter 5*), by **nosocomial**, or hospital acquired, infection of surgical wounds or burning of the skin. *Mycobacterium leprae* is the causative organism of the dreaded disease leprosy, which, although now virtually eliminated in developed countries, still affects many thousands in the developing world.

Fungi are associated with a variety of infections, particularly of the skin (*Chapter 2* and *Figure 2.11*).

3.3 INFECTIONS OF THE EYES, EARS AND CENTRAL NERVOUS SYSTEM

Eyelid infections generally involve the lid margins, eyelid glands or follicles causing styes or **hordeolums**. They are usually associated with *Staphylococcus* *aureus* infections. The conjunctiva is particularly susceptible to infection. The epithelial surface enclosed by the eyelids is a warm moist enclosed environment in which microorganisms can rapidly become established. However, microorganisms must avoid being rinsed away by tears and some pathogens, such as *Chlamydia trachomatis* (*Figure 3.3*), attach specifically to conjunctival cells. An estimated 500 million people are infected with different serotypes of *Chlamydia trachomatis*, making trachoma the most significant eye infection worldwide. The disease blinds approximately 1% of infected individuals while many others suffer visual impairment. *Chlamydia trachomatis* is transmitted by contact with contaminated flies, fingers and towels although trachoma itself results from chronic repeated infections. This is much more likely to occur in regions where restricted access to water prevents regular washing of the hands and face. Chlamydia is also a sexually transmitted disease (*Section 3.6*) and there is evidence that untreated chlamydial infections can lead to premature delivery and babies born to infected mothers can be infected in their eyes and respiratory tracts. Chlamydia is a leading cause of early infant pneumonia and conjunctivitis (pink eye) in the newborn.

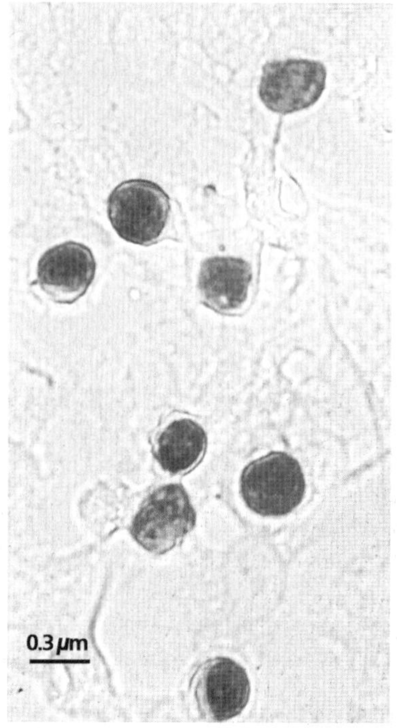

Figure 3.3 Light micrograph of *Chlamydia trachomatis*, growing in cultured eukaryotic cells, stained with iodine. Courtesy of School of Biochemistry and Microbiology, University of Leeds, UK.

A laboratory diagnosis of trachoma can be carried out using samples of conjunctival fluid or scrapings. The usual treatment is with oral or topical antibiotics, such as tetracycline or doxycycline (*Section 3.11*). *Chlamydia* infections account for only a fifth of cases of conjunctivitis; others are caused by bacteria such as *Streptococcus pneumoniae* and *Leptospira spp*.

Serious infections of the inner eye with *Pseudomonas aeruginosa* may follow trauma or after invasion by the protozoan *Toxoplasma gondii* causing chorioretinitis and possible blindness. This widespread protozoan is not a serious threat unless acquired *in utero* or when an individual is immunosuppressed, perhaps as a result of taking drugs to prevent transplant rejection (*Chapter 6*). Infection occurs by swallowing oocysts released by infected cats or by eating meat containing tissue cysts.

The eyes may also be infected by parasitic worms, for example, larval forms of the tapeworm *Echinococcus granulosus* that is transmitted by eggs passed from infected dogs. Infection by the larvae of the nematode *Toxocara canis*, which occurs naturally in the intestine of dogs, is, however, more common. An infection can lead to the detachment of the retina. River blindness in Africa and Central America is caused by the helminth *Onchocerca volvulus*. Simulium flies carrying larvae obtained from the skin of infected hosts transmit the infection. These flies develop in rivers, hence the name of the disease. River blindness is a serious infection with over 300 000 people infected worldwide. The rates of blindness, which is irreversible, may reach 40%. The usual treatment is with anthelminthic drugs.

Infections of the outer ear may cause pain or irritation and are resistant to some antibiotics. The middle ear can be colonized by viruses and bacteria from the upper respiratory tract causing acute otitis media with swelling and blockage of the Eustachian tube and may lead to deafness, though this is generally temporary. Microorganisms that cause middle ear infections include the virus that causes mumps. This may be followed by secondary infections with *Streptococcus pneumoniae* and *Haemophilus influenzae*. This is very common in children. Indeed, otitis media is the most frequent illness diagnosed in young children. The general symptoms, apart from a devastating earache, are fever, vomiting and diarrhea. The vesicles of the tympanum become dilated, with bulging of the drum itself occurring in the later stages of infection. If treatment is inadequate, then acute attacks may eventually perforate the eardrum, produce chronic discharges and defective hearing. The usual treatment is with oral antibiotics (*Section 3.11*), such as ampicillin, amoxycillin, erythromycin and cefixime.

Margin Note 3.1 *Haemophilus influenzae* type b

Haemophilus influenzae type b is the cause of 'Hib' meningitis. It is also responsible for childhood epiglottitis, causing the throat to swell alarmingly and breathing difficulties as mucus collects in the throat and fever. The condition is life threatening. *Haemophilus influenzae* type b can also cause pneumonia and other lower respiratory infections. The health risks are mainly associated with children under five years old but adults whose resistance has been weakened by sickle cell anemia (*Chapter 13*), chronic disease of the spleen, alcoholism (*Chapter 12*) or some malignancies (*Chapter 17*) are also at risk. In the developed world, the introduction of an effective vaccine in the 1980s eradicated Hib disease. However, it is still a problem with thousands of children in sub-Saharan countries affected. In 2005, it was reported that a five-year Medical Research Council led program, involving Sanofi Pasteur and the World Health Organization (WHO), of vaccinations against Hib had successfully eliminated the disease in children in the Gambia. Hopefully, this will encourage other countries to adopt a similar practice of routine Hib immunization policies.

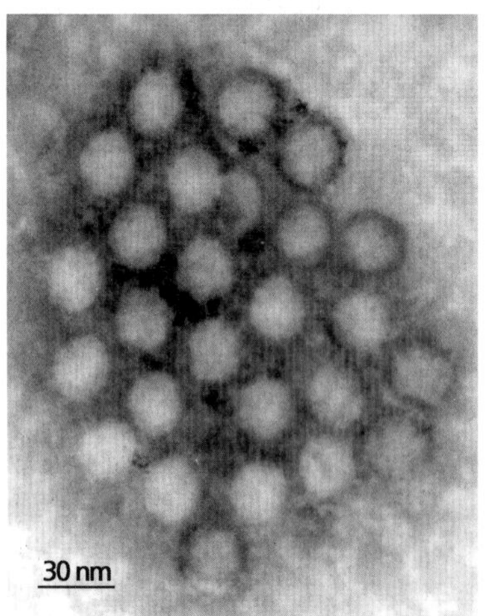

Figure 3.5 Electron micrograph of poliomyelitis virus. Courtesy of H. Cotterill, Manchester Royal Infirmary, UK.

Infections of the central nervous system (CNS) may affect the meninges, the spinal cord or the brain causing meningitis, myelitis and encephalitis respectively. More than one area may be infected simultaneously. Such infections can also become systemic infections (*Section 3.7*). Pathogens can enter these areas following head injuries, along the axons of neurons or by breaching the blood–brain or blood–cerebrospinal fluid barriers (*Figure 3.4*). The most common causes of viral meningitis are enteroviruses, such as ECHO-, Coxsackie- and, formerly, poliomyelitis viruses (*Figure 3.5*). Viral meningitis is not usually a life-threatening condition. Bacterial meningitis, in contrast, has a mortality greater than 10%. The principal causative organisms are the capsulated bacteria, *Neisseria meningitidis* (*Figure 1.3*), *Streptococcus pneumoniae* and, before the introduction of a vaccine, *Haemophilus influenzae* (*Margin Note 3.1*). Effective vaccines are also available against some serogroups of *Neisseria meningitidis* and a vaccine against *Streptococcus pneumoniae* is being tested.

Encephalomyelitis results from infections by a number of viruses or protozoa. These include some poliovirus types, *Herpes simplex*, measles virus, HIV, toga viruses, which are transmitted by arthropods, and the rabies virus that is transmitted from infected mammals. The protozoan *Toxoplasma gondii* may infect individuals with compromized immune systems while *Trypanosoma brucei* is the causative organism of African sleeping sickness.

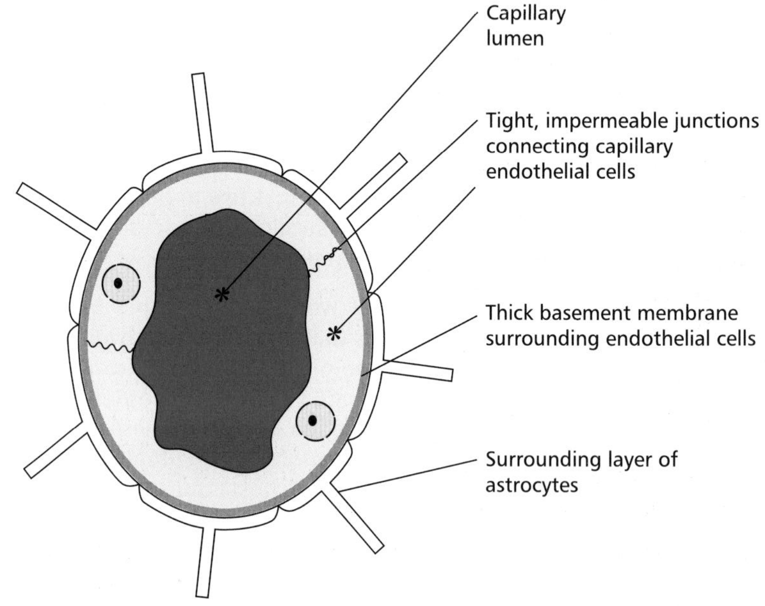

Figure 3.4 Schematic of the blood–brain barrier. Note the tightly associated endothelial cells and thick basement membrane, which prevent materials in the blood entering the brain.

3.4 INFECTIONS OF THE RESPIRATORY SYSTEM

The respiratory system is constantly exposed to inhaled microorganisms but is protected by extensive defenses. The nose filters out particles larger than 10 μm although those smaller than 5 μm may reach the bronchi and alveoli. In addition, there is a host of immune defenses including alveolar macrophages, secretory IgA antibodies, complement proteins, surfactant proteins, secreted defensins and lactoferrin (*Chapter 4*). Despite this, infections of the respiratory tract are frequent causes of illness. The World Health Organization (WHO) has reported that many hundreds of millions of patients suffer acute infections of the lower respiratory tract worldwide. *Figure 3.6* indicates the sites of a number of respiratory diseases.

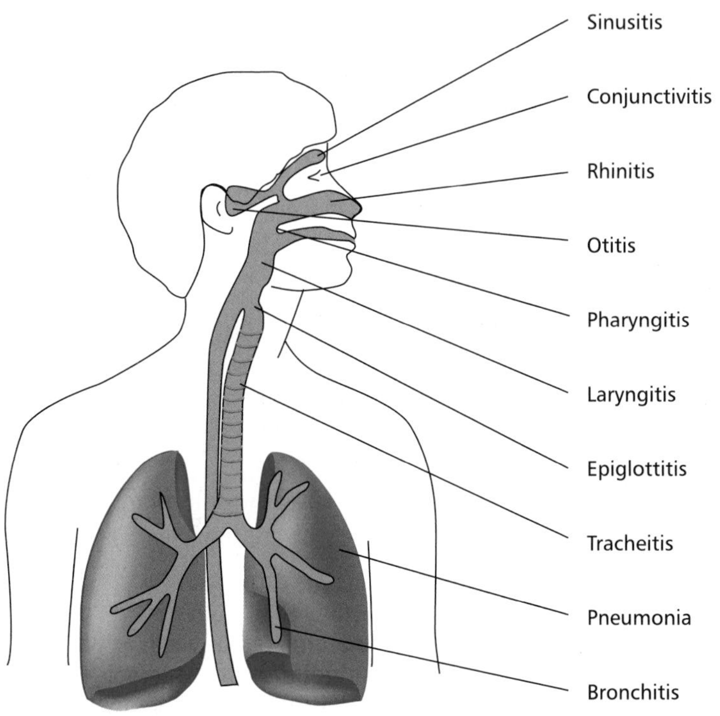

Figure 3.6 Selected infections associated with the head and respiratory system. See text for discussions.

Respiratory viruses are transmitted directly by aerosols or indirectly from contaminated surfaces. The first site of attack is, not surprisingly, the epithelium of the nose and throat. Indeed, the hundreds of corona and rhinoviruses that cause the common cold replicate at 32 to 33°C, the temperature of the mucosal surface lining the nose. The influenza viruses (*Figures 2.4* and *2.8*) infect and replicate in respiratory epithelial cells causing cellular damage. The generalized symptoms that present, such as muscular aches, malaise and anorexia, suggest the virus may spread systemically from the respiratory tract but there is no conclusive evidence for this.

The loss of ciliated and mucus producing epithelial cells impairs clearance of invading microbes and creates conditions for secondary bacterial infections of staphylococci, streptococci or *Haemophilus influenzae*. Bacterial proteases, for example the V8 protease of *Staphylococcus aureus*, can enhance the infectiveness of the influenza virus by improving virus adhesion.

Corynebacterium diphtheriae (*Figure 3.7*) and *Bordetella pertussis* (*Figure 3.8*) are obligate bacterial pathogens. *Corynebacterium diphtheriae* infects the nasopharynx and the tonsils and may lead to a lethal systemic infection affecting the heart, liver and kidneys. *Bordetella pertussis* is the causative agent of whooping cough. It adheres to the epithelial cells lining the trachea and bronchi where it interferes with ciliary action and releases toxins (*Chapter 2*) and substances that damage and kill cells and irritate the surface, causing the characteristic cough. Effective vaccines are available against both organisms although 40 million infections of whooping cough occur annually worldwide. In contrast, *Streptococcus pneumoniae, Haemophilus influenzae, Staphylococcus aureus* and *Moraxella catarrhalis*, make up to 60% of the normal bacterial population of the nasopharyngeal mucous membrane in healthy individuals. They can become opportunistic pathogens in immunosuppressed individuals

Figure 3.7 Light micrograph of *Corynebacterium diphtheriae*. Courtesy of School of Biochemistry and Microbiology, University of Leeds, UK.

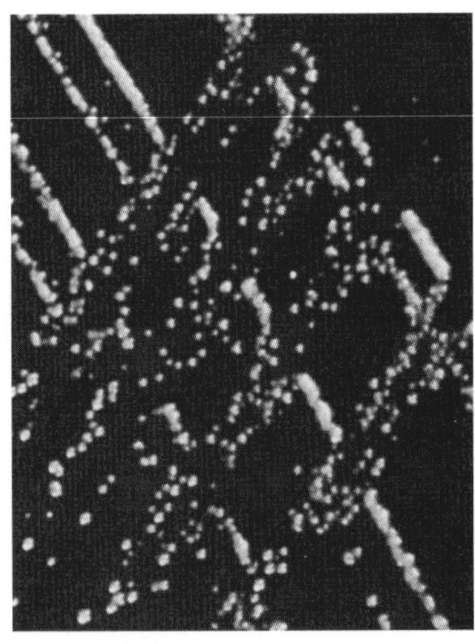

Figure 3.8 Characteristic small 'metallic' colonies of *Bordetella pertussis* growing on agar enriched with potato starch, glycerol and blood. Charcoal has been added to this medium to absorb bacterial metabolites that would inhibit the growth of the pathogen. Courtesy of School of Biochemistry and Microbiology, University of Leeds, UK.

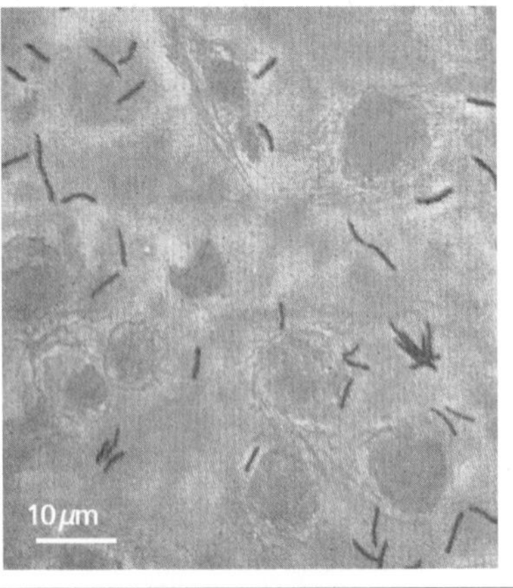

Figure 3.9 Light micrograph of *Mycobacterium tuberculosis* in a specimen of sputum. Courtesy of Public Health Image Library, Centers for Disease Control and Prevention, USA.

or following changes to the bacterium that render it increasingly virulent. The commonest form of bacterial pneumonia is lobar pneumonia caused by *Streptococcus pneumoniae* and results in a massive inflammation of one lobe of the lung. *Staphylococcus aureus* may cause bronchopneumonia, while *Haemophilus influenzae* can infect the epiglottis.

Mycobacterium tuberculosis (*Figure 3.9*) causes tuberculosis (TB) of the lung and may be considered a rather special case of bacterial infection of the lower respiratory tract. The bacteria enter the alveoli in inhaled air and are phagocytozed by macrophages where they escape being killed and multiply (*Chapter 4*). Mycobacteria can then enter the lymphatic system and invade a neighboring lymph node. The healing of local lesions leads to calcification of the lung tissues. In immunodeficient individuals, the lymph nodes and tissues may be progressively affected until eventually the mycobacteria are spread by the blood. Also with impaired immunity, dormant *Mycobacterium tuberculosis* can be reactivated causing a severe form of pneumonia.

Atypical pneumonias can result from infections with *Mycoplasma pneumoniae, Chlamydia pneumoniae* and *Legionella pneumophila.* These infections are associated with 'flu-like' symptoms, such as high temperatures and coughing, although bronchial secretions and sputum do not contain pus as would be expected of a typical bacterial lobar pneumonia.

Generally, pathogenic fungi do not produce toxins but damage tissues directly or disturb normal metabolic functions and can induce hypersensitivity responses (*Chapter 5*). Fungi can cause respiratory infections; *Aspergillus fumigatus* can invade the respiratory system and lead to one of several types of diseases. It may simply grow in the mucus of the bronchi and induce a hypersensitive state but may invade old wound cavities of the lungs, such as those resulting from TB, and grow as a solid mass called an **aspergilloma**. Aspergillosis may also result from an invasive growth in the lungs and other tissues. Generally, the infective dose of spores is extremely large although the invasive form may be secondary to other systemic diseases. Similarly, *Pneumocystis carinii* can cause a serious pneumonia (PCP) in AIDS compromized patients (*Box 3.1*). The yeast, *Candida albicans,* is also an opportunistic agent in sufferers of AIDS.

3.5 INFECTIONS OF THE GASTROINTESTINAL TRACT

All regions of the gastrointestinal tract (GIT) are subject to infection. Saliva traps and removes many pathogens and these can also be killed by stomach acid (*Chapter 11*). Unfortunately new ones are constantly introduced through breathing and eating.

Infections of the oral cavity (*Figure 3.6*) differ in type and symptoms to those of the stomach and intestines. Inflammation of oral tissues caused by fungal infection, **actinomycosis**, often occurs following injuries, such as an accidental bite to the lining of the mouth or fracture of the jaw. Immunosuppression resulting from viral infections, AIDS, cancer treatment or treatment with broad spectrum antibiotics can all allow the yeast *Candida albicans* (*Figure 3.10*) to invade and colonize the mucous membrane, eventually producing a thick layer of yeast cells called candidiasis or thrush.

Some bacteria can resist removal by saliva and become immobilized by binding to surface receptors of cells in the mouth, eventually forming biofilms and microcolonies. Oral streptococci, such as *Streptococcus sanguis* and *Streptococcus mutans* (*Figure 3.11*), secrete glycosyltransferases that mediate their adhesion to extracellular carbohydrates on tooth surfaces leading to the formation of dental plaque, which is a complex mixture of bacteria and extracellular materials. These bacteria, together with *Actinomyces* species, can cause caries by forming plaque on the tooth enamel, where they catabolize sugars to produce acid that demineralizes enamel and allows the dentine to be eroded. Abscesses of the roots of teeth can also be caused by mixed bacterial infections.

Periodontal (gum) diseases are inflammatory conditions that attack the gums, bone, and other supporting structures of the teeth. The extent of the inflammatory response depends upon the types of pathogens involved and the effectiveness of the immune response. However, they are major causes of tooth loss in adults. Gingivitis is the earliest form of periodontal disease and occurs when plaque accumulates on the teeth near the gums, which become inflamed and bleed easily. If detected and treated early, gingival tissues will return to normal without long-lasting damage. Untreated gingivitis progresses to periodontitis, which is also known as pyorrhea. Plaque hardens and extends from the gum line to the tooth root causing the gums to detach from the teeth and form pockets. Periodontal pockets create room for greater bacterial activity, particularly of facultative and obligate anaerobic bacteria leading to a progressive cycle of tissue damage until eventually the bone supporting the teeth is destroyed resulting in their loss.

Stomach and intestinal infections are caused by viruses, bacteria, protozoa and worms, all of which may be transmitted in food, contaminated drinking water or by fecal–oral contact. The need for strict personal hygiene is emphasized because these are the most frequent infections of children under five years of age. Approximately 40% of cases of diarrhea (*Chapter 11*) in children are caused by rotaviruses (*Figure 3.12*). In the very young this is potentially lethal and the WHO has estimated that out of the nearly two billion annual diarrhea diseases worldwide, three million end fatally.

Figure 3.13 indicates a number of pathogens that can infect the GIT. The acidic environment and proteolytic enzymes of the stomach kill most

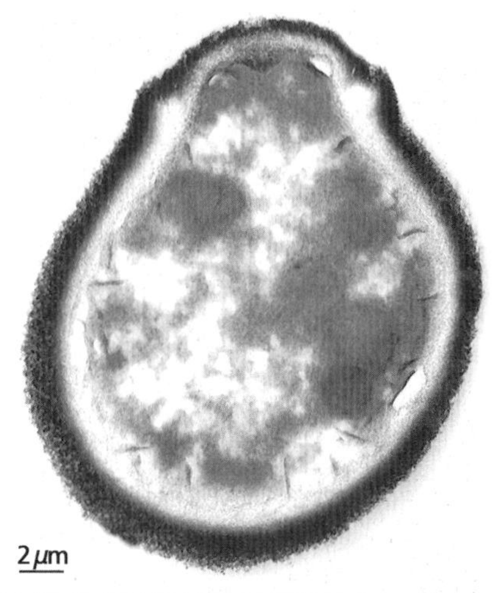

Figure 3.10 Electron micrograph of *Candida albicans*. Courtesy of H. Cotterill, Manchester Royal Infirmary, UK.

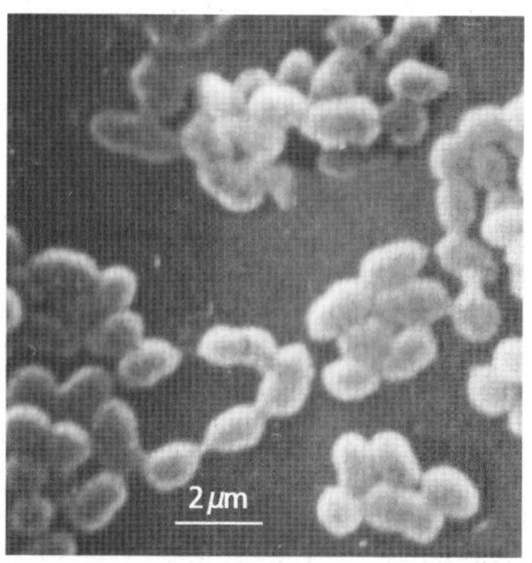

Figure 3.11 Electron micrograph of *Streptococcus mutans*. Courtesy of Professor J. Verran, School of Biology, Chemistry and Health Science, Manchester Metropolitan University, UK.

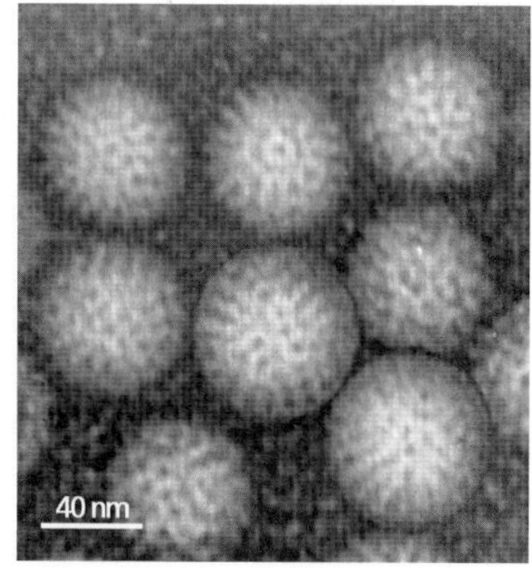

Figure 3.12 Electron micrograph of rotavirus. Courtesy of H. Cotterill, Manchester Royal Infirmary, UK.

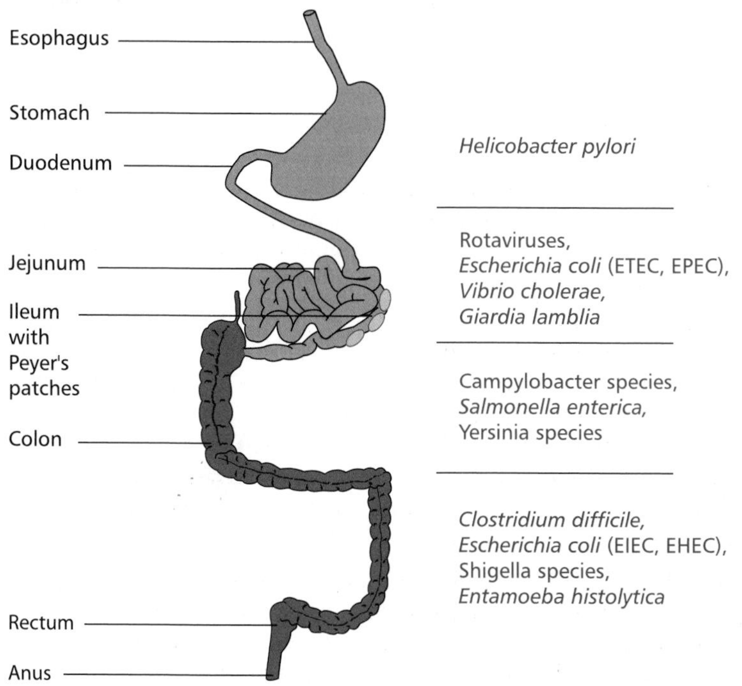

Figure 3.13 Some infectious organisms associated with the gastrointestinal tract. See text for discussions.

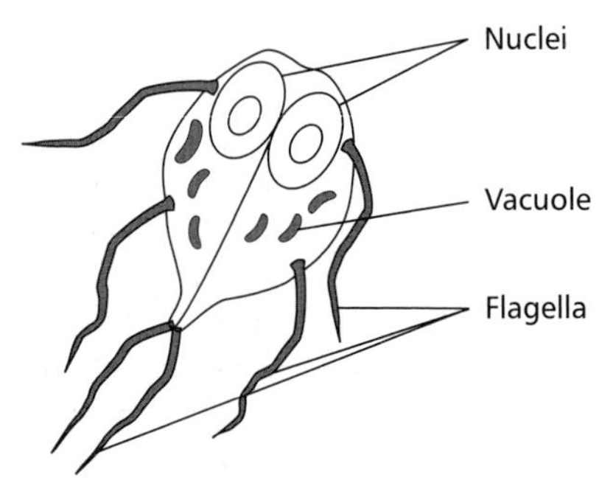

Figure 3.14 Schematic structure of *Giardia lamblia* based on several light and electron micrographs.

ingested microorganisms. However, the motile bacterium, *Helicobacter pylori* (*Chapter 11*) has specific receptors enabling it to bind to the gastric epithelium. It secretes urease, which catalyzes the hydrolysis of urea releasing ammonia that neutralizes stomach acid, and cytotoxins that damage the cells. This causes chronic inflammation of the gastric mucosal membrane and can lead to stomach and duodenal ulcers (*Chapter 11*). The partially digested food (chyme) in the stomach is made alkaline in the small intestine by secretions of the gut and pancreas and by bile salts (*Chapter 11*). In the ileum and jejunum, nonenveloped viruses, such as rotaviruses and adenoviruses, may infect enterocytes (*Chapter 11*) and damage the intestinal mucous membrane with disruption of water and electrolyte resorption. This can result in intestinal cramps, vomiting, watery diarrhea and a raised temperature. Enteropathogens, such as *Vibrio cholerae* and forms of *Escherichia coli* that are enterotoxic (ETEC) or enteropathogenic (EPEC) all have similar effects. The protozoan parasites *Giardia lamblia* and *Cryptosporidium parvum* are water-borne parasites that can infect the GIT. *Giardia lamblia* (*Figure 3.14*) frequently causes chronic disease, with watery diarrhea and, in some cases, a subfebrile temperature leading to malnutrition in children as a result of malabsorption. *Cryptosporidium parvum* can adhere to the epithelium of the small intestine and cause a shortening of the villi, which may be the cause of the diarrhea.

The lower portion of the ileum has areas of lymphoid tissue called Peyer's patches (*Figure 3.13*) composed of so-called M (microfold) cells, rather than the usual enterocytes and goblet cells (*Chapter 11*). These cells are able to translocate materials directly to the lymph follicles found beneath the mucosal surface. Invasive bacteria such as *Campylobacter jejuni* (*Figure 3.15*), Salmonellae and Yersiniae can use M cells to enter the submucosal area. Here they can multiply and destroy the adjacent epithelium, form abscesses and spread through the lymph and blood systems into the mesenteric lymph nodes, spleen and liver. The infection can also spread into the colon, causing inflammation of the colon or **colitis**. The ileum and colon can also be attacked by the bacteria *Yersinia enterocolitica*, *Salmonella enterica* and *Campylobacter jejuni* resulting in abdominal cramps, vomiting, watery, occasionally bloody, diarrhea and fever. *Shigella dysenteriae* and *Escherichia coli* pathotypes, EHEC (enterohemorrhagic) and EIEC (enteroinvasive) can cause a hemorrhagic colitis with bloody stools and subfebrile to febrile temperatures. The pathogenic protozoan *Entamoeba histolytica* is thought to infect 50 million people and kill about 100 000 per year worldwide due to amebic liver abscesses. Lastly, *Clostridium difficile*, a normal inhabitant of the gut, is an opportunistic pathogen. It is especially common in older people in hospitals and nursing homes and has been implicated in iatrogenic infections following medical interventions, such as antibiotic therapy. Infection with *Clostridium difficile* is now recognized as the major causative agent of colitis and diarrhea, which may occur following antibiotic intake and can be fatal in older patients.

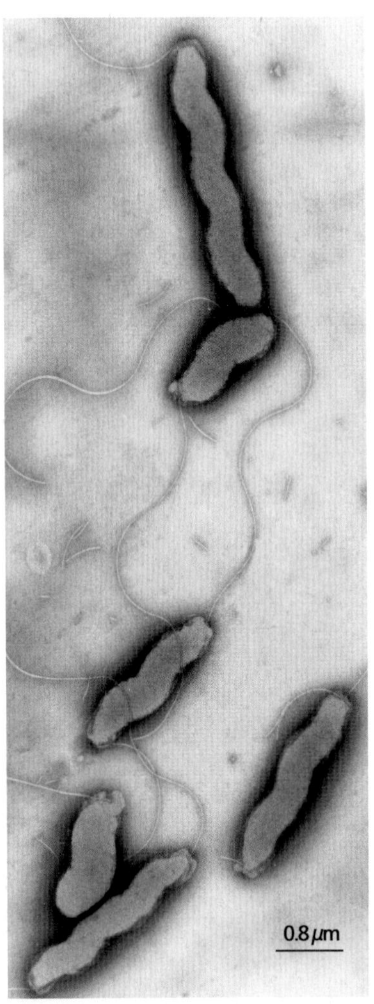

Figure 3.15 Electron micrograph of *Campylobacter jejuni*. Courtesy of Dr A. Curry, Manchester Royal Infirmary, UK.

3.6 INFECTIONS OF THE UROGENITAL SYSTEM

The urinary system and the genital systems are subject to infections by specific respective pathogens (*Figure 3.16*). Urinary tract infections (UTIs) are frequent in the developed world, with many millions of cases occurring each year. A number of factors, including diabetes mellitus (*Chapter 7*), scarring, kidney stones, use of catheters or anatomical peculiarities of the urinary tract all predispose individuals to UTIs. These originate in the perianal area and move up the urethra into the bladder causing a short-lived, acute infection

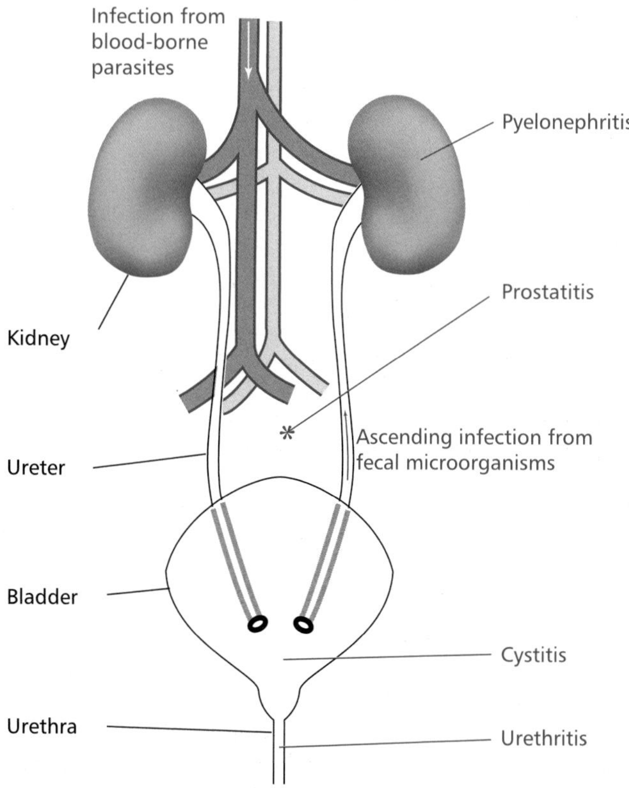

Figure 3.16 Some infections associated with the urogenital system. See text for discussions.

called cystitis. *Escherichia coli* is the most common agent causing up to 80% of the relatively uncomplicated UTIs, like cystitis. Women are more prone than males to cystitis because of their relatively short urethra and the hormonal changes associated with the menstrual cycle (*Chapter 7*). Cystitis is readily treated by increasing the uptake of fluids, particularly of acid drinks like cranberry juice, which causes increased flushing of the bladder, and by the use of antibiotics. Unfortunately, reinfections are frequent.

Pseudomonas aeruginosa (*Figure 2.1*) has been known to cause UTIs following hospitalization. Similarly, infections by *Enterococcus faecalis* and *Klebsiella pneumoniae* have followed organ transplants (*Chapter 6*). Unfortunately, these organisms often show multidrug resistance against antibiotics (*Box 3.4*). Thrush in the bladder, caused by the yeast *Candida albicans* can also occur following antibiotic treatment.

There has been a large increase in the incidence of sexually transmitted diseases (STDs) in the UK in recent years. Sexually transmitted diseases can affect the urinary system and the genital tract and are caused by a wide range of different pathogens. Human papilloma virus (HPV) can cause anogenital warts (condyloma acuminata). Strains 16 and 18 of HPV cause lesions in the cervix that are involved in the development of cervical carcinoma. Infection with *Herpes simplex* virus (HHV-2) can lead to genital herpes with painful ulcers and vesicular lesions of the genital mucous membrane. Infection with human immunodeficiency virus (*Box 3.1*) leads to acquired immunodeficiency syndrome (AIDS).

BOX 3.1 HIV and AIDS

In March 1981, physicians in New York reported eight cases of the previously rare Kaposi's sarcoma (KS), in a form that was markedly more aggressive than usual. All the affected were young gay men. Also in 1981, the Communicable Disease Centers (CDC) in Atlanta, Georgia, began to investigate an increase in requests from New York and Los Angeles for pentamidine, a drug used to treat *Pneumocystis carinii* pneumonia (PCP), an extremely serious fungal infection. In addition to PCP and/or KS, patients suffered a variety of opportunistic pathogens that eventually caused their death. The combination of infections and KS appeared to be the result of a total breakdown of the immune system and became known as Acquired Immunodeficiency Syndrome (AIDS). Subsequently, AIDS was seen in intravenous drug abusers and in a number of recipients of blood transfusions, and a viral cause was suspected.

In 1983, a virus, variously named Lymphadenopathy Virus (LAV) and Human T cell Lymphotropic Virus Type III (HTLV III), was isolated. Following international agreement the virus was renamed Human Immunodeficiency Virus (HIV) (*Figure 3.17*) in 1986. A number of strains of the virus have since emerged; HIVs I and 2 are the most prevalent. HIV is a human retrovirus, that is, its nucleic acid is RNA that on infection is transcribed into DNA by the viral enzyme reverse transcriptase (*Figure 3.18*). An electron micrograph of the virus and an illustration of its replication cycle are shown in *Figures 3.17* and *3.19* respectively. The virus infects the CD4+ Helper T lymphocyte, a key regulatory cell of the immune system (*Chapter 4*), rendering the patient susceptible to a whole range of infections with all microbial groups, from dysentery and diarrhea to pneumonia. For example, the protozoan *Cryptosporidium parvum* can cause a moderate to severe diarrhea that would soon be resolved in healthy people. In HIV infected patients, cryptosporidial diarrhea is among the commonest clinical presentations during the transition to full-blown AIDS, particularly in developed countries. The diarrhea is severe, protracted and may become life threatening. Death is commonly caused by pneumonia associated with the fungus PCP among a host of opportunistic infections. A list of infections common in AIDS patients is shown in *Table 3.2*.

Since the emergence of HIV, the virus has spread worldwide with an estimated 38 million people, including 2.3 million children, living with HIV/AIDS. It is thought that 25 million people have died of AIDS up to the end of 2005. The virus has had devastating effects on communities, particularly in sub-Saharan Africa, where about 8% of the adult population are estimated to be living with HIV/AIDS.

There is no effective treatment that completely clears the body of the virus. Drugs have been developed to inhibit replication of HIV in positive individuals to prevent them developing AIDS. These drugs target enzymes needed at different stages in the replication of the virus and include inhibitors of reverse transcriptase, for example azidothymidine (AZT), and antiproteases, such as amprenavir which prevent the virus from budding from an infected cell as shown in *Figures 1.4* and *2.7*. The combined use of a number of drugs with different actions has been extremely effective at preventing the development of AIDS but aggressive therapy can lead to problems such as HAART (*Margin Note 16.1*). Infected people can now expect to live relatively healthy lives following initial infection. However, combination therapy has to be taken for the remainder of the patient's life. The drugs have a number of side effects and are also extremely expensive. There is evidence that some multiple drug resistance is emerging amongst HIV strains. Since the virus was identified in 1983, the search for a successful vaccine for HIV has been ongoing. All the latest scientific technology has been used in this effort but, to date, no vaccine has been successful.

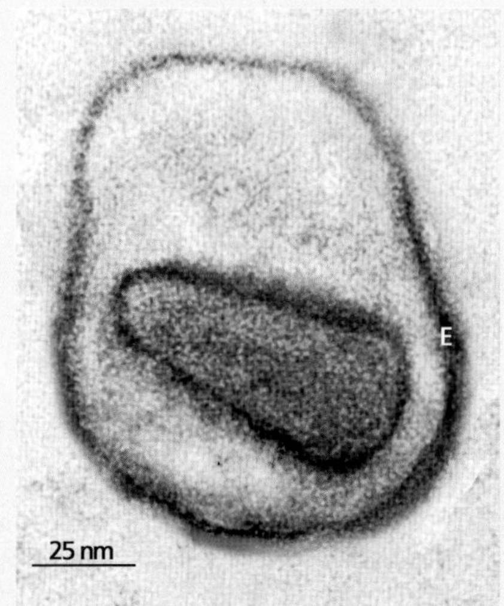

Figure 3.17 Electron micrograph of a single HIV virion. E denotes the envelope.

Disease	Caused by	Infectious agent
Pneumonia	*Pneumocystis carinii*	fungus
Tuberculosis	*Mycobacterium tuberculosis, Mycobacterium avium*	bacteria
Kaposi's sarcoma	Kaposi's sarcoma virus (HHV8)	virus
Lymphoma	*Epstein-Barr virus* (HV4)	virus
Mucocutaneous thrush	*Candida albicans*	yeast
Diarrhea and dysentery	*Cryptosporidium*	protozoan
Shingles	*Varicella zoster* (HHV3)	virus
Cryptococcosis	*Cryptococcus*	fungus

Table 3.2 Some diseases associated with AIDS

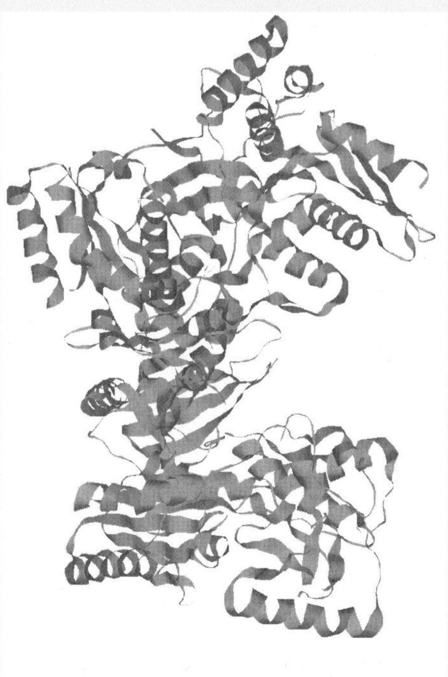

Figure 3.18 Molecular model of the HIV 2 reverse transcriptase. PDB file 1MU2.

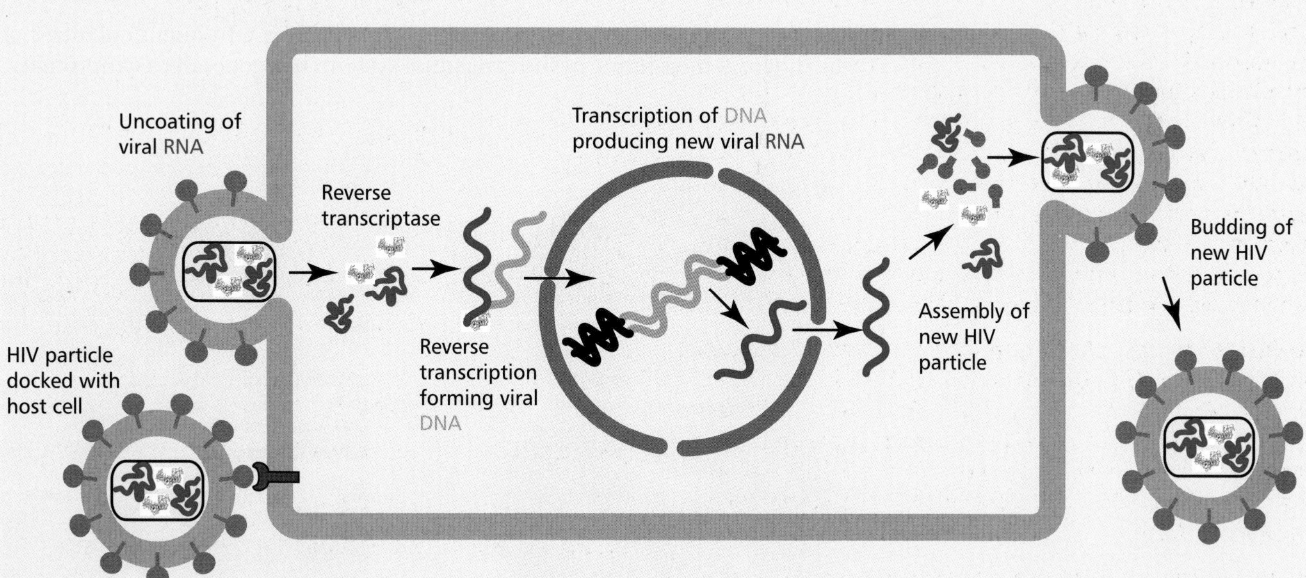

Figure 3.19 The replication cycle of HIV. See text for explanation.

Some types of *Chlamydia trachomatis* are among the commonest causes of STDs producing local inflammations of the urethra and cervix. In contrast, other types are highly invasive and infiltrate the lymphatic system leading to necrosis (lymphogranuloma venereum).

The most widely known STDs are probably gonorrhea and syphilis caused by the bacteria *Neisseria gonorrhoeae*, a Gram-negative diplococcus, and *Treponema pallidum* a spirochete (*Figure 3.20 (A)* and *(B)*) respectively. Gonorrhea, a common STD, is a pelvic inflammatory disease whose major symptoms include a purulent inflammation of the uterine cervix and urethritis. In some women, however, the infection may be relatively asymptomatic and may go unnoticed. Syphilis was thought to have originated in the Americas and been brought to Europe by sailors on the Columbian expeditions. More recent evidence suggests that it was present in the Old World long before this. *Treponema* spirochetes can enter through mucous membranes or minute abrasions in the skin during sexual acts. The infection shows three stages of pathogenesis. Initially, an ulcer called a **chancre** develops at the site of infection. The infection then spreads to nearby lymph nodes causing them to swell and harden. Secondary syphilis develops after one to three months. It is characterized by the presence of highly infectious lesions on various parts of the body. The disease may lie dormant for many years but, if not treated with antibiotics, will develop into tertiary syphilis causing inflammations of the aorta and CNS. Dementia, heart attacks and death can all result. Patients with tertiary syphilis cannot infect others with the disease. Some patients may develop benign late syphilis, which is usually rapid in onset but does respond well to treatment. It usually begins three to 10 years after infection and is characterized by the development of **gummas**. These are tumor-like growths of a rubbery consistency that are most likely to affect the skin or long bones but can also develop in the eyes, mucous membranes, throat, liver and stomach lining. However, since the use of antibiotic treatments for syphilis they are increasingly uncommon.

The protozoan *Trichomonas vaginalis* (*Figure 3.21*) is a frequent colonizer of the mucosal membrane of the urogenital system. It is generally asymptomatic

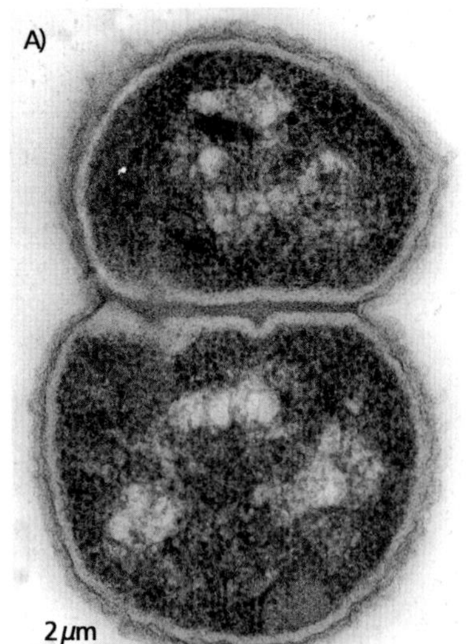

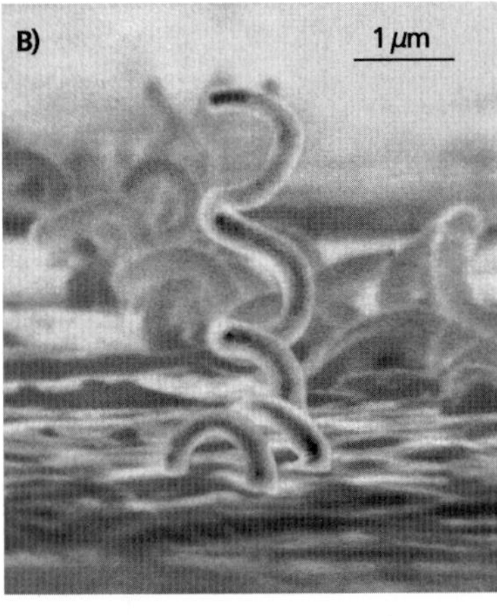

Figure 3.20 Electron micrographs of (A) *Neisseria gonorrhoea* and (B) *Treponema pallidum*. (A) Courtesy of Dr A. Curry, Manchester Royal Infirmary, UK and (B) Public Health Image Library, Centers for Disease Control and Prevention, USA.

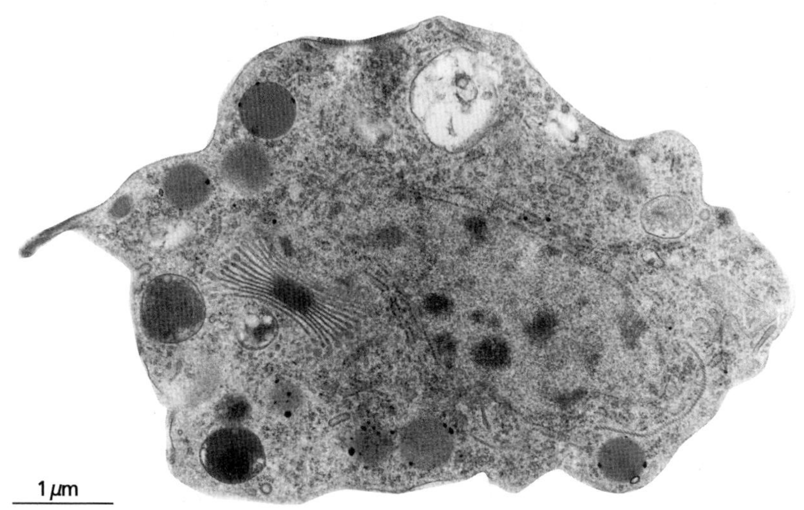

Figure 3.21 Electron micrograph of *Trichomonas vaginalis*. Courtesy of Dr A. Curry, Manchester Royal Infirmary, UK.

but an inflammatory reaction, trichomoniasis vaginitis, may result in a frothy cream-colored discharge.

A number of arthropods may be sexually transmitted. These include crab lice and the scabies mite (Figure 2.14(A)).

3.7 SEPSIS AND SYSTEMIC INFECTIONS

A relatively small number of pathogens are able to enter the body in the lymphatic or circulatory systems and produce a generalized or systemic infection involving numerous body organs: brain, bone marrow, kidneys, liver, lungs and spleen. Local skin infections, such as with *Streptococcus pyogenes* or *Staphylococcus aureus* or infections of the GIT or urogenital system with Enterobacteriaceae can unfortunately progress to acute generalized infections within only a few days, given appropriate conditions. For example, *Streptococcus pyogenes* or *Staphylococcus aureus* secrete toxins called **superantigens** (*Chapters 2* and *4*) that stimulate the release of cytokines from immune cells and produce an excessive inflammatory response called **systemic inflammatory response syndrome** (**SIRS**) consisting of four characteristic stages. In stage I, respiration, heart rate and body temperature all increase. The leukocyte count may be increased or decreased. Stage II refers to the presence of the organism in the blood of the patient; this is called **sepsis**. Normally less than 50% of cases can be identified. Stage II may be cured intrinsically by the immune system or extrinsically by the administration of antibiotics. If unsuccessful, the condition progresses to stage III: serious sepsis or multiorgan dysfunction syndrome (MODS) characterized by lactic acidosis, falling blood pressure, hypoxia and oliguria. If clinical measures fail to stabilize the patient, then stage IV, septic shock, develops with irreversible organ failure and the death of the patient in most cases.

In contrast to the possible acute developments of localized infections, generalized infections may have incubation times of up to three weeks. A classical example of such a systemic infection is typhoid fever caused by *Salmonella typhi* (*Figure 3.22*). The pathogen may be ingested and can enter the body through the tonsils in the throat and Peyer's patches in the gut. The bacterium is distributed to various organs in the lymph and blood, infects

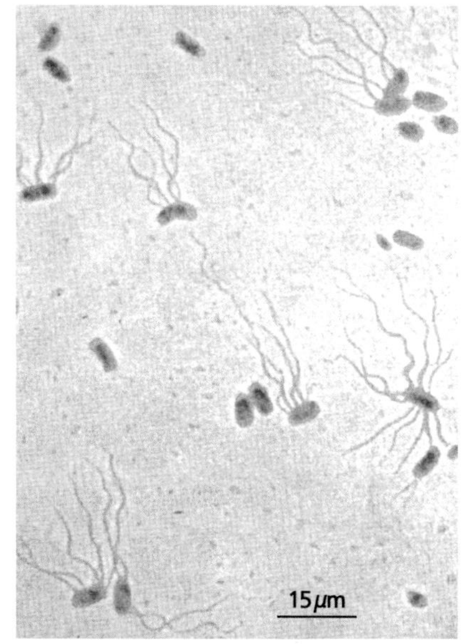

Figure 3.22 Light micrograph of *Salmonella typhi*. Courtesy of Public Health Image Library, Centers for Disease Control and Prevention, USA.

Margin Note 3.2 *Salmonella typhimurium*

Salmonella typhimurium (*Figure 3.23*), as its name suggests, causes a lethal, systemic infection that resembles typhoid fever in mice. However, in humans it does not cause as severe a disease as *Salmonella typhi* although it is a major cause of food poisoning. This rarely leads to fatalities, other than in the elderly or very young or in patients with depressed immune systems who are not treated with antibiotics. The disease is characterized by symptoms of nausea and vomiting, abdominal cramps and diarrhea that generally last up to seven days.

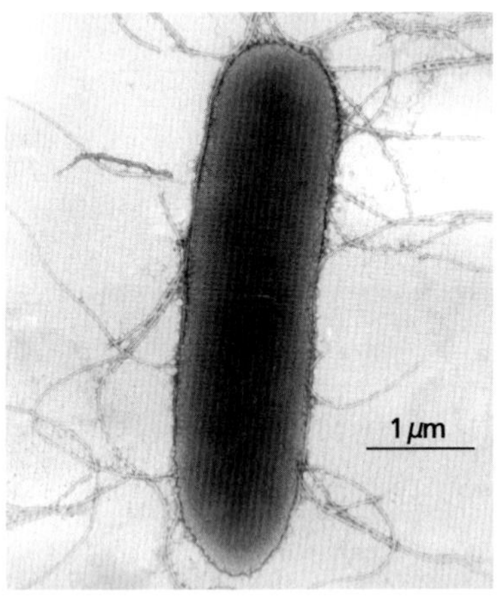

Figure 3.23 Electron micrograph of *Salmonella typhimurium*. Courtesy of Dr A. Curry, Manchester Royal Infirmary, UK.

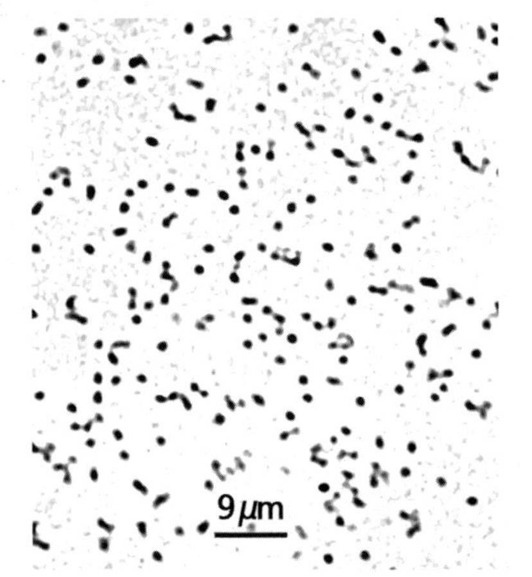

Figure 3.24 Light micrograph of *Brucella* species. Courtesy of L. Stauffer, Public Health Image Library, Centers for Disease Control and Prevention, USA.

cells and grows intracellularly. The major symptoms are fevers of up to 40°C and the development of abscesses in the intestine that may cause perforations and a fatal peritonitis. Even when the patient recovers, a relapse may occur. *Salmonella typhi* infects about 60 million patients annually, mainly in the developing world. If untreated, it can be fatal. Some recovered patients retain the pathogen in the gall bladder and excrete bacteria in their feces, becoming carriers, although remaining healthy themselves.

The zoonosis, brucellosis, caused by various species of *Brucella* (*Figure 3.24*), is also a primary systemic infection. This is a common infection of cattle (*Brucella abortus*), sheep and goats (*Brucella melitensis*) and can infect humans if they drink contaminated milk, if they inhale the bacteria into the lungs or if there is direct contact through skin lesions. People who work in direct contact with animals, such as farmers, veterinarians, shepherds, goatherds and abattoir workers, are most at risk. The pathogen colonizes and grows in the main abdominal organs and bone marrow producing periodic attacks of relapsing fever over an extended period.

3.8 INVESTIGATING INFECTIOUS DISEASES

To diagnose an infectious disease, two criteria must be satisfied. First, signs and symptoms compatible with the suspected infectious agent must be apparent. Secondly, the pathogen must be recovered from the infected site of the patient or there must be evidence of the pathogen being present at that site.

Infectious diseases can affect any organ or system and can cause a wide variety of symptoms and signs (*Chapter 1*). The clinical history and examination should aim to identify the sites of infection and the causative organism. The clinical history focuses on aspects relevant to infectious disease, such as recent travel history, food and water intake, occupational exposure, sexual activity and any use of intravenous drugs. The clinical examination involves identifying fever, skin rashes, swollen lymph nodes (lymphadenopathy) and investigations of the eyes, ears, mouth and throat. Fever is a typical symptom of infection but not all patients with fever have an infection and not all infectious diseases present with fever. Investigations of the vagina, rectum and penis are necessary in sexually transmitted diseases.

The history and clinical examination is often supported by tests to assess health and identify the organs affected. These tests include imaging techniques such as X-rays, ultrasound, computerized tomography (CT), magnetic resonance imaging and blood tests (*Chapters 18* and *13*). The blood tests involve a full blood count where eosinophilia is an important finding in parasitic infections and lymphocytosis is usually found in viral infections. The erythrocyte sedimentation rate and C-reactive protein (*Chapter 4*) levels are nonspecific tests of value in monitoring the course of infectious diseases. Other tests involve assessing liver and renal functions (*Chapters 11* and *8*), which may be disrupted by an infection.

Microbiological tests to identify the infectious agents are especially helpful. Specimens for microbiological investigations include blood, cerebrospinal fluid (CSF), feces, pus, sputum and urine (*Chapter 1*). Microbiological investigations use a variety of techniques, including culture, serological, biochemical and molecular biological tests.

CULTURE

The suspected microorganism from the patient is grown outside the body usually on or in growth media, such as nutrient agar or broth, or in selective media which support the growth of particular microorganisms, until growth is detectable. In the case of bacteria, characterization is based on their microscopic appearance and the shape, texture and color of colonies. To some extent, bacteria can be identified by their ability to grow in specific media, such as blood agar or Sabouraud's agar. Indicator media that include some substance that visibly changes as a result of the metabolic activities of particular microorganisms are also of use in identification. Fungi and mycoplasmas are cultured in similar ways to bacteria. However, they require a greater use of microscopic and colonial morphology for identification. Certain microorganisms, such as chlamydiae, and all human viruses, are intracellular pathogens and their growth requires the inoculation of cultured eukaryotic cells. When viruses infect and replicate within cultured cells, pathological changes are produced that are characteristic of particular viruses and can be used for identification purposes. Moreover, electron microscopy of supernatants from these cultures, or even on patient samples, can show the presence of particular viruses. For example, Norwalk virus (*Figure 3.25*), which causes outbreaks of vomiting and diarrhea, has been identified in stool samples of affected individuals. The diagnosis of parasitic infections, protozoa and helminths, may involve growing them in culture. More frequently, parasites are identified directly from specimens isolated from infected patients and/or indirectly by examining the cysts or eggs of the parasite.

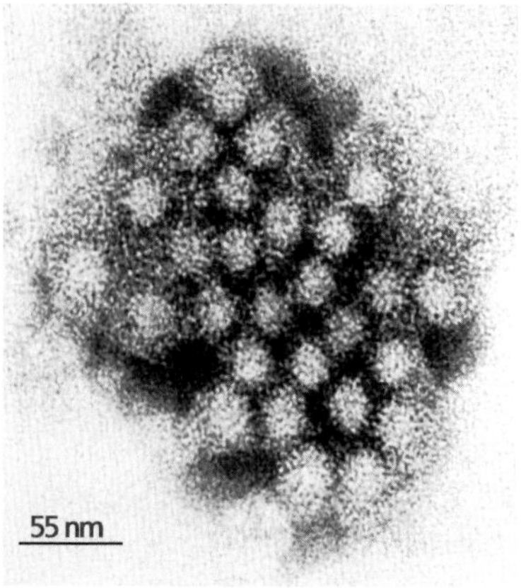

55 nm

Figure 3.25 Electron micrograph of Norwalk virus. Courtesy of Public Health Image Library, Centers for Disease Control and Prevention, USA.

SEROLOGICAL TESTS

Serological tests involve identifying infectious agents indirectly by measuring serum antibodies in the affected individual. Antibody levels against a pathogen increase in the early stages of the disease and then fall during recovery. Such tests are particularly useful in situations where it is impossible to isolate the infectious agent and are used to make a diagnosis of, say, HIV infection. It is preferable to take a blood sample early in the infection (acute serum) and 10–14 days later (convalescent serum). A fourfold or greater rise in antibody

BOX 3.2 API strips

The API system, named from the parent company *Appareils et Procédés d'Identification*, for identifying bacteria and yeasts was first developed in the 1970s. Although the tests themselves were not new, the system used standardized and miniaturized versions of the existing biochemical tests on a single strip. Each test strip contains a number of separate compartments that contain the dehydrated reagents necessary for each specific test. The test is typically performed by forming a homogeneous suspension of the microorganism to be identified in 0.85% NaCl solution. Samples of the suspension are then added to each of the wells and this also rehydrates the reagents. The organisms will produce some observable change in the wells, for example color changes due to pH differences or enzyme activities or form end products that can be identified. Any one well will, of course, give a positive (+) or negative (–) result. A number of tests are necessary to identify a species or strain and the tests on any one API strip give a profile or numerical identifier that is the sequence of positive and negative test results (*Figure 3.26*). The organisms can then be named from a codebook that correlates this sequence with the bacterial species or strain or, more usually, identifications can be made with a computerized database. Organisms can generally be identified accurately and reliably in four to 48 h, depending upon the strips used and the species of microorganism concerned.

API tests include 15 identification systems for almost all groups of bacteria and over 600 different species. These include Gram-negative and Gram-positive bacteria, such as species and subspecies of Enterobacteriaceae, bacilli, *Campylobacter*, Corynebacteria, enterococci, Listeria, micrococci, Staphylococci and Streptococci and some anaerobic bacteria and yeasts.

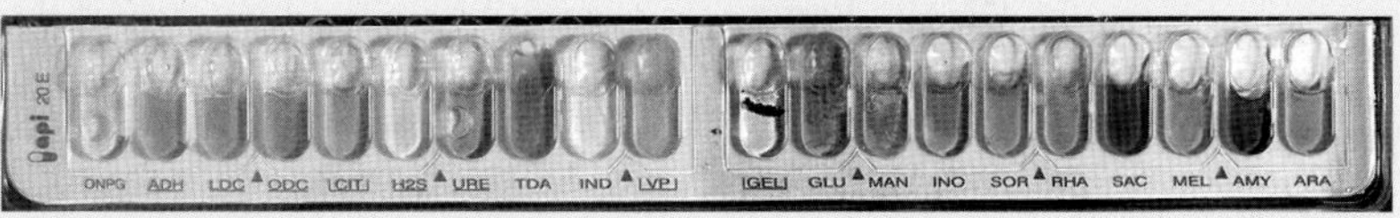

Figure 3.26 A developed API strip showing positive (darker colored) and negative (clear) results for each of the test wells.

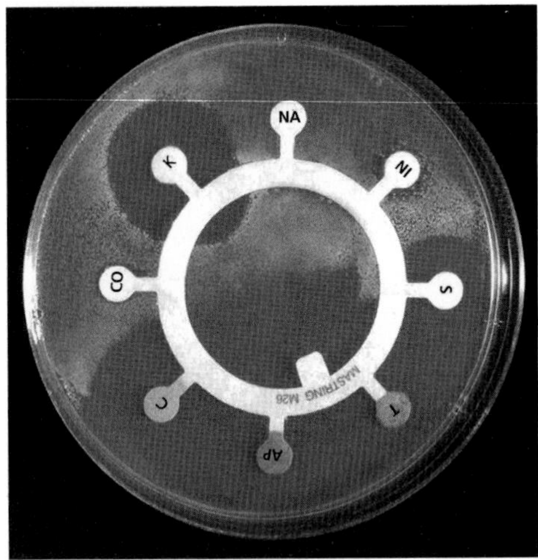

AP Ampicillin
C Chloramphenicol
CO Colistin sulfate
K Kanamycin
NA Nalidixic acid
NI Nitrofurantoin
S Streptomycin
T Tetracycline

Figure 3.27 A multidisc containing a different antibiotic in each lobe as indicated, which can be used to assess the antibiotic susceptibility or resistance of bacteria. The bacteria in this case are a strain of *Escherichia coli* that cannot grow in the area surrounding an antibiotic to which they are susceptible.

titer in the second serum is diagnostic of an acute infection. Antibodies to bacteria can be detected by their ability to agglutinate these microorganisms. Alternatively, a variety of immunotechniques are available which are outlined in *Chapter 4*.

BIOCHEMICAL AND MOLECULAR BIOLOGICAL TECHNIQUES

A variety of biochemical and molecular biological tests are available to help identify microorganisms. The simple Gram stain can immediately eliminate many possible bacteria. The abilities of some bacteria specifically to ferment some carbohydrates, use different substrates, express restricted enzyme activities or form specific products in enzyme catalyzed reactions can all aid in identification. The susceptibility of bacteria to different antibiotics (*Figure 3.27*) can also be useful and has the added utility of indicating possible therapy.

The GC ratio of the DNA of bacteria is generally expressed as a percentage 100(G + C) / (A + T + G + C) and varies from 20 to nearly 80%. Any one particular group of bacteria has a characteristic value. Specific genes from pathogenic organisms have been cloned and sequenced. That for the 16S ribosomal RNA has received considerable attention and, indeed, differences in the sequences of this gene have allowed evolutionary relationships between different groups of bacteria to be deduced. The entire genomes of a number of viruses, pathogenic bacteria and parasitic protozoa have been sequenced and the number is growing rapidly. Nucleic acid probes can be designed to detect characteristic sequences of pathogens and identify the organism in body fluids or tissues. This technique has been enhanced by the development of the **polymerase chain reaction**.

The polymerase chain reaction

The polymerase chain reaction (PCR) was devised by Mullis in 1983. He was awarded the Nobel Prize in Chemistry in 1993; the shortness of time between the two dates is indicative of the perceived importance of PCR. Indeed, it is hard to exaggerate the impact of PCR because it has revolutionized molecular biological techniques. The PCR is an elegantly simple *in vitro* method for increasing in an exponential manner the number of relatively short strands of specific DNA fragments, normally of 500 to 5kbp, although longer lengths of up to 40 kbp can also be amplified. These may be whole genes but are more usually only a fragment of one. DNA consists of two polynucleotide strands arranged together in the now familiar double helical structure. The strands run in opposite directions: one in the $3' \rightarrow 5'$ direction, the other with a $5' \rightarrow 3'$ orientation. The bases of the nucleotides of each strand associate together such that they form complementary pairs, with an adenine (A) of one strand paired with a thymine (T) of the other, and guanine (G) pairing with cytosine (C). The base pairs, and therefore the polynucleotide strands, are held by hydrogen bonds. It is the sequence of bases in the DNA strand(s) that is unique and specific to a particular gene.

The PCR is used to replicate the original DNA sample (template DNA) using a DNA dependent DNA polymerase, usually abbreviated to DNA polymerase. This enzyme copies the template DNA to produce new strands with complementary sequence. Some DNA polymerases can proofread, that is correct any mistakes in the newly formed strand to ensure the fidelity of its sequence. Crucially, DNA polymerases cannot begin the new strand *de novo*, but can only extend an existing piece of DNA. Thus two primer molecules are required to initiate the copying process. The primers are artificial oligonucleotides, short DNA strands of fewer than 50 nucleotides that are complementary to regions that flank the section of the template DNA of interest. Hence the primer determines the beginning of the region to be amplified. Primers are usually made to order by commercial suppliers who must, of course, be supplied with the required sequence.

The PCR consists of a series of cycles, each of which consists of three identical steps (*Figure 3.28*). First, the double-stranded DNA has to be heated to 90–96°C to break the hydrogen bonds that connect the two strands and allow them to separate. This denaturation is called melting. Prior to the first cycle, the DNA is often heated for an extended time, called a 'hot start', of up to 5 to 10 min to ensure that the template DNA and the primers are fully melted. In subsequent cycles, 30 s to 3 min at 94 to 96°C is normally sufficient for melting. Placing a heated lid on top of the reaction tubes or a layer of oil on the surfaces of the reaction mixtures prevents evaporation. The second step is primer annealing at temperatures of 50 to 65°C for 30 to 60 s, during which the primers bind to their complementary bases on the now single-stranded DNA templates. The primers must be present in excess of the target DNA otherwise its strands will simply rejoin. The design of the length of the primers requires careful consideration. Primer melting temperature, which must not be confused with that of the template DNA itself, increases with the length of the primer. The optimum length for a primer is generally 20 to 40 nucleotides that have melting temperatures of 60 to 75°C, although this depends upon their G/C content. If the primers are too short they will anneal at random positions on the relatively long template and result in nonspecific amplification. However, too long a primer and the melting temperature would be so high, that is above 80°C, that the DNA polymerase would have reduced

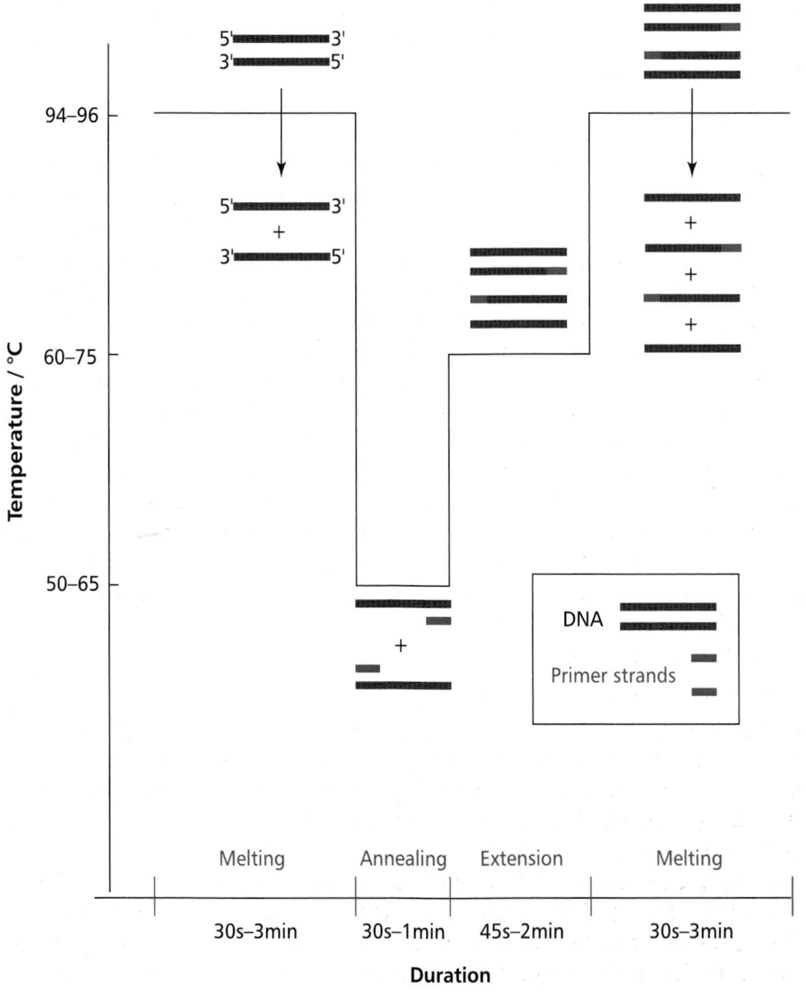

Figure 3.28 Outline of the steps associated with the cycles of the polymerase chain reaction. See text for details.

activity. The third step is the synthesis of DNA in reactions catalyzed by the polymerase and which extend the primers using the complementary strands as templates. This usually requires 45 to 120 s at 72°C. The extended portion of DNA is a complement of the template strand. Given the high temperatures of the reactions, DNA polymerases from thermophilic organisms are preferred. The *Taq* polymerase from *Thermus aquaticus* is widely used although it has the disadvantage of lacking proofreading capabilities and therefore of introducing errors (mutations) of 1 in 400 to 500 nucleotides in the new DNA. Polymerases such as *Pwo* or *Pfu*, obtained from Archea, have proofreading mechanisms that significantly reduce mutations and are used in 'long range' PCR of up to about 30 kbp.

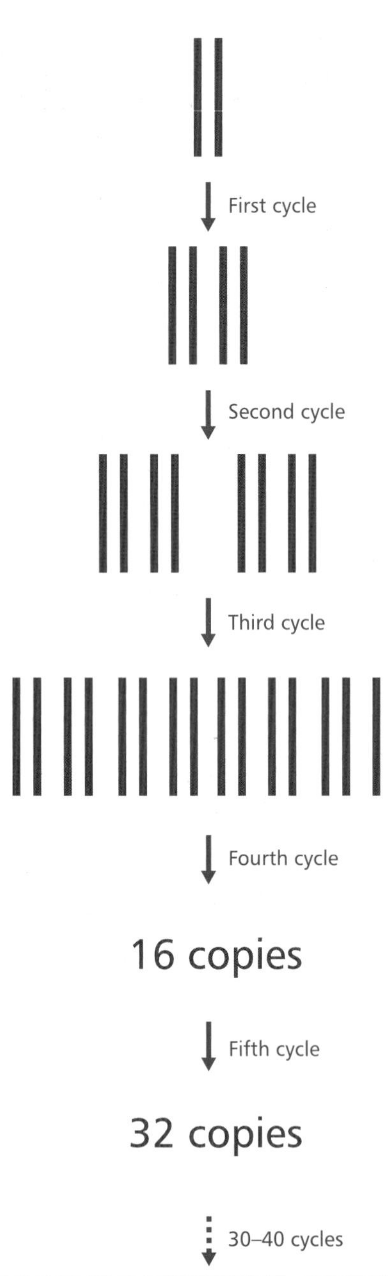

Figure 3.29 The amplification of DNA molecules during cycles of the polymerase chain reaction.

The result of the first cycle is two helices that are usually overextensions of the target sequences. Each is composed of one of the original strands plus its newly assembled complementary strand and its associated primer, for each of the original template DNA molecules. The cycle is usually repeated 20 to 30 times in identical conditions with a doubling of the amount of DNA present for every cycle (*Figure 3.29*). Hence after 30 cycles the amount of DNA in the original sample has increased over 2^{30} (10^9 fold)! A PCR experiment normally terminates with a 10 min incubation at 72°C to ensure that all of the amplified DNA molecules are fully extended by the polymerase. Note that extremely stringent conditions are necessary to prevent any unwanted DNA contaminating PCR experiments, since this would also be amplified.

The great advantages of PCR are the increase in sensitivity and its automation. Samples of DNA from even a single cell or from samples many years old can be amplified and then analyzed. The PCR reaction is automated in a thermocycler, which automatically heats and cools the reaction tubes to the appropriate temperatures, for the desired times and the required numbers of cycles.

The products of PCR are identified by determining their base sequences and/or their sizes using agarose or polyacrylamide gel electrophoresis against a known sample. The size of the product can be estimated by comparison with the electrophoretic mobility of fragments of DNA of known size (*Figure 3.30*).

The biomedical sciences apply PCR in four main areas, which are the identification of infectious disease organisms for diagnostic purpose, in the detection of variations and mutations in hereditary diseases (*Chapters 15* and *16*), detecting acquired mutations that lead to cancers (*Chapter 17*) and in tissue typing (*Chapter 6*). It is especially useful in diagnosing diseases caused by organisms that are difficult or impossible to culture. Its ability to amplify incredibly small amounts of DNA means that PCR-based tests can identify sources of infection more accurately, reliably, rapidly and cheaply than previous methods. For example, it can detect three different sexually transmitted

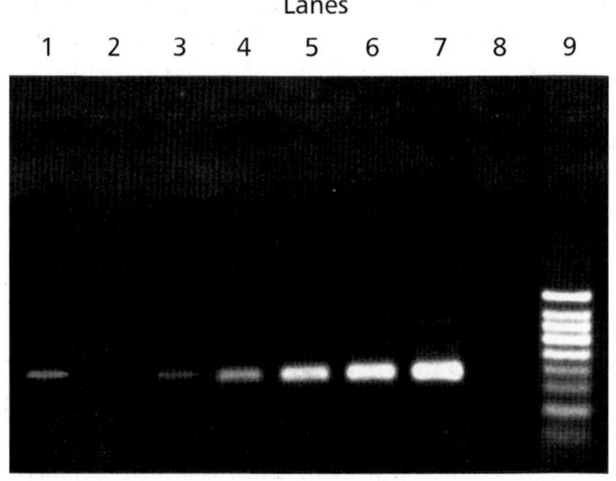

Figure 3.30 A polyacrylamide gel electrophoresis of the products of a polymerase chain reaction experiment. Lane 1 shows the DNA sample of size 390 bp (positive control) which has been amplified by 10, 20, 25, 30, 35 and 40 cycles of polymerase chain reactions in lanes 2 to 7. Lane 8 shows the results of a negative control in which the DNA sample was replaced by water and amplification did not occur. Lane 10 shows the separation of DNA molecules of known sizes between 100 and 1000 bp as a calibration marker. Courtesy of Dr Q. Wang, School of Biology, Chemistry and Health Science, Manchester Metropolitan University, UK.

disease agents, *Herpes*, papillomaviruses and *Chlamydia* on a single swab. Polymerase chain reaction tests can even distinguish the specific strains of papillomaviruses that predispose individuals to cervical cancer. Diagnostic tests are also available for the viruses involved in AIDS, viral hepatitis and viral meningitis. The PCR is the most sensitive and specific test for *Helicobacter pylori*, the main cause of stomach ulcers (*Chapter 11*). In England and Wales, 47% of cases of meningococcal infections in 2002 were diagnosed using PCR tests for meningococcal DNA in clinical samples. Bacterial infections in middle ear fluid from children suffering otitis media are detectable by PCR, indicating an active infection, even when standard culture methods fail. The Lyme disease bacterium, *Borrelia burgdorferi*, is often difficult to diagnose accurately on the basis of general symptoms but PCR can amplify its DNA in body fluids.

Margin Note 3.3 

The extreme sensitivity of PCR means it can even diagnose the diseases on old and therefore largely degraded samples. The former USA vice president and presidential candidate Hubert H. Humphrey (1911–1978) underwent tests for bladder cancer in 1967. However, these gave negative results and, untreated, he died of the disease. A urine sample taken in 1967 and a tissue sample from his cancer-ridden bladder obtained in 1976 were analyzed retrospectively with PCR amplification in 1994. The DNA of both samples showed identical mutations in the p53 gene (*Chapter 17*). This is a well-established tumor suppressing gene. Had it been possible to diagnose the cancer in 1967, then Humphrey could have received the benefits of the then current treatments and his life may have been extended. This is, of course, only one of many examples where advances in biomedical sciences have greatly improved clinical practice.

3.9 PREVENTING INFECTIOUS DISEASES

Infections are the commonest cause of human morbidity and mortality. In developing countries at the beginning of the last century diseases, such as TB, pneumonia and bacterial infections secondary to influenza, were major causes of death. However, in the developed world their impact has been lessened by public health measures such as improved housing, better sanitation and advanced social and economic conditions. Unfortunately, in developing countries infectious diseases, such as malaria, TB, respiratory and GIT infections, are still major causes of death, particularly in children.

The prevention of infectious disease is achieved by the use of vaccines (*Chapter 4*). Vaccination works by stimulating the immune system to produce antibodies against the pathogenic organism by introducing bacteria or viruses that have been rendered 'harmless' in some way. The simplest, though not necessarily the most effective, method of preparing vaccines is by killing whole microorganisms. These are then injected into the host to induce an immune response. An example of a 'killed' vaccine is that used to protect against whooping cough caused by *Bordetella pertussis*. In some instances the immune response produced against dead organisms is insufficient to induce good immunity. This is usually because killing the microorganism often involves denaturing their proteins so while the immune response recognizes the denatured proteins it does not react to the native proteins on the pathogen. To overcome this problem, live but attenuated (weakened) microorganisms may be used. These microorganisms are less virulent and, in most cases, stimulate an effective immune response in the host. This type of approach is used for the combined vaccine against measles, mumps and rubella (MMR) and for the oral vaccine against poliomyelitis (*Margin Note 3.4*).

Some patients do not develop an effective immune response to these weakened microorganisms and, unfortunately, the weakened microorganisms can become virulent again, a phenomenon known as **reversion**. To overcome this problem, selected proteins, from, for example, bacterial capsules or viral envelopes, are extracted from the microorganism and used as vaccines. These are known as subunit vaccines. Unfortunately, bacterial capsular polysaccharide is often poor at stimulating immunity and a recent development is to render the vaccine more immunogenic by attaching the polysaccharide to an immunogenic protein. Such vaccines are known as **conjugate** vaccines: examples include the most recent vaccines against *Neisseria meningitidis* serogroup C, and *Haemophilus influenzae*.

Subunit vaccines against viral proteins can now be produced more cheaply by employing recombinant DNA techniques. In such cases, nucleic acid coding for the protein in question is isolated and cloned. This DNA is then transfected into a suitable microorganism which can be cultured and induced to synthesize

Margin Note 3.4 Baby and childhood vaccination programmes

Most countries of the developed world have baby and childhood vaccination programmes that protect against a variety of potentially life-threatening conditions. In early 2006, the UK government announced it was amending its vaccination programme for children less than two years old by adding a new vaccine to its programme that protects against the Pneumococcus bacterium, which causes ear infections, pneumonia and meningitis. This would give the UK a fairly typical baby vaccination programme, as follows. At the age of two months, the baby is given a vaccine against Pneumococcus and a five-in-one vaccine that protects against Hib, diphtheria, polio, tetanus and whooping cough. At three months, a meningitis C vaccine and a five-in-one booster are administered. These are followed a month later with another five-in-one booster and boosters for meningitis C and Pneumococcus. When the baby is one year old, he or she receives a combined Hib/meningitis C vaccine, which is followed a month later with the MMR (combined mumps, measles and rubella) vaccine and another Pneumococcal booster.

the protein *in vitro*. An example is the recombinant vaccine against Hepatitis B produced in the yeast *Saccharomyces cerevisiae* that have been transfected with a gene encoding the S (surface) protein of the virus. More recently, DNA vaccines, which consist of viral genes transfected into bacterial plasmids and injected directly into muscle, have been undergoing clinical trials, although, as yet, none of these is routinely available.

For those infectious diseases where toxins (*Chapter 2*), rather than the microorganism, are responsible for the disease, vaccines may be prepared against chemically modified or heat inactivated toxins. These inactive toxins, known as **toxoids**, are then used for vaccination purposes and, indeed, this is the approach used for diphtheria and tetanus vaccines.

3.10 CONTROLLING THE SPREAD OF PATHOGENS

Controlling the spread of an infectious agent is a complex process, requiring a number of interventions, some of which will depend on the microorganism involved. Epidemiological investigations (*Chapter 1*) can help to pinpoint the source of the infection. Strict hygiene precautions, for example thorough cleaning and disinfection of contaminated materials, and hand washing, can prevent the spread of pathogens between patients. The boiling of drinking water may be necessary, depending on the source of the outbreak. Routine microbiological investigations may be carried out to ensure that the microbe is under control. For the patient, antibiotics or antiviral drugs (*Section 3.11*) may be given, while patient contacts may be offered vaccination and/or drug treatment. Examples of this multifactorial approach can be seen following outbreaks of *Neisseria meningitidis*, in schools and colleges, where all contacts with infected patients are offered antibiotics and vaccines, depending on the serotype.

3.11 TREATMENT OF INFECTIOUS DISEASES

The treatment of infectious disease is almost entirely pharmacological although nursing care is obviously necessary in many cases and surgery may sometimes be used. The development and extensive uses of antibiotics and other drugs over the last 60 years in particular has had an enormous impact in reducing the number and effects of infectious diseases. Viruses and bacteria are the most important causes of infectious diseases in developed countries, while fungi, protozoa and helminths are of increased importance in the tropical climates of developing countries.

ANTIVIRAL DRUGS

Viruses are acellular and can only replicate by utilizing the metabolic processes of their host cell. It is difficult to target viruses inside host cells without damaging both infected and normal host cells hence there are relatively few effective antiviral agents. To be fully effective, antiviral drugs should ideally inhibit viral replication but not affect the reproduction of the host cell. Unfortunately, current antiviral drugs are not fully effective as all interfere to some extent with reproduction of the host cell and so produce adverse effects. Antiviral agents do not normally directly 'kill' the virus but, rather, inhibit their replication. Therefore, they must be administered for sufficient time to allow natural immune mechanisms to eradicate all the virions present. Thus antiviral therapy may well fail in severely immunocompromized patients.

Most antiviral agents act by disrupting one of the steps in the replication cycle of the target virus. They may prevent viral internalization into the host cell. If the virion enters its host cell, other antiviral agents can interfere with the release of the viral nucleic acid from the capsid. Some antiviral agents prevent the synthesis of the viral nucleic acid or its proteins. Even if the virus is successfully synthesized, agents that interfere with its release from the host cell can prevent its dissemination. Lastly, some antivirals can help promote a more effective immune response against viral infection. *Table 3.3* lists examples of each type of antiviral agents.

Mode of action	Example of agent
Prevent internalization	γ-globulin, zanamovir
Inhibit uncoating	p133
Prevent viral nucleic acid synthesis	amantadine, gancyclovir, rimantadine, vidarabine
Inhibit viral protein synthesis	indinavir, ritonavir, saquinavir
Interfere with virion release	4-guanidino-Neu5Ac2en
Promote immune response	α-interferon

Table 3.3 The mode of action of antiviral agents with selected examples

Figure 3.31 Molecular model of HIV protease with a bound inhibitor. PDB file 1HII.

Viruses can become resistant to specific antiviral agents (*Box 3.4*). Hence a therapy combining several agents, each of which acts at a different stage in the replication cycle is likely to be a more effective treatment. For example, HIV treatment is typically a combination of the antiHIV proteases (*Figure 3.31*) amprenavir, ritonavir and the antiHIV reverse transcriptase inhibitor AZT, which is a ribonucleoside analog (*Figure 3.32* and *Box 3.1*).

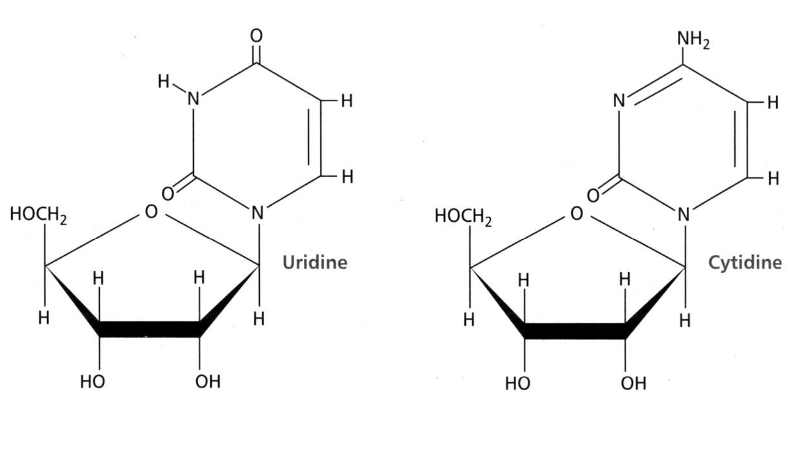

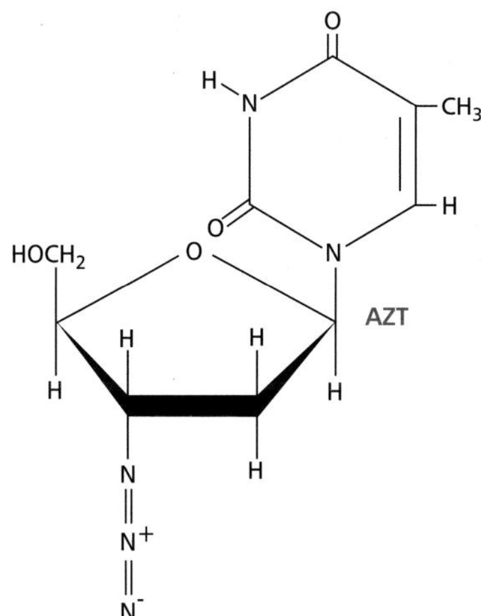

Figure 3.32 Structures of the ribonucleosides uridine and cytidine and the antiHIV drug, AZT.

ANTIBACTERIAL DRUGS

Drugs to treat bacterial diseases are generally called **antibiotics.** They are the most significant in clinical practice because bacteria are responsible for the greater proportion of infectious diseases. Most antibiotics are derived from the products of the metabolism of microorganisms, such as certain bacteria and fungi. Their actions usually rely on differences between microbial and host cells. Ideally, antibiotic drugs should kill the target bacterium, that is, they should be **bactericidal. Bacteriostatic** drugs prevent the bacteria replicating and must be administered for sufficient time to allow the immune defenses of the body to eliminate the pathogen.

Figure 3.33 Molecular model of rifampicin (a rifamycin antibiotic) shown in red, bound to an RNA polymerase. The spheres represent magnesium and zinc atoms. PDB file 1I6V.

Not all antibiotics are effective against all pathogenic bacteria. Some are only effective against Gram-negative bacteria, whereas others are effective against Gram-positive only (*see later*). Other antibiotics can be used to treat a range of both Gram-positive and negative organisms and are referred to as **broad spectrum** antibiotics. Antibacterial drugs generally act in one of four major ways: they may inhibit the synthesis of the bacterial nucleic acid, proteins or the cell wall or they may act in a variety of rather miscellaneous and, in some cases, unknown mechanisms.

Many drugs inhibit both replication and transcription at several points. These drugs fall into a number of general families. Sulphonamides inhibit the formation of folate which is essential for the synthesis of precursors of nucleic acids. Nitroimidazoles bind directly to DNA and denature its helical structure such that it is no longer a substrate for DNA-binding proteins. Clofazimine, a drug which is used to treat leprosy, also binds to DNA preventing replication and transcription, though it is not a nitroimidazole. Quinolones, nalidixic acid and norfloxin, and the synthetic antibiotic ciprofloxacin, fluoroquinolones, offloxacin, norfloxacin and others, are inhibitors of DNA topoisomerase II, an enzyme essential for the replication and transcription of DNA. The rifamycins are inhibitors of RNA polymerases (*Figure 3.33*) and suppress transcription.

Protein synthesis begins with the translation of messenger RNA molecules by ribosomes to form polypeptides. Translation is broadly similar in both bacterial and mammalian cells though there are some significant differences between the two types of cells. For instance, bacterial ribosomes consist of 30S and 50S subunits (*Figure 3.34*), whereas eukaryotic ones have larger 60S and 40S subunits. A number of the protein translation factors necessary for translation also differ. These differences are exploited by a number of antibacterial drugs.

The main antibacterial antibiotics that interfere with protein synthesis are the aminoglycosides, lincosamides, macrolides and tetracyclines. The streptogramin quinupristin-dalfopristin, a relatively newly introduced drug, also interferes with protein synthesis. Aminoglycosides, such as streptomycin and gentamicin, are bactericidal and have complex effects following irreversible binding to specific proteins of the 30S subunit. They inhibit protein synthesis by interfering with initiation, inhibiting an essential checking step called proofreading performed by the ribosome so incorrect amino acids are inserted into the polypeptide leading to the production of nonfunctional or toxic peptides, inhibit elongation and prevent ribosomes associating together as functional polyribosomes. Clindamycin and lincomycin are lincosamides. These antibiotics can be bacteriostatic or bactericidal. They interfere with the first and subsequent steps in translation within the ribosome. Macrolides may be bacteriostatic or bactericidal. Examples include the widely used erythromycin and azithromycin and clarithromycin. They prevent translocation, that is, the movement of the peptide-tRNA complex within the ribosome. The tetracyclines are a well known group of drugs and include the parent tetracycline itself, as well as other antibiotics, such as doxycycline and oxytetracycline. Their action is bacteriostatic in that they inhibit the binding of the aminoacyl-tRNA complex to the 30S subunit and so slow down translation. Streptogramins type A and B (dalfopristin and quinupristin respectively), produced by *Streptomyces pristinaepiralis*, are chemically modified to give the drug quinupristin-dalfopristin. Quinupristin and dalfopristin alone each show weak bacteriostatic activities, however, together their actions are synergistic since they both target different site/actions of the 50S ribosome of bacteria, inhibit protein synthesis and lead to the release of incomplete polypeptides. Dalfopristin binds directly within the peptidyl transferase center of the ribosome interfering with the binding of tRNA molecules and inhibiting elongation of the polypeptide, while quinupristin blocks the exit tunnel of the polypeptide from the ribosome. Quinupristin-dalfopristin was licensed for use in the UK and USA in the late 1990s for treating severe infections with Gram-positive organisms, including nosocomial pneumonia and infections related to the use of intravascular catheters. It is particularly useful for treating complicated skin infections with methicillin-resistant *Staphylococcus aureus* and *Streptococcus pyogenes* and life-threatening infections of vancomycin-resistant *Enterococcus faecium* (*Box 3.4*). Quinupristin-dalfopristin has poor activity against *Enterococcus faecalis* compared with *Enterococcus faecium*. The latter is generally a less serious pathogen and other treatments are available, even though the former is the more prevalent clinically. However, strains of *Enterococcus faecium* resistant to quinupristin-dalfopristin are being found.

Several other antibiotics that interfere with protein synthesis are also in clinical use. Fusidic acid, a bactericidal agent that is used only in the treatment of gonorrhea, inhibits translocation. Chloramphenicol is a widely known antibiotic although its action is bacteriostatic. It inhibits the formation of peptide bonds by the ribosome. Lastly, spectinomycin, used in the treatment of penicillin-resistant staphylococcal infections, prevents translocation.

Human cells, like all animal cells, lack a cell wall. This means that metabolic processes in the bacterial cell concerned with wall synthesis are excellent targets for specific antibacterial agents. The bacterial cell wall forms a protective bag around each microbial cell that prevents osmotic lysis. The wall contains layers of peptidoglycan, each of which consists of rows of amino sugars linked together by short peptides. Gram-negative bacterial cell walls have only a single layer of peptidoglycan. In contrast, those of Gram-positive organisms may have as many as 40.

Bacterial growth involves cell division and entails the breakdown of the cell wall by bacterial enzymes, followed by synthesis of new peptidoglycan. Antibiotics, such as the β-lactams, that inhibit cell wall synthesis are almost all bactericidal since their presence stops new peptidoglycan formation and therefore affects cell wall synthesis while bacterial enzymes continue to break down the existing cell wall. β-Lactams are the largest, most widely used class of antibacterial antibiotics and contain the best known antibiotics, the penicillins and cephalosporins. They are, of course, only active against growing bacterial cells. All contain a chemical structure known as a β-lactam ring (*Figure 3.35*) responsible for their antibacterial effects. β-Lactams are irreversible inhibitors of transpeptidase, the enzyme that catalyzes the cross-link between the sugar residues and peptides in the peptidoglycan layer(s). A number of non-β-lactam antibiotics also reduce the efficiency of bacterial cell wall synthesis. They inhibit a number of different enzyme catalyzed steps in the synthesis at a wide variety of sites; hence they have no common mechanism of action. Examples of these drugs include cycloserine, vancomycin, fosfomycin and isoniazid (*Figure 10.29*) among others.

A rather miscellaneous grouping of antibiotics includes the polymyxins, nitrofurantoin, pyrazinamide and metronidazole. Polymyxins are effective against Gram-negative bacteria where they disrupt the structure of the cell membrane. Nitrofurantoin may be considered a prodrug given the need for bacterial enzymes to metabolize it to an active form that is thought to damage bacterial DNA. Pyrazinamide is used to treat TB and acts by an unknown mechanism although TB treatment, in general, requires prolonged therapy with a combination of antibiotics (*see later*). Metronidazole is an effective drug against anaerobic bacteria and some protozoan parasites. The unionized form of metronidazole is readily taken up by these organisms, which possess electron transport systems able to reduce it to an active form that disrupts the helical structure of DNA, inhibiting bacterial nucleic acid

A)

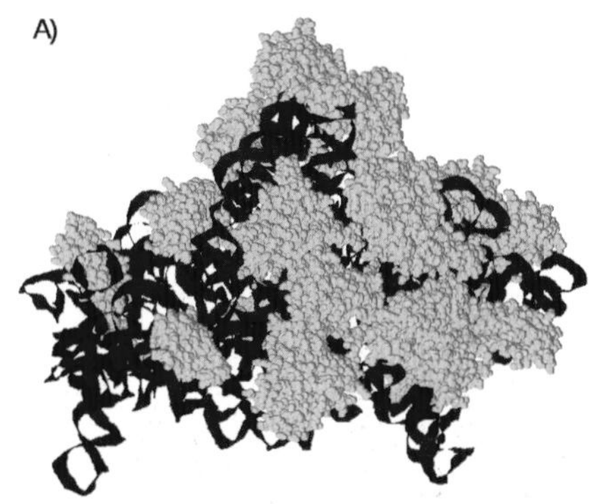

B)

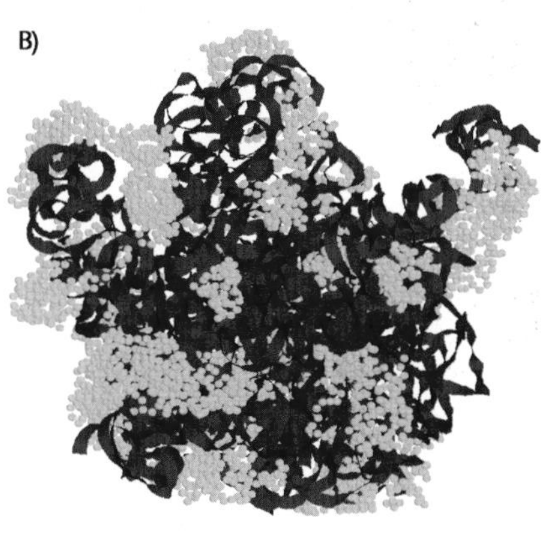

Figure 3.34 Molecular models of the (A) small and (B) large ribosomal subunits of *Escherichia coli*. PDB files 1P87 and 1P86 respectively.

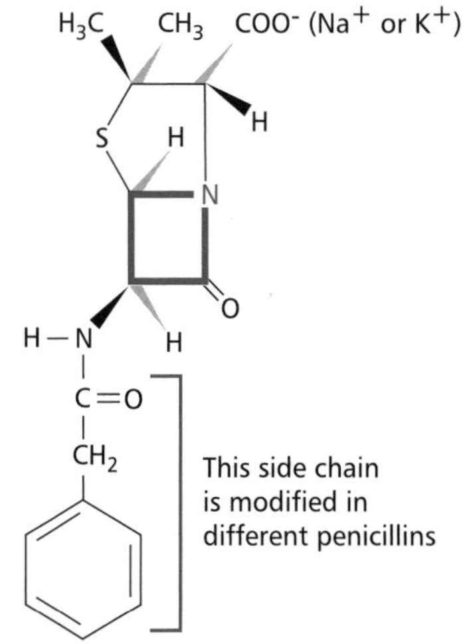

Figure 3.35 Natural penicillin. The β-lactam ring is shown in red.

Margin Note 3.5 Fleming and penicillin

The antibacterial activity of penicillin was discovered by Fleming (1881–1955) in 1928 at St Mary's Hospital in London. During this period, Fleming went on holiday and left some cultures of staphylococci on agar plates unwashed. When he returned, he noticed a fungal contaminant growing on one of the plates that was inhibiting the growth of the bacteria around it. Fleming realized that the mold was secreting a substance into the agar which was preventing the bacteria from growing. He called this substance **penicillin** after the contaminating mold, *Penicillium notatum* (*Figure 3.36*). However, Fleming was unable to isolate the substance and its purification did not occur until many years later following extensive work by Chain and Florey in Oxford during World War II. The significance of Fleming's discovery of the first antibiotic was quickly appreciated and, among many honors, Fleming was elected Fellow of the Royal Society in 1943 and knighted in 1944. In 1945 Fleming, Chain and Florey were jointly awarded the Nobel Prize in Physiology or Medicine for their discovery of penicillin and its curative effects.

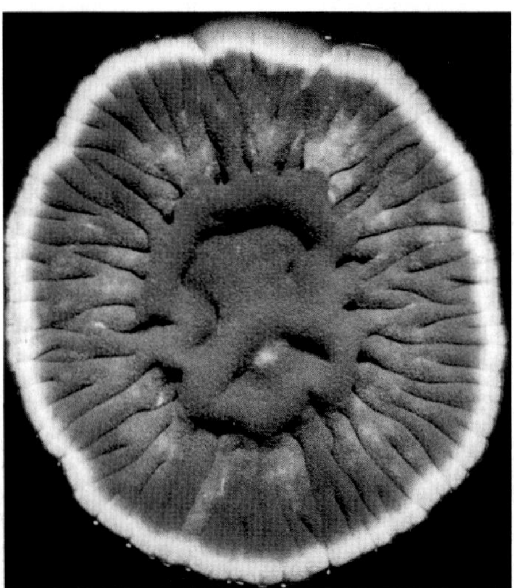

Figure 3.36 Penicillium mold growing on agar.

synthesis. Metronidazole is used alone or combined with other antibiotics to treat abscesses in the liver, pelvis, abdomen and brain caused by susceptible anaerobic bacteria and colon infections caused by *Clostridium difficile*, *Giardia* infections of the small intestine, amebic liver abscesses and amebic dysentery and *Trichomonas vaginalis* vaginal infections, and both male and female carriers of trichomonas.

Despite their specificity, antibiotics can cause toxicity in humans including allergic responses to the penicillins and sulphonamides, ear and kidney damage by aminoglycosides (*Box 3.3*), while chloramphenicol can show liver and bone marrow toxicities causing serious hematological diseases, particularly aplastic anemia. This can cause the death of rare susceptible individuals.

ANTIFUNGAL, ANTIPROTOZOAL AND ANTHELMINTHIC DRUGS

Fungi, protozoa and helminth parasites are responsible for many infections, particularly in the developing world. Given that they are all eukaryotic, then drugs to treat them are prone to act against host cells and they often have side effects. These drugs may kill the parasite or simply inhibit its growth. In the latter case, therapy must be continued for sufficient time to allow the host immune system to eradicate the organism.

Most antifungal drugs are not fungistatic but are fungicidal. One fungistatic drug, griseofulvin, inhibits intracellular transport and mitosis in fungi by interfering with the functions of their microtubules. A comparatively large number of fungicidal drugs suppress the synthesis of the essential cell membrane constituent, ergosterol. These include the allylamine antifungals, terbinafine and naftifine; the imidazoles, clotrimazole, econazole, ketoconazole and miconazole and the triazoles fluconazole and itraconazole. Ciclopiroxolamine inhibits the synthesis of fungal cell membrane proteins. The polyene antifungals, amphotericin and nystatin insert into plasma membranes of susceptible fungi. This increases the permeability of the membranes, allowing water and ions to leak and kill the parasite. Fluorocytosine inhibits the synthesis of fungal DNA.

In many cases, the precise biochemical mechanisms of antiprotozoal drugs are not known in any great detail. However, atovaquone inhibits electron transport in mitochondria. Pentamidine and isethionate interfere with the synthesis of protozoal macromolecules, while metronidazole, nifurtimox and tinidazole are thought to denature existing macromolecules. A number of other antiprotozoal therapeutic drugs are in clinical use. These variously affect protozoal enzymes, inhibit glycolysis and fatty acid oxidation or inhibit the synthesis of precursors of nucleic acids.

In general, the therapeutic bases of anthelminthic drugs are poorly understood. A number of commonly used drugs interfere with muscle contractions in the worms producing flaccid or spastic paralysis. This kills the parasite or makes it susceptible to attack by the host immune system. Paralysis is achieved by several overlapping mechanisms. Metriphonate inhibits cholinesterase leading to spastic paralysis while ivermectin potentiates inhibitory γ aminobutyric acid mediated peripheral neurotransmission and levamisole and pyrantel block nerve transmission at the neuromuscular junction. Diethylcarbamazine and piperazine cause hyperpolarization of muscle membranes. Praziquantel acts directly on muscle cells and increases the permeability of muscle membranes to Ca^{2+}.

Other anthelminthic agents act through different mechanisms. Oxamniquine interacts with helminth DNA and disrupts its structure. Niclosamide inhibits mitochondrial oxidative phosphorylation in helminth parasites. The anthelminthic agents albendazole, mebendazole and thiabendazole disrupt the microtubules of the cytoskeleton of the helminth.

BOX 3.3 Ototoxicity and Ménière's disease

Approximately 5–10% of patients treated with aminoglycosides, of which the best known examples are streptomycin and gentamicin, experience side effects, involving hearing, balance and the renal functions. **Ototoxicity**, that is, drug or chemical damage to the inner ear (*Figure 3.37*), can result in hearing loss or tinnitus and can lead to a loss of balance and feelings of dizziness. The extent of ototoxicity varies with the drug, its dose and other clinical conditions. In the majority of cases the damage is minor and reversible once medication ceases. In other cases, the extent of damage is limited, for example high-frequency hearing loss, where the damage to the ear makes it difficult to hear high pitched musical notes, but does not affect the ability to hear the spoken word or converse. In extreme cases, there may be complete and permanent deafness.

Ototoxicity is obviously undesirable; however, the ear damage associated with aminoglycosides can help some patients who suffer from Ménière's disease. Ménière's disease is named after the French physician Ménière (1799–1862). In the 1860s he theorized that attacks of vertigo, tinnitus and hearing loss arose from problems in the inner ear. The disease is an idiopathic syndrome, although some patients can identify triggers that can induce or aggravate symptoms, of endolymphatic hydrops, a condition in which abnormally large amounts of endolymph collect in the inner ear. Its symptoms are recurring episodes of hearing loss, tinnitus, rotational vertigo (a form of dizziness), nausea and a sense of pressure in the middle ear. Ménière's disease affects adults from the age of 20 years but is commonest in patients in their 40s and 50s. Given its generalized symptoms, the criteria used in diagnosing Ménière's disease are variable and estimating its incidence is difficult, although it is thought to be 0.5 to 7.5 per 1000. The incidence varies by ethnic background and is commoner in Britain and Sweden but it is known to affect black and Oriental ethnic groups.

Betahistine may be used to control Ménière's disease. Avoidance of triggers can reduce the frequency and duration of symptoms and episodes but not all episodes can be attributed to triggers. Conservative treatments for Ménière's disease involve a reduced sodium diet and diuretics to control water retention and reduce inner ear fluid pressure, and medications to reduce the vertigo, nausea/vomiting, or both during an episode. Vestibular rehabilitation therapy to help retrain the body and brain to process balance information can help with the poor balance that afflicts patients between attacks. Devices that deliver a series of low-pressure air pulses designed to displace inner ear fluid may also be used. In 20–40% of patients, conservative treatments are ineffective and a chemical labyrinthectomy may be performed with ototoxic aminoglycosides to control the vertigo associated with the affected ear but cause less damage to the hearing mechanism than some traditional treatments. Gentamicin is injected through the tympanic membrane into the middle ear from where it can diffuse into the inner ear and destroy some or all of the balance cells. In patients with Ménière's disease of both ears, streptomycin can be given intramuscularly and will have an effect on both ears. About 10% of patients require surgery, which can be used to relieve the pressure on the inner ear or to block the transfer of information from the affected ear to the brain.

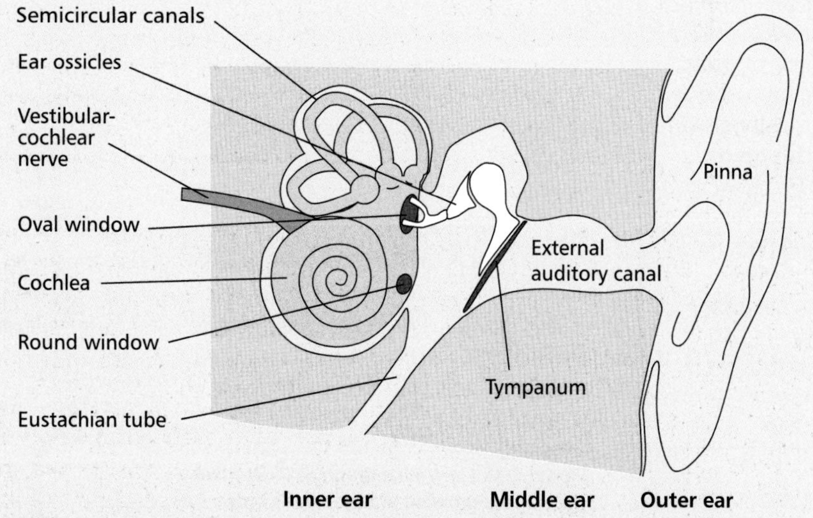

Figure 3.37 Schematic showing structure of the ear with the outer, middle and inner parts clearly separated.

Margin Note 3.6 Necrotizing fasciitis

The condition necrotizing fasciitis is an infection with the so-called 'flesh-eating bacteria', for example Group A Streptococci (GAS), most commonly *Streptococcus pyogenes*. The prevalence and incidence of this condition are both extremely rare but can begin after surgery, particularly abdominal or gynecological interventions. However, it may also develop from complications of childbirth, burns or following relatively minor traumas, for example bites and abscesses. The bacteria produce extracellular enzymes that attack soft tissues, often in an extremity, destroying muscle, fat and skin, causing an extensive necrosis of subcutaneous tissue. The lysosomal hydrolytic enzymes (*Chapter 16*) released from damaged cells of the patient may exacerbate the bacterial damage. The condition can be diagnosed by culturing the bacteria from blood samples or aspirations of pus from affected sites although surgical exploration may be necessary. Necrotizing fasciitis can have such a sudden and rapid onset that extent of destruction of soft tissues may quickly kill the patient. Hence an early diagnosis and prompt medical and surgical interventions are necessary to reduce the risk of death. Treatment often includes intravenous penicillin and clindamycin, along with aggressive surgical **debridement**, the removal of infected tissue, which can be very extensive. For those with severe illness, confinement in an intensive care unit is needed. Limb amputation may be necessary. Unfortunately, approximately 20% of patients who suffer necrotizing fasciitis die of the condition.

COMBINATION THERAPY

Combination therapy is the treatment of infections with two or more drugs usually to increase the clinical efficacy of the treatment, for example as described above for quinupristin-dalfopristin, or to minimize the development of resistant strains of the infective organism. Where the infection is of unknown origin, then multiple therapies are advisable to fight the most likely pathogens. With mixed infections involving two or more known pathogens, it is desirable to target each microorganism with one or more different antibiotics. Combination therapy may be used even if only a single infective pathogen is present as using combinations of drugs can prevent or delay the development of resistance to the drugs being used (*Box 3.4*). For example, some bacteria are resistant to β-lactams because they produce a β-lactamase that catalyzes the breakdown of the β-lactam ring. Combining an inhibitor of β-lactamase, such as clavulanic acid, with the β-lactam antibiotic helps preserve the drug *in vivo*. Other drugs commonly combined in therapeutic use are sulphamethoxazole and trimethoprim that synergistically inhibit the synthesis of folate by blocking different steps in its synthesis. The cocktail of isoniazid, rifampicin and pyrazinamide is used in the treatment of TB, while clofazimine, dapsone and rifampicin are used in combination in the therapy of leprosy.

SURGERY

Most infectious diseases can be treated using drug therapies. However, surgical intervention may be required in instances when the pathogen is resistant to available treatments or where it is the only means to contain an infection that is spreading to other areas of the body, as, for example, in gangrene caused by *Clostridium perfringens* (*Figure 3.38*) or necrotizing fasciitis (*Margin Note 3.6*). In other cases, surgery might be desirable because access to the affected site by the antimicrobial agents is limited, as in the case of some abscesses or appendectomy where surgical drainage or removal of necrotic tissue respectively can enhance the recovery process.

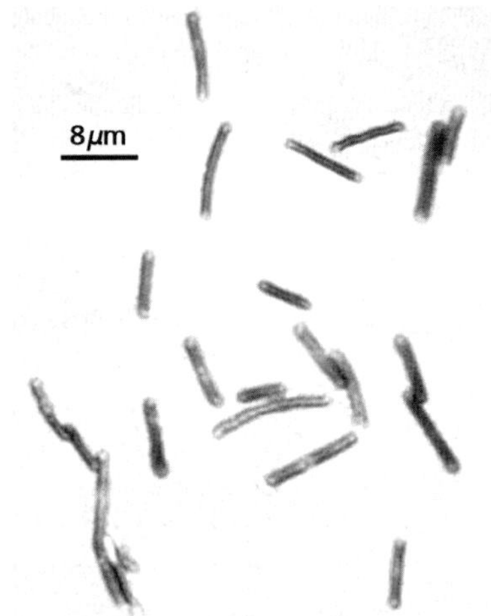

Figure 3.38 Light micrograph of *Clostridium perfringens* growing in Schaedler's broth. Courtesy of D. Stalons, Public Health Image Library, Centers for Disease Control and Prevention, USA.

BOX 3.4 Development of antidrug resistance

Drug resistance refers to the loss of effectiveness of a pharmacological agent against a pathogen. The pathogen or parasite may, of course, be initially nonsusceptible to a particular drug and this is referred to as primary resistance. Secondary or acquired resistance is that which develops over a period of time in a previously sensitive organism.

The development of secondary resistance by viruses following the chronic use of antiviral drugs is common and is normally the result of spontaneous mutations in viral genes. Many antiviral drugs target the same viral enzyme. Hence a mutation in the gene for that enzyme may render the virus resistant to several antiviral agents. In other cases, resistance to two antivirals is acquired by discrete mutations. The strain of virus that acquires resistance to one specific drug may be more susceptible to another antiviral agent. This is one reason why combination therapy involving two or more antiviral drugs maximizes the therapeutic effects, as for example, in the treatment of AIDS (*Box 3.1*). Unfortunately, given time, viral strains resistant even to combinations of antiviral drugs are likely to develop. Strains of the HIV virus resistant to the commonly used zidovudine, lamivudine and ritonavir are now present.

Like viruses, secondary or acquired resistance can also develop in previously sensitive bacteria. Indeed, the development of microorganisms that are resistant to drug treatment is well illustrated in bacteria, a number of which are now resistant to commonly used antibiotics. For example, sulphonamide resistance can be caused by a single mutation in the gene coding for dihydropterate synthase. Genes that confer resistance against a particular drug may be acquired as a result of spontaneous mutations. It may, however, occur by acquiring new genetic material from other bacteria. This genetic material may be chromosomal, most commonly transposons, or extrachromosomal, usually plasmids. Plasmids are especially significant in the transfer of antibiotic resistance between resistant and nonresistant bacteria. Further, a single plasmid may contain the genetic determinants for resistance to several drugs.

Antibiotic resistance in bacteria is achieved by four major biochemical mechanisms. There may be a decrease in the uptake or increase in the efflux of the drug by the bacterial cell. Streptococci and some anaerobe bacteria, for example, lack the electrochemical gradient across the cell membrane necessary to allow aminoglycoside entry. Some bacteria develop enzymes that are able to modify and inactivate certain drugs. The production of β-lactamase by some bacteria that renders them resistant to β-lactams has already been mentioned in the main text. However, in the 1980s it was found that *Klebsiella* spp and in some cases *Escherichia coli* were producing new β-lactamases that could hydrolyze the extended spectrum cephalosporins. These enzymes were collectively named **extended spectrum β-lactamases** (**ESBLs**).

When an antibiotic blocks a single reaction, for example, by inhibiting an enzyme, then resistance may be acquired by the bacterium acquiring mechanisms to bypass that reaction. This is the principal mechanism of resistance to sulphonamides. Bacteria may also acquire the ability to modify chemically the targets of specific drugs. Examples of the target modification include topoisomerase II, the pharmacological target of quinolones and fluoroquinolones, and ribosomal proteins targeted by aminoglycosides.

Bacteria may differ in the way they resist the actions of a drug while, in some cases, they may use more than one way of nullifying the effects of a single drug. Several clinically significant bacteria are now resistant to a number of antibiotics and, unfortunately, are sometimes spread to vulnerable patients within hospitals. An example of such a bacterium is the methicillin-resistant *Staphylococcus aureus* (MRSA). In addition to producing β-lactamase, MRSA is resistant to more than 40 β-lactams in clinical use because they contain a gene coding for an additional penicillin binding protein (PBP) which has only low affinity for the β-lactams. This allows MRSA to continue the synthesis of cell wall material during treatment with β-lactam drugs.

Enterococci are normally found in the large intestine and the female urogenital system. They can contaminate hands, equipment and the patient care environment. The recovery of enterococci from the hands of health care workers indicates that fecal-hand contact may be a major means of transmission to hospitalized patients with intravascular devices. This can lead to life-threatening bloodstream infections. Vancomycin was the drug of choice for treating such patients; however, vancomycin-resistant enterococci (VRE) were isolated in 1994. These bacteria are also resistant to many of the antibiotics previously used in treatment and patients may be affected for many months. Not surprisingly, VRE infections have become a serious health care issue. The vancomycin-resistant gene of VRE can be transmitted to other bacteria, for example *Staphylococcus aureus* and strains of this organism partially resistant to vancomycin, vancomycin-resistant *Staphylococcus aureus* (VRSA), were discovered in Japan in 1996, and in the USA and France in 1997.

Antibiotic resistance can also, of course, develop in the eukaryotic pathogens, fungi, protozoa and helminths. Resistance in the malarial parasite has been mentioned in *Chapter 2*. In all cases, the continual emergence of resistant strains of pathogens means there is a need to keep developing new pharmacological agents that are capable of treating infectious diseases.

CASE STUDY 3.1

Andrew, a 48-year-old male, presented at the Accident and Emergency Department of his local hospital. He told the examining doctor that the previous day he had experienced pain on the side of his right calf. Within 6 h, the area was septic and pus could be expressed. Over the day, the swelling and redness (inflammation) had gradually extended to cover the area from the ankle to the knee. Physical examination showed respiration, pulse, blood pressure and temperature were all within reference values. There was no obvious area of pus at the site and pus did not drain on puncturing with a 20-gauge needle. Andrew was treated with intramuscular and oral ceftriaxone. Two days later, Andrew returned with increased pain and slightly increased temperature. The center of inflammation showed an obvious pus-filled area. Culture of the pus showed the presence of clusters of Gram-positive cocci on nutrient agar and yellowish-colored colonies on 5% sheep blood agar. The area was removed by aspiration and the area excised and drained.

Questions

(a) Describe the history of this case.

(b) Why do you think antibiotic treatment alone was insufficient and that Andrew required drainage and excision at the site of infection?

CASE STUDY 3.2

Chris, a six-year-old boy, complained of a sore throat and had a temperature of 38.6°C. His mother kept him home from school and treated him with a proprietary pediatric painkiller. He slept well but awoke with similar symptoms. His mother took him to their family doctor. Physical examination showed reddening of the nasopharyngeal area and tonsils and a slight enlargement of the cervical lymph nodes. The skin was clear.

Questions

(a) What are the most likely organisms causing this disease?

(b) How could these organisms be detected?

(c) What treatment is desirable and why is this so?

CASE STUDY 3.3

After several days of recurring attacks of repeated severe coughing that left her gasping for breath and which were eventually associated with bouts of vomiting, Neha, a six-week-old baby girl, was transferred to hospital. Physical examination showed a normal trachea. A radiograph showed the chest to be clear of infection. Clinical examination showed the following data (reference values in parentheses):

- pulse of 155 min^{-1} (normal 72 for adults, but higher for babies and children)
- respiratory rate 71 min^{-1} (18 for adults, rather higher for babies and children)
- white blood cell count of 15 000 000 cm^{-3} (4 000 000 to 12 000 000 cm^{-3}).

A nasopharyngeal swab was taken. Organisms did not culture on standard media. Bacterial plaques with a mercury-like appearance were visible when charcoal blood agar was used as culture medium.

Questions

(a) What is Neha suffering from?

(b) What is the causative organism?

3.12 SUMMARY

Pathogenic microorganisms cause clinical conditions from infections of single organs or systems to systemic disease. Infections of the skin may be caused by viruses, such as papilloma and *Herpes simplex* types. Skin bacteria cause boils and abscesses and severe and chronic conditions such as leprosy. Fungi also cause a variety of skin infections, although these are not usually severe, unless the patient is immunocompromized. Severe infections of the eye include trachoma, a chlamydial disease that is the most common eye infection worldwide. Helminths, such as *Toxocara canis* can cause blindness. Viruses and bacteria may cause painful ear infections which, if recurrent, may damage hearing. Both groups of organisms are capable of infecting the CNS, sometimes resulting in encephalomyelitis by *Herpes simplex* or meningitis by *Neisseria menigitidis*. A number of viruses and bacteria cause significant respiratory infections, some of which, for example TB, are chronic. All varieties of microorganisms can infect the GIT, causing illnesses ranging from short-term gastroenteritis to stomach ulcers. An extensive number of microorganisms can infect the genital tract or are transmitted sexually. Several may cause systemic diseases.

The diagnosis of infectious disease involves a combination of laboratory tests to identify the microorganism present in specimens typically by culturing them. However, the presence of antibodies to the organisms in the patient may be used for identification and to compare levels in acute and convalescent patients. An increasing number of biochemical and molecular biological techniques have been introduced, many of which rely on identifying bacterial or viral nucleic acid in clinical specimens.

Infectious disease may be prevented by vaccination and public health measures or treatment with drugs. Depending on the infectious agent, these may be antiviral, antibacterial (antibiotics), antifungal or anthelminthic. Drugs may be given singly, or in combinations, as, for example, in the treatment of HIV patients.

QUESTIONS

1. Briefly review the causes of infections associated with the genital tract.

2. List some of the methods used in the diagnosis of infectious diseases.

3. A 19-year-old student attended a barbecue to celebrate his brother's 21st birthday. He ate a very rare steak. Two days late he began to experience pains on defecation, and to suffer from persistent diarrhea that rapidly increased in frequency and produced bloody stools. What is the student suffering from? Suggest how this condition should be treated.

4. Examine *Figure 3.27*. To which antibiotics is this strain of *Escherichia coli* resistant?

5. Compare and contrast the actions of griseofulvin and albendazole.

6. In what ways may antibiotics prevent the growth of bacterial pathogens?

7. In two words state a major way of delaying the development of antidrug resistance in pathogens.

FURTHER READING

Abraham, SN, Jonsson, A-B and Normark, S (1998) Fimbriae-mediated host-pathogen cross-talk. *Curr. Opin. Microbiol.* **1:** 75–81.

Abuhammour, W and Habte-Gaber, E (2004) Newer antifungal agents. *Indian J. Pediatr.* **71:** 253–259.

Andrews, TM (2003) Current concepts in antibiotic resistance. *Curr. Opin. Otolaryngol. Head Neck Surg.* **11:** 409–415.

Barrett, B, Ebah, L and Roberts, IS (2002) Genomic structure of capsular determinants. *Curr. Top. Microbiol. Immunol.* **1:** 137–155.

Chakraborty, C, Nandi, S and Jana, S (2005) Prion disease: a deadly disease for protein misfolding. *Curr. Pharm. Biotechnol.* **6:** 167–177.

Chambers, HF (2003) Solving staphylococcal resistance to β-lactams. *Trends Microbiol.* **11:** 135–148.

Frayha, GI, Smyth, ID, Gobert, JG and Savel, J (1997) The mechanisms of action of antiprotozoal and anthelmintic drugs in man. *Gen. Pharmacol.* **28:** 273–299.

Frieden, TR, Sterling, TR, Munsiff, SS, Watt, CJ and Dye, C (2003) Tuberculosis. *Lancet* **362:** 887–899.

Gould, D (2004) Bacterial infections: antibiotics and decontamination. *Nurs. Stand.* **18:** 38–42.

Hagan, P, Projan, S and Rosamond, J (2002) Infectious diseases: back to the future? *Trends Microbiol.* **10:** S1–S2.

Hayden, FG (2006) Antiviral resistance in influenza viruses – Implications for management and pandemic response. *N. Engl. J. Med.* **354:** 785–788.

Krcmery, VC (2005) Antifungal chemotherapeutics. *Med. Princ. Pract.* **14:** 125–135.

Lipsitich, M (2001) The rise and fall of antimicrobial resistance. *Trends Microbiol.* **9:** 438–444.

Papageorgiou, AC and Acharya, KR (2000) Microbial superantigens: from structure to function. *Trends Microbiol.* **8:** 369–375.

Prinz, C, Hafsi, N and Voland, P (2003) *Helicobacter pylori* virulence factors and the host immune response: implications for therapeutic vaccination. *Trends Microbiol.* **11:** 134–138.

Prusiner, SB (2003) *Prion Biology and Diseases.* University of California, USA.

Putman, M, van Veen, HW and Konings, WN (2000) Molecular properties of bacterial multidrug transporters. *Microbiol. Mol. Biol. Rev.* **64:** 672–693.

Razin, S, Yogev, D and Naot, Y (1998) Molecular biology and pathogenicity of mycoplasmas. *Microbiol. Mol. Biol. Rev.* **62:** 1094–1156.

Rhen, M, Eriksson, S, Clements, M, Bergström, S and Normark, J (2003) The basis of persistent bacterial infections. *Trends Microbiol.* **11:** 80–86.

Rover, JK, Vats, V, Uppal, G and Yadav, S (2001) Anthelmintics: a review. *Trop. Gastroenterol.* **22:** 180–189.

Seeger, C and Mason, WS (2000) Hepatitis B virus biology. *Microbiol. Mol. Biol. Rev.* **64:** 51–68.

Sharma, R, Sharma, CL and Kapoor, B (2005) Antibacterial resistance: current problems and possible solutions. *Indian J. Med. Sci.* **59:** 120–129.

Smith, MA (2005) Antibiotic resistance. *Nurs. Clin. North Am.* **40:** 63–75.

Stebbing, JS, Gazzard, B and Douek, DC (2004) Where does HIV live? *N. Engl. J. Med.* **350:** 1872–1880.

Stephans, RS (2003) The cellular paradigm of chlamydial pathogenesis. *Trends Microbiol.* **11:** 44–51.

White, TC, Marr, KA and Bowden, RA (1998) Clinical, cellular, and molecular factors that contribute to antifungal drug resistance. *Clin. Microbiol. Rev.* **11:** 382–400.

Yang, S and Rothman, RE (2004) PCR-based diagnostics for infectious diseases; uses, limitations, and future applications in acute-care settings. *Lancet Infect. Dis.* **4:** 337–348.

A useful web site on HIV and AIDS:

http://www.avert.org/historyi.htm

chapter 4:

THE IMMUNE SYSTEM

OBJECTIVES

After studying this chapter you should be able to:

- give an account of the role of the immune system in maintaining health;
- explain the difference between nonspecific and specific immune defenses;
- describe the cells of the immune system and outline their roles in immune defense;
- explain the roles of major proteins involved in immune defense;
- describe the major features of inflammation and the acute phase response;
- describe the major features of humoral and cell-mediated specific immunity;
- outline the use of antibodies for detecting and quantifying molecules of biomedical significance.

4.1 INTRODUCTION

The **immune system** is a set of organs, tissues, cells and molecules that protect the body from disease caused by microorganisms and multicellular parasites (*Chapter 2*). **Immunology** is the study of the immune system and how it works. Unraveling of the mechanisms of immune defense and the use of products of the immune system in investigating and treating disease has revolutionized the biomedical sciences in a way that is comparable to that of molecular biology in recent years. This chapter describes the major components of the immune system and explains how a coordinated immune response is produced against infectious agents. In addition, the use of immunological techniques in the biomedical sciences will be outlined.

4.2 TYPES OF IMMUNE DEFENSE

The immune system has to distinguish between the cells and macromolecules that make up the body that is, **self**, and those that are foreign or **nonself**. Immunological defenses are usually classified as **nonspecific** and **specific**. Nonspecific defenses constitute a first line of defense and are available immediately any foreign material, including substances such as wood splinters as well as microorganisms, enters the body. Nonspecific defenses include responses such as **inflammation**, a rapid immediate response to tissue damage, and the **acute phase response**, a relatively rapid response to infection. In specific immunological defense, cells of the immune system recognize not just individual microorganisms, but also the particular proteins or glycoproteins found on that microorganism. This type of defense may take several days to become effective, depending on whether the immune system was previously exposed to that specific microorganism but, once activated, results in a long-lasting immunity to it. This immunity can be **humoral**, in that it involves the production of antibodies and/or **cell-mediated**, which involves the production of cells that kill or recruit other cells to kill the infected cells. Humoral immunity is effective against microorganisms that do not invade cells and this includes most bacteria and multicellular parasites (*Chapter 2*). Cell-mediated immunity is effective against intracellular parasites, including viruses and some bacteria. However, these two types of specific immunity are not mutually exclusive and usually both types are activated on exposure to the infectious agent.

4.3 NONSPECIFIC DEFENSES

Microorganisms are prevented from entering the body by structural barriers, such as the skin and the mucosal membranes. These barriers are further protected by chemical secretions, for example lactic acid in sebum and hydrochloric acid secreted in the stomach. Furthermore, mucus, secreted by the mucosal membranes that line the urogenital, respiratory and gastrointestinal tracts, contains lysozyme, an antibacterial enzyme that catalyzes the hydrolysis of the peptidoglycan wall of some types of bacteria. Should microorganisms breach these barriers, other antimicrobial proteins such as complement and interferons may be activated and/or produced. Finally, there are a number of cells that have a role in the elimination of microorganisms. These cells may be found in the blood, although a number are also found in the solid tissues. Some blood cells move between the blood and the extravascular tissues, particularly during an infection.

INTERFERONS

Interferons (IFNs) are families of inducible, secretory proteins produced by eukaryotic cells in response to viral infections and other stimuli. They disrupt viral replication in neighboring healthy cells by inducing those cells to produce enzymes that inhibit replication of viral nucleic acid and the production of viral proteins. There are three major families of interferons: IFNs α, β and γ. Interferons α and β are produced by cells that have been infected with a virus. They are the predominant forms produced by leukocytes and fibroblasts respectively, although both types are produced by other viral infected cells in the body. Interferon γ is produced by cells of the specific immune system in response to any agent, whether bacterium, virus or foreign protein, which stimulates that system.

Interferons belong to a large family of proteins called **cytokines**, a general name given to proteins secreted by cells that stimulate activities in other cells after binding to receptors on their surfaces. Cytokines act at low concentrations and may stimulate different activities depending on the type of target cell. Many different cytokines are involved in the immune response and some are now used therapeutically. Interferon α has been used to treat hairy cell leukemia and Kaposi's sarcoma (*Chapter 5*), while IFN β is used in the treatment of multiple sclerosis and IFN γ has been used to treat several immune deficiency diseases (*Chapter 5*). Cytokines are also involved in the pathology of a number of diseases, particularly inflammatory disorders, such as rheumatoid arthritis (*Chapter 5*).

COMPLEMENT

Complement is the name given to a set of about 30 plasma proteins, some of which are listed in *Table 4.1* and which, when activated, can combat an infection by lysing the invading microorganisms, stimulating inflammation and promoting the uptake of the microbe by phagocytic cells. Activation of complement can be achieved in one of several ways. The **classical pathway** is initiated when antibody binds to the microorganism. This pathway may therefore take several days to become effective if specific antibody is not already present. The classical pathway for activation initially involves complement proteins C1–C4 (*Table 4.1*). An **alternative pathway** can be activated in the absence of antibody by cell wall components of bacteria, such as lipopolysaccharide. This first line of defense against microorganisms uses the complement proteins C3, and Factors B and D. Both pathways feed into a common pathway involving complement proteins C5–C9 that results in lysis of the target cell (*Figure 4.1*). Both pathways also result in the production of small peptides that induce phagocytosis and inflammation. The classical and alternative pathways are described in more detail in *Chapter 6*. Complement is also activated by C-reactive protein (CRP), and mannose-binding lectin (MBL), both of which are plasma proteins produced by the liver during the early or acute stage of an infection. Mannose-binding lectin is, as its name implies, a protein that binds to mannose residues on bacteria. This activates an associated protease that, in turn, activates proteins of the classical pathway. Thus, it provides a means of entering the classical pathway in the absence of antibody.

Complement is essential for immunological defense. Indeed, a deficiency of just a single complement protein can lead to increased susceptibility to bacterial infections. Complement is also involved in the pathology of a number of immunological disorders, including autoimmune diseases, such as rheumatoid arthritis and autoimmune hemolytic anemias (*Chapter 5*).

Protein	M_r
C1q	410 000
C1r	190 000
C1s	87 000
C2	115 000
C3	180 000
C4	210 000
C5	190 000
C6	128 000
C7	121 000
C8	163 000
C9	79 000
MBL (Mannose Binding Lectin)	200 000–700 000
MASP-II (MBP-associated serine protease II)	76 000
Factor B	93 000
Factor D	24 000
Factor H	150 000
Factor I	88 000
Factor P	220 000

Table 4.1 Complement proteins

NONSPECIFIC CELLS

Leukocytes are the white blood cells (*Chapter 13*) produced from precursor stem cells present in the bone marrow. All have immunological roles. About 80% of them are involved in nonspecific immune defense but the small lymphocytes (*Section 4.5*) are the cells of the specific immune system whose products control the numbers and activities of the nonspecific cells and so the two systems are interlinked.

All leukocytes can be classified into one of two groups, the **polymorphonuclear leukocytes (PMN)**, which have lobed nuclei and granular cytoplasm, and **mononuclear leukocytes** (**MN**) that have a more rounded nucleus (*Figure 4.2*). Polymorphonuclear leukocytes form approximately 65% of all blood leukocytes. They are classified into three groups: **neutrophils**, **basophils** and **eosinophils**.

Neutrophils, which make up around 60% of the blood leukocytes, are phagocytic cells that ingest and kill bacteria. These cells have receptors for antibodies and for the activated complement protein, C3b. Thus, neutrophils will bind readily to bacteria coated with any of these proteins, and phagocytosis is promoted (*Figure 4.3*). This phenomenon is known as **opsonization**. The killing of the

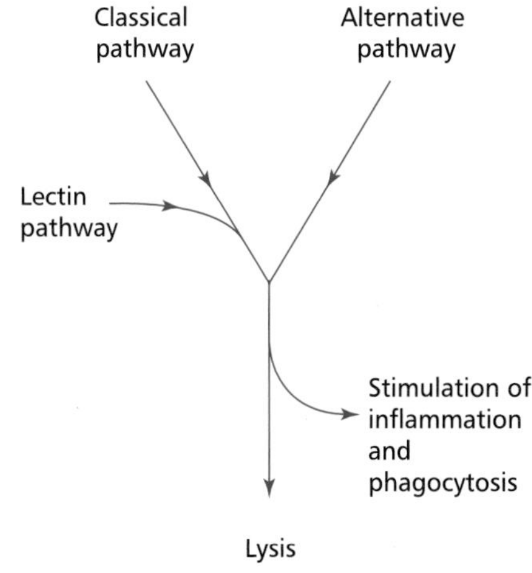

Figure 4.1 Activation of complement. See text for details.

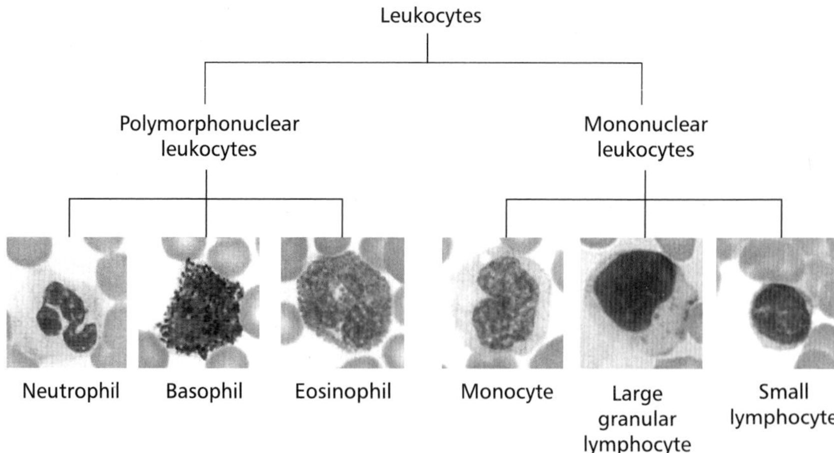

Figure 4.2 Photomicrographs of blood leukocytes. Courtesy of Drs L. Seal and S.J. Richards, School of Biology, Chemistry and Health Science, Manchester Metropolitan University, UK.

ingested bacteria is achieved through several different mechanisms, including the use of lysosomal enzymes (*Chapter 16*), production of antibacterial chemicals, such as hydrogen peroxide, hypochlorite and nitric oxide and the use of cytoplasmic proteins known as defensins that attack the membranes of the ingested microbe.

Basophils (*Figure 4.2*) are found in low numbers in the blood and usually form less than 1% of the leukocytes present. Basophils promote inflammation. They have prominent cytoplasmic granules which take up basic stains such as toluidine blue, and contain an abundance of pharmacologically active agents, such as histamine and heparin, and factors that are **chemotactic** for other PMNs (*Table 4.2*). In addition, basophils have the capacity to synthesize and secrete other mediators when appropriately stimulated. The primary, that is granular, and secondary induced mediators are necessary to promote and maintain inflammation and their inappropriate release can result in immunological disorders such as hay fever and allergic asthma (*Chapter 5*). Despite their relatively low numbers in the blood, basophils are essential for initiating inflammation. However, a similar type of cell known as a **mast cell**, which is found in solid tissues rather than blood, is of greater clinical significance. Mast cells are found throughout the body but especially in the skin, mucosal membranes and epithelia of the respiratory and gastrointestinal tracts and in the connective tissue of a variety of internal organs. Mast cells are highly granular and contain similar mediators to the basophil although the granular contents may vary according to location. Mast cells also secrete a number of cytokines, many of which are **proinflammatory**, that is they

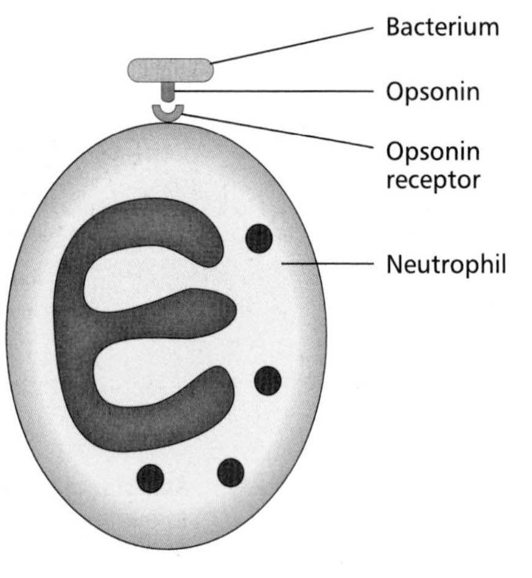

Figure 4.3 The role of opsonin (from the Greek word *opsonion*, meaning victual (food)) in triggering phagocytosis following the binding of an opsonized bacterium to a neutrophil. See text for details.

Chemical	Activity promoted
Histamine	dilates blood vessels, increases vascular permeability, promotes contraction of smooth muscle cells
Heparin	inhibits clotting of blood
Chemotactic factor for neutrophils (NCF-A)	attracts neutrophils into inflammatory area
Chemotactic factor for eosinophils (ECF-A)	attracts eosinophils into inflammatory area
Proteases	degrades basement membrane of blood vessels, promoting migration of cells associated with inflammation into that area
Serotonin	increases vascular permeability and promotes contraction of smooth muscle

Table 4.2 Some pharmacologically active chemicals in basophil granules

promote inflammation. Both mast cells and basophils have receptors for a particular class of antibody, known as immunoglobulin E (*Section 4.4*) and this has significance in the pathology of some of the most common immunological disorders, namely **allergies** (*Chapter 5*).

Like basophils, eosinophils are found in low numbers in the blood and usually constitute less than 2% of white cells. They also have prominent cytoplasmic granules but in this case they take up acidic stains, for example eosin, because they contain highly basic proteins. Though capable of phagocytosis, the major role of eosinophils is to assist in the elimination of multicellular parasites such as tapeworms and nematodes (*Chapter 2*). Eosinophils first bind to the surface of the helminth, often using antibodies and then secrete their toxic granular proteins onto its surface. Eosinophils can be attracted to areas of inflammation by chemotactic factors secreted by basophils and mast cells.

MONONUCLEAR LEUKOCYTES

The mononuclear leukocytes comprise three distinct groups of cells: **monocytes** and **large granular lymphocytes (LGLs)**, which are nonspecific cells, and **small lymphocytes** that are responsible for the specific immune response.

Monocytes (*Figure 4.2*) make up approximately 5% of the blood leukocytes. They have a characteristic indented, often horseshoe-shaped nucleus and a granular cytoplasm. Monocytes are immature cells that circulate for only a matter of hours before they enter the solid tissues and develop into **macrophages.** The spleen, lungs, liver, lymph nodes and tonsils contain especially high numbers of macrophages. Monocytes and macrophages are phagocytic cells, clearing the blood and solid tissues of microbes as well as dead or dying host cells, including neutrophils and erythrocytes. Their numbers and generalized distribution ensure they effectively clear foreign material. Macrophages have killing mechanisms similar to those found in neutrophils and can be stimulated by target cells coated with antibodies and/or complement proteins. Following phagocytosis of microorganisms, especially bacteria, they secrete a range of cytokines, including interleukins 1 (IL-1), 6 (IL-6) and 8 (IL-8) and tumor necrosis factor alpha (TNF α). Interleukin 1, IL-6 and TNF-α are proinflammatory and are responsible for initiating events in the early or acute phase of infection, while IL-8 is a chemokine that attracts neutrophils to its source.

Monocytes and macrophages also play significant roles in the specific immune response since they are able to 'process' foreign material and 'present' it to certain types of small lymphocyte in a form they can recognize. As such, they are known as **antigen presenting cells (APC)**, although they are not the only cells to carry out this activity.

Large granular lymphocytes (LGL) make up 5–10% of the blood leukocytes. These cells have rounded nuclei and a granular cytoplasm (*Figure 4.2*). Functionally, LGL represent a mixed population of cells: some are **natural killer (NK)** cells that kill virus-infected cells and some tumor cells nonspecifically. Natural killer cells bind to the target cell and release proteins, some of which perforate the target cell membrane while others induce a genetically programmed cell death called **apoptosis**. Natural killer cells are the first line of defense against viruses since they prevent their replication and spread. They may also form a defense against potential tumors by destroying some cancerous cells as they arise in the body (*Chapter 17*).

INFLAMMATION AND THE ACUTE PHASE RESPONSE

The term **inflammation** is sometimes used to describe a whole array of responses to infection and tissue damage. Acute inflammation will be used

Margin Note 4.1 Chemokines 

Chemokines are cytokines with common structural and functional features. They are all small polypeptides that are chemotactic for and/or activate different types of leukocytes. Their structure consists of 70–90 amino acid residues with a M_r between 8000 and 10 000. Nearly all belong to one of two families with four conserved cysteine residues. These families differ in the presence or absence of an amino acid between the first two cysteine residues.

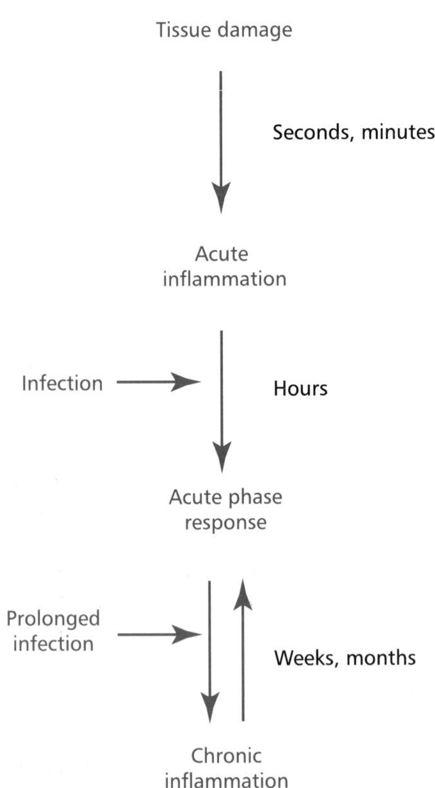

Figure 4.4 The relationship between inflammation and the acute phase response. See text for details.

here to describe the body's immediate and localized response to tissue damage. However, should an infection arise as a result of damage to tissue, acute inflammation can merge with the acute phase response, which is a whole body response that occurs within a few hours of an infection entering the body. The acute phase response involves the production of a number of chemical mediators that promote and prolong local inflammation. In addition, chronic infection can lead to chronic inflammation together with a prolonged acute phase response. This potentially confusing series of overlapping events is clarified in *Figure 4.4*.

Inflammation is a rapid local response to tissue damage, as would happen, for example, if the skin is scratched by a thorn. The response is characterized by reddening and swelling of the skin as well as sensations of heat and pain at the damaged site. Inflammation is initiated by the release of mediators, especially histamine, from mast cells in the damaged area (*Figure 4.5*). Histamine causes blood vessels to dilate and become leaky so that plasma escapes into the damaged tissue where it dilutes any noxious agents that entered at the damaged site, and helps to wash them away into the lymph (*Section 4.5*). If bacteria enter the damaged area they activate complement to release proteins that are chemotactic to neutrophils, encouraging them to move out of the blood between the lining endothelial cells.

The acute phase response occurs within hours of exposure to microorganisms. This is a systemic response involving the whole body and several organ systems. The response is brought about by cytokines, including IL-1, IL-6 and TNF α released by monocytes and macrophages. An acute phase response results in changes in the composition of the blood, including an increased neutrophil count and the appearance of, or increase in, a number of defense proteins called **acute phase proteins**. The appearance in the blood of a particular protein, C-reactive protein (CRP), which binds to bacteria and activates complement, is often used as a marker of infection. The plasma concentration of CRP prior to an acute phase response is so low as to be barely measurable; thereafter its synthesis increases 100–1000 fold.

The acute phase response is a primary defense against infection, with multiple beneficial effects. However, it has the potential to cause harm if an infection is prolonged. For example, acute phase proteins are synthesized in the liver from amino acids released by the enzyme catalyzed proteolysis of muscle tissues. Following an acute infection, this muscle protein is rapidly replaced. However, a chronic infection, such as tuberculosis, can result in severe and prolonged muscle wastage, a condition known as **cachexia** (*Chapters 10* and *17*). High body temperature, or fever, is another symptom of the acute phase response. While elevated body temperatures may inhibit the replication of bacteria, fever can be dangerous, especially in children, if the temperature becomes too high.

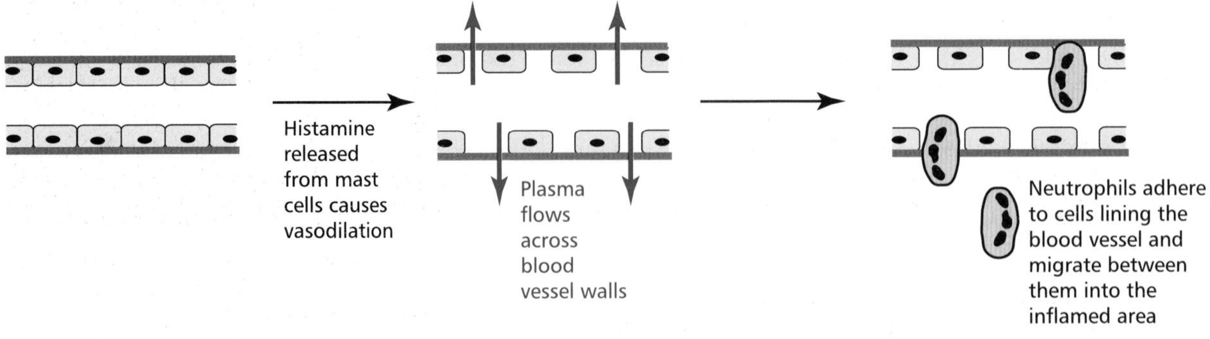

Figure 4.5 Acute inflammation as triggered by histamine release from mast cells. See text for details.

4.4 SPECIFIC IMMUNE RESPONSES

The specific immune response allows the development of true immunity to an infectious agent. Since true immunity can only develop after exposure to the virulent microorganism or a harmless vaccine derived from it, the response is often called **acquired immunity**. For example, an individual who has had measles is unlikely to suffer that disease again, even though exposure to the virus may occur subsequently even after many years. Two major features define a specific immune response. First, specific immunity is only induced towards the agent, called the **immunogen**, which stimulated it and

BOX 4.1 Vaccination

The process of protecting people from infection by deliberately exposing them to microbial components was initiated in modern times by Jenner (1749–1823) in 1796, when he showed that immunity to smallpox, which is caused by the Variola virus, was induced by the introduction into the skin of material from the crusts of cowpox lesions. This process became known as **vaccination**, from the Latin *vacca,* a cow, a term which is still used today. In this case, immunization with the cowpox virus induced immunity that cross-reacted with the variola virus, because the viruses share some molecular similarities. Cross-reactions between vaccines are quite rare and most vaccines are usually specific for only one type of microorganism.

Since Jenner's time, vaccination has been widely applied to protect people from many serious infections. Vaccination against smallpox became compulsory in several developed countries from the 1900s to the 1940s and the WHO embarked on an eradication program, using the cross-reacting Vaccinia virus (*Figure 4.6*) as a vaccine. This program was deemed successful in 1980, the last case of smallpox having been reported in 1977. With this success, the WHO also embarked on a program to eradicate polio by world-wide vaccination but this has not yet been achieved.

Traditional vaccines to induce immunity to microorganisms include the use of attenuated viruses, such as polio, measles, mumps, killed bacteria, for example whooping cough, toxoids, which are derived from bacterial toxins, for example those of tetanus and diphtheria and bacterial cell wall polysaccharides from, for example, *Neisseria meningitidis* serogroup A (*Chapter 2*). More recent developments include recombinant subunit vaccines. Here the vaccine is a microbial protein that has been produced by genetically engineered eukaryotic cells. An example of this is the vaccine for hepatitis B, which consists of a viral surface protein produced by the recombinant yeast, *Saccharomyces cerevisiae*. Other recent developments in vaccine production include polysaccharide conjugate vaccines in which a bacterial polysaccharide is conjugated to a protein, in order to stimulate a more potent immune response. Examples of polysaccharide conjugate vaccines include those for *Haemophilus influenzae* and *Neisseria meningitidis* serogroup C. Where several strains of a microorganism are known to cause disease, multivalent vaccines containing components from a number of strains may be used. The newest development in vaccine production is the DNA vaccine, which consists of plasmid DNA containing a gene coding for the microbial protein in question. This DNA is injected intramuscularly and is taken up by the muscle cells. For a limited period the gene is transcribed and translated to form the foreign protein, which stimulates the immune response ***in situ***. Several DNA vaccines are currently in clinical trial.

At present there are no successful vaccines in routine use against protozoa, including the malarial parasites (*Chapter 2*) that infect 300–500 million annually and kill two to three million people worldwide each year. Such organisms have very complex life cycles, often with secondary animal hosts and often accompanied by distinct antigenic changes, which divert the immune system. In 2005, a new vaccine against malaria was tested on a group of over 2000 children in Mozambique. This vaccine is aimed at the sporozoite form of the parasite, which is the form injected into humans by mosquitoes, and has been shown to cut the risk of developing severe malaria by 58%.

Figure 4.6 Electron micrograph of Vaccinia virus. Courtesy of North West Regional Virus Laboratory, Booth Hall Hospital, Manchester, UK.

Margin Note 4.2 Haptens

Some molecules, such as steroid hormones, are too small to stimulate a specific immune response. However, if such molecules are attached covalently to an immunogenic protein they can be recognized by the immune system and an immune response will then be mounted against them. In this case, the small molecule is called a hapten and the protein to which it is attached the carrier. The ability of the immune system to recognize small molecules when they are presented in the appropriate context has proved extremely useful as it is possible to produce antibodies to a wide range of smaller molecules. The antibodies can then be used in specific assays to determine the presence and concentration of the hapten in biological samples (*Box 4.2*). For example, assays for steroid hormones use antibodies that have been produced in animals by immunizing them with the steroid linked to a protein.

secondly, the specific immune response is more rapid following subsequent contacts with that same immunogen. It is the rapid secondary response that prevents the development of disease in the immune individual. The ability to produce a quicker response on second contact with an immunogen is known as **immunological memory**. This process is mimicked by the use of vaccines, where an attenuated or less virulent strain of the microorganism is administered to induce immunity without causing the disease.

The body's specific response to an infectious agent such as rubella virus, the cause of German measles, is to respond with two types of immunity, that is, humoral immunity and cell-mediated immunity. A specific immune response is only stimulated by an **immunogen**, which may be the whole microorganism but could merely be a protein, glycoprotein or lipoprotein. As a general rule, proteins with a M_r greater than 5000–10 000 are immunogenic. Polysaccharides are usually only weakly immunogenic, although they generally become more immunogenic when conjugated to a protein.

Cells of the specific immune system do not recognize the whole immunogen but, rather, small regions of immunogenic macromolecules known as **epitopes.** Usually, epitopes are sequences of 5–7 amino acid residues on immunogenic proteins, but they may also be short sequences of sugar residues in a polysaccharide, lipopolysaccharide or glycoprotein. A large macromolecule may have many epitopes that are recognized by the immune system. However, a protein that has some similarities with our own proteins, for example bovine serum albumin that has a similar size and structure to human albumin, will have fewer nonself regions recognizable as epitopes.

The term **antigen** is often confused with an immunogen and, indeed, some textbooks use the terms interchangeably. Here, an antigen is defined as something that will bind to the product of an immune response, such as an antibody. Nowadays the term is most commonly used when discussing immunotechniques, where an antibody is used to detect an antigen or quantify how much of it is present in a sample (*Box 4.2*).

ROLES OF HUMORAL AND CELL-MEDIATED IMMUNITY

Most immunogens stimulate both humoral and cell-mediated immunity. Antibodies are found in body fluids, including blood and lymph, and have access to extracellular organisms. They are therefore most effective at eliminating microorganisms that live outside the cells of the host. Cell-mediated immunity, on the other hand, is effective against intracellular parasites, which includes all viruses and a number of intracellular bacteria, such as *Mycobacterium tuberculosis* and *Listeria monocytogenes*. However, antibodies are useful in protecting cells in the initial stages of a viral infection and in preventing the spread of viruses from one cell to another.

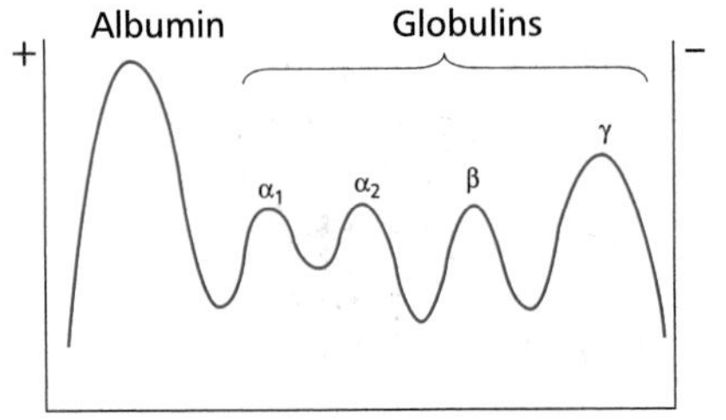

HUMORAL IMMUNITY

In a humoral response, the immune system produces specific glycoproteins called **antibodies** that are found in the blood plasma, mostly in the γ globulin fraction (*Figure 4.7*), in lymph and in body secretions such as saliva, tears, mucus and milk. Antibodies have binding sites that are complementary to the shape of an epitope. These sites allow antibodies to bind to the epitopes of the immunogen and initiate its destruction by a variety of other agents, such as complement, phagocytic cells and LGLs. Collectively, antibodies are known as **immunoglobulins (Igs)** which are heterogeneous molecules indeed because each antibody is specific for an individual epitope. Despite this heterogeneity, they can be divided into one of five major classes or isotypes, IgG, IgM, IgA, IgE and IgD, based on differences in their structures, and particularly in the amino acid sequences of their largest polypeptide subunits which are known as heavy chains (*see below*). Some of their properties are listed in *Table 4.3*.

Figure 4.7 The order of separation of serum proteins (albumin and globulins) by electrophoresis (densitometric scan).

Class	Subclasses	M_r	Heavy chains	Activates complement	Crosses placenta	Opsonization	Triggers inflammation
IgM	none	900 000	μ	yes	no	no	indirectly via complement
IgG	IgG_1 IgG_2 IgG_3 IgG_4	146 000 145 000 170 000 146 000	γ_1 γ_2 γ_3 γ_4	yes yes yes –	yes yes yes yes	yes – yes yes	indirectly via complement
IgA	IgA_1 IgA_2	160 000 to 380 000	α_1 α_2	no no	no no	yes yes	no no
IgE	none	188 000	ε	no	no	no	directly by binding to ε receptors on mast cells
IgD	none	184 000	δ	no	no	no	no

Table 4.3 Properties of immunoglobulin classes. Light chains are always either κ or λ.

Immunoglobulin G

Immunoglobulin G (IgG) is the most abundant antibody in the blood where it is found at concentrations of approximately 13.5 mg cm^{-3}. It is evenly distributed between the vascular and extravascular compartments. In humans, IgG occurs as four subclasses, all with slightly different properties and roles in the body. For example, all subclasses except IgG_4 can activate complement (*Table 4.3*). Immunoglobulin G is produced after IgM on exposure to an immunogen and is the predominant antibody produced on a second exposure, that is in the secondary response. It is also the only antibody in humans that can cross the placenta to allow maternal antibodies to protect the developing fetus from infection. On some occasions, this may cause problems as, for example, when the mother has become sensitized to fetal antigens or has an autoimmune disease (*Chapter 5*).

Immunoglobulin G is a symmetrical molecule consisting of four polypeptide chains joined by disulfide bonds as shown in *Figure 4.8(A)*. The four chains

B)

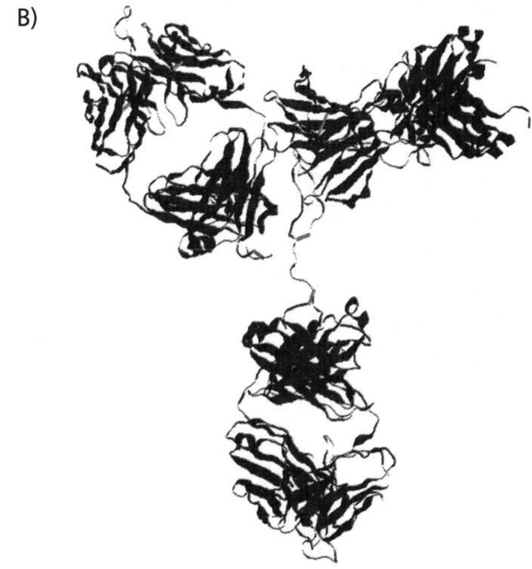

A)

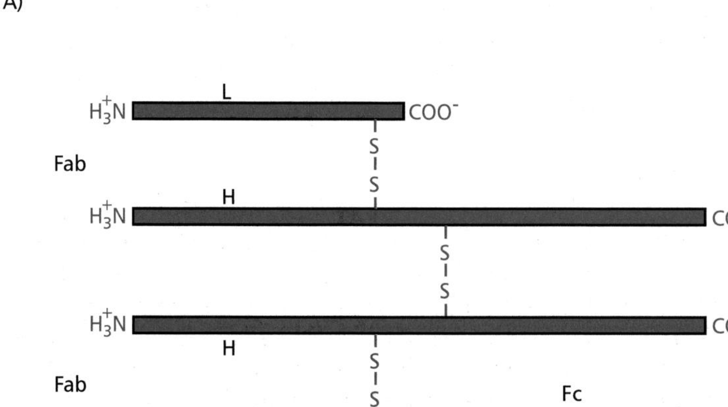

C)

Figure 4.8 The structure of IgG shown (A) diagrammatically, (B) molecular model PDB file 1HZH and (C) showing a bound antigen (black). (C) Courtesy of Dr R.S.H. Pumphrey, St Mary's Hospital, Manchester, UK.

Margin Note 4.3 Domains

The domain substructure of polypeptides is found in proteins throughout the immune system, indicating that these proteins share a common evolutionary relationship. Proteins with this domain structure are members of the immunoglobulin superfamily and include T cell receptors, MHC molecules and cell surface proteins such as Cluster of Differentiation CD4 and CD8 (*Section 4.5*).

consist of two identical heavy chains each with a M_r of 50 000 and two identical light chains each with a M_r of 25 000. Both heavy and light chains have a domain structure. A domain is a globular region made up of approximately 110 amino acid residues and stabilized by an intrachain disulfide bond (*Figure 4.9*). The heavy chain of IgG has four domains while the light chain has only two.

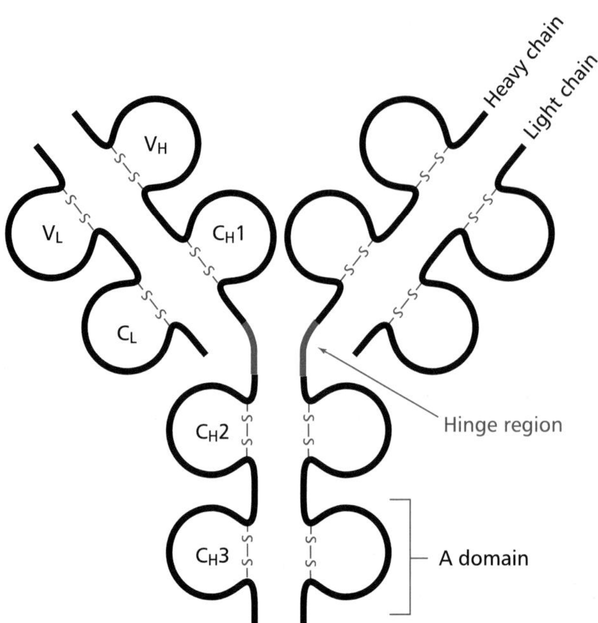

Figure 4.9 A schematic showing the domains of an immunoglobulin molecule. The interchain disulfide bonds have been omitted for clarity.

The interchain disulfide links between the two heavy chains and between the light and the heavy chains (*Figure 4.8 (A)*) produce two units, called the Fab, or fragment antigen binding, part of which can recognize and bind to an epitope (*Figure 4.8 (C)*). The remainder of the molecule, consisting only of the carboxyl terminal halves of the heavy chains is known as the Fc, or fragment crystallizable, portion. The Fc region is concerned with complement activation, placental transfer and binding to LGL and phagocytes.

Immunoglobulin M

Immunoglobulin M (IgM) is the largest of the antibodies with a M_r of about 900 000. Its concentration in the plasma is roughly 1.5 mg cm^{-3}, which is approximately 10% of the plasma immunoglobulins. Most IgM is vascular, with little present in lymph or secretions. Immunoglobulin M is always the first antibody to be produced during an immune response and is also the predominant one formed when an immunogen is encountered for the first time, that is, during a primary response (*Figure 4.10*). The structure of IgM is shown in *Figure 4.11 (A)*. Four polypeptide chains form a structure somewhat similar to that of IgG but this four-chain structure is repeated five times. The five units are joined by their Fc portions by a J chain. Thus, each molecule of IgM has ten binding sites, the largest number of any of the antibody classes. Immunoglobulin M is efficient at **agglutinating** cells or clumping them together, and is an effective activator of complement.

Immunoglobulin A

Immunoglobulin A (IgA) is found in plasma at concentrations between 0.5–3.0 mg cm^{-3} but is also the major antibody found in body secretions, including mucus, saliva and tears. Thus, it protects mucosal surfaces. In humans, most plasma IgA occurs in the familiar four chain structure, similar to IgG

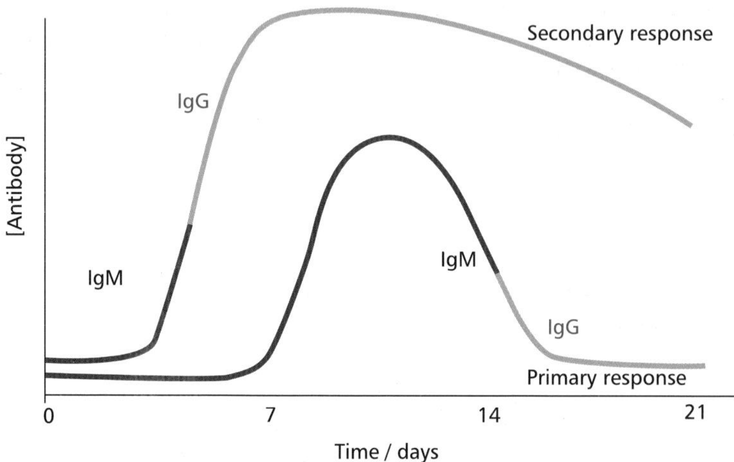

Figure 4.10 The primary and secondary antibody responses to an immunogen.

and with a M_r of 160 000. However, a proportion of plasma IgA also exists as a dimer, in which IgA molecules are joined together by a protein called the joining chain(s) (*Figure 4.11 (B)*). Secretory IgA exists in this dimeric form but is protected from enzymic attack by an additional protein known as the secretory component.

Immunoglobulin E

Immunoglobulin E (IgE) is found in plasma at a concentration of about 5×10^{-5} mg cm^{-3}. Most IgE is found bound to the surface of blood basophils and tissue mast cells, which possess a high affinity receptor for the heavy chain of this immunoglobulin. Binding to these cells prolongs the half-life of IgE from two

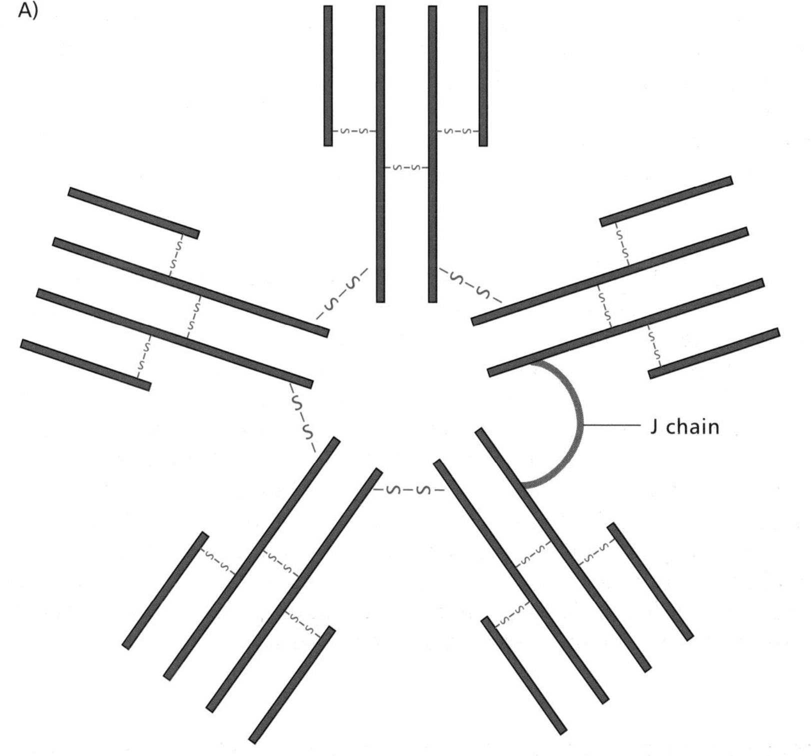

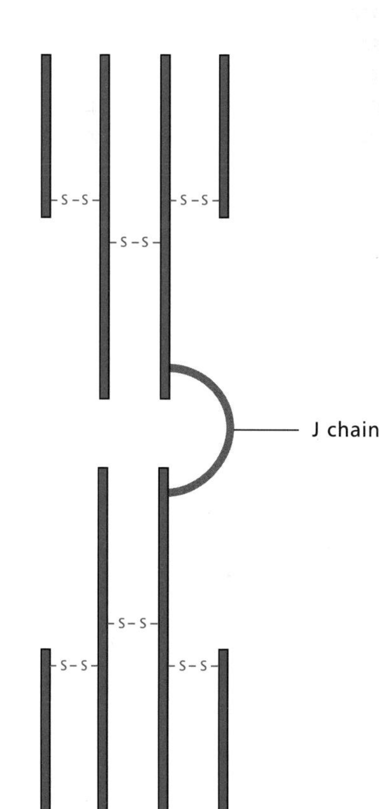

Figure 4.11 Schematics showing the structures of (A) IgM and (B) IgA molecules. J chain is the joining chain, in each case.

days to several weeks. Despite its low plasma concentrations, IgE is a potent stimulator of inflammation, since binding of epitopes to mast-cell bound antibody can trigger degranulation of these cells. The proinflammatory nature of IgE is seen in the elimination of multicellular parasites, such as tapeworms and nematodes. However, in susceptible individuals, this antibody can also trigger an inflammatory response to otherwise harmless immunogens, such as pollen, resulting in allergic reactions (Chapter 5).

Immunoglobulin D

Immunoglobulin D (IgD) is found in plasma at a concentration of about 30 μg cm^{-3}. Its role as a secreted antibody is uncertain. Hyperimmunoglobulin D syndrome (HIDS), in which IgD levels are increased, is associated with periodic fever and joint disease. However, IgD does have a role in the recognition of epitopes by cells of the specific immune system (*Section 4.6*).

The concentration of immunoglobulins shown above for all the immunoglobulins represents their mean plasma concentrations. During

BOX 4.2 Immunoassay

As well as being molecules of major significance in protecting the body from infection, antibodies are also powerful reagents for detecting and quantifying antigens. The specificity of antibodies allows biomedical scientists to measure the level of an analyte, such as a steroid hormone, in biological fluids such as plasma, which contain hundreds of other biomolecules, some of which may be very similar to the analyte being measured. The use of antibodies to quantify antigens is called **immunoassay** and immunoassays are used in all branches of the biomedical sciences. Some immunoassays are among the most sensitive assays known, detecting antigens in the range of mg cm^{-3} to pg cm^{-3}.

One of the earliest immunoassays developed is known as radial immunodiffusion (RID). This method relies on the ability of antibodies to precipitate soluble protein antigens. Antibody is incorporated evenly into an agar gel and a measured volume of antigen solution is added to wells cut into the agar. As the antigen diffuses into the agar the reaction with antibody forms a circle of precipitation (*Figure 4.12*). After allowing all the antigen to diffuse, which takes about 72 h, the diameter of this precipitin ring is measured. The antigen concentration is proportional to the square of the diameter so that the concentration of the unknown can be determined from a standard curve produced using known antigen concentrations.

Radial immunodiffusion is a simple and reliable method, which can be used, for example, to measure the concentrations of a number of serum proteins. However, it is not a sensitive method, suitable for determining concentrations in the μg cm^{-3} to mg cm^{-3} range, and takes several days before the result can be read. Nephelometry is another technique which detects precipitation,

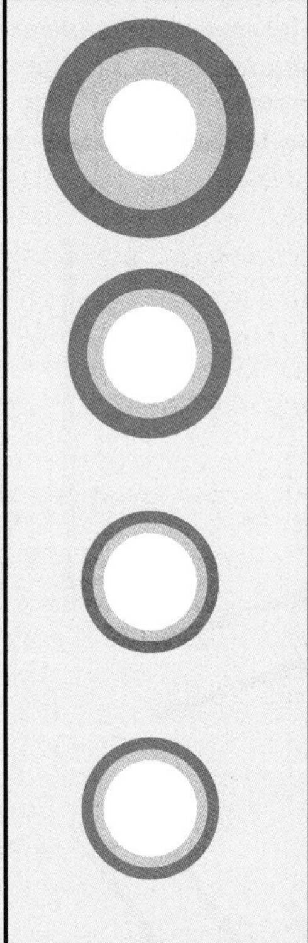

Figure 4.12 An outline of a radial immunodiffusion assay as described in the text. Note that the diameters of the precipitin rings increase with larger concentrations of antigen.

infection, concentrations increase as humoral immunity is activated. Levels significantly above the normal range for the appropriate age group are associated with myeloma, a cancer of antibody-producing cells or with infection with Epstein Barr virus. Concentrations significantly below the normal range are associated with immunodeficiency disorders (*Chapter 5*).

CELL-MEDIATED IMMUNITY

Cell-mediated immunity (CMI) involves the direct and indirect destruction of host cells infected by viruses or other intracellular parasitic microorganisms, such as rickettsias (*Chapter 2*). The direct destruction of infected cells is brought about by the production of specific **cytotoxic cells** that are capable of killing any such infected cell that induced their formation. Indirect destruction is brought about by the release of cytokines that promote destruction by macrophages and LGLs. This type of immunity forms the major defense against viral infections since the destruction of virus-infected host cells prevents the replication and spread of the virus.

Margin Note 4.4 Immunity and self damage 

Both humoral and specific immunity protect the body from infection but, sometimes, these responses can lead to tissue damage as, for example, when IgE causes allergic reactions (*Chapter 5*) or when both forms of immunity help to bring about the destruction and rejection of a transplant (*Chapter 6*).

but in solution rather than in agar. It relies on the ability of precipitates formed by an antibody reacting with a protein antigen to scatter a beam of light passed through it. Scattered light is detected at right angles to the original light source. This method is much more rapid than RID, although optically clear antibodies must be used. Other methods, which use 'labeled' reagents, are much more sensitive and can be used to measure nonprotein antigens. In addition, results from these assays can be obtained within hours rather than days. The first of these labeled reagent techniques to be developed was radioimmunoassay (RIA), which was devised in 1960, but is still used extensively. Radioimmunoassay relies on the competition between radiolabeled and unlabeled antigen for a limited amount of antibody. The more unlabeled antigen (standard or unknown) there is in a sample, the less radioactive antigen will bind to the antibody. Radioimmunoassays are extremely sensitive, measuring routinely in the pg cm^{-3} to $\mu g\ cm^{-3}$ range.

Enzyme immunoassays (EIA) rely on the use of an enzyme-labeled antibody to measure an antigen. The enzyme used is one that will convert a colorless substrate into a colored soluble product that can be measured spectrophotometrically. The most frequently used enzyme labels are horseradish peroxidase and alkaline phosphatase and the most common EIA is the enzyme-linked immunosorbent assay (ELISA). The simplest ELISA format is to allow a protein antigen to adsorb onto the wells of a plastic microtiter plate (*Figure 4.13*). An enzyme labeled antibody is then added which binds to the antigen in the wells. Wells containing a large amount of antigen will contain more antibody, and therefore more enzyme. The substrate for the enzyme is then added and, after a limited period, the reaction is stopped and the absorbance measured. There are many different adaptations of ELISA which allow the measurement of nonprotein antigens, or which allow the sensitivity to be increased to that approaching RIA.

Other labels that can be used in immunoassays include fluorescent labels and bio- and chemiluminescent labels.

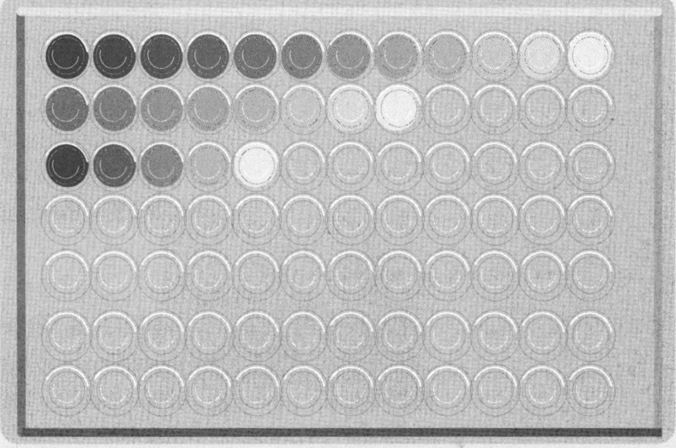

Figure 4.13 The outcome of a typical ELISA assay using three sources of antigen which have each been serially diluted across the rows of wells.

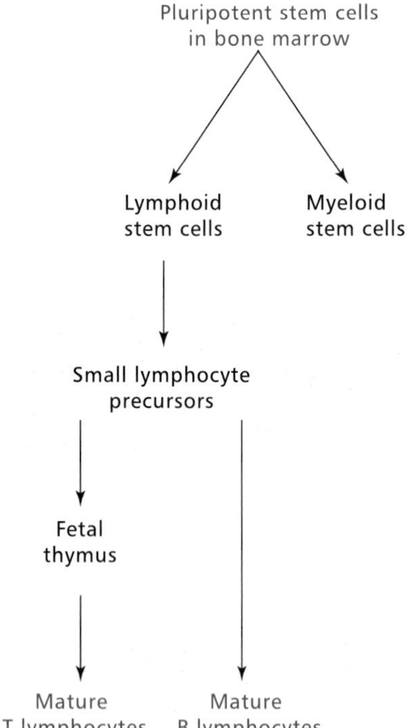

Figure 4.14 Schematic showing the development of small (B and T) lymphocytes.

4.5 SMALL LYMPHOCYTES

Small lymphocytes are the cells responsible for specific immunity (*Figure 4.2*). They make up approximately 20% of the blood leukocytes. There are two populations of small lymphocytes that mature at different sites in the body and have distinct functions (*Figure 4.14*).

The precursors of small lymphocytes originate in the bone marrow by division of **lymphoid stem cells**. Some small lymphocytes remain in the bone marrow where they mature into **B lymphocytes**. When maturation is complete, B lymphocytes have antibodies on their surface that are receptors for an individual epitope. Thus a single B lymphocyte is specific for an epitope and is capable of clonal division and of making antibody to it when stimulated appropriately by the immunogen. Thus they are responsible for humoral immunity. The second population of small lymphocytes, known as **T lymphocytes**, leave the bone marrow when immature and complete their maturation in the fetal thymus, a bilobed organ situated just above the heart. During their development in the thymus, T lymphocytes first acquire specificity for an epitope, by producing a cell surface receptor. They then mature into one of two T cell subsets. Cells of the first subset are known as **cytotoxic precursors** or **T_C cells.** When appropriately stimulated, T_C cells develop into **cytotoxic T lymphocytes (CTL)** that are capable of killing virus-infected cells. Cells of the second subset of T cells are the **helper T lymphocytes** or **T_H cells.** When stimulated by an immunogen, T_H cells develop into cytokine-secreting T_H cells that produce an array of cytokines that control the activities of both specific and nonspecific cells of the immune system. Thus T_H cells have a central role in the regulation of all immune responses.

When mature, both B and T lymphocytes are released into the circulation. However, small lymphocytes are not confined to the blood and many move into the lymphoid tissues: the spleen, lymph nodes, tonsils and the mucosa-associated lymphoid tissues found in the respiratory, gastrointestinal and urogenital tracts. Small lymphocytes constantly move between the blood and the lymphoid systems, a phenomenon known as recirculation. The route of this recirculatory process is shown in *Figure 4.15*. Lymph is the fluid that drains from the tissues into small lymphatic vessels. These merge with larger lymphatic vessels, the largest of which, the thoracic duct, delivers the lymph to the blood at its junction with the left subclavian vein. *En route* to the thoracic duct, lymph is filtered through many lymph nodes. Small lymphocytes circulating in the blood are able to move between the

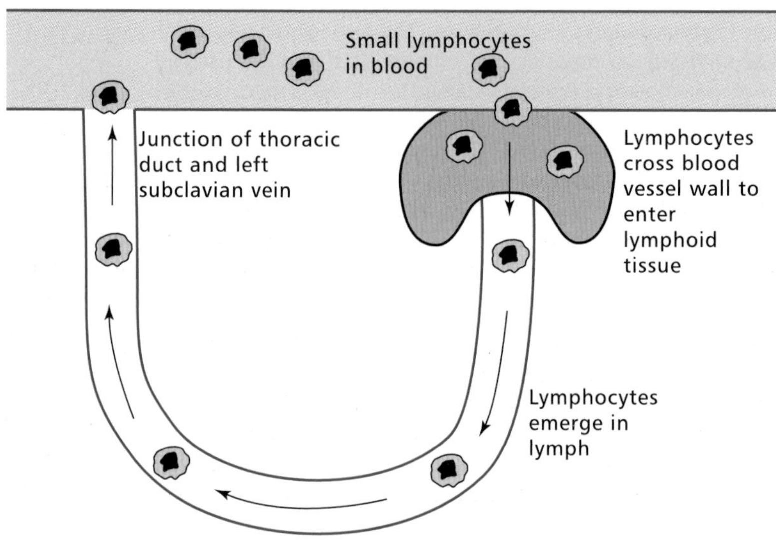

Figure 4.15 Schematic showing the recirculation of small lymphocytes between the blood and lymph systems.

endothelial cells lining the blood vessels that supply the lymph nodes. These blood vessels have a specialized endothelium which aids this process. By crossing the blood vessel wall, the small lymphocytes enter the lymph node and from this they enter the lymphatics and, eventually, return to the blood.

4.6 PRODUCTION OF A SPECIFIC IMMUNE RESPONSE

All small lymphocytes are specific for a single epitope on an individual immunogen. This means that they will respond only to that epitope and no other. The basis of this specificity depends on lymphocyte membrane proteins that act as receptors for individual epitopes. The receptors on B lymphocytes are surface immunoglobulins that have the same specificity as the antibody that the cells will subsequently secrete following appropriate stimulation. The surface antibodies on a single cell may belong to more than one class, so, for example, B lymphocytes may express both IgG and monomeric IgM, but these will have identical specificities. The T cell receptor (TCR) is composed of two polypeptide chains that form a single binding site. All the receptors on a single T cell have the same specificity.

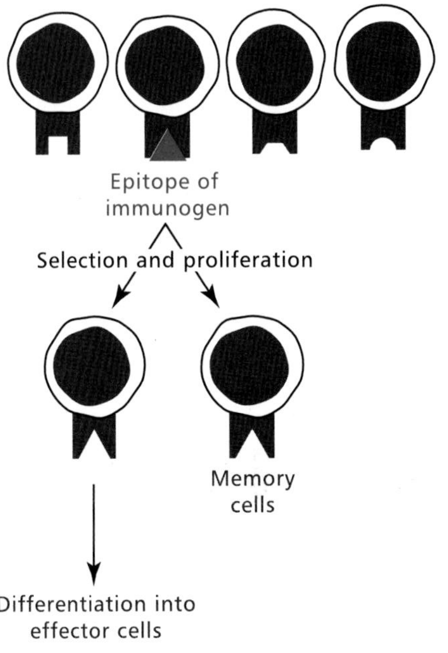

Figure 4.16 Schematic showing clonal selection in lymphocyte activation. See text for details.

Although T and B cells have different types of receptor and are involved in different aspects of the immune response, there are some similarities in the way they are activated and respond to a specific epitope. For example when both B and T lymphocytes are exposed to an appropriate epitope they enter a series of cell divisions that result in the production of a clone of cells of identical specificity (*Figure 4.16*). These cell divisions require cytokines that are produced and secreted by T_H lymphocytes once they have been appropriately stimulated. Most of the cells in the clone of small lymphocytes then differentiate into effector cells, the nature of which depends on the type of small lymphocyte that was stimulated. Not all the cells in the clone differentiate at this stage. Some remain as memory cells, awaiting the next exposure to the same immunogen when a faster and quantitatively greater response is produced.

BOX 4.3 Distinguishing T and B lymphocytes

In a standard stained blood smear, all small lymphocytes appear similar and it is not possible to distinguish between B and T lymphocytes, or between the T_C and T_H cells. However, these cells can be distinguished by staining for marker proteins in the membranes of these cells. Different markers that can be used to distinguish T and B lymphocytes are shown in *Table 4.4*.

The marker molecules can be distinguished by **immunohistochemistry**, in which a labeled antibody is used to stain the cells. One of the commonest labels used is a fluorescent molecule, such as fluorescein, which produces an apple-green colored fluorescence when irradiated with light of short wavelengths.

B lymphocytes can be stained with an anti-immunoglobulin carrying a conjugated fluorescein label. This binds to the cell surface immunoglobulin, so that the B lymphocyte fluoresces when examined with a fluorescence microscope. All mature T lymphocytes can be identified by the CD3 protein carried on their surface. Staining usually involves an indirect method that is a two-stage process. This involves incubating the lymphocytes with an unconjugated anti-CD3 antibody, which is a monoclonal antibody originating in mice (*Box 4.4*). This is then followed by the second stage involving incubation with a fluorescein-conjugated antibody to the mouse immunoglobulin bound to the CD3 proteins on the T lymphocyte surface (*Figure 4.17*). The cells can then be distinguished by their fluorescence. Similarly, T_H and T_C cells can be stained and identified using anti-CD4 and anti-CD8 antibodies respectively.

Fluorescent labeled antibody techniques, known as **immunofluorescence**, have been in use since the 1950s. They are easy to perform and reliable, although the preparations are not permanent as the fluorescent label tends to bleach, and they need to be examined soon after the test has been performed. In addition, they require the use of fluorescence microscopes although lymphocyte samples can also be quantified using a flow cytometer, which is usually available in major laboratories that deal with multiple samples on a regular basis.

BOX 4.3 Distinguishing T and B lymphocytes

Labels other than fluorochromes are available, which allow the production of permanent preparations and the use of an ordinary light microscope. For example, antibodies labeled with an enzyme, such as horseradish peroxidase or alkaline phosphatase can be located by their ability to convert a colorless substrate into an insoluble colored compound that can be seen when the cells are examined microscopically. This technique is called **enzyme immunohistochemistry** (*Figure 4.18*).

Protein	Role *in vivo*	Found on
Surface immunoglobulin	epitope receptor	B lymphocytes
CD3	signal transduction following binding of T cell receptor to epitope	all mature T lymphocytes
CD4	coreceptor molecule	helper T lymphocytes (T_H)
CD8	coreceptor molecule	cytotoxic T cell precursors (T_C) and cytotoxic T lymphocytes (CTL)

Table 4.4 Marker proteins for T and B lymphocytes

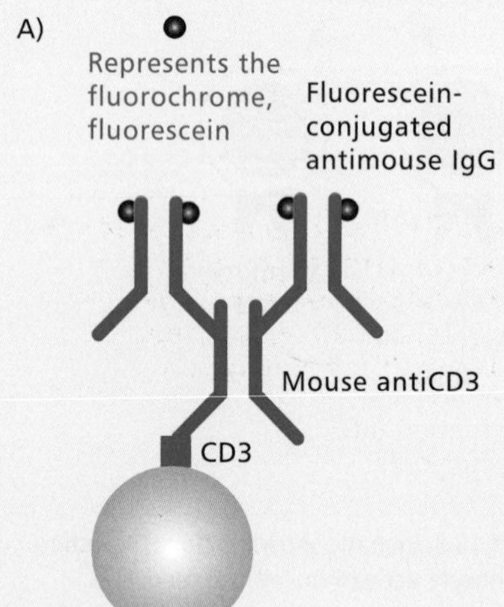

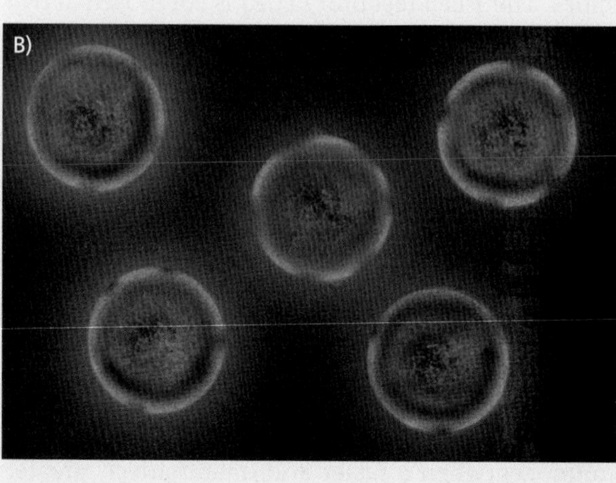

Figure 4.17 (A) Schematic showing the detection of cellular antigens, in this case CD3, by immunofluorescence as described in the text. (B) Computer generated image showing the distribution of membrane antigens as detected by immunofluorescence.

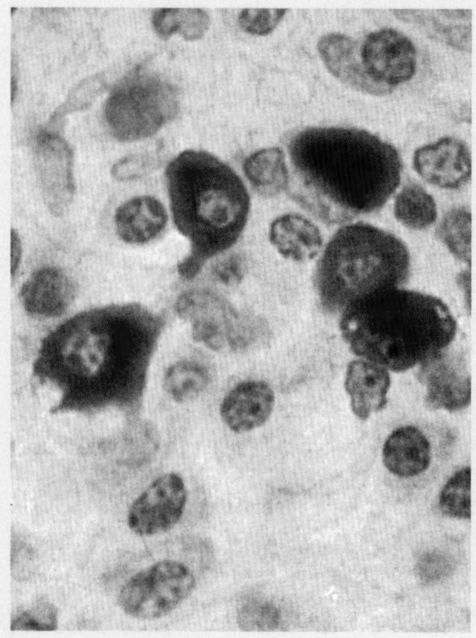

Figure 4.18 Photomicrograph showing the detection of cytoplasmic antigens following the production of a colored marker using enzyme immunohistochemistry.

HUMORAL IMMUNITY

The response of B lymphocytes to an immunogen is outlined in *Figure 4.19*. Once the receptors on B lymphocyte bind to an epitope on the immunogen, the lymphocyte is stimulated to divide and differentiate under the influence of cytokines released from T_H lymphocytes, to form a clone of antibody-secreting cells called **plasma cells** (*Figure 4.20*). Plasma cells are not found in the blood of healthy individuals. Instead, they are present in the lymph nodes and spleen, where they secrete antibodies until they die after a lifespan of a few days to several months. Antibody secreted by plasma cells in lymph nodes first appears in the lymph and then the blood, while antibody produced in the spleen moves directly to the blood.

The class of antibody that is secreted depends, in part, on which B lymphocyte was stimulated and on the cytokines that influenced its differentiation. Certain cytokines are known to favor the production of antibodies of a particular class. For example IL-4 promotes the production of IgE, and favors a response to multicellular parasites. However, a predisposition to produce IgE may also make that individual more susceptible to allergic reactions, such as hay fever and allergic asthma, as will be discussed in *Chapter 5*.

Although *Figure 4.19* illustrates a single B lymphocyte responding to a single epitope on an immunogen to give rise to a single clone of plasma cells, in reality an immunogen, such as a bacterium, contains numerous proteins each of which may have hundreds of epitopes. Thus, a humoral immune response involves the stimulation of many B lymphocytes, each of which can proliferate into a clone. Although each clone produces only a single type of antibody, called a **monoclonal antibody**, hundreds of clones are formed so that the response of the system is **polyclonal**, resulting in a heterogeneous array of antibodies appearing in the blood.

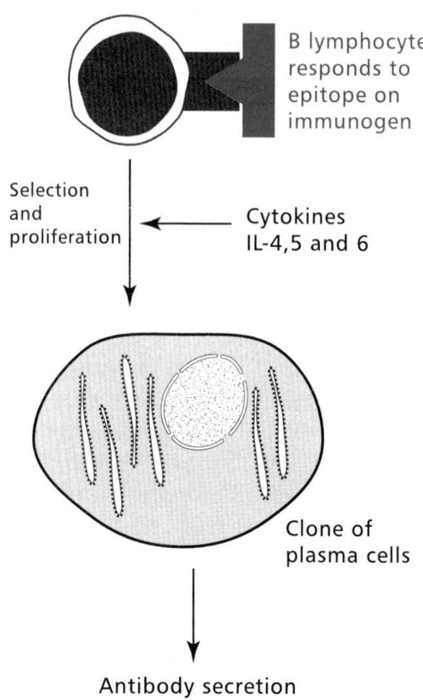

Figure 4.19 Schematic illustrating the activation of a B lymphocyte and its differentiation into a plasma cell. Note that the plasma cell has more cytoplasm, with extensive rough endoplasmic reticulum synthesizing immunoglobulins.

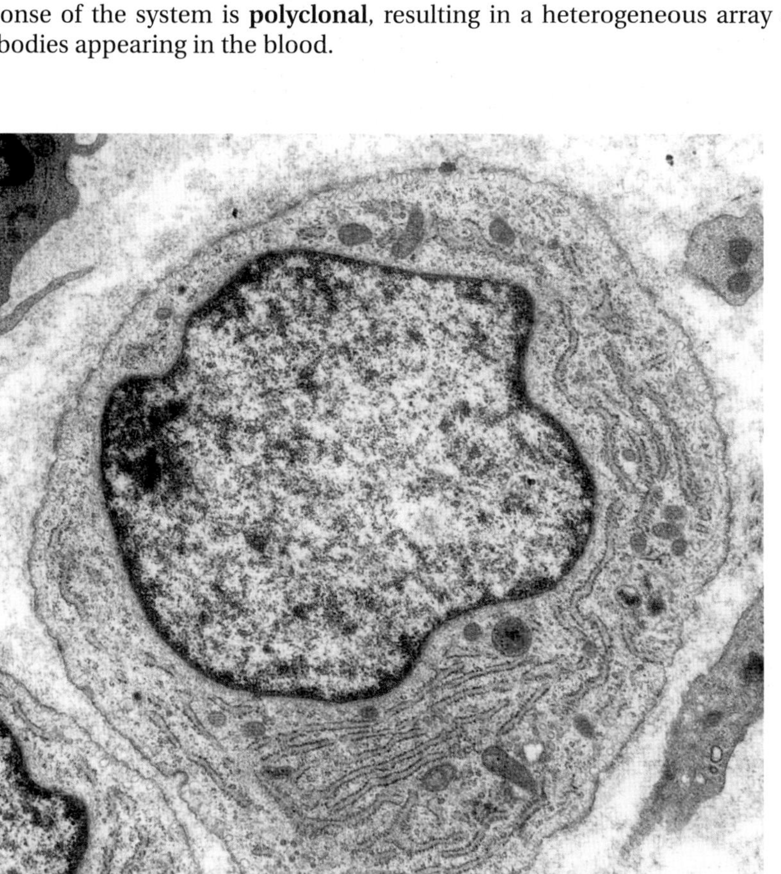

Figure 4.20 An electron micrograph of a plasma cell showing extensive rough endoplasmic reticulum.

BOX 4.4 Monoclonal antibodies

Each clone of plasma cells produces homogeneous or monoclonal antibody that is specific for a single epitope. In 1975, Kohler and Milstein, working at Cambridge, developed a technique whereby plasma cells could be immortalized producing a specified monoclonal antibody so that they could be cultured indefinitely. The technique, outlined in *Figure 4.21*, involves immunizing mice with the immunogen in question. After the required immunization protocol, the mouse spleen containing antibody-secreting plasma cells is removed and gently homogenized to form a suspension of single cells. The plasma cells are then fused with cultured cells derived from a mouse myeloma, that is a plasma cell tumor, the cells of which are immortal. The fusing agent, or fusogen, is polyethylene glycol (PEG). The resulting cells are known as hybrid myelomas, or, more frequently **hybridomas** and are, like their myeloma progenitor, essentially immortal when grown in suspension. Since the fusogen is indiscriminate in its actions, the cell suspension also contains hybrids consisting of plasma cell–plasma cell and myeloma–myeloma cell fusions. While the plasma–plasma cell fusions die within a short time in culture, the myeloma–myeloma fusions and any unfused myeloma cells have to be selectively removed, as they would quickly outgrow the hybridomas. The commonest selection procedure involves using a myeloma cell lacking hypoxanthine guanine phosphoribosyl transferase (HGPRT) activity and growing the resulting hybridomas in medium containing hypoxanthine, aminopterin and thymidine (HAT medium) for a period after hybridization. Aminopterin inhibits dihydrofolate reductase (DHFR), which is essential for the synthesis of DNA. Cells that possess HGPRT can overcome this block by using HGPRT and thymidine kinase (TK), as long as they are supplied with hypoxanthine and thymidine. Thus, hybridoma cells, which have HGPRT, supplied by the plasma cells, and TK activities survive in HAT medium but myeloma cells do not.

A suspension of fused cells produced from a single mouse spleen may contain millions of hybridoma cells derived from different clones. To produce monoclonal antibodies, individual hybridoma cells need to be isolated and grown individually. Isolation is achieved by diluting the cell suspension to such a degree that there is a high degree of certainty that aliquots will contain only a single hybridoma cell. Such aliquots can then be grown on in culture to produce a clone, which will secrete a monoclonal antibody.

Monoclonal antibodies have widespread uses in the diagnosis and treatment of disease. They can be used in immunoassays (*Box 4.2*) to measure the concentrations of biomolecules in clinical samples. For example, commercial monoclonal antibodies to hormones such as thyroxine, estrogen, and testosterone can be used to confirm suspected hormonal deficiencies (*Chapter 7*). In addition, monoclonal antibodies to cancer associated antigens can be used to screen for cancers or to monitor the treatment of cancers. An example of such an antigen is the prostate-specific antigen (PSA) which is elevated in the blood of patients with benign hyperplasia of the prostate gland, prostatitis and tumors of the prostate gland (*Chapter 17*).

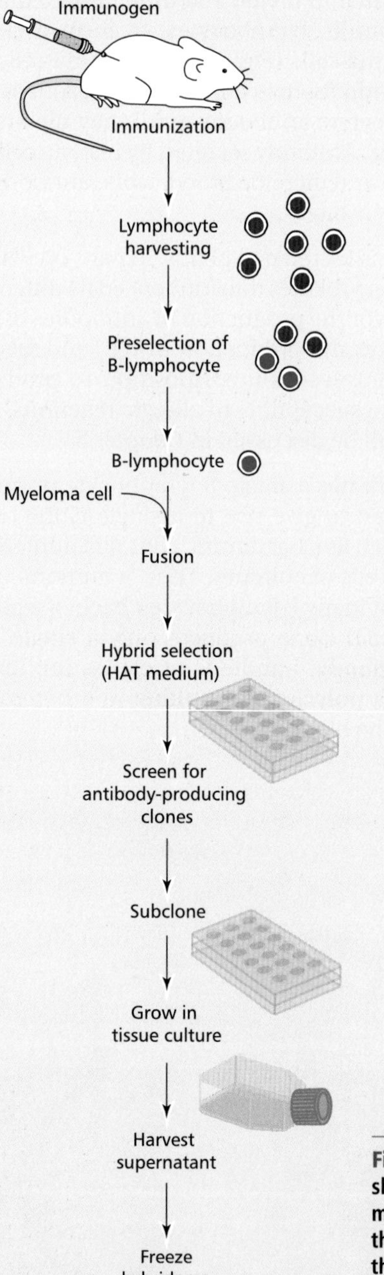

Figure 4.21 Schematic showing the production of monoclonal antibodies from the immunization of a mouse to the culture and storage of the hybridoma cells for future use.

Monoclonal antibodies have been used in clinical trials for treating autoimmune diseases and a variety of malignancies. Since most monoclonal antibodies are mouse immunoglobulins, they need to be 'humanized' by linking the mouse Fab region to a human Fc portion so that they are less likely to be recognized as foreign protein by the immune system when injected. The monoclonal antibody, MRA, is a humanized antibody to the human IL-6 receptor, which began clinical trials in 2005 for the treatment of systemic lupus erythematosus (SLE), an autoimmune disease (*Chapter 5*).

Activation of T lymphocytes

The manner in which T lymphocytes recognize foreign epitopes is rather more complex than that used by their B lymphocyte counterparts, as the receptors on T cells are unable to bind to epitopes on 'native' proteins. Instead, peptides derived from the foreign protein are 'presented' to them on the surfaces of other host cells, bound to membrane proteins encoded by a genetic region called the **Major Histocompatibility Complex** (MHC). In humans, this genetic region is found on chromosome 6. The proteins encoded by the MHC include two classes of membrane proteins. Class I proteins are found on all nucleated cells in the body. They consist of a single polypeptide, the α chain, which is associated with a smaller protein called β_2 microglobulin (β_2M) that is not encoded within the MHC. Class II proteins are found on only a few cell types and are made up of two polypeptides, α and β, both encoded within the MHC. Both Class I and II proteins have grooves that can bind a foreign peptide in an extended form (*Figures 4.22* and *4.23*). The peptide-binding groove is formed by the α_1 and α_2 domains of the Class I molecule and the α_1 and β_1 domains of the Class II molecule, while these structures are supported, in the Class I molecule by the α_3 domain and the β_2M, and in the Class II molecules by the α_2 and β_2 domains.

The requirement to have foreign peptides presented by different MHC molecules to T_C and T_H cells can be explained by looking at their roles *in vivo.*

Activation of T_C cells

The role of T_C cells is, ultimately, to destroy virus-infected cells. Since all nucleated cells are susceptible to such infections, MHC encoded Class I molecules are required by these cells to present endogenously produced viral peptides to T_C cells. Thus, viral protein produced within the cytoplasm of an infected cell may be hydrolyzed to produce short peptides, about 8–12 amino acid residues in length. These peptides are then transported across the endoplasmic reticulum where they become attached to Class I proteins and the complex transported to the surface of the cell membrane in Golgi vesicles.

A)

α chain

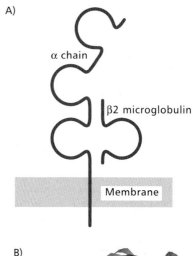

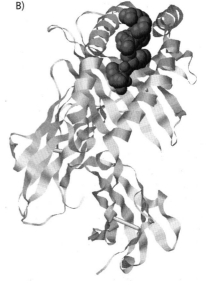

β2 microglobulin

Membrane

B)

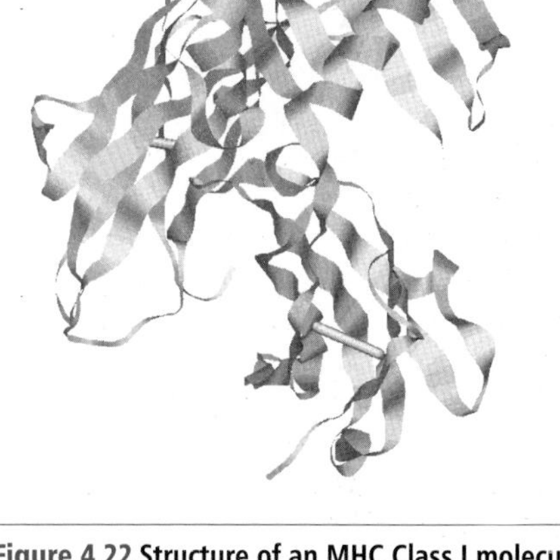

Figure 4.22 Structure of an MHC Class I molecule shown (A) diagrammatically and (B) molecular model showing a bound peptide (red) PDB file 1KJV.

A)

α chain

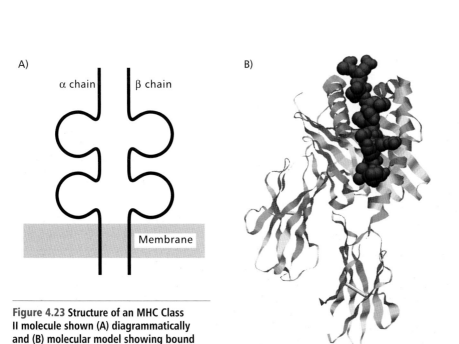

β chain

Membrane

B)

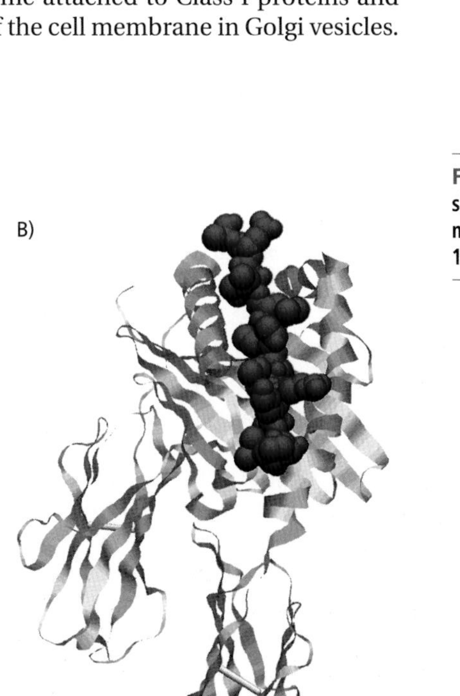

Figure 4.23 Structure of an MHC Class II molecule shown (A) diagrammatically and (B) molecular model showing bound peptide (red) PDB file 1MUJ.

The T_C cell is activated when the T cell receptor recognizes the complex of MHC Class I protein with its bound virus derived peptide and binds to it (*Figure 4.24 (A)* and *(B)*). An accessory protein, CD8, is also involved in this interaction as it also binds to the MHC protein. Once bound, the T_C cell is stimulated into a series of mitotic divisions to form a clone of cells, which then differentiate into cytotoxic T lymphocytes (CTL). These cells resemble T_C cells but their cytoplasm is more granular, owing to the presence of vesicles containing cytotoxic proteins called granzymes and perforins. The CTL binds to a virus infected cell using the same mechanism as the T_c cell, and then releases these cytotoxic proteins which destroy the target cell. Cytotoxic T lymphocytes also release IL-2 and IFN γ which stimulate NK cells and macrophages to kill virus infected cells.

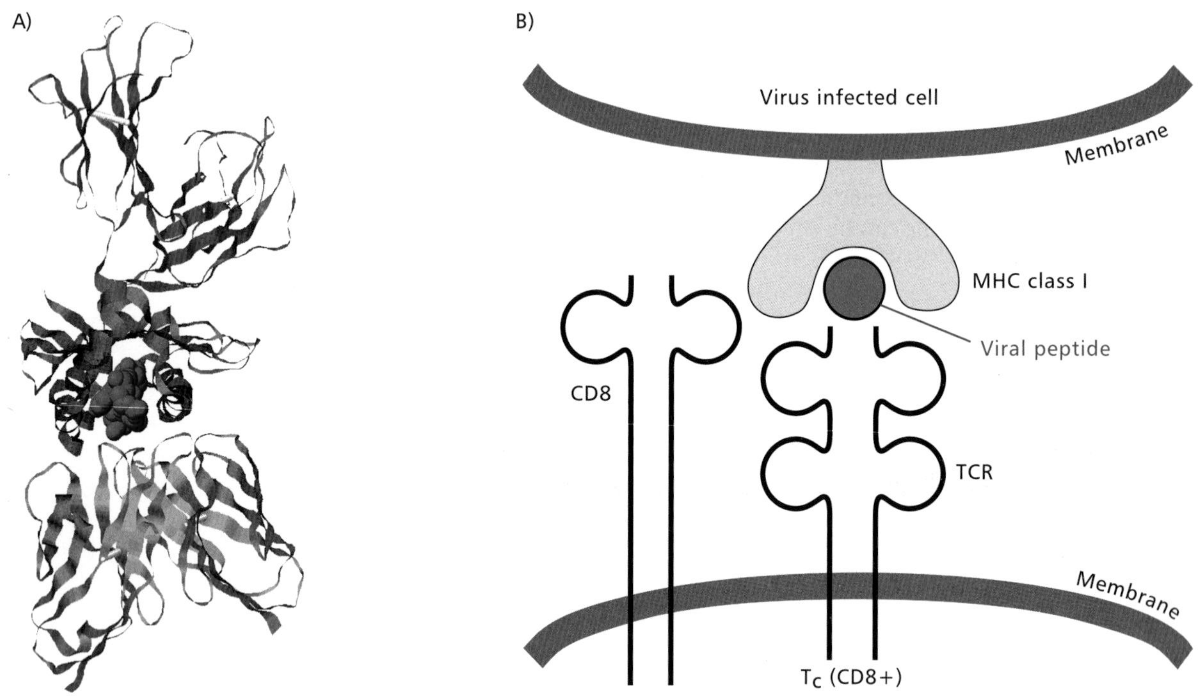

Figure 4.24 (A) Molecular model showing the recognition of a MHC Class I-peptide complex by a T_c cell receptor. PDB file 1NAM. (B) A schematic showing the interactions between a virus-infected cell and a T_c, as described in the text. TCR, T cell receptor.

Activation of T_H cells

When stimulated by an immunogen, T_H cells secrete cytokines required for specific and nonspecific immune responses. Thus they are required for an effective immune response to the whole range of infectious agents, from viruses to multicellular parasites. In order to activate a T_H cell, an immunogen has to be taken up by an **antigen presenting cell (APC)**. Here, its proteins are hydrolyzed within endocytic vesicles and the resulting peptides of 12–19 amino acid residues are attached to MHC encoded Class II proteins and transported to the cell membrane. The MHC Class II protein and foreign peptide complex is recognized and bound by a TCR (*Figure 4.25*). The coreceptor protein, CD4, also binds to the MHC protein. Several different types of cells can act as APCs, including monocytes, macrophages, dendritic cells in the lymphoid tissues and Langerhans cells of skin. However, under certain conditions, other cells, including epithelial cells, can be induced to express MHC Class II proteins and it has been proposed that such cells may initiate autoimmune reactions (*Chapter 5*).

When a T_H cell is stimulated by the MHC class II protein–peptide complex on the surface of an APC, it is stimulated to proliferate and differentiate into cells that actively secrete cytokines. The cytokines include growth and differentiation factors that stimulate B and T cells, hemopoietic factors, as well as factors that stimulate mast cells, macrophages and eosinophils.

There are two subsets of T_H cells, called T_H1 and T_H2. They differ in the profile of cytokines they secrete as shown in *Table 4.5*. Cytokines produced by T_H1 favor the development of cell-mediated immunity, which is used to destroy intracellular parasites such as viruses. Those produced by T_H2 cells stimulate humoral immunity to extracellular parasites, including most bacteria and helminth worms and flukes. In addition, cytokines produced by T_H1 inhibit T_H2 cells and *vice versa*.

Margin Note 4.5 Dendritic cells 

Dendritic cells are cells with characteristic membrane processes that resemble the dendrites of nerve cells. The processes give the cells a large surface area for presenting antigens to T_H cells. Dendritic cells are found in the lymphoid tissues and are highly efficient at antigen presentation. Langerhans cells in skin resemble dendritic cells in having multiple processes. They are involved in taking up antigens that have entered the skin before migrating to lymph nodes where they present the antigen to T lymphocytes.

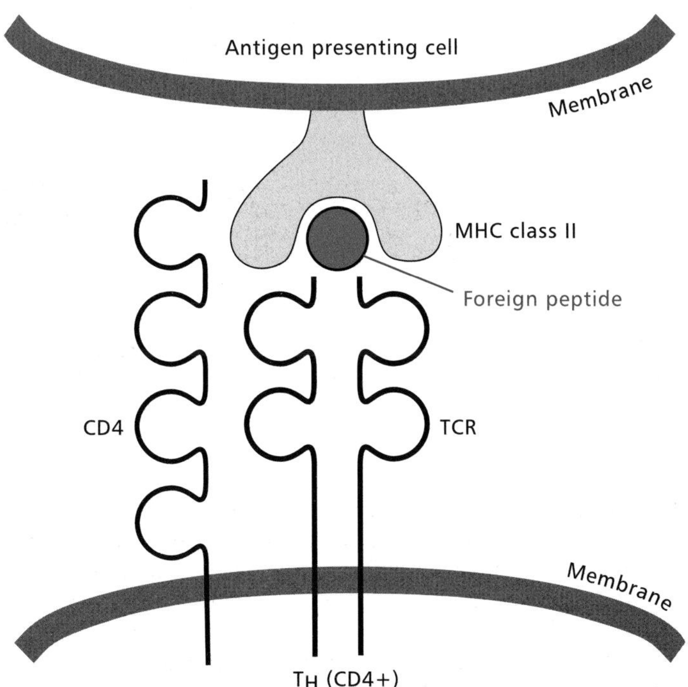

Figure 4.25 Schematic showing the activation of a T_H cell by an antigen presenting cell as described in text. TCR, T cell receptor.

Cytokine	Roles *in vivo*	T_H1 or T_H2 cytokine
IL-2	stimulates growth of T lymphocytes; increases activity of NK cells	T_H1
IL-3	stimulates hemopoiesis	T_H1 and T_H2
IL-4	promotes growth and differentiation of B lymphocytes; promotes growth of T_H2 cells; growth factor for mast cells	T_H2
IL-5	promotes growth and differentiation of B lymphocytes and eosinophils	T_H2
IL-6	promotes acute phase response; stimulates differentiation of B lymphocytes	T_H2
IL-9	activates mast cells	T_H2
IL-10	inhibits production of T_H1 cells	T_H2
IL-13	growth and differentiation of B lymphocytes	T_H2
IL-14	B cell growth factor	
Interferon γ	activates macrophages and NK cells; inhibits T_H2 cells	T_H1
TNF-β	promotes acute phase response; cytotoxin	T_H1
Granulocyte-Monocyte Colony Stimulating Factor (GM-CSF)	stimulates hemopoiesis	T_H1 and T_H2

Table 4.5 A range of T_H cytokines

Margin Note 4.6 Superantigens 

Superantigens such as Toxic Shock Syndrome Toxin cause massive stimulation of T_H cells by linking the T cell receptor to the MHC Class II protein outside the peptide binding groove (*Figure 4.26*). Since this does not depend on the TCR being specific to the antigen, many more T cells are stimulated than usual. The high level of cytokine release leads to shock with severe clinical consequences (*Chapter 3*).

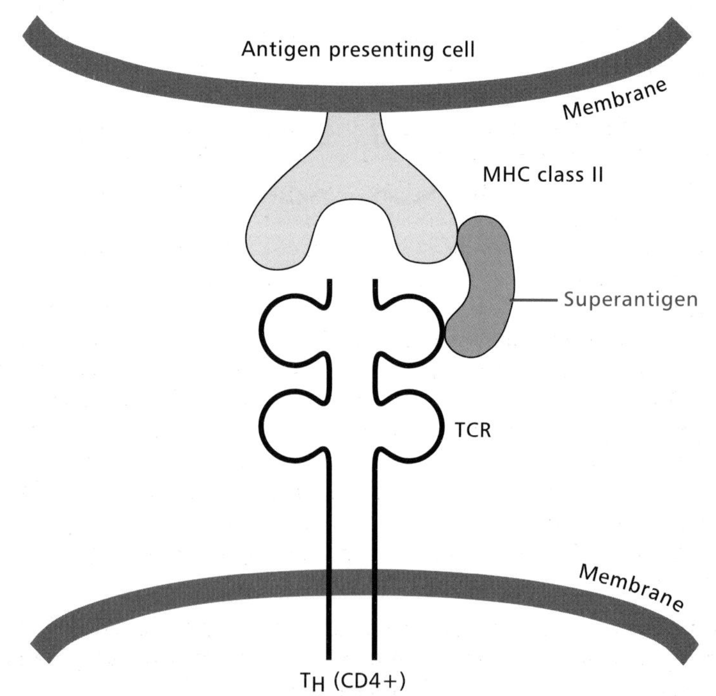

Figure 4.26 A schematic showing the linking of an MHC Class II molecule to a TCR by a superantigen as described in text.

CASE STUDY 4.1

Maria is a 28-year-old biology teacher who suspected she might be pregnant as it was now eight weeks since the start of her last period. Her pregnancy was confirmed when she carried out a home pregnancy test. Maria and her partner were delighted and began to make plans for the future. However, Maria became concerned when, during the eighth week of her pregnancy, several children in one of her classes developed German measles (rubella). Maria knew that, if she developed rubella, there was a high risk that the baby would be harmed. Unfortunately, she could not remember if and when she had ever been immunized against rubella. She consulted her family doctor who took a blood sample and had it tested for antibodies. The test revealed that Maria's blood was positive for IgM antibodies specific for rubella but no IgG antibodies specific for rubella were detected.

Questions

(a) What are the consequences of this result for the unborn baby?

(b) Why would the consequences be different if IgG antibodies specific for rubella had been detected?

(c) What counseling and/or advice would you recommend to Maria?

CASE STUDY 4.2

Alfred is a 70-year-old man who is suspected of having a myeloma, that is a plasma cell tumor. Suggest tests which could be carried out to confirm the diagnosis. Assuming the plasma cell was producing IgG, suggest an assay which could be used to measure the level of IgG in his blood.

4.7 SUMMARY

The basis of the actions of the immune system is its ability to distinguish self from nonself. It defends the body in a variety of nonspecific and specific ways. Nonspecific defenses include structural barriers and complement. Specific defenses are the development of immune responses against infectious agents. An effective immune response results from the complex interaction of nonspecific and specific cells. The nonspecific cells include the monocytes, large granular lymphocytes and polymorphonuclear leukocytes found in the blood. The specific cells are the small lymphocytes found in the blood and lymphoid tissues. Nonspecific responses include inflammation and the acute phase response, while specific responses include the production of antibodies, known as humoral immunity, and the production of cytotoxic cells, in the process known as cell-mediated immunity. Humoral immunity is effective at dealing with extracellular bacteria and multicellular parasites, while cell-mediated immunity is effective at killing cells infected with a virus. Small lymphocytes are highly specific since they bear cell surface receptors for epitopes found on foreign proteins. Small lymphocytes belong to one of two major subsets, the B lymphocytes responsible for humoral immunity and the T lymphocytes, some of which, the T_C cells, can develop into cytotoxic cells, while others, the T_H cells, regulate immune responses through the secretion of cytokines.

Antibodies are glycoproteins which are highly specific for epitopes. This specificity has enabled them to be used in the detection and quantitation of antigens, in highly sensitive techniques such as ELISA.

QUESTIONS

1. Which of the following is **NOT** part of the nonspecific defense against invading microorganisms?

   a) skin;
   b) mucus;
   c) antibodies;
   d) lysozyme;
   e) lactic acid.

2. Which of the following is **NOT** found in the blood?

   a) macrophage;
   b) monocyte;
   c) neutrophil;
   d) small lymphocyte;
   e) large granular lymphocyte.

3. Which of the following statements best describes complement?

   a) a defense protein found in sweat;
   b) a group of proteins which results in the lysis of bacteria;
   c) a cytokine produced by macrophages;
   d) an antibody which lyses bacteria;
   e) a protein found in the cell wall of bacteria.

4. Pair up the cells in column one with the appropriate cell surface molecule in column two.

Column 1	Column 2
a) All mature T lymphocytes	1) CD4
b) All B lymphocytes	2) CD8
c) Macrophages	3) CD3
d) Helper T lymphocytes	4) Surface immunoglobulin
e) Cytotoxic T lymphocytes	5) Receptors for complement

5. Which immunoglobulin class in each case best fits the following description?

   a) the only antibody that crosses the placenta;
   b) the most abundant antibody in the blood;
   c) also known as the secretory antibody;
   d) produced in response to parasitic worms;
   e) the antibody with the largest M_r.

6. Why is it unnecessary for erythrocytes to have MHC encoded Class I proteins?

FURTHER READING

Alonso, PL, Sacarial, J, Aponte, JJ *et al.* (2004) Efficacy of the RTS,S/AS02A vaccine against *Plasmodium falciparum* infection and disease in young African children: randomised controlled trial. *Lancet* **364:** 1411–1420.

Carroll, MC (2004) The complement system in regulation of adaptive immunity. *Nature Immunol.* **5:** 981–986.

Ceciliani, F, Giordano, A and Spagnolo, V (2002) The systemic reaction during inflammation: the acute phase proteins. *Protein Pept. Lett.* **9:** 211–223.

Chapel, H, Haeney, M, Misbah, S and Snowden, N (1999) *Essentials of Clinical Immunology*, 4th edn. Blackwell Science, Oxford, UK.

Helm, T (2004) Basic immunology: a primer. *Minn. Med.* **87:** 40–44.

Hyatt, MA (2002) *Microscopy, Immunohistochemistry, and Antigen Retrieval Methods for Light and Electron Microscopy.* Kluwer Academic/Plenum Publishers, New York, USA.

Jack, DL and Turner, MW (2003) Anti-microbial activities of mannose binding lectin. *Biochem. Soc. Trans.* **31:** 753–757.

Jackson, DC, Purcell, AW, Fitzmaurice, CJ, Zeng1, W and Hart, DNJ (2002) The central role played by peptides in the immune response and the design of peptide-based vaccines against infectious diseases and cancer. *Curr. Drug Targets* **3:** 175–196.

Kärre, K and Colonna, M (Editors) (1998) Specificity, function, and development of NK cells: NK cells: the effector arm of innate immunity. *Curr. Top. Microbiol. Immunol.* **230**, contains a number of interesting articles.

King, DJ (1998) *Applications and Engineering of Monoclonal Antibodies.* Taylor and Francis, London, UK.

Laroux, FS (2004) Mechanisms of inflammation: the good, the bad and the ugly. *Front. Biosci.* **9:** 3156–3162.

Pestka, S, Krause, CD and Walter, MR (2004) Interferons, interferon-like cytokines, and their receptors. *Immunol. Rev.* **202:** 8–32.

Price, CP and Newman, DJ (2001) *Principles and Practice of Immunoassay.* Macmillan, UK.

Todd, I and Spickett, G (2005) *Lecture Notes on Immunology,* 5th edn., Blackwell Science, Oxford, UK.

Vladutiu, AO (2000) Immunoglobulin D: Properties, measurement and clinical relevance. *Curr. Diagn. Lab. Immunol.* **7:** 131–140.

chapter 5:

DISORDERS OF THE IMMUNE SYSTEM

OBJECTIVES

After studying this chapter you should be able to:

- describe the features and explain the clinical consequences of immunodeficiency diseases;
- describe the features and explain the clinical consequences of autoimmune diseases;
- describe the features and explain the clinical consequences of immunological hypersensitivity;
- outline the ways in which immunological disorders are investigated and treated.

5.1 INTRODUCTION

A fully functioning immune system is essential for maintaining health (*Chapter 4*) but, like all body systems, the immune system itself is subject to many clinical disorders. Immunological disease can occur for one of three reasons. Firstly, **immunodeficiency diseases** occur in individuals whose immune systems are inadequate, perhaps due to the absence or malfunction of essential components, making them susceptible to infectious disease and even to certain types of cancer. Secondly, **autoimmune diseases** can occur as a result of the immune system mounting an immune response against 'self' components. Finally, **immunological hypersensitivity** occurs when immune responses, often mounted against seemingly innocuous immunogens, result in tissue damage. Indeed, immunological hypersensitivities are among the commonest of immunological disorders. This chapter will examine the variety, etiologies, diagnoses and treatments of these three groups of immunological disorders.

5.2 IMMUNODEFICIENCY DISEASES

Immunodeficiency diseases result from a failure of one or more components of the immune system and may involve the absence or malfunction of organs, cells or proteins of the immune system. Where the immunodeficiency results from a direct defect within the immune system, the disease is classified as a **primary immunodeficiency (PID)**. All PIDs are inherited or have a genetic component. Most, but not all, primary deficiencies are congenital, that is are present from birth, although some do not manifest themselves until later in life. **Secondary immunodeficiencies (SIDs)** arise as a consequence of other conditions. For example, some viruses are associated with an immunosuppression, which may be transient, as for example in measles, or permanent, as in HIV infection. Conditions that can give rise to secondary immunodeficiency are listed in *Table 5.1*. This chapter will concentrate on the primary immunodeficiencies.

Contributing factor	Comments
Malnutrition	a proper functioning immune system requires a supply of essential nutrients (*Chapter 10*)
Physical trauma	for example extensive burns and surgery are associated with an immunosuppression
Immunosuppressive drugs	drugs given to prevent transplant rejection (*Chapter 6*); chemotherapeutic therapy for cancer (*Chapter 17*)
Infectious diseases	for example HIV (*Chapter 2*); measles; cytomegalovirus
Hematologic disease	for example myeloma (*Chapter 4*); leukemia (*Chapter 17*)
Immaturity of immune system	for example in premature infants
Aging of the immune system	all body systems start to malfunction with aging, the immune system is no different (*Chapter 18*)
Metabolic and hereditary disorders	for example diabetes mellitus (*Chapter 7*)
Stress	stress hormones, such as cortisol (*Chapter 7*) can be immunosuppressive

Table 5.1 Some factors contributing to secondary immunodeficiency

Patients with immunodeficiency diseases invariably suffer infections more frequently and with increased severity although the type of infection depends to a large extent on the nature of the immune deficit. Some general rules to guide diagnoses are given in *Table 5.2*. Other types of disorder may also arise as a consequence of a PID. So, for example, deficiencies of some complement proteins may lead to systemic lupus erythematosus (SLE), an autoimmune disorder (*Section 5.2*). Certain types of cancer, such as non-Hodgkins lymphoma, a tumor of the lymph glands, and Kaposi's sarcoma, a tumor derived from blood vessels and which frequently shows in the skin, are also more frequent in the immunodeficient patient, because these tumors are linked to infections with certain viruses (*Chapter 17*).

PRIMARY IMMUNODEFICIENCY DISEASE

Primary immunodeficiencies are classified according to the site of the immunological defect. The 100 or so inherited PIDs were classified into eight groups or tables (*Table 5.3*) at a meeting of the International Union of Immunological Societies in 2003.

Deficiencies of:	Consequences
T lymphocytes	viral and fungal infections (for example herpes simplex; *Pneumocystis carinii*) opportunistic infections such as *Candida albicans* infections with intracellular bacteria, e.g. Mycobacteria Kaposi's sarcoma (a tumor derived from blood vessels)
Antibodies	bacterial infections causing recurrent chest infections and boils in the skin
Phagocytic cells	bacterial infections causing boils and abscesses
Complement	recurrent meningococcal meningitis bacterial pneumonia , septicemia systemic lupus erythematosus (SLE, an autoimmune disorder)

Table 5.2 Diseases associated with PIDs of different origins

Table	Deficiencies in:	*Examples
I	T and B lymphocytes	severe combined immunodeficiency (SCID)
II	antibody deficiencies	X-linked agammaglobulinemia
III	other well-defined immunodeficiency syndromes	Wiskott Aldridge syndrome DiGeorge anomaly
IV	diseases of immune dysregulation	Chediak-Higashi syndrome
V	congenital deficiencies of phagocyte numbers and/or function	X-linked neutropenia/myelodysplasia leukocyte adhesion deficiencies (LADs) types I and II chronic granulomatous disease
VI	defects in innate immunity	mannose binding lectin (MBL) deficiency
VII	autoinflammatory disorders	hyper IgD syndrome
VIII	complement deficiencies	C4 deficiency C2 deficiency C1 inhibitor deficiency

* examples only, many more disorders are known

Table 5.3 Classification of PIDs

Combined T and B deficiencies

Deficiencies that affect both B and T lymphocyte (*Chapter 4*) numbers and/or their functions are life threatening. Such deficiencies are termed **severe combined immunodeficiency (SCID)**. The term SCID represents a group of disorders associated with more than 20 different mutations and with a frequency of between one in 50 000 and one in 500 000 births. As some forms of SCID are inherited in an X-linked fashion (*Chapter 15*), more boys than girls are affected, with the male : female ratio of approximately 3 : 1.

Depending on the mutation involved, T and B cells may both be decreased (T^-B^- SCID) or B cells numbers may be normal or increased (T^-B^+SCID). In the latter, the absence of T cells renders B cells functionally inactive, owing to the need for the cytokines produced by T_H cells for antibody production as described in *Chapter 4*. In either case, the disorder becomes apparent in the first few weeks

BOX 5.1 Laboratory tests for lymphocyte function

It is theoretically easy to test for lymphocyte function *in vivo* by immunizing a patient with immunogens known to stimulate cell-mediated or humoral immunity and monitoring the response of their immune systems. A test for humoral immunity would involve measuring the levels of specific antibody *in vitro*. Tests for cell-mediated immunity might involve injecting a small amount of immunogen subcutaneously and looking for delayed **hypersensitivity** (*Section 5.4*). However, such tests are not necessarily appropriate or, indeed, ethical, especially for sick babies. Fortunately, there are tests that can be used to measure the function of lymphocytes *in vitro*. For example, small lymphocytes can be incubated with antigens to which the patient has been exposed. If the child is immune, the small lymphocytes will respond by starting to proliferate. The amount of cell division can be measured by determining the amount of radiolabeled thymidine incorporated into the DNA of the dividing cells. Of course, an antigen will only stimulate those cells which bear specific receptors, perhaps only one in 100 000 cells and so the test may not be sufficiently sensitive. However, some proteins and other types of molecules, such as lipopolysaccharide, act as **mitogens** and stimulate many lymphocytes to divide and are known as polyclonal activators because they stimulate many different clones of lymphocytes. Mitogens are examples of **lectins**, which stimulate lymphocyte proliferation by binding to carbohydrate residues on the lymphocyte membrane. Lectins are frequently of plant origin (*Table 5.4*) although some are derived from other groups of organisms. It is also possible to stimulate lymphocytes to divide by incubating them with antibodies directed at cell surface receptors. Thus, an antibody to CD3 will stimulate all mature T lymphocytes to divide.

A typical response of T lymphocytes to the lectin phytohemagglutinin (PHA) from runner beans in a healthy individual is shown in *Figure 5.1*. A failure to respond in this way is indicative of a lymphocyte defect. In patients with SCID, the proliferative response is less than 10% of control values.

Mitogen	Derived from	Small lymphocytes stimulated
Phytohemagglutinin (PHA)	*Phaseolus vulgaris*	T lymphocytes
Concanavalin A (ConA)	*Canavalia ensiformis*	T lymphocytes
Poke weed mitogen (PWM)	*Phytolacha americana*	B lymphocytes
Lipopolysaccharide (LPS)	*Escherichia coli*	B lymphocytes
AntiCD3	monoclonal antibody originating from mice	T lymphocytes
Anti-immunoglobulin	monoclonal antibody originating from mice	B lymphocytes

Table 5.4 Polyclonal activators of small lymphocytes

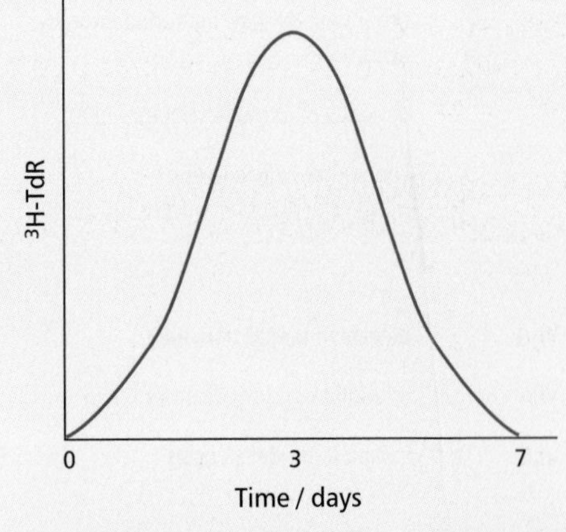

Figure 5.1 The stimulation of T lymphocytes with phytohemagglutinin (PHA), which stimulates all T cells to divide as measured by the uptake of tritiated thymidine (^{3}H-TdR) into dividing cells.

or months of life with the mean age at diagnosis being 6.5 months. The disease is characterized by chronic viral and fungal infections. Chronic diarrhea and oral Candida infections are common and in the presence of other infections, the child fails to thrive. Affected infants may suffer generalized viral infections if given live viral vaccines, such as the MMR and polio vaccines. Laboratory tests show fewer than 3000 per mm^3 circulating lymphocytes (where the reference range for an infant of six months is between 4000 and 13 500 per mm^3). The lymphocytes present are functionally inactive and do not respond *in vitro* to known **mitogens** (*Box 5.1*). Chest X-rays show an abnormally small or absent thymus.

Approximately 20% of T$^-$B$^-$ SCID arises from mutations in the gene encoding adenosine deaminase (ADA). The absence of this enzyme results in the accumulation of metabolites, such as ADP, GTP and dATP, which are toxic to small lymphocytes. A similar syndrome arises from deficiency of the purine nucleoside phosphorylase (PNP) that results in an accumulation of dGTP. Approximately 40% of cases of T^-B^+ SCID are X-linked and arise from mutations in a polypeptide that forms part of the receptor for several interleukins (ILs), so that affected lymphocytes are unable to respond to interleukin signals.

If not diagnosed early and treated appropriately, children with SCID usually die from infections in the first few years of life. Management of this condition includes administering antiviral, antibacterial and antifungal drugs and measures must be taken to avoid infection. Keeping such infants in totally aseptic conditions is neither feasible nor ethical, since this would preclude direct contact with other humans. Their immune system may be restored with a bone marrow transplant. This can lead to long-term survival but is not without danger, chiefly graft versus host disease (GVHD; *Chapter 6*). Gene therapy has been attempted with some ADA deficient patients and, in a few cases, has been reported to be successful.

Antibody deficiencies

The normal concentration ranges for the five immunoglobulin classes, IgM, IgG, IgA, IgE and IgD (*Chapter 4*), in adults are shown in *Table 5.5*. Deficiencies involving immunoglobulins of all classes are commonly referred to as **agammaglobulinemias** or **hypogammaglobulinemias**, depending on the level of deficiency. However, with some disorders there may be a selective deficiency of a single immunoglobulin class, as in selective IgA deficiency, or a dysregulation, where some antibody classes are reduced while others are increased.

Immunoglobulin	Serum concentration in adults/g dm^{-3}
IgM	0.5–2.0
IgG	7.2–19.0
IgA	0.8–5.0
IgD	trace
IgE	trace

Table 5.5 Normal adult levels of serum immunoglobulins

Transient hypogammaglobulinemia (TH) occurs when the start of production of IgG in very young children is delayed. It possibly arises when the maturation of helper T lymphocytes is itself delayed. During pregnancy, IgG is the only immunoglobulin to cross the placenta from mother to fetus. Thus a newborn baby has adult levels of IgG, most of which is maternally derived (*Figure 5.2*).

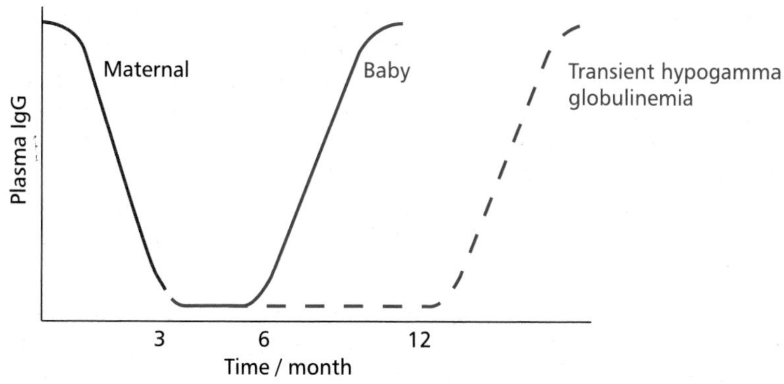

Figure 5.2 Transient hypogammaglobulinemia (TH). The production of IgG in the baby normally starts between three and six months of age by which time the maternal IgG in the baby has almost disappeared. In TH, the production of IgG by the baby is delayed.

During the first three months postpartum, the maternally derived IgG is catabolized and steadily disappears from the baby's circulation. Between three and six months, serum IgG levels may be quite low, after which the levels begin to increase, and should normally attain 'adult' levels at about 12–18 months of age. In infants with transient hypogammaglobulinemia, the production of IgG is delayed considerably, sometimes for as long as two years. During this time the child is susceptible to recurrent infections with **pyogenic** (pus producing) bacteria and antibiotics must be administered. The incidence of TH has been estimated as 23 to 61 per million births.

Common variable immunodeficiency (CVID) is the commonest primary immunodeficiency involving all classes of antibody. It is a heterogeneous group of disorders and includes a range of phenotypes. Many patients are not diagnosed until early adulthood. Most patients show low levels of IgG and IgA, with near normal or 50% of normal levels of IgM and normal lymphocyte counts. The latter allows CVID to be distinguished from other antibody deficiencies such as X-linked agammaglobulinemia (*see below*). Some patients also have impaired cell-mediated immunity (*Chapter 4*). The incidence has been estimated at one in 10 000–50 000.

The etiology of CVID is unknown and the majority of cases are sporadic. The B lymphocytes are immature and, when stimulated, do not differentiate into antibody-secreting plasma cells following the binding of an antigen (*Chapter 4*), owing to defects in their cell surface receptors or signal transduction mechanisms (*Chapter 7*). However, it is possible that in some patients there may be other defects, such as mutations in immunoglobulin regulatory genes. In addition, many CVID patients have defects in CD4+ T lymphocytes so that the T and B cell interactions required for antibody production are impaired.

Patients with CVID present with recurrent pyogenic infections, especially of the respiratory tract, commonly involving *Streptococcus pneumoniae, Staphylococcus aureus, Haemophilus influenzae* and *Moraxella catarrhalis*. This can lead to **bronchiectasis**, in which the bronchi and bronchioles are abnormally dilated. Those patients who also have some impairment in cell-mediated immunity also suffer infections with mycobacteria and the fungus *Pneumocystis carinii*. Such patients also suffer recurrent severe infections with herpes simplex and herpes zoster (*Chapter 3*), and may develop viral illness when immunized with live viral vaccines.

The diagnosis of CVID relies on the measurement of immunoglobulins, including specific antibodies to common vaccines, and ruling out other immunodeficiencies, such as X-linked agammaglobulinemia. Its treatment usually involves replacement therapy with pooled immunoglobulins obtained from healthy donors. All pooled immunoglobulin preparations are treated to inactivate viruses, such as hepatitis virus or HIV, which may be present. These preparations are available commercially and need to be given intravenously every three to four weeks to maintain plasma levels and protect against infections. The dose depends on the weight of the patient and is usually 400–600 mg kg^{-1}. Alternatively, weekly subcutaneous administration of lower doses, which can be done at home, may be more convenient. Intramuscular injection, which allows a greater volume to be administered than can be delivered subcutaneously, can also be given.

X-linked agammaglobulinemia (XLA) or Bruton's disease is caused by a deficiency in Bruton's tyrosine kinase (Btk), which is required for the maturation of preB cells in the bone marrow to form B lymphocytes. The deficiency of Btk is due to one or more of 300 different mutations in the *BTK* gene located on the X chromosome. B lymphocytes are therefore absent from the circulation and plasma cells are not present in the spleen and lymph nodes. Tonsils and adenoids may be absent, as demonstrated by radiography. However, circulating T lymphocytes are normal. Serum immunoglobulin levels are extremely low and all classes of immunoglobulin are affected. This is a rare inherited disorder with an incidence of about one in 250 000 males. Patients present with recurrent pyogenic infections, which occur from around three to nine months of age, when maternally derived IgG is low. Infections encountered may result in pneumonia, otitis media, meningitis and diarrhea.

It is essential that patients with XLA are diagnosed as early as possible so that replacement immunoglobulin therapy can begin. Infections are treated with antibiotics. The prognosis for children diagnosed before the age of five years is good, with patients often surviving to middle age. Tests for mutations of the *BTK* gene are available, which allows for genetic counseling of affected females and prenatal diagnosis of fetal cells obtained by chorionic villus sampling or amniocentesis (*Chapter 15*), with the possibility of a therapeutic abortion.

Selective IgA deficiency, as its name implies, affects only a single class of immunoglobulin. Many IgA deficient individuals are asymptomatic, with the condition only being detected during investigation of other disorders. In contrast, other patients with selective IgA deficiency suffer recurrent infections, typically ear infections, sinusitis and pneumonia. A high proportion of sufferers also develop autoantibodies that are directed against a variety of self antigens and approximately a third present with autoimmune diseases, such as systemic lupus erythematosus (SLE) (*Section 5.3*). It is not known which features determine the severity of the disease. Selective IgA deficiency is a relatively common disorder with an incidence of one in 500 to 700 Caucasians, although the frequency is much lower in other ethnic groups.

A patient who presents with a history of recurrent infection, chronic diarrhea, and autoimmune disease should be suspected of having a selective IgA deficiency. This can be confirmed by measuring serum immunoglobulin concentrations. Values of IgA below 0.07 mg dm^{-3}, while other immunoglobulins are normal, would confirm the deficiency. Treatment of selective IgA deficiency normally involves using antibiotics to treat bacterial infections and replacement therapy is not usually necessary. If the disease presents with autoimmunity then anti-inflammatory drugs, such as corticosteroids, may be given. The prognosis is good, with patients living normal lifespans. However, approximately 10% of patients with a selective IgA deficiency also have a deficiency of the IgG_2 subclass, which is usually produced in response to polysaccharide antigens. Patients with both defects suffer more severe bacterial infections, especially with encapsulated bacteria. Immunoglobulin replacement therapy may be appropriate in these cases.

The DiGeorge anomaly and the Wiskott Aldridge syndrome

The third group of PIDs (*Table 5.3*) contains a number of well-defined immunodeficiency syndromes, of which the DiGeorge anomaly and the Wiskott Aldridge syndrome are well-known examples.

The DiGeorge anomaly (DGA) is a developmental disorder involving organs that develop from the third and fourth pharyngeal pouches of the embryo. It is associated with a deletion or partial monosomy of chromosome 22 (*Chapter 15*) that results in a range of defects. Several different patterns of inheritance have been reported, including autosomal dominant and autosomal recessive. Its incidence has been estimated to be between one in 20 000 to 66 000, depending on the country.

The DGA is characterized by facial abnormalities, hypoparathyroidism and hypocalcemia with symptoms of convulsions and tetany, congenital heart disease that may be so severe as to be life threatening, and a small under-developed or sometimes absent thymus that results in a profound immuno-deficiency. Patients suffer severe and recurring viral and fungal infections.

Indeed, the immunological defects are the second commonest cause, after heart conditions, of death in DGA patients. The number of circulating T lymphocytes is severely reduced leading to defects in cell-mediated immunity. T cell proliferative responses to mitogens vary in DGA patients, such that they can be classified either as partial or complete. In the former, proliferation is reduced but in the latter it is completely absent. The absence of helper T lymphocytes reduces antibody production, so that antibacterial immunity may also be compromised, even though the number of circulating B lymphocytes is normal.

A diagnosis of DGA is based on the cardiac malformations, hypoparathyroidism resulting in hypocalcemia and a small or absent thymus. T lymphocytes in the circulation are reduced and the proliferative response to mitogens is impaired. Fluorescence *in situ* hybridization (FISH) has been used to detect deletions in chromosome 22 in the majority of patients (*Margin Note 5.1*). Other syndromes, without any apparent genetic link, but which have known environmental causes, bear some resemblance to DGA. One example is fetal alcohol syndrome, which results from prolonged exposure to alcohol during fetal development. Children with fetal alcohol syndrome also show the characteristic facial features associated with DGA.

Margin Note 5.1 Fluorescence *in situ* hybridization 

Fluorescence *in situ* hybridization (FISH) relies on the ability of fluorochrome-labeled DNA probes to hybridize with complementary DNA in tissue sections. The hybridized probe can be seen as fluorescent 'spots' in the nuclei of target interphase cells and can be located to specific chromosomes when applied to cells in metaphase (*Chapters 17* and *18*).

Attempts have been made to treat the immunological deficit in DGA with thymus transplants, (*Chapter 6*) although results have been variable. The associated hypocalcemia is treated with calcium and vitamin D supplements, while cardiac malformations must be rectified surgically. The prognosis for patients with DGA is variable and depends mostly on the degree of cardiovascular abnormality. For patients with severe cardiac problems it is poor, with a mortality rate of over 80% at the age of six months.

The Wiskott Aldridge syndrome (WAS) arises from mutation in the *WAS* gene, which was identified on the short arm of the X chromosome in 1994. The gene codes for the cytoskeletal protein sialophorin, found in lymphocytes and platelets, that is involved in the assembly of actin filaments. The incidence of WAS is approximately one per 250 000 male births.

The syndrome is characterized by decreased levels of IgM but often with increased production of IgE and IgA. In the early stages, T and B cell numbers in the blood are normal. Since IgM is the prevalent antibody in immune responses to bacterial polysaccharides, there is an increased incidence of infections with encapsulated bacteria. Sufferers may also develop eczema. Blood platelets are small, short-lived and reduced in number, leading to thrombocytopenia and increased bleeding times which may prove fatal (*Chapter 13*). As WAS progresses, there is a loss of both humoral and cell-mediated immunity and, along with severe infections, there is also an increase in leukemia and lymphoid tumors.

The treatment for WAS includes antibiotics for infections and platelet transfusions to prevent bleeding. Immunoglobulin replacement therapy may also be given to provide some protection against infection. Bone marrow transplants (*Chapter 6*) have been successful in some cases. Unfortunately the prognosis for WAS sufferers is poor, with death commonly occurring before the age of four years usually from severe infection and bleeding. Genetic counseling is recommended for women who have had a child with WAS. Detection of the abnormal gene in cells obtained by chorionic villus sampling or amniocentesis allows a prenatal diagnosis, with the possibility of terminating the pregnancy if the fetus is found to be affected.

The Chediak-Higashi syndrome

Chediak-Higashi syndrome (CHS) is a rare autosomal recessive disorder first described in 1943. It is sometimes classified as a phagocytic defect.

However, even though phagocytic cells, such as neutrophils and monocytes, are defective, Natural Killer (NK) cells, which form the first line of defence against viruses, are also abnormal. The syndrome arises from a mutated form of the *CHS1* gene located on chromosome 1. The gene product is involved in the intracellular transport of proteins and the synthesis of storage granules in certain cells. The mutation results in neutrophils with abnormally large or 'giant' lysosomes (*Chapter 16*) and platelets (*Chapter 13*) with abnormal dense bodies. In addition, melanocytes, the pigment-producing cells of the skin, contain larger than normal melanosomes; the pigment storing organelles.

Chediak-Higashi syndrome presents as an immune deficiency that leads to recurrent bacterial infections, most commonly with *Staphylococcus aureus, Streptococcus pyogenes,* and *Pneumococcus* spp and viral infections, such as with Epstein-Barr virus, that frequently result in tumors of the lymph nodes (**lymphomas**). Leukocyte counts reveal neutropenia and abnormal neutrophils that do not respond to chemotactic molecules, for example activated complement proteins, and which fail to kill ingested bacteria. Infants suffer recurrent skin infections, which may result in ulcers and abscesses. The abnormalities of melanocytes means that patients are deficient in skin pigment and have blond hair and translucent blue eyes. Infants also bruise very easily, due to defective platelets. In addition, they suffer progressive neurological dysfunction, with abnormal gait, mental retardation and peripheral neuropathy. If the child survives beyond the first decade this may lead to Parkinsonism and/or dementia. Morbidity and mortality are high in CHS, with infants frequently dying before the age of 10 years, usually from overwhelming pyogenic infections attributable to poor neutrophil function.

The recommended treatment for children with CHS is a bone marrow transplant to correct the immune deficit. However, success has been variable and this treatment has no effect on the lack of pigmentation since melanocytes do not arise from bone marrow. Antibiotics are given to treat bacterial infections and antiviral drugs, such as acyclovir or interferon α, to limit infection with the Epstein-Barr virus. Patients with lymphoma are given anticancer drugs, such as vincristine.

Phagocytic defects

Phagocytic cells, such as monocytes, macrophages and neutrophils, form part of the nonspecific immune defense. These cells kill ingested bacteria using several different mechanisms as described in *Chapter 4*. A defect in any of these mechanisms can lead to increased incidences of infections. Thus, patients may be severely compromized by defective phagocytes, even if their B and T cell populations and functions are normal. Some examples of phagocytic defects are chronic granulomatous disease and leukocyte adhesion deficiency.

Chronic granulomatous disease (CGD) is named from the granulomatous inflammatory nodules that present on the skin and in the gastrointestinal and genitourinary tracts. It is an inherited disorder of phagocytic cells characterized by their inability to generate the reactive oxygen intermediates needed to produce bactericidal compounds, such as hydrogen peroxide. The formation of reactive oxygen intermediates is dependent on NADPH oxidase activity. This enzyme is composed of four subunits and a defect in any one of them can result in CGD. Approximately 65% of all CGD cases is due to a defect in the *CYBB* gene located on the X chromosome which encodes cytochrome b_{245}. The genes for the other subunits are located on autosomal chromosomes and females and males are equally affected. The incidence of CGD is estimated to be about one in 200 000 to 250 000.

Sufferers of CGD usually present before the age of five years. Skin infections, pneumonia, gastroenteritis, perianal abscesses are common. Abscesses on internal organs, such as the lungs, spleen and liver, may also be present. The

Margin Note 5.2 Hyper IgD syndrome 

Hyper IgD and periodic fever syndrome (HIDS) is a rare disease which was first reported in 1984. Patients suffer recurrent attacks of fever and symptoms of inflammation and an acute phase response from about the age of one year. These attacks may last up to six days and may be triggered by surgery, trauma or vaccination. Clinical findings include, usually, elevated levels of serum IgD sometimes with higher levels of IgA. The disease is inherited as an autosomal recessive trait and sufferers have been shown to have a defective gene for mevalonate kinase. Most sufferers are of Western European origin, with the majority being Dutch or French. The mevalonate kinase enzyme is involved in the metabolism of cholesterol, although how this deficiency is related to the inflammatory condition is not known.

small amounts of hydrogen peroxide produced by CGD patients makes them resistant to catalase negative bacteria. However, catalase positive bacteria, by definition produce catalase, which catalyzes the degradation of hydrogen peroxide; hence these types of bacteria give rise to infections in CGD sufferers. Pneumonia is generally associated with fungal infections; and disseminated fungal disease also occurs.

A diagnosis of CGD takes into account the recurrent infections of early onset, granulomas, hepatosplenomegaly, that is enlarged liver and spleen, and lymphadenopathy. Laboratory investigations include the nitroblue tetrazolium (NBT) test to determine the activity of the NADPH oxidase. In neutrophils with normal levels of enzyme, the pale yellow NBT is reduced to a blue colored compound as NADPH is oxidized, and can be observed in the cytoplasm. Patients with CGD are treated with high doses of antibiotics over long periods of time. This treatment also helps to dispel the granulomas. Abscesses may need to be drained. Bone marrow transplantation has been used successfully to treat some patients.

Leukocyte adhesion deficiency (LAD) occurs in two forms, but both are caused by the failure of leukocytes to express cell adhesion molecules essential for their movement through blood vessel walls during inflammation. Thus phagocytes are unable to enter inflamed tissues and remove bacteria. In LAD I, patients do not express the integrin, CD18 on neutrophils, macrophages and lymphocytes that allows them to bind to endothelial cells lining the blood vessels. In addition, CD18 is the receptor for C3b, which is an opsonin for phagocytic cells and crucial molecule of the complement pathway. Patients suffer localized bacterial infections that may become life threatening. In LAD II leukocytes fail to express ligands for other cell adhesion molecules, namely E and P selectins. Binding of leukocytes to these ligands allows them to roll along the endothelial cell surfaces before crossing into the tissues. Both LAD I and LAD II are autosomal recessive disorders. While LAD I affects all ethnic groups, LAD II has only been reported in people of Middle Eastern origin.

Patients with LAD I suffer localized bacterial infections that may become life threatening. Children do not usually survive beyond two years of age unless they have a bone marrow transplant. Patients with LAD II also suffer repeated infections as well as severe growth and mental retardations. Blood counts from patients with either form of LAD show a leukocytosis, that is, a white blood cell count in excess of 20×10^9 dm^{-3} in the absence of infection, compared to normal values of $4–11 \times 10^9$ dm^{-3} (*Chapter 13*). Both diseases may be diagnosed by flow cytometry, to assess the presence of the cell adhesion molecule on blood leukocytes. Leukocyte adhesion deficiency I has been treated successfully with bone marrow transplantation (*Chapter 6*).

Complement deficiencies

The role of complement in immune defense was outlined in *Chapter 4*. Here it will be described in more detail. The activation of complement results in the:

- lysis of bacteria;
- stimulation of the inflammatory response;
- promotion of phagocytosis;
- clearance of immune complexes.

The value of these roles cannot be overestimated. Complement may be activated by one of three pathways: the classical pathway, which is activated by IgG or IgM, following binding to an antigen; the alternative pathway, which is stimulated principally by components of the cell walls of bacteria and yeasts; and the lectin pathway, which is initiated by the binding of mannose binding lectin (MBL) to bacteria. Complement proteins C1, 2, 3 and 4 are involved in classical activation, while Factors B, D, H, I and P are involved in the alternative pathway activation. Activation of the lectin pathway requires MBL, MBL-associated serine proteases (MASP) 1 and 2 and C2, C3 and C4. All three pathways can lead to the production of membrane attack complexes (MACs), which are composed of C5b, C6, C7, C8 and C9. These cause lysis of the pathogen; in addition, fragments of activated complement proteins, such as C3a, C4a, C5a, promote inflammation by binding to mast cells and causing them to degranulate and release inflammatory chemicals, such as histamine (*Chapter 4*). These chemicals may also promote the chemotaxis of neutrophils out of the inflamed vessels. Other complement proteins, such as C1q, C3b and C567 act as opsonins. The absence of any of these proteins can severely compromise health. Complement is potentially very inflammatory once activated and there are a number of regulatory molecules which either prevent unregulated activation of complement, such as C1 inhibitor (C1INH), or which prevent damage to innocent bystander cells, such as Decay Accelerating Factor (DAF).

Inherited deficiencies of each complement protein have been described (*Table 5.6*). Deficiencies in several of the early classical pathway proteins result in an increased incidence of immune complex disorders, showing the role of complement in helping to remove immune complexes from the circulation. Over 90% of patients with a homozygous C1q deficiency and 10% of patients with a homozygous C2 deficiency develop SLE (*Section 5.3*). A deficiency of MBL is associated with increased incidence of infections with, for example, *Pseudomonas aeruginosa*. Newborn babies and infants are particularly at risk from this deficiency, indicating the importance of this pathway in protecting the young from infection.

Complement protein	Comments	Effects reported
C1q, C1r, C1s	autosomal recessive	increase in immune complex disorders, e.g. SLE; vasculitis* increase in pyogenic infections (Gram positive)
C4	autosomal recessive	increase in immune complex disorders, e.g. SLE; vasculitis * increase in pyogenic infections (Gram positive)
C2	autosomal recessive	increase in immune complex disorders, e.g. SLE; vasculitis * increase in pyogenic infections (Gram positive)
C3	autosomal recessive	frequent and severe bacterial infections resulting in pneumonia, septicemia and meningitis increase in immune complex disorders
C5, C6, C7, C8	autosomal recessive	recurrent neisserial infections (meningitis, gonorrhea)
C9		asymptomatic
Factor D, P		recurrent neisserial infections (meningitis; gonorrhea)
C1INH	autosomal dominant	hereditary angioedema
MBL	autosomal dominant or recessive	increased susceptibility to a variety of extracellular pathogens
Factor H	autosomal dominant	leads to depletion of C3 and symptoms similar to C3 deficiency

*inflammation of blood vessels

Table 5.6 Some complement deficiencies

Deficiencies in the classical and the alternative pathways can be detected in the laboratory by carrying out the CH_{50} and AP_{50} tests respectively. The CH_{50} test measures the ability of the patient's serum to lyse antibody-coated sheep erythrocytes, while the AP_{50} determines its ability to lyse these uncoated cells in rabbits. Results are expressed in terms of the ability of the serum to induce lysis of 50% of the target erythrocytes. Individual complement proteins can be measured by immunoassay, using specific antibodies. Thus C3 and C4 can be measured by nephelometry, single radial immunodiffusion or by ELISA (*Chapter 4*).

Hereditary angioedema (HAE) results from faults in the activity of the complement regulator, C1 inhibitor (C1INH). This protein normally functions to prevent the overactivation of the first part of the classical pathway by inhibiting C1r and C1s. The lack of the inhibitor leads to consumption of C4 and C2. Two types of HAE occur. Type I results from reduced levels of C1INH while in Type II the inhibitor is present but nonfunctional. The condition is an autosomal dominant disorder, which has an incidence of one in 50 000 to 150 000, with 85% of cases being Type I.

The disorder presents as noninflammatory and painless swellings of the skin, especially that of the limbs, which is often precipitated by physical trauma and anxiety. Abdominal pain is caused by the involvement of internal organs, such as the stomach, bladder and intestines. Severe edema of the larynx can cause death. Treatment during attacks is to administer fresh frozen plasma or commercially available C1-INH. Prophylactic treatment with danazol, an androgen, has been shown to stimulate the synthesis of C1-INH, but prolonged treatment may result in unpleasant side effects, such as virilization in women and suppression of testosterone production in men (*Chapter 7*).

5.3 AUTOIMMUNE DISORDERS

The macromolecules of the body are potentially highly immunogenic but, fortunately, immune systems do not usually mount immune responses against them. In fact, we are 'tolerant' to 'self' (*Box 5.2*). Whatever mechanisms lie behind the induction and maintenance of tolerance, it is clear that a number of disorders arise when these mechanisms fail and the immune system starts to attack self antigens. Failures of immunological tolerance lie behind the development of **autoimmune disease**. Autoimmune disease affects 5 to 7% of the population and autoimmune disorders are debilitating, chronic and painful.

CLASSIFICATION OF AUTOIMMUNE DISORDERS

Autoimmune disorders are often classified according to whether they are organ-specific, affecting only one organ or are systemic that is, affecting multiple organ systems (*Table 5.7*). In addition, destruction of cells and tissues can be brought about by autoantibodies and/or cell-mediated immunity. For example, in multiple sclerosis (MS) patients produce antibodies against myelin, the fatty material surrounding the axons of nerves. In addition, MS patients have T_H and T_C lymphocytes in their blood and cerebrospinal fluid, which are specific for myelin protein. Thus, humoral and cell-mediated autoimmunity may contribute to the demyelination of nerves in MS patients. In some instances, autoantibodies can block or stimulate a cell receptor. Myasthenia gravis is an example of the former, while Graves disease (*Chapter 7*) is an example of the latter. Like most classification schemes, that of autoimmune disorders is not perfect. For example, Goodpasture's syndrome directly affects both kidneys and lungs while MS exerts systemic effects by attacking one type of tissue only.

Type of disorder	Example	Effect of disorder	Autoantibodies present	Autoreactive T cells
Organ specific	autoimmune hemolytic anemia	destruction of erythrocytes (*Chapter 13*)	antibodies to erythrocyte antigens	
	autoimmune thyroiditis	hypothyroidism (*Chapter 7*)	antibodies to thyroglobulin and thyroid peroxidase	T_H1 cells specific for thyroid antigens
	Addison's disease	adrenal insufficiency (*Chapter 7*)	antibodies to cytoplasmic antigens of cells of adrenal cortex	infiltration of adrenal cortex with autoreactive T cells
	type 1 diabetes mellitus	destruction of insulin-producing cells in pancreas; serious metabolic disturbances	antibodies to islet cells found in classical juvenile form	infiltration of pancreas with autoreactive T cells
	Goodpasture's syndrome	progressive kidney and lung damage	antibodies to basement membrane antigens of kidney and lung	
	Graves disease	hyperthyroidism (*Chapter 7*)	antibodies to thyroid stimulating hormone receptors	destruction of thyroid cells by autoreactive T lymphocytes
	myasthenia gravis	progressive muscle weakness	antibodies to acetyl choline receptors on muscle cells	
	pernicious anemia	failure to absorb vitamin B_{12} in the stomach (*Chapters 11* and *13*)	antibodies to intrinsic factor	
Systemic disease	rheumatoid arthritis (RA)	inflammatory disorder affecting joints, skin and internal organs	antibodies to IgG (rheumatoid factor)	infiltration of joints with autoreactive T lymphocytes
	systemic lupus erythematosus (SLE)	inflammatory disorder affecting multiple organ systems	antibodies to DNA, chromatin and histones; rheumatoid factor in some individuals	evidence of T cell reactivity in some of the many organs affected
	multiple sclerosis (MS)	inflammatory disorder affecting central nervous system	antibodies to myelin basic protein	destruction of myelin membrane by autoreactive T lymphocytes

Table 5.7 Some examples of autoimmune disease

Many, but not all, autoimmune disorders, affect a preponderance of female patients, with three times as many females as males presenting with autoimmune diseases. The reasons for this gender bias are unclear but may be related to sex hormone levels. Many autoimmune disorders show a link with the type of MHC antigens that are present on cells. In humans, the MHC is known as the HLA system. The links between HLA type and different diseases is described in *Chapter 6*. So, for example, patients with Goodpasture's syndrome have a higher incidence of HLA-DR2 than the healthy population.

AUTOIMMUNE DISORDERS AFFECTING ENDOCRINE GLANDS

Autoimmune disorders of the thyroid gland are among the most common autoimmune disorders. Some will be discussed in *Chapter 7*. Autoimmune thyroiditis, also known as Hashimoto's thyroiditis, results in hypothyroidism and myxedema. The disease presents, typically, in women of middle age who are overweight, lethargic, constantly feel cold, are constipated and have coarse, dry hair and skin. The thyroid is swollen with a generally painless goiter, but which has a rubbery consistency when palpated. A biopsy of the thyroid shows infiltration with both CD8+ and CD4+ T lymphocytes that progressively destroy the thyroid gland. Patients also have antibodies to thyroglobulin and to thyroid peroxidase, which can be determined by

BOX 5.2 How to recognize self

There are a number of theories to explain the development of immunological tolerance and it may indeed be that several different mechanisms are involved. It has been shown in experimental systems that rodents become tolerant to potential immunogens if they were exposed to them during fetal development. Thus mice exposed to foreign proteins *in utero* do not mount immune responses against these immunogens when adults. The suggestion that lymphocytes exposed to epitopes, including self epitopes, during fetal development are selectively removed or deleted from the immune system explains the experimental induction of tolerance, and there is certainly evidence that this happens to developing T lymphocytes in the thymus (*Figure 5.3*). However, this does not explain the development of tolerance to immunogens which are not present in the fetus but which are expressed in the adult. It may be that some immunogens are kept anatomically separate from the immune system during life, to avoid potentially immunogenic self proteins inducing an immune response. For instance, it has been shown that when rabbits are injected with lens protein they make antibodies that then bind to the lenses of their own eyes. Another example is seen in vasectomized men who may start to make antisperm antibodies, presumably because the sperm they continue to produce become exposed to their immune systems following the operation. Finally, there is evidence that some types of T lymphocytes can suppress immune responses against self antigens. These T lymphocytes have, in the past, been called suppressor T lymphocytes and were thought to belong to the CD8+ subset. However, it has been shown that both CD8+ and CD4+ cells can have suppressor activity by producing inhibitory cytokines, such as IFN γ and IL-10 respectively.

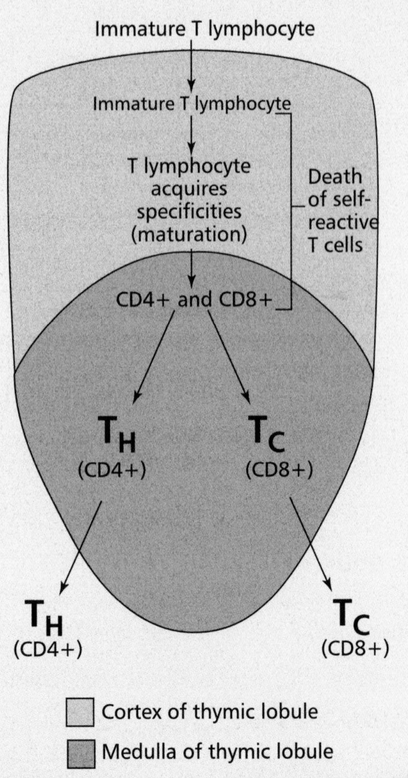

Figure 5.3 Schematic showing the development of T lymphocytes. Immature T lymphocytes enter the thymic lobule and mature as they progress through the thymic lobule into helper and cytotoxic subsets. In the cortex, cells that recognize self epitopes are deleted.

ELISA (*Chapter 4*). Immunohistochemical techniques (*Chapter 4*) can show the patient's serum to have antibodies that bind to microsomal antigens in sections of normal thyroid (*Figure 5.4*). Patients may be given thyroxine to treat the myxedema, and thyroidectomy may be required. The prognosis for patients with Hashimoto's thyroiditis is good.

In Graves disease, patients suffer symptoms of thyrotoxicosis: are thin, have a high resting pulse rate, constantly feel hot, have bulging eyes, or **exophthalmos**, due to growth of tissue around the orbit of the eye and may suffer diarrhea and general agitation. They have nodules in the thyroid that are foci of infiltrating T lymphocytes. Low levels of antibodies to thyroid microsomal antigens are seen in the plasma. However, more than 90% of patients have antibodies to the thyroid stimulating hormone (TSH) receptor on the surface of thyroid cells. These antibodies bind to the receptor and stimulate the production of thyroid hormone (*Figure 5.5*). This production is not regulated by the usual negative feedback mechanisms (*Chapter 7*) leading to the disease symptoms. Graves disease may be treated successfully by destruction of thyroid tissue. This can be achieved by its surgical removal or by giving the patient radioactive iodine that becomes concentrated in the thyroid. When women with Graves disease are pregnant, autoantibodies to TSH cross the placenta and the baby is born with thyrotoxicosis. Urgent treatment is required but, in time, the baby recovers as its levels of maternally derived antibodies drop.

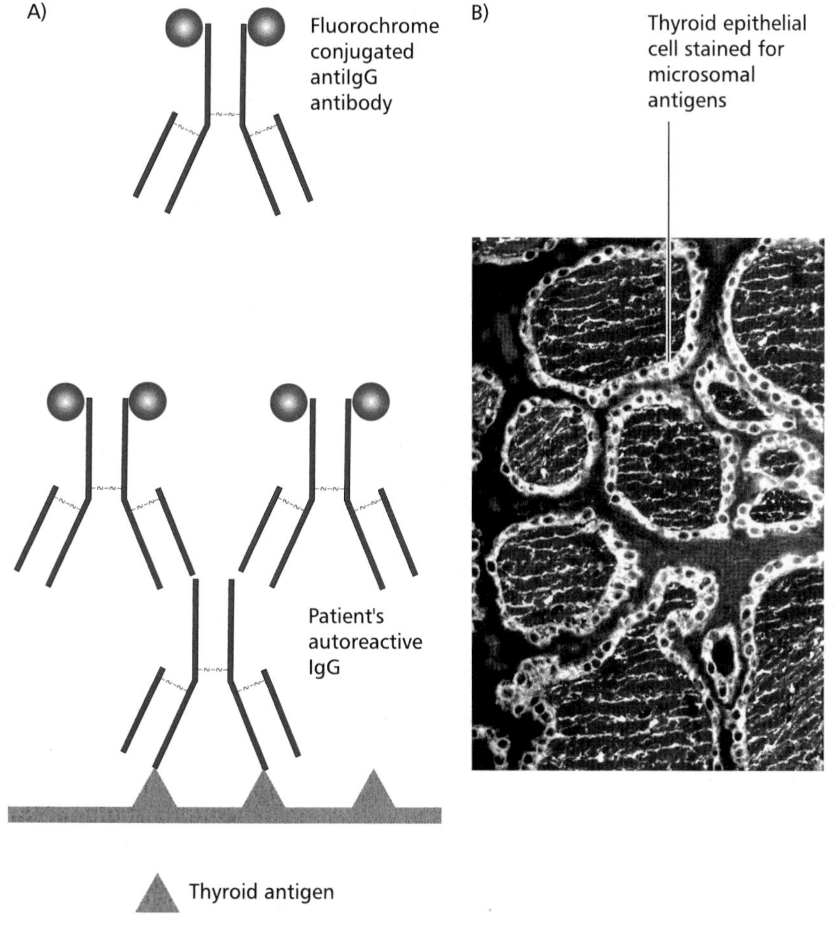

Figure 5.4 Immunofluorescence in autoimmune thyroiditis. (A) Schematic showing the basis of indirect immunofluorescence staining of thyroid antigens by a patient's serum; (B) Photomicrograph showing thyroid microsomal antigens stained with a patient's serum by indirect immunofluorescence. Courtesy of EUROIMMUN AG, Germany.

Both insulin dependent diabetes mellitus Type 1 and Addison's disease are discussed extensively in *Chapter 7*. The former is caused by an autoimmune destruction of the insulin-producing cells in the pancreas; the latter by autoimmune damage to the adrenal cortex. Both diseases are fatal unless treated by replacing the missing hormones.

ANTIGLOMERULAR BASEMENT MEMBRANE DISEASE

Antiglomerular basement membrane diseases are characterized by autoantibodies to the glomerular basement membrane (antiGBM). They include Goodpasture's syndrome and Goodpasture's disease. The former disorder shows glomerulonephritis, pulmonary hemorrhage and the presence of circulating antibodies to glomerular basement membrane; the latter is similar, but without the lung involvement. Both diseases are now included under the more general heading of antiGBM disease. Tissue damage is caused by antiGBM antibodies binding to the glomerular basement membrane and activating complement. Complement-mediated inflammation then ensues. The symptoms of glomerulonephritis include proteinuria and hematuria and erythrocyte casts (*Margin Note 5.3*) are seen.

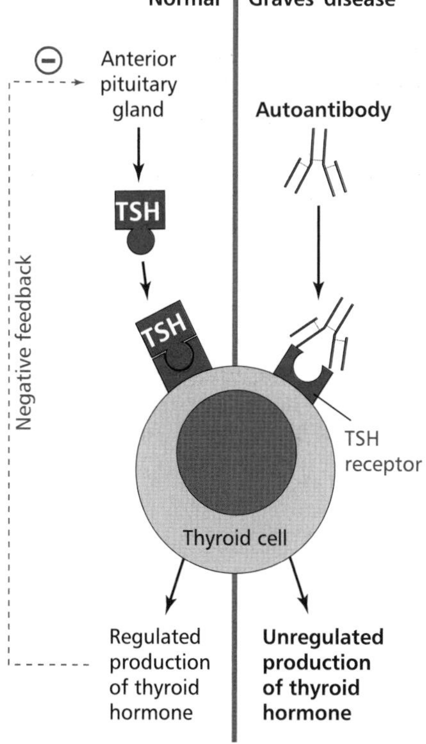

Figure 5.5 Schematic showing how the binding of an autoantibody to the TSH receptor leads to Graves disease (right-hand side). The left-hand side shows how the production of thyroid hormone is normally regulated by a feedback mechanism.

The binding of antibodies to alveolar membranes causes **hemoptysis**, that is, the coughing up of blood from the lungs and about 40% of patients experience chest pain. Hemorrhaging from the lungs may eventually lead to respiratory failure. There is some suggestion that the binding of antibodies to the alveolar basement membranes is facilitated by exposure to organic solvents, which increase the permeability of the alveolar capillaries. The incidence of antiGBM disease is rare, of the order of 0.5 cases per million in the UK. Unlike most

Margin Note 5.3 Erythrocyte casts

Erythrocyte casts are produced in the kidney when the cells become trapped in hyaline, which deposits around the kidney tubules. Healthy individuals will have a few plain hyaline casts in their urine, but finding casts containing erythrocytes or leukocytes is indicative of kidney disease.

autoimmune disease, it shows a predominance in males, with younger males generally presenting with both lung and kidney involvement.

Antiglomerular basement membrane disease can be diagnosed from linear deposits of antiGBM antibodies, which can be visualized by immunofluorescence on a kidney biopsy. An early diagnosis is essential and treatment must be started immediately. Therapy involves removal of circulating antibodies by plasmapheresis and the administration of immunosuppressive drugs (*Chapter 6*). The mortality rate for antiGBM disease is improving, and is currently at 10%, although most patients develop end stage renal disease. In the past, the disease was invariably fatal.

MYASTHENIA GRAVIS

Myasthenia gravis (MG) is an autoimmune disorder in which patients produce antibodies to acetylcholine receptors at the neuromuscular junction of striated muscle. The antibodies block the receptors so that they fail to respond to acetylcholine (*Figure 5.6*). This results in intermittent but progressive weakness of skeletal muscles, including those for breathing and the facial muscles involved in chewing, swallowing, talking and eye movements. The latter can lead to double vision and an inability to raise the eyelids, a condition known as **ptosis**. Difficulty with respiration may lead to inadequate intake of air and an inability to clear secretions from the respiratory tract. The incidence of pneumonia is increased in these patients. Approximately 75% of patients with MG also have thymic abnormalities such as hyperplasia and thymoma. The incidence of MG has been quoted as up to 14 in 100 000, with a female to male ratio of about 3:2.

The problems associated with movement of eye muscles are often the first sign of MG. The presence of autoantibodies may be confirmed by indirect immunofluorescence tests and the levels are a useful measure of disease progression. However, autoantibodies may not be detected in patients where the disease is confined to the facial muscles. Patients are treated with immunosuppressive drugs and cholinesterase inhibitors and with plasmapheresis (*Chapter 6*) to remove the autoantibodies. The condition is improved in the majority of patients by thymectomy. The mortality rate in

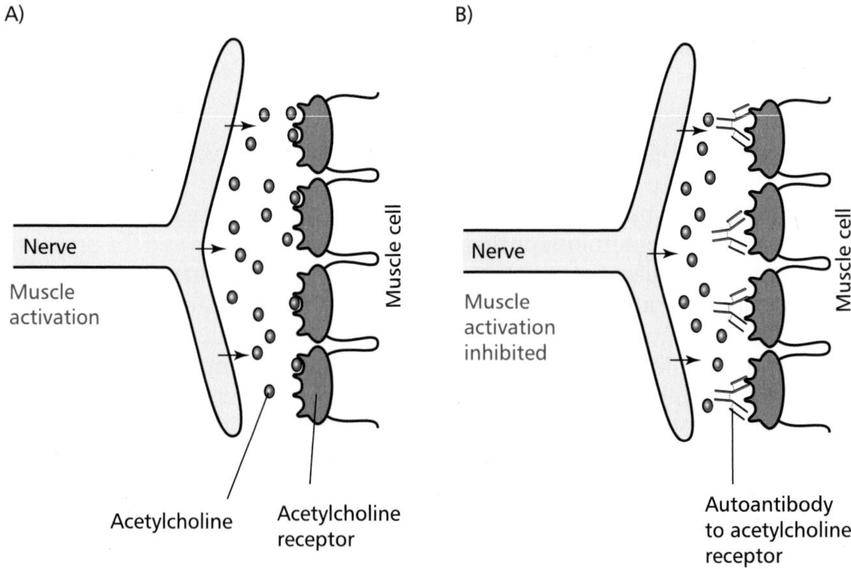

Figure 5.6 Schematics showing (A) the normal stimulation of muscle cells by the binding of acetylcholine to receptors on muscle cells and (B) the blocking of acetylcholine receptors by autoantibodies, which leads to myasthenia gravis.

MG patients is currently around 10%, which is a significant improvement on previous decades.

RHEUMATOID ARTHRITIS

Rheumatoid arthritis (RA) is a chronic, painful and debilitating condition characterized by symmetrical arthritis and radiological changes to the bone. The revised classification of rheumatoid arthritis (1988) is shown in *Table 5.8*. Autoantibodies are present in the plasma of patients with RA. The commonest, occurring in 70% of cases, is an IgM class antibody directed against IgG, called **rheumatoid factor** (**RF**). However, RFs belonging to the IgA and IgG classes have also been detected. The presence of RF causes large amounts of immune complexes to be formed, since IgG is present at relatively high concentrations in the blood. The complexes may adhere to blood vessel walls, activating complement and initiating an inflammatory reaction. While a minority of patients suffer a single episode of joint inflammation with long-term remission, most have a progressive illness characterized by intermittent 'flares'. In periods of active joint inflammation, the affected joints (*Figure 5.7*) are painful, swollen, red and warm to the touch; all characteristics of the inflammation within them. The presence of RF and subsequent inflammatory disease does not, however, adequately explain the pathogenesis of RA. Cell-mediated immunity is known to be heavily involved in joint destruction. The synovial membranes of affected joints are infiltrated with small lymphocytes, especially T_H1 cells, monocytes and macrophages so that the membranes themselves become thickened. Activated macrophages within the synovial fluid produce cytokines, such as IL-1 and tumor necrosis factor α (TNF α), which mediate erosion of bone. The accumulation of inflammatory neutrophils within the synovial fluid also contributes to the damage to the cartilage. Patients may also suffer inflammation of blood vessels or vasculitis and about 20% have subcutaneous rheumatoid nodules, often on the elbows and forearms but which may also occur in internal organs. The nodules consist of a mass of monocytes, lymphocytes and plasma cells surrounding a necrotic core, and probably represent the progression of vasculitis.

The etiology of RA remains unknown, despite numerous infectious agents having been implicated over the years. However, RA remains one of the most common autoimmune disorders, with an incidence of one to two per 100. The female to male ratio is approximately 3 : 1 and the disease manifests maximally between the ages of 40 and 60, although juvenile forms also exist.

Rheumatoid factor can be detected in plasma or serum by using the Rose-Waaler test, which determines the ability of the serum to agglutinate sheep

Criteria for diagnosis
1. Stiffness of the joints in the morning
2. Arthritis in three or more joints
3. Arthritis of the joints in the hand
4. Symmetrical arthritis
5. Rheumatoid nodules
6. Serum rheumatoid factor
7. Radiological changes to the bone

Table 5.8 1988 Revised classification for rheumatoid arthritis. To be diagnosed with RA, the patient must have four or more of these symptoms and symptoms one to four for at least six weeks.

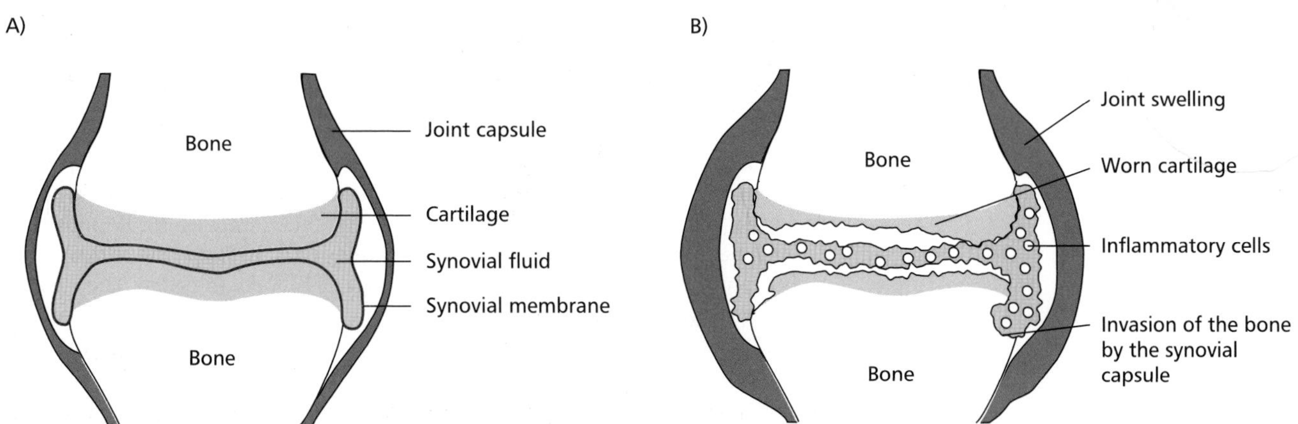

Figure 5.7 **Schematics showing (A) a normal synovial joint and (B) the characteristic changes associated with rheumatoid arthritis.**

erythrocytes sensitized with specific IgG antisheep erythrocyte antibody (*Figure 5.8*). Alternatively, latex particles nonspecifically coated with IgG can be used. A smaller proportion of individuals, around 40%, with RA also have **antinuclear factors** and these are also seen in patients with systemic lupus erythematosus (*see below*). Treatment of RA is with immunosuppressive agents such as methotrexate (*Chapter 6*) and anti-inflammatory drugs, including steroidal anti-inflammatory drugs (SAIDs), such as corticosteroids, and the nonsteroidal anti-inflammatory drugs, for example aspirin. New treatments aimed at blocking the effect of TNF α have been trialled. These involve either the infusion of a monoclonal antibody to TNF α, or the administration of soluble receptors for TNF α. In the latter case, the soluble receptors bind to TNF and prevent this inflammatory cytokine from binding to receptors on cells. The prognosis for patients with very severe disease is poor in terms of five-year survival. However, even with less severe disease, the condition is painful and debilitating, particularly during periods of active disease. The long-term use of immunosuppressive drugs leads to an increased susceptibility to infection.

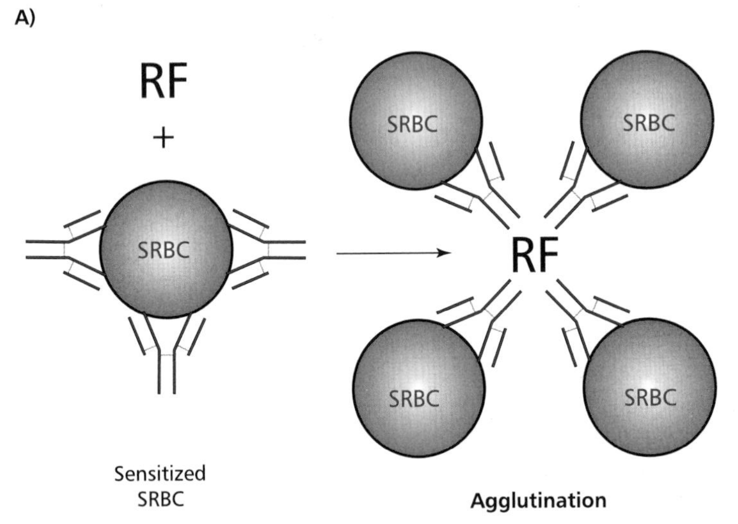

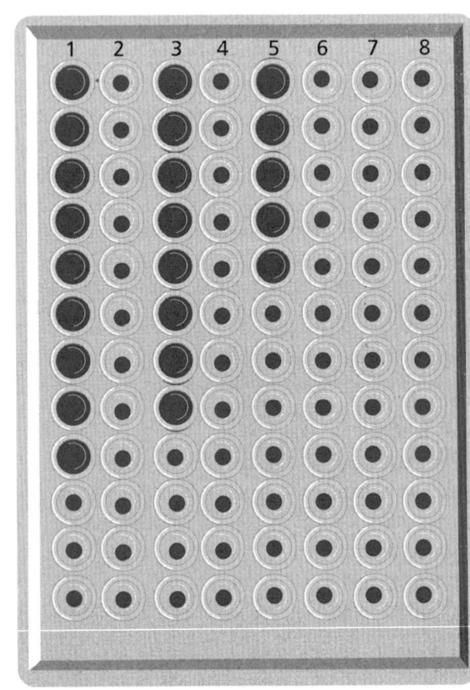

Figure 5.8 (A) A schematic illustrating the Rose-Waaler test for rheumatoid factor, which measures the ability of a patient's serum to agglutinate sensitized sheep red blood cells (SRBCs) as described in the text. (B) Shows the result of the test on several sera. Patients 1, 3 and 5 show positive agglutination at high dilutions, indicating the presence of rheumatoid factor.

SYSTEMIC LUPUS ERYTHEMATOSUS

Systemic lupus erythematosus (SLE) is an autoimmune disease in which about 95% of patients have antinuclear antibodies (ANAs) in their plasma. The deposition of immune complexes leads to widespread inflammation that affects many organs systems within the body. Clinical features of the disease are summarized in *Table 5.9*, the commonest presentations are arthritis and skin rash, particularly the butterfly rash of the face (*Figure 5.9*). Renal disease occurs in approximately half of affected individuals, with nephritis developing early on in the disease. Although the etiology of SLE is uncertain, the systemic nature of the disease is linked to the type of autoantibodies present. The disease primarily affects women of reproductive age, although much younger and much older patients have been seen. The female to male ratio is about 4 : 1, although in younger patients this gender bias does not occur.

Unfortunately different patients have different patterns of symptoms and their variety in this multisystem disorder has, in the past, presented

Symmetrical arthritis involving any joint
Arthralgia (bone pain)
Erythematous rash; butterfly rash on face (*Figure 5.9*)
Mucosal ulcerations
Pleurisy
Pericarditis
Renal involvement
Fever
Nervous system involvement (psychosis, depression, convulsions, migraine)
Heart disease
Eye involvement (retinal vasculitis; corneal ulceration)
Gastrointestinal ulcers
Pancreatitis
Hepatitis
Sjogren's disease (involving autoimmune destruction of lacrimal and salivary glands)
Hemolytic anemia

Table 5.9 Clinical manifestations of SLE

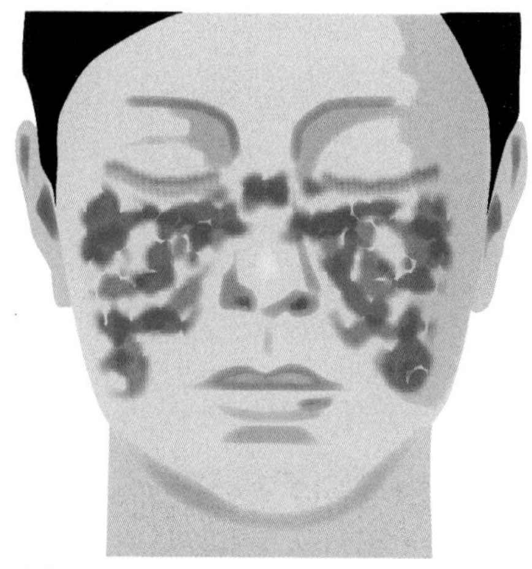

Figure 5.9 Schematic illustrating the area of face typically covered with a 'butterfly' rash in systemic lupus erythematosus.

difficulties with diagnosis, so it must take full account of the range of presentations. Diagnosis can be helped by demonstrating the presence of ANAs in samples of plasma, using indirect immunofluorescence on cultured Hep-2 cells. The commonest pattern seen is a diffuse staining throughout the nucleus due to antibodies against chromatin. It is also possible to detect different patterns of fluorescence which are indicative of antibodies to different nuclear antigens and which can help in diagnosis or in assessing prognosis (*Table 5.10*). Autoantibodies against **extractable nuclear antigens (ENA)** can also be detected by RIA or ELISA (*Chapter 4*). Positive tests for ANAs require further investigations, for example with tests for antidouble-stranded dsDNA (dsDNA) antibodies. The presence in the blood of so-called LE cells, which are neutrophils containing phagocytozed nuclei and resemble large multinucleate cells, is also indicative of SLE. Other laboratory tests used in the diagnosis of SLE include those listed in *Table 5.11*. Rheumatoid

Pattern of staining by indirect immunofluorescence (IDID)	Target antigen	Use
Homogeneous ANA	chromatin	screening but present in some normals
	double strandedDNA	specific for SLE
Speckled ANA (coarse)	Sm*	specific for SLE
Speckled ANA (coarse)	U1-RNP	found in SLE and mixed connective tissue disease

*Antibodies to the Sm (Smith) antigen were first discovered in the serum of a patient with SLE called Stephanie Smith. Antibodies to the Sm antigen bind to molecules in the nucleus called small nuclear riboproteins.

Table 5.10 Antinuclear antibodies in SLE

Laboratory test	Reasons
Full blood counts	to detect anemia, low platelet counts (*Chapter 13*)
Creatinine and electrolytes	to detect damage to kidneys (*Chapter 8*)
Erythrocyte sedimentation rate (*Chapter 13*)	increases in SLE
Urine tests	to detect proteinuria (*Chapter 8*)
Complement levels	C2 or C4 deficiency predisposes to SLE; inflammatory process in active SLE lowers complement levels

Table 5.11 Some laboratory tests for SLE

factor and the presence of antibodies against cardiolipin may also be present, although these are not specific for SLE.

Systemic lupus erythematosus is treated with immunosuppressive drugs, such as azathioprine or cyclosporine (*Chapter 6*) although the use of such drugs in patients prone to kidney disease needs careful monitoring. In addition, patients maintained on immunotherapy are more susceptible to infectious diseases. The prognosis for sufferers of SLE has improved greatly over the last 50 years because the disease is now diagnosed earlier. For example, in the 1950s, most patients died within 10 years of diagnosis, whereas today around 90% are alive 10 years after diagnosis.

5.4 IMMUNOLOGICAL HYPERSENSITIVITY

Immunological hypersensitivities are disorders in which the immune response to a foreign immunogen results in tissue damage. The term **allergy**, which originally meant 'altered reactivity', is sometimes used synonymously with hypersensitivity. In fact, allergies are only some specific types of hypersensitivity. Although the word **hypersensitivity** implies an overreaction, in fact, there is nothing essentially abnormal about the immune response in these cases. It may simply be the extent and nature of the exposure to the immunogen that results in the damage. Indeed, immunogens that provoke hypersensitivities are often 'harmless' in that they are not necessarily infectious agents. Immunological hypersensitivities represent the most common group of immunological disorders and, collectively, affect approximately 10% of the population.

In 1963, Gell and Coombs classified hypersensitivities into four 'types' based on the part of the immune system that caused the tissue damage (*Table 5.12*). This classification scheme is still used today, although not all hypersensitivities belong exclusively to one type. For example, immunological reactions to drugs can be both Types I and III, while intrinsic allergic alveolitis has components of Types III and IV. In addition, expansion of the classification to include further types has been suggested. This chapter retains the original classification.

Margin Note 5.4 Atopy 

People who suffer Type I hypersensitivities are frequently referred to as being atopic. The term atopy refers to an inherited tendency to develop an allergic condition typified by rhinitis, asthma and eczema, that is Type I hypersensitivity. There is a definite genetic predisposition to atopy, with different members of the families sometimes showing different manifestations.

TYPE I HYPERSENSITIVITY

Type I hypersensitivity is also referred to as **immediate**, because its effects are apparent within eight h of exposure to an immunogen. The effects can be fairly trivial, as in hay fever, or life threatening, as in atopic (*Margin Note 5.4*) or allergic asthma or anaphylactic shock. The term **allergy** is often used for Type I, although this is also used for some of the other types. Immunogens that cause allergies are often referred to as **allergens** and this is the term that will be used here.

Type	Names	Examples of disorder	Immune system component involved
I	immediate; anaphylactic	hay fever; allergic asthma; food allergies; anaphylactic shock	IgE
II	cytotoxic	transfusion reactions; hemolytic disease of the newborn (HDN)	IgM, IgG and complement
III	complex-mediated	intrinsic allergic alveolitis; serum sickness	antigen/antibody complexes (usually IgG)
IV	delayed-type hypersensitivity (DTH)	Mantoux reaction; contact hypersensitivity	sensitized CD4+ T lymphocytes

Table 5.12 The Gell and Coombs classification of immunological hypersensitivity

The different manifestations (*Table 5.13*) of Type I depend on the degree of previous exposure to the allergen and also on the route of exposure. The underlying cause is the production of IgE in response to an allergen. This type of antibody stimulates inflammatory responses that are aimed at eliminating parasitic worms. Atopic individuals produce IgE in response to allergens that, in nonallergic individuals, would stimulate the production of IgG. The tissue mast cells and blood basophils have receptors for the Fc region of IgE, so that IgE binds to the surface of these cells. The more sensitized an individual, the more their mast cells are coated with IgE. Further exposure to the same allergen results in the cross-linking of mast-cell bound IgE by the allergen (*Figure 5.10*). This triggers an explosive degranulation of the mast cell that releases pharmacologically active mediators, including histamine, which causes vasodilation, smooth muscle contraction and mucus secretion, depending on where they are released. In addition the subsequent release of further mediators, for example leukotrienes and prostaglandins, which are synthesized at the mast cell membrane potentiate inflammation and smooth muscle contraction. This response which evolved as a defense against multicellular parasites, causes the characteristic symptoms of the hypersensitivity.

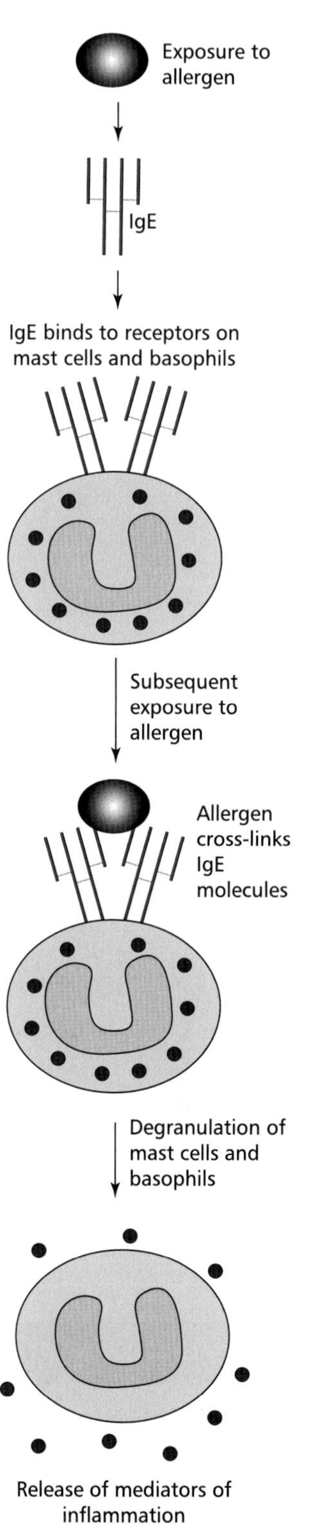

Figure 5.10 Schematic illustrating how exposure to an allergen can lead to degranulation of mast cells and basophils and Type I hypersensitivity.

Disorder	Effects
Allergic rhinitis (hay fever)	seen in upper respiratory tract; excess mucus; sneezing and wheezing
Atopic eczema	extensive and very itchy rash in skin; a common manifestation in atopic children
Food allergies	skin rash (hives); gastrointestinal 'cramps' and diarrhea; may sometimes result in anaphylactic shock
Allergic asthma	severe inflammation in respiratory tract; severe respiratory distress
Anaphylactic shock	sudden drop in blood pressure; respiratory distress; skin rash; gastrointestinal 'cramps' and diarrhea; may result in death within an hour of exposure
Drug allergies, e.g. penicillin, sulphonamides, salicylates	may be trivial (as in skin rash) or severe, as in anaphylactic shock

Table 5.13 Manifestations of Type I hypersensitivity

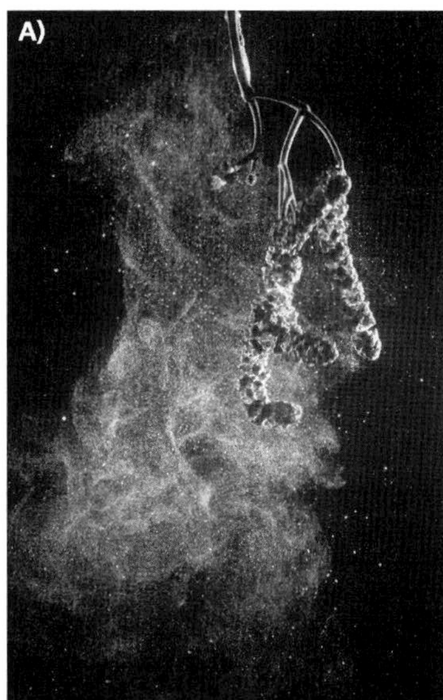

Figure 5.11 Photographs showing the release of pollen grains from (A) hazel, *Corylus avellana* and (B) Timothy grass, *Phleum pratense.*

Type I hypersensitivity is the most common form of the immunological hypersensitivities and is estimated to affect between one in 5–10 individuals, with conditions, such as allergic rhinitis, allergic asthma and food allergies.

Allergic rhinitis

Allergic rhinitis or hay fever is extremely common and its effects are well known. Seasonal allergic rhinitis typically presents in the spring and summer and is generally brought on by pollen to which the individual has been sensitized. Some individuals show sensitivity to many types of pollen (*Figure 5.11*), including well-known 'culprits' such as ragweed, while others are allergic to a single type of tree pollen. Individuals who suffer allergic rhinitis all year round are most likely to be allergic to other types of allergens, such as the dead skin cells sloughed from household pets, or the feces of house dust mites (*Figure 5.12*) which thrive in warm, carpeted dwellings. Although allergic rhinitis is not life threatening, it can be very debilitating.

Allergic asthma

Allergic asthma is a serious and potentially life-threatening condition brought on by air-borne allergens, such as those which trigger hay fever. In patients with allergic asthma, sensitization to a range of air-borne allergens stimulates inflammation in the respiratory tract, narrowing the airways and eventually leading to hyperreactivity of the muscles of the bronchial tree. The condition is very distressing particularly during an asthmatic attack, when severe respiratory distress may require emergency treatment. The incidence of asthma is increasing in children for which various explanations have been put forward. For example, it may be that increasing air pollution is predisposing patients to more frequent asthmatic attacks by increasing the hyperreactivity of the bronchial muscles.

Food allergies

Allergies to food are also seen in atopic individuals. Normally, the allergy will present as a skin rash, possibly with diarrhea, within an hour of consumption of the particular food involved. Foods known to cause allergies include eggs, shellfish, mushrooms and strawberries. Recently, much attention has focused on the extent of severe allergic reactions to nuts, especially peanuts. This may be due to the increased use of peanut paste as a thickening agent in the preparation of processed foods. Individuals who are allergic to peanuts may suffer rapid and life-threatening allergic reactions and are advised to avoid processed food unless the ingredients are clearly labeled. It has also been known for individuals to suffer severe reactions to certain fruits, such as papaya, though this is rare.

BOX 5.3 Anaphylactic shock

Anaphylactic shock is a life-threatening reaction to an allergen that may begin within minutes of exposure. Anaphylactic shock is often due to intramuscular or intravenous exposure to an immunogen as, for example, in a penicillin sensitive individual who has been given intramuscular or intravenous penicillin to treat a bacterial infection. People who are highly sensitized to bee venom may also develop anaphylactic shock following a bee sting. Anaphylactic shock may also be brought about by sensitization to latex found, for example, in rubber gloves used in hospitals, and this appears to be an increasing problem. In addition, food allergies in highly sensitized individuals occasionally result in anaphylactic shock. Typically, those affected suffer extreme respiratory distress accompanied by hypotension and may also develop a skin rash (urticaria) and diarrhea. The condition arises from a systemic release of mast cell mediators. The extensive vasodilation causes a sudden massive drop in blood pressure. The effects of mediators, in particular leukotrienes, on respiratory muscles also lead to breathing difficulties. Anaphylactic shock is a medical emergency and must be treated immediately with, for example, rapid intramuscular injection of adrenalin to relax the smooth muscle tissues.

Management of Type I hypersensitivity

The most appropriate strategy for treating Type I hypersensitivity is to identify the allergen and to avoid it. Laboratory tests to identify the allergen in question may involve skin testing (*Box 11.5*). Extracts of common allergens are injected intradermally. In sensitive individuals, the causative allergen will produce a 'wheal and flare' skin reaction within 20 min. The 'wheal' is a raised red lump, while the flare is the red inflamed area that surrounds it. Investigations of Type I hypersensitivity include the radioallergosorbent test (RAST) which measures the level of allergen-specific IgE in the blood. This involves incubating samples of serum from a patient with the potential allergen immobilized on a solid support, any IgE that becomes bound to the allergen is then detected by the addition of ^{125}I-labeled anti-IgE antibodies.

When complete avoidance of the allergen is not feasible, drugs may be used to control the symptoms. These include antihistamines, such as brompheniramine maleate and loratadine, anti-inflammatory drugs, such as corticosteroids, and 'Intal' which prevents mast cell degranulation.

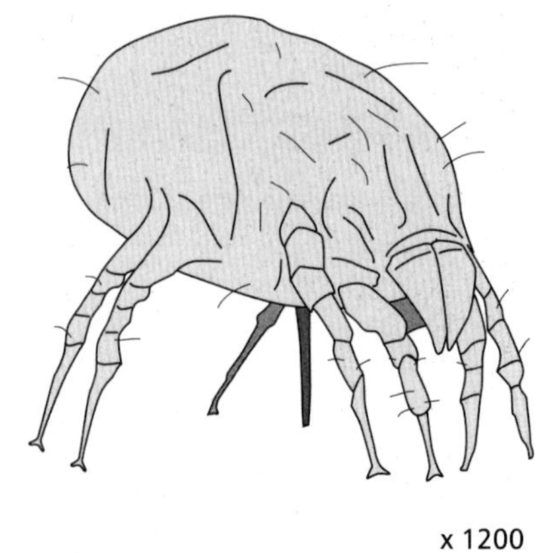

Figure 5.12 The house dust mite.

TYPE II HYPERSENSITIVITY

Type II or cytotoxic hypersensitivity refers to those situations in which antibody activates complement causing tissue damage. Examples include transfusion reactions to mismatched blood, and hemolytic disease of the newborn (HDN). In addition, autoimmune reactions that involve lysis of cells can be included here. Autoimmune diseases have already been discussed above and transfusion and HDN are described in *Chapter 6*, hence only a brief discussion of the causes and consequences of the transfusion of ABO incompatible blood will be given here.

Transfusion reactions

The ABO blood group system contains four blood groups: A, B, AB, or O, according to the types of antigens found on erythrocyte membranes (*Chapter 6*). In the plasma, there are also antibodies to the blood group antigens that are not present on the erythrocyte membranes. Thus, individuals with blood group A have anti-B antibodies in their plasma, while those of blood group B have anti-A (*Figure 5.13*). These antibodies are known as **isohemagglutinins**, and usually belong to the IgM class, which are efficient activators of complement. If a blood group A individual is transfused with group B blood, then antibodies from the donated blood will bind to recipient erythrocytes, activate complement and cause their lysis. Similarly, antibodies in the recipient will lyse the donated erythrocytes. The sudden and simultaneous lysis of cells leads to kidney failure and death.

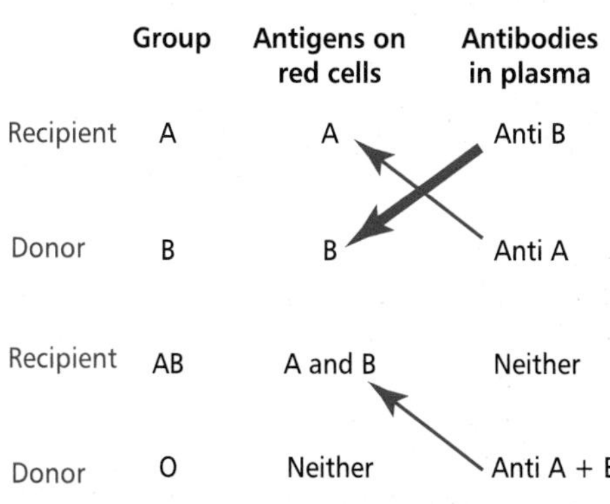

Figure 5.13 The antigens and antibodies of the ABO blood group system. The results of only two incompatible transfusions are highlighted, with the thickness of the arrows indicating the relative amounts of agglutination (*Chapter 6*).

TYPE III HYPERSENSITIVITY

Type III or complex-mediated hypersensitivity is brought about by immune complexes that usually involve antibodies to soluble antigens. Immune complexes can be harmful because they activate complement, triggering inflammation and the influx of neutrophils into an area. Over a period of time, this can cause tissue damage, principally due to lytic enzymes released by dying neutrophils. As the size of immune complexes varies, depending on the relative proportions of antigen and antibody, the clinical consequences may vary. For example, in autoimmune diseases such as RA (*Section 5.3*) immune complexes between rheumatoid factor and IgG are produced in antigen excess. These complexes are small and soluble and travel in the circulation. They may adhere to the insides of blood vessels, triggering vasculitis, or terminate in the kidney and cause nephritis. In intrinsic allergic alveolitis, immune complexes produced in the alveoli are large, and precipitate in the lungs, causing alveolitis. The name 'intrinsic allergic alveolitis' covers a number of

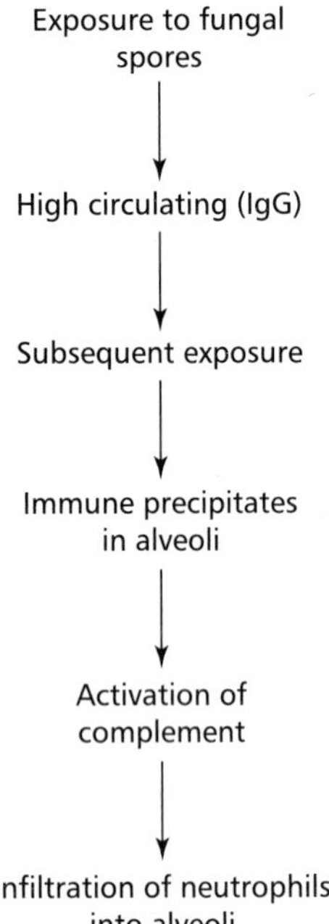

Figure 5.14 A long-term exposure to spores can cause the production of high levels of circulating antibody (IgG) which can precipitate spore antigens in the alveoli and activate complement leading to Farmer's lung.

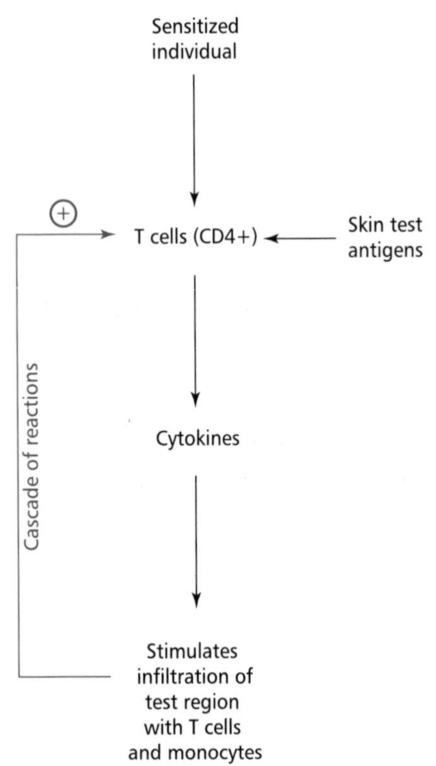

Figure 5.15 The Mantoux test for Type IV hypersensitivity. See text for details.

occupational diseases, in which workers are exposed to air-borne immunogens. A typical example of this disorder is Farmer's lung (*Figure 5.14*). This disorder typically begins in winter time, when the farmer develops a cough. Since this is not an unusual occurrence, the disease may go unrecognized for several years. However, the disease is progressive and, if untreated, will progress to emphysema. The disease is related to the farmer moving hay to feed livestock during the winter. Ascomycete fungi grow well in the warm damp conditions at the center of a haystack, and a cloud of spores are released when the hay is moved. The farmer inhales the spores and, over months and years, develops high levels of circulating IgG to immunogenic molecules which have been leached from the spores. With high levels of IgG in the blood supply to the alveoli, further exposure to spore antigens causes large immune complexes to precipitate in the lungs, setting up inflammation in the alveoli.

Successive winters may result in immune complex-mediated damage to the lungs, with fibrosis and loss of gas exchange capacity. Although Farmer's lung is classified as an antibody-mediated hypersensitivity, it is now recognized that cell-mediated immunity is also involved and that damage caused by specific T cells also contributes to the disease. It is essential that Farmer's lung be diagnosed early to avoid permanent damage to lung tissue. Treatment in the early stages may simply involve avoidance of the antigen, although corticosteroids may also be used to treat the inflammatory reaction.

TYPE IV DELAYED TYPE HYPERSENSITIVITY

Delayed-type hypersensitivity (DTH) requires more than 18 h after exposure to an immunogen for the symptoms to become apparent. This type of hypersensitivity is caused by T lymphocytes rather than antibody. The DTH reaction is typified by the Mantoux reaction. The vaccine to protect against tuberculosis (TB) consists of an attenuated form of Bacillus Calmette Guérin (BCG) which is a strain of *Mycobacterium bovis* (*Chapter 2*). Before being given the vaccine, individuals are skin tested to see if they are already sensitized by infection or previous vaccination. This involves injecting an extract of mycobacteria, called **purified protein derivative (PPD)** intradermally. If an individual is sensitized, then after about 18–24 h, the injection site becomes swollen and red. The swelling increases for around 48 h then subsides slowly so that this 'positive' reaction is still visible after several weeks. An individual who gives a positive Mantoux test may then be investigated to ensure that the positive result was due to prior sensitization rather than active disease.

The swelling is caused by small lymphocytes and monocytes infiltrating into the area. Initially, sensitized CD4+ T cells respond to PPD by releasing cytokines (*Figure 5.15*) that attract and retain monocytes at the site and induce inflammation, allowing the entry of more CD4+ cells. Thus a cascade reaction occurs, producing a slow but progressive swelling at the site of the injection. The reaction subsides once the phagocytic monocytes have removed all the PPD.

Delayed type hypersensitivity is also seen in **contact allergies** to a number of chemicals, including certain biological stains, some hair dyes, nickel salts in cheap jewellery, mercuric salts in some tattoo dyes, fluorodinitrobenzene and some plant biochemicals, such as urushiol in poison ivy. Typically, once sensitized, an individual will develop dermatitis approximately 18 to 24 h after further skin contact with the same chemical. Skin sensitizing chemicals are not typical immunogens, since they are neither proteins nor large molecules. However, it appears that sensitization involves their chemical binding to skin proteins to form a hapten-protein conjugate (*Chapter 4*). Langerhans cells, which are antigen presenting cells in the skin, process the 'new' antigen and present it to helper T lymphocytes, which become sensitized. Once this has happened, any further contact with the chemical will promote cytokine release from T cells, producing a cascade similar to that of the Mantoux reaction.

Laboratory workers need to be aware of the skin sensitizing nature of many of the chemicals used in biomedical science and to conduct risk assessments for their use, since sensitization, once induced, is irreversible. The only treatment for Type IV hypersensitivy is to avoid the allergen, although corticosteroid creams may give some relief during an episode of dermatitis.

CASE STUDY 5.1

George is a nine-month-old boy who has suffered repeated bouts of upper respiratory tract infections since the age of four months. He was admitted to hospital with bacterial pneumonia. A routine examination of his throat showed his tonsils to be much smaller than usual in boys of his age, especially those who have had numerous throat infections. George has no siblings.

Questions

(a) What diagnoses might plausibly be suggested?

(b) How could these suggestions be confirmed?

(c) If confirmed, what types of treatment would be beneficial?

CASE STUDY 5.2

Peter is a teacher at a primary school. In his spare time he keeps bees and sells the honey he produces at local country fairs. At one such fair he was allowing potential customers to sample his honey. The stall became very attractive to some nearby wasps, and, in trying to keep the wasps away, Peter was stung. With a few minutes he collapsed with obvious severe breathing problems and a generalized rash.

Questions

(a) What is the most likely cause of Peter's collapse?

(b) How should he be treated?

CASE STUDY 5.3

Jane is a 24-year-old student studying for a PhD. Her first two years went well and she was hard-working and dedicated. In the last year, however, her attendance at university declined and her work suffered. Her supervisor was worried about her constant tiredness and weight gain, and, although the university is well heated, Jane was always complaining about the cold. Her supervisor advised her to consult her physician who notices that Jane has a goiter.

Questions

(a) What is the likely cause of Jane's tiredness?

(b) What clinical tests are appropriate to Jane?

5.5 SUMMARY

Disorders of the immune system include immunodeficiencies, autoimmune diseases and hypersensitivities. The lack of a component of the immune system renders the sufferer much more susceptible to infectious disease and some forms of cancer. These diseases may be treated by giving antibiotics and antiviral drugs to treat infections and, where possible, to replace the missing component by administration of antibodies or by a bone marrow transplant. Autoimmune diseases occur when the mechanisms for preventing self-reactivity fail. These diseases can be organ specific or systemic. They can be treated with immunosuppressive drugs, although this renders the individual more susceptible to infection. Hypersensitivities are classified into four types depending on the underlying cause. They are best treated by avoiding contact with the immunogen to which the individual is sensitized.

QUESTIONS

1. Which one of the following statements is **CORRECT**?
   a) Rheumatoid factor is only present in patients with rheumatoid arthritis.
   b) All patients with rheumatoid arthritis have rheumatoid factor.
   c) Antibodies to double stranded DNA are diagnostic for SLE.
   d) Graves disease is characterized by antibodies to the acetylcholine receptor.
   e) Goodpasture's disease is a systemic disorder.

2. State whether the following statements are TRUE or are FALSE.
   a) Bruton's agammaglobulinemia is an autosomal recessive condition.
   b) Complement deficiencies result in increased neisserial infections.
   c) Deficiencies in T cells cause an increased susceptibility to viral infections.
   d) T cell deficiencies are corrected with infusions of plasma.
   e) Immunodeficiency may lead to increased risk of some cancers.

3) Which of the following statements is **INCORRECT**?
   a) In autoimmune disease generally, more women than men are affected.
   b) Transient hypogammaglobulinemia refers to a brief antibody deficiency in pregnant women.
   c) Deficiencies of C9 are asymptomatic.
   d) The classical pathway for complement action depends on the presence of antibodies.
   e) In SCID, B lymphocyte levels may be normal or decreased.

4. List three ways in which the body prevents autoimmune reactions.

5. Arrange the following two lists into their most appropriate pairings

Selective IgA deficiency	Is due to IgE
Farmer's lung	Is caused by T lymphocytes
Chediak Higashi disease	May be asymptomatic
Anaphylactic shock	Affects melanocytes
Delayed hypersensitivity	Is caused by immune complexes

FURTHER READING

Chapel, H, Haeney, M, Misbah, S and Snowden, N (1999) *Essentials of Clinical Immunology*, 4th edn. Blackwell Science Ltd, Oxford.

Chinen, J and Shearer, WT (2003) Basic and clinical immunology. *J. Allergy Clin. Immunol.* **111:** S813–S818.

Conley, ME, Rohrer, J and Minegishi, Y (2000) X-linked agammaglobulinemia. *Clin. Rev. Allergy Immunol.* **19:** 183–204.

Di Renzo, M, Pasqui, AL and Auteri, A (2004) Common variable immuno-deficiency; a review. *Clin. Exp. Immunol.* **3:** 211–217.

Etzioni, A (2003) Immune deficiency and autoimmunity. *Autoimmun. Rev.* **2:** 364–369.

Fischer, A, Hacein-Bey, S, Le Diest, F *et al.* (2001) Gene therapy for human severe combined immunodeficiencies. *Immunity* **15:** 1–4.

Goldsby, RA, Kindt, TJ, Osborne, BA and Kuby, J (2003) *Immunology*, 5th edn. W.H. Freeman and Co., New York, USA.

Heyworth, PG, Cross, AR and Curnutte, JT (2003) Chronic granulomatous disease. *Curr. Opin. Immunol.* **15:** 578–584.

Ismail, AA and Snowden, N (1999) Autoantibodies and specific serum proteins in the diagnosis of rheumatological disorders. *Ann. Clin. Biochem.* **36:** 565–578.

Lekstrom-Himes, JA and Gallin, JI (2000) Advances in Immunology: immunodeficiency diseases caused by defects in phagocytes. *New Engl. J. Med.* **343:** 1703–1714.

Mancias, C, Infante, AJ and Kamani, NR (1999) Matched unrelated donor bone marrow transplantation in leukocyte adhesion deficiency. *Bone Marrow Transplant* **24:** 1261–1263.

Notarangelo, L, Casanova, J-L, Fischer, A, Puck, J, Rosen, F, Seger, R and Geha, R (2004) Primary Immunodeficiency disease: an update. *J. Allergy Clin. Immunol.* **114:** 677–687.

Todd, I and Spickett, G (2005) *Lecture notes: Immunology, 5th edn,* Blackwell Science, Oxford.

Turner, MW (2003) The role of mannose-binding lectin in health and disease. *Molec. Immunol.* **40:** 423–429.

Wen, L, Atkinson, JP and Giclas, PC (2004) Clinical and laboratory evaluation of complement deficiency. *J. Allergy Clin. Immunol.* **113:** 585–593.

Useful web sites:

C. Lucy Park *Common Variable Immunodeficiency* http://www.emedicine.com (accessed May 2005)

http://www.hopkins-arthritis.som.jhmi.edu/rheumatoid/tnf.html (accessed June 2005)

http://emedicine.com (for up to date articles on all the diseases mentioned in this chapter)

chapter 6:

TRANSFUSION AND TRANSPLANTATION

OBJECTIVES

After studying this chapter you should be able to:

- outline the role of the biomedical scientist in the transfusion of blood and blood products;
- describe how blood group systems are named;
- describe the characteristics of the ABO and Rh blood group systems;
- outline the ways in which blood groups are determined;
- discuss some of the adverse effects of transfusion;
- outline how tissue typing is carried out in the laboratory;
- outline the role of the immune system in the rejection of transplants;
- discuss the significance of MHC antigens in determining the rejection of a transplant;
- outline the uses of some of the drugs used in immunosuppression;
- describe situations that give rise to graft versus host reactions;
- explain the uses of bone marrow and stem cell transplants.

6.1 INTRODUCTION

Blood transfusion is the transfer of blood or blood products into the bloodstream of a patient who has lost blood due to injury, disease or an operation. The amount and type of blood or component transfused depends on the needs of the patient. Transfusions of blood and blood products are routine and are generally safe therapeutic procedures that are rarely associated with adverse reactions. However, it is only in the last hundred or so years that the foundations of these procedures have been established. Similarly, the transplantation of organs and solid tissues from one individual to another to replace diseased or nonfunctional tissue is now well established. However, it is little more than 50 years since the first successful clinical transplant of a kidney in 1954 paved the way for the range of tissue transplants currently available. Advances in the transplantation of bone marrow and stem cells have brought the two fields of transfusion and transplantation closer together.

This chapter will explain the role of the transfusion scientist and will discuss the biochemical and genetic bases of a number of blood group systems and the methods used to determine blood groups. In addition, the adverse reactions of transfusion and the consequences to a fetus of antibodies against fetal erythrocytes being transmitted across the placenta will also be outlined. The evidence for the immunological basis for the rejection of tissue transplants will also be examined, together with the genetics of the HLA system, a series of genes encoding proteins that stimulate rejection of tissues (*Chapter 4*). The chapter will also consider how the rejection of tissues can be prevented by immunosuppressive therapy and will review the set of circumstances which result in **graft versus host disease** (**GVHD**), as well as considering the consequences of GVHD for the relatively new treatments which involve the transplantation of stem cells.

6.2 BLOOD AND BLOOD PRODUCTS FOR TRANSFUSION

The role of the biomedical scientist in the transfusion laboratory is to ensure that the blood and blood products being transfused into a patient are safe. To ensure safety, the blood is tested to determine its blood group and to check that it is not contaminated with harmful microorganisms. In addition, checks are made to ensure that the transfused blood does not contain antibodies that will destroy the erythrocytes of the recipient and cause death.

Blood transfusions are required to replace blood lost as a result of accident or surgery. Surgical procedures which require transfusions include the transplantation of organs, such as the liver and heart, where significant bleeding may occur. Blood may also be given to treat certain diseases, such as anemia. Plasma may also be transfused to treat badly burned patients who have lost significant amounts of fluid or in the treatment of bleeding disorders. Plasma products, such as Factor VIII, to treat hemophilia (*Chapter 13*), or immunoglobulins, to treat certain immunodeficiency disorders (*Chapter 5*), may also be given. Platelet concentrates are also used to treat bleeding disorders (*Figure 6.1*).

In the UK, blood containing leukocytes is no longer transfused for a number of reasons, as shown in *Table 6.1*. Sensitization to Major Histocompatibility Complex (MHC) antigens, which are present on blood leukocytes but not erythrocytes, may have consequences if the recipient later requires a transplant (*Section 6.11*) and GVHD may have a fatal outcome in immunosuppressed individuals. As a consequence, leukocytes are removed from blood, usually within a few hours of collection. This involves filtering the blood through leukocyte-specific filters, which trap the leukocytes but not the smaller erythrocytes or platelets. Such a process is called **leukodepletion** and it reduces the leukocyte count to less than 5×10^6 dm^{-3}. The number of leukocytes left in blood can be assessed by counting in a hemocytometer, or by using a flow cytometer (*Box 6.1*).

Reason	Examples
Possibility of transmitting infectious agents in leukocytes	human immunodeficiency virus (HIV) cytomegalovirus (CMV) prion proteins responsible for variant Creutzfeldt-Jakob disease (vCJD)*
Possibility of immunizing against leukocyte antigens	sensitization to major histocompatibility antigens
Increased risk of graft versus host disease (GVHD) with whole blood	in immunosuppressed or immunodeficient individuals
Possibility of inducing febrile reactions post-transfusion (nonhemolytic febrile reactions)	caused by cytokines released from leukocytes in storage

*In 2003 a recipient of blood taken from a healthy donor who developed vCJD, also died of vCJD 6 years after transfusion. Other evidence since that time suggests that it is possible to transmit vCJD with leukocytes (see also *Chapters 2* and *15*).

Table 6.1 Reasons for leukodepleting blood before transfusion

6.3 THE DISCOVERY OF THE BLOOD GROUP SYSTEMS

It has been known since the seventeenth century that the transfusion of blood between individuals could have rapid and fatal consequences. Fortunately, in 1900 Landsteiner (1868–1943) discovered that individuals could be classified into different groups depending on the characteristics of their erythrocytes and the presence of specific antibodies in their plasma to erythrocyte antigens. These discoveries laid the foundations for the routine and safe therapeutic transfusion of blood. Landsteiner drew blood from a number of individuals and separated the erythrocytes from the plasma. He then mixed together all possible combinations of erythrocytes and plasma from these individuals together and showed that only certain combinations resulted in the clumping or **agglutination** of the erythrocytes (*Figure 6.2*). These patterns of agglutination showed that there were different blood groups, which Landsteiner named A, B and O. In 1902 von Decastelo (1872–1960) and Sturli (1873–1964) discovered a fourth blood group which he called AB. It became clear that fatal blood transfusions resulted from incompatible blood being transfused and

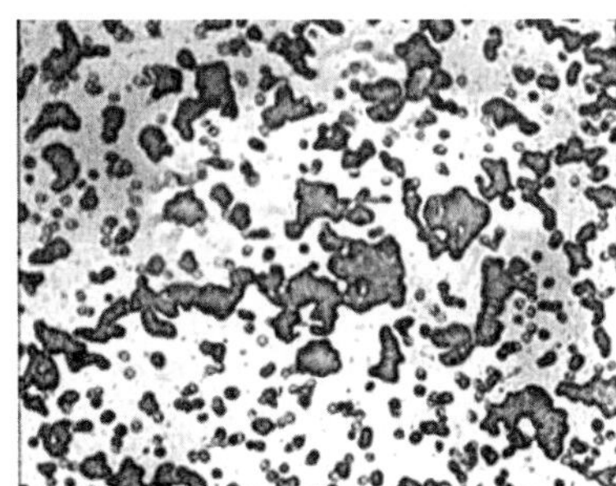

Figure 6.2 The agglutination of erythrocytes by antierythrocyte antibodies.

Figure 6.1 Some blood products: (A) plasma (B) erythrocytes and (C) platelets. Courtesy of the Manchester Blood Transfusion Service, UK.

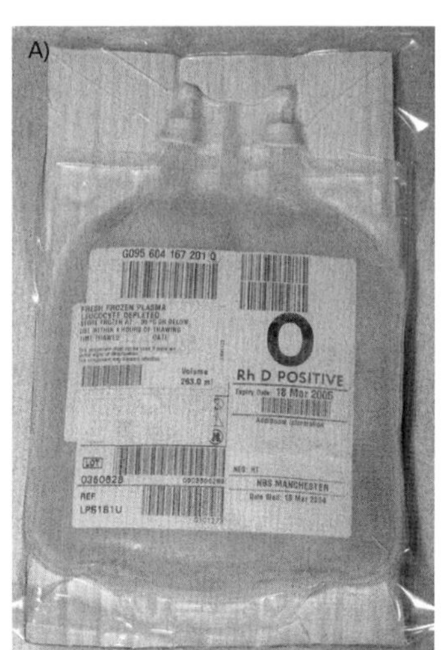

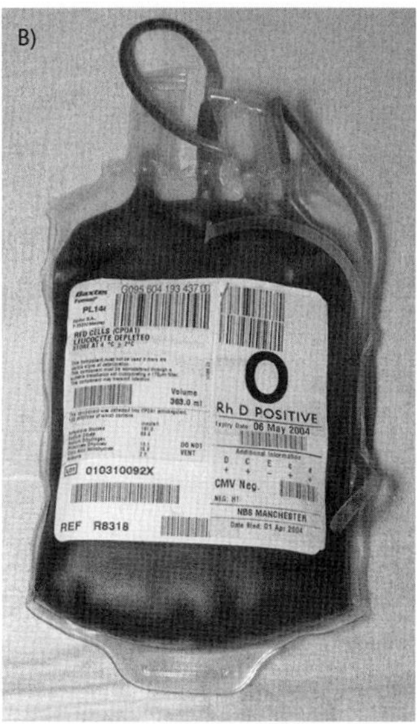

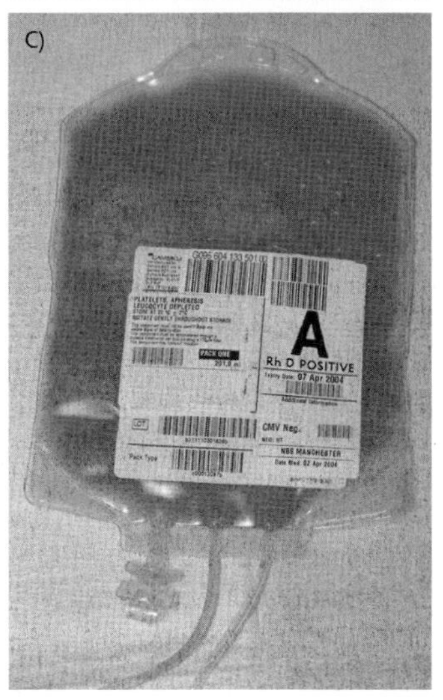

BOX 6.1 The flow cytometer

The flow cytometer (*Figure 6.3*) is an instrument that can analyze several properties of cells simultaneously in mixed populations. The cells to be analyzed are passed as single cells in a stream past a laser light source. This is usually achieved by having a sheath of fluid passing through an orifice of 50–300 μm. The sample is injected into the sheath fluid as it passes through the orifice. With the right sheath fluid flow rate, the sample and the fluid do not mix (*Figure 6.4*). As the cells are illuminated by the laser beam some of the light is scattered. The scattered light is detected simultaneously by two detectors. One measures side scatter, that is the light deflected 90° from the incident beam. The other detects the light scattered in a forward direction up to 10° from the incident beam. This is the forward scatter. The intensity of the forward and side scatters are related to the size and shape of the cells. The forward scatter is sensitive to the surface characteristics of the cell, whereas side scatter is more sensitive to the granularity of a cell. Thus a mixed population of cells can be analyzed on the basis of these measurements. In addition

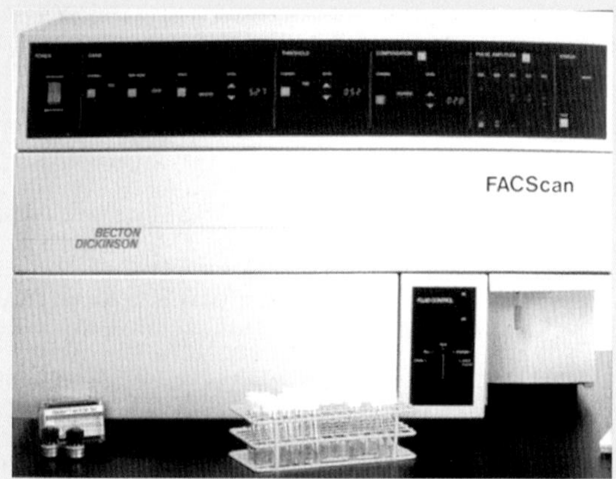

Figure 6.3 A flow cytometer. Obtained from http://flowcyt.cyto.purdue.edu/flowcyt/educate/photos/flowware/fwarepre.htm

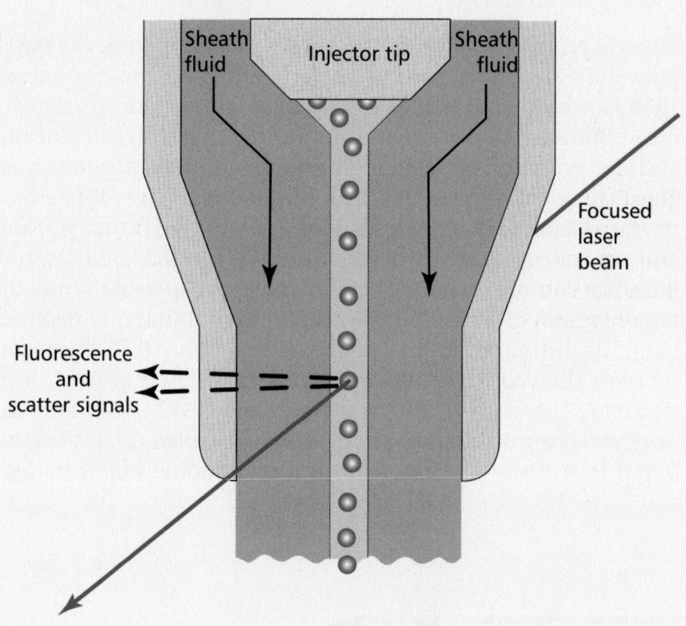

Figure 6.4 Schematic of a Flow Cell. See main text for details.

that these procedures were successful when the blood transfused was of an identical blood group. The discovery of the ABO blood group system led to Landsteiner receiving the Nobel Prize for Physiology and Medicine in 1930. Landsteiner later discovered other blood group systems, including the Rh system. Since then, numerous other systems have been discovered, as shown in to forward and side scatter, cells can be stained with fluorescent antibodies to proteins that are characteristic of a cell population. For example, in a mixed population of lymphocytes, the T lymphocytes (*Chapter 4*) could be stained with a fluorescent antibody to the CD3 marker. This would allow T lymphocytes to be distinguished from B lymphocytes, which do not have this protein. As the cells are illuminated in the laser beam the T lymphocytes will fluoresce, and this fluorescence is picked up by further detectors, using appropriate filters. The intensity of the fluorescence is related to the amount of CD3 on the surface of the cell (*Figure 6.5*).

It is also possible to distinguish different populations simultaneously if they are stained with antibodies to unique marker proteins and if the antibodies are conjugated to different fluorescent molecules. Thus, helper T lymphocytes (T_H cells) can be stained with a fluorochrome labeled antibody to CD4. At the same time all the T cells can be stained with an antibody to CD3 which is conjugated to a different fluorochrome. Analysis by flow cytometry would reveal four populations of small lymphocytes: those which are CD3– and CD4–, presumably B_C lymphocytes, those that are CD3+ and CD4–, presumably T_C lymphocytes, those that are CD3– and CD4+, a very minor population, and those that are CD3+ and CD4+, that is, the T_H cells (*Figure 6.6*).

The flow cytometer is used in many aspects of pathology science. In transfusion it can be used, for example, to measure the number of fetal erythrocytes in the maternal circulation following a placental bleed. In transplantation it may be used to evaluate the results of a cross match (*Section 6.11*).

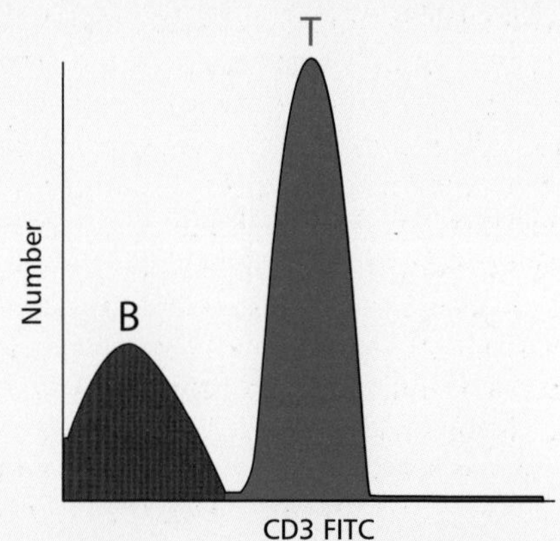

Figure 6.5 Diagram to show how T and B lymphocytes can be distinguished in a flow cytometer.

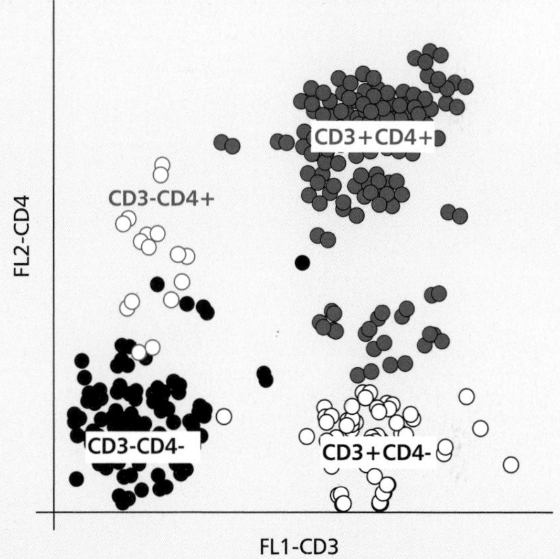

Figure 6.6 Schematic to show how the double staining of lymphocytes can be used to distinguish several populations of cells in the flow cytometer. The cells are represented by the 'dots' on the diagram. See main text for details.

Table 6.2. These blood groups systems have been assigned numbers by the International Society for Blood Transfusion (ISBT) and these, together with the conventional abbreviations, are shown. There is insufficient space in this chapter to discuss all blood groups and only those of greatest clinical significance will be discussed.

ISBT Number	Name	Symbol
001	ABO	ABO
002	MNS	MNS
003	P	P
004	Rh	RH
005	Lutheran	LU
006	Kell	KEL
007	Lewis	LE
008	Duffy	FY
009	Kidd	JK
010	Diego	DI
011	Cartwright	YT
012	XG	XG
013	Scianna	SC
014	Dombrock	DO
015	Colton	CO
016	Landsteiner-Weiner	LW
017	Chido/Rodgers	CH/RG
018	Hh	H
019	Kx	XK
020	Gerbich	GE
021	Cromer	CROM
022	Knops	KN
023	Indian	IN
024	Ok	OK
025	Raph	RAPH
026	John Milton Hagen	JMH

Table 6.2 ISBT Human blood group systems

6.4 THE ABO BLOOD GROUP SYSTEM (ISBT 001)

The ABO blood group system classifies people into one of four major blood groups: A, B, AB and O, according to the different type of antigen present on the surfaces of their erythrocytes. The frequencies of these blood groups vary between different populations (*Table 6.3*).

Phenotype	Antigen	Caucasian	Indian	Black	Overall in UK
A	A	44	24	27	43
B	B	9	38	19	9
AB	A and B	3	15	3	3
O	H	44	22	49	45

Table 6.3 Frequency (as %) of A, B, AB and O blood groups in different ethnic groups and populations

THE ANTIGENS OF THE ABO SYSTEM

The antigens that determine ABO blood groups are oligosaccharide constituents of cell surface glycolipids and glycoprotein. These sugars are added to an existing chain of oligosaccharides which protrudes from the erythrocyte membrane (*Figure 6.7*). The *H* gene, located on chromosome 19, encodes an enzyme, L-fucosyl transferase, which adds L-fucose to the terminal galactose, to form the H antigen. In blood group O individuals, the H antigen is found on their erythrocytes and also on a variety of other cells. People of blood group A and AB possess the *A* gene, encoded on chromosome 9. This gene encodes *N*-acetylgalactosaminyl transferase which adds *N*-acetylgalactosamine to the terminal galactose of the H antigen, to form the A antigen. This antigen is found on the erythrocytes of individuals of blood group A and AB. The *B* gene, which is allelic to the *A* gene, encodes D-galactosyl transferase which adds D-galactose to the terminal galactose of the H antigen, to form the B antigen. This antigen is found on the erythrocytes of blood groups B and AB individuals. The *O* gene, also allelic to the *A* and *B* genes, does not appear to produce a protein, that is, it is a silent gene. The *A*, *B* and *O* genes are inherited in a Mendelian fashion (*Chapter 15*) and A and B are codominant. The genotypes that determine the different phenotypes in the ABO system are shown in *Table 6.4*.

Genotype	Comment	Blood Group (phenotype)
AA	The gene encoding the 'A' glycosyl transferase is present on both copies of chromosome 9	A
AO	The gene encoding the 'A' glycosyl transferase is present only on a single copy of chromosome 9	A
BB	The gene encoding the 'B' glycosyl transferase is present on each copy of chromosome 9	B
BO	The gene encoding the 'B' glycosyl transferase is present on only a single copy of chromosome 9	B
AB	One copy of chromosome 9 has the gene encoding the 'A' enzyme while the other has the gene encoding the 'B' enzyme	AB
OO	Neither the 'A' gene nor the 'B' gene is present on either chromosome 9	O

Table 6.4 Genotypes and phenotypes of the ABO system

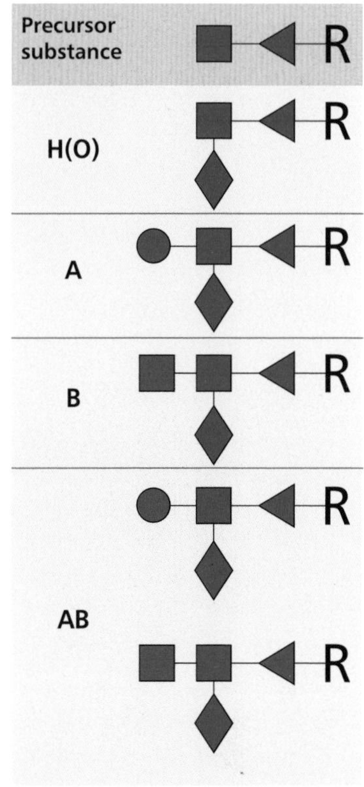

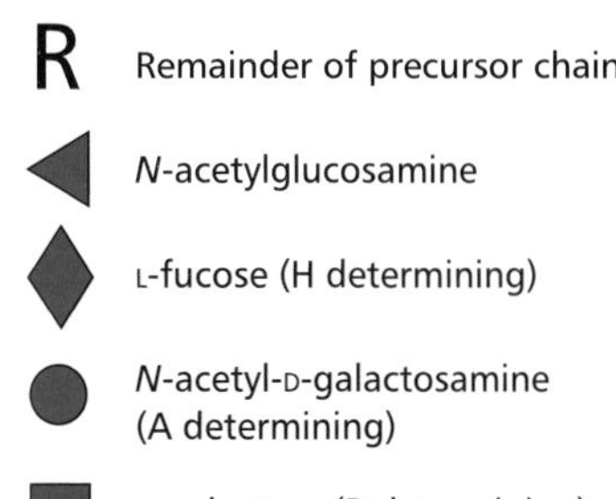

Figure 6.7 Schematic of the antigens of the ABO blood group system. See main text for details.

Approximately 75% of Caucasians attach A, B and H antigens (depending on their blood group) to soluble proteins and secrete them into saliva and other body fluids, and are known as **secretors**. Secretor status is dependent on the expression of a fucosyl transferase, which is encoded by the *SECRETOR* gene on chromosome 19. In the past, the presence or absence of secreted ABH antigens has been used in forensic science for the purpose of identification.

ANTIBODIES OF THE ABO BLOOD GROUP SYSTEM

Transfusion of ABO incompatible blood causes problems because individuals possess plasma antibodies to the complementary antigens. Thus, people of blood group A have antibodies to the B antigen, those with blood group B have antibodies to the A antigen (*Table 6.5*). People of blood group AB have neither antibody, while those of blood group O possess both. These antibodies will agglutinate erythrocytes of the complementary group. The pattern of agglutination when serum and erythrocytes of different ABO types are mixed *in vitro* is shown in *Table 6.6*.

Blood group	Antibodies in plasma
A	anti-B
B	anti-A
AB	neither anti-A nor anti-B
O	anti-A and anti-B

Table 6.5 Antibodies of the ABO system

	Serum			
Cells	**A**	**B**	**AB**	**O**
A	negative	positive	negative	positive
B	positive	negative	negative	positive
AB	positive	positive	negative	positive
O	negative	negative	negative	negative

Table 6.6 Patterns of hemagglutination when sera and cells of different ABO types are mixed

Antibodies directed against the ABO antigens generally belong to the IgM class though IgG antibodies (*Chapter 4*) may also be found. IgM antibodies are pentameric molecules, which are efficient agglutinators of cells and activators of complement. These antibodies, which are sometimes called isohemagglutinins because they agglutinate erythrocytes, are sometimes also termed 'natural antibodies' because they are present even though people have never been 'immunized' with blood of an inappropriate group. In fact, the blood group antigens are also found on some common bacteria and the antibodies that are produced following infection are capable of cross-reacting with blood group antigens. Of course, people do not make antibodies to antigens that they themselves express, because of immunological tolerance (*Chapter 5*).

CONSEQUENCES OF ABO INCOMPATIBLE TRANSFUSIONS

An example of an incompatible blood transfusion would be when a patient of blood group A, the recipient, is given blood from a donor who is blood group B. In this case the anti-B antibodies present in the plasma of the recipient bind to the donated erythrocytes and activate complement. This results in the simultaneous lysis of billions of erythrocytes in the donated blood. At the same time, the limited amount of anti-A antibodies in the donated blood binds to the recipient's erythrocytes and hemolyzes those too. The lysis of so many erythrocytes, called **acute intravascular hemolysis**, releases so much hemoglobin that acute renal failure and shock result. This has serious clinical consequences and, indeed, is fatal in approximately 10% of cases. In addition to the problems caused by the release of hemoglobin, the fragments of erythrocyte membrane released may also initiate the blood clotting systems, leading to disseminated intravascular coagulation (DIC). The consequences of an incompatible transfusion may vary according to the blood group involved. For example, antibodies to some of the other blood group systems may result in a delayed, extravascular hemolysis. Slower destruction of the donated cells may lead to a decreasing hemoglobin concentration, with the patient suffering a fever and general malaise.

The administration of an incompatible donation is most often due to errors postdonation, and is rarely due to mismatching of bloods.

6.5 THE Rh BLOOD GROUP SYSTEM (ISBT 004)

The Rh blood group system divides people into Rh positive and Rh negative groups depending on whether or not their erythrocytes carry the Rh antigen. Landsteiner and Wiener discovered this system in 1940. They showed that antisera raised in guinea pigs against erythrocytes from rhesus monkeys reacted with 85% of Caucasian blood donors in New York. The Rh system of blood group antigens is often described as if it is a single antigen. However, it consists of a complex series of antigens, which are specified by two genes: *RHD* and *RHCE*. The former encodes the RhD protein which expresses the D antigen while the latter encodes the RhCcEe protein which carries either the C or c antigen together with the E or e antigen. At one time it was thought that another antigen, termed the 'd' antigen, was present when the D antigen was absent. It is now recognized that the d antigen does not exist. However the term is still used to indicate the D negative phenotype.

An individual person inherits a set of genes from each parent, with the possible haplotypes (haploid genotypes) being shown in *Table 6.7*, where the symbol '*d*' is used to express a lack of the *RHD* gene. Under the Fisher system for nomenclature, each Rh haplotype is assigned a code. The commonest genotypes are shown in *Table 6.8*. If an individual has the D antigen, they are said to be RhD positive. Thus, amongst the most common genotypes the only

Haplotype	Fisher system code
DCe	R1
DcE	R2
DCE	Rz
Dce	Ro
dCe	r′
dcE	r′′
dCE	r^y
dce	r

Table 6.7 Haplotypes of the Rh blood group system

Genotype	Frequency in Caucasians / %	Rh status
DCe/dce	33	positive
DcE/dce	11	positive
Dce/dce	2	positive
DCe/DCe	18	positive
DcE/DcE	2	positive
DCe/DcE	14	positive
DCe/Dce	2	positive
dce/dce	15	negative

Table 6.8 Most common genotypes of the Rh blood group system (in Caucasians)

individuals who are RhD negative are those with the genotype dce/dce (or rr using the Fisher code) and these constitute 15% of Caucasians. RhD negative individuals do not normally have antibodies to the D antigen. However, they can become sensitized if transfused with blood from an Rh positive individual. For example, a person with the genotype dce/dce who is transfused with blood from a DCE/DCE individual may make antibodies to C, D and E antigens, although anti-D antibodies are the most common.

HEMOLYTIC DISEASE OF THE NEWBORN

Hemolytic disease of the newborn (HDN) is a serious disease characterized by anemia, splenomegaly, hepatomegaly and edema. The condition is caused by the transfer of maternal anti-erythrocyte antibodies across the placenta. The condition may arise in the offspring of women who are RhD negative. Such women may become sensitized to RhD antigen when they give birth to an RhD positive child. At birth, some of the baby's blood can enter the maternal circulation and the mother will respond by making anti-D antibodies. This does not have any clinical consequence for the child itself, but may cause problems during a subsequent pregnancy, if that fetus is RhD positive. Antibodies to Rh antigens are clinically significant during pregnancy because they are of the IgG class. Thus they can cross the placenta and bind to the fetal erythrocytes. Although Rh antibodies are IgG, they do not appear to activate complement. Instead, the antibody-coated erythrocytes bind to receptors for the Fc portion of IgG (*Chapter 4*) on monocytes and macrophages, a phenomenon known as **immune adherence**. This occurs when blood is traveling through the spleen and liver. The uptake and subsequent destruction of antibody-coated erythrocytes in these organs is known as **extravascular hemolysis**.

The consequences for the second or subsequent child depend on the extent of maternal sensitization and, hence, the amount of circulating antibody. Destruction of fetal erythrocytes *in utero* may lead to fetal anemia and hyperbilirubinemia, an excess of bilirubin in the blood (*Chapter 13*). The concentrations of serum bilirubin indicate the degree of hemolysis. Babies born with significant levels of bilirubin will suffer **kernicterus** or brain damage due to the build up of the lipid soluble bilirubin in the brain. The baby will need phototherapy, which helps to breakdown the bilirubin (*Chapters 11* and *13*).

If the anemia is severe, the fetus may die of heart failure or *hydrops fetalis*, which is an extreme edema of the entire body of the fetus. This may lead to spontaneous abortion. A longer term child may be stillborn. The considerably enlarged spleen and liver in a child born alive with HDN is associated with erythrocyte destruction in these organs. The baby may also have a facial rash indicative of hemorrhages due to impaired platelet function.

The commonest cause of HDN is antibody to the RhD antigen although it may also be caused by antibodies to other blood group antigens, for example, antibodies to blood group A or B antigens, if the antibodies are of the IgG class. In addition, antibodies to the Rhc antigen, and to the Kell blood group antigen (*Section 6.6*), may also be involved.

All pregnant women in the UK and other developed countries now have their ABO and Rh status checked at their initial hospital booking, which is usually at 12–16 weeks of pregnancy. They are also checked for anti-D, anti-c or anti-Kell (*Section 6.6*) and, if these are present, the concentrations will be monitored regularly throughout the second trimester of pregnancy. If the levels of clinically significant antibodies start to rise, clinical intervention may be necessary. Concentrations of anti-D below 4 international units (IU) cm^{-3} are unlikely to cause HDN, those between 4–15 IU cm^{-3} have a moderate risk of HDN, while values above 15 IU cm^{-3} are associated with a high risk of HDN.

Margin Note 6.1 HDN and the ABO system 

Hemolytic disease of the newborn does not occur within the ABO system if Anti-A and Anti-B are of the IgM class. This is because IgM does not cross the placenta. In addition, a woman who is blood group A, RhD negative, is unlikely to become sensitized to a fetus who is blood group B, RhD positive because, when the fetal red cells enter her circulation following the birth, her anti-B antibodies will destroy the fetal erythrocytes before sensitization to RhD can occur.

HDN prophylaxis

Since the 1970s, a prophylactic treatment to prevent Rh sensitization has been available which has greatly reduced the incidence of HDN due to RhD sensitization. The treatment involves the intramuscular injection of at least 500 IU of anti-D immunoglobulin within 72 h of the birth of a RhD positive baby. The administered antibodies bind to any erythrocytes from the baby which have entered the maternal circulation and destroy them, preventing the mother from making antibodies.

In 2002, the National Institute for Clinical Excellence (NICE) in the UK recommended that antenatal anti-D prophylaxis should be routinely offered to any pregnant Rh negative woman who has not made anti-D antibodies to prevent sensitization predelivery caused by, for example, a placental bleed.

Direct antiglobulin testing and Kleihauer testing

The direct antiglobulin test (DAT), formerly known as the Direct Coombs test, is undertaken to see if maternal antibodies are present on the baby's erythrocytes. If they are present, then a sample of the erythrocytes from the baby will be agglutinated by an antibody to IgG, known as antihuman globulin (*Figure 6.8*).

The Kleihauer test uses the Kleihauer-Betke stain and is a method of assessing the volume of fetal blood that has entered the maternal circulation. In most cases the volume that has entered will be less than 4 cm^3, and 500 IU of Anti-D is sufficient to remove the erythrocytes in this volume. However, for less than 1% of women, the volume of fetal blood is larger and the mother consequently needs more than this amount of anti-D. The Kleihauer test is carried out on a sample of maternal blood 2 h after delivery. The principle of the test is that the hemoglobin in adult erythrocytes can be eluted with acid, whereas the hemoglobin of fetal erythrocytes is resistant to acid elution. A smear of maternal blood is placed in a solution of hematoxylin and hydrochloric acid, pH 1.5, and then is counterstained with eosin. The maternal red cells appear as pale 'ghosts' whereas the fetal erythrocytes stain pink with eosin, while the leukocytes stain blue (*Figure 6.9*). The ratio of fetal to maternal cells is an indicator of the volume of blood that has entered the circulation. This test can also be carried out during pregnancy if a placental bleed is suspected.

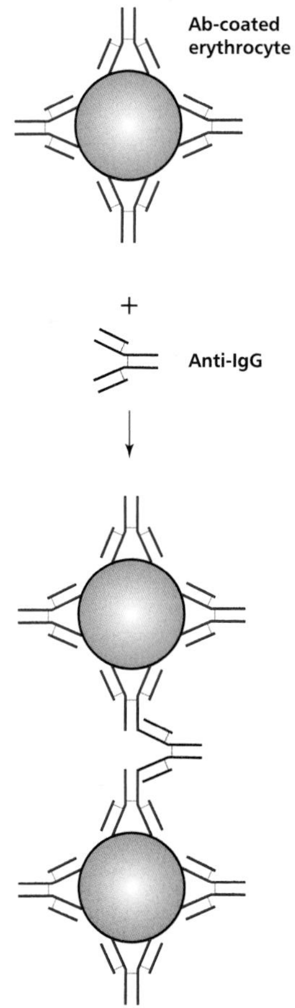

Figure 6.8 Schematic to show the direct antiglobulin test. See main text for details.

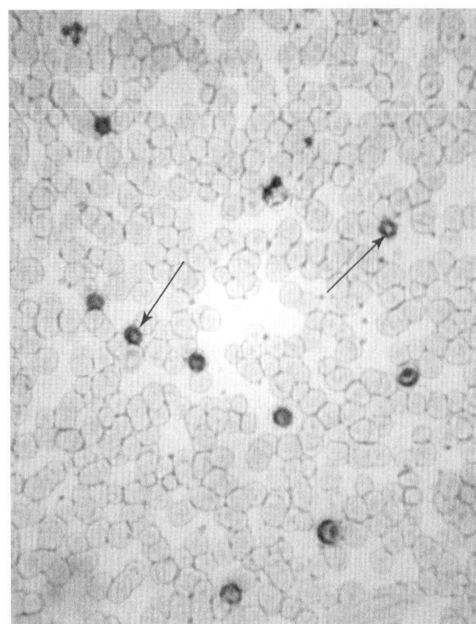

Figure 6.9 Results of Kleihauer test. The darker staining erythrocytes are fetal cells.

6.6 OTHER BLOOD GROUP SYSTEMS

The Lewis blood group system (ISBT 007, symbol Le) is related the Lewis antigens Le^a and Le^b present on erythrocytes. However, these antigens are not integral parts of the membrane but are soluble plasma proteins which become reversibly adsorbed onto erythrocyte membranes. The levels of bound antigens therefore vary, although the erythrocytes of children 2 years old and above have approximately adult levels.

The Le^a and Le^b antigens are not the products of different forms of a single gene, but arise from different actions of a fucosyl transferase that attaches fucose residues to an oligosaccharide known as type-1 precursor oligosaccharide. If the fucose is added to a subterminal position it produces the Le^a antigen, whereas attachment to the terminal position gives the Le^b antigen. Approximately 72% of white populations are Le(a–b+), that is, they lack the Le^a but have the Le^b, 22% are Le(a+b–), and 6% lack both antigens. Antibodies to the Le antigens are usually of the IgM class and, as such, do not cause HDN since they do not cross the placenta.

The Duffy system of blood group antigens (ISBT 008, symbol FY) is comprised of six antigens, of which Fy^a and Fy^b are the most significant in transfusion reactions. These antigens are expressed on an erythrocyte membrane glycoprotein and also form the site of attachment for malarial parasites (*Chapter 2*). Thus a Fy(a–b–) individual, who does not express the blood group antigens, has a selective advantage in a malarial area. Indeed, 68% of blacks of African descent are of this phenotype, which is rare in whites. Antibodies to Duffy antigens belong to the IgG class and may cause HDN.

The Kidd antigens (ISBT 009, symbol JK) are expressed on a membrane glycoprotein, which is associated with urea transport. The Jk^a and Jk^b antigens result from the expression of a codominant pair of genes. Approximately 27% of whites, and 57% of blacks are Jk(a+ b–) while 50% of whites and 34% of blacks are Jk(a+ b+). The Jk(a–b+) phenotype is found in 23% of whites and only 9% of blacks, while the Jk(a–b–) phenotype is rare in both populations. Antibodies to Jk^a and Jk^b belong to the IgM or the IgG class and may cause a mild form of HDN.

The Kell blood group system (ISBT 006, symbol KEL) is formed from 24 antigens expressed on a glycoprotein of the erythrocyte membrane. The antigen K (formerly Kell) is highly immunogenic and IgM or IgG antibodies to it are common in transfused patients. Similarly, antibodies to its allele, designated *k* (formerly Cellano), can also cause HDN, although specific antibodies are rare in transfused patients.

6.7 LABORATORY DETERMINATION OF BLOOD GROUPS

Traditional methods for determining blood groups rely on the agglutination of erythrocytes by antibodies, usually referred to as hemagglutination. Hemagglutination can be carried out on glass microscopy slides or in microtiter plates in which agglutination patterns are easily distinguished from the settling of erythrocytes. Recent years have seen increasing use of the Diamed typing system to detect hemagglutination. This is a system which uses monoclonal typing antibodies, distributed in a gel, contained in individual tubes set in plastic 'cards'. Cells are added to the antibodies and the cards are centrifuged. Where agglutination has occurred, the agglutinates remain on top of the gel, whereas nonagglutinated cells settle through the gel to the bottom (*Figure 6.10*). Most transfusion laboratories now use gel technology for blood grouping and compatibility testing.

Whichever technique is used, a blood group, such as the ABO grouping, is determined by incubating the individual's erythrocytes with antibodies to known antigens (anti-A and anti-B in this case) and also mixing the individual's plasma with erythrocytes of known A, B, AB or O blood groups. The pattern of hemagglutination shown will enable the determination of the blood group.

Hemagglutination occurs when antibodies to an erythrocyte antigen cross-link the cells, forming visible aggregates. The extent of hemagglutination depends on the temperature, pH and the ionic strength of the medium. Agglutination is favored in low ionic strength saline (LISS). Erythrocytes have a net electronegative charge and repulsive forces normally keep them about 20 nm apart. When antibodies bind to the erythrocyte, the reduced surface charge allows the cells to agglutinate. This is most effectively achieved with IgM antibodies, which can cause direct agglutination of erythrocytes. To obtain a direct agglutination with IgG antibodies, it is usually necessary to include bovine serum albumin in the medium, which masks the charges on the erythrocytes and allows them to come closer together. Another method of reducing the negative charge is to use proteolytic enzymes to remove surface proteins that carry the charge. The enzyme can be added to the erythrocytes prior to the addition of the antibody, or all the components can be added together. Polycationic polymers such as polybrene will also reduce the negative charge on erythrocytes.

The antiglobulin test *(Section 6.5)* uses the ability of antihuman globulin (AHG) to agglutinate erythrocytes coated with nonagglutinating erythrocyte-specific IgG. This can be used to detect erythrocytes already coated with anti-erythrocyte IgG in the direct antiglobulin test (DAT), or can be used on cells which have been incubated with antibody *in vitro* (see *Figure 6.8*).

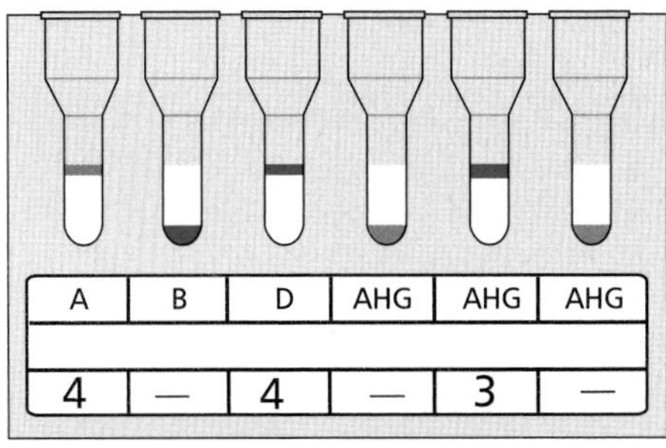

Figure 6.10 The Diamed gel card system for determination of blood group. Agglutinated cells do not penetrate the gel. The blood group indicated here is A Rh+, as indicated by the first three tubes where agglutination has occurred in tubes 1 and 3. AHG is antihuman globulin (*Section 6.5*).

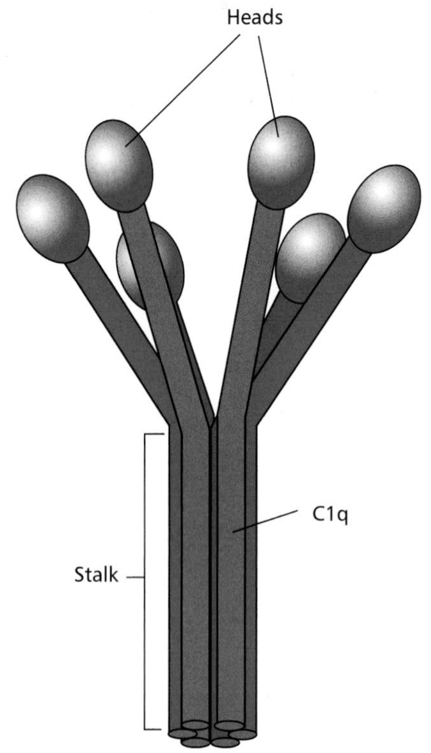

Figure 6.11 A diagram of the C1q molecule to show the six sites ('heads'), each of which can bind to an Fc region of an IgG or an IgM molecule.

6.8 ROLE OF COMPLEMENT IN TRANSFUSION REACTIONS

In *Chapter 4*, the 'beneficial' role of complement in eliminating immunogens was discussed. The activation of complement is also involved in some forms of immunological hypersensitivity and can cause some of the problems associated with autoimmune disease (*Chapter 5*). However, given that it amplifies the actions of antibodies, complement can cause many of the problems associated with transfusion reactions. Thus, complement causes lysis of sensitized erythrocytes, that is, erythrocytes coated with anti-erythrocyte antibodies.

The classical pathway for complement activation is initiated when IgG or IgM binds to an epitope, in this case on the erythrocyte's membrane. The antibody could be IgM, as is usually the case with Anti-A or Anti-B, or IgG, as is the case with antibodies to Rh antigens. The binding of antibody to the epitope induces a conformational change in the Fc region of IgG or IgM, allowing the binding of C1 protein. The C1 is comprised of three loosely associated proteins called C1q, C1r and C1s. The C1q is a large protein and has several binding sites allowing it to bind to multiple Fc regions of antibodies (*Figure 6.11*). It requires at least two of these sites to bind to adjacent Fc regions on the surface of a cell to activate complement. For this reason IgM, which has several Fc regions per molecule, is more efficient than IgG at activating complement. Indeed, a single molecule of IgM can activate complement, whereas it takes about 1000 molecules of IgG to achieve the density required for activation. The binding of C1q activates C1r and this, in turn, activates C1s which acquires proteolytic activity (*Figure 6.12*). C1s has two substrates: C4 and C2 each of which are hydrolyzed to two fragments: C4a, C4b and C2a and C2b respectively. Proteins C4b and C2a are the larger fragments in each case and they combine to form a new proteolytic enzyme, C4b2a. A single enzyme molecule can generate a number of product molecules. Thus for a limited number of antibody molecules, many molecules of C4b2a are formed, because enzymic steps allow amplification to occur. C4b2a is the classical pathway **C3 convertase** (*Chapter 4*) which cleaves C3 into two fragments: a larger C3b molecule and a C3a. The former binds to the target cell membrane where it may bind a molecule of the C3 convertase to form a **C5 convertase**, which catalyzes the hydrolysis of C5 into C5a and C5b. C5b binds to the cell membrane and forms a site for the build up of the **Membrane Attack Complex** (**MAC**). This is a large, cylindrical, hydrophobic structure constructed from single molecules of C5b, C6, C7, C8 and several molecules of C9. When it inserts into the membrane it forms a pore of approximately 10 nm diameter. Since amplification has occurred at each enzymic step, the target cell membrane may be covered with MACs (*Figure 6.13*). The MACs allow small ions to equilibrate across the cell membrane, increasing the osmotic pressure within the cell so that water moves across the membrane into the cell causing it to lyze. *In vitro* this can be seen as a sudden clearing of the cloudy suspension of erythrocytes. *In vivo*, several regulatory proteins may prevent direct lysis of the erythrocytes. Instead, the cells are lyzed by phagocytic cells that have receptors for C3b and other complement proteins on their membranes. *Table 6.9* lists some of the receptors involved in the clearance of sensitized erythrocytes. In the transfusion laboratory it is much easier to look for complement proteins on erythrocytes than to look for antibodies, since a small amount of antibody may result in large amounts of complement on the cells. Thus, the presence of complement proteins on cells is used as an indicator of the presence of complement binding antibodies.

It is essential for the transfusion scientist to be able to detect hemolytic antibodies. Such antibodies may be present due to transfusion reactions, to HDN or they may be autoantibodies to erythrocytes, as happens in autoimmune hemolytic anemia. The presence of relevant antibodies may be detected in

Receptor	Distribution on cells involved in clearance of sensitized erythrocytes	Protein bound
CR1	erythrocytes, neutrophils, eosinophils, monocytes, macrophages	C3b, C4b, C3bi (an inactive form of C3b)
CR3	neutrophils, large granular lymphocytes, macrophages	C3bi
CR4	neutrophils, monocytes, macrophages	C3bi

Table 6.9 Complement receptors

samples of serum by hemolysis *in vitro* using the complement activity in the serum or they may be detected by examining the surface of erythrocytes for activated complement proteins. The presence of complement in serum diminishes with storage, therefore it is recommended that samples of sera should be stored at –20°C to retain activity if they are to be used for hemolysis determination. Further, some anticoagulants, such as EDTA, inhibit complement, which may be significant if plasma rather than serum is available. Other sera may have 'anticomplementary' activity due to the presence of a denatured form of complement known as complementoid.

Receptors for C3b are found on all the major types of phagocytic cells and also on erythrocytes themselves, so that even these cells have a role in the clearance of immune complexes from the blood. Antigen–antibody complexes coated in C3b bind to erythrocytes and are removed by macrophages in the spleen and liver.

In vivo, other complement proteins trigger inflammatory reactions. For example, C3a, C4a and C5a cause blood basophils and tissue mast cells to degranulate. The mediators released stimulate inflammation (*Chapter 4*), which may have consequences in a patient who has anti-erythrocyte antibodies. In addition, both C3a and C5a are chemotactic factors for neutrophils and promote a build up of these cells, which may itself lead to clinical problems.

The alternative pathway for complement activation is a positive feedback loop which is usually initiated by microorganisms such as bacteria and yeasts. However, feedback may utilize C3b, produced in the classical pathway, and amplify the amount of C3b produced. The positive feedback loop is controlled to prevent an overproduction of C3b. One regulatory step binds C3b to a plasma protein called Factor H and in the bound form is inactivated by Factor I, which converts it to C3bi, a form that can no longer enter the amplification loop. C3b may then be degraded into smaller fragments, C3dg and C3d, which may remain bound to the erythrocyte membrane. The presence of natural regulators means that many antibodies that are potentially able to lyze erythrocytes are unable to do so *in vitro* and the transfusion scientist may look for the presence of C3d on erythrocytes to determine whether antibodies to them are present.

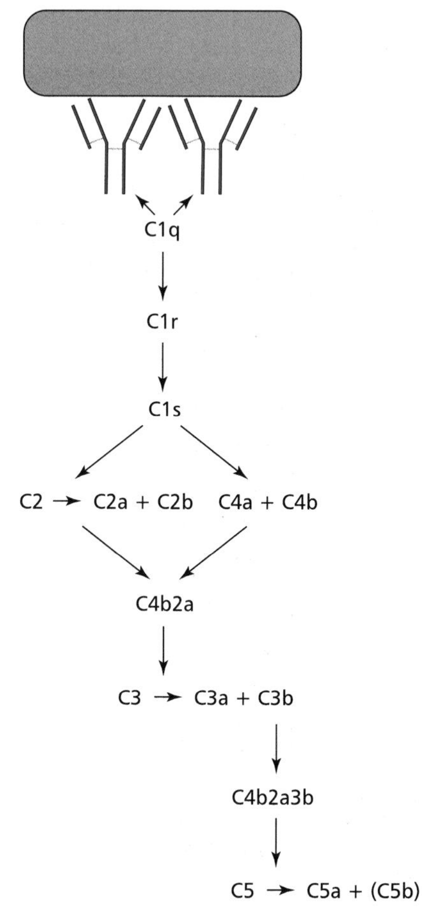

Figure 6.12 The classical pathway for complement activation. See main text for details.

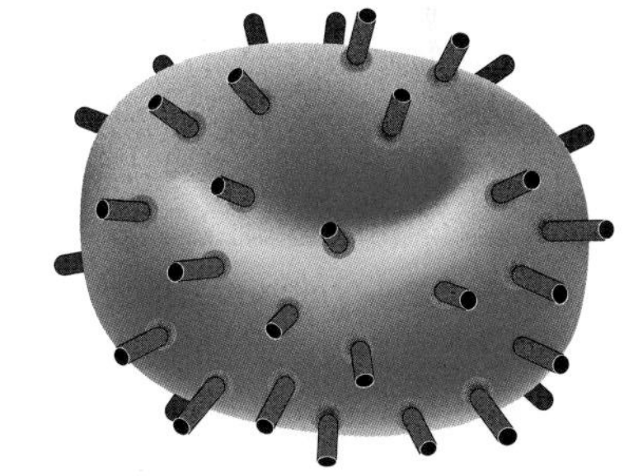

Figure 6.13 Schematic to show an erythrocyte covered with membrane attack complexes (MACs) inserted into the cell membrane.

6.9 HAZARDS OF TRANSFUSION

One hazard of transfusion is a hemolytic transfusion reaction (HTR) if preexisting antibodies are present in the recipient. This may result in acute intravascular hemolysis, as in ABO incompatibility, or in delayed extravascular hemolysis, as with several of the other blood group systems. Acute intravascular hemolysis has serious clinical consequences and, indeed, may be fatal. With delayed extravascular hemolysis the patient may suffer fever and general malaise as the donated erythrocytes are destroyed and the hemoglobin levels

Margin Note 6.2 The Serious Hazards of Transfusion group 

In the UK, adverse reactions to transfusion are reported to the Serious Hazards of Transfusion (SHOT) group based at the Manchester Blood Centre. Here, the data are collected and an annual report is produced. This process enables trends to be recognized, and recommendations on safe practice in transfusion to be made.

fall. To prevent either occurrence, the recipient should be screened before the transfusion for the presence of anti-erythrocyte antibodies.

The transfusion of leukocytes may also lead to adverse reactions, such as **nonhemolytic febrile transfusion reactions**. Such patients exhibit flushing, fever, rigors and hypotension. These may be caused by the reaction between antibodies and leukocyte antigens in the recipient, resulting in the lysis of donated leukocytes and release of cytokines from them. In addition, activated complement proteins cause the release of histamine from basophils, triggering inflammatory reactions. As all blood is leukodepleted before transfusion, such reactions are rare.

TRANSFUSION RELATED ACUTE LUNG INJURY

Transfusion related acute lung injury (TRALI) is a life-threatening complication of transfusion. It presents as a rapidly progressing and severe respiratory failure with diffuse damage to alveolar cells and the filling of alveolar spaces with fluid. Histological examinations reveal an infiltration of the alveoli with neutrophils and monocytes, indicative of an acute inflammatory reaction.

The symptoms of TRALI include dyspnea, cyanosis, hypotension, fever and pulmonary edema, which usually occur within 6 h of transfusion. The condition is thought to arise from the interaction of leukocytes and specific antileukocyte antibodies, usually in the donor plasma, though occasionally in that of the recipient. The most likely antibodies are those to the Major Histocompatibility (MHC) antigens, which are commonest in women who have had multiple pregnancies (*Section 6.11*) and in males and females who have received multiple transfusions. In addition, antibodies to neutrophil antigens have also been implicated.

TRALI has been reported to occur after transfusion of fresh frozen plasma, platelets, whole blood and concentrated erythrocytes. In the UK in 2003, 36 cases of suspected TRALI were reported to SHOT. Nine patients died, seven possibly, and one definitely due to transfusion. Plasma-rich components were implicated in 20/21 cases in which there was proven leukocyte incompatibility between the patients and the donor. In 2004, however, 23 suspected cases were reported in the UK. Of these, 13 were highly likely to be TRALI and were linked to fresh frozen plasma, in six cases, platelets in four, erythrocytes in two and whole blood in one. The incidence of TRALI has since decreased in the UK, due to the processing of fresh frozen plasma from untransfused male donors who have tested negative for antileukocyte antibodies.

IRON OVERLOAD

Patients who receive many transfusions over a long period of time may develop iron overload. The excess of iron, for which there is no excretory route, from transfused blood may cause tissue damage, especially to the liver, heart and endocrine glands. Signs of iron overload can be detected after 10–20 transfusions and the condition may be fatal if left untreated. Patients should be treated with a chelating agent such as deferrioxamine mesylate to remove unwanted iron (*Chapter 13*).

ALLOIMMUNIZATION

Patients who receive regular erythrocyte transfusions may become immunized to other blood group antigens present on the ABO compatible cells. This may have consequences for future transfusions. If patients receive whole blood, rather than leukodepleted blood, they may become immunized to the MHC antigens on the foreign leukocytes. This may become clinically significant if in the future they require an organ transplant (*Section 6.11*). However, the increasing use of leukodepleted blood, in which the leukocytes have been removed prior to transfusion, prevents the immunization of recipient to these antigens.

GRAFT VERSUS HOST DISEASE

Graft versus host disease (GVHD) is a potentially fatal condition which is associated with the transfusion of whole blood or blood products, such as packed erythrocytes or platelets, which contain residual lymphocytes. It usually results from the transfusion of leukocytes into an immunodeficient patient or when blood is transfused into neonates or premature babies. The small lymphocytes present in the donated blood recognize the antigens of the recipient as foreign and mount an immune response against them. The donor lymphocytes proliferate in the patient and attack tissues, causing enlargement of the spleen and liver, diarrhea and an extensive skin rash. Acute GVHD may be fatal and for this reason it is recommended that products such as packed erythrocytes be irradiated prior to use to prevent residual small lymphocytes from reacting to the host antigens. Graft versus host reaction can also be a consequence of bone marrow transplantation (*Section 6.14*). Frozen plasma is safe in this respect, since freezing destroys leukocytes. Transfusion associated GVHD is not usually linked to AIDS.

INFECTIONS

One potential hazard of transfusion is infection with microorganisms present in the donor. In the past, blood transfusions have spread infections such as HIV (*Chapter 4*) and hepatitis C (*Chapter 11*) to the patient. For this reason there is now extensive screening of blood donors (*Section 6.10*).

Margin Note 6.3 Plasmapheresis and leukapheresis 

Different components of the blood can be isolated as the donation is occurring by sending the blood through a cell separator. A catheter is first inserted into the vein of the donor. Blood entering the catheter is transferred to the cell separator where it is centrifuged at a speed necessary to collect the required component. The blood, minus the component that has been removed, is returned to the donor via a catheter inserted into a vein in the other arm. **Plasmapheresis** is the collection of plasma, with the return of the erythrocytes and leukocytes to the donor. **Leukapheresis** is the collection of leukocytes, with plasma and erythrocytes being returned. Similarly, **plateletpheresis** involves the collection of platelets, with all other components being returned. The term **apheresis** is a general term that covers the collection of specific blood components in this way.

6.10 SCREENING OF BLOOD DONORS

Transfusion scientists must be assured that the process of transfusing blood poses minimal risk to the patient and donor alike. Aside from the obvious need for blood to be carefully matched to avoid a possibly fatal transfusion reaction, it is essential that donors are carefully screened to avoid those who are ill or who may be harmed by giving blood, or whose blood poses a health risk because, for example, it is contaminated with certain viruses even though the donor shows no signs of ill health.

In the UK, blood is taken from healthy donors aged between 17 and 70 and is a voluntary and unpaid activity. Potential donors who are excluded from donation include individuals with HIV or hepatitis viral infections, as well as individuals who are at risk of becoming HIV and/or hepatitis virus positive, for example prostitutes, drug abusers who inject themselves with drugs, and individuals who have had sex with men or women living in Africa (*Table 6.10*). In addition, people with low hemoglobin levels (below 135 g dm^{-3} and 125 g dm^{-3} for men and women respectively), those who have had infectious diseases such as a cold or sore throat within the last seven days or viral infections such as measles, mumps, rubella, chickenpox, shingles or herpes simplex cold sores within the last three weeks. Other reasons for exclusion include the recent use of therapeutic drugs, for example aspirin, antibiotics, antihistamines and antidepressants.

All donated blood is screened for a variety of infectious agent as shown in *Table 6.11*. Some tests are mandatory while others are optional. Optional tests such as those for cytomegalovirus (CMV) are used when the blood is to be transfused into immunocompromized individuals.

Category	Examples
Recent history of clinical intervention	had a minor operation within the last month had a major operation within the last six months had a child or a miscarriage within the last 12 months had a local anesthetic for dental treatment within the last two days had a general anesthetic for dental treatment within the last month is currently receiving medical treatment or under medical investigation
Donor is currently unwell	has a hemoglobin level below 135 g dm^{-3} (male) or 125 g dm^{-3} (female) is currently taking aspirin or has taken antibiotics, antihistamines or antidepressants within the last seven days
Donor has recently been unwell/ is unwell	recently had chicken pox, shingles, measles, mumps, rubella, herpes simplex cold sores within the last three weeks had a sore throat, cold, cough in the last seven days is known to be infected with HIV or hepatitis viruses
Donor is in a high risk category for HIV and/or hepatitis viruses	has injected recreational drugs at any time since 1977 had sex with men or women living in Africa at any time since 1977 is a prostitute

Table 6.10 Reasons for exclusion from donating blood

Mandatory tests	Optional tests
Hepatitis B virus (HBV)	cytomegalovirus (CMV)
Hepatitis C virus (HCV)	malaria (*Chapter 3*)
Human immunodeficiency viruses (*Chapter 3*)	
Syphilis (*Chapter 3*)	

Table 6.11 Mandatory and optional tests for infectious agents in donated blood

BOX 6.2 Artificial blood

The potential risks arising from the transfusion of blood have led to research into the development of an artificial blood or blood substitute that would alleviate these risks. In addition, the supply of blood in some regions of the world is too small to make its availability reliable. For an artificial material to be a 'good' blood substitute it should contain a material that will carry appropriate amounts of oxygen around the body. In addition, the fluid carrier must be isotonic to prevent cell lysis due to osmotic differences. It must also be able to withstand sterilization to prevent the risk of infection. Two classes of artificial blood substitute have been investigated. One is based on the use of solutions of modified hemoglobin, the other has involved the use of products based on perfluorocarbons.

MODIFIED HEMOGLOBINS

Originally, these substitutes were based on free hemoglobin extracted from erythrocytes. However, it became apparent that free hemoglobin is associated with renal toxicity. Hemoglobin inside the erythrocyte forms tetrameric molecules whereas outside the erythrocyte it forms toxic dimers. Cross-linking the tetramers or processing them to polymers prevents dimer formation. However, success has been limited. The use of hemoglobin-based solutions has been hampered by severe side effects including hypertension and vasoconstriction. One cross-linked hemoglobin product was withdrawn from a phase II clinical trial in 1998 because of unacceptably high mortality compared with standard treatments. However, other attempts may prove to be more useful.

6.11 TRANSPLANTATION OF SOLID TISSUES AND ORGANS

The transplantation of solid tissues from one individual to another for therapeutic purposes is a fairly routine medical procedure. Interest in transplantation mainly arose during the Second World War during attempts to treat badly burned airmen using skin grafts from unrelated donors. Indeed, it was from experiments with skin grafting in rodents that the role of the immune system in the rejection of transplants was identified. Moreover, once this role was recognized, the search for drugs that could prevent rejection became more focused. The first successful kidney transplant was performed by Murray (1919–) in 1954 while the first human heart transplant, by Barnard (1922–2001), was performed in 1967.

Once technical problems associated with the transplantation of whole organs were overcome, the major problem associated with transplanting organs was, and remains, rejection. This is caused by the immune system of the recipient recognizing the cells of the donated organ as foreign and mounting an immune response against them. Unless steps are taken to prevent it, transplant rejection is inevitable, unless the donor and the recipient are identical twins. However, organ transplantation is now remarkably advanced compared with initial attempts. Indeed, a range of transplants are routinely performed (*Table 6.12*).

IMMUNOLOGICAL REJECTION OF TRANSPLANTS

The rejection of a transplant is due to the immune system of the recipient recognizing the donated cells as foreign, that is nonself. Thus, the greater the genetic disparity between the donor and recipient, the greater the chances of rejection. The commonest type of clinical transplant is called an **allograft**, that is, a graft between two genetically nonidentical people. However, **isografts**, in which pieces of tissue are transplanted from one site to another on the same individual are also routinely undertaken, for example in skin grafting to treat burns. Isografts are not rejected since the donor and recipient are, of course, genetically identical. Occasionally, the graft may come from another species, such as those occasions when a baboon heart has been transplanted into a human. Such grafts are known as **xenogeneic** transplants. Xenogeneic transplants are also subject to rejection,

In 2003, trials at the Karolinska hospital in Stockholm used an artificial blood to treat patients. The product was said to transport oxygen through the body 'better than real blood'. It is thought that the substitute was based on hemoglobin solutions. In October 2005, the American Food and Drug Administration (FDA) was reported to be conducting a trial with patients in Kansas. Patients traveling by emergency ambulance to hospitals in four counties in the state were being given an artificial blood substitute called PolyHeme, rather than saline, to treat severe bleeding.

PERFLUOROCARBON-BASED SOLUTIONS

Perfluorocarbons (PCFs) are compounds in which fluoride and bromide atoms are attached to an inert carbon chain. They are able to dissolve large quantities of oxygen and have been shown to improve the oxygenation of tissues, even at low doses. However, PCFs are immiscible with water and must be administered as emulsions. They have been investigated over a considerable period. In 1966 mice were found to survive a 10-min immersion in an oxygen saturated PCF liquid and were able to breathe atmospheric oxygen when removed from the trial. Several products based on PCFs have been the subjects of clinical trials, although patients also have to breathe 70–100% oxygen and flu-like symptoms have been reported. The use of PFCs to enhance artificially the performance of athletes has also been reported.

Tissue organ transplanted	Examples of clinical conditions
Bone marrow	immune deficiency; in treatment of leukemia
Cornea	types of blindness
Heart	heart failure
Heart and lungs	cystic fibrosis
Kidney	end-stage renal failure
Liver	cirrhosis
Pancreas	type 1 diabetes mellitus
Skin	treatment of burns

Table 6.12 Tissue and organ transplants

owing to the increased genetic disparity between the donor and the recipient. Other more common examples of xenogeneic grafts involve the use of porcine and bovine heart valves to replace diseased human ones. In the latter case the valves are treated with glutaraldehyde to stiffen them and to mask their antigenic determinants, and these are often referred to as bioprosthetic valves. There is much debate about proposals to breed genetically engineered **transgenic** pigs to carry some human antigens on their cells to provide transplants for humans. Apart from the ethics of breeding animals purely to supply organs for humans, there is also the risk of the transfer of an unknown virus from pigs to humans, with possible disastrous consequences.

CAUSES OF GRAFT REJECTION

The transplanted cells of an allograft carry histocompatibility antigens on their membranes, which are recognized as foreign by the cells of the immune system. These histocompatibility antigens are found on proteins that are encoded by the **Major Histocompatibility Complex** (MHC) which, in humans, is also known as the Human Leukocyte Antigen (HLA) complex (*Chapter 4*). The MHC complex is a genetic region that encodes several classes of proteins, some of which are membrane proteins. Class I MHC proteins (MHC I) are found on the membranes of all nucleated cells and are involved in the recognition of virus-infected cells by the precursors of cytotoxic T lymphocytes (T_C cells). Class II MHC proteins (MHC II) are found on the membranes of antigen presenting cells such as macrophages and are involved in the recognition by T_H lymphocytes of foreign proteins on the surface of antigen presenting cells. It is relatively minor differences between the amino acid sequences of the MHC molecules of the donor and of the recipient that lead to the rejection of the transplanted tissue.

The rejection of an allograft usually takes place a few weeks after the transplant unless immunosuppressive treatments are given. Small lymphocytes recognize the transplanted cells as foreign. An **acute rejection** is caused by T lymphocytes that infiltrate the graft. The presence of T lymphocytes and monocytes in the infiltrate is a strong indicator of cell-mediated immunity. Both T_H and T_C cells are involved in graft rejection. The T_C cells develop into cytotoxic T lymphocytes directed against the foreign histocompatibility antigens of the grafted cells and are able to destroy cells of the grafts directly, or indirectly by producing cytokines that attract monocytes and macrophages. The T_H lymphocytes respond by producing cytokines that activate a variety of nonspecific cells to attack the graft.

Sometimes rejection may take place within hours or minutes of transplantation, once the tissue has become revascularized. This is known as **hyperacute rejection** and is due to antibodies against graft antigens being already present in the plasma of the recipient. These antibodies bind to the graft cells and activate complement leading to the rapid destruction of donor cells. Hyperacute rejection can also be brought about if the recipient already has antibody to MHC antigens present on the graft. These antibodies may be present for a number of reasons. For example, women who have had a number of pregnancies often have antibodies to the MHC antigens on the fetus, which were inherited from the father. Secondly, patients who have had a number of blood transfusions may become immunized to the MHC antigens on residual leukocytes present in the transfusion. Thirdly, patients who have had a previous transplant and rejected it will almost certainly have antibodies to any foreign MHC antigens that were present on that graft. Finally, antibodies to blood group antigens can also cause hyperacute rejection if they are already present in the recipient. For example, the blood group A, B and H antigens are present on the endothelial cells lining blood vessels. If a recipient of blood group A is given a transplant, for example a kidney from a person of blood group B, the anti-B antibodies in the plasma of the recipient will attack the endothelial cells of the graft and activate complement causing destruction of the graft. For this reason, transplants are no longer carried out against a major blood group barrier. Given that pre-existing antibodies can cause a rapid rejection of a graft it is essential to know which potential recipients have such antibodies. Hence, a cross match is performed in which serum from the recipient is incubated with cells from the donor. If the donor cells are killed in the presence of recipient serum and complement, the transplant will not be undertaken and another potential recipient will be sought.

A **chronic rejection** takes place months or years after transplantation and is brought about by a combination of cell-mediated and humoral mechanisms.

6.12 THE HLA SYSTEM

The HLA system of genes is found on the short arm of chromosome 6. This region encodes MHC proteins. The structures of MHC I and II were discussed in *Chapter 4* in the context of their roles in the immune response. Here their involvement in triggering rejection will be emphasized. Molecules of MHC I consist of a single polypeptide encoded by the MHC that is always associated with a smaller protein, β_2M, encoded outside the MHC. However, MHC II proteins consist of two polypeptides, α and β, both of which are encoded by the MHC region.

Figure 6.14 illustrates the structure of the HLA-region although it has been greatly simplified to aid understanding. The HLA complex contains a number of genetic loci, including those that encode different types of Class I proteins. Thus, the HLA-A, B and C regions contain genes that encode HLA-A, B and C proteins respectively. These are all found on nucleated cells and are distinct types of proteins and not allelic forms of each other. However, there *are* allelic forms of each of the *HLA-A, B* and *C* genes, and these encode HLA proteins, each of which have small differences in their amino acid sequences. Each of the Class I genes is **polyallelic**, which means that many different alleles exist although, of course, each individual only expresses a maximum of two alleles at each locus, given chromosomes occur in pairs (*Chapter 15*). Moreover, the alleles are codominant so that each nucleated cell expresses two different alleles of the *HLA-A* gene, as well as two different forms of the *HLA-B* and of the *HLA-C* gene. A large number of allelic forms of each gene exist (*Table 6.13*).

The HLA complex in humans is one of the most highly polymorphic systems known. Given that each individual has two of each of these alleles, and that the allelic forms are codominant, it can be seen that the chances of two unrelated

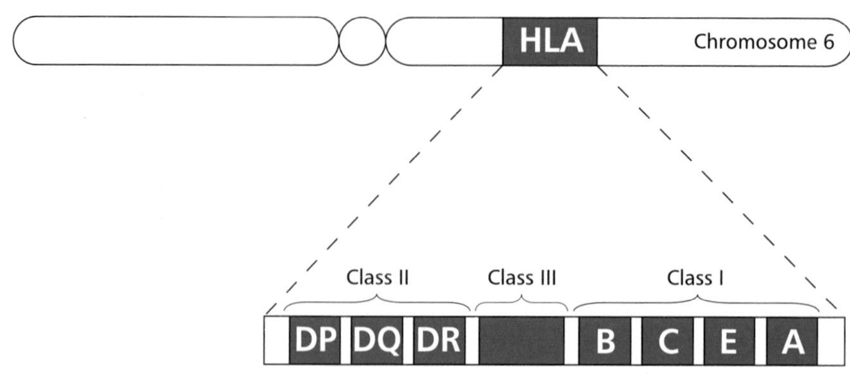

Figure 6.14 Schematic to show the structure of the HLA system on chromosome six.

individuals having the same 'set' of HLA genes is very small, making the possibility of finding 'good' transplant matches, other than identical twins, low.

In the HLA complex, the Class II region is located nearest to the centromere, which contains the DP, DQ and DR loci. Within each of these loci there are genes encoding the α and β chains of the class II molecules. The situation is rather more complex than for Class I because each locus may contain more than one gene that encodes the α and β chains. For example, the HLA-DR region contains three or four genes for the HLA-β chain. All of the β gene products may be expressed in a single cell, making the degree of variation much higher than that of class I proteins. Like Class I, the Class II region also displays a high degree of polymorphism (*Table 6.14*).

HLA TYPING

Human Leukocyte Antigen typing is the process whereby the HLA antigens present on the cells of the recipient and the potential donor are determined. By determining the HLA types of potential transplant recipients, and storing these details on computer databases it is possible to match the donor organ, when one becomes available, to the most appropriate recipient. The degree

Gene	Number of alleles	Number of proteins
HLA-A	372	348
HLA-B	661	580
HLA-C	190	153

Table 6.13 Genes and alleles of Class I genes of the HLA complex

Gene	Codes for	Number of alleles
HLA-DRB	β chain of HLA-DR	249
HLA-DQA1	α chain of HLA-DQ	20
HLA-DQB1	β chain of HLA-DQ	36
HLA-DPA1	α chain of HLA-DP	13
HLA–DPB1	β chain of HLA-DP	82

Table 6.14 Alleles of the Class II HLA genes

of match required depends on the organ being transplanted, and is much more stringent for bone marrow transplants than for solid organ transplants. The transplantation of bone marrow is used to treat a number of conditions, including immunodeficiency and some kinds of cancer (*Chapter 17*). Transplantation of bone marrow presents particular difficulties because bone marrow contains immunocompetent cells that can lead to fatal GVHD.

The traditional method for HLA typing is a serological technique that uses antibodies to known HLA antigens. This method is still used but is being superseded by molecular biological techniques. The serological method is the lymphocytotoxicity assay. In this assay, the cells to be typed are peripheral blood mononuclear cells (PBMC) that are readily obtained from whole blood. For typing Class II HLA antigens it is necessary to use purified B lymphocytes, since resting T lymphocytes do not express Class II molecules. Aliquots of the cells to be typed are pipetted into the wells of 96-well trays, known as Teresaki plates, after the inventor of this test. Antibodies to individual HLA antigens are added to each well and will bind to the lymphocytes if they express the appropriate antigen. The addition of complement to all the wells results in the death of cells with bound antibody. Viability stains, such as the fluorescent acridine orange and ethidium bromide are then used to reveal wells containing dead cells. Acridine orange enters living cells and stains the nuclei green, while ethidium bromide enters dead cells and stains the nuclei red (*Figure 6.15*). The anti-HLA antibodies used in these tests may be obtained from people already sensitized to HLA antigens or they may be monoclonal antibodies with specificities for known HLA antigens. Rabbit serum is used as a source of complement. A test is scored as strongly positive when more than 50% of the cells in a well are killed.

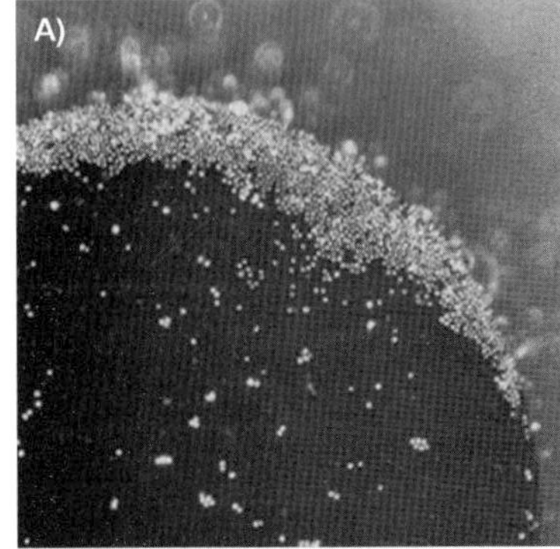

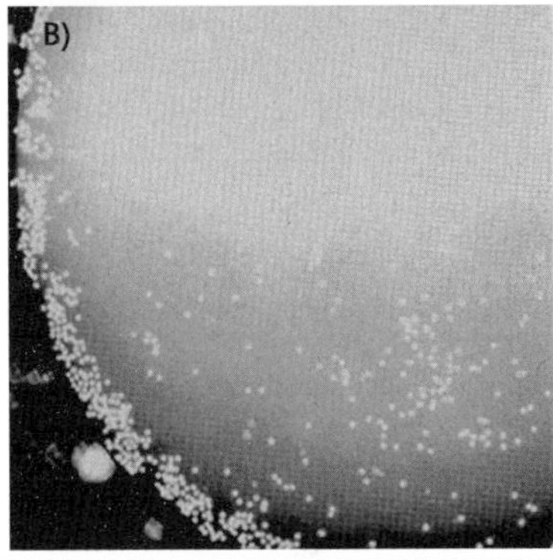

Figure 6.15 Serological test to determine HLA type. (A) A negative result, that is, no cell death and (B) a positive result in which cells have been killed as indicated by different flourescent colors. Courtesy of the Transplantation Laboratory, Manchester Royal Infirmary, UK.

Molecular biology techniques

The main molecular biological technique used in histocompatibility testing is the polymerase chain reaction (PCR; *Chapter 3*) in which samples of DNA from the individual being typed is amplified many times over. Methods that use PCR require very little DNA and do not require living cells. For example, they can be carried out on whole blood that has been stored frozen. In the sequence specific primer (SSP) assay the primers used in the PCR are those specific for individual HLA alleles. The DNA will only be amplified in those mixtures that contain a probe complementary to that of the DNA being typed (*Figure 6.16*). Another method, called a sequence specific oligonucleotide probe (SSOP) assay, amplifies all the DNA by PCR and then uses sequence specific oligonucleotide probes to identify the products. In some laboratories the oligonucleotide probes are attached to microspheres. Incubation of the microspheres with the PCR product allows the latter to bind to any complementary HLA sequence. The microspheres, with bound DNA, are then analyzed in a flow cytometer to identify target HLA sequences.

The serological cross match

A serological cross match is carried out in order to detect the presence of antibodies to graft antigens in the serum of a potential recipient. Serum from the potential recipient is added to PBMC from the donor. Complement is added and the viability of the cells tested as previously. If donor cells are killed then

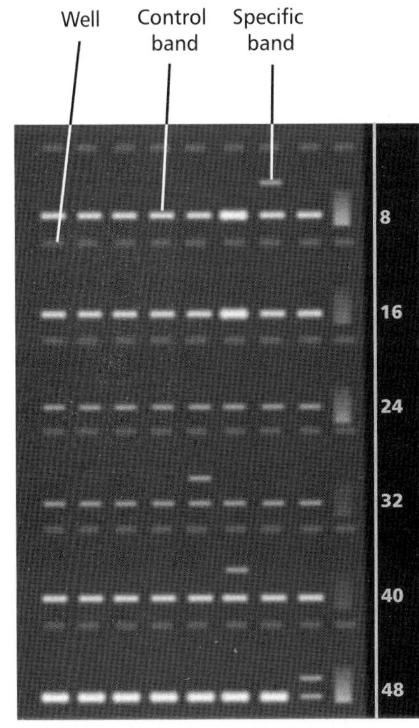

Figure 6.16 Sequence Specific Primer Assay to determine HLA type. The specific band shows that the individual being typed is positive for that specific HLA gene. The controls ensure that the individual's DNA has been amplified correctly. The numbers down the right-hand side are reference values to indicate the appropriate sizes of the bands. Courtesy of the Transplantation Laboratory, Manchester Royal Infirmary, UK.

it is probable that the recipient already has antibodies against graft antigens and the donor is unlikely to be a suitable match.

Matching donor and recipient

Individuals inherit a set of HLA genes from each parent. Thus, siblings are more likely to have a closer match than unrelated donors and, occasionally, people have donated a healthy kidney to a sibling with renal failure. Other sources of kidneys for transplantation include cadavers, often people who have been killed in accidents. The HLA types of potential recipients are stored in computer databases so that when a kidney donor becomes available, they can be HLA typed and the kidneys given to those recipients whose HLA antigens match as closely as possible.

Retrospective studies on the outcome of kidney transplantation have shown that matching the donor to the recipient improves graft survival. Thus, in the most favorable match, that is no mismatches of the HLA-DR, HLA-A and HLA-B alleles graft survival is superior to those with no mismatch of HLA-DR and only one mismatch of HLA-A and/or HLA-B alleles. These also gave significantly better graft survival than any other mismatched graft.

6.13 IMMUNOSUPPRESSION

All patients who receive an allograft are liable to reject the transplant, even if the recipient and host are closely matched for histocompatibility antigens. This is because relatively few of the HLA antigens are tested for, and complete matches are rare. Thus, all patients who receive an allograft have to take immunosuppressive drugs to prevent rejection resulting from an immune response. Immunosuppressive treatments fall into a number of categories. The first generation of immunosuppressive drugs were used to prevent lymphocytes from proliferating (*Table 6.15*). As these drugs act by inhibiting cell division, they are also used in the treatment of cancer. Their actions are described in more detail in *Chapter 17*. Corticosteroids, such as cortisol, are also immunosuppressive agents but act principally by suppressing inflammation. They are still used, often in combination with other drugs such as methotrexate. All the first generation drugs produce a 'blanket' immunosuppression and prevent all immune responses. This makes the patient more susceptible to infections of all kinds, but especially to opportunistic infections caused by such organisms as *Candida albicans*. Immunosuppressed patients are also more susceptible to the types of cancers associated with viruses, including lymphoma, associated with the Epstein-Barr virus (EBV) and Kaposi's sarcoma, associated with the Kaposi's sarcoma associated herpes virus (KSV). First generation immunosuppressive treatments also have significant toxicity, because they affect all dividing cells, including those of the bone marrow and of the GIT. Some, such as methotrexate, also show liver toxicity.

Type of treatment	Mechanisms of action	Examples
Purine analogs	incorporated into DNA during the process of DNA synthesis; prevent further DNA elongation	azathioprine; mercaptopurine
Folic acid antagonists	prevent the action of dehydrofolate reductase, an enzyme required for the synthesis of purines and pyrimidines	methotrexate; aminopterin
Alkylating agents	become incorporated into developing DNA and cross-links DNA strands, preventing further replication	cyclophosphamide

Table 6.15 First generation immunosuppressive drugs

The second generation of immunosuppressive treatments target T lymphocytes rather than B cells. These included the use of antilymphocyte globulin, an antibody raised against human T lymphocytes. The third generation of immunosuppressive treatments are much more selective in their action and target only those T cells which have been activated by an antigen. Cyclosporin A, a cyclic peptide (*Figure 6.17*) derived from the soil fungus *Tolypocladium inflatum gams*, is most commonly used. The peptide shows considerable immunosuppressive action, without the bone marrow toxicity demonstrated by other drugs. Tacrolimus, a macrolide antibiotic (*Figure 6.18*) derived from *Streptomyces tsukubaensis*, is similar to cyclosporin A in action. It is more powerful than cyclosporin A but also has more side effects. Mycophenolate mofetil is an example of a prodrug, which is converted in the body to mycophenolic acid, another powerful immunosuppressive agent.

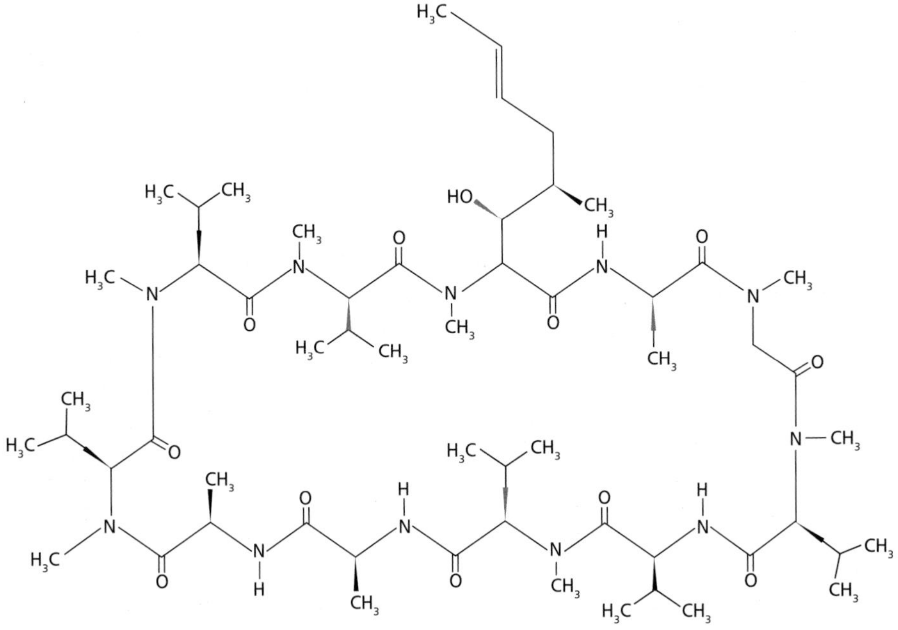

Figure 6.17 Cyclosporin A.

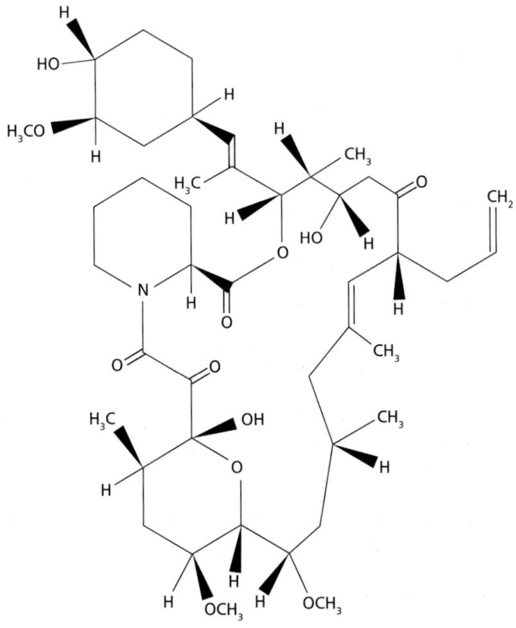

Figure 6.18 Tacrolimus.

6.14 HEMOPOIETIC STEM CELL TRANSPLANTATION

The transplantation of hemopoietic stem cells is used to correct some immunodeficiency diseases (*Chapter 5*) and during the treatment for some types of cancer, such as leukemias and lymphomas (*Chapter 17*). Patients with total bone marrow failure, such as in aplastic anemia, or partial failure, as seen in some of the hemoglobinopathies, may also benefit from a stem cell transplant (SCT).

Hemopoietic stem cells for transplantation may be derived from bone marrow, from peripheral blood, and from umbilical cord blood and SCTs may be autologous, **syngeneic** or **allogeneic**. In an **autologous** transplant the patient's own stem cells are harvested prior to a treatment in which their own stem cells are destroyed, as, for example, in patients who receive radiation treatment or high dose chemotherapy to treat leukemia. The stem cells are stored in liquid nitrogen until the transplant can take place, when they are thawed and reinfused.

A SCT between identical twins is called a syngeneic transplant. In contrast, an allogeneic SCT involves donors and recipients who are genetically nonidentical. The donor could be a sibling or a potential donor identified in a bone marrow registry (*Margin Note 6.4*). Donors who are related to the recipient are more likely to be compatible. There is a much greater need for close HLA matching of the donor and recipient involved in bone marrow transplants (BMTs), as GVHD can be a major hazard of this type of transplantation and can occur between seven and 30 days post-transplantation. In the acute form of GVHD epithelial cells in the skin and those lining the intestine are attacked by the sensitized T cells originating in the graft. Patients present with a severe maculopapular skin rash and sloughing of the intestinal epithelium may lead to diarrhea. Splenomegaly and hepatomegaly occur, as these organs become attacked by T lymphocytes and the patient may become jaundiced. Acute GVHD is frequently fatal. A chronic form with similar symptoms may occur over a longer time frame. The patients may suffer frequent secondary infections.

Margin Note 6.4 Bone marrow databases 

Potential bone marrow donors are always required to increase the number of HLA types available. The National Marrow Donor Program in the USA maintains an international registry of stem cell donors. Similarly, in the UK the two main bone marrow registries are the British Bone Marrow registry and the Anthony Nolan Bone Marrow registry.

One method of avoiding GVHD is to remove T lymphocytes from SCTs prior to giving the graft. This process, known as T cell depletion involves the use of an antibody to T cells. Once GVHD develops it must be treated with immunosuppressive drugs.

BONE MARROW TRANSPLANTATION

The first successful bone marrow transplant (BMT) was achieved in 1968. Bone marrow contains the hemopoietic stem cells that give rise to all the formed elements of the blood (*Chapter 13*). In bone marrow transplantation (BMT), the donor receives a general or local anesthetic and the marrow is harvested using a needle inserted through the skin over the pelvic bone and into the bone cavity. The process takes approximately one hour to harvest sufficient quantity; usually a minimum of 3×10^8 nucleated bone marrow cells containing approximately 2×10^6 stem cells per kilogram of recipient body weight is required. At this stage the marrow may be infused into the recipient or it may require further processing. The marrow is treated to remove blood and plasma, especially if there is a disparity between the ABO blood group of the donor and the recipient. Fragments of bone are removed and the marrow may also be depleted of T cells to reduce the risk of GVHD prior to intravenous transfusion into the recipient. The recipient may be given antibiotics because, until the marrow is engrafted and starts to produce blood cells, the patient is at risk of infection. They may also be given platelet and erythrocyte transfusions to prevent bleeding and anemia. A patient receiving a bone marrow transplant may show adverse side effects, such as nausea, fatigue, hair loss and loss of appetite.

PERIPHERAL BLOOD STEM CELL TRANSPLANTATION

Peripheral blood stem cell transplantation (PBSCT) is the most common form of SCT. Collection of stem cells from peripheral blood is easier both for the collector and the donor. In addition, engraftment of PBSCT is often more rapid than with bone marrow. The donor is treated with granulocyte colony stimulating factor (G-CSF) to increase the number of stem cells in the blood. The stem cells are then obtained from the donor by leukapheresis (*Margin Note 6.3*). Leukapheresis may take several hours to complete and more than one session may be needed, with stem cells being stored frozen between donations.

UMBILICAL CORD STEM CELL TRANSPLANTATION

Stem cells may also be obtained from umbilical cord blood, of course with the permission of the family involved. Following birth, blood derived from the baby is obtained from the umbilical cord and the placenta. Since only a small amount of blood is retrieved in this way, the collected stem cells are typically used to treat children.

IDENTIFICATION OF HEMOPOIETIC STEM CELLS

Hemopoietic stem cells have a marker protein called CD34 that may be used in their identification. Thus, the numbers of CD34 positive cells in a preparation can be assessed if a sample is stained with a fluorescent antibody to CD34. The cells can be estimated either by using a fluorescence microscope or by flow cytometry (*Box 6.1* and *Figure 6.19*). Both peripheral and cord blood may be further processed to obtain the stem cells. For example, an anti-CD34 antibody linked to magnetic particles will bind to CD34+ cells, which can then be purified using a magnet.

Storage of transplant material

Harvested stem cells may be stored for 2–3 days in a refrigerator at 4°C. This may be required, for example, if more than one harvesting procedure is needed or if a patient has to undergo radiation or chemotherapy for cancer treatment (*Chapter 17*) prior to receiving the graft. If longer term storage is required, the stem cells may be stored in liquid nitrogen vapour at –176°C. Prior to storage, the cryopreservative dimethylsulfoxide (DMSO) is added to prevent ice crystal formation, which would destroy the cells. A programmed freezer allows the cells to be cooled at the optimum rate for cell survival, which is normally approximately 1°C per min.

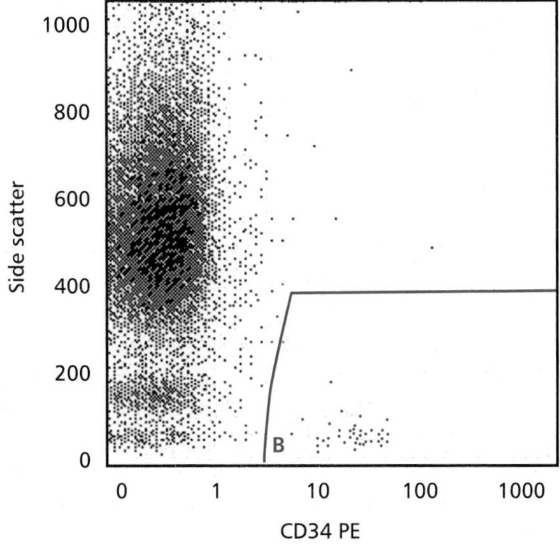

Figure 6.19 Flow cytometric characterization of CD34+ stem cells, shown in the gated Section B, in cord blood. Courtesy of Dr T.F. Carr, Royal Manchester Children's Hospital, UK.

Post-transplant support

Patients who have just undergone BMT or PBSCT transplantation are highly susceptible to infections. If transfusions of blood products are required, for example for anemia, these products should be free of CMV, a virus which is a latent infection in some individuals. In an immunosuppressed individual, CMV can cause serious illness.

CASE STUDY 6.1

Marie is a 31-year-old florist who is blood group O RhD negative and is seven months pregnant. The father of her baby is blood group A RhD positive. Marie is worried about the pregnancy, as she had several miscarriages, each of which occurred before three months of gestation. Despite these miscarriages, Marie has never become sensitized to the D and has no anti-D antibodies in her circulation.

Questions

(a) What are the chances of her baby developing hemolytic disease of the newborn?

(b) Is Marie likely to become sensitized to the D antigen after the birth of this child?

CASE STUDY 6.2

John is a 50-year-old school teacher who has recently been treated for acute myeloid leukemia. His consultant has recommended that he receives aggressive chemotherapy followed by a stem cell transplant. John is likely to die without this treatment. He has no living siblings who could donate bone marrow and both his parents are dead.

Questions

(a) What would be the best approach to treating John?

(b) Is a bone marrow transplant feasible?

CASE STUDY 6.3

Michael is a patient who was taken to the accident and emergency unit of his local hospital following a stabbing incident. Michael had lost a lot of blood and required an immediate transfusion. He was given compatible leukodepleted blood and fresh frozen plasma. However, he suffered acute respiratory distress approximately 4 h after the transfusion. He had hypotension and cyanosis and a temperature of 39.5°C. Examination of the chest showed signs of fluid in his lungs.

Question

What is the likely cause of Michael's respiratory problems?

6.15 SUMMARY

Transfusion of blood and blood products is a routine and safe clinical procedure, which rarely causes harm. Transfusion has been greatly facilitated by knowledge of the range of blood group antigens and the conditions under which antibodies to blood group antigens can cause problems. The screening of donors, for example for HIV or hepatitis viruses, has also increased the safety of the procedure. The transfusion of incompatible blood can cause the death of a recipient but this is a rare occurrence. Transfusion laboratories are involved in ensuring that compatible blood is given to patients and that all aspects of this procedure are safe. In addition, laboratories monitor the presence of antibodies in pregnant women that might cause problems for a developing fetus.

Transplantation of solid tissue has progressed considerably since the first successful kidney transplant and a wide range of tissues and organs are now transplanted routinely. The rejection of transplants is minimized by a careful matching of donor and recipient and by the administration of immunosuppressive drugs. Bone marrow and other forms of stem cell transplant carry the risk of GVHD, although this is decreased if good HLA matches are achieved and if the patients are carefully monitored post-transplant.

Questions

1. Which of the following combinations of plasma and erythrocytes would result in agglutination?
   a) Group A erythrocytes and group B plasma;
   b) Group O erythrocytes and group O plasma;
   c) Group B erythrocytes and group O plasma;
   d) Group B erythrocytes and group B plasma;
   e) Group AB erythrocytes and group A plasma.

2. Which of the following statement(s) is (are) **TRUE**?
   a) MHC II proteins are found on all nucleated cells.
   b) MHC I proteins are found on antigen presenting cells.
   c) Complement is a protein complex which lyses bacteria.
   d) The classical pathway for complement activation is initiated by IgA.
   e) An ABO incompatible transfusion may be fatal.

3. Which of the following conditions would mean that a blood donor should be removed from the register of donors?
   a) previous infection with hepatitis virus;
   b) current infection with HIV;
   c) pregnancy;
   d) donor aged 78 years;
   e) intravenous drug abuse.

4. List the possible causes of a potential transplant patient having anti-HLA antibodies in their plasma.

5. Give one reason why PBSCT is preferable to bone marrow transplantation.

FURTHER READING

Avent, ND and Reid, ME (2000) The Rh blood group system: a review. *Blood* **95:** 375–387.

Basara, N, Blau, WI, Kiehl, MG, Schmetzer, B, Bischoff, M, Kirsten, D, Günzelmann, S and Fauser, AA (2000) Mycophenolate mofetil for the prophylaxis of acute GVHD in HLA-mismatched bone marrow transplant patients. *Clin. Transplant.* **14(2):** 121–126.

Daniels, GL, Cartron, JP, Fletcher, A, Garratty, G, Henry, S, Jorgenson, J, Judd, WJ, Levene, C, Lin, M, Lomas-Francis, C, Moulds, JJ, Moulds, JM, Overbeeke, M, Reid, ME, Rouger, P, Scott, M, Sistonen, P, Smart, E, Tani, Y, Wendel, S and Zelinski, T (2003) ISBT Committee on Terminology for Red Cell Surface Antigens. Vancouver Report. *Vox Sang* **84:** 244–247.

Dyer, P and Middleton, D (eds) (1993) *Histocompatibility testing: a practical approach.* IRL Press, Oxford.

Edgar, J and David, M (2006) *Master medicine immunology: A core text with self-assessment.* Churchill Livingstone, Edinburgh.

Elmaagacli, AH, Peceny, R, Steckel, N, Trenschel, R, Ottinger, H, Gross-Wilde, H, Schaefer, UW and Beelen, DW (2003) Outcome of transplantation of highly purified blood CD34+ cells with T-cell add-back compared with unmanipulated bone marrow or peripheral blood stem cells from HLA-identical sibling donors in patients with first chronic phase myeloid leukaemia. *Blood* **101:** 446–453.

Ginns, LC, Cosimi, AB and Morris, PJ (1999) *Immunosuppression in Transplantation.* Blackwell Science, Malden, MA, USA.

Hill, B (2004) Transfusion science: Aiming for safety. *Med. Lab. World* http://www.mlwmagazine.com

Hughes-Jones, NC (2002) Historical review: Red cell agglutination: the first description by Creite (1869) and further observations made by Landois (1875) and Landsteiner (1901). *Br. J. Haematol.* **119:** 889–893.

Kjellstrom, BT (2003) Blood substitutes: where do we stand today? *J. Int. Med.* **253:** 495-497.

Kopko, M and Holland, PV (1999) Transfusion related acute lung injury. *Br. J. Haematol.* **105:** 322–329.

Lee, AH and Reid, ME (2000) ABO Blood group system: a review of molecular aspects. In *Immunohematology* **16:** 1–6 Special Millennium Issue.

Llewellyn, CA, Hewitt, PE, Knight, RSG, Amar, K, Cousens, S, Mackenzie, J and Will, RG (2004) Possible transmission of variant Creutzfeldt-Jakob disease by blood transfusion. *Lancet* **363:** 417–421.

Lomas-Francis, C and Reid, ME (2000) The Rh blood group system: the first 60 years of discovery. In *Immunohematology* **16:** 7–17 Special Millennium Issue.

Palfi, M, Berg, S, Ernerudh, J and Berlin, G (2001) A randomised controlled trial of transfusion-related acute lung injury: is plasma from multiparous blood donors dangerous? *Transfusion* **41:** 317–322.

Schwartz, HP (2003) Historical Review: Karl Landsteiner and his major contributions to haematology. *Br. J. Haematol.* **121:** 556–565.

Useful web sites:

http://www.anthonynolan.com

http://www.cyto.purdue.edu/ (an excellent website with all you need to know about flow cytometry)

http://www.shot-uk.org

http://www.transfusionguidelines.org.

chapter 7:

DISORDERS OF THE ENDOCRINE SYSTEM

OBJECTIVES

After studying this chapter you should be able to:

- list the different types of hormones and give examples;
- outline the structure and roles of the endocrine system;
- describe general mechanisms underlying hormone actions;
- explain how disorders of the endocrine system occur;
- discuss the causes, investigation and management of some endocrine disorders.

7.1 INTRODUCTION

The endocrine system is one of two major control systems in the body, the other being the nervous system, that help control the activities of the body. It consists of a number of ductless glands (*Figure 7.1*) that produce hormones. Hormones are molecules that circulate in the blood and excite or inhibit the metabolic activity of target tissues or organs. These responses maintain and regulate body functions, such as growth and development, responses to stress and injury, reproduction, homeostasis and energy metabolism (*Figure 7.2*).

Hormones can be divided into three chemical groups: amines, peptides and proteins, and steroids (*Table 7.1*). Many amine hormones, such as adrenaline (epinephrine) and those produced by the thyroid gland, are derivatives of tyrosine. The majority of hormones are peptides and proteins, examples being insulin and growth hormone. A number of protein hormones, for example thyroid stimulating hormone, are glycoproteins in that they have carbohydrate groups covalently attached to them. All steroid hormones are derivatives of cholesterol and include cortisol and testosterone. *Figure 7.3* shows examples of each type of hormones.

Class	Examples	Associated glands
Amines	adrenaline noradrenaline thyroid hormones T_3 and T_4	adrenal medulla thyroid gland
Peptides and proteins	insulin, growth hormone	islets of Langerhans anterior pituitary
Steroids	cortisol aldosterone testosterone	adrenal cortex

Table 7.1 A structural classification of hormones with selected examples

Figure 7.1 The major endocrine glands of the body.

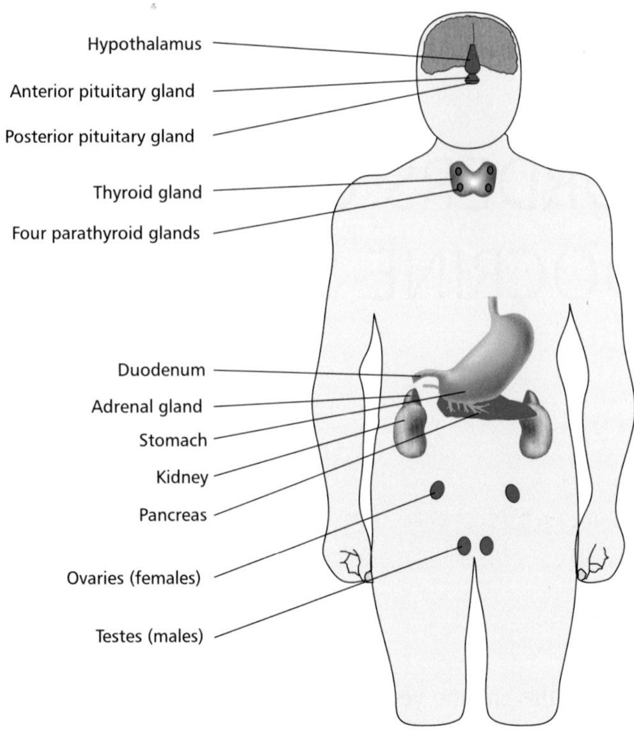

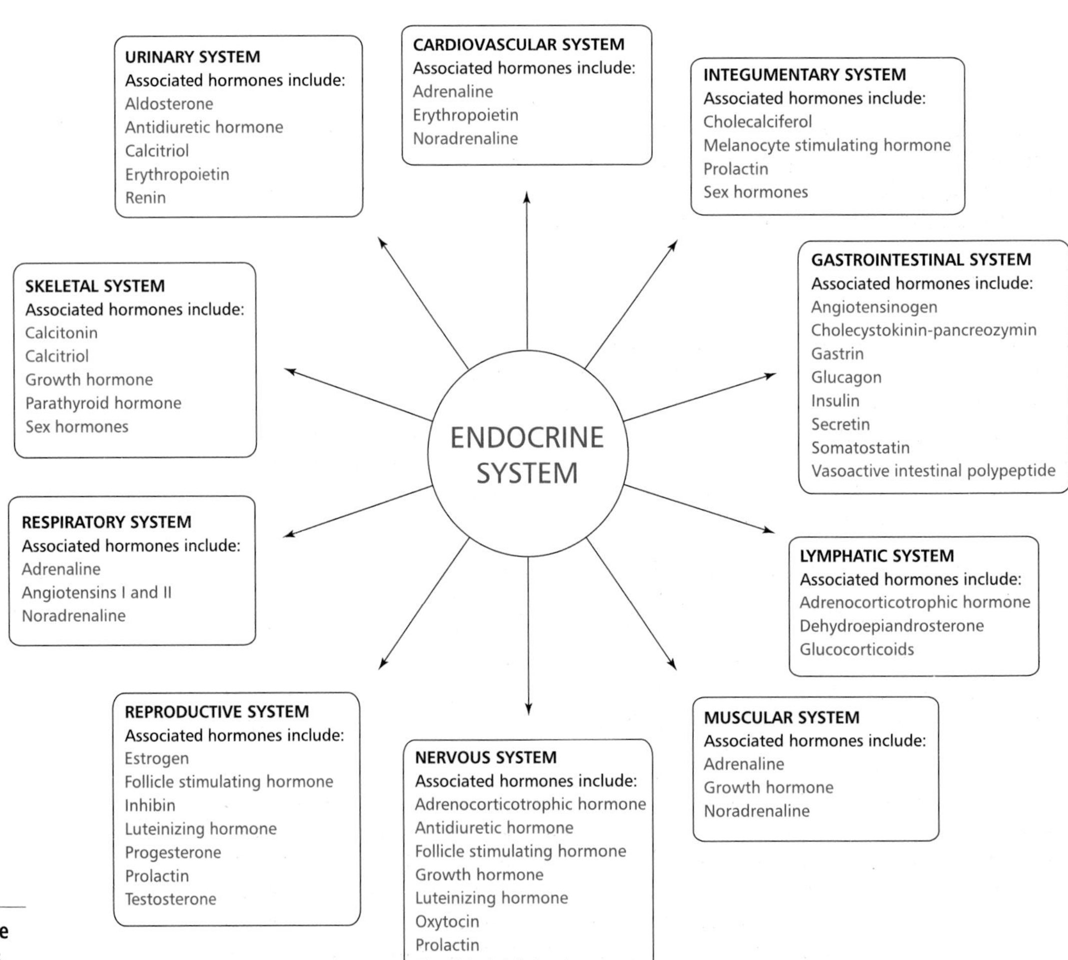

URINARY SYSTEM
Associated hormones include:
Aldosterone
Antidiuretic hormone
Calcitriol
Erythropoietin
Renin
CARDIOVASCULAR SYSTEM
Associated hormones include:
Adrenaline
Erythropoietin
Noradrenaline
INTEGUMENTARY SYSTEM
Associated hormones include:
Cholecalciferol
Melanocyte stimulating hormone
Prolactin
Sex hormones
SKELETAL SYSTEM
Associated hormones include:
Calcitonin
Calcitriol
Growth hormone
Parathyroid hormone
Sex hormones
GASTROINTESTINAL SYSTEM
Associated hormones include:
Angiotensinogen
Cholecystokinin-pancreozymin
Gastrin
Glucagon
Insulin
Secretin
Somatostatin
Vasoactive intestinal polypeptide
ENDOCRINE SYSTEM
RESPIRATORY SYSTEM
Associated hormones include:
Adrenaline
Angiotensins I and II
Noradrenaline
LYMPHATIC SYSTEM
Associated hormones include:
Adrenocorticotrophic hormone
Dehydroepiandrosterone
Glucocorticoids
REPRODUCTIVE SYSTEM
Associated hormones include:
Estrogen
Follicle stimulating hormone
Inhibin
Luteinizing hormone
Progesterone
Prolactin
Testosterone
MUSCULAR SYSTEM
Associated hormones include:
Adrenaline
Growth hormone
Noradrenaline
NERVOUS SYSTEM
Associated hormones include:
Adrenocorticotrophic hormone
Antidiuretic hormone
Follicle stimulating hormone
Growth hormone
Luteinizing hormone
Oxytocin
Prolactin
Thyroid stimulating hormone

Figure 7.2 Effects of endocrine glands on other body systems.

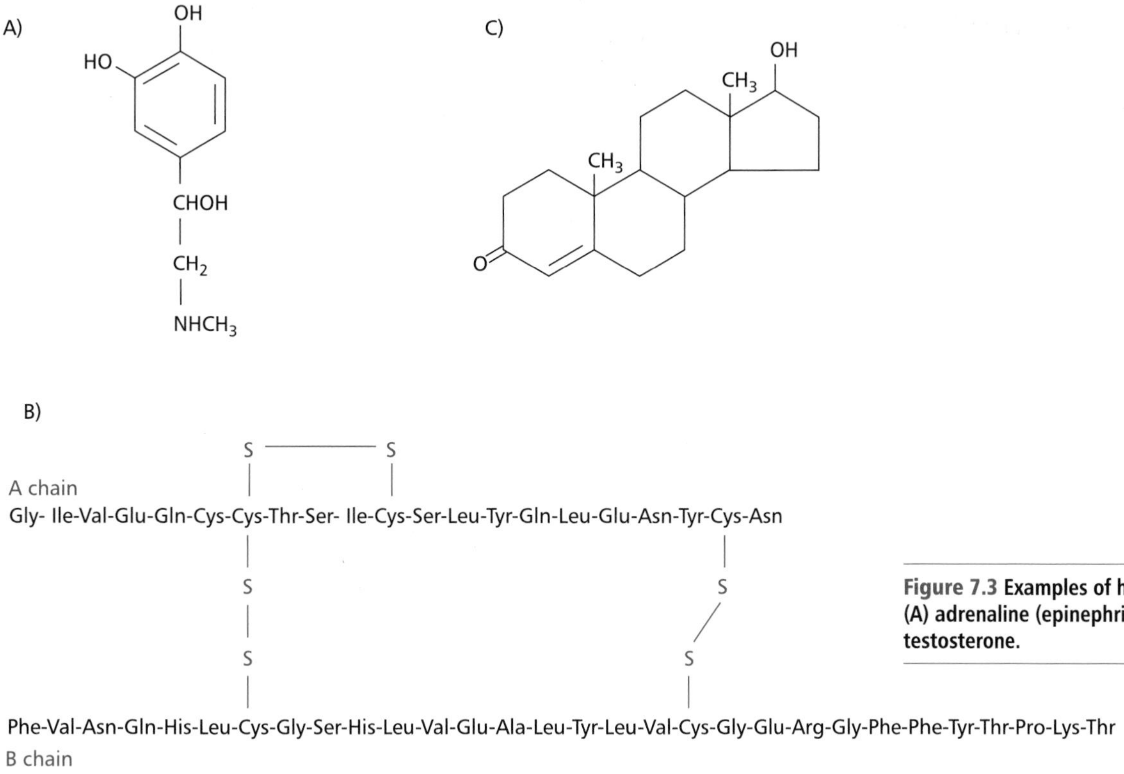

A)

OH
HO
CHOH
CH_2
$NHCH_3$

C)

OH
CH_3
CH_3
O

B)

A chain
Gly- Ile-Val-Glu-Gln-Cys-Cys-Thr-Ser- Ile-Cys-Ser-Leu-Tyr-Gln-Leu-Glu-Asn-Tyr-Cys-Asn

Phe-Val-Asn-Gln-His-Leu-Cys-Gly-Ser-His-Leu-Val-Glu-Ala-Leu-Tyr-Leu-Val-Cys-Gly-Glu-Arg-Gly-Phe-Phe-Tyr-Thr-Pro-Lys-Thr
B chain

Figure 7.3 Examples of hormones. (A) adrenaline (epinephrine), (B) insulin and (C) testosterone.

7.2 HORMONE PRODUCTION

Amine hormones include adrenaline, noradrenaline (also called epinephrine and norepinephrine respectively) and thyroid hormones. Their synthesis involves a series of enzyme-catalyzed reactions in the cytoplasm of endocrine cells. For example, thyroid hormones (*Section 7.7*) are synthesized by iodination of tyrosine residues in the protein thyroglobulin found in the thyroid. Most peptide hormones are synthesized as large inactive prohormones, which are subsequently cleaved by enzymes to produce the active hormone. Sometimes a number of hormones may be derived from the same prohormone. Steroid hormones are synthesized by a sequence of enzymatic reactions using cholesterol as a common precursor. The enzymes responsible for conversion of cholesterol to hormone are located in the smooth endoplasmic reticulum and mitochondria of cells. The presence or absence of particular enzymes determines the type of steroid hormone synthesized by that specific cell.

Amine hormones, such as adrenaline and noradrenaline, are stored in secretory granules within the cytoplasm, but thyroid hormones are stored within the thyroid follicles as components of thyroglobulin. Peptide hormones are usually stored in membrane-bound vesicles within the cytoplasm of the endocrine cell. Steroid hormones are not usually stored but are released upon synthesis. However, lipid droplets may be found in the cytoplasm containing precursor material for these hormones.

Hormones are released in response to nervous, hormonal or metabolic stimuli. Hormones stored in granules are released when the granules move and fuse with the plasma membrane. Some hormones, for example thyroxine, are released continuously whereas others show diurnal variation and their release varies during the day. For example, cortisol shows diurnal variation with levels being high in the morning but low at night. The concentrations of hormones in the plasma must be kept within narrow ranges for optimum function. A

BOX 7.1 The pineal gland

The pineal gland or body is an appendage of the posterior end of the roof of the third ventricle in the brain (*Figure 7.4*). It is shaped like a pinecone, hence its name, weighs 100 to 150 mg and is only 8 to 10 mm long. It is composed largely of pinealocytes and glial cells. In older animals, the pineal often contains calcium deposits, sometimes referred to as brain sand! Descartes (1596–1650) regarded the pineal gland as 'the seat of the soul', because he believed it was unique in the human brain in being the only structure not duplicated on the right and left sides. This observation is not strictly true, however, since it is finely divided into two hemispheres that can be observed with a microscope.

The physiological function of the pineal gland in humans is poorly understood. However, pinealocytes are known to produce the hormone, melatonin (5-methoxy-*N*-acetyltryptamine), from tryptophan by acetylating serotonin (*Figure 7.5*). Melatonin has a relatively simple structure yet communicates information about environmental lighting to physiological systems of the body. This light-transducing ability has led some to call the pineal gland the 'third eye'. Melatonin, therefore, helps regulate the internal 'clock' of the body affecting biological rhythms and, in particular, patterns of sleeping and waking and possibly the onset of puberty. It may also function as a free radical scavenger and reduce oxidative damage. Most melatonin is synthesized during the night and can be secreted directly into the blood because the capillaries of the pineal gland are permeable and do not form part of the normal blood–brain barrier (*Figure 3.4*). In a normal environment in healthy humans the release of melatonin usually starts at 21.00 to 22.00 h and ends between 07.00 to 09.00 h, with peak levels of 60 to 70 pg cm^{-3} occurring in the plasma of adults between 02.00 and 04.00 h. The activity of the usual rate-limiting enzyme, serotonin-*N*-acetyltransferase (NAT), increases 7–150 fold during peak production. In daylight the activity of NAT, and therefore melatonin production, is reduced. The rhythm is endogenously synchronized to 24 h by the suprachiasmatic nucleus of the hypothalamus, an area of the brain known to coordinate biological clock signals, but can be entrained to values between 23 and 27 h primarily by the environmental light–dark cycle acting through the retina. When the retina is exposed to light, nervous impulses are relayed to the suprachiasmatic nucleus. Nerve fibers from the hypothalamus descend to the spinal cord, from where postganglionic fibers ascend back to the pineal gland. Thus the pineal gland can measure day length and adjust its secretion of melatonin appropriately.

Small cysts are commonly seen within the pineal region although their discovery is frequently incidental to radiographic investigations. However, they are benign and nonprogressive and should be treated conservatively. The pineal gland is also subject to numerous types of malignant tumors, for example teratomas, germinomas, choriocarcinomas, endodermal sinus tumors, mixed germ cell tumors, pineoblastoma and pineocytoma, and gliomas. Fortunately, all are rare and collectively account for less than 1% of intracranial space-occupying lesions. With the exception of parenchymal cell tumors, pineoblastoma and pineocytoma, they occur mainly in patients below the age of 20 years. Germinomas and teratomas occur predominantly in males. The commonest symptoms are secondary to hydrocephalus, such as headaches, vomiting, and drowsiness, with visual problems, diabetes insipidus, and reproductive abnormalities. In children, pineal tumors of the region are often associated with abnormal pubertal development. Some evidence suggests that precocious puberty is due to the production of human chorionic gonadotrophin (hCG) by germ cell tumors of the pineal gland. Delayed puberty has also been associated with pineal tumors.

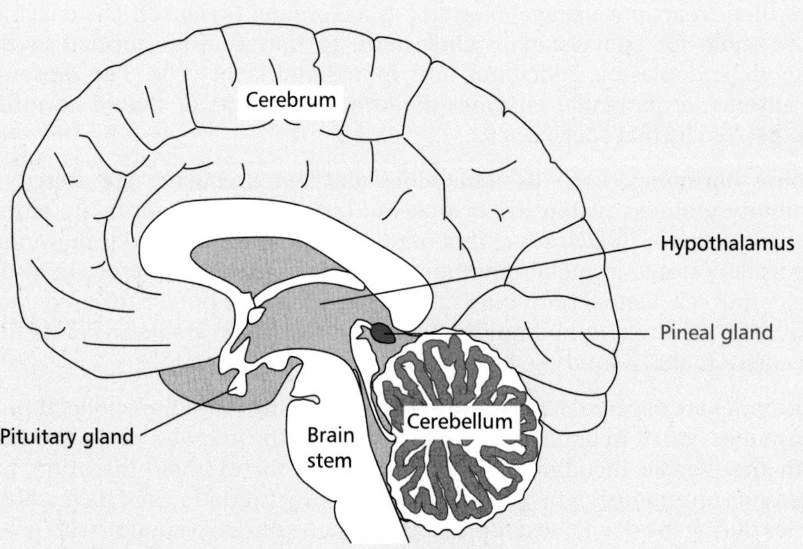

Figure 7.4 Schematic showing the location and structure of the pineal gland.

BOX 7.1 continued

Tryptophan

Tryptophan hydroxylase

5-Hydroxytryptophan (5-HTP)

5-HTP decarboxylase

Serotonin

Serotonin-*N*-acetyltransferase (NAT)

N-Acetylserotonin

Hydroxyindole-*O*-methyl transferase (HIOMT)

Melatonin

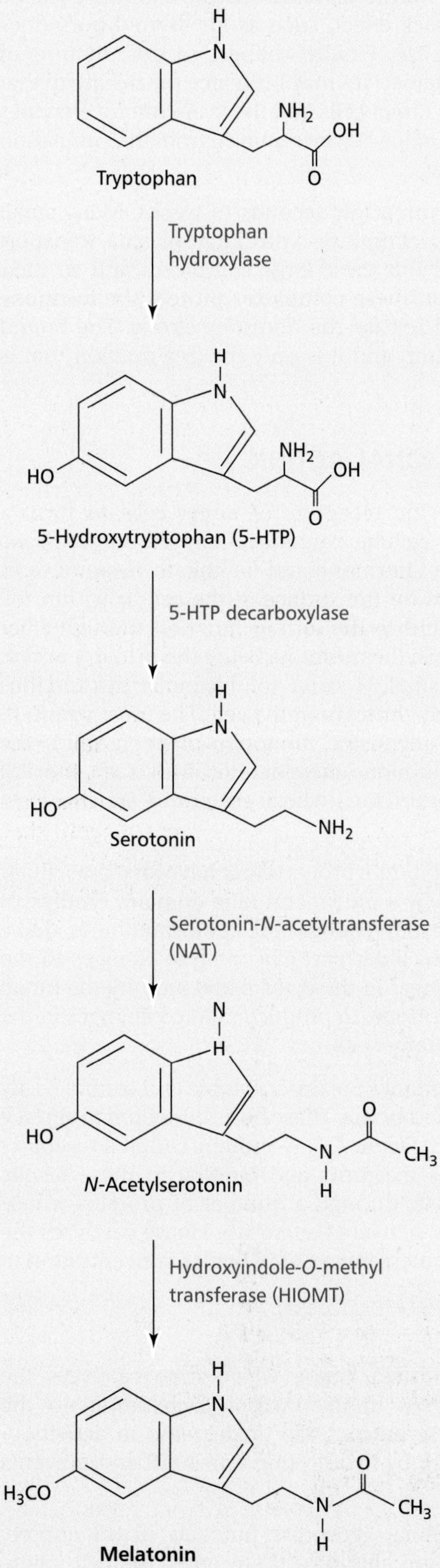

Figure 7.5 The synthesis of melatonin.

Pineal tumors can be diagnosed from the symptoms, physical examination and tests for tumor markers in samples of cerebrospinal fluid (CSF), such as α fetoprotein and hCG, together with CSF cytology. Computer aided tomography (CAT) or magnetic resonance imaging (MRI) (*Chapter 18*) can aid in differential diagnosis. Germinomas respond well to radiation therapy, whereas surgery is usually the choice for other types. There have been significant improvements in prognosis, with 62% surviving germinomas for over five years, although survival is only 14% for other malignant tumors.

Given that melatonin affects normal sleep patterns, there is interest in using it, possibly combined with **phototherapy**, to treat sleep disorders. Phototherapy, more properly called full spectrum bright light therapy, is a repeated exposure to 3 to 6 h of artificial light that is as bright as sunlight. The use of melatonin in shift workers, who often find it difficult to adjust to working at night and sleeping during the day, unfortunately has not shown promise. Melatonin therapy does not appear to help their condition and appears not to be as effective as phototherapy. However, melatonin therapy appears to be modestly beneficial in elderly insomniacs who have lower concentrations compared with matched noninsomniacs. Jet lag also involves the disruption of circadian rhythms where melatonin therapy seems to be of use. During long distance air travel, taking melatonin close to the bedtime of the destination appears to reduce the symptoms associated with jet lag. The greatest benefits occur when jet lag would be expected to be greatest, that is, on journeys that cross many time zones.

The production and secretion of melatonin is related to the length of the night: the longer the night, the more produced over a longer secretion time. Also, melatonin profiles show a seasonal phase, with an earlier secretion in summer than in winter. The condition **seasonally affective disorder (SAD)** or **winter depression** is characterized by changes of moods and eating and sleeping patterns. It appears to develop in people living at high latitudes in winter when sunlight is lacking. The increased release of melatonin in winter has been suggested as a possible cause of SAD but at present there is no consensus as to its causes. The assumption that melatonin duration is a seasonal signal in humans led to the treatment of SAD with bright light in an attempt to induce advancement of the melatonin rhythm. It appears to be somewhat more efficient when given in the morning, although there is a large placebo effect. However, other mechanisms are also possible; indeed, many pharmacological antidepressant treatments stimulate melatonin secretion.

number of factors control hormone production by the endocrine glands. The secretion of pituitary hormones is under the influence of peptides released from the hypothalamus and, this in turn, is influenced by signals from the central nervous system (CNS). Most hormones released from endocrine glands are controlled by a negative feedback effect, such as for thyroid hormones (*Section 7.7*) and cortisol (*Section 7.9*). Finally, changes in the amounts of materials regulated by hormones themselves may influence the release of that hormone, as is the case for insulin. Target cells and the liver contain enzymes that degrade hormones. Hormones of low M_r are removed from the circulation by the kidneys and excreted in urine.

The half-lives of hormones vary from a few seconds to weeks. Many small or water insoluble hormones form complexes with large plasma transport proteins. The kidneys cannot filter out these large complexes and so their rapid loss is prevented. In addition, these complexes protect the hormone from degradation by enzymes and release the hormone slowly. The bound and free hormones are in equilibrium and it is only the free fraction that is biologically active.

7.3 MECHANISMS OF HORMONAL ACTION

Hormones act by binding to specific receptors of target cells to form a complex (*Figure 7.6*) that elicits a cellular response. Only the target tissue will express the receptor for a given hormone and be able to respond to it. Hormone receptors may be located on the surface of the cell or within the cell respectively. Hormones that bind to the former function through what are called second messenger systems, the hormone being the primary or first messenger. Second messengers are small M_r water-soluble molecules and ions that are generally able to move freely throughout the cell. The most common secondary messengers are cyclic adenosine monophosphate (cAMP), the structurally related cyclic guanosine monophosphate (cGMP), Ca^{2+}, inositol triphosphate (IP_3), and diacylglycerol (DAG) whose structures are shown in *Figure 7.7*.

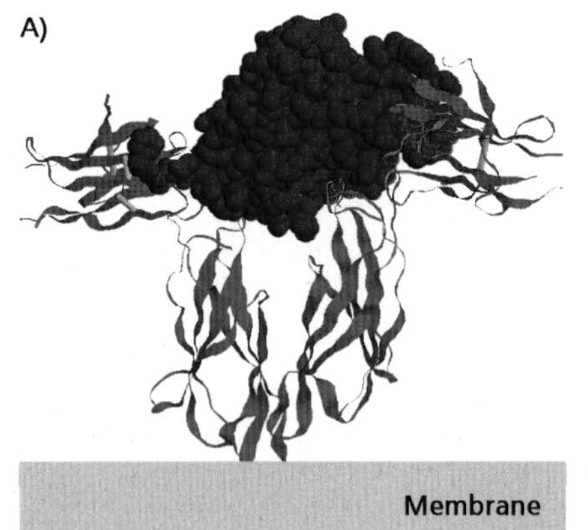

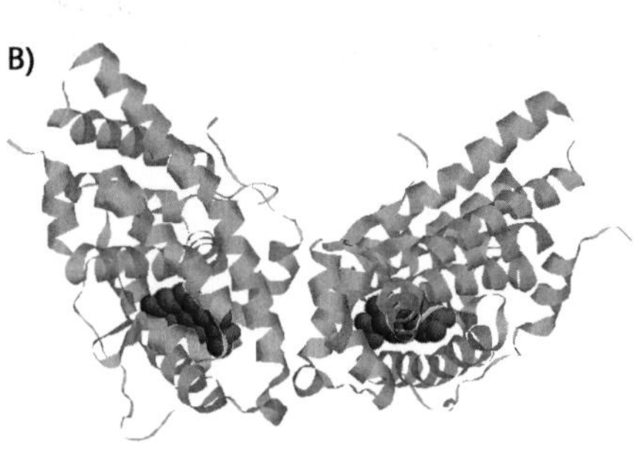

Figure 7.6 (A) Molecular model of human growth hormone (red) bound to the extracellular portion of its receptor (black). PDB file 3HHR. The gray bar represents the surface of the target cell. (B) Molecular model of a dimer of the steroid hormone progesterone (red) bound to its intracellular receptor. PDB file 1A28.

Extracellular receptors are transmembrane proteins that have an extracellular and an intracellular portion joined by a transmembrane domain. Binding of the hormone to the extracellular portion changes the conformation or shape of the complex, such that the intracellular part can catalyze changes to the concentration of the second messenger in the cytosol and amplify the initial signal, that is the binding of the hormone, to produce marked changes in the activities of existing proteins in the target cell.

Amine and peptide and protein hormones are water soluble and cannot easily cross the lipid layer of the plasma membrane. These hormones bind to surface receptors on the plasma membrane (*Figure 7.8*). G-protein-coupled receptors are the most common cell surface receptors and binding to these results in the activation of adenylate kinase, through a number of proteins whose conformations are changed in turn. Activated adenylate kinase catalyzes the conversion of ATP to the secondary messenger cAMP, whose concentration in the cytoplasm therefore increases.

$$ATP + H_2O \longrightarrow cAMP + PP_i$$

Cyclic AMP, in turn, stimulates protein kinase which then catalyzes the phosphorylation of specific enzymes in the cytosol. Depending on the enzyme, phosphorylation can cause an increase or decrease in activity. A phosphodiesterase inactivates cAMP by hydrolyzing it to AMP and prevents its accumulation in the cytoplasm.

Hormones that recognize intracellular receptors function in an entirely different fashion. Such hormones are able directly to enter the cell where

A)

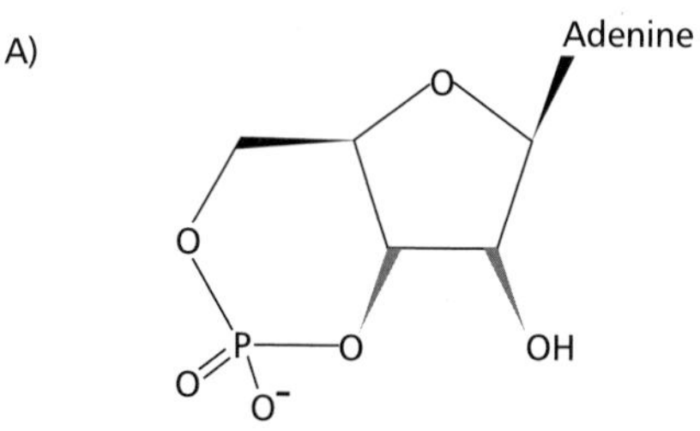

B)

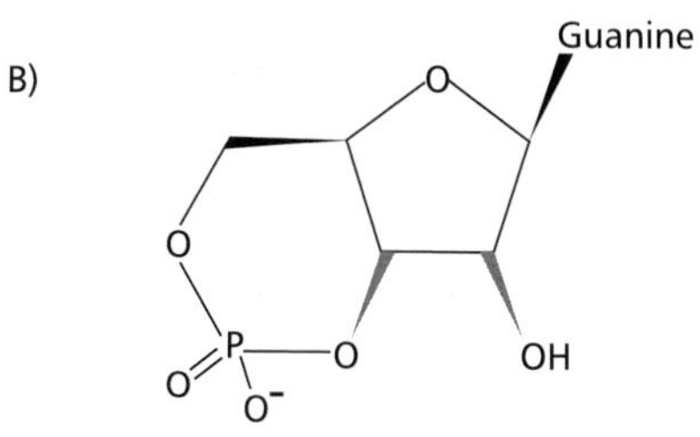

C)

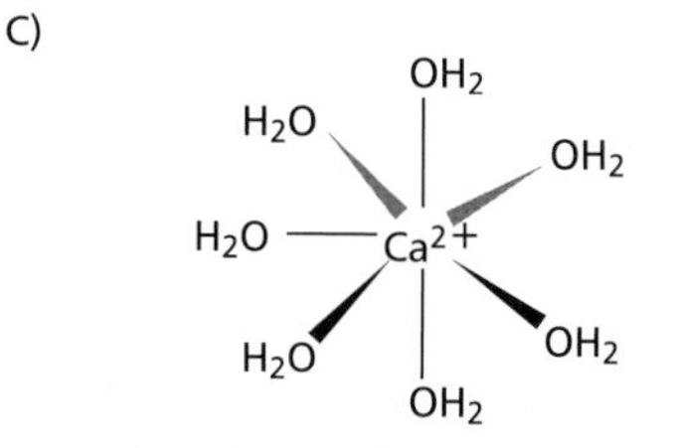

D)

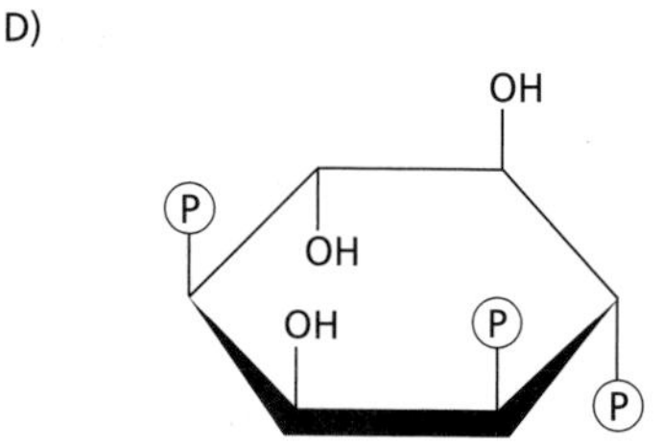

E)

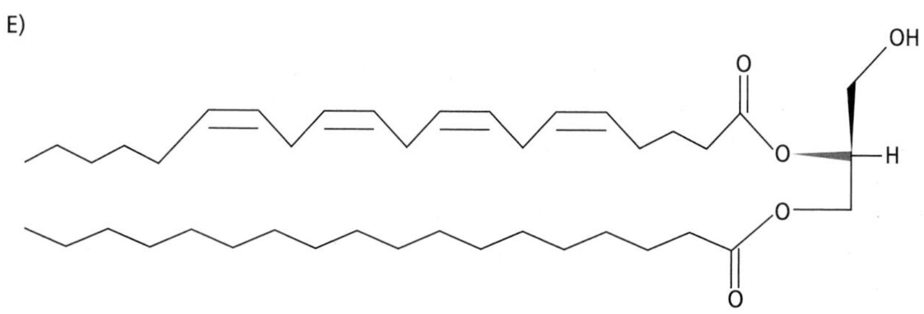

Figure 7.7 Examples of hormone secondary messengers. (A) cAMP, (B) cGMP, (C) a hydrated Ca^{2+}, (D) inositol triphosphate (IP_3) and (E) a diacylglycerol (DAG).

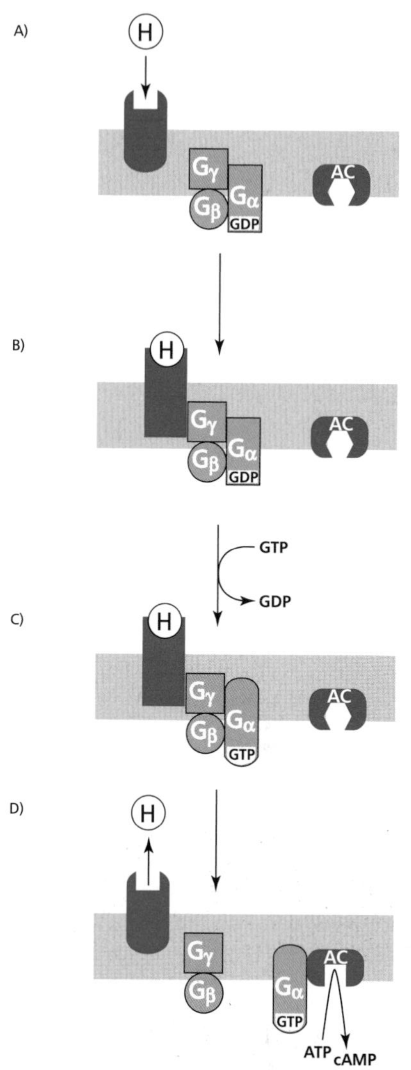

Figure 7.8 Mechanism of action of a hormone that acts through the second messenger, cAMP. Note the conformational changes that are shown schematically as (A) the hormone binds to its receptor, (B) and (C) to the trimeric G protein as GDP and GTP are exchanged and to the (D) adenylate cyclase (AC) as it is activated by the Gα-GTP complex.

Margin Note 7.1 G proteins

G proteins are so called because they bind the guanyl nucleotides, GDP and GTP. They are a diverse group, ranging from small soluble monomers to large multisubunit membrane proteins. Their functions are also wide ranging, from the activities associated with numerous hormones, to sense perception and the nucleocytoplasmic transport described in *Chapter 16*. However, all are GTPases that can hydrolyze their bound GTP to GDP

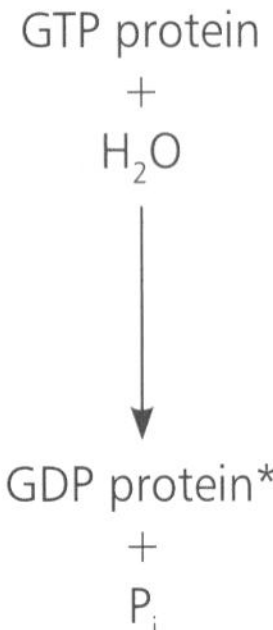

The GDP portion can be exchanged for a GTP. This alteration of the nucleotide between the GTP–GDP forms allows the protein to switch between two conformations (*Figure 7.9*). The form with a bound GTP is active (functional); the other has GDP bound and is nonactive. Thus G proteins provide on–off switches in physiological systems as they oscillate between the two forms.

*indicates a change in the conformation of the protein (see also *Figure 7.9*)

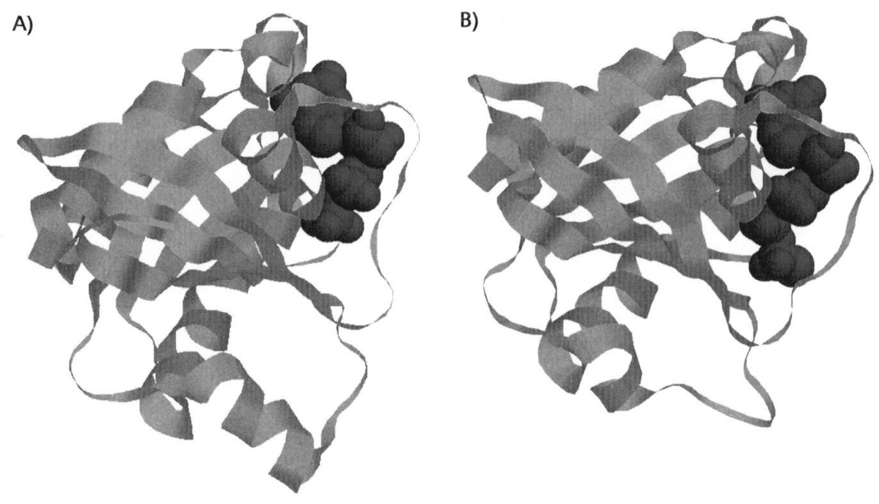

Figure 7.9 A small soluble G protein with bound (A) GDP and (B) GTP. In both cases the nucleotide is shown in red. Note the difference in conformation between the two forms. PDB files 1OIV and 1OIW respectively.

they regulate the synthesis of proteins. For example, steroid hormones are hydrophobic and lipid soluble and so diffuse directly through the plasma membrane into the cytoplasm of target cells. Thyroid hormones enter their target cells by facilitated diffusion. In the cytoplasm, steroid and thyroid hormones bind to intracellular receptors forming hormone–receptor complexes (*Figure 7.10*). The complex then interacts with the DNA of specific genes in the nucleus to switch their transcription on or off, allowing or preventing the production of appropriate mRNA molecules respectively. Thus the cell's production of proteins, such as enzymes, is regulated to produce a physiological response to the hormone.

7.4 CAUSES OF ENDOCRINE DISORDERS

Endocrine disorders arise because a disruption to the endocrine system causes decreased (hypofunction) or increased (hyperfunction) hormonal activity or resistance to hormone action. There may be defects in synthesis of the hormone due to an inherited deficiency of an enzyme required for its synthesis. Inappropriate stimuli may impair the release of the hormone or certain drugs may stimulate hormone release. Defects in the negative feedback mechanism may cause abnormal hormone secretion. Faulty inactivation or excretion of hormones in liver or renal diseases respectively can increase hormone levels. Excessive hormone secretion can occur ectopically from a nonendocrine source, such as a tumor. Even if correctly synthesized and released, the target tissues may not recognize the hormone because of a lack of receptors or because the receptors themselves are nonfunctional. Disorders will also occur if the target cells do recognize the hormone but there is a defect in the secondary messenger system responsible for converting the hormonal signal to a physiological action.

In some autoimmune diseases (*Chapter 5*), antibodies are produced that stimulate or destroy endocrine glands, as in Grave's disease and autoimmune thyroiditis respectively. The various causes underlying endocrine disorders are summarized in *Figure 7.11*. A considerable number of endocrine disorders have been described including disorders associated with the pituitary, thyroid, pancreas, adrenal glands and the reproductive systems.

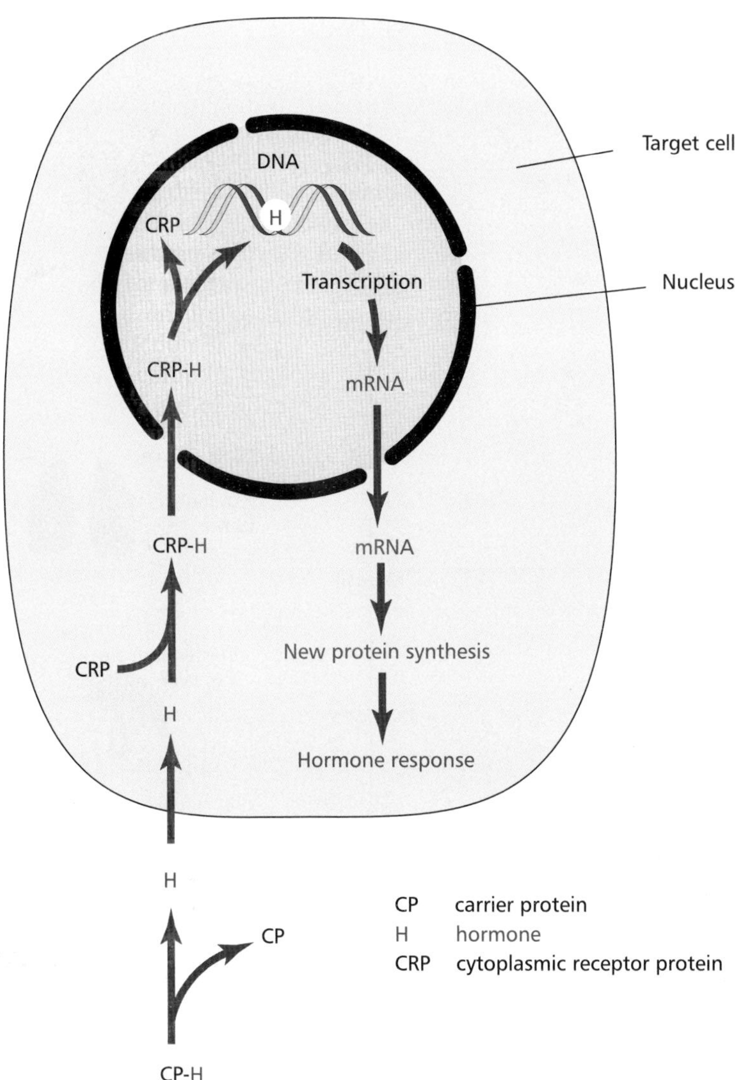

Figure 7.10 Mechanism of action of a hormone that binds to an intracellular receptor. See text for details.

7.5 THE PITUITARY GLAND

The pituitary gland is often referred to as 'the master gland' given that its secretions regulate the activities of many of the other hormone-producing glands. Despite its crucial role, it weighs only about 0.5 g. The pituitary is found in a bony cavity at the base of the skull and is connected to the hypothalamus by a pituitary stalk composed of blood vessels and nerve fibers. It is composed of two lobes, an anterior pituitary or adenohypophysis and a posterior pituitary or neurohypophysis (*Figure 7.12*).

The anterior pituitary secretes a number of hormones (*Figure 7.13*) that are regulated by the release of peptides from the hypothalamus with stimulatory or inhibitory effects on the anterior pituitary. The principal peptides released by the hypothalamus reach the anterior pituitary by a portal blood circulation. These peptides include thyrotrophin releasing hormone (TRH), growth hormone releasing hormone (GHRH), gonadotrophin releasing hormone (GnRH), corticotrophin releasing hormone (CRH) and dopamine. With the exception of dopamine, all are stimulatory, controlling the release of thyroid stimulating hormone (TSH), growth hormone (GH), follicle stimulating hormone (FSH) together with luteinizing hormone (LH) and adrenocorticotrophic hormone (ACTH) respectively. Thyroid stimulating hormone targets the thyroid gland causing it to release thyroid hormones.

Figure 7.11 Overview of the causes of endocrine disorders.

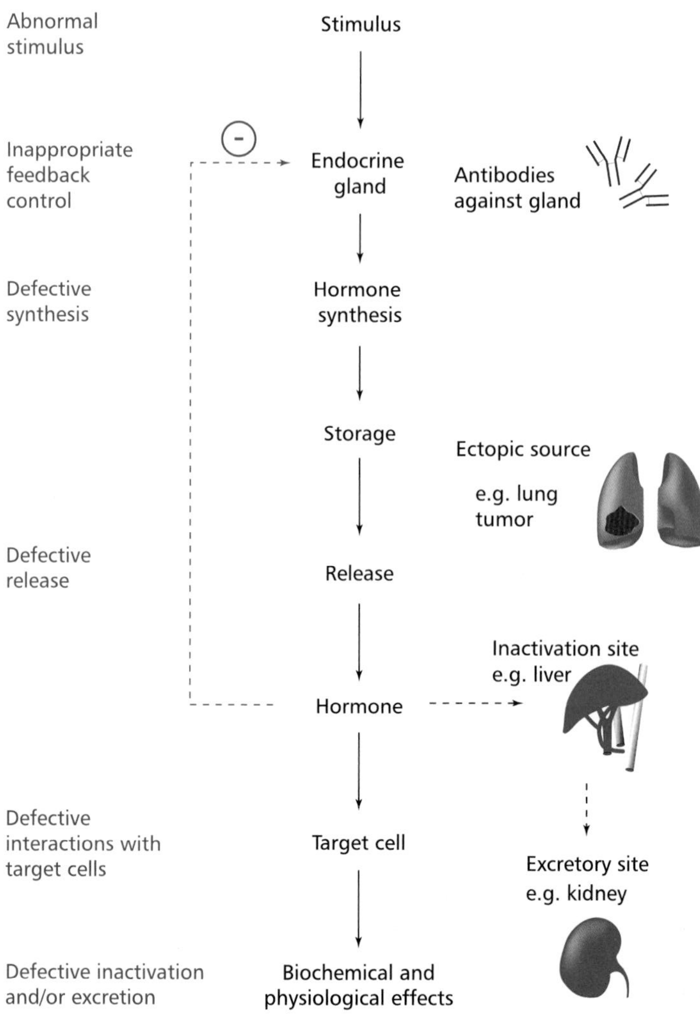

Figure 7.12 Structure of the pituitary gland, with the secreted hormones indicated.

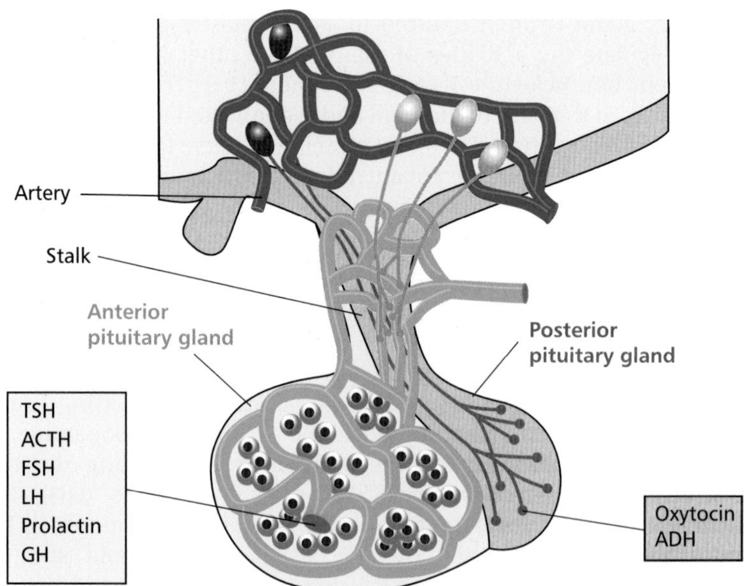

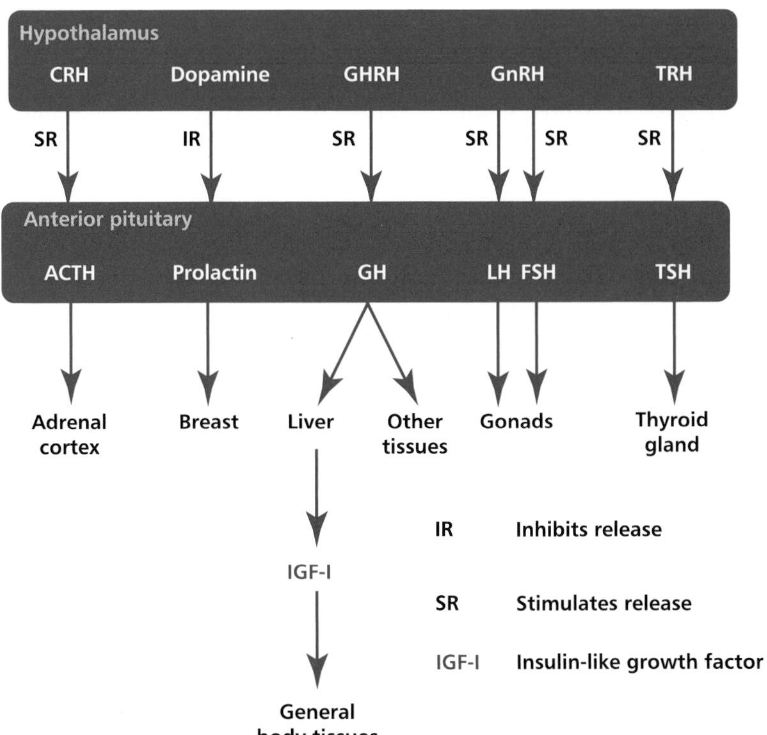

Figure 7.13 Hypothalamic factors that regulate anterior pituitary function.

Growth hormone acts on general body tissues to promote growth and development. Follicle stimulating hormone and LH, collectively referred to as gonadotrophins, act on the testes and ovaries and are essential for reproduction. Adrenocorticotrophic hormone acts on the adrenal cortex and stimulates release of cortisol. Prolactin stimulates the female mammary glands to control lactation. The release of prolactin is inhibited by dopamine.

The posterior pituitary is composed of a collection of nerve fibers originating in the hypothalamus. It secretes two main hormones, antidiuretic hormone (ADH) or vasopressin, and oxytocin although both are made in the hypothalamus and reach the posterior pituitary in the nerve fibers of the stalk. The former stimulates the kidneys to conserve water (*Chapter 8*), whereas the latter promotes uterine contractions during childbirth and stimulates the release of milk in breast feeding.

DISORDERS OF PITUITARY FUNCTION

The majority of disorders in pituitary function are caused by tumors of the gland although some pituitary diseases can lead to underproduction of its hormones. Approximately 80% of pituitary tumors are functional, that is they cause an excessive secretion of hormones. The most common secrete prolactin, and account for 50% of functional tumors. The incidence of tumors affecting other hormones is 15% GH, 10% ACTH, 4% FSH/LH with less than 1% promoting TSH secretion. Large pituitary tumors may also exert pressure on nerves, causing headaches and visual disturbances. Radiological investigations such as X-rays, computer-aided tomography (CAT) scans and magnetic resonance imaging (MRI), outlined in *Chapter 18*, are important in locating tumors and estimating their sizes.

Prolactin secreting tumors, prolactinomas, cause hyperprolactinemia that, in turn, can lead to infertility in both males and females. Nonprolactin secreting tumors or pituitary stalk section by surgery, which block the dopaminergic inhibition of prolactin secretion, also result in hyperprolactinemia. Hyperprolactinemia abolishes menstruation (**amenorrhea**) and causes an inappropriate release of breast milk (**galactorrhea**) and impotence and breast development (**gynecomastia**) in males. Other causes of hyperprolactinemia include drugs, for example phenothiazines, that block dopamine receptors, or methyldopa that reduces the level of dopamine in the brain.

Investigating a patient with a possible prolactinoma includes assessing the concentrations of prolactin in the plasma (*Table 7.2*) following stimulation with TRH, although this test is not commonly used in most hospitals. In addition to TRH, prolactin is also secreted in response to stress and estrogens. Patients with a prolactinoma have plasma prolactin concentrations in excess of 2000 mU dm^{-3}. These high values are generally not affected by TRH stimulation in individuals with prolactinomas. The first line of treatment is with a dopamine antagonist, such as bromocriptine, although surgical removal of the tumor may be necessary in cases which do not respond to drug therapy.

Disorders of oxytocin are uncommon and have little clinical significance. However, ADH release is essential for life and disorders of its release are well recognized. The release of ADH is stimulated by increased osmolality of the plasma and a decrease in blood volume detected by hypothalamic osmoreceptors and cardiac baroreceptors respectively. The role of ADH in fluid regulation is outlined in *Chapter 8*. A decreased output of ADH gives rise to diabetes insipidus, characterized by excessive production of dilute urine (polyuria). Patients are constantly thirsty (**polydipsia**), have hypernatremia and a plasma osmolality in excess of 295 mmol kg^{-1}.

[Prolactin] / mU dm^{-3}	Interpretation
Males < 381	normal
Females < 629	normal
Males 381–1000	repeat test; does not usually indicate a serious problem
Females 629–1000	repeat test; does not usually indicate a serious problem
1000–2000	repeat test; the increased [prolactin] may be secondary to stress, drug use, hypothalamic disorders, acromegaly, primary hyperthyroidism or chronic renal failure
2000–4000	possible microprolactinoma or a hypothalamic disorder
4000–6000	prolactinoma probably present, although the possibilities of a hypothalamic disorder or pregnancy should be investigated
> 6000	virtually always indicates the presence of a macroprolactinoma

Table 7.2 Interpretation of serum prolactin concentrations in the diagnosis of hyperprolactinemia

Cranial diabetes insipidus may be caused by brain tumors, meningitis, trauma and following surgery, but can be idiopathic. Nephrogenic diabetes insipidus occurs when the kidneys fail to respond to ADH. The lack of response may be caused by drugs, such as lithium, chronic renal disease, hypercalcemia or it may be congenital. Patients with suspected diabetes insipidus are investigated by performing a fluid deprivation test in which the patient is deprived of fluid intake for a period of 8 h. In normal individuals, this results in concentrated urine with a plasma osmolality below 295 mmol kg^{-1}. However, in patients with diabetes insipidus, the urine does not become concentrated and the plasma osmolality increases. At the end of the 8 h period, the patient is allowed to drink water and given desmopressin, a synthetic analog of ADH, after which the urine is collected hourly for a further 4 h. In cranial diabetes insipidus the urine becomes concentrated, but with nephrogenic diabetes insipidus this does not occur as the kidneys are insensitive to ADH (or desmopressin in this case). Hence the test discriminates between cranial and nephrogenic diabetes insipidus.

Patients suffering from diabetes insipidus require access to rehydrating fluids but, in each case, the underlying cause must be treated. Patients with cranial diabetes insipidus are often given desmopressin in a nasal spray or chloropropamide which increases renal sensitivity to ADH. The use of the latter drug requires careful monitoring since it can lead to hypoglycemia. Individuals with nephrogenic diabetes insipidus do not respond to analogs of ADH and often there is no suitable treatment. Especial attention to adequate water intake is essential.

A number of patients present with hypopituitarism, which is a failure to secrete one or more pituitary hormones, although this is a relatively uncommon complaint. Hypopituitarism may result from a tumor, infarction, infections, trauma affecting the pituitary or may be secondary to disorders of the hypothalamus. The clinical presentation of hypopituitarism often depends on the age of the affected individual. A decreased release of GH is often an early feature, leading to dwarfism in children (*Section 7.6*). Inadequate secretion of gonadotrophins may cause amenorrhea (see above) and infertility in adult females, and loss of secondary sexual characteristics in males. Elderly patients with hypopituitarism may complain of symptoms, such as hypoglycemia and hypothermia, relating to ACTH and TSH deficiencies respectively. In most cases, GH and gonadotrophin deficiencies tend to present before that of ACTH. Hyposecretion is assessed by stimulatory tests where the ability of the anterior pituitary to secrete the hormone in question is assessed following stimulation of the patient with the hypothalamic peptide or its analog. A failure to respond would suggest hypopituitarism.

7.6 GROWTH HORMONE DISORDERS

Growth hormone (GH) or somatotrophin promotes linear growth and the maintenance of tissues by stimulating the uptake of amino acids by cells, protein synthesis, increasing blood glucose concentration and fat metabolism and promoting epiphyseal bone growth. These effects are mediated by locally acting effectors called somatomedins that are synthesized by many tissues but particularly liver. Somatomedins stimulate cell proliferation and/or differentiation. They include insulin-like growth factors-I and II (IGF-I and IGF-II). Insulin-like growth factor-I is the most significant physiologically and, indeed, its concentration correlates with that of GH.

Growth hormone is a polypeptide 191 amino acid residues long (*Figure 7.14 (A)*). Approximately 70% of plasma GH is bound to growth hormone binding protein. Growth hormone is synthesized in the anterior pituitary gland in

A)

B)

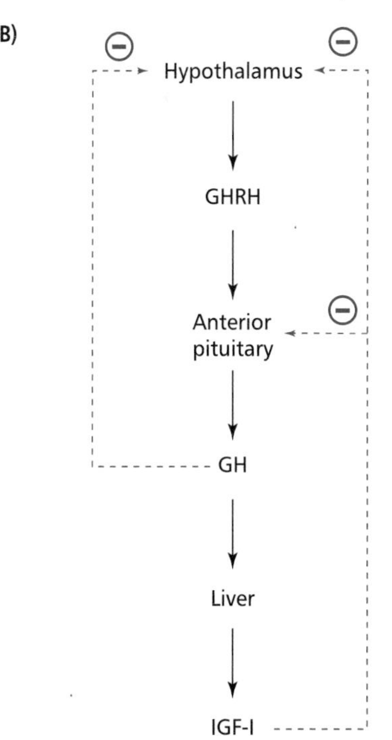

Figure 7.14 (A) Structure of human growth hormone. PDB file 1HGU. (B) The regulation of the secretion of GH.

an inactive form called preprogrowth hormone that is hydrolyzed in several enzyme-catalyzed reactions to give active GH prior to its secretion. A number of factors, such as sleep, amino acids, exercise and stress stimulate GHRH release from the hypothalamus and that, in turn, stimulates GH secretion. Hyperglycemia stimulates the secretion of somatostatin from the hypothalamus and this inhibits the secretion of GH. Increasing concentrations of serum GH and IGF-I exert a negative feedback effect that prevents further release of GH (*Figure 7.14 (B)*). The rate of adult secretion varies but is generally about 1.4 mg daily and occurs in pulses with the largest amounts being released during sleep.

The clinical features resulting from an excess or deficiency of GH depend on the age of the person. A deficiency during childhood leads to a stunted growth called dwarfism (*Figure 7.15 (A)*) and therefore requires early detection. However, GH deficiency is a rare cause of dwarfism and other causes, for example thyroid deficiency or inadequate nutrition, need to be excluded first. The commonest causes of GH deficiency are nonendocrine tumors that affect the pituitary gland or hypothalamus. Growth hormone deficiency may also be a consequence of generalized pituitary disease or a congenital defect leading to a deficient production of GHRH. Whatever the cause, the major clinical feature is stunted growth, well below that expected for a child of comparable age, with a short height, immature face and skeleton as revealed by radiological investigations. Clinical signs of other anterior pituitary hormone deficiencies may be evident.

The most common cause of excessive GH release is a GH secreting pituitary tumor or adenoma (*Chapter 17*). Although these are benign, they are not subject to normal control and continually release large amounts of GH. The

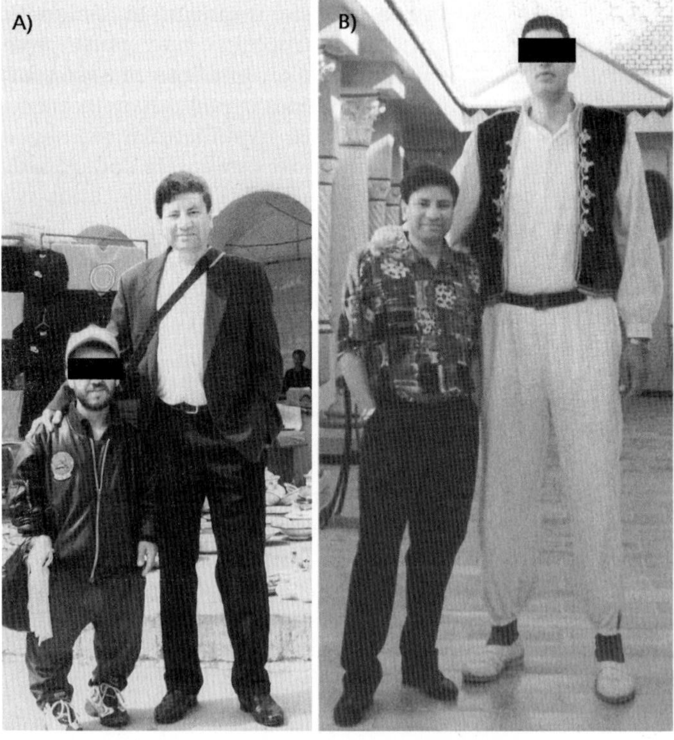

Figure 7.15 Photographs showing one of the authors beside (A) a dwarf and (B) a person with combined gigantism and acromegaly.

causes of these tumors are unknown but a genetic basis has been suggested. Ectopic GH secretion is extremely rare but has been reported in patients with bronchial carcinoma.

Excessive GH release causes gigantism in children and acromegaly in adults. Children with GH excess grow as much as 6 inches per year to abnormal heights, often in excess of 8 feet. Muscle weakness is seen in longstanding cases. Acromegaly has an insidious onset and may take years for its clinical features, enlargement of bones of hands, feet and face, thickening of soft tissues causing coarse facial features, enlarged tongue and lips, prognathism or protruding jaw, increased sweating and enlargement of internal organs, such as liver, spleen and heart, to become apparent. Additionally, acromegalics suffer from paresthesia of the hands and feet due to entrapment of nerves by thickened bone and subcutaneous tissue, headaches/vision disorders due to the growing pituitary tumor and sensory nerve entrapment, impaired glucose tolerance or diabetes mellitus and increased incidences of coronary heart disease and stroke (*Chapter 14*). Individuals affected by excessive GH secretion throughout life show features of both gigantism and acromegaly (*Figure 7.15 (B)*). The prognosis for both gigantism and acromegaly depends on how far the disorder has advanced. Gigantism is rarely life threatening and prognosis is usually good. However, an individual with advanced acromegaly will develop serious complications, such as coronary heart disease, cerebrovascular disease and diabetes mellitus.

DIAGNOSIS AND TREATMENT OF GROWTH HORMONE DISORDERS

A variety of tests are used to assess GH deficiency. Random measurements of serum GH are of limited value due to fluctuations in plasma GH levels in normal individuals. Urinary GH excretion is low in deficiency but obtaining an accurately timed collection of urine is difficult. Most tests rely on demonstrating that the hormone does not increase in concentration following a stimulus. Growth hormone increases after exercise and this has been used as a preliminary screening test. In the exercise test, the patient is subjected to hard physical exercise until they have a pulse rate greater than 150 beats per min (*Chapter 14*). Blood is collected at 0, 2 and 20 min after stopping exercise. In normal individuals, the plasma concentration of GH increases by 20 mU dm^{-3} above the initial value. Growth hormone release increases during sleep, hence high values in a nocturnal sample may exclude deficiency. Blood samples are collected using a venous catheter at 30 min intervals for 3 to 4 h after the onset of sleep. A peak of at least 10 mU dm^{-3} occurs in normal individuals but not in patients with GH deficiency. Clonidine is a potent stimulator of GH secretion and is used in a definitive test for GH deficiency. Growth hormone from genetically engineered sources is used in treatment but must be continued until longitudinal growth is completed. In cases where deficiency is due to low levels of GHRH, analogs of this peptide, for example hexarelin, have been used.

A diagnosis of excess GH is made on clinical grounds supported by biochemical and radiological investigations. Photographs taken of the patient when younger are particularly useful when making a diagnosis. Basal serum concentrations of GH are increased in GH excess but, because release is influenced by so many factors, the result of a single sample is unreliable.

The concentration of IGF-I in serum is raised in patients with acromegaly and is of diagnostic significance. The oral glucose tolerance test (OGTT) is used to confirm a diagnosis of acromegaly and is similar to that used for diagnosis of diabetes mellitus (*Section 7.8*) except that levels of plasma GH are also determined. In a healthy individual, the glucose load suppresses GH release to below 2 mU dm^{-3} by stimulating the release of somatostatin. In patients with acromegaly, this suppression is not seen.

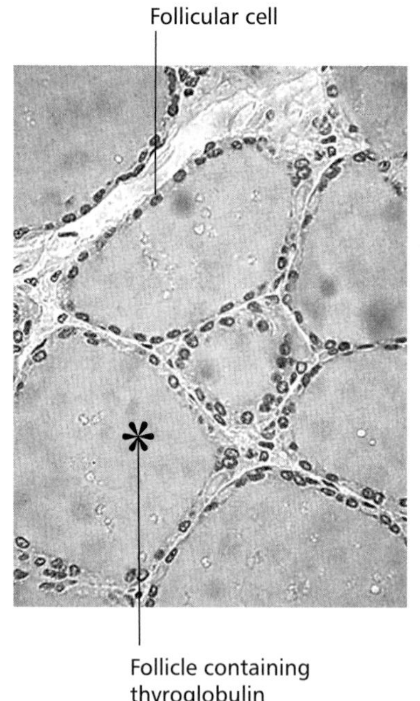

Figure 7.16 The histological structure of the thyroid gland. Courtesy of Dr A.L. Bell, University of New England, College of Osteopathic Medicine, USA.

The management of GH excess cannot reverse any clinical changes that have taken place prior to treatment. However, management is important because it improves survival and reduces deaths due to heart disease or stroke. Growth hormone secreting tumors are treated in one of three ways, the clinical decision depending on severity of disease, the age of the patient and the response to treatment. Surgical removal of the pituitary tumor may be attempted depending on its size and location with steps taken to minimize damage to the anterior pituitary. If GH levels do not normalize, other treatments must be considered. Radiotherapy is usually performed over a 4–6 week period by external irradiation using a cobalt source. Its effects are slow and, in some cases, hyposecretion of other pituitary hormones may occur. Drugs, such as bromocriptine, suppress GH release in acromegalics and are often used in patients too old to undergo surgery or radiotherapy. Somatostatin analogs, for example octeotride, have also been used to inhibit GH release. Some patients require all three modes of treatment but are not always successfully treated. To detect recurrence, serum IGF-I needs to be monitored at regular intervals.

7.7 THYROID HORMONE DISORDERS

The thyroid gland is the largest endocrine gland in the body, weighing about 20 g. It is a bilobular organ that consists of microscopic spherical follicles with secretory cells (*Figure 7.16*) that synthesize the thyroid hormones. Thyroid hormones (*Figure 7.17 (A)* and *(B)*) consist of triiodothyronine (T_3) and thyroxine (T_4). Triiodothyronine is the most active, four times that of T_4, and has a half-life of 1.5 days compared with 9 days for T_4. However in most tissues, particular the liver, T_4 can be readily converted to T_3. More than 99% of T_3 and T_4 are transported in the serum as complexes: 70% to thyroxine binding globulin (TBG) and about 20% to albumin and around 10% to prealbumin. The remaining small free portion is the metabolically active fraction.

Figure 7.17 The major thyroid hormones (A) T_3, (B) T_4 and (C) the regulation of the secretion of T_4 and T_3 by a negative feedback mechanism.

A)

I, I, I; HO–(ring)–O–(ring)–CH_2—CH(—NH_2)—COOH

B)

I, I, I, I; HO–(ring)–O–(ring)–CH_2—CH(—NH_2)—COOH

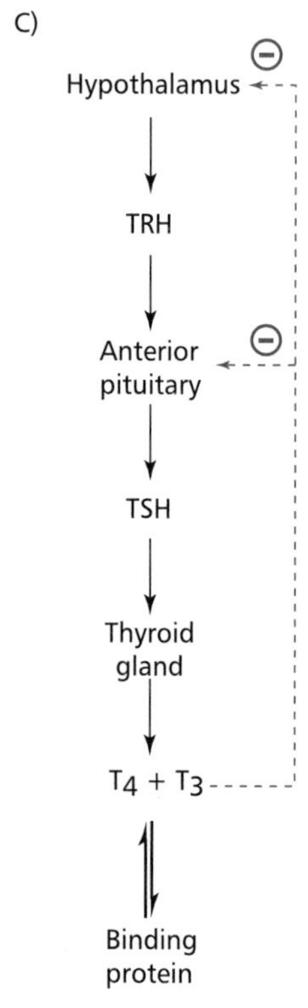

The effects of thyroid hormones are to increase heat production, oxygen consumption, the metabolism of proteins, fats and carbohydrates and to promote normal growth. They are also necessary for the normal functioning of the CNS. Levels of plasma T_4 and T_3 are regulated by the release of thyroid stimulating hormone (TSH) from the anterior pituitary that, in turn, is controlled by the release of thyrotrophin releasing hormone (TRH) from the hypothalamus. Increasing concentrations of free T_4 and T_3 inhibit further release of TSH and TRH by a negative feedback effect (*Figure 7.17 (C)*).

The most serious disorder of thyroid function is hyperthyroidism caused by an excessive production of thyroid hormones. The clinical syndrome resulting from hyperthyroidism is thyrotoxicosis (*Figure 7.18*). Its clinical features are weight loss, sweating, heat intolerance, anxiety, hyperkinesis, increased appetite, osteoporosis, menorrhagia, tachycardia and pretibial edema.

The commonest cause of hyperthyroidism is Grave's disease, which can occur at any age but particularly in 20- to 40-year-old females. Patients with Grave's disease suffer from **exophthalmos** or protrusion of the eyeballs, in addition to clinical features of hyperthyroidism. It is an autoimmune disease (*Chapter 5*), characterized by the presence of thyroid stimulating antibodies in the blood that bind to TSH receptors in thyroid cells and stimulate them in a similar manner to TSH. Toxic nodules are the second main cause of hyperthyroidism and tend to be found in elderly patients. They may occur singly or as multiples in a nodular goiter and are autonomous (self-governing) secretors of thyroid hormones. An excessive intake of thyroxine in individuals who are treated for hypothyroidism can cause hyperthyroidism. Rare causes of hyperthyroidism include ectopic thyroid tissue and tumors that secrete TSH although the latter are very uncommon.

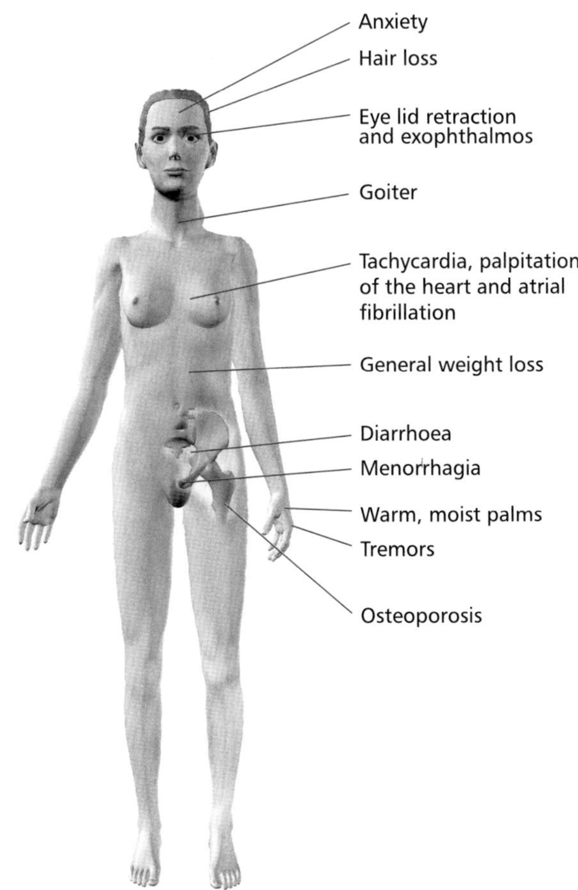

Figure 7.18 **Schematic to show the clinical features of thyrotoxicosis.**

Hypothyroidism (myxodema) is most common in women of 30–60 years of age (*Figure 7.19*). Its clinical features include psychosis, diminished sweating, hypokinesis, weight gain, muscle weakness, constipation, dry, cold skin and dry hair, ischemic heart disease, bradycardia and menstrual irregularities. The primary cause of hypothyroidism is a defect in the secretion of thyroid hormones by the thyroid gland. In, for example, Hashimoto's thyroiditis, an autoimmune disease, there is destruction of thyroid tissue by antibodies (*Chapters 4* and *5*) produced against the thyroid. Hypothyroidism may also occur after surgery (postthyroidectomy) and following treatment with antithyroid drugs. Congenital hypothyroidism occurs because of the failure of the thyroid gland to develop normally during embryonic growth. If untreated, this condition results in **cretinism** where the child suffers from mental retardation, muscle weakness, short stature, neurological signs and is often dumb and mute. Hypothyroidism may also arise due to iodine deficiency in certain parts of the world. Secondary causes of hypothyroidism are linked to the pituitary or the hypothalamus with defective secretions of TSH and TRH respectively.

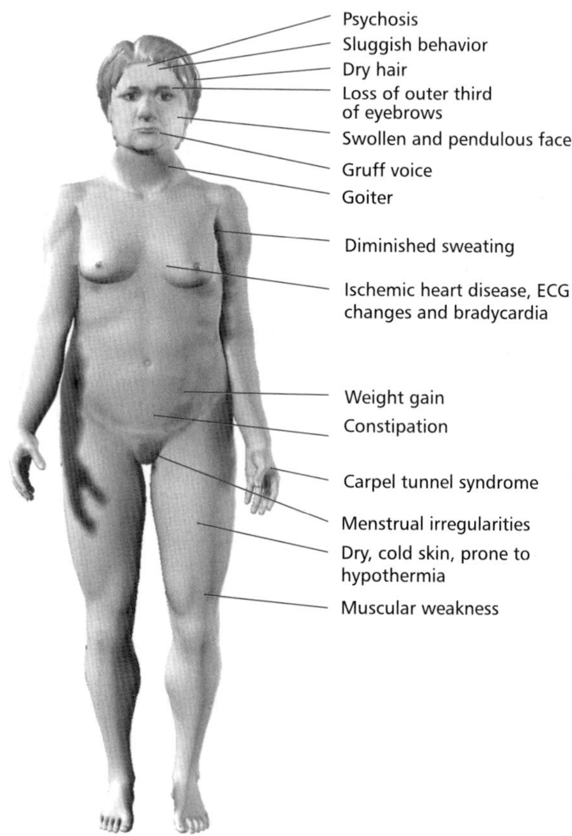

Figure 7.19 Schematic to show the clinical features of myxodema.

DIAGNOSIS AND TREATMENT OF THYROID DISORDERS

Investigating thyroid abnormalities involves measurements of serum TSH as a first line test for thyroid function. Some laboratories also include measurement of free T_4 and/or free or total T_3 in their first line screen. Sensitive assays for measurement of TSH are readily available. Thyroid stimulating hormone levels are increased in primary hypothyroidism, normal in euthyroid individuals (those without thyroid disease) and low or undetectable in hyperthyroidism.

Total T_4 and T_3 measurements were once used widely in pathology laboratories but have the disadvantage that their values are dependent on plasma TBG levels, which can give misleading results. When TBG levels increase, for example in pregnancy or in females receiving estrogen-containing oral contraceptives, then total T_4 and total T_3 are also increased even though the individual is not hyperthyroid. Decreases in TBG concentrations occur in malnutrition, protein loss, severe illness and malabsorption (*Chapters 10* and *11*) and causes a reduction in total T_4 and T_3. There has been some controversy on the validity of free thyroid hormone measurements, but most laboratories now determine free T_4 and T_3, rather than the total concentration. Free T_4 and particularly free T_3 are usually increased in hyperthyroidism. Free T_4 is low in hypothyroidism and is the preferred measurement for its detection because free T_3 can be normal in hypothyroidism due to an increase in its peripheral formation from T_4. In a few patients with hyperthyroidism, free T_4 is within the reference range but free T_3 is increased and TSH is nearly always undetectable. This form of hyperthyroidism is referred to as T_3 toxicosis.

In any systemic illness, such as myocardial infarction, fever or liver disease, the normal metabolism of thyroid hormones is disturbed, reducing the concentrations of T_4 and T_3 in the plasma because T_4 is converted to an inactive isomer called reverse T_3 or rT_3 (*Figure 7.20*) and T_3 is not replenished from T_4. Thyroid stimulating hormone levels may be normal or reduced and concentrations of TBG, albumin and prealbumin may also decline. Patients may have reduced T_4, T_3 and TSH, although there is no thyroid dysfunction. For this reason, thyroid function tests should not be performed on sick patients until they recover. *Table 7.3* outlines the results of tests used in thyroid disorders. The TRH test is rarely used now in the diagnosis of thyroid disease. It is almost exclusively used in the diagnosis of patients with pituitary disease and to assess the capacity of the pituitary to secrete TSH. The patient is given 200 μg of TRH intravenously and the serum TSH is measured after 0, 20 and 60 min. A normal response involves a three- to fivefold increase in TSH above the basal level. A slow rise in TSH (where the 60-min concentration is greater than the one at 20 min) together with low basal levels of TSH and thyroid hormones suggests hypothalamic disease, while a lack of response is suggestive of pituitary hypothyroidism or hyperthyroidism (*Figure 7.21*).

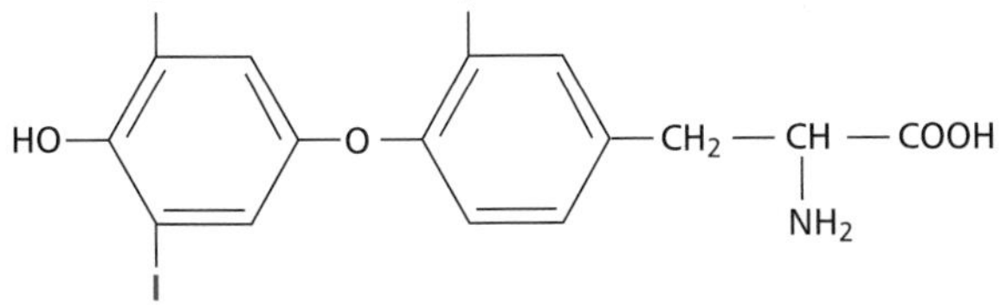

Figure 7.20 **Reverse T_3 (rT_3). Compare its structure with those shown in *Figure 7.17*.**

Test	Hyperthyroidism	Hypothyroidism	Developing hypothyroidism	T_3 toxicosis	Nonthyroidal illness
TSH	decreased	increased	increased	increased	decreased or normal
Free T_4	increased	decreased	normal	normal	decreased
Free T_3	considerable increase	decreased or normal	decreased or normal	considerable increase	decreased

Table 7.3 Interpretation of results for thyroid function tests

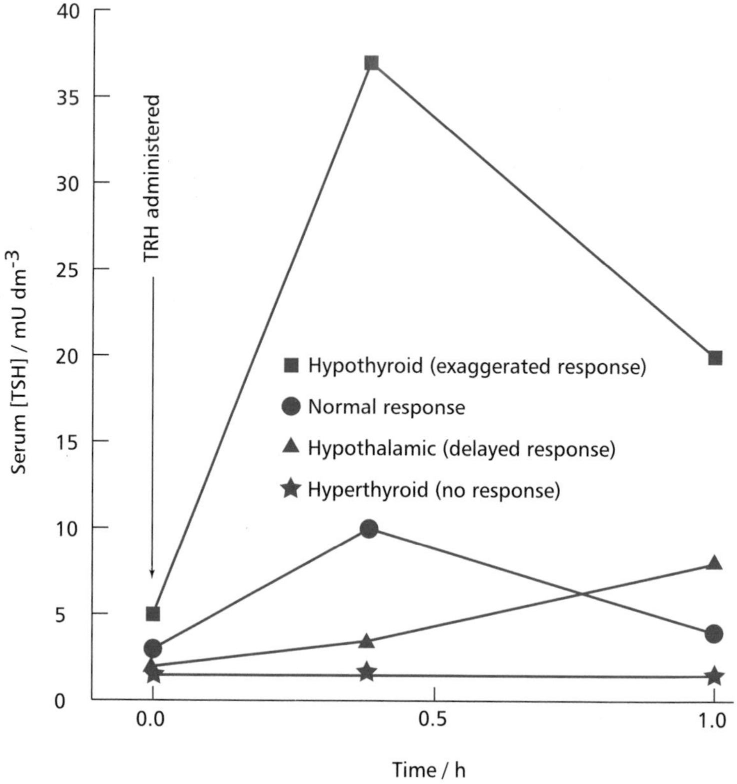

Figure 7.21 Potential results from a TRH test. See text for details.

Other techniques for investigating thyroid function include administration of isotopes, such as ^{99m}Tc-pertechnetate, and determining their distribution using a camera that detects γ radiation. This technique distinguishes between active and inactive thyroid nodules and can distinguish between Grave's disease, multinodular goiter or an adenoma affecting the thyroid gland. A thyroid biopsy involves aspirating tissue from the affected region of the thyroid using a syringe and fine needle. The collected cells are examined microscopically for evidence of thyroid nodules. Often thyroid disease has an autoimmune basis and measurement of antibodies can aid diagnosis. Antiperoxidase and antithyroglobulin antibodies are found in patients affected by Hashimoto's thyroiditis and often in those with Grave's disease. However, the detection of antibodies is not always diagnostic, as low levels of these antibodies can occur in older people who are euthyroid.

The management of hyperthyroidism includes using antithyroid drugs, for example, carbimazole. This is useful in young patients and acts by reducing the production of thyroid hormones. Other forms of treatment include radioiodine therapy with ^{131}I, although this is normally used in older patients, and partial or complete surgical removal of the thyroid gland (thyroidectomy). Some patients develop hypothyroidism, often as a consequence of treatment, and may have to be placed on thyroxine therapy. In this regard it is important to monitor TSH levels to detect developing hypothyroidism. Management of hypothyroidism involves replacement therapy with thyroxine, often for life. Thyroxine is readily available, safe and inexpensive. The treatment is monitored at regular intervals by measurement of serum concentrations of TSH to ensure they are kept within its reference range.

7.8 REGULATION OF BLOOD GLUCOSE

Adequate concentrations of glucose in the blood are necessary for brain cells as they cannot metabolize substances other than glucose and ketone bodies as energy sources nor can they store or synthesize glucose. After a meal, any released glucose is absorbed by the gastrointestinal tract (*Chapter 11*) enters the bloodstream and is delivered to the peripheral tissues where it may be metabolized to allow ATP production. Surplus glucose is converted to glycogen and stored in the liver and skeletal muscles or converted to triacylglycerols and stored in adipose tissue. During fasting, the liver produces glucose by glycogenolysis or gluconeogenesis and this is used to maintain blood glucose concentration (*Figure 7.22*).

The concentration of glucose in plasma is regulated by the hormones insulin and glucagon. Insulin is synthesized as preproinsulin in the β cells of the islets of Langerhans in the pancreas but during its secretion is enzymatically converted to active insulin (*Figures 7.3 (B)* and *7.23*). Insulin has a number of functions. It inhibits glycogenolysis, gluconeogenesis, lipolysis, ketogenesis

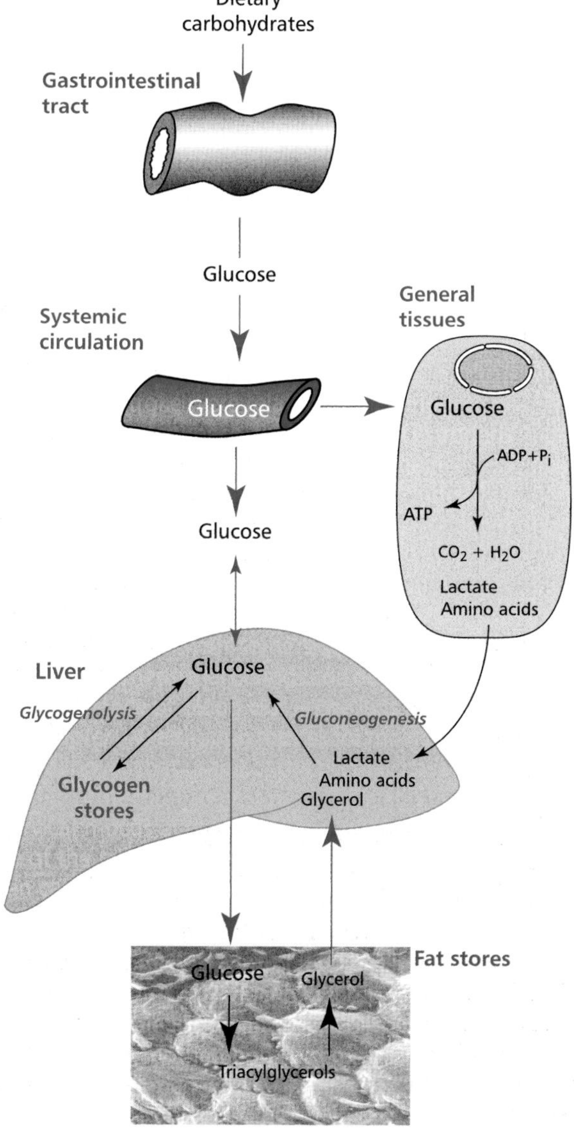

Figure 7.22 Overview of glucose homeostasis. See text for details.

Margin Note 7.2 Sanger and sequences

Sanger determined the complete sequences of the amino acid residues, the primary structures, of both chains of insulin in the early 1950s. This was the first unequivocal demonstration that proteins have strictly defined primary structures. With the techniques of chemistry and molecular biology available at that time, this was an incredible achievement and Sanger was awarded the 1958 Nobel Prize in Chemistry. Nowadays, sequencing a protein as small as insulin would be a trivial task. The sequences of smallish peptides are easily determined by mass spectrometry and it is relatively easy to determine the base sequence of a gene (DNA) and then interpret it in terms of the amino acid order of the encoded protein. Indeed, complete genome sequences of numerous organisms, including those of human beings and a number of human pathogens and parasites, are known and are widely available in various websites. Incredibly, the major method for sequencing DNA, the dideoxy method, was also devised by Sanger in 1977, when he and coworkers published the complete sequence of the genome of the virus, φX174. Sanger was awarded the Nobel Prize in Chemistry in 1980 for this work. Sanger thus belongs to the tiny elite group of people to have received two Nobel Prizes.

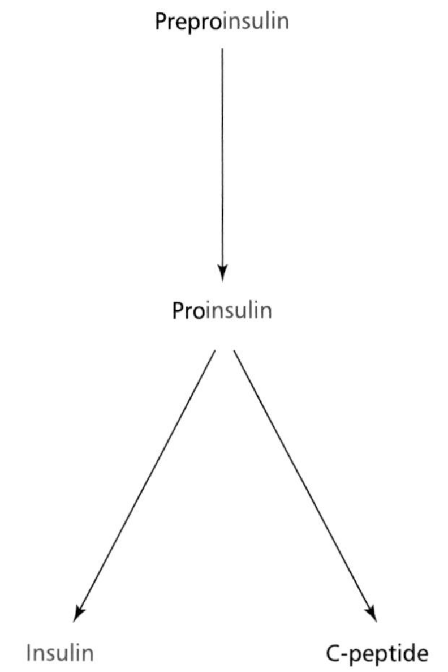

Figure 7.23 The conversion of preproinsulin to active insulin. See also *Figure 7.3 (B).*

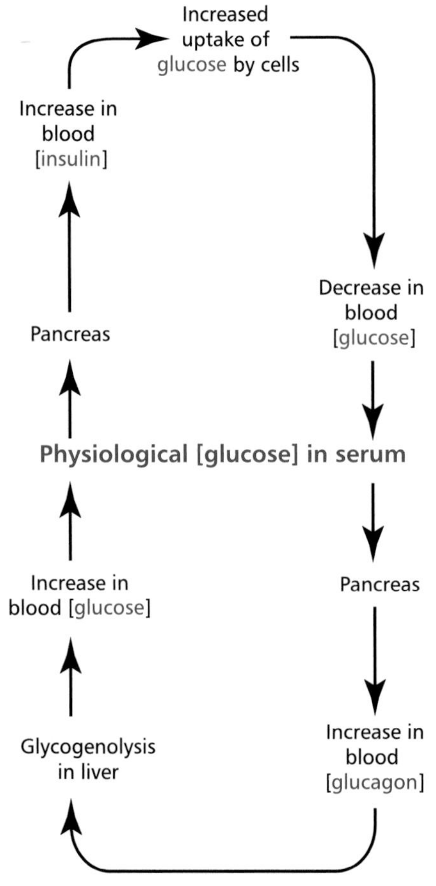

Figure 7.24 Regulation of blood glucose concentration by insulin and glucagon.

and proteolysis and stimulates glucose uptake by muscle and adipose tissues, glycolysis, glycogenesis, protein synthesis and uptake of K^+ and P_i. Glucagon is released by the α cells of the pancreas. Its effects are antagonistic to those of insulin. An increase in blood glucose stimulates the pancreas to produce insulin which, in turn, promotes the uptake and utilization of glucose by cells lowering its concentration. A reduction in blood glucose stimulates release of glucagon that promotes glycogenolysis in the liver thereby increasing blood glucose levels (*Figure 7.24*). Disorders of insulin release or activity can cause an increase in blood glucose, hyperglycemia, or its reduction, hypoglycemia.

DIABETES MELLITUS

Diabetes mellitus is a syndrome characterized by hyperglycemia due to an absolute or relative deficiency of, and/or resistance to, insulin. This is the commonest endocrine disorder, affecting about 2% of the world's population. Diabetes can be primary when caused directly by malfunction of one or more of the systems regulating blood glucose concentration, or secondary as a result of another disease. Primary diabetes is divided into types 1 and 2. In type 1 diabetes, also known as insulin dependent diabetes mellitus (IDDM), there is a decrease or absence of insulin production. It occurs in 15% of all diabetics and typically presents acutely during childhood or adolescence, although it can occur at any age. Patients have marked weight loss and ketoacidosis (*see below*) can occur readily. Type 1 diabetes mellitus is an autoimmune disease (*Chapter 5*) and antibodies that react with β cells of the islets of Langerhans in the pancreas have been demonstrated in over 90% of patients. It also has a strong association with certain histocompatibility antigens such as HLA-DR3, DR4 and certain DQ alleles (*Chapters 4* and *6*). Many cases of type 1 diabetes may develop after a viral infection, such as with Coxsackie B, which initiates an autoimmune reaction that destroys the β cells of the pancreas. Type 2 diabetes is also known as noninsulin dependent diabetes mellitus (NIDDM), where insulin secretion tends to be normal or even elevated. It accounts for about 85% of all cases of diabetes, has a gradual onset and tends to occur in middle-aged and elderly individuals. Patients are less likely to develop ketoacidosis. The etiology of type 2 diabetes is still unclear but has a strong association with obesity. The disease may arise because an abnormal insulin, not recognized by its receptor, is produced, lack of insulin receptors, the presence of defective receptors or by a defective secondary messenger system linking the insulin receptor to the glucose transporter in the plasma membrane. Type 2 diabetes has a strong familial incidence. For example, if an identical twin develops type 2 diabetes there is a strong likelihood that the other twin will become diabetic.

Secondary diabetes mellitus is uncommon and is a consequence of other disorders that involve excess secretion of hormones antagonistic to insulin, such as cortisol in Cushing's syndrome and GH in acromegaly. Damage to the pancreas following, for example, chronic pancreatitis or pancreatic surgery, may result in secondary diabetes.

Patients with diabetes mellitus suffer from a number of symptoms, including polydipsia, the production of large volumes of urine (**polyuria**), unexplained weight loss, blurred vision, tiredness and an increased susceptibility to infections. Diabetic patients are also susceptible to acute and chronic complications.

Acute complications of diabetes mellitus include diabetic ketoacidosis (DKA), hyperosmolar nonketotic (HONK) coma and hypoglycemia. Diabetic ketoacidosis occurs most commonly in patients with uncontrolled type 1 diabetes mellitus as a result of their failure to comply with insulin therapy or it can also be precipitated by infections, such as the common cold, when the body responds by releasing more glucose into the bloodstream and by reducing the action of insulin. The pathogenesis of DKA is outlined in *Figure 7.25*. In DKA, glucose uptake by cells decreases whereas gluconeogenesis and glycogenolysis are both stimulated, causing severe hyperglycemia. Increased breakdown of proteins occurs and the released amino acids enter gluconeogenesis or are degraded to form urea. As a result of insulin deficiency, there is a shift of both K^+ and P_i from the intracellular to the extracellular compartments causing hyperkalemia and hyperphosphatemia respectively (*Chapter 8*). The insulin deficiency also stimulates lipolysis producing free fatty acids (FFAs) and glycerol. In the liver, FFAs are converted to acetyl CoA which is then converted to acetoacetate and finally acetone and β-hydroxybutyrate. Acetoacetate, acetone and β-hydroxybutyrate are called **ketone bodies** even though β-hydroxybutyrate is not a ketone. The liver cannot utilize ketone bodies and they accumulate in the blood (**ketonemia**) and may be excreted in the urine (**ketonuria**). The ketone bodies in the blood are moderately strong acids, $H^+Ketone^-$, and react with buffers, such as $NaHCO_3$ (*Chapter 9*),

$$H^+Ketone^- + NaHCO_3 \longrightarrow Na^+Ketone^- + H_2CO_3$$

decreasing the concentration of hydrogen carbonate but increasing that of carbonic acid producing an acidosis. The carbonic acid dissociates to CO_2 and H_2O and the blood PCO_2 increases. Thus the lung ventilation rate increases as the body attempts to remove the excess CO_2. The ketonemia is believed to be responsible for the abdominal pain, vomiting and acidosis associated with DKA. Severe hyperglycemia in DKA exerts a high osmotic pressure causing water to move out of the cells leading to cellular dehydration. Blood volume rises and the kidneys respond with polyuria. If blood glucose levels exceed the renal threshold, glucose is lost in the urine causing glycosuria. Eventually an osmotic diuresis may occur with loss of water and electrolytes. The blood volume therefore declines further (**hypovolemia**) reducing the glomerular

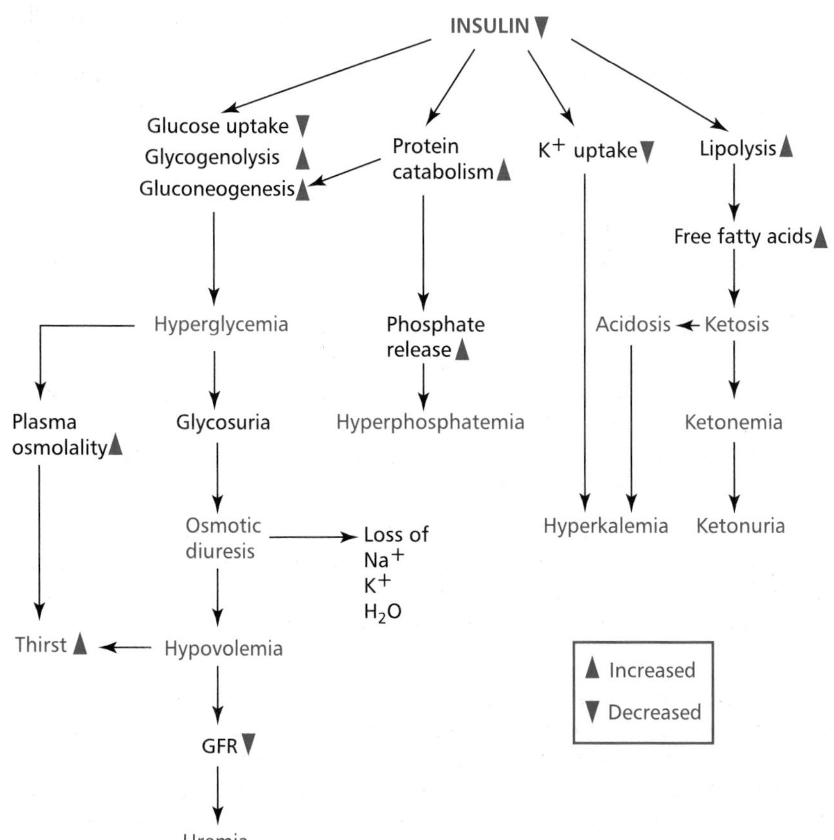

Figure 7.25 An overview of the factors causing diabetic ketoacidosis (DKA). See text for details.

filtration rate (GFR) and causing uremia. Dehydration and hypovolemia may stimulate the thirst center and the patient may suffer polydipsia.

Diabetic ketoacidosis is a medical emergency and is fatal in about 10% of adults and 5% of children if untreated. Death in untreated cases of DKA is due to poor tissue perfusion, acidosis and cardiovascular failure. The following approaches are taken to correct the metabolic disturbance. Isotonic saline and insulin are administered to correct the dehydration and hyperglycemia respectively. Following administration of insulin, K^+ enters the cells and this may cause hypokalemia therefore it may be necessary to administer K^+ supplements. In the past, severe cases of acidosis were treated by infusion of hydrogen carbonate although this approach is rarely adopted nowadays. Finally, if the DKA was precipitated by an infection, then it is necessary to identify and treat this infection.

Patients with uncontrolled type 2 diabetes mellitus may enter a HONK coma. This tends to occur in the elderly and develops over a period of days or even weeks. In these patients, the levels of insulin are sufficient to prevent ketosis but they often have severe hyperglycemia, characterized by blood glucose concentration above 35 and often in excess of 50 mmol dm^{-3}, together with severe dehydration and high serum osmolality. A HONK coma is usually precipitated by severe illness, diuretics, dehydration and glucocorticoid therapy. Its treatment is similar to that for DKA, except that the patient is rehydrated at a slower rate and insulin requirements are lower than those for a patient with DKA.

The chronic complications of diabetes mellitus include retinopathy, cataract, atherosclerosis, nephropathy and neuropathy. Diabetics with poor blood glucose control are most susceptible to these chronic complications. It is generally accepted that increased protein glycation and the accumulation of tissue advanced glycation end products (*Chapter 18*) are involved in the pathogenesis of chronic complications.

Diagnosis and treatment of diabetes mellitus

Investigation of diabetes is made on the basis of clinical features and laboratory investigations. A preliminary screening test may identify the presence of urinary glucose, although this is not diagnostic of diabetes mellitus. A patient presenting with symptoms of diabetes mellitus must have a venous blood specimen taken and its glucose concentration determined. A patient is diagnosed as diabetic if the fasting plasma glucose concentration is equal to or greater than 7.0 or the random concentration is greater than 11.1 mmol dm^{-3}. Diabetes mellitus is excluded if the fasting or random plasma glucose concentrations are less than 6.1 or 7.8 mmol dm^{-3} respectively. If the individual under investigation lacks the typical symptoms of diabetes then diagnosis cannot be confirmed by a single glucose determination but reconfirmed by at least one additional positive test on another day or investigated using the oral glucose tolerance test (OGTT). During the OGTT, the patient is kept on a normal diet for three days prior to the test and then fasts overnight prior to the test. A basal (fasting) venous blood sample is taken for glucose determination before the patient drinks 75 g of anhydrous glucose dissolved in a small volume of water. Blood specimens are collected after one and two hours and plasma glucose determined. Plasma glucose values greater than or equal to 7.0 mmol dm^{-3} for the basal sample or 11.1 mmol dm^{-3} for the 2 h samples are diagnostic of diabetes mellitus. Individuals with plasma glucose concentrations less than 7.0 for the basal sample or between 7.8 and 11.1 mmol dm^{-3} for the 2h samples are categorized as having impaired glucose tolerance (IGT). This group has an increased risk of developing cardiovascular disease (*Chapter 14*). Patients with plasma glucose concentrations of 6.1 to 7.0 mmol dm^{-3} for the basal samples or less than 7.8 mmol dm^{-3} for the 2 h samples are categorized as having impaired fasting glucose (IFG) and are at risk of developing diabetes. The values given relate to venous plasma samples and are different from those for whole blood samples.

The management of diabetes mellitus aims to provide relief from symptoms and reduce the chances of developing acute and chronic complications. This includes educating the patient that diabetes is a life-long disease and affected individuals must be responsible for their own treatment. Regular clinical and laboratory assessment of the patient is required to ensure that treatment is effective, to detect early signs of treatable complications so as to reduce their progression and ensure compliance with treatment. Management involves the dietary restriction of simple sugars and of saturated fats and cholesterol and the use of complex carbohydrate and fibers. Dietary control is often accompanied by use of injected insulin or oral hypoglycemic drugs, such as sulfonylureas, in patients with type 1 and type 2 diabetes respectively. Occasionally, it may be necessary to use insulin in patients with type 2 diabetes to control blood glucose effectively. Hypoglycemic drugs act by increasing the sensitivity of β cells to glucose therefore stimulating insulin release or by increasing sensitivity of target cells to insulin. Both effects will reduce blood glucose levels. Some hypoglycemic drugs act to reduce the absorption of glucose by the GIT or reduce glycogenolysis in the liver. Diabetic patients on treatment are monitored regularly to ensure that blood glucose is kept in control. Most patients measure their own blood glucose at home regularly using kit methods based on reagent strips and a portable glucose meter (*Figure 7.26*) and adjust insulin dosage according to needs, perhaps following a change in diet, during illness or after exercise. The amount of glycated hemoglobin (*Figure 1.14*) in a patient is determined regularly to assess therapy compliance because its presence is an indicator of average glycemia over the previous 6–8 weeks. The amount of glycated hemoglobin tends to be less than 6% in nondiabetics but may exceed 10% in uncontrolled diabetes. Diabetic patients with high values have poor blood glucose control and their treatment or compliance must be reviewed.

HYPOGLYCEMIA

Hypoglycemia is defined as a blood glucose concentration less than 2.2 mmol dm^{-3} in a random specimen collected into a tube containing an inhibitor of glycolysis. Hypoglycemia occurs because of an imbalance between glucose intake, endogenous glucose production and glucose

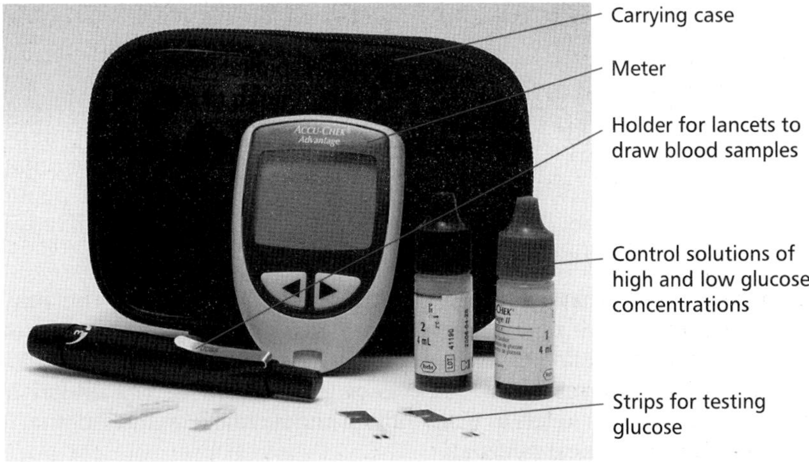

Figure 7.26 A blood glucose meter.

utilization. Its clinical features are mainly due to abnormal function of the CNS (neuroglycopenia) and can be acute or chronic. Acute features include tiredness, confusion, hunger, dizziness, blurred vision, convulsions, coma, anxiety, profuse sweating and tachycardia. The typical signs and symptoms are more likely to occur if blood glucose falls rapidly: in children and young adults the symptoms often present when the glucose concentration is less than 2.2 whereas neonates develop symptoms only when it is less than 1.5 mmol dm^{-3}. Acute hypoglycemia is usually associated with diabetics who have taken too much insulin. Its chronic features include personality changes, memory loss and dementia and are generally observed in patients with insulin secreting tumors.

Most cases of hypoglycemia occur in patients with type 1 diabetes mellitus because of insufficient carbohydrate, too high a dose of insulin, inappropriate use of hypoglycemic drugs, excessive alcohol intake and strenuous exercise. Hypoglycemia can occur during fasting if the individual has an insulin secreting tumor or insulinoma which is a primary tumor of the β cells of the islets of Langerhans. These tumors produce excessive amounts of insulin or secrete insulin when it is not required. Nonpancreatic tumors, especially carcinomas of the liver and sarcomas, may cause hypoglycemia by increasing cellular uptake of glucose but this is unlikely to be the sole cause. Many of these tumors secrete IGF-II which has insulin-like effects and is capable of causing hypoglycemia.

Hypoglycemia in children is particularly dangerous because of the high risk of permanent brain damage. Its diagnosis is especially necessary in the first few months of life. A fetus exposed to maternal hyperglycemia will have pancreatic islet cell hyperplasia and elevated insulin levels. Following birth, the neonate has hyperinsulinism and may develop hypoglycemia now the high glucose supply from the mother has been removed. In addition, these babies are larger than average in size since insulin promotes growth.

Diagnosis and treatment of hypoglycemia

An initial assessment of a patient with frequent episodes of hypoglycemia involves a thorough clinical evaluation with an emphasis on the patient's drug history, relationship of symptoms to meals, presence/history of endocrine disease and an investigation of a possible nonpancreatic tumor. Laboratory tests can confirm the diagnosis of hypoglycemia by demonstrating a low blood glucose concentration. To confirm that the clinical features are due to hypoglycemia, the patient can be given glucose by mouth or parenterally as appropriate. Symptoms that are due to acute neuroglycopenia resolve immediately whereas those due to chronic neuroglycopenia often persist. Measurement of plasma insulin concentration can help in the diagnosis or exclusion of an insulinoma. An insulinoma is likely if the patient has fasting hypoglycemia together with high serum insulin levels that are greater than 10 mU dm^{-3} and raised C-peptide levels (*Figure 7.23*). Insulin secretion in an insulin-treated diabetic cannot be determined for obvious reasons. However, insulin and the C-peptide are secreted by islet cells in equimolar amounts and therefore a measurement of C-peptide together with insulin can differentiate between hypoglycemia due to an insulinoma, which will have a high C-peptide concentration, from that due to exogenous insulin, which will have relatively low amounts of C-peptide.

Hypoglycemia can be fatal and patients must be treated urgently. The aim in comatose patients is to rapidly correct the hypoglycemia using intravenous dextrose or intramuscular glucagon. If a patient is conscious and can swallow, then they are given sweet drinks, sweets or glucose tablets. The underlying cause of hypoglycemia needs to be identified and rectified; an insulinoma, for example, requires surgical removal.

7.9 DISORDERS OF THE ADRENAL CORTEX

Adrenal glands each weigh about 5 g and are found on the upper surfaces of the kidneys. They consist of an outer cortex and an inner medulla. The cortex consists of three layers: the zona glomerulosa, zona fasciculata and zona reticularis (*Figure 7.27*). The adrenal cortex is essential for life since it produces three groups of steroid hormones. The glucocorticoids, such as cortisol, and adrenal androgens, for example testosterone, are produced by the zona reticularis and zona fasciculata and mineralocorticoids, such as aldosterone, from the zona glomerulosa. Adrenal cortex cells have many low density lipoprotein receptors on their surfaces enabling them to take up precursor cholesterol rapidly from the plasma.

Cortisol (*Figure 7.28 (A)*) is released in response to adrenocorticotrophic hormone (ACTH) from the anterior pituitary that, in turn, is controlled by the release of corticotrophin releasing factor (CRF) from the hypothalamus. Cortisol exerts a negative feedback effect on the anterior pituitary and hypothalamus (*Figure 7.28 (B)*). The secretion of cortisol shows a diurnal variation: highest in the morning, lowest at night. Cortisol stimulates an increase in protein catabolism, hepatic glycogenolysis and gluconeogenesis and a redistribution of adipose tissue but suppresses inflammation. About 90% of blood cortisol is bound to a cortisol binding globulin called transcortin whilst the remaining 10% is free.

The major mineralocorticoid, aldosterone (*Figure 7.29 (A)*), is released in response to hypotension, Na^+ depletion or hyperkalemia (*Chapter 8*). A group of cells in the kidneys called juxtaglomerular cells detect a fall in blood pressure and secrete renin which circulates in the blood and catalyzes the conversion of the plasma protein angiotensinogen to angiotensin I. A converting enzyme in the lungs converts angiotensin I to angiotensin II that stimulates the release of aldosterone from the adrenal cortex and ADH from the posterior pituitary (*Figure 7.29 (B)*). Aldosterone stimulates the retention of Na^+ in exchange for K^+ and H^+ in the kidney tubules increasing the osmolality of the ECF and the retention of water raising the blood pressure or the ECF volume back to

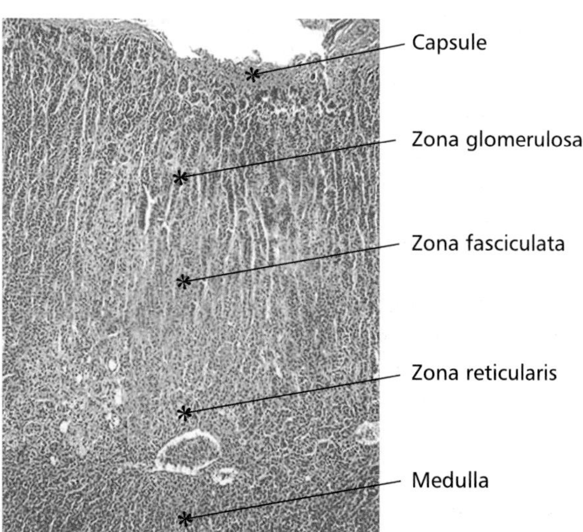

Figure 7.27 The histological structure of the adrenal cortex. Courtesy of Dr A.L. Bell, University of New England, College of Osteopathic Medicine, USA.

A)

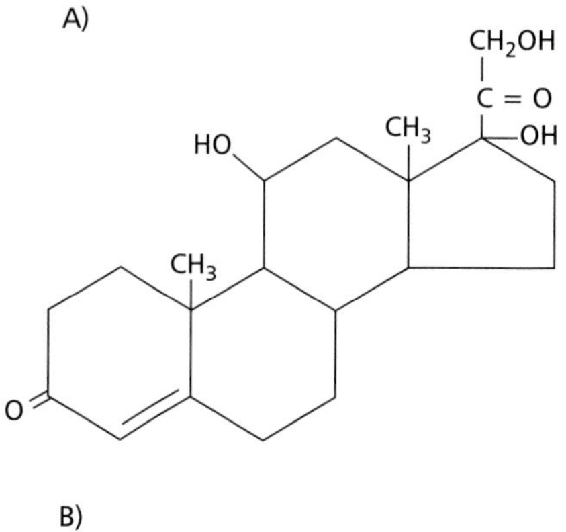

B)

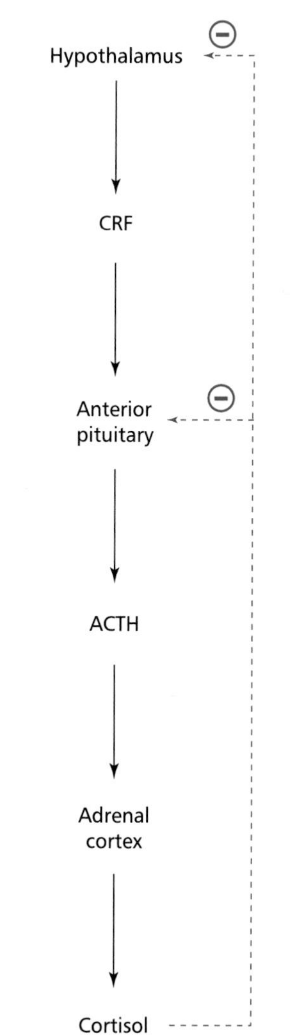

Figure 7.28 (A) Structure of cortisol and (B) the regulation of its secretion.

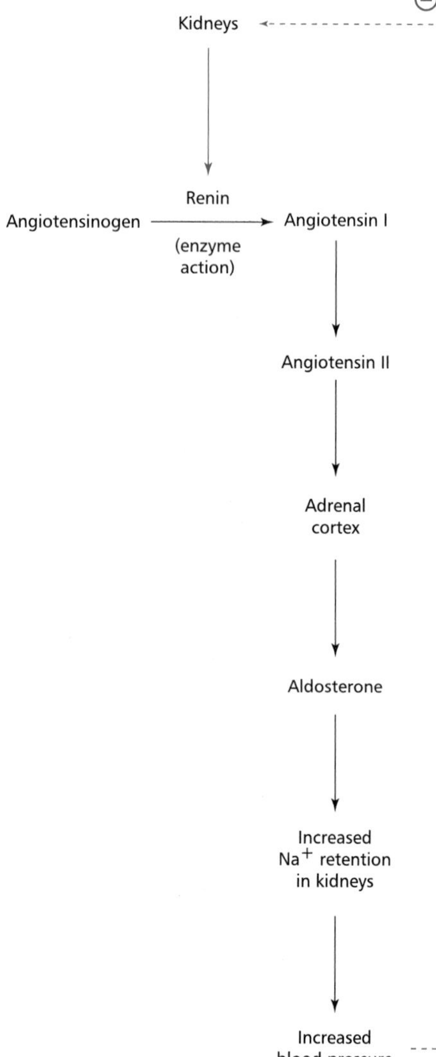

Figure 7.29 (A) Structure of aldosterone and (B) the regulation of its secretion.

physiological values. Angiotensin II also stimulates thirst that helps to increase blood pressure. The adrenal cortex secretes androgens, such as testosterone, dehydroepiandrosterone sulfate (DHEAS) and androstenedione. Testosterone is the hormone that stimulates the growth and development of the male characteristics (*Section 7.10*).

ADDISON'S DISEASE

Adrenal hypofunction or Addison's disease is a rare condition but is simple to treat once diagnosed. It can arise from one of a number of causes: an autoimmune destruction of the adrenal cortex, as a response to tuberculosis (TB), amyloidosis, hemochromatosis, following adrenalectomy or hypothalamic or pituitary diseases referred to as secondary adrenal insufficiency. Addison's disease is characterized by a deficiency of glucocorticoids and mineralocorticoids. Its clinical features include weakness, lethargy, anorexia, nausea, vomiting, weight loss, hypotension, skin pigmentation, hypoglycemia and depression (*Figure 7.30*). In the first few months, symptoms are usually vague with only lethargy, weakness and weight loss presenting as a result of glucocorticoid deficiency. Later, patients start to vomit and have abdominal pain. The lack of mineralocorticoids leads to an excessive loss of Na^+ and therefore hypotension is common in these patients. Plasma ACTH is increased because of pituitary response to low levels of cortisol given the lack of a negative feedback mechanism. Adrenocorticotrophic hormone can stimulate melanocytes in skin to produce the melanin, hence pigmentation is a common feature in Addison's disease.

Diagnosis and treatment of Addison's disease

A clinical suspicion of Addison's disease can be confirmed by demonstrating hyperkalemia with hyponatremia (*Chapter 8*). The plasma cortisol is usually, but not always, low in these patients. A high concentration of ACTH coupled with a low concentration of cortisol is indicative of Addison's disease whereas low cortisol and ACTH values are suggestive of secondary adrenal insufficiency. The situation can be resolved using complex biochemical tests with the analog of ACTH, synacthen. The short synacthen test involves an intramuscular injection of 0.25 mg of synacthen. The concentration of plasma cortisol is measured within 30 min. If it rises by at least 200 or to a value greater than 550 nmol dm^{-3}, then Addison's disease is unlikely. If this is not the case, then it is appropriate to proceed to the long synacthen test which involves an intramuscular injection of 1 mg of ACTH daily for three days. On the fourth day, the short synacthen test is performed and the serum concentration of cortisol is measured. If this is less than 200 nmol dm^{-3} with no increase following administration of synacthen there is primary adrenal failure and the patient is suffering from Addison's disease. If, however, there is an incremental increase of at least 200 nmol dm^{-3} above the baseline, then the decreased output of cortisol from adrenal gland is secondary and due to a deficiency of ACTH caused by a hypothalamic or pituitary disorder. Once Addison's disease is diagnosed, it is necessary to ascertain its cause. A number of laboratories test for the presence of antibodies against the adrenal glands to see if there is an autoimmune cause. A plain abdominal X-ray may be useful in that it can detect calcification of adrenal glands as a result of TB (*Chapter 3*).

The conventional therapy for Addison's disease involves treatment with steroids, such as hydrocortisone and fludrocortisone, which possess glucocorticoid and mineralocorticoid activities respectively. If a patient is left untreated he or she will eventually experience an adrenal crisis precipitated by stress, bacterial infection, trauma or surgery, which is a medical emergency. Typical clinical features of a crisis are abdominal pain, vomiting, hypotension together with hyponatremia, hypoglycemia and hyperkalemia. Its treatment involves administering saline infusions to correct the hypotension, fluid and salt losses and intravenous steroids to correct glucocorticoid and mineralocorticoid deficiencies. The precipitating factors, such as bacterial infections, require identification and appropriate treatment.

CUSHING'S SYNDROME

Adrenal hyperfunction can cause Cushing's syndrome and arise from a number of causes. The commonest cause is a pituitary lesion secreting high levels of ACTH (referred to as Cushing's *disease*). Other causes include ectopic production of ACTH from a carcinoma of the lungs, or excessive production of cortisol from an adrenal adenoma or carcinoma and iatrogenic causes, such as corticosteroid or ACTH treatment. The major clinical features (*Figure 7.31*) include truncal obesity, thinning of skin, excessive bruising, poor wound healing, purple striae on the abdomen and thighs, muscle weakness and wasting, **hirsutism**, the development of increased body hair on the face, chest upper back and abdomen in females (especially in adrenal carcinoma), hypertension, amenorrhea and psychiatric disturbances. Excess of cortisol has a mineralocorticoid effect leading to the retention of Na^+ and water producing hypertension. Hypokalemia may also occur because of an excessive loss of K^+ (*Chapter 8*). Excess cortisol increases blood glucose levels and some of these patients may have diabetes mellitus. The clinical features are due to increased cortisol production and, partly, to excessive androgen release.

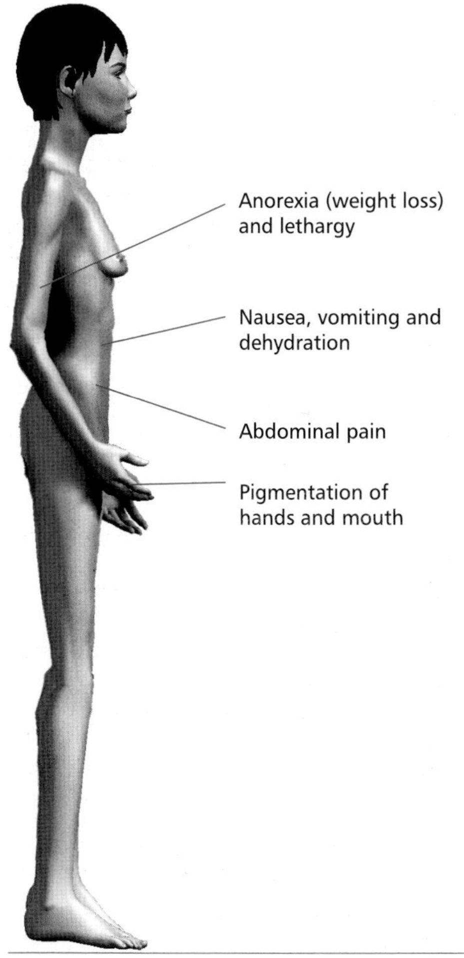

Figure 7.30 Schematic to show the clinical features of Addison's disease.

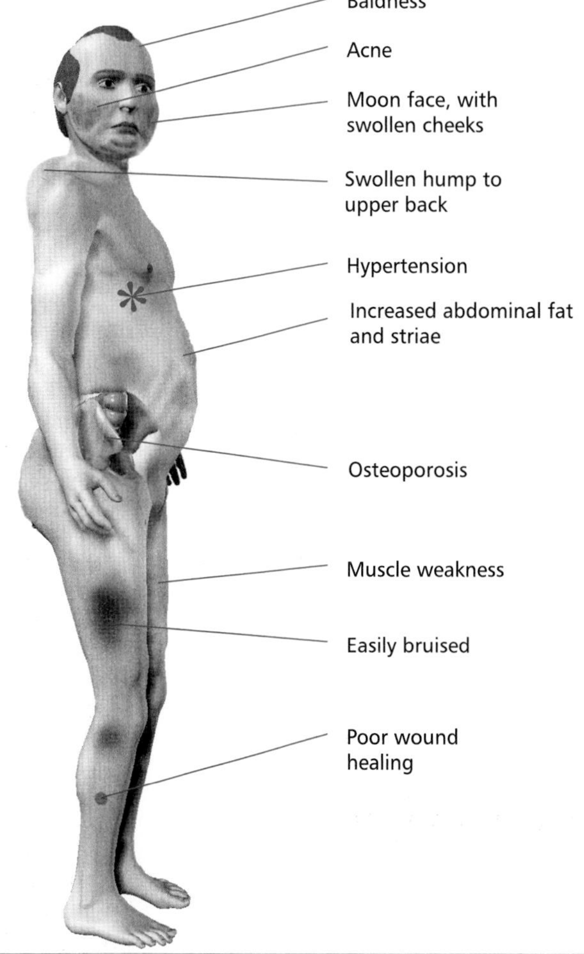

Figure 7.31 Schematic to show the clinical features of Cushing's syndrome.

Diagnosis and treatment of Cushing's syndrome

A clinical suspicion of the syndrome is supported by hypokalemia and alkalosis, high urinary free cortisol, which is normally less than 300 nmol per 24 h, and loss of the usual diurnal rhythm of cortisol secretion. Initial screening criteria are followed by tests using the cortisol analog, dexamethasone, which is not detected by the usual methods of measuring cortisol. Dexamethasone, however, suppresses ACTH production and cortisol secretion in normal people. The low dose dexamethasone test involves giving 1 mg of dexamethasone at night. A blood specimen is taken for cortisol measurement the following morning. A failure of dexamethasone to suppress cortisol release is suggestive of Cushing's syndrome or disease. To distinguish between the two, a high dose dexamethasone test may be used or the concentration of ACTH in plasma measured. The high dose dexamethasone test consists of administering 2 mg of dexamethasone every 6 h for a period of 48 h. The concentration of cortisol in the plasma is then measured at 09.00 h on the morning following the last dose. In Cushing's syndrome, due to excessive secretion of cortisol by an adrenal tumor or in response to an ectopic source of ACTH, suppression of cortisol does not occur. In Cushing's disease, caused by a pituitary lesion secreting ACTH, the concentration of cortisol is suppressed to less than 50% of its value prior to the test. Plasma ACTH levels are raised in patients with Cushing's disease and ectopic ACTH production, but are low in patients who have an adrenal tumor that secretes cortisol.

The management and treatment of Cushing's syndrome depends upon its cause. Drugs, such as metyrapone, that inhibit the synthesis of cortisol may be used. Tumors can be removed surgically.

CONGENITAL ADRENAL HYPERPLASIA

Congenital adrenal hyperplasia (CAH) is an autosomal recessive condition (*Chapter 15*) characterized by an abnormal biosynthesis of steroid hormones in the adrenal glands. The clinical features depend on whether cortisol and/or aldosterone or androgens are involved. The commonest cause of CAH is a complete or partial deficiency of 21-hydroxylase activity (*Figure 7.32*) that accounts for 95% of all cases, and occurs with an incidence of 1 in 12 000 newborn babies in the UK. A deficiency of 21-hydroxylase activity blocks the synthesis of cortisol and, as a consequence, negative feedback to the anterior pituitary is diminished. The anterior pituitary increases its secretion of ACTH, which causes hyperplasia of the adrenal glands. The substrate, 17α-hydroxyprogesterone accumulates and stimulates the production of adrenal androgens. A deficiency in 21-hydroxylase activity is complete in approximately one-third of CAH patients. Less common forms of CAH are characterized by deficiencies of other enzymes in the synthetic pathway.

Females affected with CAH are born with ambiguous genitalia, although in cases of partial enzyme deficiency this may not be apparent until early adulthood where the female presents with hirsutism, amenorrhea and infertility. Males with CAH present with premature development of the male secondary sexual characteristics called pseudoprecocious puberty. In individuals with a complete deficiency of 21-hydroxylase activity, aldosterone production is also inhibited and these individuals usually present shortly after birth with a life-threatening condition characterized by excessive salt and water loss.

Diagnosis and treatment of congenital adrenal hyperplasia

Diagnosis of CAH due to 21-hydroxylase deficiency is made by detecting increased concentrations of 17α-hydroxyprogesterone in the baby's blood at least two days following birth. Maternal 17α-hydroxyprogesterone may still be present at two days postparturition, hence the need for the delay. The affected

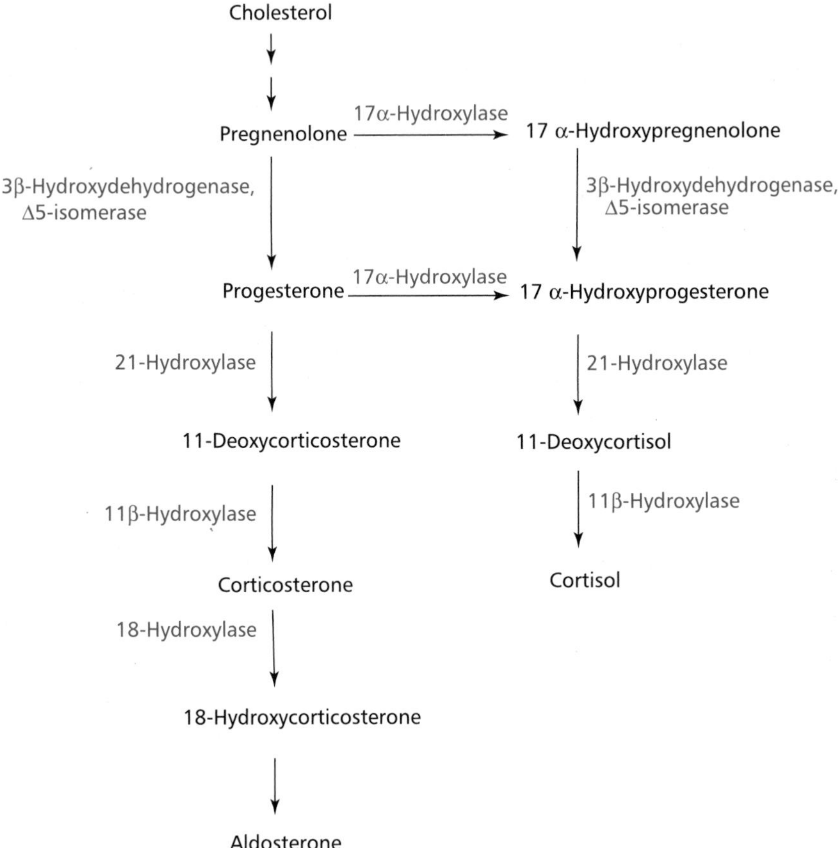

Figure 7.32 The synthesis of cortisol and aldosterone.

individuals are treated with cortisol and, if necessary, aldosterone. The treatment should reduce ACTH secretion and therefore excessive androgen production. The treatment requires monitoring by regular measurements of plasma 17α-hydroxyprogesterone.

CONN'S SYNDROME

Conn's syndrome, also called primary hyperaldosteronism, is characterized by an increased production of aldosterone. In 80% of cases this is due to an adrenal adenoma. Other causes include hypertrophy of the zona glomerulosa of the adrenal cortex and adrenal carcinoma, although the latter is extremely rare. The excessive secretion of aldosterone leads to the increased retention of Na^+ and loss of K^+ by the kidneys. Most of the clinical features, muscle weakness, tetany, paresthesiae, polydipsia and polyuria, are due to the hypokalemia (*Chapter 8*) whereas the excessive Na^+ retention causes hypertension.

Diagnosis and treatment of Conn's syndrome

Investigating the possibility of Conn's syndrome involves determining the concentrations of Na^+ and K^+ in the serum and urine of the patient. Plasma Na^+ can be high, slightly increased or normal, whereas that of K^+ is always reduced. Measurements of plasma aldosterone and renin in patients with Conn's syndrome often show a high aldosterone concentration and a low renin activity. Conn's syndrome is managed by treating its underlying cause. Often this is a tumor that must be removed surgically. Other approaches involve using diuretics, such as spironolactone, an antagonist to aldosterone that helps to control hypertension. This is particularly useful when the cause of Conn's syndrome is adrenal hyperplasia.

BOX 7.2 Pheochromocytoma

Adrenaline (epinephrine) and noradrenaline (norepinephrine) are catecholamine hormones secreted by the adrenal medulla. They stimulate increases in heart, metabolic and breathing rates, the depth of breathing and dilate blood vessels to muscles and increase the metabolism of carbohydrate stores to prepare the body for 'fright, flight or fight'.

Pheochromocytomas are tumors of the adrenal medulla that cause uncontrolled and irregular secretion of adrenaline and noradrenaline. Such tumors are usually benign, that is they do not metastasize (*Chapter 17*), and are rare causes of hypertension (*Chapter 14*) accounting for some 0.5% of such cases.

Adrenaline and noradrenaline are metabolized to metadrenaline and normetadrenaline respectively by catechol-*O*-methyl-transferase (COMT). Metadrenaline and normetadrenaline are converted to 4-hydroxy-3-methoxymandelic acid (HMMA), also known as vanillylmandelic acid (VMA), by monoamine oxidase (MAO). Metadrenaline and normetadrenaline and HMMA are all excreted in the urine (*Figure 7.33*).

Patients affected by pheochromocytomas present with headaches, pallor, tremors, increased heart rate and palpitations, sweating and abdominal discomfort. Pheochromocytoma leads to high levels of catecholamines in the plasma. However, a diagnosis based on measurement of plasma levels is difficult, as they are subject to variation by a number of factors including posture. Screening tests include measuring and demonstrating increased concentrations of urinary HMMA or metadrenaline and normetadrenaline (*Figure 7.33*). The reference range for urine HMMA is less than 35 μmol per 24 h whereas those for metadrenaline and normetadrenaline are less than 0.1 and less than 0.57 μmol per 24 h respectively. In cases where the results are equivocal, the pentolinium test is conducted. Patients are given 2.5 mg of pentolinium by intravenous injection. A blood specimen is collected before and 15 min after the injection. Pentolinium is a drug that inhibits release of catecholamines in normal individuals but not in those suffering from pheochromocytoma.

The treatment of pheochromocytoma involves surgical removal of the tumor. However, it is necessary to block the action of excess adrenaline and noradrenaline before any surgery by administration of phenoxybenzamine. This is because during the operation, the sudden release of large amounts of catecholamines results in high blood pressure which, in turn, could cause a heart attack or brain hemorrhage.

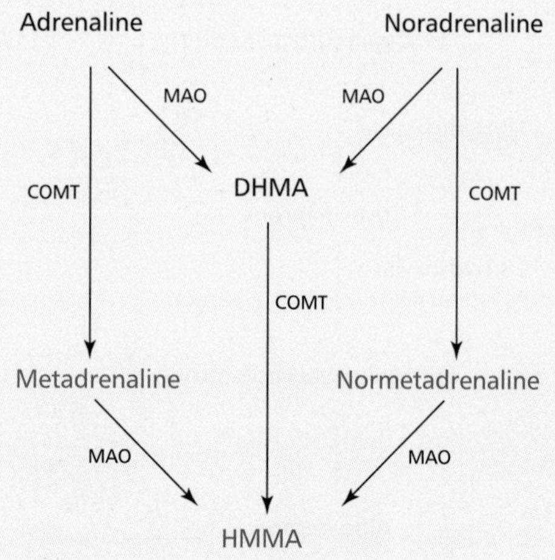

Figure 7.33 Metabolism of adrenaline and noradrenaline.

7.10 REPRODUCTIVE HORMONES

The male and female reproductive systems (*Figure 7.34 (A)* and *(B)*) produce and secrete a number of sex hormones and are responsible for the maturation of germ cells, the production of gametes and, in the female, the fertilization of the ovum and its subsequent growth and development. The testes produce male gametes or spermatozoa ('sperms') that mature and are stored in the epididymis and vas deferens. Testes are composed of lobules with up to three seminiferous tubules containing cells undergoing spermatogenesis. These cells are supported and nourished by Sertoli cells. Spermatogenesis involves meiosis and produces haploid sperm (*Chapter 15*) as outlined in *Figure 7.35*. Each sperm has a head and a tail consisting of a midpiece and flagellum. The midpiece contains mitochondria that provide energy for the locomotory movements of the flagellum. The head contains a nucleus and

A)

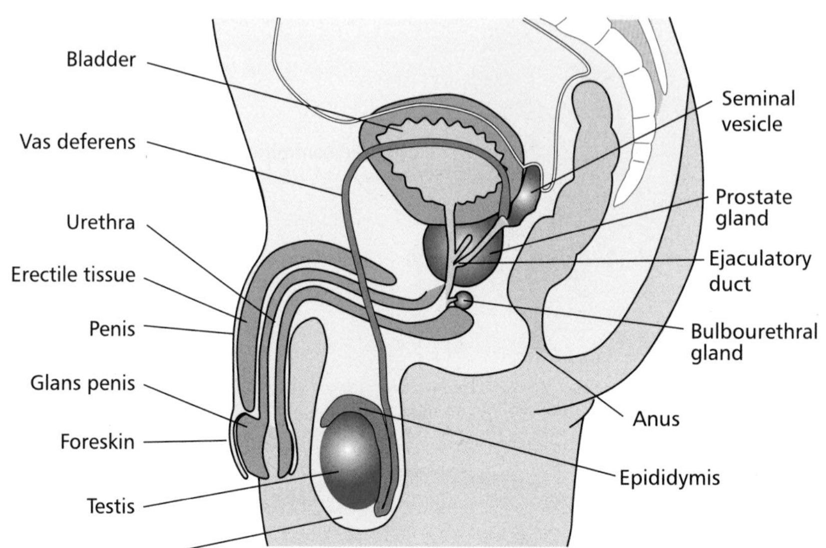

B)

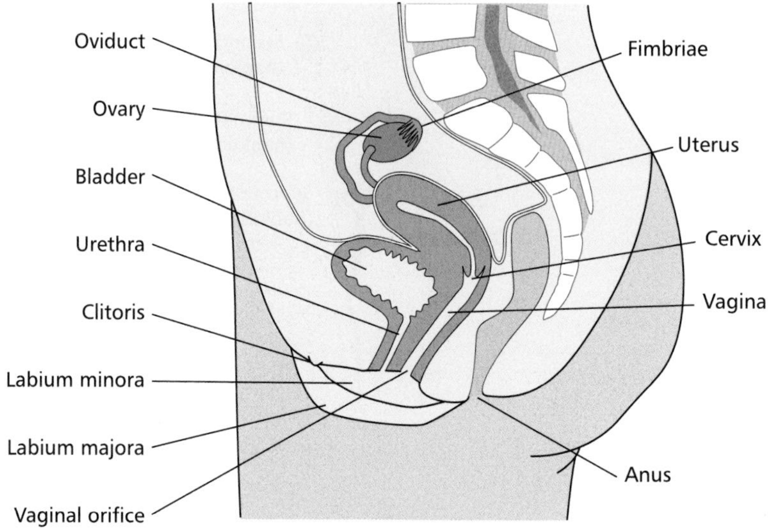

Figure 7.34 Overviews of the (A) male and (B) female reproductive systems.

is covered by a cap called an acrosome, which contains enzymes required to penetrate the ovum or egg. The production of ova, female gametes, begins in the ovaries by a process called oogenesis (*Figure 7.36*). Primordial germ cells in the outer germinal epithelium divide by mitosis to form a diploid primary oocyte that becomes surrounded by follicle cells to produce primary follicles. These migrate into the center of the ovary. As many as two million primary follicles are present at birth and remain dormant until puberty. Approximately 400 primary follicles mature over the lifetime of a female until follicle development ceases at the menopause. The primary follicle matures to form a secondary follicle. During this development, the primary oocyte divides by meiosis but this is arrested and forms a haploid secondary oocyte, which is the precursor of the ovum, and a small polar body. In an adult fertile female, the nucleus of a secondary oocyte begins the second meiotic division at each monthly ovulation but progresses only to metaphase, when

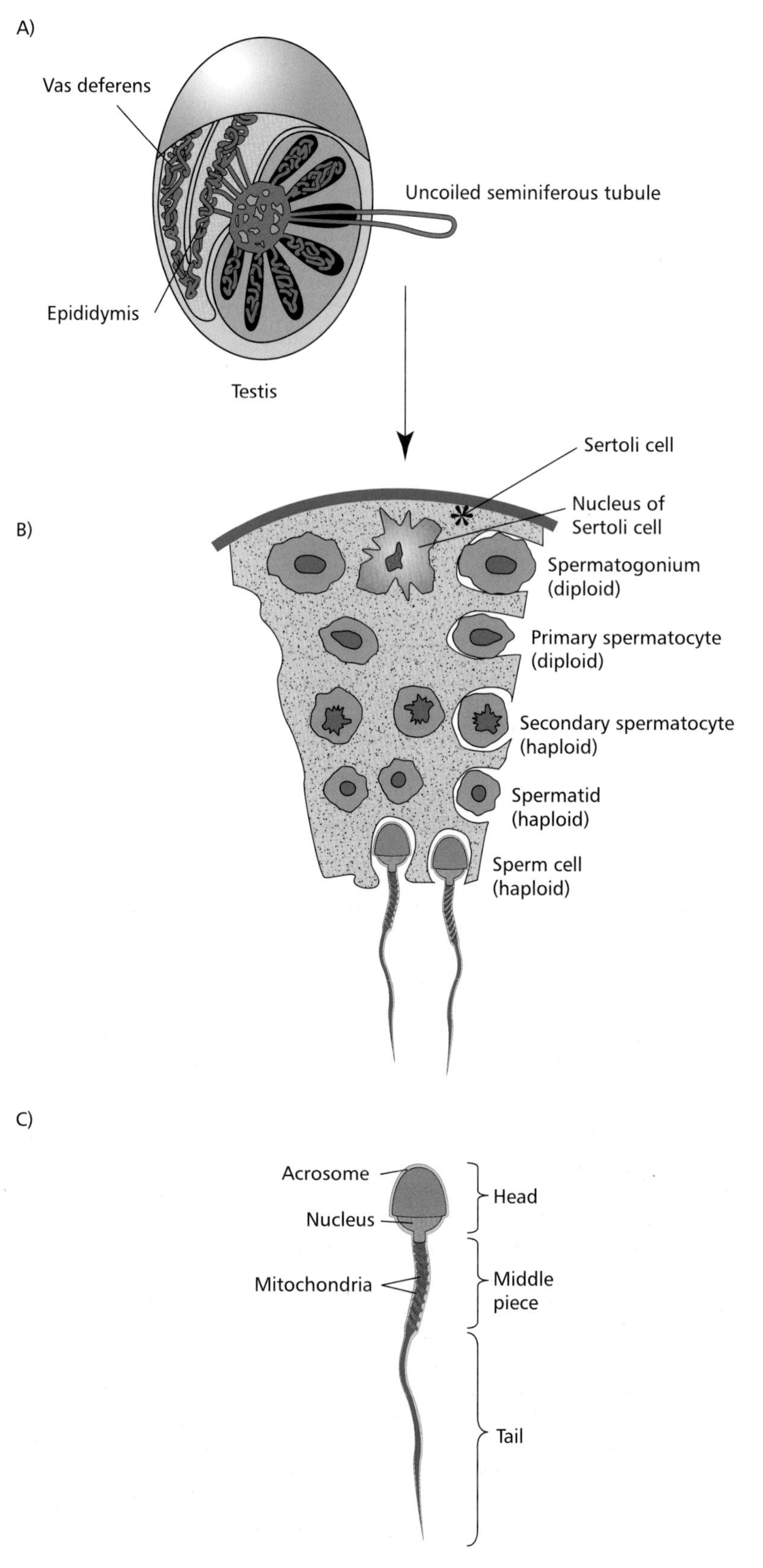

Figure 7.35 (A) Schematic of a testis. (B) Overview of spermatogenensis. (C) Schematic of a sperm. See text for details.

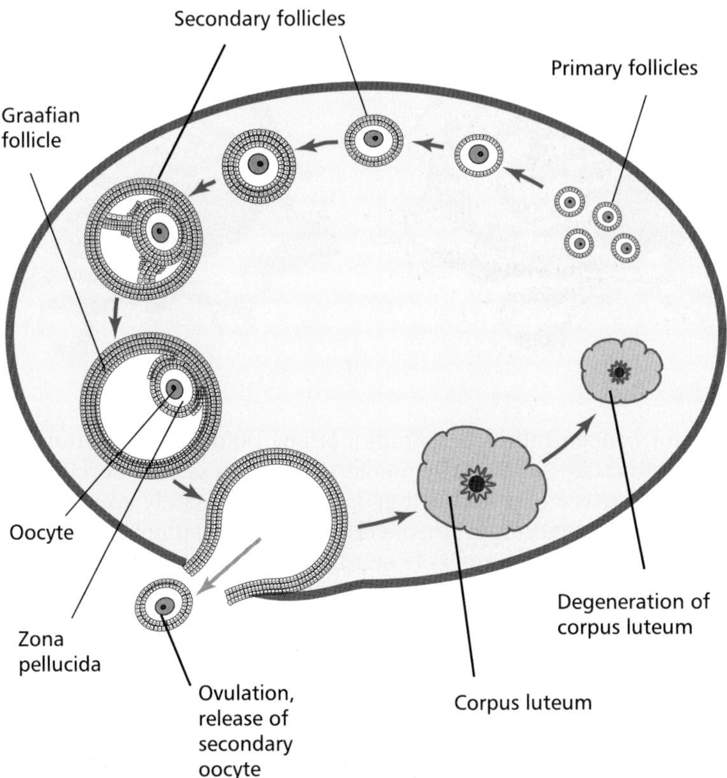

Figure 7.36 Overview of oogenesis and release of oocyte. See text for details.

division again ceases. The second meiotic division of the secondary oocyte is completed at fertilization.

In males and females, the hypothalamus secretes gonadotrophin releasing hormone (GnRH) which regulates secretion of LH and FSH from the basophil cells of the anterior pituitary. Secretion of GnRH, LH and FSH occurs in pulses. Follicle stimulating hormone and LH act cooperatively to stimulate the ovaries and testes to secrete sex hormones and to develop germ cells.

The testes are stimulated by LH to release testosterone from their Leydig cells (*Figure 7.37*). Testosterone is the principal androgen and its secretion inhibits further release of LH by a negative feedback mechanism. Follicle stimulating hormone and testosterone are required by Sertoli cells in the basement membrane of the seminiferous tubules to produce inhibin that, in turn, inhibits the secretion of FSH by negative feedback (*Figure 7.37*). Testosterone is also required for sexual differentiation, the development of secondary sexual characteristics and spermatogenesis. It is transported in the blood, bound to sex hormone binding globulin, SHBG (*Figure 7.38 (A)*) and to a lesser extent albumin. Typically, only its free fraction is metabolically active. Testosterone enters target cells and is converted to the potent androgen, dihydrotestosterone by the enzyme 5α-reductase. Testosterone is also found in the plasma of normal females, half of which is secreted by the ovaries, the remainder arising from the peripheral conversion of androstenedione and DHEAS, both of which are secreted by the adrenal cortex.

The ovaries produce estrogens, of which estradiol is required for the development of female secondary sexual characteristics and normal menstruation. Circulating estradiol is bound mostly to SHBG (*Figure 7.38 (B)*), although the blood concentration of estradiol varies widely with the menstrual cycle.

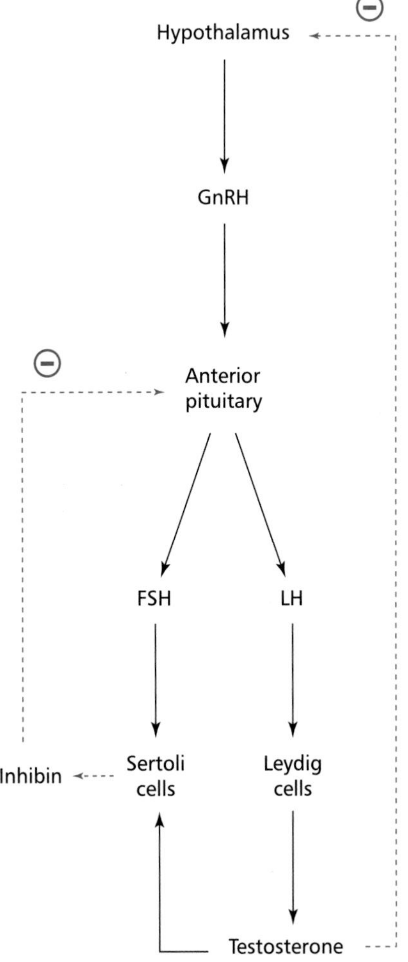

Figure 7.37 The regulation of testosterone (*Figure 7.3 (C)*) secretion.

Figure 7.38 Molecular models of (A) the active form of testosterone and (B) estradiol shown in red bound to SHBG. PDB files 1KDM and 1LHU respectively.

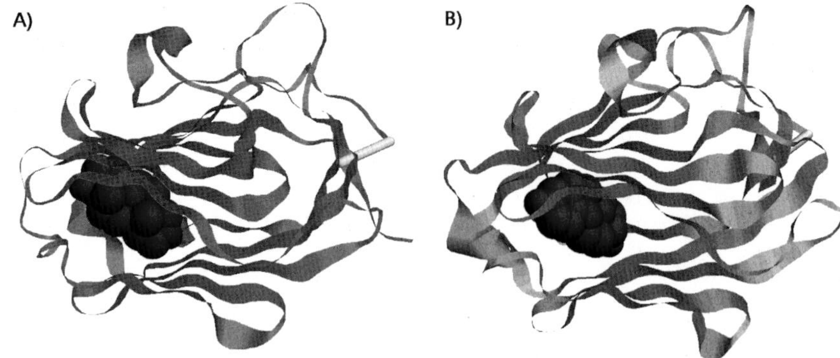

The plasma concentration of estradiol is low before puberty but increases rapidly and fluctuates during the menstrual cycle, a series of cyclical changes in the ovary, uterus and pituitary that occur approximately every 28 days until the menopause. Variations in plasma hormones in the menstrual cycle (*Figure 7.40*) depend on interactions between the hypothalamus, anterior pituitary and ovaries. Follicle stimulating hormone is released at the beginning of the cycle and increases growth of the follicles in the ovaries. Estradiol production increases the sensitivity of the pituitary to GnRH but decreases its secretion by the hypothalamus. The release of estradiol gradually increases and a follicle matures during the first half of the cycle. At the start of each cycle, about 20 secondary follicles enlarge and begin to secrete estrogen and the hormone inhibin, and a cavity filled with follicular fluid forms around their ova. This is referred to as antrum formation. By about the sixth day, one of the secondary follicles in an ovary has outgrown the others and becomes the dominant follicle. Its secretion of estrogen and inhibin decreases the secretion of FSH

BOX 7.3 Oral contraceptive pill

Oral contraceptives (*Figure 7.39*) known colloquially as *the pill*, contain a synthetic version of estrogen, called ethinylestradiol and the synthetic version of progesterone, progestogen. The estrogen prevents ovulation taking place, whereas progestogen acts on the pituitary gland to block the normal physiological control of the menstrual cycle. Progestogen alters the lining of the uterus so that it is unsuitable for implantation and increases the viscosity of the mucus in the cervix, so that conception is less likely even if ovulation does occur. Oral contraceptive pills are taken daily for three weeks and then stopped for a week during menstruation.

Pincus (1903–1967) began development of the contraceptive pill in 1950. Within a few years, clinical trials on 6000 women began in Puerto Rico and Haiti. The first commercially available contraceptive pill was introduced in 1960 after it was discovered that Mexican yam (*Pachyrhizus erosus*) was a cheap natural source of the hormone precursors required to make the pill. Over 60 million women worldwide use the pill with about three million in the UK. Early contraceptive pills contained between 100–175 μg of estrogen and 100 mg of progestogen. However, shortly after introduction of the pill, some concern was expressed about their side effects. These included an increased disposition to blood clots, heart attacks and strokes, although the risks involved were still relatively small. Studies by 1969 showed that the increased

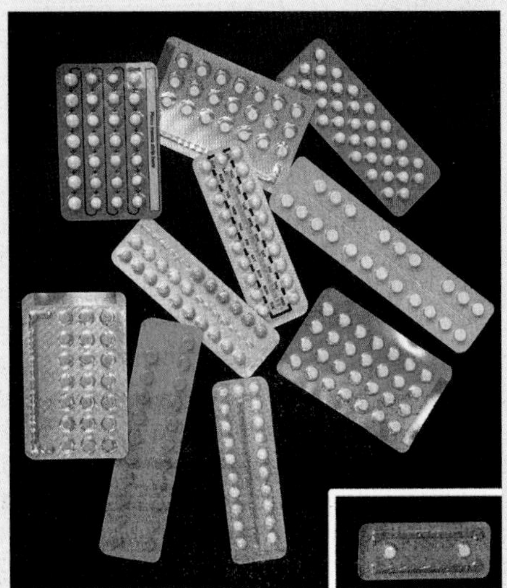

Figure 7.39 Examples of some types of oral contraceptive pills. The inset shows a packet of morning after pills. Courtesy of the Young Person's Sexual Health Clinic, Brook Advisory Center, Manchester, UK.

(*Figures 7.36* and *7.40*), leading to the regression of the other follicles in an apoptotic process to form atretic follicles. It is uncertain how only the one follicle becomes dominant, but appears to be related to its ability to secrete the estrogen, needed for its maturation under the influence of LH. Maturation involves the dominant secondary follicle accumulating fluid filled cavities that eventually enlarge to the point where they are called a Graafian follicle. Ovulation occurs each month when a Graafian follicle ruptures to release the oocyte, now usually called an ovum, into the Fallopian tube. The ovum is transported along the tube by ciliary action. The portion of the follicle remaining in the ovary develops into a corpus luteum. If fertilization does not occur, this degenerates within 10 days or so.

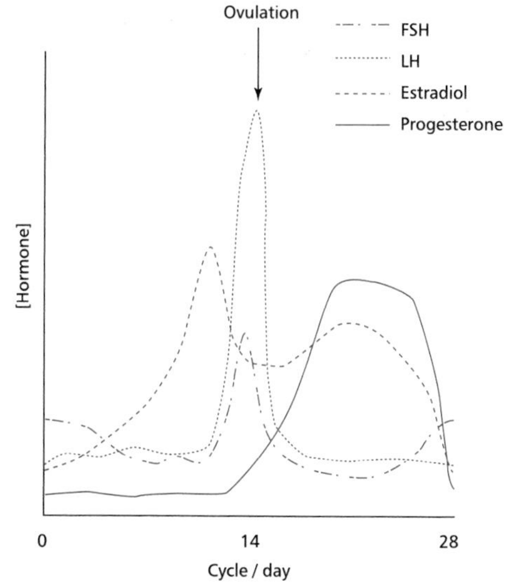

Figure 7.40 Hormonal changes in the menstrual cycle. See text for details.

Following copulation, the sperm are propelled through the vas deferens by muscular contractions into the urethra. The sperm are suspended in liquid semen produced by the seminal vesicles, prostate and bulbourethral glands. Semen contains nutrients, which activates and increases the motility of sperm, and is alkaline to counteract the acidity of the vagina. The ruptured follicle develops into the corpus luteum, which secretes progesterone and estradiol and stimulates the development of the endometrium for implantation. Fertilization of the egg to form a zygote usually takes place in the Fallopian tubes and the developing embryo is transported to the uterus by ciliary action and muscular contractions. The zygote begins a series of mitotic divisions to form a developing embryo that embeds into the endometrium lining the uterus and undergoes further development to produce a fetus and eventually a neonate in 9 months. Fertilization ensures that the corpus luteum does not degenerate but begins to produce a number of sex hormones, together with those produced by the gonads and anterior pituitary.

Following the menopause, plasma levels of estradiol decline despite the high levels of gonadotrophins and ovulation ceases.

risks of heart attacks and stroke were related to the amounts of estrogen in the pills. As a consequence, the amounts of estrogens used have decreased over the years and by 2006 contained less than one third of that in earlier contraceptive pills. Indeed, a reduction in the risk of heart disease and stroke has been detected in females on modern versions of the pill. Initially, there was some concern that usage of the pill increased the likelihood of cancers of the breast and cervix, but clinical data have cast doubt on this. Using the pill is still associated with some side effects, such as nausea, bleeding between menstrual periods and depression. The pill does increase the chance of blood clot in the legs (deep vein thrombosis, DVT, *Chapter 14*) although the risk for most pill users is very low. It is now known that usage of the pill may have many benefits, for example protection against pelvic infection because the thickened cervical mucus acts as a barrier to bacteria. Also, long-term usage of the pill has been reported to reduce the risk of certain ovarian cancers and can prevent ectopic pregnancies.

The development of the contraceptive pill has been a remarkable achievement as it allows women to control their fertility in a safe and effective manner. Since its introduction, the pill has had a tremendous impact on female liberty and has aided the process of making pregnancy and motherhood a choice for women. This, particularly in the developed world, has given them greater choices in marriage, work, love and lifestyle.

There is also a 'morning after pill', also known as the 'postcoital pill'. This pill is used by women to reduce the chances of pregnancy following unprotected sexual intercourse. It contains the active ingredient levonorgestrel, which is a synthetic derivative of progesterone. The precise mechanism of action of this pill is still unclear but it is believed to act by preventing ovulation, fertilization and implantation of the fetus. The whole process from fertilization to implantation in the womb can take up to three days, so the morning after pill can prevent pregnancy occurring for up to 72 h after intercourse. This pill is more effective the earlier it is taken after intercourse and it is estimated that 85% of pregnancies would be prevented if the morning after pill was taken within 72 h of sexual intercourse.

There is considerable interest in a contraceptive pill designed for use by men. This pill contains desogestrel as well as testosterone. This combination blocks the production of sperm while maintaining male characteristics and sex drive. As with the female contraceptive pill, it must be taken daily. In preliminary studies, the male pill reduced sperm counts to zero and is expected to be more effective than the female pill or the condom. According to the Food and Drug Administration (FDA) of the USA, the condom has a failure rate of about 14% under typical conditions, while the failure rate of the female pill is less than 1%. The male pill appears to be 100% effective.

DISORDERS OF SEX HORMONES

Disorders associated with male sex hormones include **hypogonadism** and gynecomastia. In the former, there is deficient sperm production and decreased testosterone secretion. Hypogonadism can occur because of testicular disease, referred to as primary or hypergonadotrophic hypogonadism, or to a defect in the hypothalamus or pituitary gland leading to secondary or hypogonadotrophic hypogonadism. In the latter, there may be a deficiency of both gonadotrophins or only of LH. The causes of primary hypogonadism are varied. They include congenital defects, such as Klinefelter syndrome (*Chapter 15*), a deficiency of 5α-reductase activity, testicular agenesis (failure of the testes to develop), acquired defects due to testicular infections, for example mumps or cytotoxic drugs, trauma or irradiation. The causes of secondary hypogonadism include pituitary tumors and hypothalamic disorders such as Kallmann's syndrome. The treatment of hypogonadism is usually directed at the underlying cause. Testosterone is given in cases of testosterone deficiency. However, if fertility is required then gonadotrophins may have to be administered.

Gynecomastia, or breast development in males, is usually related to a disturbance in the balance of estrogens to androgens. In puberty, it occurs in approximately 50% of normal boys due to a temporary increase in the secretion of estrogens. Other than at puberty, the condition is pathological. It may arise because of decreased androgen activity in hypogonadism or because of increased estrogen production from various endocrine tumors. Such tumors may secrete large quantities of estrogens or may secrete hCG that stimulates estrogen production. Some drugs possess estrogen or antiandrogen activity and their use can lead to gynecomastia.

Disorders associated with female sex hormones include amenorrhea, **oligomenorrhea**, infrequent menstruation, and **virilism**, with the development of hirsutism, muscle mass, deepening of the voice and male psychological characteristics. Amenorrhea can be primary where menstruation fails to occur by the age of 16 years, or secondary, where menstruation stops for three months or more after normal menstruation has been established and before menopause. The other clinical features of amenorrhea include hirsutism, acne, menstrual cycle disturbances and obesity, although these features vary in their severity and prevalence. A common reason for secondary amenorrhea in females is pregnancy. This condition must be excluded before other possible causes, for example stress, severe weight loss, polycystic ovary syndrome (PCOS, *see below*), gonad dysgenesis, such as in Turner syndrome (*Chapter 15*), the decrease in gonadotrophin secretion associated with some tumors, hyperprolactinemia and congenital adrenal hyperplasia, are investigated. The causes of amenorrhea are investigated by measuring the concentrations of FSH, LH and prolactin in plasma as outlined in *Figure 7.41*. A high value for FSH indicates ovarian failure. One for prolactin suggests hyperprolactinemia and requires further investigations to confirm this diagnosis. If, however, the values for FSH, LH and prolactin are normal, further tests to investigate pituitary or hypothalamic diseases are necessary. The management of amenorrhea is aimed at treating its underlying cause.

Patients with virilism present with enlargement of the clitoris, deepening of the voice, atrophy of the breasts and hirsutism. Hair growth is not only excessive but shows a male-like distribution. The cause of virilism is increased androgen secretion although abnormally low levels of SHBG can also increase the free testosterone fraction. In some cases virilism occurs because of an increased sensitivity to androgens by target cells. Its causes include PCOS, androgen secreting tumors, congenital adrenal hyperplasia, Cushing's syndrome and may be iatrogenic following treatment with androgens and progesterone. Its commonest cause is PCOS, characterized by multiple cysts in the ovaries that arise from follicles that have failed to ovulate. The ovaries secrete large amounts of androgens although why this is so is unclear. Many

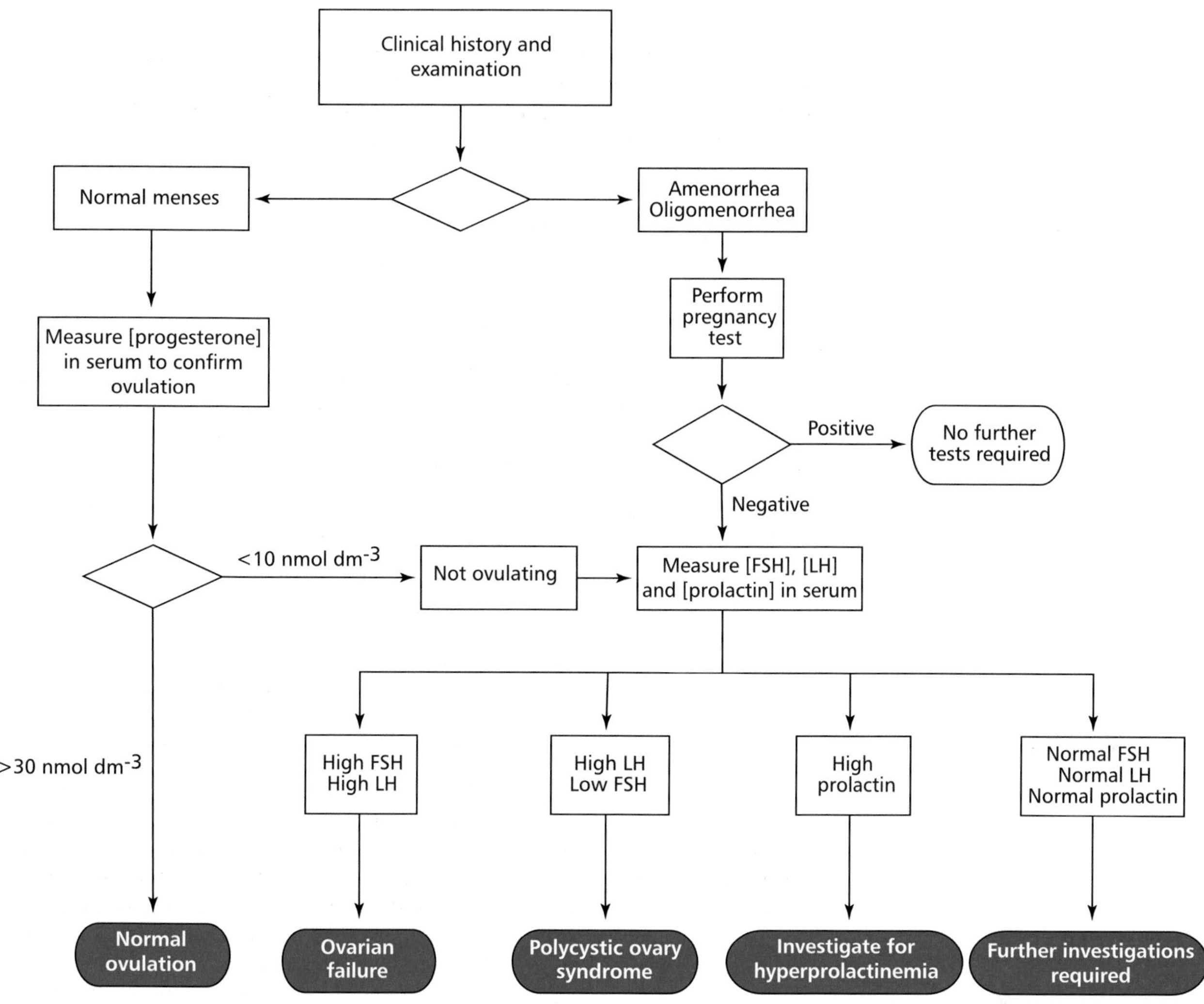

Figure 7.41 Overview of the investigation of amenorrhea.

patients with PCOS suffer from acne, obesity, type 2 diabetes mellitus and may be infertile. A diagnosis of PCOS is made on clinical grounds and assisted by ultrasonography. Plasma LH is often increased in these patients but may be normal, whereas the concentration of testosterone is increased. Treatment of severe PCOS may involve using the antiandrogen drug, cyproterone but this leads to infertility. If fertility is required, then clomiphene, an antiestrogen drug, may be used and can induce ovulation in 75% of cases.

Infertility

Infertility is defined as a failure to conceive despite regular unprotected sexual intercourse for one year. Female infertility may be due to failure to ovulate, obstruction of the Fallopian tubes or to diseases of the uterine lining. In females, failures to ovulate due to hyperprolactinemia or to hypothalamic-pituitary dysfunction are responsible for 20% of cases of infertility. Damage to the Fallopian tubes may also be a cause of female infertility. Male infertility is usually due to decreased numbers or motility of sperm (oligospermia) or complete absence of sperm (azoospermia). Infertility due to endocrine dysfunction occurs only rarely in males. The normal volume of ejaculate is

2 to 5 cm^3 and contains 40 to 500 million sperm. A sperm count of less than 20 million per cm^3 is believed to be subnormal and causes 25% of infertility cases. In general, the lower the total sperm number the greater the chances of infertility although men with low counts have been known to father children. The motility of sperm is also necessary for fertilization to take place and at least 60% of sperm should have a normal shape and be mobile with beating flagella.

Investigating infertility involves a clinical and laboratory evaluation of both partners. Women should be investigated for regular menstruation. If it is regular, then ovulation is probably occurring. The concentration of progesterone in plasma is also indicative and should exceed 30 nmol dm^{-3} on day 21 of the menstrual cycle, values below 10 nmol dm^{-3} are suggestive of abnormal ovulation. If ovulation is confirmed then it may be necessary to examine cervical mucus following intercourse to determine the presence of motile, normal shaped sperm.

Microscopic examination of samples of ejaculate will indicate whether the sperm produced are motile and normal in shape or show abnormalities, such as malformed heads and twin sperm formed by a failure in development. Sperm counts will show if adequate numbers of sperm are produced. A low sperm count may be further investigated by measuring the concentrations of testosterone, FSH and LH in the plasma to compare with their reference ranges of 9–30 nmol dm^{-3}, 2–10 U dm^{-3} and 2–10 U dm^{-3} respectively. In some cases, a biopsy of the testes may by necessary. Investigation of infertility in males is outlined in *Figure 7.42*.

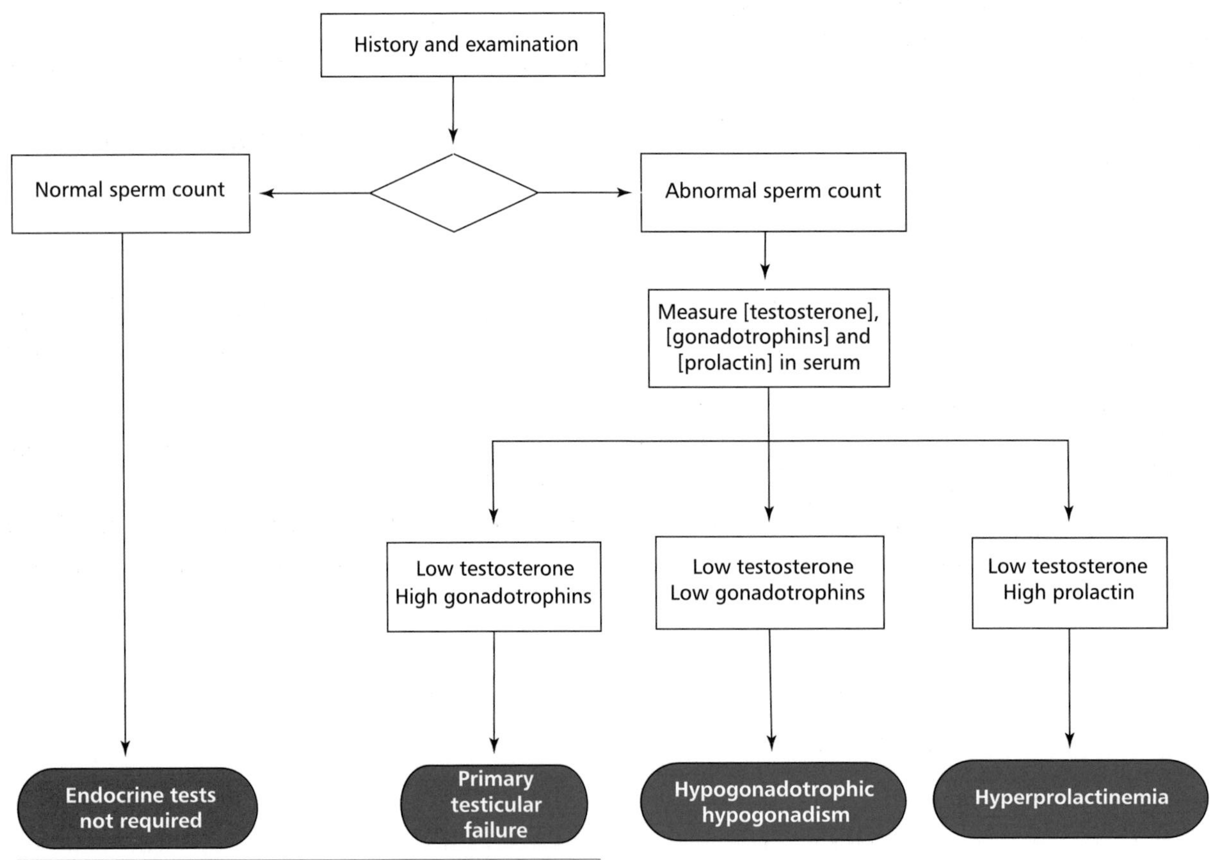

Figure 7.42 Overview of the investigation of infertility in males.

CASE STUDY 7.1

Sarah, a 46-year-old woman, was referred to hospital because of her excessive thirst and frequent urination over the preceding two months. Her thirst was severe and she had to drink water every few hours. An analysis of her plasma and urine yielded the following results (reference ranges are given in parentheses):

Plasma:	Na^+	139	(136–146 mmol dm^{-3})
	K^+	4.2	(3.7–5.1 mmol dm^{-3})
	Creatinine	105	(65–121 μmol dm^{-3})
	Osmolality	296	(282–296 mmol kg^{-1})
Urine:	Glucose		negative
	Osmolality	90	(up to 1400 mmol/kg)

A fluid deprivation test was performed and the following results were obtained for the plasma and urine osmolality:

Plasma:	Osmolality	310	(282–296 mmol kg^{-1})
Urine:	Osmolality	225	(up to 1400 mmol kg^{-1})

She was allowed to drink water on completion of the test and given a dose of desmopressin after which her urine osmolality increased to 620 mmol kg^{-1}.

Questions

(a) What could be the causes of Sarah's symptoms?

(b) Account for the plasma and urine data.

(c) Discuss the need for and results of the fluid deprivation test.

CASE STUDY 7.2

Ian, a 53-year-old man went to see his family doctor complaining of weakness. Thyroid function tests were requested to exclude the possibility of hypothyroidism. These tests showed the following results (reference ranges are given in parentheses):

TSH	7.5 mU dm^{-3}	(0.2–4.0 mU dm^{-3})
Free T_4	14.5 pmol dm^{-3}	(12–25 pmol dm^{-3})

Questions

(a) What is the most appropriate clinical explanation of these results?

(b) What other signs and symptoms should be looked for?

(c) How should Ian's progress be monitored?

CASE STUDY 7.3

Amelia, a 20-year-old student, had not been feeling well for three weeks. She was admitted to her local hospital with a history of weakness and vomiting over the previous three days. On admission she was unconscious. Clinical tests on serum yielded the following data (reference ranges are given in parentheses):

Na^+	113 mmol dm^{-3}	(137–144 mmol dm^{-3})
K^+	5.8 mmol dm^{-3}	(3.3–4.2 mmol dm^{-3})
Urea	30 mmol dm^{-3}	(2.6–6.5 mmol dm^{-3})
pH	7.3	(7.35–7.45)

Questions

(a) What is the most likely explanation of these results?

(b) What is the probable cause of Amelia's condition?

(c) Having identified her condition, how should Amelia be treated?

CASE STUDY 7.4

An 18-year-old girl, Jaclyn, was admitted to hospital in a coma. She presented with a deep sighing respiration and her breath smelt of acetone. She had vomited earlier. On presentation, she had a low blood pressure and a high pulse rate of 120 per min. The following results were obtained (reference ranges are given in parentheses):

Plasma:

Na^+	136 mmol dm^{-3}	(132–144 mmol dm^{-3})
K^+	5.7 mmol dm^{-3}	(3.2–4.8 mmol dm^{-3})
Urea	15 mmol dm^{-3}	(3.0–8.0 mmol dm^{-3})
Glucose	31.0 mmol dm^{-3}	(3.0–5.5 mmol dm^{-3})
Osmolality	371 mmol dm^{-3}	(282–295 mmol dm^{-3})

Arterial blood:

pH	7.09	(7.35–7.45)
PCO_2	2.7 KPa	(4.7–6.0 KPa)
HCO_3^-	11 mmol dm^{-3}	(24–34 mmol dm^{-3})

Urine:

Ketones	positive	negative

Questions

(a) Explain these results.

(b) How should Jaclyn be treated?

CASE STUDY 7.5

Rachel, a 32-year-old teacher, was suffering from irregular periods and acne. She was examined by her doctor and found to be overweight and hirsute. She was referred for hospital tests, which yielded the following data (reference ranges are given in parentheses):

Plasma:

Testosterone	3.9	(1.1–3.3 nmol dm^{-3})
LH	14	(2.0–10 U dm^{-3})
FSH	5.3	(2.0–8.0 U dm^{-3})

Questions

(a) What can be inferred from Rachel's symptoms?

(b) What is the most likely diagnosis?

7.11 SUMMARY

The endocrine glands and the hormones they produce control many of the activities of the body. Hormones are released into the blood and travel to their sites of action where they bind to cell surface or intracellular receptors to initiate activities. Hormones belong to several categories of molecules including amines, peptides, proteins and steroids. Endocrine disorders may arise from damage to the endocrine gland, causing hypo- or hypersecretion of the hormone, or some failure of the hormone–receptor interactions. Disorders of the pituitary gland may arise from pituitary tumors, such as prolactinomas, with excessive secretion of hormones. Abnormal GH secretion leads to dwarfism or acromegaly. Several thyroid disorders arise as a consequence of autoimmune conditions which can cause hypo- or hyperactivities. However, the commonest endocrine disorders involve insulin resistance or its absence leading to dysregulation of blood glucose concentrations. Disorders of the adrenal gland can result in Addison's disease or Cushing's syndrome. Dysfunctions associated with reproductive hormones can cause a number of clinical conditions, including gynecomastia in men and disruption of menstruation, infertility, and virilism in women.

The diagnosis of endocrine disorders usually involves measurement of the hormone in question and replacement for deficiencies. Disorders due to hypersecretion from endocrine tumors may be treated by surgical excision of the tumor.

QUESTIONS

1. Pituitary tumors can secrete various hormones. Which of the following hormones is the most commonly secreted?
   a) thyroid stimulating hormone;
   b) follicle stimulating hormone;
   c) luteinizing hormone;
   d) prolactin;
   e) growth hormone.

2. Which of the following may cause hyperthyroidism?
   a) Grave's disease;
   b) Hashimoto's thyroiditis;
   c) thyroidectomy;
   d) antithyroid drugs;
   e) diabetes insipidus.

3. Which of the following can cause acromegaly?
   a) diabetes mellitus;
   b) pituitary adenoma;
   c) excessive ingestion of growth hormone;
   d) bronchial carcinoma;
   e) excessive secretion of prolactin.

4. Which one of the following is **NOT** a cause of Cushing's syndrome?
   a) ectopic ACTH production;
   b) adrenal adenoma;
   c) corticosteroid therapy;
   d) autoimmune destruction of the thyroid gland;
   e) all of the above.

5. Ketone bodies are produced during diabetic ketoacidosis as a result of an increase in which of the following?
   a) protein breakdown;
   b) urea production;
   c) insulin release;
   d) lipolysis;
   e) K^+.

6. Estradiol is required for?
   a) the development of male secondary sexual characteristics;
   b) spermatogenesis;
   c) the development of female secondary sexual characteristics;
   d) the secretion of breast milk;
   e) stimulating the endometrium to receive fertilized egg.

7. Sajida, a 27-year-old, found over a 2-month period that she was expressing breast milk. She was not pregnant and, indeed, had not had any children. She was on medication with analgesic and antidopaminergic drugs for migraines attacks. An analysis of her serum gave a prolactin value of 1200 (reference range: 70–395 mU dm^{-3}).

    a) What is the most likely diagnosis?
    b) What could have caused such a condition?
    c) What should her doctor recommend?

8. Graham, a 50-year-old man, visited his doctor for the first time in many years complaining of blurred vision and frequent headaches. The doctor noticed changes in the facial features characteristic of acromegaly. An oral glucose tolerance test including measurements of GH was performed.

Time / min	[Glucose] / mmol dm^{-3}	[GH] / mU dm^{-3}
0	6.3	47
30	7.3	55
60	8.6	52
90	9.5	64
120	8.8	53
150	7.5	49
180	6.8	49

Reference range for fasting glucose 3.0–5.5 mmol dm^{-3}

Reference range for GH following glucose load < 2 mU dm^{-3}

Plot a graph to illustrate these data. Do these findings support the diagnosis of acromegaly?

9. Tabulate the differences between types 1 and 2 diabetes mellitus.

10. Give four functions of human GH.

11. Why is it necessary to screen newborn babies to detect congenital hypothyroidism?

FURTHER READING

Ahmed, N (2005) Advanced glycation endproducts – role in pathology of diabetic complications. *Diabetes Res. Clin. Pract.* **67:** 3-21.

Ali, I and Dawber, R (2004) Hirsutism: diagnosis and management. *Hosp. Med.* **65:** 293–297.

Ben-Shlomo, A and Melmed, S (2001) Acromegaly. *Endocrinol. Metab. Clin. North Am.* **30:** 565–583.

Boscaro, M, Barzon, L, Fallo, F and Sonino, N (2001) Cushing's syndrome. *Lancet* **357:** 783–791.

Carroll, MF, Burge, MR and Schade, DS (2003) Severe hypoglycemia in adults. *Rev. Endoc. Metab. Disord.* **4:** 149–157.

Cornia, PB and Anawalt, BD (2004) Male hormonal contraception. *Expert Opin. Emerg. Drugs* **9:** 335–344.

Dattani, M and Preece, M (2004) Growth hormone deficiency and related disorders: insights into causation, diagnosis and treatment. *Lancet* **363:** 1977–1987.

Ehrmann, DA (2005) Polycystic ovary syndrome. *N. Engl. J. Med.* **352:** 1223–1236.

Erxheimer, A and Waterhouse, J (2003) The prevention and treatment of jet lag. *BMJ* **326:** 296–297.

Holdaway, IM (2004) Treatment of acromegaly. *Horm. Res.* **62:** 79–92.

Kaufman, FR (2003) Type I diabetes mellitus. *Pediatr. Rev.* **24:** 291–299.

Korc, M (2004) Update on diabetes mellitus. *Dis. Markers* **20:** 161–165.

Lee, KC and Kraus, WL (2001) Nuclear receptors, coactivators and chromatin: new approaches, new insights. *Trends Endocrinol. Metab.* **12:** 191–197.

Manger, WM and Eisenhofer, G (2004) Phaeochromocytoma: Diagnosis and management update. *Curr. Hypertens. Rep.* **6:** 477–484.

Nadar, S, Lip, GYH and Beevers, DG (2003) Primary hyperaldosteronism. *Ann. Clin. Biochem.* **40:** 439–452.

Roberts, CGP and Ladenson, PW (2004) Hypothyroidism. *Lancet* **363:** 793–803.

Speiser, PW and White, PC (2003) Congenital adrenal hyperplasia. *N. Engl. J. Med.* **349:** 776–788.

Stumvoll, M, Goldstein, BJ and van Haeften, TW (2005) Type 2 diabetes: principles of pathogenesis and therapy. *Lancet* **365:** 1333–1346.

Sulak, PJ (2004) Oral contraceptive update: new agents and regimens. *J. Fam. Pract.* **Suppl**: S5–S12.

Taylor, A (2003) ABC of subfertility: Making a diagnosis. *BMJ* **327:** 494–497.

Ten, S, New, M and Maclaren, N (2001) Addison's disease 2001. *J. Clin. Endocrinol. Metab.* **86:** 2909–2933.

Topliss, DJ and Eastman, CJ (2004) Diagnosis and management of hyperthyroidism and hypothyroidism. *Med. J. Aust.* **180:** 186–193.

Trachtenbarg, DE (2005) Diabetic ketoacidosis. *Am. Fam. Physician* **71:** 1705–1714.

Verbalis, JG (2003) Diabetes insipidus. *Rev. Endocr. Metab. Disord.* **4:** 177–185.

Verhelst, J and Abs, R (2003) Hyperprolactinemia: pathophysiology and management. *Treat. Endocrinol.* **2:** 23–32.

Williams, M (1998) Disorders of the adrenal gland. *Semin. Preoper. Nurs.* **7:** 179–185.

chapter 8:

DISORDERS OF WATER, ELECTROLYTES AND URATE BALANCES

OBJECTIVES

After studying this chapter you should be able to:

- describe the structure of the kidney;
- list the functions of the kidney;
- describe renal function tests;
- describe the types of renal failure;
- explain the causes and clinical consequences of selected disorders of water, electrolyte and urate homeostases;
- describe the investigation of selected disorders of water, electrolyte and urate homeostases;
- discuss the management and treatment of such disorders.

8.1 INTRODUCTION

Homeostasis is the maintenance of a stable internal environment within the body. This stability is necessary for optimum functioning of proteins, particularly enzymes, cells, tissues, organs and systems. Many substances have to be maintained at appropriate concentrations, including water, electrolytes, such as Na^+, K^+, Ca^{2+}, Mg^{2+} and P_i, and the acid–base components H^+ and HCO_3^-. In addition, waste products, such as urea and urate, must be kept below toxic levels. Normally biochemical and physiological mechanisms regulate and control the concentrations of all these components and, in general, homeostatic controls involve negative feedback mechanisms. A receptor detects unacceptable levels of a particular substance under homeostatic control and sends a signal to a regulatory center that initiates a response that corrects the imbalance and returns conditions to a physiologically acceptable state (*Figure 8.1*). Once normality returns, the receptor is no longer stimulated and the center ceases to respond. Disorders of homeostasis can occur, often as a result of failures in the control mechanisms or because of damage to the regulatory center by external agents.

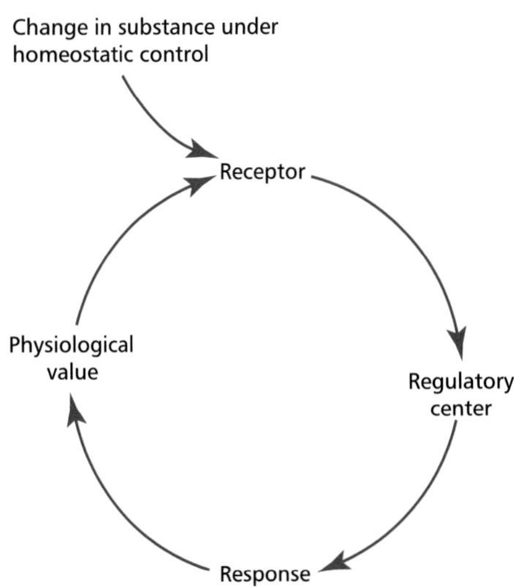

Figure 8.1 A generalized feedback mechanism. See text for details.

Within the body, the kidneys regulate water, electrolyte concentrations, including acid–base balance (*Chapter 9*), and excrete nitrogenous wastes, for example urate. Disorders of water, electrolyte and acid–base homeostasis and urate excretion account for a large number of investigations that are carried out in hospital laboratories.

8.2 KIDNEYS

The functions of the kidney include maintaining the volume of the plasma and its concentrations of electrolytes, such as Na^+ and K^+, and keeping its pH within normal physiological ranges. This is largely achieved by varying the amounts of water and salts excreted, the removal of excess H^+ (*Chapter 9*) and the regeneration of HCO_3^-. Kidneys also excrete waste products, such as urea, urate and creatinine, and produce the enzyme rennin and the hormones erythropoietin and calcitriol (also called 1α,25-dihydroxycholecalciferol, 1α,25DHCC). These control blood pressure, stimulate the production of erythrocytes by the bone marrow (*Chapter 13*) and regulate the absorption of Ca^{2+} by the gastrointestinal tract (GIT) (*Chapter 11*) respectively. Kidneys also synthesize prostaglandins and degrade hormones, such as insulin.

Each kidney is composed of an outer fibrous capsule, a cortex, a middle medulla and an inner pelvis region (*Figure 8.2A*). The tough capsule surrounding each kidney offers protection against trauma and prevents the entry of bacteria (*Chapter 3*). Kidneys are composed of about a million functional units called **nephrons** each composed of a tuft of capillaries called a glomerulus and a tubule. Tubules have four different regions: the proximal tubule, the loop of Henle, the distal tubule and the collecting duct (*Figure 8.2B*). The cortex

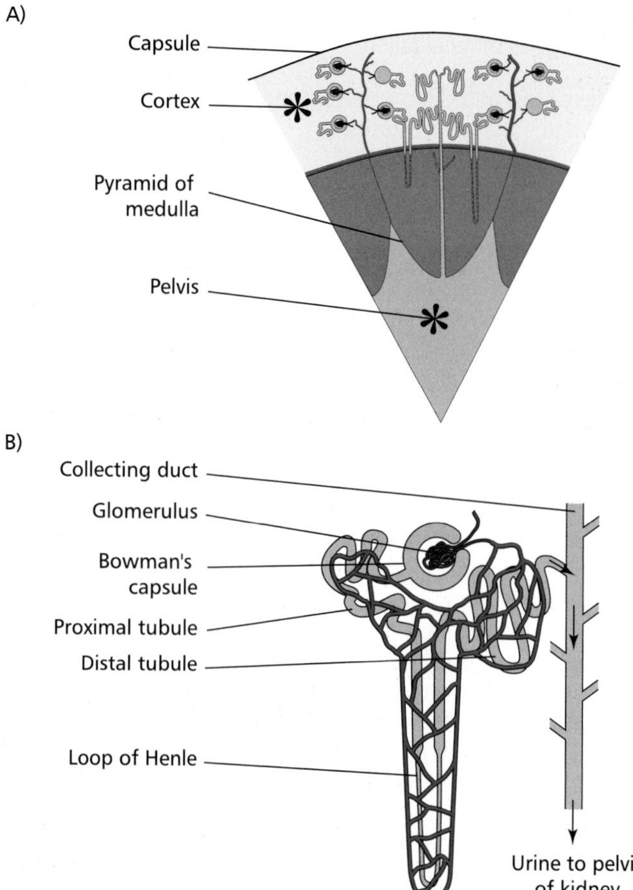

Figure 8.2 (A) Schematic showing the structure of the kidney. (B) The structural relationships between the glomerulus, proximal tubule, loop of Henle and distal tubule (nephron).

contains the Bowman's capsules and their glomeruli. The medulla is composed of the tubules arranged in pyramids that lead into calyces that, in turn, lead into the pelvis of the kidney. The pelvis drains into the ureters that take urine from the kidneys to the bladder.

Kidneys have a rich blood supply and normally receive about 25% of the cardiac output. The renal artery supplies blood at a high pressure to afferent arterioles that supply the glomerular capillaries. Glomerular capillaries drain into efferent arterioles that, in turn, divide to form a capillary network covering the nephron (*Figure 8.2B*). Blood eventually leaves each kidney in a renal vein. Blood is separated from the lumen of the tubules by three layers: the capillary endothelial cells, a basement membrane and specialized epithelial cells of the Bowman's capsule, called podocytes (*Figure 8.3*). Openings between the extensions of the podocytes are called fenestra. The basement membrane contains negatively charged glycoproteins that give the basement membrane an overall negative charge. The hydrostatic pressure of blood in the glomerulus is high at 10 kPa because of its direct route from the heart and because the diameters of afferent arterioles supplying the glomeruli are less than the efferent arterioles collecting blood from them. This forces the plasma to filter through the layers into the lumen of the capsule. The hydrostatic pressure of blood in the glomerulus is opposed by osmotic pressure of 4 kPa generated by its plasma proteins and a back pressure of 2.7 kPa exerted by the filtrate in the Bowman's capsule. Thus the effective pressure, P*eff*, is:

$$P_{eff} = 10 - (4 + 2.7) = 3.3 \text{ kPa}$$

Water and small molecules are passively filtered into the Bowman's capsule leaving blood cells and plasma proteins in the capillary. Particles with a M_r less than 5000, such as electrolytes, sugars, amino acid, urea and some small polypeptides and proteins pass freely from the plasma through the glomerular wall into the lumen of the capsule. Substances with M_r up to 68 000, can penetrate to some extent but larger molecules, such as proteins above 68 000 are excluded because of their size and to some extent their charge, given that most plasma proteins are negatively charged at pH 7.4. Hence the initial filtrate in the capsule lumen has a composition similar to that of plasma except that it is largely free of protein. Most of the filtrate is reabsorbed as it passes along the nephron. The proximal tubule is responsible for bulk reabsorption of filtrate while the distal tubule is important for fine tuning its composition depending upon the needs of the body. Normally all the glucose, amino acids, K^+ and HCO_3^- and about 75% of the Na^+ are absorbed by energy dependent mechanisms. The reabsorption of water occurs passively and follows Na^+ reabsorption. Approximately 90% of the filtered Na^+ and 80% of water is reabsorbed in the distal tubule. More Na^+ is reabsorbed in the distal tubule by the cells exchanging it for K^+ and H^+. This exchange is controlled by

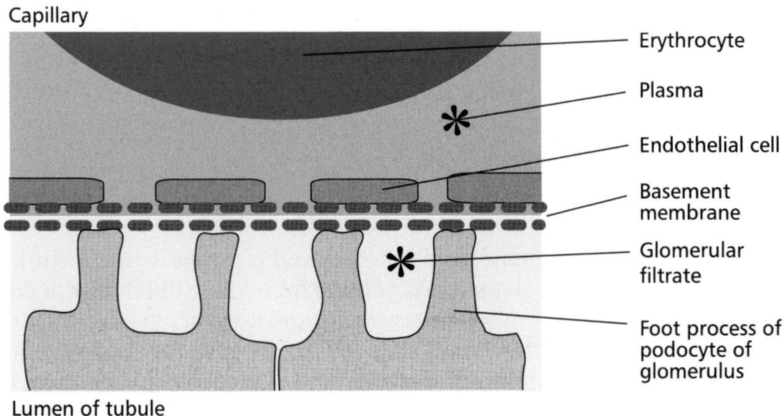

Figure 8.3 Schematic of the filtration unit of the glomerulus. See text for details.

aldosterone (*Chapter 7*). Between 6 and 12% of the filtered urate is excreted by the kidneys with the remainder being reabsorbed in the proximal convoluted tubule. Tubular fluid passes into collecting ducts that extend into the renal medulla and discharge urine into the renal pelvis. About 1 to 2 dm^3 of urine is produced per day depending on the amount of fluid intake with larger volumes being produced after increased intake of water.

RENAL FUNCTION TESTS

Renal function tests are used to detect the presence of renal diseases and assess their progress. They are, however, of little use in determining the causes of renal disease. The most widely used test is to measure the glomerular filtration rate (GFR), that is, the rate of filtrate formation by the kidneys. The value of the GFR depends on the net pressure across the glomerular membrane, the physical nature of the membrane and the surface area of the membrane that reflects the number of functioning glomeruli. All three factors can change as a result of disease and this will be reflected in the value of the GFR. In adults, the GFR is about 120 cm^3 per minute although it is related to body size, being higher in men than women. The GFR is also affected by age and declines in the elderly.

Measuring the GFR

The GFR is determined by measuring the concentration of a substance in the urine and plasma that is known to be completely filtered from the plasma at the glomerulus. This substance must not be reabsorbed nor secreted by renal tubules and must remain at a constant concentration in the plasma throughout the period of urine collection. It should also be possible to measure the concentration of this substance in the plasma and urine both conveniently and reliably. Inulin and creatinine have been used to assess GFR using the equation:

$$\mathbf{GFR = (U_c \times V) / P_c}$$

where U_c is the concentration of substance in urine, P_c is the concentration of substance in plasma, V is rate of formation of urine in cm^3 per min giving the GFR units of $cm^3\ min^{-1}$.

Creatinine is derived from creatine phosphate in the muscle and the amount produced daily is relatively constant. An estimate of creatinine clearance can be made by a determination of the creatinine concentration in the plasma (*Figure 1.17*) and the creatinine content in a 24-h urine collection. Normal creatinine clearance in adults is between 115 and 125 $cm^3\ min^{-1}$. Reliable measurements of creatinine clearance are often difficult because of the need to obtain a complete and accurately timed urine sample.

Measurements of the concentration of serum creatinine may be used to assess renal function and they are easier to determine than creatinine clearance values. The concentration of creatinine in serum increases with deteriorating renal function but this test lacks sensitivity. For example, the GFR must fall to less than 50% of the original value before there is a significant increase in serum creatinine. This means that a normal serum creatinine value does not necessarily exclude the presence of renal disease.

RENAL FAILURE

Renal failure is the cessation of renal function and it can be acute or chronic. In acute renal failure there is rapid loss of renal function within hours or days, although the condition is potentially reversible and normal renal function can be regained. The deterioration is sudden, with increases in the concentrations of urea, creatinine and H^+ in serum. Patients with acute renal failure often, but not always, present with **oliguria**, where there is less than 400 cm^3 of urine passed per day. Indeed, patients are sometimes **anuric** and do not pass any urine at all. Chronic renal failure is the gradual, progressive deterioration of kidney function. As kidney function declines, there is accumulation of waste products that eventually reach toxic levels in the blood and may affect other organs.

Acute renal failure

Acute renal failure can be categorized as prerenal, where the loss in renal function is due to a decrease in renal blood flow, postrenal, where the loss is due to an obstruction of the urinary tract, or intrinsic, where the loss is due to damage to the kidney itself.

Prerenal kidney failure can occur because of a decreased plasma volume following blood loss, burns, prolonged diarrhea or vomiting, decreased cardiac output or occlusion of the renal artery. Whatever the reason, prerenal acute renal failure results in a low GFR and decreased blood flow to kidneys. Aldosterone and antidiuretic hormone (*Section 8.3* and *Chapter 7*) release is stimulated by the low blood pressure and the kidneys respond by producing smaller volumes of concentrated urine. The biochemical indicators of prerenal acute renal failure include increased amounts of urea and creatinine in serum due to the low GFR, metabolic acidosis (*Chapter 9*), because of an inability of the kidneys to excrete H^+, and hyperkalemia (*Section 8.5*) because of the low GFR and acidosis. Postrenal kidney damage can be the consequence of a blockage of the urinary tract by, for example, renal calculi, kidney stones or neoplasms. These obstructions increase the hydrostatic pressure that opposes glomerular filtration. If this persists for a sufficiently long time it can cause **intrinsic** renal damage. If pre- or postrenal acute failures are not corrected, patients can develop intrinsic renal damage. A variety of conditions cause intrinsic acute renal failure. These include nephrotoxins, for example drugs such as aminoglycosides and analgesics, septic shock (*Chapters 2* and *4*), a low cardiac output (*Chapter 14*), burns or crush injuries and renal diseases, such as glomerulonephritis. Glomerulonephritis is inflammation of the renal cortex which affects the filtration mechanism of the kidney. It may develop following an infection (*Chapters 2, 3* and *5*).

Three phases occur in acute renal failure. The first is an oliguric phase with a low urine output. The second is a diuretic phase where the urine output increases while the third is a recovery phase when normal function returns. The oliguric phase is characterized by increased concentrations of K^+, urea, creatinine and H^+ in the serum. If the patient survives the oliguric phase, then urine output increases after a few days when the diuretic phase starts. The GFR increases during the diuretic phase and as the output of urine increases the amounts of urea and creatinine in the serum gradually fall. Tubular function may still be abnormal in this phase so that the acidosis may still persist. In the recovery phase, tubular cells regenerate and tubular function is restored to normal. The concentrations of urea and creatinine in serum decrease and K^+ returns to normal levels as the GFR improves.

The management of acute renal failure includes correction of prerenal factors if they are present. This could be achieved, for example by increasing the extracellular fluid volume by administering fluids. Biochemical monitoring of creatinine and K^+ is required and dialysis may be necessary when K^+ concentrations are high or when severe acidosis is present. The cause of the renal failure should be identified and treated wherever possible.

Chronic renal failure

Many diseases, such as glomerulonephritis, diabetes mellitus, hypertension and polycystic kidney disease, can lead to irreversible renal damage. All these conditions effectively decrease the number of functioning nephrons. Patients may remain asymptomatic until the GFR falls below 15 cm^3 min^{-1}. Chronic renal failure progresses to end-stage renal failure (ESRF) where dialysis or a kidney transplant is necessary for survival. Metabolic features of ESRF include the impairment of urinary concentration and dilution, abnormal electrolyte and H^+ homeostasis, the retention of waste products and decreased syntheses of erythropoietin and calcitriol. Chronic renal failure is accompanied by increases in the concentrations of K^+, urea, creatinine, P_i and H^+ and decreased Ca^{2+} in the plasma.

Patients with chronic renal failure often present with a number of clinical features including neurological symptoms, such as lethargy, growth failure, myopathy, anorexia, nausea, vomiting, anemia, hypertension, *nocturia* and impotence. The causes of many of these are not known but are probably due to the retention of toxins that cannot be excreted.

In some cases, it is possible to delay the progression of the disease by treating its cause. A number of measures may be taken to alleviate symptoms before dialysis becomes necessary. These include careful matching of water and Na^+ intake with their losses. High blood K^+ is controlled with oral ion exchange resins given as their Ca^{2+} or Na^+ salts (*Section 8.5*) whereas a high blood P_i is controlled by oral aluminum or magnesium salts that precipitate ingested phosphate in the GIT. A restriction of dietary protein may reduce the formation of nitrogenous waste.

8.3 DISORDERS OF WATER HOMEOSTASIS

Water is necessary to maintain the volumes of body compartments, for excretion of waste products and as a medium in which biochemical reactions occur. Water intake is variable and can depend, to some extent, on social habits but is supplied in the diet, from food as well as water and as a product of oxidative metabolism. Its loss is variable although an almost fixed amount, called the insensible loss, occurs from the GIT, skin and lungs. An average 70 kg man has 42 dm^3 of water distributed between various body compartments (*Figure 8.4*). Water accounts for 60% of body weight in men but only 55% in women given they have a higher proportion of fat. In disease, patients can be **dehydrated**, where water loss caused by vomiting and diarrhea exceeds gain, or **overhydrated**, with an accumulation of water in body compartments. The clinical features of dehydration and over hydration are listed in *Table 8.1*. A reduced extracellular fluid (ECF) volume causes a decline in blood circulation with decreased excretion of wastes and reduced oxygen and nutrient supply to the cells. Humans deprived of fluid intake die after a few days because the reduced total body fluid leads to a circulatory collapse.

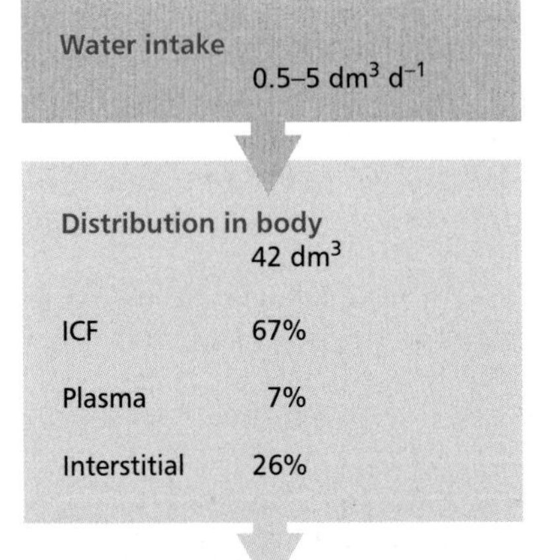

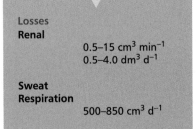

Figure 8.4 The distribution of body water. (ICF, intracellular fluid.)

Clinical feature affected	Dehydration	Overhydration
Pulse	increased	normal
Blood pressure	decreased	normal/increased
Skin turgor	decreased	increased
Eyeballs	soft/sunken	normal
Mucous membranes	dry	normal
Urine output	decreased	normal/decreased
Consciousness	decreased	decreased

Table 8.1 Clinical features affected in patients suffering dehydration and overhydration

The kidneys regulate water balance by varying the output of urine from 0.5 to 15 cm^3 min^{-1} to match water intake. When there is an excess of water, the kidneys lose water rapidly but in times of shortage it is conserved. The total body water is therefore kept constant. Water loss from the kidneys can be regulated by the hormone arginine vasopressin also called antidiuretic hormone (ADH). Antidiuretic hormone acts by altering the permeability to water of the collecting ducts in the kidneys. Osmoreceptor cells in the hypothalamus detect an increase or decrease in osmolality between the intracellular fluid (ICF) and ECF. An increase in the osmolality of the ECF stimulates the receptors and these, in turn, stimulate the release of ADH from the posterior pituitary gland (*Chapter 7*). Antidiuretic hormone then stimulates the kidneys to retain water and produce a more concentrated urine. The retention of water helps return the osmolality of the ECF back to normal. If the osmolality of the ECF is low, the osmoreceptors are not stimulated and ADH is not released. This results in water loss from the kidneys in dilute urine. The loss of water helps to increase the osmolality of the ECF back to normal values. A low blood or ECF volume can be detected by baroreceptors in the aortic arch and carotid sinus (*Chapter 14*). These receptors also stimulate a release of ADH and, indeed, this mechanism can override the release of ADH by osmolality to maintain blood volume and therefore circulation. Antidiuretic hormone interacts with a second hormone, aldosterone to maintain the normal volume and concentration of the ECF. Aldosterone, a steroid hormone, is produced by the adrenal cortex (*Chapter 7*) and released in response to a low ECF volume or blood pressure. It stimulates retention of Na^+ together with water in the kidneys returning the ECF volume back to normal.

There are distinctive signs and symptoms associated with loss of water from body compartments. For example, loss of water from the ICF results in cell dysfunction that presents clinically as confusion, lethargy and coma. Loss of water from the ECF decreases blood pressure, leading to renal shutdown and shock. A reduction in total body water (ICF and ECF) produces a combination of both effects.

All body fluids contain electrolytes (*Table 8.2*). The regulation of water content by ADH helps to maintain normal electrolyte concentrations within the body. The concentration of Na^+ and K^+ in the ICF and ECF are maintained largely by the activity of the plasma membrane Na^+/K^+-ATPase (*Chapter 11*). This enzyme acts as an energy-dependent pump that expels Na^+ from the cell in exchange for an intake of K^+ to maintain both at physiological concentrations. The concentrations of these ions are maintained within narrow ranges and, since water can flow freely through most membranes, the concentrations of Na^+ and K^+ are responsible for maintaining the appropriate osmolalities of these compartments. The movement of water from one compartment to another is mainly responsible for determining their volumes.

Homeostatic mechanisms exist to minimize changes in body water and electrolyte composition and are particularly important in maintaining the

	Intracellular Fluid	Extracellular Fluid
K^+ / mmol dm^{-3}	110	4
Na^+ / mmol dm^{-3}	10	135
Cl^- / mmol dm^{-3}	5	100
HCO_3^- / mmol dm^{-3}	15	28
PO_4^{2-} / mmol dm^{-3}	31	1

Table 8.2 Typical compositions of the ICF and ECF

Margin Note 8.1 Osmolality and the movement of water

Osmolality is the concentration of particles (molecules or ions) dissolved in a fluid. Compartments of cells and, indeed, the body in general are divided by selectively permeable membranes. Most biological membranes are freely permeable to water. If the osmolality is not the same on both sides of the membrane, then there will be a net movement of water from the side of low to high osmolality since the osmolality of a body compartment is proportional to the concentration of a compartment expressed as mmoles of solute per kg of water. In humans, the osmolality of serum is approximately 285 mmol kg^{-1}. Serum osmolality can be measured directly using an osmometer or, if the concentration of solute particles is known, it can be calculated in units of mmol kg^{-1} using the expression:

$$\text{Osmolality} = 2[Na^+] + 2[K^+] + [\text{glucose}] + [\text{urea}]$$

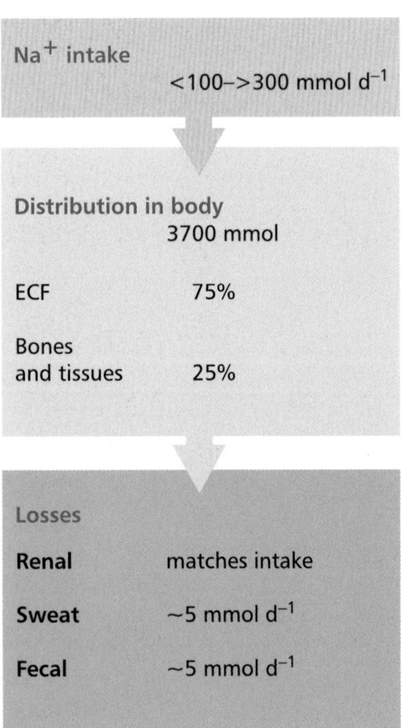

Figure 8.5 The distribution of body Na^+.

volume of the ECF. Water will remain in the extracellular compartment only if its osmolality is sufficiently high.

The assessment of fluid and electrolyte disorders in patients is a significant workload in the hospital pathology laboratory. In most cases, clinical tests to determine the concentrations of electrolytes in blood must be interpreted in conjunction with a clinical examination which involves taking the patient's clinical history, looking for signs and symptoms of hydration or dehydration and assessing kidney function.

8.4 DISORDERS OF Na^+ HOMEOSTASIS

Sodium ions are significant constituents of tissues, including bone, they control the volume of ECF and are required for normal neuromuscular functions. The intake of Na^+ is variable, from less than 100 mmol to more than 300 mmol day^{-1}. Losses are also variable, but renal loss is normally matched to intake. Small amounts of Na^+ are lost via skin and in feces and, under some circumstances, the GIT (*Chapter 11*) can be a major route of Na^+ loss, as in diarrhea.

The average 70 kg man contains 3700 mmol of Na^+ (*Figure 8.5*), of which 75% is found in the ECF. Hyponatremia and hypernatremia refer to serum concentrations of Na^+ below and above the reference range of 135–145 mmol dm^{-3}. Hyponatremia is caused by an excessive retention of water or the loss of Na^+, these two conditions resulting in different clinical features. The retention of water produces behavioral disturbances, headaches, confusion, convulsions and eventually coma. The symptoms associated with excessive loss of Na^+ are weakness, apathy, dizziness, weight loss and hypotension. Hyponatremia due to water retention is the more common. Water may be retained with or without an increase in total body Na^+. The former produces an **edema**, giving an edematous hyponatremia, whereas the latter results in nonedematous hyponatremia.

Edema is the excessive accumulation of fluid in interstitial compartments of the body, resulting from an increase in the concentration of Na^+ in the ECF. It results in swelling, which may be localized in, for example, legs and ankles (*Figure 8.6*) but can be more general in the chest cavity, abdomen and lungs. The major causes of edematous hyponatremia are heart failure, nephrotic syndrome and liver disease. All three reduce blood volume and stimulate aldosterone secretion, which, in turn, stimulates the retention of Na^+. The reduced blood volume also stimulates release of ADH from the posterior pituitary. Both result in more water than Na^+ being retained, giving rise to hyponatremia. Nephrotic syndrome leads to a loss of blood proteins to the urine, reduced concentrations of albumin leading to edema. The commonest cause of nephrotic syndrome is renal damage by diseases such as glomerulonephritis (*Chapter 3*). The treatment of edematous hyponatremia is aimed at its underlying cause, for example heart failure, kidney or liver diseases, and at removing the excess water and Na^+ using diuretics, and restricting water intake.

Nonedematous hyponatremia, the result of water overload without an increase in total body Na^+, is due to a decreased excretion of water from the syndrome of inappropriate secretion of ADH (SIADH), a severe renal failure or an increased intake by compulsive drinking or excessive parenteral fluid. The SIADH is a common finding in clinical practice. Patients present with reduced plasma osmolality, normal kidney function and a low output of urine. This syndrome is associated with many conditions, including malignancies, for example carcinoma of the lungs or bowel (*Chapter 17*), infections, such as pneumonia and tuberculosis (*Chapter 3*), trauma following, for example abdominal surgery, or it may be induced with drugs, such as chlorpropamide. All these conditions result in SIADH with water retention and a low urinary output. The excessive water is distributed between ICF and ECF and so the clinical signs of water overload, edema, may be mild or absent. Treatment of SIADH is to reduce water intake to less than 750 cm^3 day^{-1} and to correct its underlying cause.

Hyponatremia from loss of Na^+ decreases the total body Na^+ of patients. The losses may occur from vomiting, diarrhea, kidneys (aldosterone deficiency), the effects of drugs, such as spironolactone, or a decreased dietary intake of Na^+, although this is very rare. The loss of Na^+ is always accompanied by water loss; as the volume of the ECF decreases the release of ADH is stimulated and the increased reabsorption of water produces hyponatremia. The decreased volume of ECF means that the patient presents with the clinical symptoms of dehydration. Treatment is aimed at correcting the Na^+ losses with intravenous infusions of 0.9% NaCl, and treating the underlying cause, for example steroid therapy for aldosterone deficiency as in Addison's disease (*Chapter 7*).

Hypernatremia is caused by water depletion, water and Na^+ depletion with the loss of water predominating, or to an excess of Na^+ (*Figure 8.7*). Its clinical features are variable but, in general, patients present with muscular weakness, hypertension, intense thirst (**polydipsia**) and polyuria. If fluid loss occurs, the features associated with dehydration may be present. However, if Na^+ is in excess, raised blood pressure or edema may be seen.

Water depletion results from a decreased intake, such as in comatose patients, infants or the elderly. The body conserves water by producing a low volume of concentrated urine. Increased water losses can also occur in diseases such as diabetes insipidus that result in large quantities of dilute urine (*Chapter 7*). Hypernatremia with water and Na^+ depletion occurs only if relatively more water than Na^+ is lost. It is commonly caused in children by excessive sweating or diarrhea. Patients respond by producing low volumes of concentrated urine. The condition may also occur during osmotic diuresis in patients with diabetes mellitus (*Chapter 7*) where both water and Na^+ are lost, together with other electrolytes in large volumes of dilute urine, producing hypernatremia and a decreased ECF. An excess of Na^+ in the ECF is caused by an increased intake or decreased excretion of Na^+. The intake may be oral, for example salt tablets or seawater, or parenteral as in the treatment of Conn's or Cushing's syndromes (*Chapter 7*). Both disorders produce dilute urine due to retention of Na^+ by the kidneys.

Hypernatremia is treated by oral administration of water. If this is not possible, then 5% dextrose is administered parenterally. If hypernatremia is due to an excessive Na^+ intake, measures to remove it must be considered.

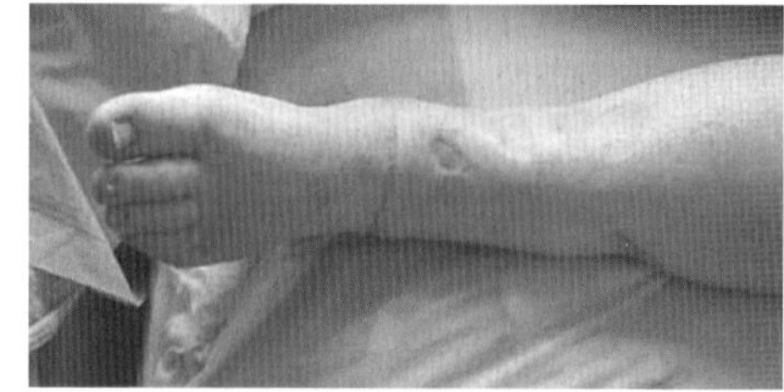

Figure 8.6 A massive edema of the lower limb. Courtesy of Charlie Goldberg, M.D., medicine.ucsd.edu/clinicalmed.

TYPICAL NUTRITIONAL VALUES

	Per pack (25g)	Per 100g
ENERGY	545 kJ 131 kcal	2180 kJ 525 kcal
PROTEIN	1.6 g	6.5 g
CARBOHYDRATE of which Sugars	12.5 g 0.1 g	50.0 g 0.5 g
FAT of which Saturates	8.3 g 2.5 g	33.0 g 10.0 g
FIBRE	1.0 g	4.0 g
SODIUM* *Equivalent as Salt	0.3 g 0.8 g	1.2 g 3.0 g
Per 25g BAG	131 CALORIES	8.3g FAT

THIS IS A PACK FROM

Figure 8.7 Many highly processed foods and snacks, such as potato crisps, contain high levels of Na^+ as its chloride salt (NaCl).

8.5 DISORDERS OF K^+ HOMEOSTASIS

Potassium ions are necessary to maintain cell volume, for the optimal activities of a number of enzymes, and to maintain the resting potential of cell membranes and therefore neuromuscular functions, especially in the heart (*Chapter 14*). The intake of K^+ varies between 30 and 100 mmol day^{-1} and losses are equally variable. The kidneys excrete most ingested K^+ with a smaller amount being eliminated by the GIT. A high concentration of plasma K^+ stimulates the release of aldosterone (*Chapter 7*) that, in turn, increases the renal excretion of K^+. Gastrointestinal losses can be significant during vomiting and diarrhea. Only very small amounts of K^+ are lost in sweat. The average 70 kg human contains about 3600 mmol of K^+ (*Figure 8.8*), almost all being found in the ICF. Values for the concentration of K^+ in the serum below and above the reference range of 3.4 to 4.9 mmol dm^{-3} are called hypokalemia and hyperkalemia respectively. Hyperkalemia is the more common clinical condition.

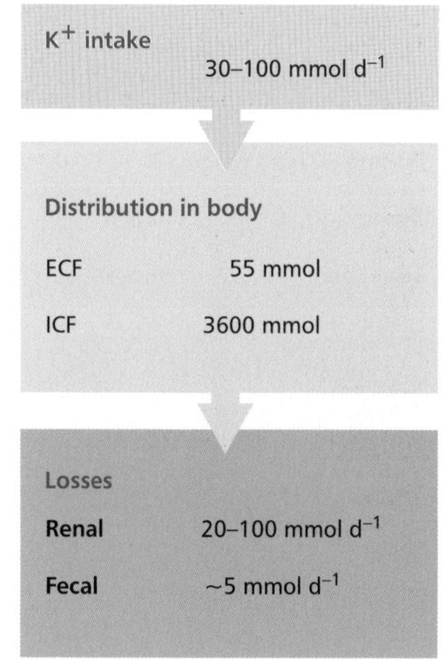

Figure 8.8 The distribution of body K^+.

Unlike Na^+, the plasma K^+ concentration does not vary significantly with water loss or overload. However, hyperkalemia must be identified because concentrations of serum K^+ above 7 mmol dm^{-3} can result in cardiac arrest and death. Renal failure, acidosis, aldosterone deficiency, damage to cells and an excess intake of K^+ can all cause hyperkalemia. In renal failure, the kidneys are unable to excrete K^+ because of the low GFR. Further, acidosis, a common feature of renal failure, leads to hyperkalemia because the low pH of the ECF means that K^+ moves out of cells in exchange for H^+, to return the pH to reference values. A deficiency of aldosterone, such as in Addison's disease where the kidneys lose their ability to excrete K^+, can result in hyperkalemia. The destruction of cells during trauma can release large amounts of K^+ causing hyperkalemia. Lastly, an excessive oral or parenteral intake of K^+ is a rare cause of hyperkalemia. The treatment of hyperkalemia includes infusion of insulin and glucose to promote the entry of K^+ into cells. Severe hyperkalemia may require dialysis.

Hypokalemia is clinically significant, giving rise to muscular weakness and cardiac arrhythmias, hence patients often present with breathlessness and chest pain. The causes of hypokalemia include increased K^+ losses from the GIT or kidneys, alkalosis, certain clinical disorders, some drugs, or a decreased K^+ intake. Excessive losses from the GIT can occur during vomiting and diarrhea. Hypokalemia occurs in alkalosis because the pH of the ECF is high and H^+ moves from the ICF to the ECF as part of the buffering process, while K^+ moves in the opposite direction leading to hypokalemia. A number of disorders, for example, Cushing's and Conn's syndromes, are associated with increased cortisol and aldosterone production respectively. Both hormones have mineralocorticoid activity and stimulate the renal retention of Na^+ in exchange for K^+ causing hypokalemia. Drugs, such as carbenoxolone used to treat gastric ulcers (*Chapter 11*), can cause hypokalemia because of their mineralocorticoid activity. Decreases in oral or parenteral intakes of K^+ are rare but can lead to hypokalemia. Patients with hypokalemia are treated with oral K^+ salts. Severe hypokalemia may require intravenous infusions of K^+.

Ca^{2+} intake	
Food	25 mmol d^{-1}
Distribution in body	
ECF	22.5 mmol
Plasma	9 mmol
Bone	24 750 mmol
Exchanges	
ECF – Bone	
Exchange	500 mmol d^{-1}
Bone formation	7.5 mmol d^{-1}
Bone resorption	7.5 mmol d^{-1}
ECF – Kidney	
Glomerular filtration	240 mmol d^{-1}
Reabsorption	234 mmol d^{-1}
Plasma – Gut	
Absorption	12 mmol d^{-1}
Secretion	6 mmol d^{-1}
Losses	
Renal	6 mmol d^{-1}
Skin	0.3 mmol d^{-1}
Fecal	19 mmol d^{-1}

Figure 8.9 The distribution of body Ca^{2+}.

8.6 DISORDERS OF Ca^{2+} HOMEOSTASIS

Calcium is required for bone and teeth structure, the release of neurotransmitters and initiation of muscle contraction, as a cofactor for coagulation factors (*Chapter 13*), some enzyme activities and it also acts as an intracellular second messenger for a number of hormones (*Chapter 7*).

The normal dietary intake of Ca^{2+} of about 25 mmol day^{-1} is supplemented by the reabsorption of Ca^{2+} from gastrointestinal secretions. Approximately 19 mmol of Ca^{2+} is lost in the feces daily. The kidneys normally filter about 240 mmol of Ca^{2+} daily but, as most of this is reabsorbed by the tubules, normal renal loss of Ca^{2+} is only about 6 mmol per day (*Figure 8.9*). Calcium is the most abundant mineral in the body and the average adult contains approximately 1 kg or 25 000 mmol of Ca^{2+}. Approximately 99% of Ca^{2+} is present in the bone. About 500 mmol of Ca^{2+} is exchanged daily between bone and the ECF. The ECF contains about 22.5 mmol of Ca^{2+}, of which 9.0 mmol is present in the plasma. Approximately 47% of Ca^{2+} in plasma occurs as free ionized Ca^{2+}, 46% is protein bound and 7% is complexed with citrate or phosphate. Only free Ca^{2+} is physiologically active and its plasma concentration is controlled by homeostatic mechanisms involving the hormones parathyroid hormone (PTH), calcitriol and calcitonin (*Figure 8.10*). Parathyroid hormone is secreted by the parathyroid glands in response to a fall in the concentration of plasma ionized Ca^{2+} and *vice versa*. It stimulates the release of Ca^{2+} from bone, a process called **bone resorption**, and a decreased reabsorption of HCO_3^- by the kidneys that produces an acidosis, which helps to increase plasma ionized Ca^{2+} and stimulates the synthesis of calcitriol from cholecalciferol in the liver.

This hormone is also formed in the skin by the action of ultraviolet light on 7-dehydrocholesterol. Calcitriol increases Ca^{2+} and P_i absorption from the GIT and increases bone resorption. The physiological function of calcitonin remains unclear but it is known to reduce the concentration of Ca^{2+} in plasma by inhibiting both bone resorption and the renal reabsorption of Ca^{2+}.

The serum reference range for total Ca^{2+} is 2.20–2.60 mmol dm^{-3} and for free 1.20–1.37 mmol dm^{-3}. Values above and below these are called hypercalcemia and hypocalcemia respectively.

The renal damage associated with hypercalcemia is its most serious consequence. Hypercalcemia may suppress neuromuscular excitability causing constipation and abdominal pain and affect the CNS, resulting in depression, nausea and anorexia. The nausea may cause vomiting and therefore dehydration. Calcium can stimulate gastrin and therefore gastric acid secretion and so hypercalcemia may be associated with peptic ulcers (*Chapter 11*). Hypercalcemia may cause arrhythmias and in severe cases may result in cardiac arrest (*Chapter 14*). The commonest causes of hypercalcemia are malignant disease or primary hyperparathyroidism. Less common causes include thyrotoxicosis, vitamin D intoxication, thiazide diuretics and familial hypocalciuric hypercalcemia. Rare causes are tuberculosis, sarcoidosis, acromegaly, milk-alkali syndrome and idiopathic hypercalcemia of infancy.

Cancerous tumors of the lungs stimulate an increase in plasma Ca^{2+} by producing a PTH related protein (PTHrp) that resembles the structure of PTH (*Figure 8.11*). Cytokines and prostaglandins released by tumors that have metastasized to the bones, may lead to increased resorption of Ca^{2}. Primary hyperparathyroidism occurs most commonly due to a parathyroid adenoma, which is a benign tumor, and only rarely due to a parathyroid carcinoma. It affects both men and women at any age but is most common in postmenopausal women. In primary hyperparathyroidism, there is excessive PTH secretion that causes hypercalcemia and sometimes hypophosphatemia (*Section 8.7*), which increases bone turnover particularly of the metaphyses (*Figure 8.12*). Thyroid hormones have no direct effect on Ca^{2+} homeostasis but can cause increased bone turnover by increasing osteoclastic activity and giving rise to mild hypercalcemia during thyrotoxicosis. An excessive iatrogenic or accidental ingestion of vitamin D or thiazide diuretics that interfere with renal Ca^{2+} loss can also cause hypercalcemia.

Familial hypocalciuric hypercalcemia is a recently recognized autosomal dominant (*Chapter 15*) condition that develops from childhood. It is

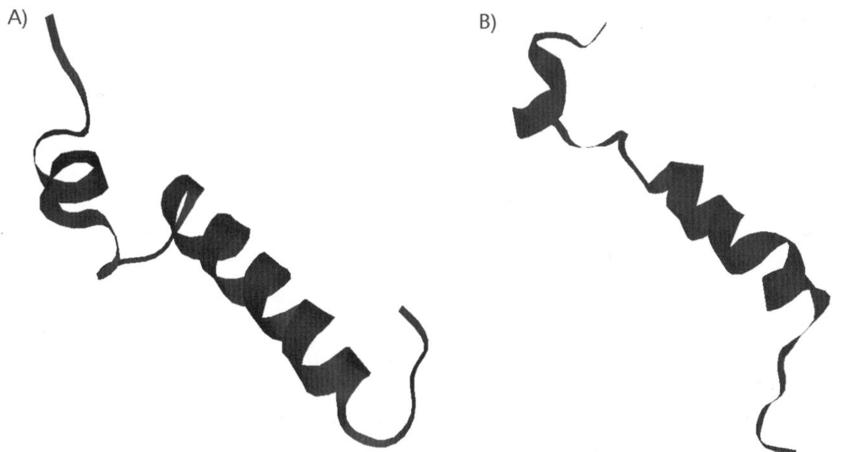

Figure 8.11 Molecular models of (A) PTH and (B) PTHrp. PDB files 1BWX and 1BZG respectively. Note the similarity in overall structures.

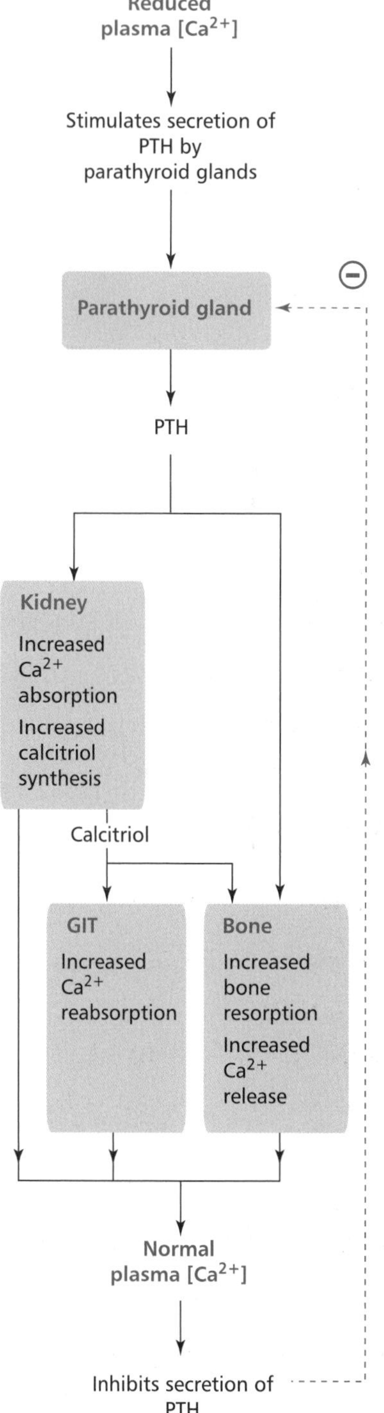

Figure 8.10 The hormonal regulation of the concentration of plasma Ca^{2+}.

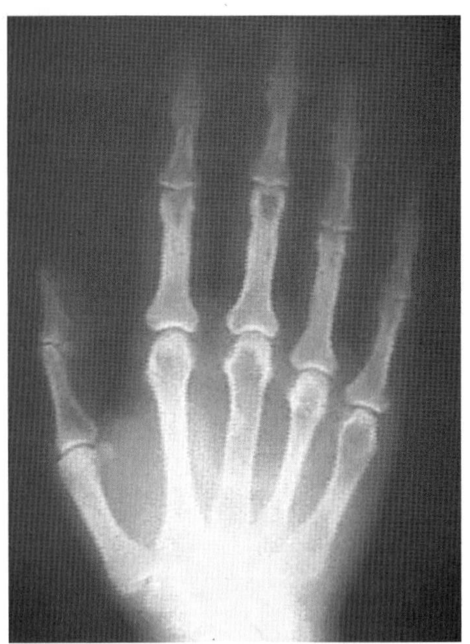

Figure 8.12 An isotope scan of hand and finger bones showing sites in the fingers with a decreased opacity where greater bone turnover has occurred because of increased Ca^{2+} resorption. Courtesy of Dr I. Maddison, London South Bank University, UK.

characterized by chronic hypercalcemia but is usually asymptomatic, with normal levels of PTH and no parathyroid adenoma. The mechanism underlying this condition is unknown. Both sarcoidosis and tuberculosis are granulomatous diseases. In these conditions, hypercalcemia occurs as there is increased production of calcitriol by macrophages in the granulomas. Hypercalcemia is occasionally seen in acromegaly, probably due to stimulation of calcitriol production by excess growth hormone. Hypercalcemia may occur in people who ingest large amounts of milk together with alkali antacids, such as HCO_3^-, to relieve symptoms of peptic ulceration. An alkalosis occurs that is believed to reduce renal Ca^{2+} excretion although the precise mechanism is still unclear. This milk-alkali syndrome is very rare as antacid treatment of peptic ulcers has been replaced by drugs that inhibit gastric acid secretion. The condition idiopathic hypercalcemia of infancy is associated with hypercalcemia because of an increased sensitivity to vitamin D in bone and the GIT but the precise mechanism underlying this hypercalcemia is unknown.

Patients who present with hypercalcemia are investigated for malignancy or primary hyperparathyroidism as this accounts for up to 90% of cases. If both malignancy and primary hyperparathyroidism are excluded, other causes must be considered and investigated (*Figure 8.13* and *Box 8.1*). A number

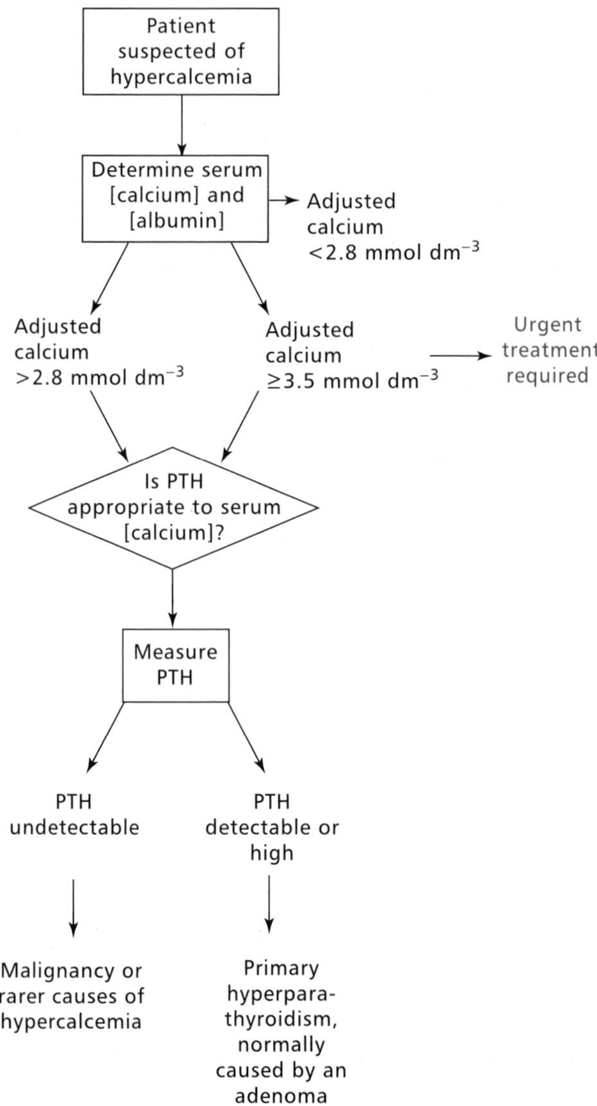

Figure 8.13 Overview of the clinical investigation of hypercalcemia. ***See also Box 8.1.***

BOX 8.1 Albumin and Ca^{2+}

Changes in the concentration of albumin (*Figure 8.14(A)*) in plasma affect the total Ca^{2+} concentration but not the free Ca^{2+} content. A high plasma albumin concentration gives rise to a high total plasma Ca^{2+} concentration and, conversely, a low plasma albumin concentration produces a low total plasma Ca^{2+} concentration (*Figure 8.14(B)*). The determination of total plasma or serum Ca^{2+} concentration can be misleading since it is affected by conditions that change the concentration of albumin. The effect of these changes can be overcome by measuring free Ca^{2+}. However, its determination is difficult and expensive and the usual solution is to calculate an adjusted value for total Ca^{2+} using the following formulae if the [albumin] is given in g dm^{-3} and $[Ca^{2+}]$ in mmol dm^{-3}.

When [albumin] is less than 40 g dm^{-3} then:

$$\text{adjusted } [Ca^{2+}] = \text{measured concentration of total } Ca^{2+} + 0.02\,(40 - [\text{albumin}])$$

When [albumin] is more than 45 g dm^{-3} then:

$$\text{adjusted } [Ca^{2+}] = \text{measured concentration of total } Ca^{2+} - 0.02\,([\text{albumin}] - 45)$$

Since the free Ca^{2+} concentration is unaffected by variations in the concentration of albumin, clinical symptoms are not manifested. However, free Ca^{2+} competes with H^+ for negatively charged binding sites on albumin and changes in the concentration of free Ca^{2+} may occur in acute acid–base disorders, with clinical consequences (*Figure 8.15*). These do not, however, affect the total plasma Ca^{2+} concentration. In alkalosis, more H^+ dissociates from albumin allowing increased amounts of free Ca^{2+} to bind to it. This has the effect of increasing the protein bound Ca^{2+} fraction but at the expense of decreasing the free plasma Ca^{2+} leading to hypocalcemia. In acidosis, increasing amounts of H^+ bind to albumin as Ca^{2+} dissociates from it, decreasing the protein bound Ca^{2+} and increasing free plasma Ca^{2+} producing hypercalcemia.

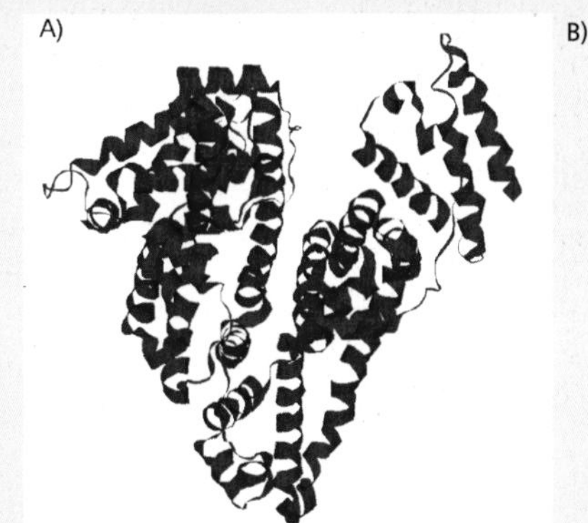

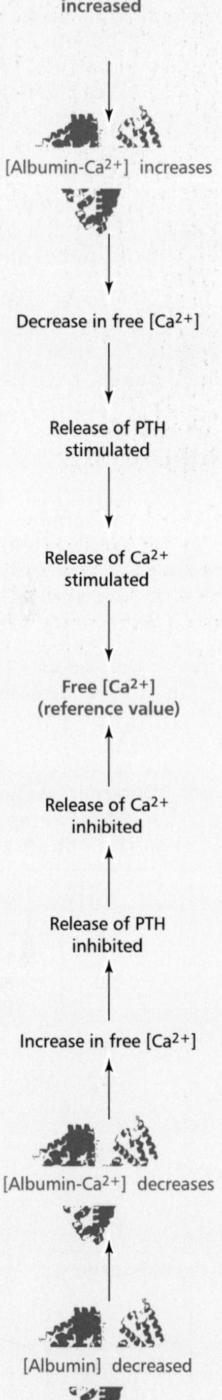

Figure 8.14 (A) Molecular model of human serum albumin. PDB file 1E78. (B) The effect of changes in the concentration of albumin on the concentration of total plasma Ca^{2+}.

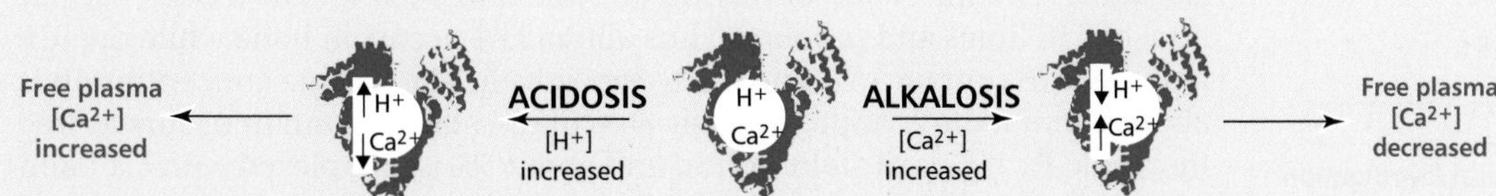

Figure 8.15 Effect of acidosis and alkalosis on the concentration of free Ca^{2+} in the plasma.

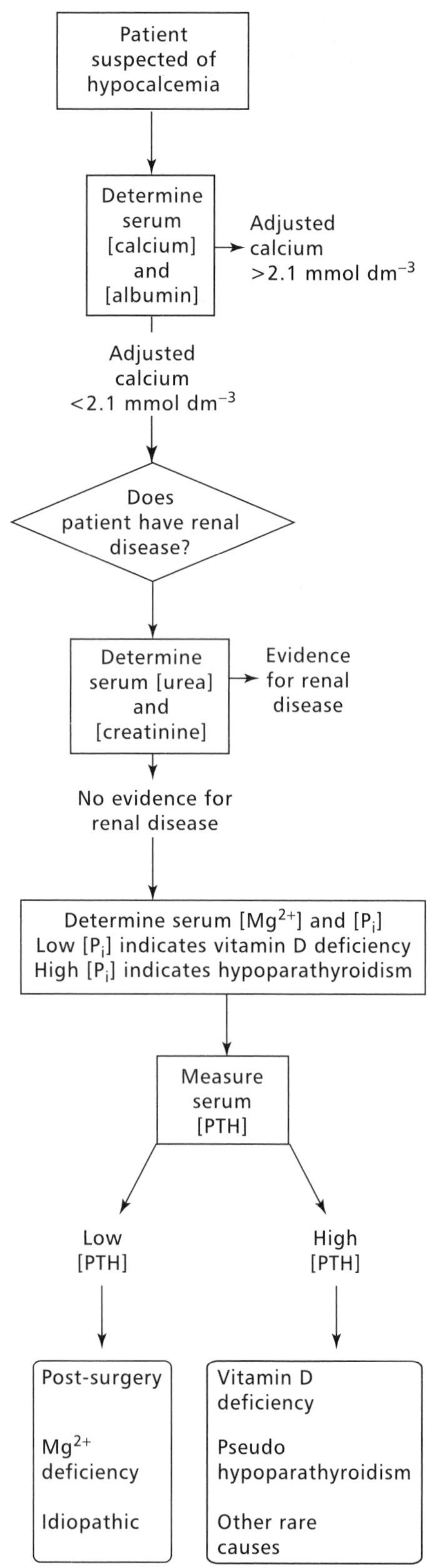

Figure 8.16 Overview of the clinical investigation of hypocalcemia.

of approaches are taken to managing hypercalcemia. The underlying cause should be treated wherever possible. Intravenous saline may be administered in dehydrated patients to restore the glomerular filtration rate and enhance Ca^{2+} loss and hydration. Drugs, such as frusemide, inhibit renal reabsorption of Ca^{2+} and promote its excretion while bisphosphonates lower Ca^{2+} levels by inhibiting bone resorption. In very severe cases, dialysis or emergency parathyroidectomy may be necessary. In some cases, an artefactual hypocalcemia may be reported when blood samples are erroneously collected into tubes containing ethylene diaminetetraacetic acid (EDTA). This anticoagulant is a chelator of Ca^{2+} and its use will lead to low values for Ca^{2+} concentrations.

The clinical effects of hypocalcemia include behavioral disturbances, paresthesiae, tetany, convulsions and cataracts. Its major causes are renal failure, Mg^{2+} and vitamin D deficiencies, hypoparathyroidism and pseudohypoparathyroidism. Chronic renal failure may decrease the reabsorption of Ca^{2+} by decreasing the synthesis of calcitriol leading to hypocalcemia. This may lead to bone disease because the increased output of PTH arising from the hypocalcemia can increase osteoclast activity. Magnesium ions are required for PTH secretion and its action and a deficiency produces hypocalcemia. A deficiency in vitamin D may arise from a poor diet, malabsorption (*Chapters 10* and *11*) or inadequate exposure to sunlight leading to an inadequate absorption of Ca^{2+} from food. Hypoparathyroidism, or a reduced activity of the parathyroid glands with decreased production of PTH, results in hypocalcemia. The condition can be congenital, where there is an absence of the parathyroid glands, or acquired hypoparathyroidism that may be idiopathic, or caused by autoimmune conditions or surgery, for example thyroidectomy. In pseudohypoparathyroidism, there is excessive PTH secretion because target tissues fail to respond to the hormone, producing a persistent hypocalcemia. This condition is more common in males than females and patients present with skeletal abnormalities including short stature, mental retardation, cataracts and testicular atrophy.

The investigation of hypocalcemia is outlined in *Figure 8.16*. The underlying cause of hypocalcemia should be treated wherever possible. Magnesium supplements may be prescribed in hypocalcemia due to Mg^{2+} deficiency, whereas calcitriol and its precursors may be prescribed in vitamin D deficiency. Oral Ca^{2+} supplements are prescribed in mild cases of hypocalcemia.

8.7 DISORDERS OF PHOSPHATE HOMEOSTASIS

Phosphate (P_i) combines with Ca^{2+} to form hydroxyapatite, the mineral component of bone and teeth and is also required for some enzymic activities, oxidative phosphorylation and the synthesis of 2,3-bisphosphoglycerate that regulates the dissociation of oxyhemoglobin (*Chapter 13*), the excretion of H^+ (*Chapter 9*) and for cell membrane integrity. The daily intake of P_i is about 40 mmol. The kidneys lose approximately 26 mmol daily and 14 mmol are lost in feces. The total body content of P_i in the average male is over 20000 mmol (*Figure 8.17*) with 17000 occurring in bone and 3000 in soft tissues, largely attached to lipids and proteins. Thus about 85% occurs in bone while the ICF and the ECF contain 15% and 0.1% respectively. The plasma concentration is about 1 mmol dm^{-3}. Approximately 80% of the plasma content occurs as free inorganic P_i, 15% is protein-bound and about 5% is complexed with Ca^{2+} and Mg^{2+}. Parathyroid hormone (*Figure 8.11*) and the hormone, calcitriol, control the homeostasis of P_i; the former decreasing the reabsorption by the kidneys and reducing its plasma concentration, the latter stimulating P_i absorption in the GIT and increasing the concentration.

The reference range for total serum P_i is 0.8–1.4 mmol dm^{-3} but a higher reference range applies in infancy and childhood. Hyperphosphatemia and hypophosphatemia are used to describe concentrations above and below the reference range respectively. Hypophosphatemia causes more damage than hyperphosphatemia but, fortunately, is less common.

Hyperphosphatemia may cause metastatic calcification, for example the deposition of calcium phosphate in soft tissues as the excess P_i precipitates with Ca^{2+} and causes hypocalcemia and tetany in affected patients. The commonest cause of hyperphosphatemia is renal failure where the GFR and P_i excretion decline. Hypoparathyroidism reduces renal excretion of P_i giving rise to hyperphosphatemia. In diabetic ketoacidosis (*Chapter 7*), a deficiency of insulin prevents the uptake of P_i by cells leading to hyperphosphatemia. Other causes are an increased intake of P_i or its release from damaged cells in intravascular hemolysis. Indeed, any condition where there is increased turnover of cells, for example following treatment of malignant disease with chemotherapy, results in release of P_i during cell destruction. Excessive intake, either oral or intravenous, is a rare cause and is more likely when there is also renal failure as in pseudohypoparathyroidism where there is resistance by the kidneys to PTH that decreases their excretion of P_i. A delay in the separation of plasma or serum from blood before analysis for P_i or hemolysis of a blood sample prior to its analysis can indicate artefactual hyperphosphatemia but this does not reflect the true clinical situation.

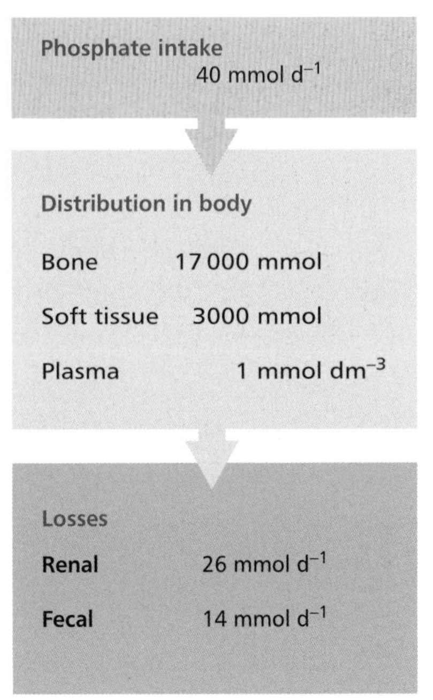

Figure 8.17 The distribution of body P_i.

A number of biochemical tests are useful when investigating hyperphosphatemia. These include determining the concentrations of P_i, Ca^{2+}, urea and creatinine in serum and the concentration of P_i in urine. The following strategy has proved useful in investigating obscure causes of hyperphosphatemia. First, it is necessary to exclude artefactual causes. Secondly, serum concentrations of creatinine and urea should be determined to exclude renal failure. If the serum concentration of Ca^{2+} is normal or above reference values, vitamin D intoxication or untreated diabetes mellitus should be considered. Thirdly, if the plasma or serum concentration of Ca^{2+} is low, then hypoparathyroidism should be investigated. Finally, if the urinary concentration of P_i is low, then hypoparathyroidism is, again, a consideration, whereas a high urinary concentration indicates increased intake, malignancy or intravascular hemolysis. Patients with hyperphosphatemia are managed by treating the underlying cause wherever possible. The oral intake of aluminum, Ca^{2+} and Mg^{2+} salts may be used as these can bind P_i in the GIT reducing its absorption.

The clinical features of hypophosphatemia include paresthesiae, ataxia, coma, osteomalacia and muscle weakness. There may be increased susceptibility to infection possibly due to defective phagocytosis. The causes of hypophosphatemia are varied. Vitamin D deficiency results in a decreased synthesis of calcitriol and therefore decreased P_i absorption in the GIT. Increased renal loss of P_i may occur in primary hyperparathyroidism where increased secretion of PTH causes excessive renal loss of P_i. Certain diuretics that increase renal loss of P_i can cause hypophosphatemia. It may also occur during the recovery phase of diabetic ketoacidosis when patients are administered insulin, which promotes cellular uptake of P_i. Total body P_i may be depleted as a consequence of osmotic diuresis. There are a number of rare causes of hypophosphatemia. These include an inadequate dietary intake usually associated with parenteral nutrition, or when agents, such as aluminum hydroxide are used as antacids and prevent its absorption in the GIT, and in chronic alcoholics who have a complex and multifactorial condition with poor diet and reduced GIT absorption (*Chapter 11*).

Determination of the serum concentrations of P_i and Ca^{2+} and the urinary concentration of P_i are useful in investigating hypophosphatemia. The following strategy may be used when its cause is not obvious. First, exclude causes such as alkalosis and chronic alcoholism. Secondly, a reduced urinary P_i suggests decreased dietary or parenteral intakes or increased cellular uptake, for example in insulin therapy. Thirdly, if the urinary concentration of P_i is above its reference range then excessive renal losses are occurring and the concentration of Ca^{2+} in the plasma or serum should be determined. If this is increased, then primary hyperparathyroidism or malignancy may be present. If, however, the concentration is low or normal, renal defects or inappropriate diuretic therapy are considerations. Hypophosphatemia should be managed by treating the underlying cause wherever possible. In some situations it may be necessary to administer oral or parenteral P_i.

8.8 DISORDERS OF Mg^{2+} HOMEOSTASIS

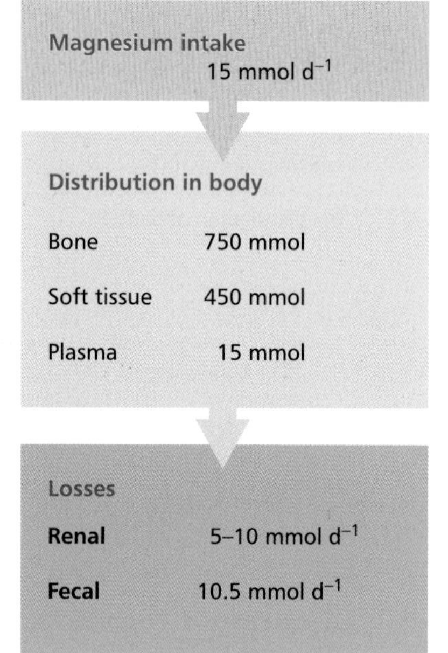

Figure 8.18 The distribution of body Mg^{2+}.

Magnesium is required to maintain the structures of ribosomes, nucleic acids and numerous proteins and acts as a cofactor for over 300 enzymes, including those involved in energy metabolism and protein synthesis. It is also required for normal cell permeability and neuromuscular functions. The usual dietary intake of Mg^{2+} is about 15 mmol day^{-1} and approximately 30% of this is absorbed in the GIT, the rest is lost in the feces. The adult human body contains over 1200 mmol of Mg^{2+} (*Figure 8.18*). Approximately 750 mmol is found in bone and about 450 mmol in muscle and soft tissues. The ECF contains only 15 mmol. Approximately 55% of plasma Mg^{2+} occurs as free ionized Mg^{2+}, 32% is protein-bound and 13% complexed with P_i or citrate.

The kidneys lose 5 to10 mmol of Mg^{2+} daily but losses are adjusted to control Mg^{2+} homeostasis. An increased dietary intake of Mg^{2+} results in increased renal loss and *vice versa*. This is achieved principally by adjusting the reabsorption of Mg^{2+} by cells of the proximal tubules and loop of Henle. A number of factors influence the rate of excretion of Mg^{2+} including hypercalcemia and hypophosphatemia (*Section 8.6*) that decrease renal reabsorption and PTH, which stimulates renal retention.

The reference range for serum Mg^{2+} is 0.8–1.2 mmol dm^{-3}. Hypo- and hypermagnesemia refer to concentrations below and above the reference range respectively. Note that measurements of the concentration of Mg^{2+} in plasma or serum are unreliable indicators of its body status since only 1% of body Mg^{2+} occurs in the ECF.

The clinical effects of hypomagnesemia are similar to those seen in hypocalcemia and include tetany, muscle weakness, convulsions and cardiac arrhythmias. These effects are related to the role of Mg^{2+} in neuromuscular function. The causes of hypomagnesemia include decreased intake as in starvation (*Chapter 11*), poorly managed parenteral nutrition or malabsorption. Increased losses of Mg^{2+} as in osmotic diuresis in diabetics (*Chapter 7*), diuretic therapy, hyperaldosteronism and excessive losses from the GIT in prolonged diarrhea, GIT fistula and laxative abuse can also cause hypomagnesemia. The use of anticancer drugs (*Chapter 17*), such as cisplatinum, can damage the kidneys and prevent the renal reabsorption of Mg^{2+}. In alcoholism (*Chapter 12*), hypomagnesemia is believed to occur due to increased renal excretion, inadequate dietary intake, vomiting and diarrhea.

In many cases, the cause of hypomagnesemia is determined by clinical examination. However, measuring the urinary Mg^{2+} may be useful as the amount of Mg^{2+} excreted per day decreases with decreased intake. If hypomagnesemia occurs with increased renal excretion then losses are likely to be due to renal damage. Hypercalcemia may increase renal Mg^{2+} excretion causing hypomagnesemia but hypocalcemia may occur in hypomagnesemia due to hypoparathyroidism.

In hypomagnesemia, the underlying cause should be treated wherever possible. Oral Mg^{2+} supplements may be adequate for mild cases but severe Mg^{2+} deficiency together with malabsorption may require intravenous infusions of Mg^{2+}.

The clinical effects of hypermagnesemia are also largely related to the role of Mg^{2+} in neuromuscular activities and include muscular weakness, respiratory paralysis and, in very severe cases, cardiac arrest. Acute or chronic renal failures are the commonest causes of hypermagnesemia; others include its release from damaged cells from, for example, crush injuries. Mild hypermagnesemia may occur in mineralocorticoid deficiency, as in Addison's disease. In rare cases, hypermagnesemia may occur from an increased oral or parenteral intake of Mg^{2+} or from the use of Mg^{2+} containing antacids or laxatives. When this does occur, it is usually combined with renal failure. The management of hypermagnesemia involves treating the underlying cause wherever possible. Hypermagnesemia due to renal failure may require dialysis.

8.9 DISORDERS OF URATE METABOLISM

In humans, the end product of the metabolism of the purines, adenine and guanine is urate (*Figure 8.19*). There are three sources of purines namely diet, the breakdown of endogenous nucleotides and nucleic acids and *de novo* synthesis. Most dietary nucleic acids are ingested in the form of nucleoproteins from which urate is produced by the GIT (*Chapter 11*). The degradation and *de novo* synthesis of purines are linked (*Figure 8.20*). The body urate pool, and therefore plasma concentration, depends upon the relative rates of urate formation and excretion. Both the kidneys and the GIT excrete urate with renal excretion accounting for approximately 66% of the total. Almost all the urate is filtered at the glomerulus but most is reabsorbed by the proximal tubule. However, both reabsorption and secretion occur in the distal tubule, so that the net effect is to excrete about 10% of the urate. Urate secreted into the GIT is metabolized to CO_2 and NH_3 by bacterial action or **uricolysis**.

The reference range for serum urate is 0.1 to 0.4 mmol dm^{-3}. However, there is a wide variation in the concentration of urate in plasma or serum even in health. Plasma urate concentration tends to be higher in males than females, is highest in obese individuals, those from affluent social classes and those with a high protein and alcohol intake. Thus hyperuricemia is defined as a concentration greater than 0.42 mmol dm^{-3} in men and more than 0.36 mmol dm^{-3} in women.

Hyperuricemia may arise as a result of increased production of uric acid or decreased excretion or both. Excessive synthesis may occur because of a defective synthetic metabolic pathway, stimulation of *de novo* purine synthesis by alcohol or by increased nucleic acid turnover, as in malignant disease, or the use of cytotoxic drugs. An excessive dietary intake of purines will also produce hyperuricemia. A decreased urate excretion may be due to a reduced GFR giving rise to hyperuricemia. Increased proximal tubular reabsorption and decreased distal tubular secretion of urate have similar effects. Lactate and β-hydroxybutyrate compete with urate for excretion by the distal tubule. Therefore lactic acidosis or ketosis (*Chapter 7*) are often associated with hyperuricemia. Some drugs, for example low doses of aspirin, can inhibit the distal tubular secretion of urate causing hyperuricemia.

A)

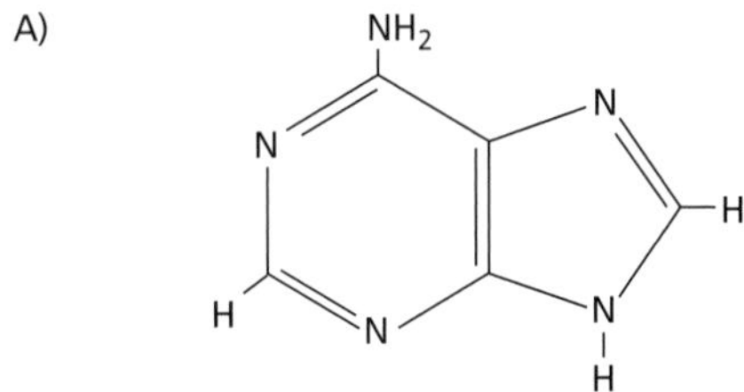

B)

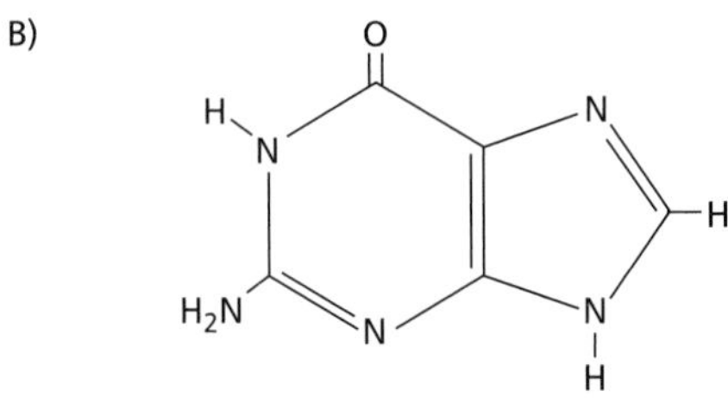

C)

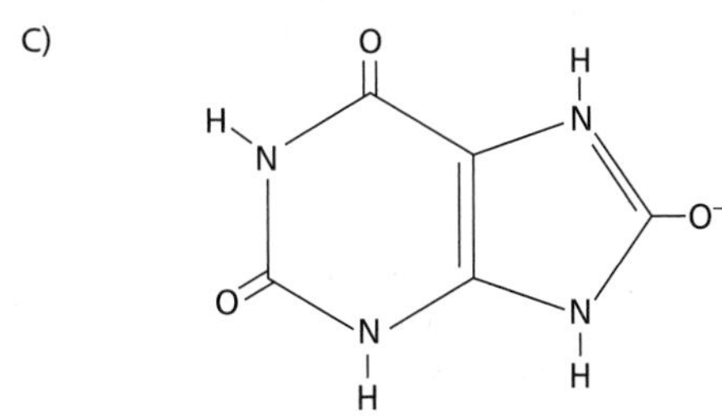

Figure 8.19 (A) Adenine, (B) guanine and (C) urate.

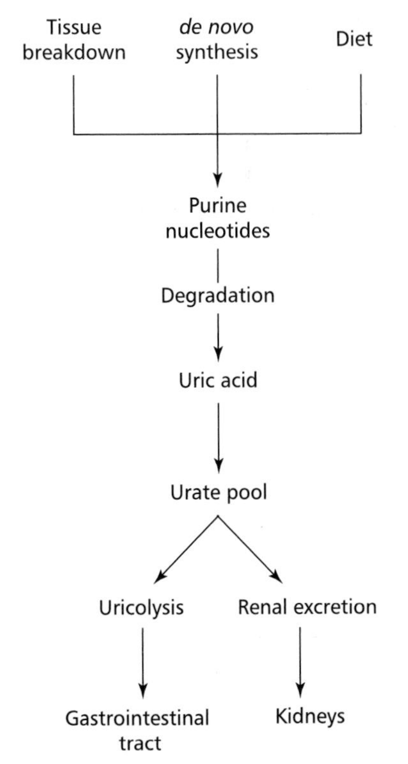

Figure 8.20 The synthesis and degradation of purine bases.

Urate has low solubility and the ECF easily becomes saturated at concentrations just above the upper limit of the reference range. There is a tendency for crystalline monosodium urate to form in people with hyperuricemia, giving rise to gout. Crystals of monosodium urate (*Figure 8.21(A)*) tend to form in cartilage and synovial fluid of joints and particularly those of the big toe causing gout (*Figure 8.21(B)*). The crystals are phagocytosed by neutrophil leukocytes and may cause damage to lysosomal membranes within these cells (*Chapter 16*). As a consequence, lysosomal contents are released, causing damage to both leukocytes and surrounding tissues with an associated inflammatory response (*Chapter 4*). Gout may be primary, with no known cause, or secondary, as a consequence of another disorder. Primary gout is characterized by recurrent attacks of arthritis. It is more common in men than women. The metabolic defect in patients is unknown but a number of abnormalities may be responsible for the overproduction of urate and therefore increased urinary urate output. In many patients there is a combined defect of urate overproduction together with its impaired renal excretion. Patients with primary gout often have deposits of urate in their soft tissues and some can develop renal stones composed of urate salts. The risk that a normal person will develop gout varies with their urate concentration. The annual incidence of gout in men is low, about 0.1%, when the urate concentration is less than 0.42 mmol dm^{-3}. This increases to 0.6% when the concentration is 0.42–0.54 mmol dm^{-3} and 5% when urate concentrations are greater than 0.54 mmol dm^{-3}. The reason for the onset of acute attacks in gout is unclear since a sharp rise in the concentration of urate is not usually demonstrable.

Secondary gout is rare but can arise from a number of other disorders including myeloproliferative disorders (*Chapter 17*) such as polycythemia vera, where the hyperuricemia is due to an increased cell turnover, the use of cytotoxic drug therapy that increases cell destruction and the breakdown of nucleic acids, and psoriasis with its increased turnover of skin cells.

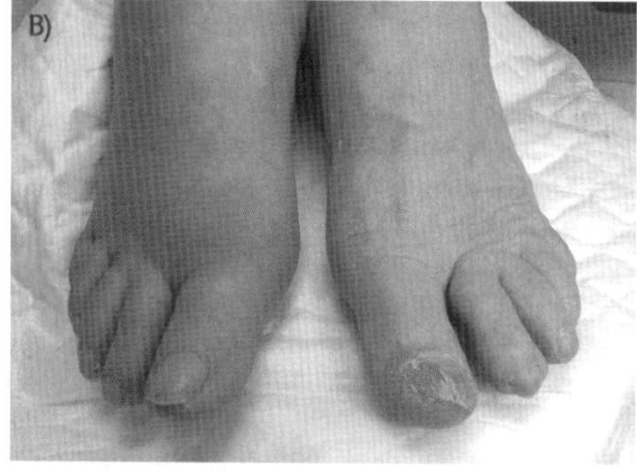

Figure 8.21 (A) Crystals of monosodium urate viewed using polarized light. (B) Gout of the right big toe showing diffuse swelling and inflammation centered where the toes join the foot but also extending over much of the foot. Courtesy of Charlie Goldberg, M.D., medicine.ucsd.edu/clinicalmed.

A diagnosis of gout is made on clinical grounds, a demonstration of hyperuricemia and a satisfactory response to uricosuric drugs. A high plasma urate concentration does not always mean that the patient has gout, that is high plasma urate concentration makes the diagnosis of gout more likely, whereas a consistently low plasma urate concentration excludes the diagnosis. To confirm diagnosis it is necessary to aspirate the joint fluid during an acute attack. The finding of urate crystals, 2–10 μm long and needle shaped, within neutrophils will confirm the diagnosis.

Anti-inflammatory drugs, such as indomethacin, are used to treat acute attacks of gout but have no effect on the hyperuricemia which is treated with a diet low in protein and alcohol. Urate lowering drugs, for example allopurinol, that prevent the formation of urate and decrease *de novo* synthesis of purines, are used in long-term treatment or when plasma urate levels are persistently higher than 0.6 mmol dm^{-3}.

Hypouricemia, where the concentration of urate in serum is below the reference range, is uncommon and not of clinical significance. Its occurrence is due to a decreased urate synthesis, as in congenital xanthine oxidase deficiency and severe liver disease, or to increased excretion of urate as seen in renal tubular disorders, such as the Fanconi syndrome. Hypouricemia may also result from excessive use of drugs such as allopurinol.

CASE STUDY 8.1

Ted was admitted to hospital following a car accident. The following results were obtained on a serum specimen three days later. Reference ranges are given in parentheses.

Urea	45 mmol dm^{-3}	(2.4–6.5 mmol dm^{-3})
Na^+	132 mmol dm^{-3}	(133–145 mmol dm^{-3})
K^+	6.9 mmol dm^{-3}	(3.4–4.8 mmol dm^{-3})
HCO_3^-	13 mmol dm^{-3}	(21–28 mmol dm^{-3})
Osmolality	332 mmol kg^{-1}	(280–290 mmol kg^{-1})

Questions

(a) What may be the primary cause of these results?

(b) Suggest reasons why Ted shows hyperkalemia and low serum HCO_3^-.

CASE STUDY 8.2

Arnie, a 25-year-old man, presented with a history of severe diarrhea, abdominal pain, weight loss, cramp in the arms and legs and tetany. He had suffered several previous episodes of diarrhea and abdominal pain. His serum was investigated and yielded the following results. Reference ranges are given in parentheses.

Na^+	140 mmol dm^{-3}	(133–145 mmol dm^{-3})
K^+	3.3 mmol dm^{-3}	(3.4–4.8 mmol dm^{-3})
Urea	5.8 mmol dm^{-3}	(2.4–6.5 mmol dm^{-3})
Total Ca^{2+}	2.40 mmol dm^{-3}	(2.15–2.46 mmol dm^{-3})
Phosphate	1.0 mmol dm^{-3}	(0.8–1.44 mmol dm^{-3})
Albumin	42 g dm^{-3}	(38–48 g dm^{-3})
Mg^{2+}	0.39 mmol dm^{-3}	(0.8–1.2 mmol dm^{-3})

Questions

(a) Explain the significance of these results.

(b) Suggest ways in which Arnold should be treated.

CASE STUDY 8.3

John, a 58-year-old obese lecturer woke in the middle of the night with severe pain in his large toe which was hot, swollen and red. The pain was so intense he could not place his foot on the floor. John had been to a dinner party the night before. In the morning he visited the local hospital where a blood sample was taken and analyzed for serum urate. Reference ranges are given in parentheses.

Serum urate 0.81 mmol dm^{-3} (0.1–0.4 mmol dm^{-3})

Questions

(a) What is the most likely diagnosis?

(b) What further investigations should be performed?

(c) How should John be treated?

8.10 SUMMARY

It is essential to maintain appropriate levels of water and electrolytes in the body so that metabolic reactions can function effectively. The stable environment within the body is maintained by homeostatic mechanisms, which return levels to normal, following a shift in equilibrium. In addition, waste products such as urate need to be removed to prevent toxicity. The kidneys help to maintain the balance of water, electrolytes and waste products and a number of renal function tests are available to assess their function in cases of suspected renal failure. Dehydration, possibly as a result of gastrointestinal disease causing diarrhea and vomiting, is a severe, life-threatening condition. Disorders of electrolyte balance can involve a lack or excess of the electrolyte in question. Thus distinct syndromes can occur with disorders affecting the levels of Na^+, K^+, Ca^{2+}, Mg^{2+} and P_i. Disorders of urate metabolism may result in high levels of urate in the blood, leading to gout and renal stones, whereas low levels of urate in the blood are rare.

QUESTIONS

1. Robin was rescued from a raft at sea. He had been without food or water for several days.

   a) What will have happened to Robin's body compartments?

   b) Should he have drunk seawater to survive?

2. Jane, an 80-year-old woman who lives alone was suffering from a urinary tract infection and had little food or water for several days. She was found in a drowsy confused state by her neighbors and taken to hospital. An analysis of her serum gave Na^+ and urea of 160 and 20 mmol dm^{-3} respectively. Glucose and K^+ concentrations were within their references ranges. Account for Jane's symptoms and test results.

3. Hyperkalemia may be caused by which one of the following?

   a) hemolysis;

   b) delayed separation of plasma;

   c) increased intake of K^+ supplements;

   d) renal failure;

   e) all of the above.

4. What is the adjusted serum Ca^{2+} concentration for a patient with a serum albumin concentration of 29 g dm^{-3} and a total serum Ca^{2+} of 1.78 mmol dm^{-3}?

5. Which one of the following does NOT cause hyperuricemia?
   a) excessive dietary purine intake;
   b) malignant disease;
   c) chronic renal failure;
   d) hypomagnesemia;
   e) psoriasis.

6. Alice, a 48-year-old woman, was treated by parathyroidectomy for her hyperparathyroidism. Unfortunately, she developed hypocalcemia and was placed on vitamin D therapy although the hypocalcemia proved difficult to control. Eventually, Alice was seen in outpatients complaining of feeling unwell and vomiting. Her blood results showed her to be hypercalcemic but other values were within their reference ranges. Suggest the most plausible reason for Alice's hypercalcemia.

7. Explain why hypercalcemia is often associated with malignancy.

FURTHER READING

Andreoli, TE (2000) Water: normal balance, hyponatraemia and hypernatraemia. *Ren. Fail.* **22:** 711–735.

Ariyan, CE and Sosa, JA (2004) Assessment and management of patients with abnormal calcium. *Crit. Care Med.* **32:** S146–S154.

Costa, J, Crausman, RS and Weinberg, MS (2004) Acute and chronic renal failure. *J. Am. Podiatr. Med. Assoc.* **94:** 168–176.

DiMeglio, LA, White, KE and Econs, MJ (2000) Disorders of phosphate metabolism. *Endocrinol. Metab. Clin. North Am.* **29:** 591–609.

Fall, PJ (2000) Hyponatraemia and hypernatraemia: A systematic approach to causes and their correction. *Postgrad. Med.* **107:** 75–82.

Fukagawa, M and Kurokawa, K (2002) Calcium homeostasis and imbalance. *Nephron* **92:** 41–45.

Gennari, FJ (2002) Disorders of potassium homeostasis: Hypokalaemia and hyperkalaemia. *Crit. Care Clin.* **18:** 273–288.

Inzucchi, SE (2004) Understanding hypercalcaemia. Its metabolic basis, signs and symptoms. *Postgrad. Med.* **115:** 73–76.

Kapoor, M and Chan, GZ (2001) Fluid and electrolyte abnormalities. *Crit. Care Clin.* **17:** 503–529.

Lamb, EJ, Tomson, CRV and Roderick, PJ (2005) Estimating kidney function in adults using formulae. *Ann. Clin. Biochem.* **42:** 321–345.

Molitoris, BA (1999) Acute renal failure. *Drugs Today* **35:** 659–666.

Pascual, E and Pedraz, T (2004) Gout. *Curr. Opin. Rheumatol.* **16:** 282–286.

Rott, KT and Agudelo, CA (2003) Gout. *J. Am. Med. Assoc.* **289:** 2857–2860.

Topf, JM and Murray, PT (2003) Hypomagnesemia and hypermagnesemia. *Rev. Endocr. Metab. Disord.* **4:** 195–206.

Touyz, RM (2004) Magnesium in clinical medicine. *Front. Biosci.* **1:** 1278–1293.

Yu, HT (2003) Progression of chronic renal failure. *Arch. Intern. Med.* **163:** 1417–1429.

chapter 9:

DISORDERS OF ACID–BASE BALANCE

OBJECTIVES

After studying this chapter you should be able to:

- describe how hydrogen ions are transported by the blood;
- outline the role and operations of the blood buffering systems;
- calculate the concentrations of ingredients used to prepare buffered solutions of defined pH and concentration;
- explain the role of kidney tubule cells in excreting hydrogen ions and regenerating plasma hydrogen carbonate;
- explain the causes of disorders of acid–base balance;
- describe the strategies for investigating disorders of acid–base balance;
- outline the management of disorders of acid–base balance.

9.1 INTRODUCTION

The concentration of hydrogen ions, H^+, in the blood is kept within a narrow reference range to give the blood a pH of approximately 7.4. The body possesses physiological and biochemical mechanisms that maintain this pH by removing excess H^+ and carbon dioxide produced during metabolism (*Figure 9.1*). These activities are vital for normal bodily functions and are performed by the renal and respiratory systems respectively. Failure to maintain the

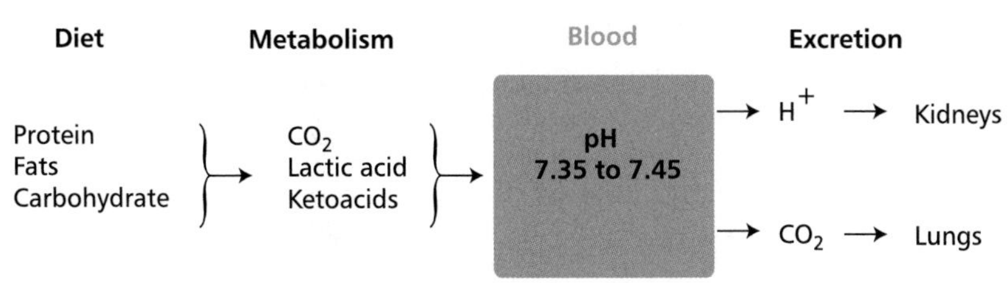

Figure 9.1 Overview of the production, transport and excretion of CO_2 and H^+.

Figure 9.2 Molecular model of carbonic anhydrase. The red sphere represents a Zn^{2+} in the active site. PDB file 2CBD.

acid–base balance at an appropriate value will give rise either to an acidosis, with a blood pH below the reference range, or an alkalosis with the pH above it. Different types of acidoses and alkaloses produce specific characteristic clinical features. Once a specific acid–base disorder has been identified, a clinical strategy must be adopted to manage the symptoms and to treat the underlying cause(s).

9.2 THE PRODUCTION AND TRANSPORT OF CARBON DIOXIDE

Body tissues produce about 20 moles of CO_2 per day during oxidative metabolism. The CO_2 diffuses from the cells into the extracellular fluid (ECF), that is the blood and tissue fluid, and eventually enters the plasma in quantities with the potential to form enough carbonic acid to disturb its pH. However, in normal circumstances this does not occur because the CO_2 is transported to the lungs and excreted. During transport, a substantial proportion of the CO_2 enters the erythrocytes by diffusion. Within the erythrocytes, a small proportion of the CO_2 remains dissolved or combines with proteins, mainly hemoglobin, to form carbamino compounds:

$$\text{Protein-NH}_2 + CO_2 \rightleftharpoons \text{Protein-NH-COO}^- + H^+$$

The major portion, however, combines with water to produce carbonic acid in a reaction catalyzed by carbonic anhydrase (*Figure 9.2*):

$$CO_2 + H_2O \xrightleftharpoons{\text{Carbonic anhydrase}} H_2CO_3$$

Carbonic acid dissociates to H^+ and hydrogen carbonate (HCO_3^-, 'bicarbonate')

$$H_2CO_3 \rightleftharpoons H^+ + HCO_3^-$$

Figure 9.3 shows how H^+ are removed from solution when they react with oxyhemoglobin (HbO_8) and promote the release of its oxygen to the tissues and forms protonated hemoglobin ('H^+Hb'). The HCO_3^- formed diffuses down its electrochemical gradient out of the erythrocytes to the plasma in exchange for Cl^-, thus maintaining the electrochemical equilibrium of the erythrocyte. The exchange of HCO_3^- for Cl^- is normally called the **chloride shift**. Since both ions are charged, neither would pass freely across biological membranes, however, an anion exchanger protein facilitates their transport. This exchanger is a membrane protein that forms a pore through the membrane allowing the cotransport of the ions across the membrane. Given that the ions move in opposite directions, the anion exchanger or cotransporter is said to be an antiporter. The concentration of HCO_3^- in the plasma is normally kept between 21–28 mmol dm^{-3}.

In the lungs, the partial pressure of oxygen is high while that of carbon dioxide is low. Thus oxygen enters the erythrocytes forming oxyhemoglobin, releasing the bound H^+ and promoting the reverse of the events that occur in other body tissues (*Figure 9.3*). Thus, H^+ associates with HCO_3^- to produce carbonic acid which then breaks down to carbon dioxide and water. The water enters the large body pool of water while the CO_2 leaves the erythrocytes and is excreted on exhalation.

These events provide an interesting confirmation that enzymes catalyze reactions in either direction depending upon the position of equilibrium. Thus carbonic anhydrase promotes the formation of carbonic acid in most body tissues where the concentration of CO_2 is relatively high. However, in the lungs, where the concentration of CO_2 is reduced, the enzyme catalyzes the formation of CO_2 and H_2O from carbonic acid.

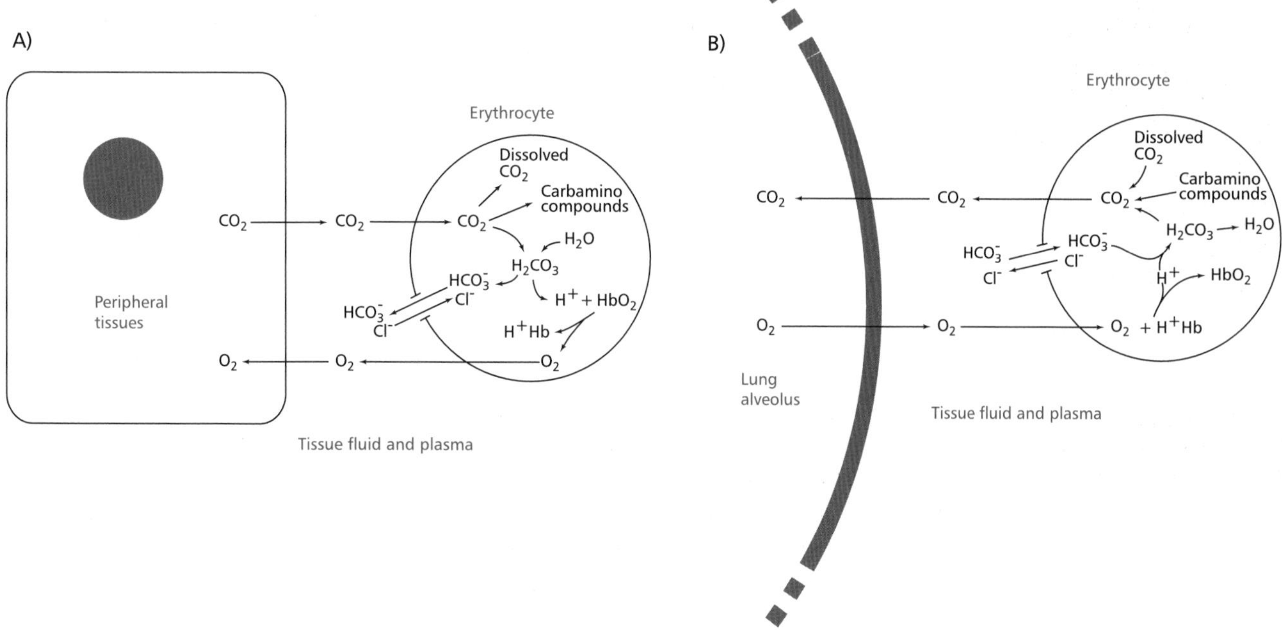

Figure 9.3 (A) Uptake by erythrocytes of CO_2 with the release of O_2. Note the Cl^- shift. See text for details. **(B) The excretion of CO_2 and uptake of O_2 at the lung epithelium.** See text for details.

9.3 BUFFERING AND THE EXCRETION OF H^+

About 60 mmol of H^+ are produced each day from the oxidation of sulfur-containing amino acid residues or from incomplete metabolic activities, such as anerobic glucose metabolism or ketone body formation (*Chapter 7*). If all the H^+ were released into the approximately 14 dm^3 of ECF, the concentration of H^+ would be 4 mmol dm^{-3} or about 100 000 times more acidic than normal. In reality, the concentration of H^+ is kept within the narrow limits of 40 ± 5 nmol dm^{-3} to maintain the appropriate body physiological pH of 7.4 ± 0.05. This pH is necessary for normal physiological functions and is maintained by temporary buffering systems that resist changes to the pH of the plasma until the excessive H^+ are excreted by the kidneys (*Chapter 8*).

When H^+ are released by cells, the ECF is buffered by the hydrogen carbonate–carbonic acid buffer system (*Box 9.1*):

$$H^+ + HCO_3^- \rightleftharpoons H_2CO_3 \rightleftharpoons CO_2 + H_2O$$

Other buffering systems, such as hemoglobin in the erythrocytes, also make significant contributions as described in *Section 9.2*. If the concentrations of H^+ and HCO_3^- reach equilibrium, buffering would become ineffective. However, in the case of the hydrogen carbonate–carbonic acid system this is usually prevented from occurring by the breakdown of carbonic acid to CO_2 and water. The formation of carbonic acid from H^+ and HCO_3^- is a rapid reaction. Its potentially slow breakdown to CO_2 and H_2O is accelerated by carbonic anhydrase in the erythrocytes and kidneys and the removal of carbon dioxide at the lungs prevents the system from reaching equilibrium.

Buffering by the hydrogen carbonate–carbonic acid system removes H^+ from the ECF but at the expense of HCO_3^-. The ECF contains relatively large amounts of HCO_3^-, for example, its concentration is usually about 24 mmol dm^{-3}. If, for any reason, the amount of H^+ produced increases, then the concentration of HCO_3^- will decrease as the hydrogen carbonate–carbonic acid buffering system operates. Any excess H^+ must be excreted from the body by the kidneys

BOX 9.1 Relationship between H^+, PCO_2 and HCO_3^-

The pH of a solution is defined as:

$$pH = -\log [H^+]$$

The pH scale (*Figure 9.4*) ranges from 0 to 14 and describes concentrations of H^+ of 10^0 (or 1) to 10^{-14} mol dm^{-3}. A pH of 7 is neutral; values below this are increasingly acidic and those above it increase in alkalinity.

A buffered solution is one that resists changes to its pH when relatively small amounts of acid or alkali are added to it. In organisms, the most significant buffers are protein molecules. However, buffered solutions can be prepared in the laboratory,

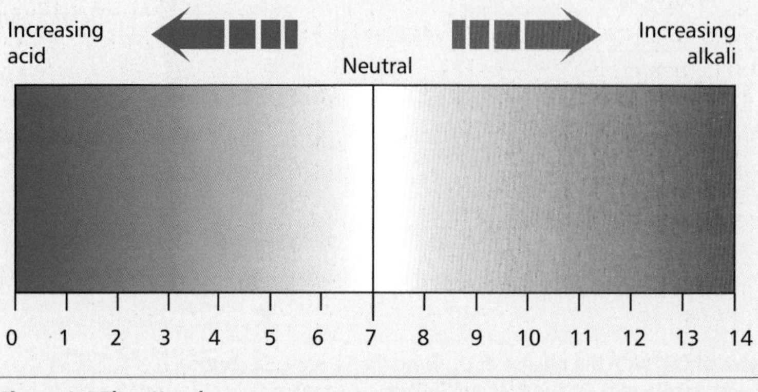

Figure 9.4 The pH scale.

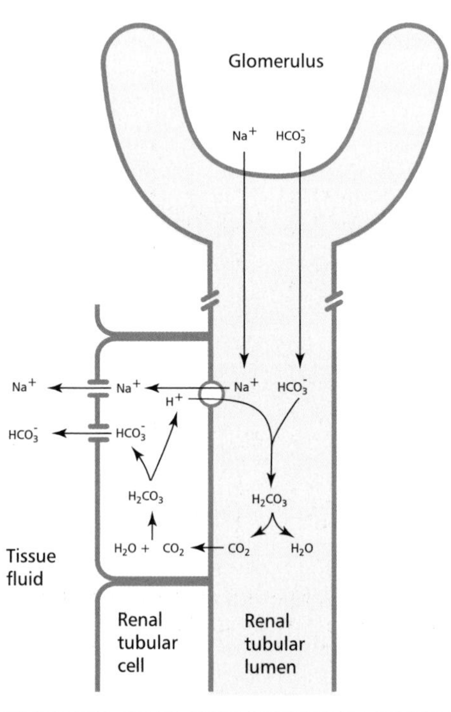

Figure 9.5 The indirect reabsorption of HCO_3^- by kidney tubule cells. See text for details.

(*Chapter 8*) to keep the ECF and tissues at an appropriate pH. In addition, the HCO_3^- used in buffering must be regenerated to return its concentration in the plasma to normal values, otherwise the body will become depleted of HCO_3^- and buffering capacity. Two separate mechanisms operate in the kidneys to recover the HCO_3^- initially removed from the blood by filtration at the glomerulus and to regenerate that used in buffering. The first mechanism is the HCO_3^- recovery system while the second regenerates the HCO_3^- (*Figures 9.5* and *9.6*).

In health, virtually all of the HCO_3^- is reabsorbed from the kidney tubule lumen. The operation of the system depends upon the tubule cells being polarized, that is, their luminal and basal surfaces differ in composition and permeability (*Figure 9.5*). In this manner, they resemble the enterocytes that line the absorptive surface of the gastroinstinal tract (*Chapter 11*). Direct reabsorption of HCO_3^- from the renal tubular fluid cannot occur because the luminal surfaces of renal tubular cells are impermeable to HCO_3^-. However, the concentration of CO_2 within the tubular cells is maintained at a relatively high value and so carbonic anhydrase catalyzes the formation of carbonic acid. The acid dissociates to HCO_3^- and H^+. The continuous formation of HCO_3^- and H^+ within the tubule cells is promoted by their removal. The HCO_3^- is transported across the basal membrane of the cell into the interstitial fluid and then into the capillaries. In contrast, H^+ is exchanged for Na^+ across the luminal membrane and enters the lumen of the kidney tubule. A membrane protein called the sodium bicarbonate cotransporter 1 (NBC 1) present in the luminal cell membrane facilitates the exchange of ions. Within the lumen, the H^+ combine with HCO_3^- to form carbonic acid. The acid breaks down spontaneously to CO_2 and H_2O in the proximal tubule, but carbonic anhydrase activity on the luminal surfaces of the cells speeds up the reaction in the distal tubule. The CO_2 can enter the tubule cell across its luminal membrane and so the HCO_3^- is recovered, indirectly, as CO_2. Approximately 80% of the filtered HCO_3^- in the proximal tubule is recovered by this mechanism. However, there usually by combining a weak acid with the salt of that acid. The pH of these solutions can be calculated using the Henderson-Hasselbalch equation:

$$pH = pK_a + \log [\text{base}] / [\text{acid}]$$

It is important to note that buffering is only effective at pH values equal to the $pK_a \pm 1$.

The buffering systems of the body do not excrete excess H^+ but temporarily remove them from free solution preventing excessive changes in pH. The effect of any H^+ produced by the body is neutralized largely by the hydrogen carbonate–carbonic acid buffer system.

For the hydrogen carbonate–carbonic acid buffer system:

$$H^+ + HCO_3^- \rightleftharpoons H_2CO_3$$

Therefore:

$$pH = pK_a + \log [HCO_3^-] / [H_2CO_3]$$

In plasma, H_2CO_3 breaks down to release carbon dioxide and water:

$$H_2CO_3 \rightleftharpoons CO_2 + H_2O$$

Since the concentration of H_2CO_3 is directly proportional to the partial pressure of CO_2 (PCO_2), it follows that:

$$[H_2CO_3^-] = PCO_2 \times 0.225$$

where 0.225 is the solubility constant of CO_2. Hence the Henderson-Hasselbalch equation can be rewritten as:

$$pH = pK_a + \log [HCO_3^-] / PCO_2 \times 0.225$$

It follows that the concentration of H^+ is directly proportional to the ratio, $PCO_2 / [HCO_3^-]$. Thus the concentration of H^+ in the blood varies as the concentration of HCO_3^- and the PCO_2 change: an increase in H^+ occurs when there is an increase in PCO_2 or a decline in HCO_3^-; while a decrease in H^+ will occur when the PCO_2 decreases or HCO_3^- increases.

has not been a net loss of H^+ and the HCO_3^- used in buffering has not been regenerated. The HCO_3^- is regenerated when carbonic acid is formed in the luminal cell as described above (*Figure 9.6*). Again, H^+ are exchanged for Na^+ and enter the lumen. Here the H^+ react with phosphate (HPO_4^{2-}) and ammonia (NH_3) to give $H_2PO_4^-$ and NH_4^+ respectively. Ammonia is a significant urinary buffer produced by the deamination of glutamine in the renal tubular cells in a reaction catalyzed by glutaminase (*Figure 9.7*). The ammonia formed can readily diffuse across cell membranes but NH_4^+ cannot enter the cells by passive reabsorption. Thus the NH_4^+, and the $H_2PO_4^-$, are excreted in the urine. For every H^+ excreted as NH_4^+ and $H_2PO_4^-$, a single HCO_3^- is formed in the tubule cell and secreted across the basal surface to the interstitial fluid and then into the blood. Hence the HCO_3^- concentration of the ECF is regenerated.

The synthesis of glutaminase is induced in states of chronic acidosis (*Section 9.4*) allowing an increase in the production of ammonia and an increased excretion of H^+ as NH_4^+.

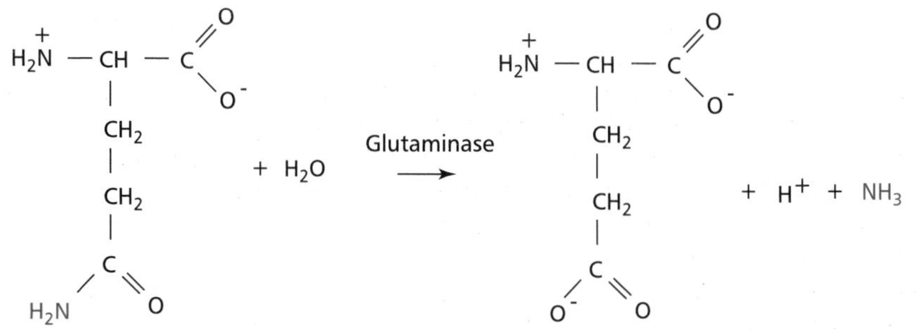

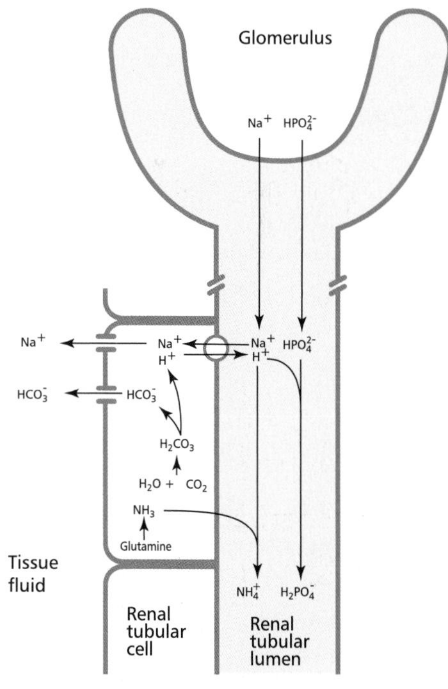

Figure 9.6 The regeneration of HCO_3^- by kidney tubule cells. See text for details.

$$H_2\overset{+}{N}-CH(COO^-)-CH_2-CH_2-C(=O)NH_2 + H_2O \xrightarrow{\text{Glutaminase}} H_2\overset{+}{N}-CH(COO^-)-CH_2-CH_2-COO^- + H^+ + NH_3$$

Figure 9.7 The formation of ammonia by the deamination of glutamine catalyzed by glutaminase

9.4 TYPES OF ACID–BASE DISORDERS

Disorders of acid–base balance are either acidoses or alkaloses. In an acidosis, there is accumulation of H^+ in the blood and its pH falls below the reference range. In an alkalosis there is a depletion of H^+ and therefore the blood has a pH above its reference range. Acid–base disorders can be further divided into two groups depending on their causes. If the abnormal pH occurs because of a metabolic or renal dysfunction, it is referred to as a **metabolic** acid–base disorder. When the abnormal pH is due to lung dysfunction, then it is a **respiratory** acid–base disorder. Physiological mechanisms that attempt to return the pH back to values within the reference range are referred to as **compensation**. Metabolic disorders cause a change in the concentration of HCO_3^- in the blood but respiratory disorders cause a change in its PCO_2 (*Table 9.1*). In any acid–base disorder, the pH of the blood depends on the severity of the primary disturbance and the amount of compensation that has occurred.

Margin Note 9.1 Fistula 

A **fistula** is an abnormal passage from a cavity or tube to another cavity or free surface.

Metabolic acidosis and alkalosis are the results of decreases and increases, respectively, in the concentration of HCO_3^-. These could be caused by the production of ketone bodies during diabetic ketoacidosis (*Chapter 7*) or from the loss of HCO_3^- from a duodenal fistula. Respiratory acidosis is associated with an increased PCO_2 whereas respiratory alkalosis occurs when the PCO_2 is decreased. For example, an impairment of respiratory function can increase the PCO_2 in the blood while hyperventilation would decrease it.

Compensation of acid–base disorders occurs by two major mechanisms: renal compensation and respiratory compensation. Renal compensation occurs when a respiratory disorder impairs lung function. The body attempts to adjust the pH of blood back to within its reference range by increasing the excretion of H^+ by the kidneys. Respiratory compensation is necessary when there is a metabolic acid–base disorder and involves changes in the ventilation of the lungs. Renal compensation is a relatively slow mechanism while respiratory compensation is much quicker to take effect. An acid–base disorder is said to be fully compensated if the compensatory mechanism returns the pH of the blood back to its reference range. However, compensation is usually partial and the pH remains outside the reference range.

METABOLIC ACID–BASE DISORDERS

Metabolic acid–base disorders lead to an accumulation or a loss of H^+ resulting in changes in the concentration of HCO_3^- in the blood. The direct loss or gain of HCO_3^- will also cause a metabolic acid–base disorder. Thus metabolic disorders are recognized by investigating the concentration of HCO_3^- in the blood. Respiratory compensation occurs quickly, often within hours, and patients will show some change in blood PCO_2 because of hypo- or hyperventilation.

Metabolic acidosis may arise from an increase in the amount of H^+ formed or a decrease in the concentration of HCO_3^-. Diabetic ketoacidosis (*Chapter 7*),

Primary disorder	Effect	Compensatory response
Respiratory acidosis	increased PCO_2	increased $[HCO_3^-]$
Respiratory alkalosis	decreased PCO_2	decreased $[HCO_3^-]$
Metabolic acidosis	decreased $[HCO_3^-]$	decreased PCO_2
Metabolic alkalosis	increased $[HCO_3^-]$	increased PCO_2

Table 9.1 Types of acid–base disorders and their compensatory mechanisms

lactic acidosis, poisoning with, for example, salicylate, methanol, ethylene glycol (*Chapter 12*) or the condition known as inherited organic acidosis all increase the production of H^+. In contrast, a decreased excretion of H^+ as in renal tubular acidosis, acute and chronic renal failure (*Chapter 8*), the use of inhibitors of carbonic anhydrase or a deficiency of mineralocorticoid, such as aldosterone (*Figure 9.8*) all increase blood H^+ content. Acid ingestion, as in acid poisoning, the excessive intake of amino acids by infusion, the direct loss of HCO_3^- by diarrhea or a pancreatic fistula can all reduce the concentration of HCO_3^- in the blood.

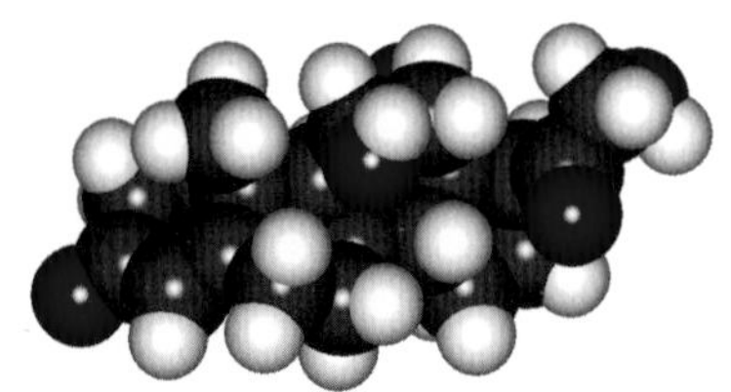

Figure 9.8 Computer generated model of aldosterone.

The clinical effects of metabolic acidosis include hyperventilation, where the increased H^+ concentration acts as a rapid and powerful stimulant of the respiratory center leading to a deep sighing breathing called **Kussmaul respiration**. The patients may also present with neuromuscular irritability that can cause cardiac arrhythmias. Cardiac arrest is more likely in the presence of hyperkalemia (*Chapters 8* and *14*). Eventually metabolic acidosis can depress the activities of the central nervous system and this can progress to coma and even death. Patients with metabolic acidosis are managed by treating the underlying cause and this usually resolves the acid–base disorder. In severe cases, the patients may be administered HCO_3^- intravenously to correct the acidosis.

Metabolic alkalosis may occur as a consequence of gastrointestinal loss of H^+ following vomiting and gastric aspiration or from excessive renal loss of H^+ in Conn's and Cushing's syndromes (*Chapter 7*). Some clinical treatments, such as the use of carbenoxolone, an anti-inflammatory drug used to treat ulcers, and thiazide diuretic drugs that reduce blood pressure by promoting the secretion of urine and by K^+ depletion, can also result in this condition. Finally, the administration of alkali, including alkali ingestion, and inappropriate treatment for acidosis can also cause metabolic alkalosis.

The clinical effects of metabolic alkalosis include hypoventilation that is a consequence of the low H^+ concentration. It is often accompanied by mental confusion and eventually coma. Patients may also suffer from paresthesia. Other effects of metabolic acidosis include tetany and muscle cramps that arise due to a decrease in the concentration of unbound Ca^{2+} in the plasma (*Chapter 8*) arising from the alkalosis. Metabolic alkalosis is usually managed by treating its underlying cause.

RESPIRATORY ACID–BASE DISORDERS

In respiratory acid–base disorders, the primary disturbance is caused by a change in the partial pressure of arterial CO_2. Respiratory disorders are related to a defect in the rate of ventilation of lungs or the exchange of gases across the alveolar membrane. The changes in PCO_2 (*Box 9.1*) alter the concentrations of carbonic acid in the blood, which, in turn, dissociates to HCO_3^- and H^+.

Some causes of respiratory acidosis are shown in *Table 9.2*. In general, obstruction of the airways by disease, or inhibition of the respiratory center in the brain by disease, trauma or drugs can cause respiratory acidosis.

Respiratory acidosis may be acute or chronic. Acute conditions occur within minutes or hours. It is usually the low PO_2 (**hypoxemia**) that is more dangerous than the high PCO_2 (**hypercapnia**). Further, renal compensation is slow, taking two or three days to become effective, so respiratory acidosis is usually uncompensated. Alveolar hypoventilation is usually the most common reason for acute respiratory acidosis. Hypoventilation increases the arterial PCO_2 and so the concentration of H^+ also rises quickly. The high PCO_2 and associated low PO_2 can cause coma and eventually death if untreated. Causes of acute respiratory acidosis include choking, bronchopneumonia and acute exacerbation of asthma.

Conditions giving rise to respiratory acidosis	Examples
Chronic obstructive airway diseases	bronchitis, emphysema
Obstruction of airways due to bronchospasms	asthma
Inhibition of respiratory center	anesthetics; sedatives
Cerebral damage	accidental trauma; stroke; tumors
Neuromuscular disease	poliomyelitis; tetanus; Guillain Barré syndrome
Pulmonary disease	fibrosis; pneumonia; respiratory distress syndrome
Sleep apnea	obesity

Table 9.2 Some conditions giving rise to respiratory acidosis

Chronic respiratory acidosis is, again, usually due to a decline in alveolar ventilation. However, this is normally a well established condition and subject to maximum renal compensation. Long-standing conditions responsible for chronic respiratory disorders include chronic bronchitis and emphysema.

The high PCO_2 is believed to be responsible for the clinical features of respiratory acidosis, such as peripheral vasodilatation and headaches. The acidosis can cause central nervous system depression leading to a coma. The treatment of respiratory acidosis is to improve alveolar ventilation, lowering the PCO_2 and increasing the PO_2. In chronic respiratory acidosis, it is usually not possible to treat the underlying cause and treatment is aimed at maximizing alveolar ventilation by using physiotherapy or bronchodilators.

Respiratory alkalosis is less common than respiratory acidosis. However, it is often an acute condition due to hyperventilation. Often renal compensation does not occur.

The clinical effects of respiratory alkalosis include confusion, headaches, dizziness and coma. Respiratory alkalosis may be caused by hypoxia, increased respiration or pulmonary disease. Hypoxia is associated with high altitudes, severe anemia and pulmonary disease. Increased respiration may result from the use of respiratory stimulants, such as salicylates, from primary hyperventilation syndrome, artificial hyperventilation, and pulmonary diseases, such as pulmonary edema and embolisms. The treatment of respiratory alkalosis is aimed at removing its underlying cause as this usually resolves the acid–base disturbance.

MIXED ACID–BASE DISORDERS

Sometimes patients may present with more than one acid–base disorder and this is known as a **mixed acid–base disorder**. These may present as (i) severe acidemia, that is a low blood pH, (ii) with a normal or near normal pH or (iii) with alkalemia, that is, a high blood pH. Whatever the underlying cause, all mixed acid–base disorders are associated with abnormal levels of blood PCO_2 and HCO_3^-.

For example, a patient with chronic bronchitis may also have renal failure. Both these disorders increase the concentration of H^+ in the blood. Chronic bronchitis leads to respiratory acidosis while the renal failure causes metabolic acidosis. This patient will therefore present with a mixed acid–base disorder with a high blood PCO_2 and H^+ concentration but a low concentration of HCO_3^-. In some cases, however, the two disorders in a mixed acid–base disorder can be antagonistic, that is, have opposing effects on the concentration of H^+ in blood. In this case the blood H^+ concentration may be near normal although the PCO_2 and HCO_3^- concentration will both be abnormal. For example, a patient with salicylate poisoning may have a metabolic acidosis together with a respiratory alkalosis. Patients may also present with metabolic and respiratory alkaloses. This could occur in someone with congestive cardiac failure who is on diuretic therapy. The former will cause a respiratory alkalosis and the latter a metabolic alkalosis. Such individuals will usually have a high blood pH and increased HCO_3^- but the PCO_2 will be decreased.

9.5 INVESTIGATING ACID–BASE DISORDERS

An investigation of an acid–base disorder involves three stages. The first stage involves identifying whether the patient has an acidosis or an alkalosis. The second stage is to determine whether the acid–base disorder is metabolic or respiratory in nature while the third stage involves determining the degree of compensation.

Acid–base disorders are investigated as outlined in *Figure 9.9*. An arterial blood specimen is collected and its H^+ concentration (pH) and PCO_2 measured. The blood must be collected from an artery into a syringe containing an

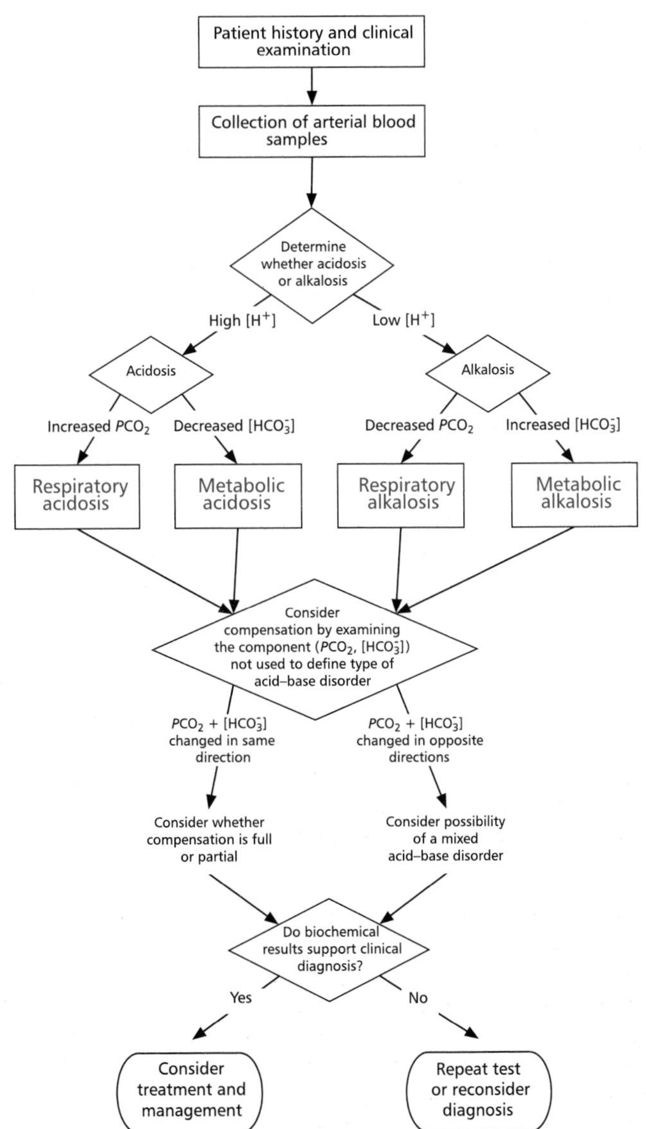

Figure 9.9 Outline of how disorders of acid–base balance are investigated. See text for details.

anticoagulant, such as heparin, and transported to the hospital laboratory at 4°C. Care must be taken that air does not enter any blood samples during collection. The concentration of H^+ and PCO_2 level are measured directly using an automated analyzer that is also programmed to calculate the corresponding concentration of HCO_3^- (*Figure 9.10*) according to the Henderson-Hasselbalch equation. An acidosis or alkalosis can be identified from the pH, that is the concentration of H^+, in the blood. An examination of the levels of PCO_2 and HCO_3^- shows whether the identified disorder is metabolic or respiratory in origin and indicates to what extent compensation is occurring.

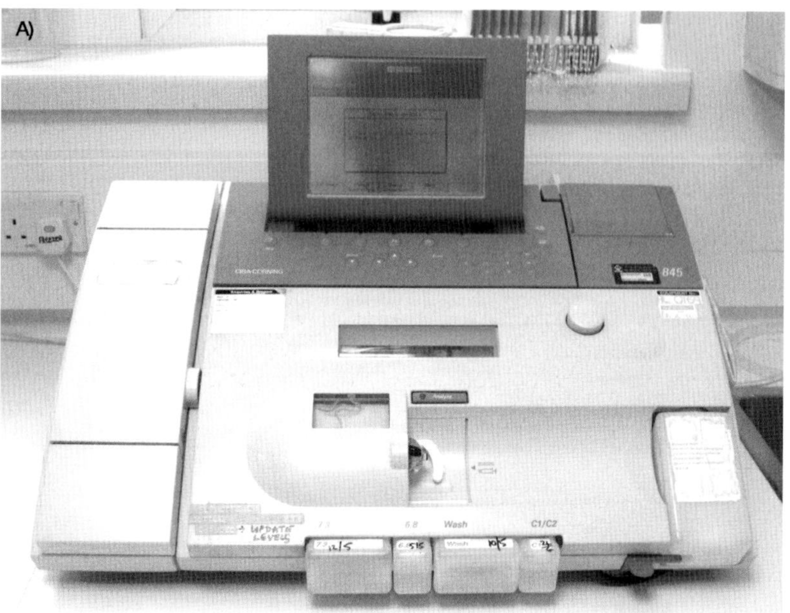

B)

PATIENT SAMPLE REPORT
SYSTEM 845-1008

Sequence no 17688
Accession no 111
Source

Patient ID
Birthdate
Age

14 MAY 2004 11:50
Analysis Date 02 MAR 2004
Analysis Time 18:53
Draw Date
Draw Time
Operator ID 1

Sex
Physician ID
Location

SYRINGE SAMPLE
ACID/BASE 37°C

		Units	Reference Range
pH	7.431		(7.360 - 7.440)
pCO2	5.17	kPa	(4.53 - 6.00)
pO2	14.58	kPa	(12.00 - 14.67)
HCO3-act	25.2	mmol/L	
BEvt	0.9	mmol/L	

Figure 9.10 (A) An automated analyzer for determining blood pH and PCO_2 and programmed to calculate the corresponding concentration of HCO_3^-. (B) A sample of a read out from the automated analyzer. Courtesy of the Department of Clinical Biochemistry, Manchester Royal Infirmary, UK.

CASE STUDY 9.1

John is a young man with a history of dyspepsia (severe indigestion) and excessive alcohol intake. He was admitted into his local hospital following 24 hours of vomiting. An arterial blood specimen was taken and analyzed for a suspected acid–base disorder. The following results were obtained (reference ranges are indicated in parentheses).

$[H^+]$	28 nmol dm^{-3}	(35–45 nmol dm^{-3})
PCO_2	7.2 kPa	(4.4–5.6 kPa)
$[HCO_3^-]$	48 mmol dm^{-3}	(21–28 mmol dm^{-3})

Question

Does John have an acid–base disorder and, if so, what type of acid–base disorder is present?

CASE STUDY 9.2

Tom was admitted to hospital following an acute attack of asthma. A blood specimen was taken and analyzed giving the following results (reference ranges are shown in parentheses).

$[H^+]$	24 nmol dm^{-3}	(34–44 nmol dm^{-3})
PCO_2	2.5 kPa	(4.1–5.6 kPa)
$[HCO_3^-]$	22 mmol dm^{-3}	(21–28 mmol dm^{-3})

Question

Does Tom have an acid–base disorder and, if so, what type?

CASE STUDY 9.3

Terry, a 62 year factory worker has suffered with vomiting and diarrhea for the last two weeks. When examined by his doctor, he was dehydrated and his breathing was deep and noisy. An arterial blood specimen was taken and analyzed for blood gases (reference ranges are shown in parentheses).

$[H^+]$	65 nmol dm^{-3}	(34–44 nmol dm^{-3})
PCO_2	2.9 kPa	(4.1–5.6 kPa)
$[HCO_3^-]$	9 mmol dm^{-3}	(21–28 mmol dm^{-3})

Questions

(a) Does Terry have an acid–base disorder?

(b) Is there any compensation?

9.6 SUMMARY

The pH of the blood is maintained at 7.4 within normal limits by physiological mechanisms that remove the H^+ and CO_2 produced by metabolism. Excess CO_2 is transported by the blood to the lungs and excreted. In the blood, the dissolved CO_2 is converted to hydrogen carbonate in a reaction catalyzed by carbonic anhydrase. Buffering systems such as the hydrogen carbonate–carbonic acid system and hemoglobin serve to maintain the physiological pH in health. Acid–base disorders result in the blood pH being lower (acidosis) or higher (alkalosis) than normal. Such disorders may be metabolic or respiratory in nature and can be fatal if untreated, although there are physiological mechanisms for at least a partial compensation of the disorder. Treatment of most acid–base disorders involves treatment of the underlying causes.

QUESTIONS

1. Approximately how many molecules of CO_2 are produced daily by oxidative metabolism? Avogadro's number is 6.02×10^{23}.

   a) 12×10^{24}

   b) 1.2×10^{23}

   c) 1.2×10^{26}

   d) 1.2×10^{25}

   e) 1.2×10^{24}

2. (a) How much greater is the $[H^+]$ in a solution of pH 3 than one of pH 6?

   (b) What is the pH of a buffer prepared from 100 cm^3 of 0.1 mol dm^{-3} ethanoic acid ('acetic acid') solution (pK_a of ethanoic acid = 4.76) and 75 cm^3 of a 0.2 mol dm^{-3} sodium ethanoate solution?

3. A blood analysis of a patient gives values for PCO_2 of 5.0 kPa (4.4–5.6 kPa) and a pH of 7.56. The pK_a for the carbonic acid–hydrogen carbonate system is 6.10. (a) Calculate the corresponding $[HCO_3^-]$. (b) Is the patient in an acidosis or alkalosis? (c) If so, is this metabolic or respiratory in origin?

4. The following blood gas results were obtained for a patient (reference ranges are shown in parentheses):

$[H^+]$	50 nmol dm^{-3}	(35–45 nmol dm^{-3})
PCO_2	11.5 kPa	(4.4–5.6 kPa)
$[HCO_3^-]$	34 mmol dm^{-3}	(21–28 mmol dm^{-3})

   Which of the following conditions match most closely with these results?

   a) diabetic ketoacidosis;

   b) laboratory transcription error;

   c) chronic obstructive airways disease;

   d) pyloric stenosis;

   e) none of the above.

FURTHER READING

Allen, K (2005) Four-step method of interpreting arterial blood gas analysis. *Nurs. Times* **10:** 42–45.

Fall, PJ (2000) A stepwise approach to acid–base disorders. Practical patient evaluation for metabolic acidosis and other conditions. *Postgrad. Med.* **107:** 249–258.

Gluck, SL (1998) Acid–base. *Lancet* **352:** 474–479.

Kraut, JA and Madias, NE (2001) Approach to patients with acid–base disorders. *Respir. Care* **46:** 392–403.

Soleimani, M and Burnham, CE (2000) Physiologic and molecular aspects of the Na^+: HCO_3^- cotransporter in health and disease processes. *Kidney Int.* **57:** 371–384.

Soleimani, M and Burnham, CE (2001) Na^+: HCO_3^- cotransporters (NBC): Cloning and characterization. *J. Membr. Biol.* **183:** 71–84.

Whittier, WL and Rutecki, GW (2004) Primer on clinical acid–base problem solving. *Dis. Mon.* **50:** 122–162.

Williamson, JC (1995) Acid–base disorders: classification and management strategies. *Am. Fam. Physician* **52:** 584–590.

chapter 10:

DIET AND DISEASE

OBJECTIVES

After studying this chapter you should be able to:

- define the term *nutrient*;
- outline the types and roles of nutrients;
- list some of the common nutritional disorders;
- explain the causes of some common nutritional disorders;
- review the general methods for investigating nutritional disorders;
- discuss the treatment and management of some common nutritional disorders.

10.1 INTRODUCTION

Nutrition is concerned with food and how the body uses it. All nutrients must be ingested, and most of them digested, before they can be assimilated and used by the body. Carbohydrates, lipids and proteins are nutrients that are required in comparatively large amounts. Others, such as vitamins, minerals and trace elements, are also necessary but are required in much smaller quantities. Water is also necessary to support life. Nutrients have a variety of metabolic roles and are needed for normal growth, development and the maintenance of health. The body can make some nutrients from others that must be supplied in the diet; the former are **nonessential**, the latter **essential nutrients**. Specific daily nutritional requirements are determined by a number of factors and, indeed, vary throughout the life of any individual. The major factors concerned include the age, sex, physical activity and the general well-being of the person concerned and, in the case of females, menstruation, pregnancy and lactation. Good nutrition is essential for health and the prevention of diseases. Nutritional disorders may arise from a deficiency or, in some cases, an excess of some component of the diet.

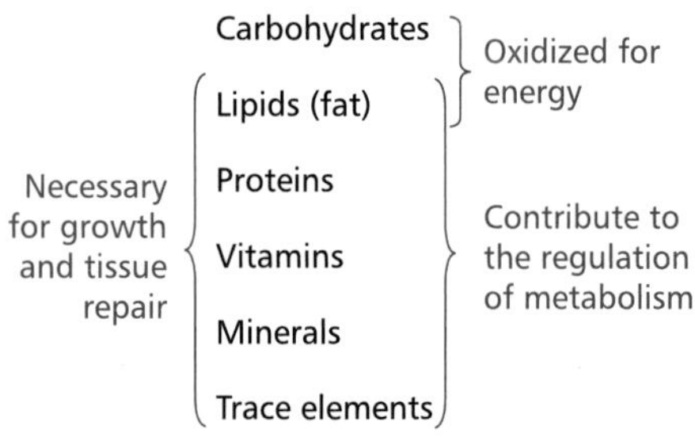

Figure 10.1 Overview of the general roles of nutrients.

10.2 DIET AND NUTRITION

Food is obviously needed to sustain life but a balanced diet is vital for good health. General evidence for good nutrition is a well developed body with an ideal weight (*Section 10.3*), good skin, including hair, and muscle tone, and physical and mental alertness. The food eaten by an individual constitutes the diet and it should supply appropriate amounts and proportions of a variety of substances, usually called nutrients. Some nutrients, the macronutrients, such as carbohydrates, lipids and proteins, are required in amounts that are large compared with the micronutrients, such as vitamins, minerals and trace elements, which are needed in much smaller quantities. Nutrients are needed to sustain a number of activities within the body if good health is to be maintained. They supply the raw materials required for growth and maintenance of tissues, include substances that help regulate metabolism, for example the cofactors and coenzymes necessary for enzyme activities and provide the energy to drive metabolism and maintain homeothermy. Nutrients vary in their contributions to these roles and there is also overlap between their contributions to each (*Figure 10.1*).

NUTRIENTS AND ENERGY

Diets are often described in terms of their energy content, despite the fact that they need to supply materials for needs other than just their energy provision (*Figure 10.1*). It is possible for a diet to supply an adequate amount of energy but be deficient in some essential raw materials, such as vitamins and minerals needed by the individual. A **balanced diet** is one that supplies adequate energy distributed appropriately between carbohydrates, lipids, and proteins and contains the necessary amounts of vitamins, minerals, trace elements, water and nondigestible fiber.

Nutritional energy is usually measured in joules or kilojoules. The number of joules required by any one person needs to be matched to their energy output (*Table 10.1*). The term **recommended daily amount (RDA)** was originally used by the UK Department of Health in 1979 to define the amounts of certain nutrients needed by different groups within the population.

Group	Approximate daily energy requirements /kJ required per kg desirable body weight
Babies up to one year old	500
Children one to 10 years old	335–420
Male children 11–15 years old	270
Female children 11–15 years old	145
Sedentary males and most females	117
Physically active males and females	125
Males and females over 55 years old	117
Pregnant females (first trimester)	117–134
Pregnant females (second and third trimesters)	150–159
Lactating females	150–159

Table 10.1 Energy required to maintain body functions. For example, a physically active man weighing 70 kg requires $70 \times 125 = 875$ kJ (or 2080 kcalories) daily.

Unfortunately, RDAs were often used inappropriately to assess the dietary needs of individuals. Accordingly, in 1991, they were replaced with **dietary reference values (DRVs)** that are guidance values as to the amounts of energy and nutrients that should be ingested, rather than exact recommendations. Dietary reference values indicate the amount of energy or amount of an individual nutrient required by a group of a certain age range and sex for good health. The value of any DRV is given as a daily intake. However, in practice DRVs must be determined from the energy and nutrient intakes averaged from several days eating, because food intake and appetite vary from day to day. Groups for which DRVs have been set include seven groups of infants of both sexes from birth to age 10 years, four groups each for males and females between the ages of 11 and 50-plus, with extra groups for women who are pregnant or are lactating. The value of a DRV applies to healthy people, since patients who are unwell for whatever reason may have differing nutritional requirements. Dietary terms that are related to DRVs are the **estimated average requirement (EAR)**, **reference nutrient intake (RNI)** and **lower reference nutrient intake (LRNI)**. The EAR is the mean amount of energy or nutrient needed by a population, while the RNI is an amount of a nutrient that is greater than the dietary needs of 97.5% of that group. The LRNI is the amount of a nutrient sufficient to meet the requirements of 2.5% of a population. Note that most people will need more than the LRNI. *Figure 10.2* shows how EAR, RNI and LRNI are related. Finally, the **safe intake** is the amount judged to be sufficient for the needs of most people and anything below this level could be undesirable. This arbitrary value is given when there is simply insufficient evidence to establish reliable values for EAR, RNI or LRNI.

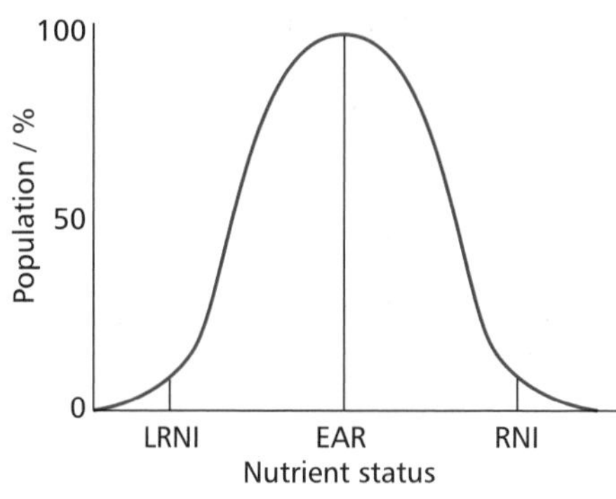

Figure 10.2 The relationships between EAR, RNI and LRNI.

CARBOHYDRATES

Carbohydrates have the general formula $(CH_2O)_n$. The main dietary carbohydrates (*Figure 10.3*) are mono- and disaccharide sugars, for example fructose, glucose, lactose and sucrose, and polysaccharides, mainly starch, and these are usually the major suppliers of energy. Current recommendations from the WHO suggest that 55% of dietary energy should be in the form of carbohydrates. Most of the carbohydrates should be in the form of starch with no more than 10–15% of energy intake in the form of sugars. However, dietary fiber or roughage largely consists of cellulose, a polysaccharide that cannot be digested and absorbed, although it may be metabolized by bacteria in the large intestine (*Chapter 11*). Foods rich in fiber include wholemeal cereals, flour, root vegetables, nuts and fruits. Dietary fiber stimulates peristalsis and protects against constipation and is known to reduce blood cholesterol and glucose and the incidence of colorectal cancers (*Chapter 17*). Communities that consume high fiber diets have relatively low incidences of these cancers.

LIPIDS

Dietary lipids are the fats and oils, both of which contain fatty acids that are concentrated sources of energy. Fats are also needed as carriers for the fat-soluble vitamins A, D, E and K. Fatty acids may be saturated or unsaturated, that is, they lack or contain double bonds respectively. The only difference between fats and oils is in their melting points which are related to their degree of unsaturation and chain length: oils are liquid at room temperatures, indicating a higher degree of unsaturation than fats which are solid at these temperatures. Dietary fats (*Figure 10.3*) consist largely of triacylglycerols together with small amounts of phospholipids and cholesterol. The WHO has recommended that total fats should not supply more than 30% of the energy intake of the diet.

Essential fatty acids

Humans lack the enzymes necessary for synthesizing certain unsaturated fatty acids necessary for health and these are **essential fatty acids (EFAs)**.

Carbohydrates
bread
cereals
pasta
peas
potatoes
rice

Fats
butter
cheese
cooking oils
cream
lard
margarine
milk
nuts
oily fish
suet

Proteins
cheese
eggs
fish
meat
milk
potatoes
pulses

Figure 10.3 A selection of carbohydrate, fat and protein rich foods. Some foods, of course, contain more than one of these macronutrients.

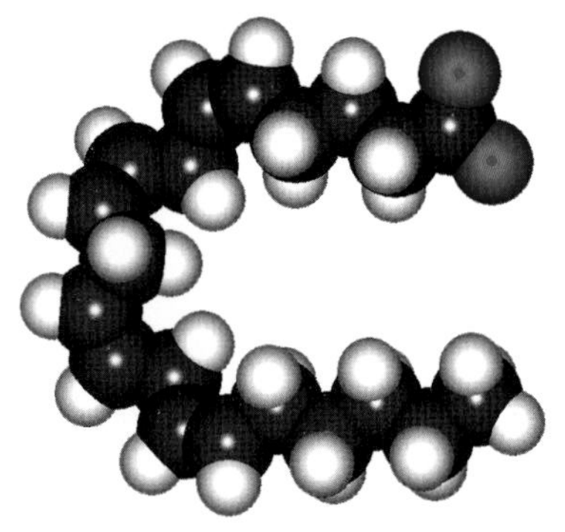

Figure 10.4 A computer generated model of arachidonic acid. Oxygen atoms are shown in red, carbon in black and hydrogen in gray.

A number of fatty acids cannot be synthesized by the body including linoleic, linolenic and arachidonic acid (*Figure 10.4*) and are precursors of many biologically active and clinically relevant molecules, such as the eicosanoids (prostaglandins, thromboxanes, prostacyclins and leukotrienes). These molecules act like hormones and mediate a wide range of physiological activities affecting, for example inflammatory responses, blood pressure and clotting, reproductive activities and the sleep–wake cycle. Their actions are local, affecting only cells near their sites of production, and they act at low concentrations via second messengers (*Chapter 7*).

Humans, like all mammals, lack the enzymes needed to form a double bond beyond C-9 or within the terminal seven carbon atoms of a fatty acid (*Figure 10.5*). Rather than the strict chemical convention of numbering carbon atoms in fatty acids from the carboxyl group, the double bonds in fatty acids are often numbered from the terminal or ω carbon atom, giving rise to three families of fatty acids with their first double bond occurring at positions ω-3, ω-6 and ω-9. These three families cannot be metabolically interconverted. The term omega fatty acid has entered everyday English and the structures of a number of examples are shown in *Figure 10.6*. Linoleic acid is an ω-6 acid. Two forms of linolenic acid occur, a ω-3 type called α linolenic acid (ALA) and the ω-6 form, γ linolenic acid (GLA). Other ω-3 EFAs are eicosapentaenoic acid (EPA) and

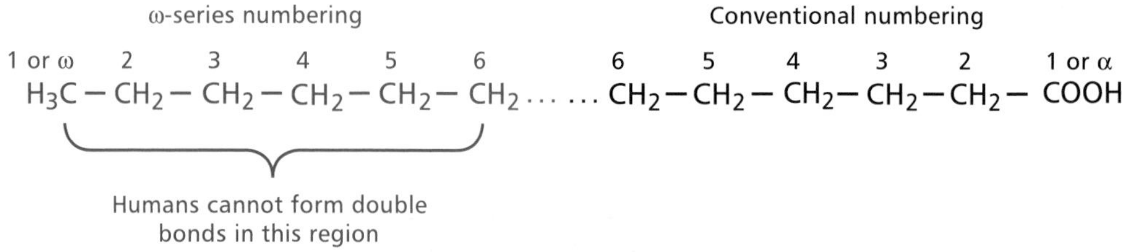

ω-series numbering | Conventional numbering

1 or ω 2 3 4 5 6 ... 6 5 4 3 2 1 or α

$H_3C-CH_2-CH_2-CH_2-CH_2-CH_2\ldots\ldots CH_2-CH_2-CH_2-CH_2-CH_2-COOH$

Humans cannot form double bonds in this region

Figure 10.5 The numbering and terminal unsaturation of fatty acids.

ω-6 series

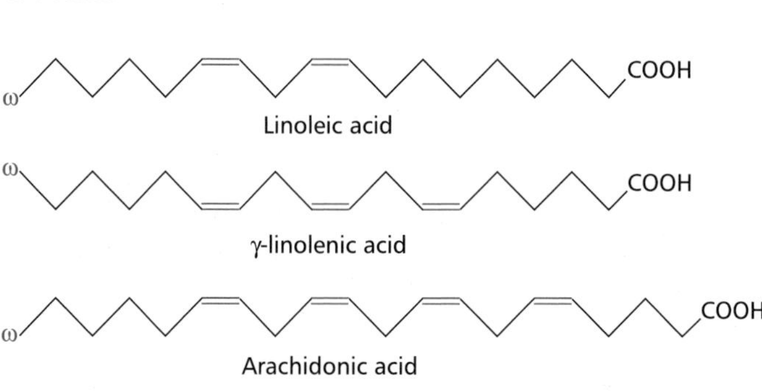

ω-3 series

ω COOH

α-linolenic acid

ω COOH

Eicosapentaenoic acid (EPA)

ω COOH

Docosahexaenoic acid (DHA)

Figure 10.6 The structures of some unsaturated fatty acids.

docosahexaenoic acid (DHA). Omega-3 and ω-6 fatty acids are EFAs because they are not synthesized by the body but must be obtained in the diet. Seeds and vegetable oils are excellent dietary sources of ω-6 EFAs and the body is able to convert linoleic acid to GLA and arachidonic acid. Many vegetable oils contain only low amounts of ALA since this is normally present in the chloroplasts of plants and may only be a significant dietary constituent if green leafy vegetables are consumed. However, ALA can be used to synthesize EPA and DHA by body tissues. Rich sources of the ω-3 EFAs, EPA and DHA are in the oils from the muscles and skins of a number of cold deepwater fishes, such as herring, mackerel, salmon, sardines and tuna (*Figure 10.7*). Cod, which live in similar environments, store fat in the liver: hence cod liver oil is also an excellent source of EPA and DHA.

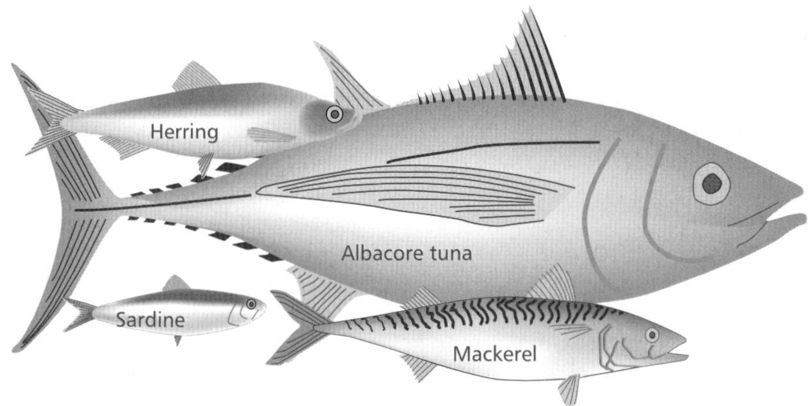

Figure 10.7 A selection of fish rich in essential fatty acids.

A dietary deficiency of EFAs leads to a dry scaly skin subject to erythema, poor healing of wounds and hair loss and a failure to thrive in infants. The first associations between ω fatty acids and health came from studying Greenland Inuit (Eskimo) and inhabitants of fishing villages in Japan and people of Okinawa. These groups have much lower incidences of diseases, such as coronary heart disease (CHD, *Chapter 14*), rheumatoid arthritis (*Chapters 5* and *18*) and diabetes mellitus (*Chapter 7*), than their European or USA counterparts, even though their diets are high in fat from eating seal and fish. However, these organisms are rich in ω-3 fatty acids that have been shown subsequently to provide significant health benefits. The metabolites of ω-3 fatty acids decrease platelet function, reduce the risk from sudden death caused by cardiac arrhythmias and slow the progress of atherosclerosis. Modest decreases in blood pressure also occur with high intakes of ω-3 fatty acids. Some studies have indicated that ω-3 fatty acids inhibit the synthesis of very low density lipoprotein (VLDL) and triacylglycerols (TAGs) and so decrease their concentration in plasma, particularly in patients suffering from hypertriglyceridemia (*Chapters 13* and *14*). Furthermore, they have been associated with reducing morning stiffness and the number of tender joints in patients with rheumatoid arthritis. There is, however, no established recommended daily intake for ω-3s although it has been suggested that 1–2% of the total daily energy intake should be in the form of EFAs. Human adults are thought to require up to 10 g of linolenic acid daily.

The eicosanoids formed from ω-6 fatty acids, although produced by many of the same enzymes as those from ω-3 acids, have different physiological functions and effects. For example, the ω-3 group is antithrombic and anti-inflammatory, both processes promoted by ω-6 acid derived metabolites. Thus a balancing of the intakes of ω-3 and ω-6 fatty acids in the diet is thought to be of importance. The diet typical of the developed world is far richer in ω-6 than ω-3 fatty acids; it has been suggested that ratios of

Essential	Nonessential
His	Ala
Ile	Arg[1]
Leu	Asn
Lys	Asp
Met[2]	Cys
Phe[3]	Glu
Thr	Gln
Trp	Gly
Val	Pro
	Ser
	Tyr

[1]Arg can be synthesized by tissues but not in sufficient amounts to support growth in children.
[2]Met is required in substantial dietary quantities to form Cys if this is not ingested in adequate amounts.
[3]Increased quantities of Phe are needed to form Tyr if this is not ingested in adequate quantities.

Table 10.2 Essential and nonessential amino acids

ω-3:ω-6 of 2:1 or 1:1 are a healthier balance. The American Heart Association and the UK Food Standards Agency recommends that people without a history of CHD should consume two servings of fish weekly and those with known CHD should eat one serving of fish daily. More than these amounts are not recommended because, unfortunately, significant amounts of environmental contaminants, such as methylmercury, polychlorinated biphenyls and dioxins, are concentrated in some species of fish. For the same reason, young children and women who may become pregnant, who are pregnant or who are breastfeeding should avoid eating excessive amounts of fish and shellfish. Omega-3 fatty acids exert a dose related effect on bleeding time and it has been suggested that care be applied to their excessive intake, particularly when combined with anticoagulant medications, such as warfarin or heparin (*Chapter 13*). In general, however, their ingestion is well tolerated, although side effects can include a fishy aftertaste and gastrointestinal tract (GIT) disturbances, for example nausea, bloating, belching, all of which appear to be dose dependent.

PROTEINS

Dietary proteins (*Figure 10.3*) are necessary to supply the amino acids needed for growth and the general repair and maintenance of tissues. A daily intake of about 65 g and 50 g of dietary protein is required in males and females respectively, which provides about 10–15% of the total energy in a balanced diet, although only about 5% of body energy comes from their catabolism under normal circumstances. Protein must be obtained from a variety of sources to supply all essential amino acids. Humans are unable to synthesize nine of the 20 amino acids found in proteins that have codons in the genetic code and these are therefore essential dietary constituents (*Table 10.2*).

However, the nonessential amino acids can be synthesized if the supply of the essential ones is adequate. The 'quality' of dietary protein is important and protein intake needs to be varied, particularly as some plant proteins lack one or more of the essential amino acids.

WATER

Water is vital for life because it helps stabilize the structures of large molecules, such as proteins and starch, functions as a medium for most biochemical reactions, acts as a solvent for electrolytes, glucose, vitamins, minerals and numerous other small molecules and transports nutrients to, and waste products away from, cells as well as around the whole body. The GIT (*Chapter 11*) secretes some 7–9 dm^3 of water that aids the digestion and absorption of nutrients. To maintain these functions, loss of body water should match intake to ensure water balance (*Chapter 8*). It is recommended that two to three dm^3 of water should be consumed daily, of which about 60% should be liquid water and the rest obtained from seemingly solid foods. The homeostasis of water has been described in *Chapter 8*.

Margin Note 10.1 Selenocysteine 

The amino acid selenocysteine (*Figure 10.8*) is a component of a number of selenoproteins, including the enzymes glutathione peroxidase, tetraiodothyronine 5′ deiodinase, thioredoxin reductase, formate dehydrogenase, glycine reductase and a number of hydrogenases. Selenocysteine is specified by the codon UGA that would normally function as a termination codon in protein synthesis. However, in the presence of a selenocysteine insertion sequence element (SECIS) in the mRNA, UGA specifies selenocysteine. Characteristic nucleotide sequence and base-pairing patterns in the SECISs form secondary structures that lead to the change in codon specificity.

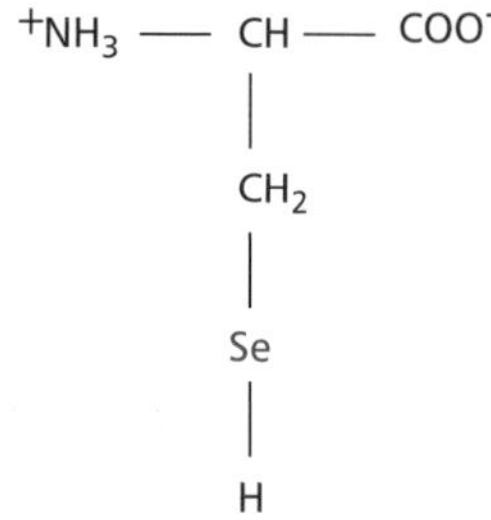

Figure 10.8 The structure of selenocysteine.

VITAMINS

Vitamins are organic substances that the body cannot synthesize or can make only from chemically closely related compounds. They are needed in the diet but only in relatively small amounts (*Table 10.3*).

Vitamins may be classified as water soluble, such as the B group and vitamin C (*Figure 10.9*) and fat soluble, such as the vitamins A, D, E and K. Some of the B vitamins and, in general, each fat soluble vitamin comprise a group of closely related compounds called **vitamers.** In such cases, the name of the vitamin is used as a collective descriptor. Vitamins have specific biochemical roles and are essential for normal metabolism, growth and good health.

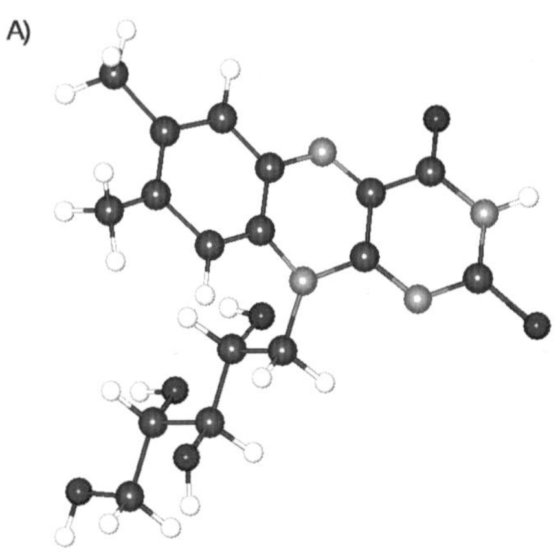

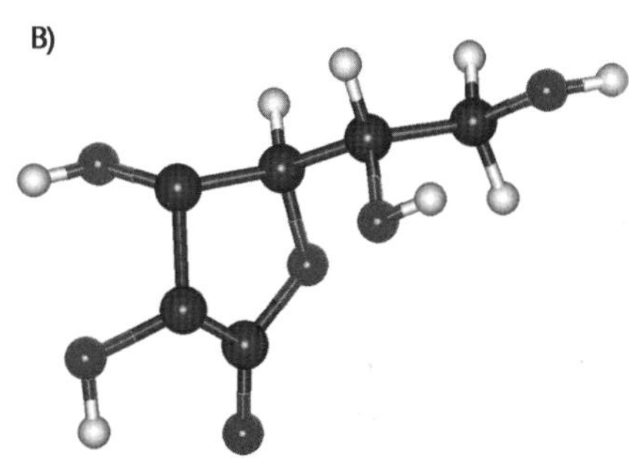

Figure 10.9 Computer generated models of the water soluble vitamins (A) B_2 and (B) C. Oxygen atoms are shown in red, carbon in black, nitrogen in gray and hydrogen are colorless.

Vitamin or related compounds	Recommended intake / mg day^{-1}	Sources
Vitamin B_1 Thiamin	1.0 in males 0.8 in females	yeast, eggs, milk, cheese and cereals
Vitamin B_2 Riboflavin	1.3 in males 1.1 in females	liver, milk, cereals, mushrooms and eggs
Niacin or nicotinic acid Nicotinamide	17 in males 13 in females	meat, fish, pulses, cereals and synthesized endogenously from tryptophan
Vitamin B_5 Pantothenic acid	3–7 is normally provided in diet and appears more than adequate	meats, offal, eggs, green vegetables, whole grains
Vitamin B_6 Pyridoxine Pyridoxal Pyridoxamine	1.4 in males 1.2 in female	vegetables, fruit, cereals, eggs, milk and nuts
Vitamin H Biotin	0.01–0.2 in adults	eggs, pulses, liver, yeast, milk, dried fruit and can be synthesized by intestinal bacteria in the GIT
Folic acid	0.2 in adults 0.4 in pregnancy	liver, spinach, cabbage, peas, bananas and oranges
Vitamin B_{12} Cobalamin	0.0015 in adults	liver, egg and milk
Vitamin C Ascorbic acid	40 in adults	green vegetables and citrus fruits
Vitamin A Retinol Retinal β carotene	0.7 in males 0.6 in females	liver, fish oils, eggs, cheese and can be synthesized from carotene in the diet
Vitamin D Cholecalciferol Ergocalciferol	0.01 in pregnancy	liver and dairy products and endogenous synthesis in the skin if there is adequate exposure to sunlight.
Vitamin E Tocopherol Tocotrienols	4 in males 3 in females	vegetable oils, nuts and wheat germ
Vitamin K Phylloquinone, Menaquinones	0.001 in adults	leafy vegetables, cheese or synthesized by colon bacteria

Table 10.3 Recommended intakes of vitamins and some of their sources

Vitamin B_1 or thiamin is an essential component of the coenzyme or prosthetic group, thiamin pyrophosphate (TPP, *Figure 10.10*). This is necessary for the actions of some enzymes, for example transketolase activity of the pentose phosphate pathway for oxidizing glucose and oxidative decarboxylations catalyzed by pyruvate and 2-oxoglutarate dehydrogenases in carbohydrate metabolism. Thiamin is therefore necessary for the metabolic formation of ATP, the major energy carrier in metabolism, and NADPH. Thiamin pyrophosphate is also known to function in nerve conductance. Vitamin B_2 or riboflavin is also required to form coenzymes or prosthetic groups, in this case flavin mononucleotide (FMN, *Figure 10.11 (A)*) and flavin dinucleotide (FAD, *Figure 10.11 (B)*), which function as electron carriers in flavoproteins. The flavoproteins are extremely numerous and include renal L-amino acid oxidase, NADH reductase, 2-hydroxyacid oxidase (FMN) and D- and L-amino acid oxidases, succinate dehydrogenase and glutathione reductases (FAD). Hence riboflavin is essential to many oxidation and reduction reactions in, for example, the TCA cycle, electron transport and the oxidation of fatty acids in mitochondria.

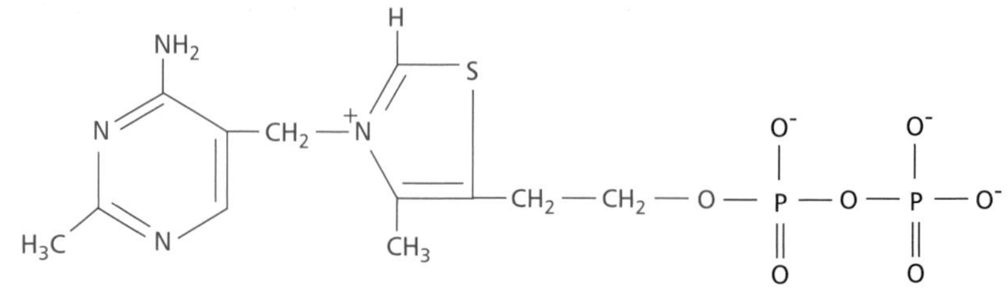

Figure 10.10 **The structure of TPP.**

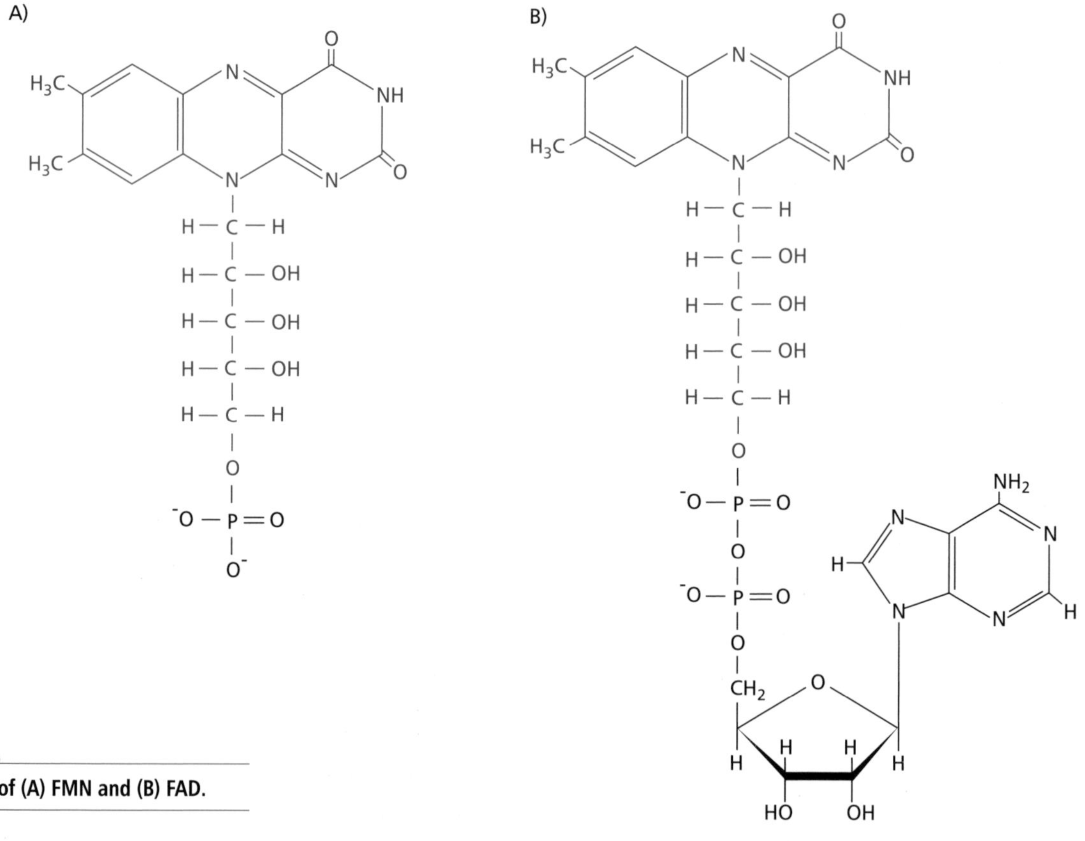

Figure 10.11 **The structures of (A) FMN and (B) FAD.**

Nicotinic acid and nicotinamide are vitamers of niacin, which strictly is not a vitamin since limited amounts of it can be synthesized from tryptophan. Niacin is required to form coenzymes NAD^+ and $NADP^+$ (*Figure 10.12*) which are electron and hydrogen carriers. The former is crucial in electron transport associated with oxidative phosphorylation and ATP formation, the reduced form of NADPH is essential to the biosynthesis of, for example sugars, lipids, amino acids and nucleotides.

Vitamin B_5 or pantothenic acid is an essential part of coenzyme A (*Figure 10.13*), which is the major carrier of metabolically active acyl (fatty acid residues) in metabolism. Thus it is essential to many of the reactions involved in the oxidation of lipids and in the synthesis of lipids including steroid hormones, some neurotransmitters and hemoglobin.

The vitamers pyridoxine, pyridoxal and pyridoxamine all have vitamin B_6 activity (*Figure 10.14*), forming pyridoxal 5-phosphate, which is a cofactor for a number of enzymes. These include glutamate decarboxylase that catalyzes the formation of γ aminobutyric acid, a neurotransmitter of the central nervous system, and enzymes that catalyze transamination and decarboxylation reactions of amino acids. Vitamin B_6 is therefore essential for the synthesis of nonessential amino acids and in the catabolism of amino acids. Pyridoxal 5-phosphate is also a cofactor for glycogen phosphorylase of liver and muscle tissues and helps regulate the actions of steroid hormones (*Chapter 7*) by participating in the dissociation of hormone–receptor complexes from DNA.

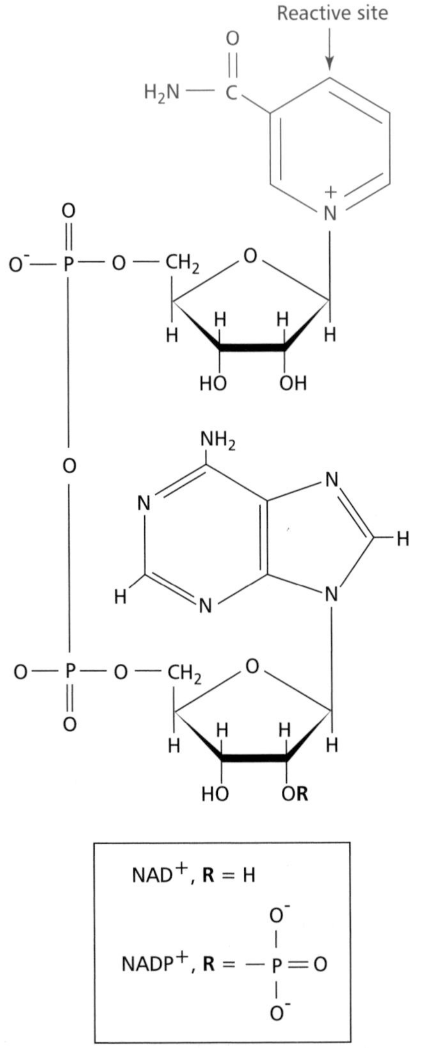

NAD^+, **R** = H

$NADP^+$, **R** = $-PO_3^{2-}$

Figure 10.12 The structures of NAD^+ and $NADP^+$.

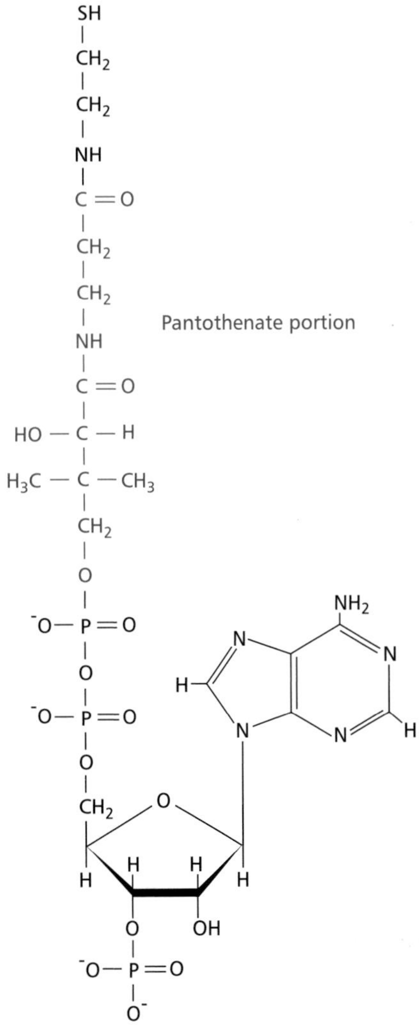

Figure 10.13 The structure of coenzyme A.

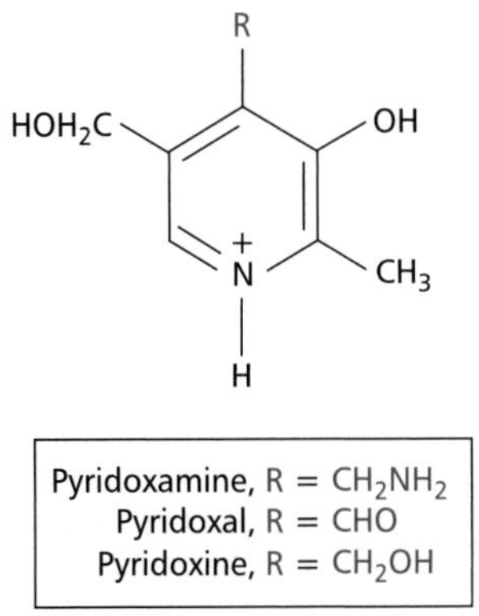

Pyridoxamine, R = CH_2NH_2
Pyridoxal, R = CHO
Pyridoxine, R = CH_2OH

Figure 10.14 The structures of vitamin B_6.

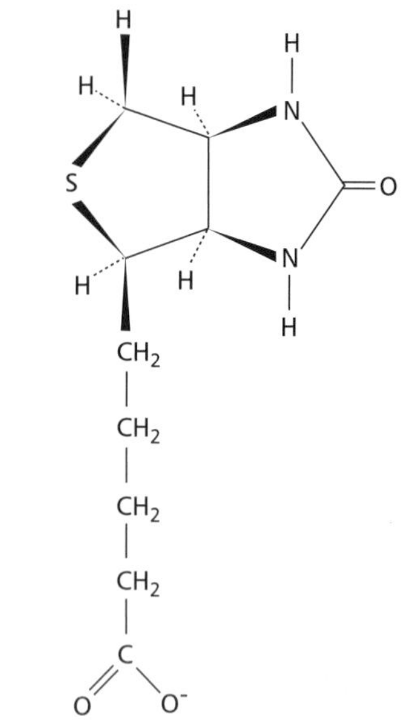

Figure 10.15 The structure of vitamin H (biotin).

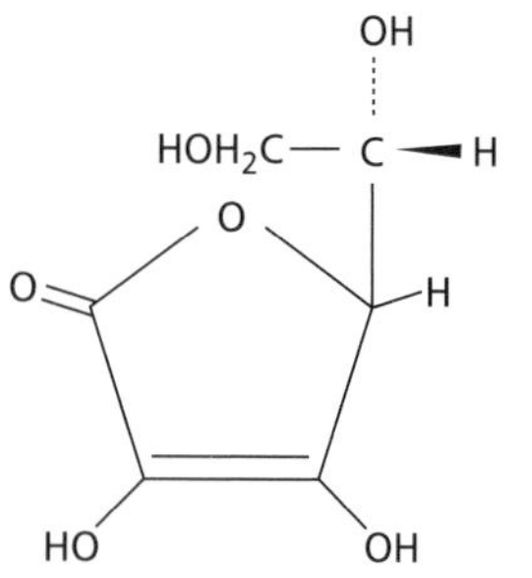

Figure 10.16 The structure of vitamin C (ascorbic acid).

Vitamin B_{12} or cobalamin is an unusual molecule in that it contains an organometallic bond between cobalt and carbon (*Figure 13.18 (A)*). A close relationship exists between the functions of vitamin B_{12} and folic acid and, to some extent at least, they depend on each other for activation. Organic one carbon groups, for example methyl (CH_3–), methylene (–CH_2-), methenyl (–CH=), formyl (–CHO), formate (–COO^-) and formino (–CHNH), are generally toxic. In metabolism, they are bound to carriers derived from vitamin B_{12} and folic acid, which allows them to be converted to different oxidation states and used in a variety of different reactions in a nontoxic manner. These reactions are necessary for the catabolism of some amino acids, for the formation of a number of proteins and the synthesis of purine and pyrimidine bases and therefore nucleotides and nucleic acids. Unlike vitamin B_{12} and folic acid that carry organic one carbon compounds, vitamin H or biotin (*Figure 10.15*) is required to form the prosthetic group that carries CO_2 in a number of enzymes. These include acetyl CoA carboxylase and pyruvate decarboxylase which are key enzymes of fatty acid synthesis and gluconeogenesis, the production of glucose from noncarbohydrate precursors.

Vitamin C or ascorbic acid (*Figure 10.16*) is required to reduce metal ions in a number of enzymes following catalysis. Prolyl and lysyl hydroxylases contain Fe^{2+}(II) which is oxidized to Fe^{3+}(III) during hydroxylation reactions involved in cross-linking collagen molecules, which adds strength to connective tissues. Ascorbate reduces their iron back to the ferrous state, regenerating an active enzyme. Similarly, the Cu^{2+}(II) in enzymes involved in synthesis of catecholamine hormones (*Chapter 7*) is returned to the Cu^{+}(I) state following the oxidation of the copper during catalysis. The antioxidant properties of ascorbic acid, in association with vitamin E, help protect lipids in the cell membranes and blood lipoproteins from oxidative damage. It also enhances the absorption of iron and regulates the absorption of copper from the GIT.

Several vitamers, retinol, retinaldehyde or retinal and retinoic acid, show vitamin A activities. Retinol (*Figure 10.17 (A)*) can be metabolically converted to retinaldehyde, which, in turn can be oxidized to retinoic acid. In addition, the provitamin A carotenoids, for example β carotene (*Figure 10.17 (B)*), can be converted to active forms in the liver. Retinoic acid helps regulate the proliferation and development of cells in a tissue specific manner that resembles the actions of steroid hormones (*Chapter 7*). It binds to nuclear receptors, which then interact with DNA and activate specific genes. Vitamin A is associated with the development of epithelial cells, such as the skin and the mucosal membranes that cover internal and external surfaces of the body and have numerous functions, for example, as structural barriers that

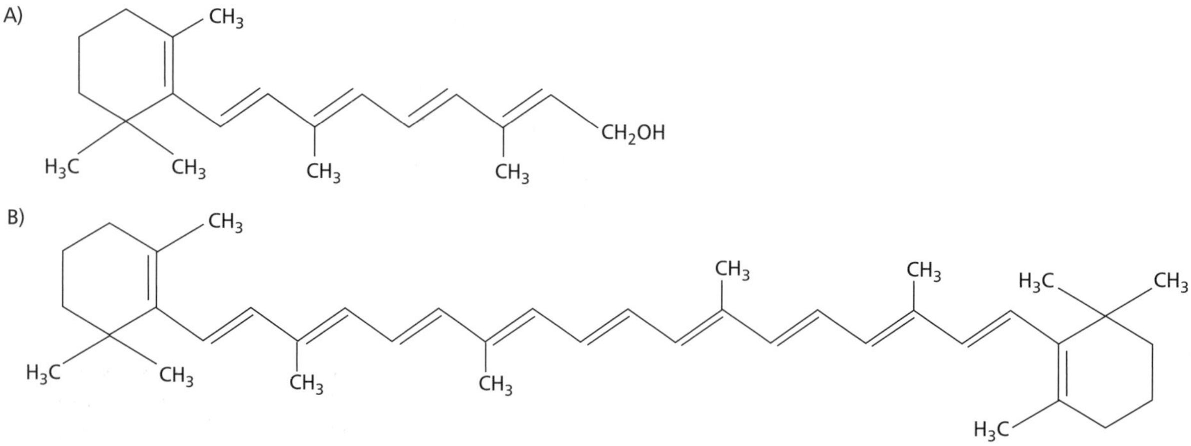

Figure 10.17 The structures of (A) vitamin A (retinol) and (B) β carotene.

prevent microorganisms from entering the body (*Chapter 4*). Retinaldehyde is necessary for vision and functions as the prosthetic group (visual pigments) of opsin proteins in light-sensitive retinal cells. Vitamin A is a weak antioxidant that can protect against free radical damage. There is some evidence that β carotene reduces the incidence of cardiovascular disease and some forms of cancer.

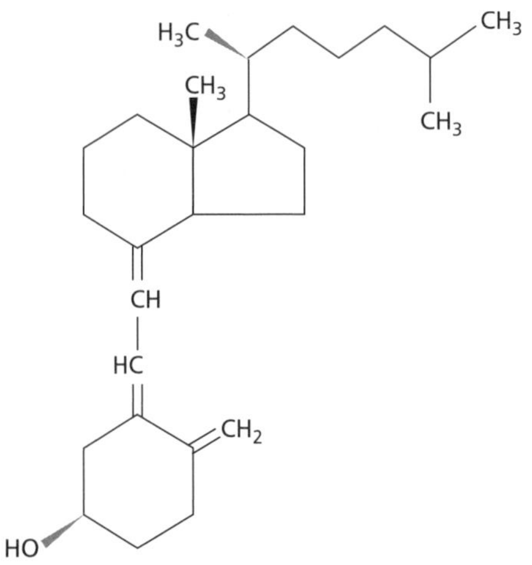

The usual dietary form of vitamin D is cholecalciferol (*Figure 10.18*), although this is not strictly a vitamin since it can be formed by ultraviolet irradiation of the skin from 7-dehydrocholesterol. Foods are sometimes fortified with the synthetic ergocalciferol, which has the same biological activity as cholecalciferol. Enzyme-catalyzed hydroxylations yield the active metabolites 1α, 25 dihydroxycholecalciferol (*Chapter 8*) and calcitriol respectively. Vitamin D mainly functions in the homeostasis of calcium as described in *Chapter 8*.

Figure 10.18 **The structure of vitamin D (cholecalciferol).**

Vitamin E is generally used to describe the tocopherols and tocotrienols that comprise a number of vitamers of differing biological potencies, of which the most active is α tocopherol (*Figure 10.19 (A)* and *(B)*). Like vitamins A and D, vitamin E also has a role in regulating gene expression, although a receptor has yet to be found, and also in signal transduction. A major function generally ascribed to vitamin E is to protect cellular membranes against free radicals (described in *Chapters 12* and *18*) and prevent the oxidation of plasma lipoproteins, especially low density lipoproteins (*Chapter 14*). The corresponding reduction of the vitamin produces a relatively unreactive and therefore less damaging tocopheroxyl radical. This is also relatively long lived and so persists sufficiently long to be reoxidized back to the active form by vitamin C or glutathione peroxidases. However, the utility of this mechanism has been challenged and it has been suggested that the antioxidant role is more restricted and arises from an inhibition of NADPH oxidase and so reduces the production of radicals, such as superoxide.

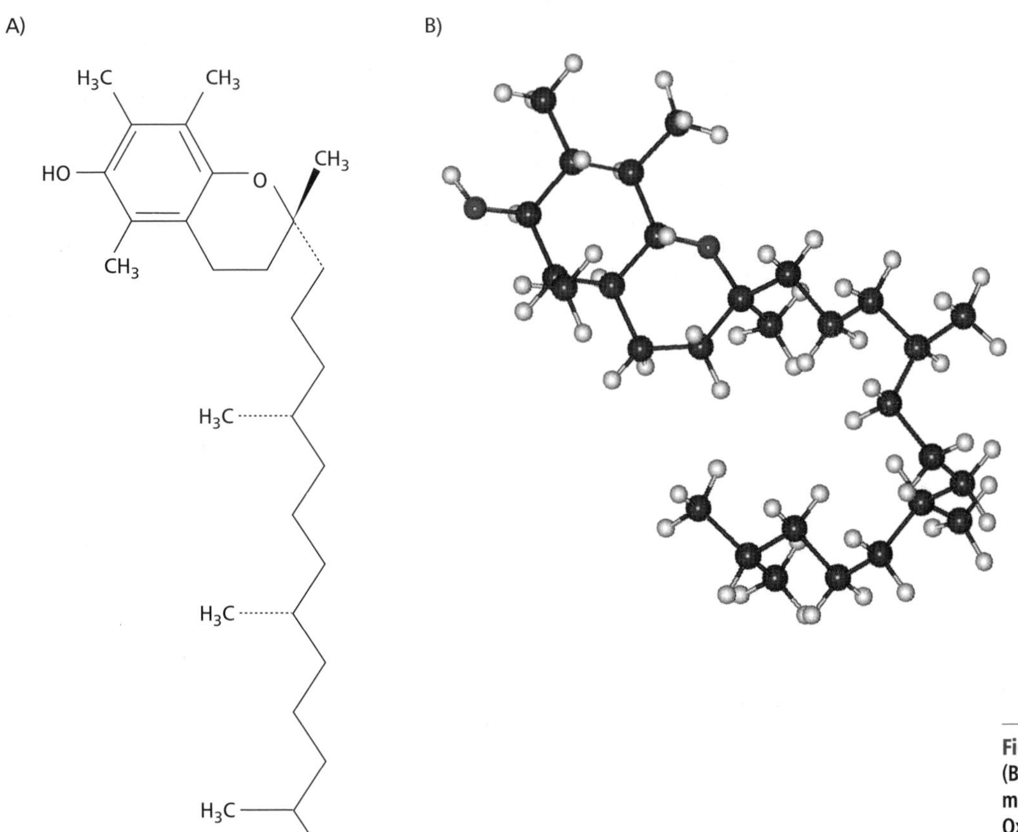

Figure 10.19 **(A) The structure and (B) a computer generated molecular model of vitamin E (α tocopherol). Oxygen atoms are shown in red, carbon in black and hydrogen in gray.**

A number of compounds are possible vitamers of vitamin K, including phylloquinone and the menaquinones. The first is the normal dietary source, while the menaquinones are a group of compounds with similar structures synthesized by GIT bacteria. Phylloquinone (*Figure 10.20*) is metabolically necessary for the conversion of glutamate residues in some proteins to γ carboxyglutamates. This is necessary for the synthesis of some blood clotting factors, which are described in *Chapter 13*, and some proteins of the bone matrix. The ability of menaquinones to function as vitamin K is unclear; they may partially satisfy the human requirements for vitamin K but their contribution is probably much less than previously thought.

Figure 10.20 The structure of vitamin K (phylloquinone).

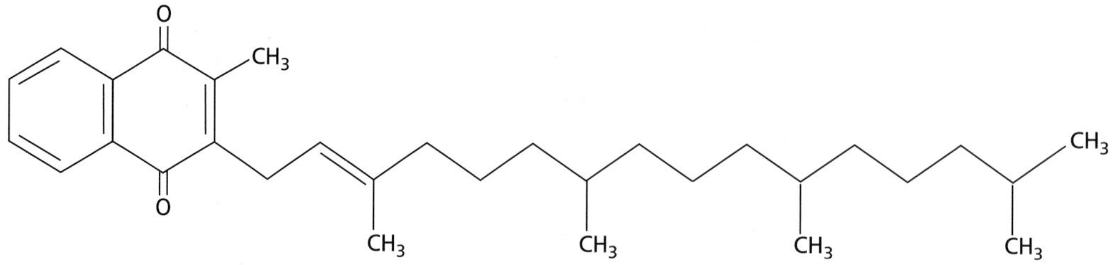

O
CH_3
CH_3
O
CH_3 CH_3 CH_3 CH_3

MINERALS AND TRACE ELEMENTS

Minerals and trace elements are inorganic dietary substances required to maintain good health. Minerals include calcium, magnesium, sodium, potassium, phosphate, chloride and sulfate. They are present in the body in amounts larger than 5 g and some are required in dietary quantities of more than 100 mg per day and are provided by a variety of foods (*Table 10.4*).

Both groups have diverse functions. Minerals play roles in promoting growth and are important constituents of body tissues such as bone, teeth, hair, skin and nails, and as cofactors in some enzymes and other proteins. Sodium, potassium and chloride are required to maintain the electrolyte and osmotic composition of intra- and extracellular fluids, generate electrochemical gradients across plasma membranes and for nerve conductance and muscle contraction. Calcium is an essential component of bone and teeth, is required for muscle contraction and is a second messenger for some hormones and neurotransmitters. Magnesium is a cofactor for many enzymes, especially those utilizing nucleotides involved in energy metabolism and nucleic acid synthesis. Like calcium, phosphate is needed for bone and teeth formation and is also a component of nucleic acids and phospholipids. It activates a number of enzymes, especially some involved in energy metabolism and is

Mineral	Recommended daily intake / mg day^{-1}	Sources
Sodium and chloride	< 6000 in adults	milk, cheese, salt
Potassium	3500 in adults	bananas, vegetables, nuts, fish, pulses, poultry
Calcium	700 in adults	milk, cheese, vegetables, nuts, fish
Magnesium	300 in males 270 in females	vegetables, nuts, bread, fish, meat
Phosphate	550 in adults	meat, fish, poultry, rice, oats
Sulfate	balanced diet adequate	milk, meat, legumes, eggs

Table 10.4 Recommended intakes and sources of minerals

an important renal buffer. Sulfate is required for the synthesis of cartilage and other components of the extracellular matrix.

Compared with minerals, trace elements are required in much smaller quantities, but like them are supplied by a variety of foods (*Table 10.5*). Trace elements include iron, cobalt, copper, molybdenum, chromium, manganese, zinc, selenium, iodide and fluoride. They are present in the body at concentrations less than 100 parts per million and are required in milligrams or even micrograms per day and a number are toxic in excess.

Trace elements have specific and diverse functions. Chromium helps maintain blood glucose concentration by acting as a cofactor for insulin activity. The role of cobalt as a component of vitamin B_{12}, has already been mentioned. Copper is also an essential cofactor for a number of enzymes, including those involved in collagen and elastin synthesis and some redox proteins. Copper is also required for iron absorption and metabolism and hemoglobin synthesis. Fluoride is necessary for the 'hardening' of bone and teeth. All the thyroid hormones contain iodine as described in *Chapter 7*. Iron is a component of the prosthetic group, heme found in hemoglobin and myoglobin, where it maintains its oxidation state and binds a dioxygen molecule. Iron is also found in the heme of cytochromes and in nonheme iron proteins involved in electron transfer, where, of course, its oxidation state does alter. Manganese is essential for the activities of a number of enzymes. For example, pyruvate carboxylase and phosphoenolpyruvate carboxykinase function in gluconeogenesis; arginase is a key enzyme of the urea cycle that detoxifies ammonia produced during amino acid metabolism and superoxide dismutase is a major antioxidant defence. Enzymes that contain molybdenum are common and catalyze several reactions in purine metabolism, for example xanthine oxidase, and maintains the synthesis of sex hormones. Selenium is a cofactor in, for example, glutathione peroxidases, and zinc in carbonic anhydrase and RNA polymerase.

Trace element	Recommended intake /mg day^{-1}	Sources
Chromium	0.025 in adults	meats, cheese, water, whole grains, lentils, spices
Cobalt	0.0015 in adults	beef, eggs, offal, fish, milk products, nuts, vegetables, cereals
Copper	1.2 in adults	offal, fish, nuts
Fluoride	diet generally adequate	present in drinking water in many areas of the world
Iodide	0.14 in adults	fish, milk, eggs , cereals, iodized salt
Iron	8.7 in males 14.8 in females	liver, meat, eggs, beans, nuts, dried fruit, brown rice, green vegetables and fortified cereals
Manganese	diet generally adequate	bananas, egg yolk, bread, nuts, cereals, vegetables, tea
Molybdenum	diet generally adequate	whole grains, leguminous vegetables, meats
Selenium	0.075 in males 0.06 in females	fish, meat, offal, Brazil nuts and eggs
Zinc	5.5 to 9.5 in males 4.0 to 7.0 in females	meat, offal, eggs, fish, cereals, milk, cheese, spinach, beans

Table 10.5 Recommended intakes and sources of trace elements

BOX 10.1 Inherited disorders of copper metabolism

Menkes or kinky hair disease is a neurodegenerative disease of impaired copper transport that results in extreme tissue copper deficiency and was first described in 1962. It is a sex-linked inherited disorder that only affects male children (*Chapter 15*). Babies born with classic Menkes disease appear normal at birth and symptoms typically begin about three months later. The disease is characterized by a failure to thrive, psychomotor deterioration, seizures, hypothermia and strikingly peculiar hair, which is characteristically kinky, stubby, tangled, sparse or steely and easily broken and is often white, ivory or gray colored. Brain problems, such as blood clots at the base of the brain (subdural hematoma) and/or rupture or thrombosis of arteries in the brain may occur. Recurrent respiratory and urinary tract infections are common and weakened bones (osteoporosis) may result in fractures. Menkes disease has an incidence of 1 in 50 000 to 250 000.

The condition arises from mutations in the *ATP7A* gene found on the X chromosome, the product of which is a copper transporting adenosine triphosphatase (ATPase). Transport of dietary copper from intestinal cells is impaired, leading to the low serum copper levels and poor incorporation of copper into cuproproteins. Copper accumulates in excessive amounts in the liver, but is deficient in most other tissues and impairs the functions of a number of essential enzymes, including tyrosinase, monoamine oxidase, cytochrome *c* oxidase, lysyl oxidase and ascorbic acid oxidase. Low activities of tyrosinase leads to depigmentation of hair, while monoamine oxidase results in kinky hair. Impaired lysyl oxidase leads to defects in elastin and collagen, resulting in vascular weaknesses and cytochrome *c* oxidase malfunction results in hypothermia. A defective ascorbate oxidase leads to skeletal demineralization and osteoporosis.

A diagnosis of Menkes disease in neonates could be based on high concentrations of copper levels in the placenta. However, this would be unusual, given a general absence of signs and symptoms at this age. After six weeks, low serum levels of copper and the copper binding protein, ceruloplasmin, are indicative of the condition. A microscopic examination of hair samples shows characteristic Menkes abnormalities and a skin biopsy can be used to assess copper metabolism. There is no effective treatment for Menkes disease and prognosis is poor. Milder forms of the disease respond to intramuscular injections of copper, but the severe form does not show much change. Other treatments are focused on relieving the symptoms. Most patients die within the first decade of life, however, survival to the late 20s has been reported. Genetic screening of the individual's family can identify carriers and provide guidance for counseling on recurrence risks.

Wilson's disease is characterized by an excessive accumulation of copper, although patients have normal or even low concentrations of copper in their blood and increased urinary excretion, and was first described in 1912. It is an autosomal

Margin Note 10.2 Kwashiorkor in the developed world 

Kwashiorkor is rarely seen in developed countries except in neglected children, the very elderly or as a consequence of nutritional ignorance. For example, cases of kwashiorkor have been reported in the USA in a small number of children fed somewhat idiosyncratic diets. Most of these cases were not associated with poverty and about half were associated with perceived or presumed food allergies. A significant portion of the food of some of these children was rice-based beverages, popularly referred to as 'rice milk', which contain less than half the amount of protein of breast milk. Other patients were fed brown rice emulsion, goat's milk or atole, a liquid emulsion of barley, water, and sugar.

10.3 NUTRITIONAL DISORDERS

Nutritional disorders can arise from a deficiency or excess of nutrients and may affect growth or cause specific diseases and even death. They are proving to be major health issues. Nutritional disorders can arise from an inadequate intake of food, such as protein-energy malnutrition, or an excessive intake, for example obesity, or due to an inadequate amount of a specific dietary nutrient, such as scurvy in vitamin C deficiency.

PROTEIN-ENERGY MALNUTRITION

Protein-energy malnutrition (PEM) is the name for a spectrum of disorders that arise due to lack of food. Despite the name, affected individuals may not be suffering from a lack of protein, however, because of their deficiency of total energy, dietary proteins that would normally be used for tissue repair or growth are used as a fuel. In addition, vitamin and mineral intakes are usually inadequate. In the developed world, PEM is rare and is usually associated with solitary elderly patients who are malnourished or with children suffering from neglect. The clinical features of PEM vary depending upon the severity, from the merely underweight to two major conditions, **marasmus** and **kwashiorkor** (*Margin Note 10.2* and *Table 10.6*). Protein-energy malnutrition can also be life threatening in that it increases susceptibility to infectious diseases that would not normally be lethal.

Marasmus is a chronic disorder that develops over a period of months to years and is caused by an inadequate energy intake (*Figure 10.21*). It occurs recessive condition caused by mutations to the *ATP7B* gene located on chromosome 13 that, like *ATP7A*, encodes a copper transporting ATPase. Indeed, the two proteins are 55% homologous in amino acid sequence. However, *ATP7A* is predominantly expressed in the placenta, GIT and blood–brain barrier, *ATP7B* expression is mainly hepatic. In contrast to *ATP7A*, mutations lead to an accumulation of copper in the liver and brain and damage to other tissues also. The first symptoms shown by about 50% of patients with Wilson's disease are swelling and tenderness of the liver and sometimes fever; this resembles more common disorders, such as viral hepatitis. Abnormal levels of circulating liver enzyme activities can indicate serious liver damage, which can progress to cirrhosis. Other symptoms may include jaundice, abdominal swelling and abdominal pain. An extremely severe hepatitis called fulminant hepatitis may occur in about 5% of patients with jaundice, fluid leaking into the abdomen, low blood protein content and clotting abnormalities, swelling of the brain and damage to erythrocytes causing hemolytic anemia. There may also be difficulties with speech, swallowing and coordination resulting in trembling, an unsteady walk and writing problems. A characteristic brown pigmentation in the cornea of the eye, called Kayser-Fleischer rings may occur. Decreased renal functions and osteoporosis may occur prematurely in some patients. Associated psychiatric problems include severe insomnia, poor concentration, depression and suicidal impulses. Wilson's disease, though more common than Menkes disease, is still rare with an incidence of 1 in 30 000.

Wilson's disease is diagnosed by a combination of blood and urine tests, eye examination and liver biopsy. A decrease in serum ceruloplasmin is seen in 95% of patients and an increased urinary excretion of copper is present in most, but not all, symptomatic patients. Kayser-Fleischer rings are visible with a slit-lamp examination performed by an optometrist or ophthalmologist in about 50% of patients presenting with liver disease. The definitive test is the demonstration of a high copper content in liver tissue obtained by biopsy.

Long-term maintenance therapy of Wilson's disease is possible with D-penicillamine, trientine or zinc treatments. Penicillamine and trientine chelates copper, leading to an increased urinary excretion and reduced tissue levels. Zinc, with vitamin B_6 (pyridoxine) supplements, reduces copper absorption and promotes its loss from the GIT. Foods high in copper, such as shellfish, nuts, liver, chocolate and mushrooms should be avoided, as should alcohol. Medical therapies must be continued for life. Patients with acute liver failure or those with advanced liver disease who do not respond to medical therapy should be considered for liver transplantation, which effectively cures the condition and results in long-term survival of approximately 80%. Genetic testing can be used to assist counseling and to screen siblings of an identified patient.

Condition	Expected weight for age/%	Edema
Underweight	60–80	not present
Marasmus	< 60	not present
Kwashiorkor	60–80	present
Marasmic kwashiorkor	< 60	present

Table 10.6 Distinguishing features of different types of protein-energy malnutrition

in epidemics due to famine and is endemic in many areas of Africa, Asia and South America, and in patients with long-term illnesses, such as chronic pulmonary disease and anorexia nervosa (*see later*). Children with marasmus fail to thrive, are emaciated and lack subcutaneous fat. Cachexia, muscle wastage associated with some chronic infections, such as tuberculosis (*Chapter 4*), or the severe and prolonged weight loss seen in some cancers (*Chapter 17*) produces similar clinical features to marasmus but the etiologies are different.

The name, **kwashiorkor** is derived from one of the Kwa languages of Ghana and means 'the one who is displaced' and reflects the development of a nutritional condition in children typically three to five years old who have been abruptly weaned when a new sibling is born (*Figure 10.22*). Kwashiorkor

Figure 10.21 An Ethiopian child of the early 1980s showing signs typical of marasmus. Courtesy of Catholic Fund for Overseas Development, London.

is common in parts of the world affected by famine and where there is poor education and knowledge of nutrition. In the original description of kwashiorkor in children of the Gambia reported in 1932, Williams (1894–1992) implied that a deficiency of protein was its major cause, even when the energy input was adequate. Since then, a number of explanations for the development of kwashiorkor have been proposed. Environmental toxins, such as aflatoxins from moldy foods, general conditions of overcrowding and poverty, a lack of other key nutrients and high rates of disease have all been implicated. A combined protein and energy deficiency, although important, is not the key factor and it is generally accepted that the condition is likely to be due to deficiency of one of several nutrients, including copper, selenium, zinc and the vitamins folic acid, C, A, β carotene and E that are associated with oxidative stress management. It is likely that infectious diseases are the precipitating factor because children with reduced antioxidant status exposed to the stress of an infection are most liable to develop kwashiorkor. Further, kwashiorkor usually occurs following infectious conditions, such as diarrhea, or diseases, for example measles, indicating that its causes are not purely nutritional. Invading pathogens trigger a macrophage respiratory burst (*Chapters 4* and *5*) that considerably increases the total free radical load of the patient and this may be the start of events that result in kwashiorkor. Unfortunately, its causes are still not fully known; siblings in the same household and on the same type of diet may develop marasmus or kwashiorkor. However, lower concentrations of the antioxidants β carotene, glutathione and vitamin E are observed in children suffering from kwashiorkor than in those affected by marasmus; both of course, have reduced levels compared with healthy children.

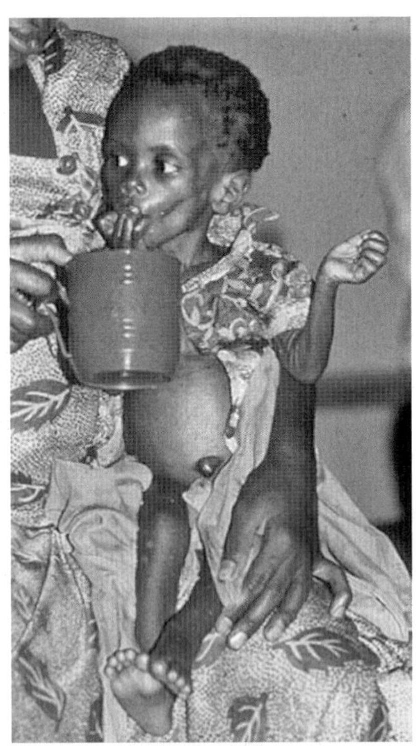

Figure 10.22 Child showing severe signs of kwashiorkor. Courtesy of Catholic Fund for Overseas Development, London.

A significant feature of kwashiorkor is a large protruding abdomen due to edema and an enlarged liver. The edema is traditionally thought to occur because an inadequate intake of protein leads to a reduced plasma albumin concentration, which in turn causes edema, although electrolyte disturbances, such as potassium deficiency and sodium retention, are also thought to play a role. The hepatomegaly occurs from a large infusion of fats into the liver; the cause of which is unknown. Early features of kwashiorkor include fatigue, irritability, lethargy, stunted growth, muscular wasting, edema in the lower limbs and impaired neurological development. The skin may be affected by dermatitis and have areas of hypo- or hyperpigmentation and thin, brittle, light colored hair that is easily pluckable. The patients may also present with delayed wound healing and anemia. If untreated, kwashiorkor can result in shock, coma and death, with a mortality rate as high as 60%. In the long term, it can lead to impaired physical and mental development.

Margin Note 10.3 BMI and IBW 

The generally accepted way to assess the weight of patients is to determine their **body mass index (BMI)** where

BMI = weight (kg) / height (m)^2

Health risks associated with weight are lowest for a BMI of 20–25 kg m^{-2}. A BMI of less than 18.5 kg m^{-2} is underweight, one of greater than 30 kg m^{-2} is defined as obese. **Ideal body weight (IBW)** is another index that can be useful in assessing the nutritional status of a patient. This index is defined differently for men and women. The IBW of a man 5 feet tall is 106 pounds and this increases by an additional six pounds for each inch over the height of 5 feet. For a woman, the IBW at 5 feet tall is 100 pounds and this increases by five extra pounds for each additional inch.

OBESITY

Obesity is diagnosed when the patient's BMI is greater than 30 kg m^{-2} (*Margin Note 10.3*). It is characterized by an excess of fat in the body, particularly under the skin and is generally recognizable when a person is 20% above their IBW. Environmental and genetic factors can determine body weight, but overeating combined with lack of exercise are its usual causes (*Box 10.2*). Very rarely, obesity may be secondary to endocrine disease (*Chapter 7*), for example hypothyroidism that decreases energy requirements, Cushing's syndrome where the distribution of body fat is altered, and some hypothalamic disorders that are associated with overeating.

BOX 10.2 Leptin and *obese/OBESE*

In 1994, the **obese* gene was discovered in mice. Its expression in adipocytes produces the hormone leptin, a polypeptide of 146 amino acid residues (*Figure 10.23*), which is secreted in amounts that are proportional to the mass of body fat. Among the effects of leptin are to decrease adiposity or total body fat and to pass through the blood–brain barrier and act on the hypothalamus to inhibit eating. It also stimulates the oxidation of fatty stores and prevents lipid accumulating in nonadipose tissues and so protects the individual against some of the adverse effects of fat accumulation, for example coronary artery disease. During periods when more energy is used than ingested, that is the starved state, adipose tissue is metabolized and the amount of leptin secreted declines. Hence, the mouse is stimulated to eat more. When more energy is consumed than used, the converse holds. Mutations to the gene or its absence in mice can lead to excessive overeating and morbid obesity (*Figure 10.24 (A)* and *(B)*). Injections of synthetic leptin to obese mice cause a weight loss. These findings led to great excitement that obesity in humans could be a genetic phenomenon and controllable by leptin therapy. However, although the physiological effects of leptin are similar in humans and mice, mutations in the **OBESE* gene have been discovered in only a handful of obese humans. Two of these patients were members of the same highly consanguineous family and weighed 29 and 86 kg at the ages of two and eight years respectively. Their *OBESE* genes had a deletion mutation of a single guanine nucleotide at codon 133 leading to a biologically inactive leptin. Injections of leptin relieved their symptoms. However, most obese people, in fact, secrete leptin in amounts that exceed its production in thinner people because they have a greater mass of fat. Unfortunately, they appear to be less sensitive to its effects.

*The convention is to write mice genes in lower case italic letters and humans in upper case italics.

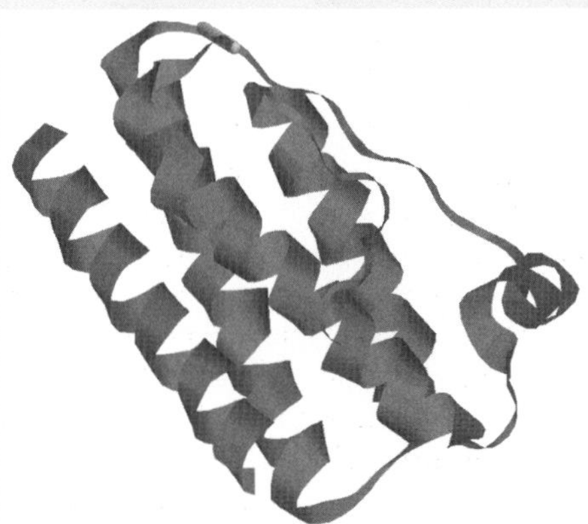

Figure 10.23 Molecular model of a leptin molecule. PDB file 1AX8.

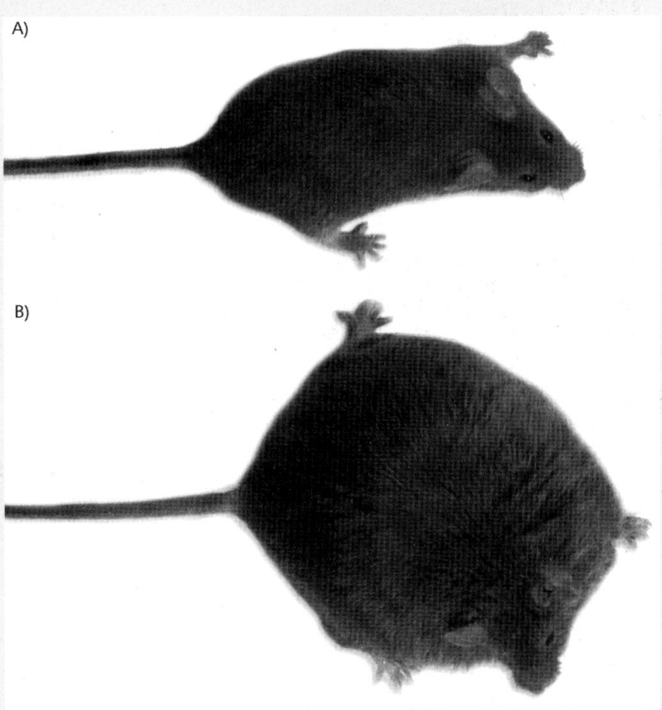

Figure 10.24 (A) Normal and (B) genetically obese mice.

A number of diseases occur more frequently in obese individuals (*Figure 10.25*), including type 2 diabetes mellitus (*Chapter 7*), coronary heart disease, hypertension (*Chapter 14*), cholelithiasis (*Chapter 11*) and osteoarthritis (*Chapter 18*). Not surprisingly, mortality rates are also greater as body weight increases. Some of the complications of obesity are listed in the *Table 10.7*.

The increasing incidence of obesity in many developed countries (*Figure 10.26 (A)*), and particularly in children of the USA and UK (*Figure 10.26 (B)*), is of concern.

Figure 10.25 Schematic indicating the characteristic shape of an obese person with some of the associated complications indicated.

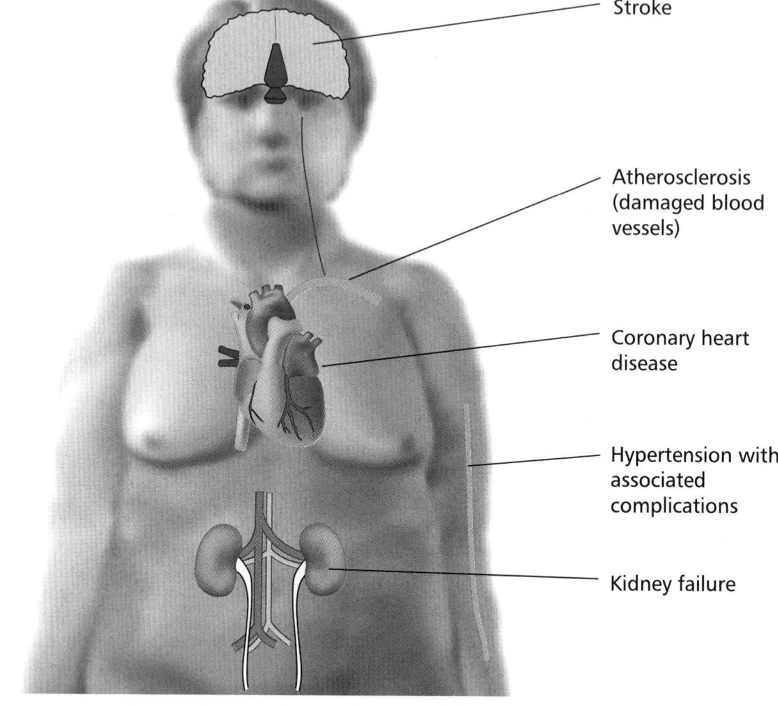

Disease	Pathophysiology
Cholelithiasis	increased cholesterol, bile stasis
Diabetes mellitus	insulin resistance
Hypertension	obesity, inappropriate lifestyle
Osteoarthritis	increased wear and tear of joints
Respiratory disorder	impaired lung ventilation
Vascular disease	hypertension, dyslipidemia, diabetes

Table 10.7 Complications associated with obesity

A)
Men
Women
Obesity / %
25
20
15
10
5
1980 1991 1997 2005
Year

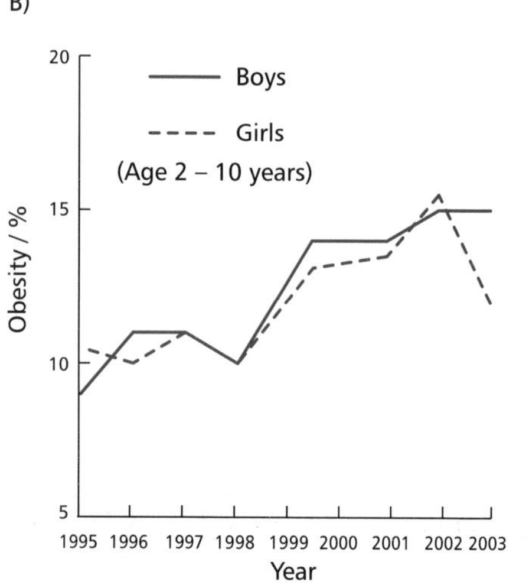

Figure 10.26 The increasing incidences of obesity in (A) adults and (B) children in the UK. Data from a variety of sources.

BOX 10.3 The Atkins Diet

The Atkins diet is a high protein, high fat and low carbohydrate slimming diet (*Figure 10.27*) introduced by Atkins (1930–2003) in 1972. The Atkins diet has received much publicity, and is one of the more popular of those diets in which carbohydrate intake is restricted. Individuals on the Atkins diet exclude most cereal-based foods, beans, fruits and starchy vegetables. Instead they eat generous amounts of meat, poultry, eggs, cream and butter. Carbohydrates promote insulin production that, in turn, causes weight gain because of the growth promoting effects of insulin (*Chapter 7*). A low carbohydrate diet will therefore reduce insulin production leading to a loss of weight. When the intake of carbohydrates is reduced, the body responds quickly by utilizing stored glycogen for energy, reducing body weight as the glycogen and its associated water are lost. Muscle, lean body mass, is not depleted because of the high intake of dietary proteins but body fats are used for energy. The oxidation of fats produces ketone bodies (ketosis) that are lost in the urine, although this condition is much less severe than the ketoacidosis of diabetes mellitus (*Chapter 7*). A number of tissues, for example brain, utilize glucose in preference to other fuels and this can be produced by gluconeogenesis in the liver using amino acids when the dietary intake of carbohydrate is low. The glucose 6-phosphate supplied from gluconeogenesis also partially replenishes glycogen stores (*Figure 10.28*). The high protein and fat intakes and the circulating ketone bodies have satiating effects that suppress appetite, reduce food intake and aid weight loss.

The program for a typical Atkins diet consists of four phases: induction, ongoing weight loss (OWL), premaintenance and a final or lifetime maintenance phase. In the induction phase, carbohydrate intake is limited to 20 g daily and is associated with the greatest weight loss, which can be as high as 3–4 kg per week. In the OWL phase, carbohydrate intake is increased by 5 g per day each week, with the goal of finding the intake of carbohydrates that will maintain weight reduction. The OWL phase continues until weight is within 4.5 kg of the set target. In the premaintenance phase, carbohydrate intake is further increased with the aim of finding the maximum amount of carbohydrate intake that can be consumed each day without causing a weight gain. The final or lifetime maintenance phase continues the dietary plan of the previous phase and also aims to avoid a return to previous dietary habits.

It has been shown that the weight losses in the first six months on the Atkins diet are greater than conventional diets but there is little difference between them over a one-year period. Supporters of the Atkins diet claim that the ketosis produced during low carbohydrate diets is a safe and natural condition necessary for weight loss. Surprisingly given the high fat intake, some studies have indicated that individuals on the Atkins diet have higher levels of HDL, the so-called 'good' cholesterol and lower levels of triacylglycerols in their blood and that levels of LDL cholesterol (*Chapter 14*) and glucose remain unchanged. Some studies have shown these reductions are attributable to the initial weight loss and concentrations often return to levels that are higher than before the diet. Critics have also expressed concerns that diets high in saturated fats may increase the risk of cardiovascular diseases. The ketosis can result in bad breath, tiredness, weakness, dizziness, insomnia and nausea. Constipation may also occur as a consequence of avoiding high-fiber foods such as fruit, vegetables, rice and cereals. There are also concerns that the unbalanced nature of the diet may lead to nutritional deficiencies with undesirable effects. For example, poor intakes of calcium may increase the risk of developing osteoporosis. The high protein intake can also produce acidosis that can leach calcium from bones and increase the likelihood of kidney stones leading to osteoporosis and kidney problems respectively. The poor intakes of antioxidants, found in fruit and vegetables, increase susceptibility to heart disease and cancers. Whether the Atkins diet should be recommended for weight loss is a controversial issue. Long-term studies are required to assess nutritional status, body composition and cardiovascular risk factors in individuals on the diet.

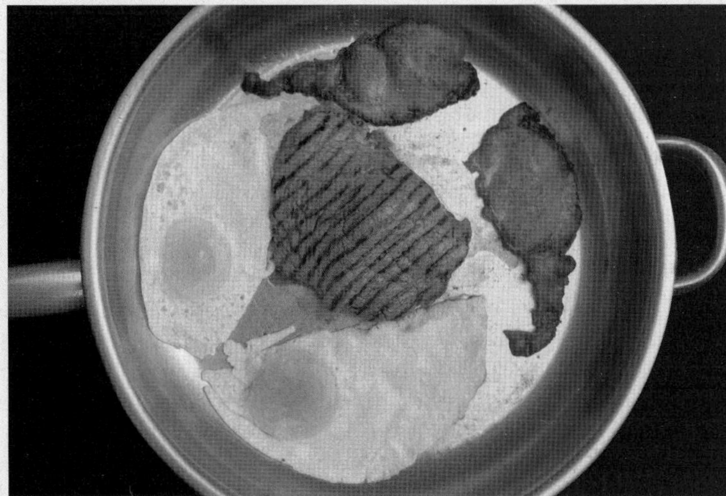

Figure 10.27 An example of a meal on the Atkins diet!

BOX 10.3 *continued*

Figure 10.28 An overview of how dietary proteins can be metabolized to support gluconeogenesis to provide glucose during a low dietary intake of carbohydrates.

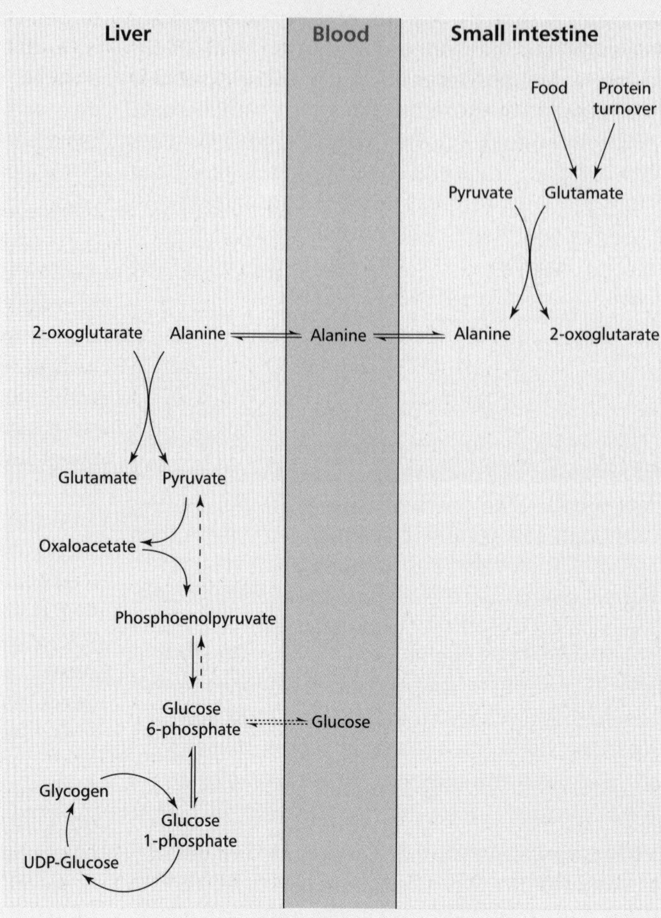

ANOREXIA NERVOSA AND BULIMIA NERVOSA

Anorexia nervosa (AN) and bulimia nervosa (BN) are both disorders of eating behavior and body weight regulation. Anorexia is simply a lack of appetite; bulimia is derived, in part, from the Greek word *limos* meaning hunger. Both have received considerable publicity in recent years although they were reported many centuries ago. The prevalence rate for AN in young females is estimated to be 0.3% and for BN 1% and 0.1% in young women and young men respectively. The incidence in the general population is much lower with the incidences of AN and BN being approximately 8 and 12 cases per 100 000 respectively.

Anorexia nervosa is an extreme refusal or reluctance to eat and associated psychological problems, leading to a severe weight loss. Compulsive exercising and the abuse of laxatives and diuretics often compound the reduced input of dietary energy. Patients are therefore normally extremely hungry and are obsessed with food but they avoid eating, especially carbohydrates. The signs and symptoms of AN include a body weight at least 15% below the recommended weight, a 'wasted' appearance with reduced muscle mass, and swelling of the joints. In younger females, puberty is delayed; an older female is likely to become amenorrhetic and infertile because her body weight reduces to less than 45 kg and/or fat content becomes less than 22% of body weight. The skin is dry, hair thin and the nails brittle. Constipation and decreased heart rate and blood pressure are common. If left untreated, long-term damage to the skeleton and cardiac systems are likely and death can result from starvation, cardiac arrest or other complications. Sufferers of AN are often of above average intelligence but generally have a grossly dysmorphic view of their own bodies, seeing themselves as obese, even though they 'know' they are underweight. It is unclear as to what causes this view. It has been suggested that overreactions to relatively mild obesity, peer or social pressures regarding an ideal human shape or a wish to delay the onset of menarche may all be linked to the condition.

Bulimia nervosa is characterized by episodes of excessive or 'binge' eating that induce feelings of guilt such that sufferers induce vomiting to void the food. This cycle of eating and induced vomiting can be repeated many times. As with anorexics, the use of laxatives, diuretics and dieting pills may be abused. The condition tends to affect older patients than those with anorexia. The signs and symptoms of BN include puffy cheeks due to enlarged salivary glands and often severely damaged tooth enamel because of the excessive vomiting. Electrolyte imbalance, such as loss of potassium, can cause health problems and increases the risk of cardiac arrest (*Chapters 8* and *13*). However, patients generally manage to maintain their weight at an appropriate value and so the condition may not be noticed and can remain undetected for many years.

DISORDERS OF VITAMIN NUTRITION

An inadequate dietary intake of a vitamin, its impaired absorption, or insufficient utilization of an adequate intake, increased dietary requirements, for example in pregnancy, without a corresponding increased intake or an increased excretion of a vitamin give rise to **hypovitaminoses**. In many cases, the symptoms of a hypovitaminosis can be correlated with the known functions of the vitamin (*Section 10.2*), although in other cases they are rather generalized. Hypovitaminoses often develop over an extended period. Initially there is depletion of body stores with a biochemical impairment, that is a subclinical deficiency. This eventually results in an overt deficiency with frank signs and symptoms and is usually accompanied by other evidence of malnutrition, for example PEM. A covert deficiency does not present with clinical features under normal conditions, but any trauma or stress may precipitate the hypovitaminosis. Starving individuals will suffer from multiple vitamin deficiencies. At the other extreme, an excess of the vitamin can be toxic and may result in a **hypervitaminosis**.

Hypovitaminoses

A deficiency of vitamin B_1 or thiamin in some developing countries is common because of the high consumption of foods, such as polished rice, where the vitamin is lost during milling, and in chronic alcoholics, who often have a poor diet. The consequences of a deficiency are depression, irritability, defective memory, peripheral neuropathy and beriberi. Beriberi, which literally means 'I cannot, I cannot' in Singhalese, occurs in two forms, which affect different body systems. Dry beriberi mainly affects the nervous system, wet beriberi the heart and circulation; both types usually occur in the same patient but one set of symptoms predominates. Patients with the dry form may present late, with polyneuropathy and Wernicke-Korsakoff syndrome (*Box 10.4*). The polyneuropathy is characterized initially by heavy, stiff legs, then weakness, numbness and paresthesia and absent ankle jerks. Later stages involve the trunk and arms. Wet beriberi, also known as shoshin, from the Japanese for acute heart damage, is less common and characterized by edema. Symptoms appear rapidly with acute heart failure in addition to the polyneuropathy. It is highly fatal and known to cause sudden deaths in young migrant laborers in Asia whose diet consists of white rice.

Vitamin B_2 (riboflavin) deficiency is rare in the developed world and usually only seen in alcoholics who normally have diets lacking other nutrients or

BOX 10.4 Wernicke-Korsakoff syndrome or cerebral beriberi

Wernicke-Korsakoff syndrome or cerebral beriberi, also known as Wernicke's disease or Korsakoff psychosis, is a brain disorder with the loss of specific functions. The term Gayet disease is applied when lesions are more extensive than those in the Wernicke type. The condition is primarily due to a deficiency of vitamin B_1 (thiamin) or secondary to alcoholism and/or starvation. Even in alcoholics on a balanced diet, which is very unusual (*Chapter 12*), the heavy drinking interferes with the absorption of thiamin from the GIT. The syndrome includes two separate conditions. The first, Wernicke's encephalopathy, involves damage to nerves of the central and peripheral nervous systems. Korsakoff syndrome, or Korsakoff psychosis, begins as Wernicke's symptoms diminish and involves damage to areas of the brain involved with memory. Patients often attempt to hide the memory loss by confabulating, that is telling detailed and credible stories that are untrue, although these are not usually deliberate attempts at deception because the patient often believes them to be factual. Other symptoms include double vision and other eye abnormalities, a loss of muscle coordination, unsteady gait and hallucinations. Many of these symptoms are indicative of alcohol withdrawal and may also be present or develop even when Wernicke-Korsakoff syndrome is not present. Other disorders related to the abuse of alcohol may also be apparent. Incidence and prevalence data for Wernicke-Korsakoff syndrome are poorly reported although one study has indicated a prevalence of 1 to 3%.

Abstinence from alcohol or moderate intake together with adequate nutrition reduces the risk of developing Wernicke-Korsakoff syndrome. Thiamin supplements and a good diet may also help prevent the condition occurring in a heavy drinker but not if damage has already occurred. The syndrome is diagnosed following an examination of the nervous and muscular systems that demonstrates polyneuropathy or damage to multiple nerves and decreased or abnormal reflexes. Testing of gait and coordination can indicate damage to the parts of

suffer malabsorption. However, it is a major problem globally affecting some 200 million people, especially children, in developing countries. Other susceptible people include the elderly with a poor diet and regular laxative users. The main metabolic effects are on lipid metabolism, which leads to clinical features of weakness, dermatitis, glossitis, insomnia and sensitivity to bright light, normochromic and normocytic anemia. A deficiency of vitamin B_2 is rarely fatal because it is present in most foods, which provide adequate amounts to maintain function, and there is efficient reutilization of the vitamin during the turnover of flavoproteins, so that little is lost.

A deficiency of niacin causes pellagra, where the patients present with weight loss, anemia, dementia, dermatitis and diarrhea. Pellagra can be primary when the diet is deficient in nicotinic acid, such as when maize is the staple, but it can be secondary, where other diseases interfere with absorption. Causes of secondary pellagra include prolonged diarrhea, cirrhosis of the liver, alcoholism (**Chapter 12**) and use of isoniazid (*Figure 10.29*) to treat TB. Isoniazid reacts with vitamin B_6 to form an inactive hydrazone and can lead to a deficiency.

A deficiency of vitamin B_5 (pantothenic acid) is rare but has been induced experimentally in animals by feeding a diet virtually devoid of the vitamin or by giving metabolic antagonists, such as ω methylpantothenic acid. Symptoms included irritability, fatigue, malaise, GIT problems, muscle cramps and paresthesia. Historically, pantothenic acid deficiency has been implicated in the 'burning feet' syndrome experienced by severely malnourished prisoners of war.

A deficiency of vitamin B_6 impairs the synthesis of γ aminobutyric acid and amino acid metabolism and may be implicated in the development of some cancers (*Chapter 17*). Clinical features associated with a deficiency of vitamin B_6 include peripheral neuropathy. However, deficiencies are uncommon because it is widely available. Cases have been reported in infants fed overheated and inadequately fortified milk formula who developed severe symptoms: irritability, **opisthotonos**, which is an arching of the head, neck and spine backwards, and convulsions that were relieved by vitamin B_6 supplements.

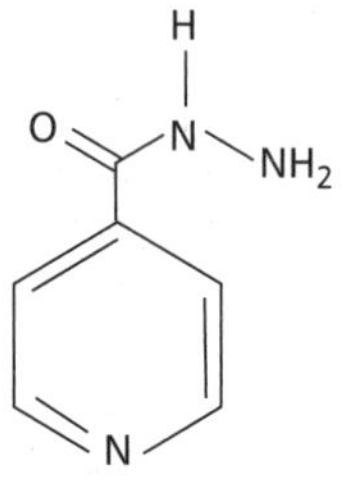

Figure 10.29 Isoniazid.

the brain that control muscle coordination. Muscles may be weak and atrophied and abnormalities of eye movement may be present. The patient may present as cachectic. Clinical tests should demonstrate low serum vitamin B_1 concentration and low erythrocyte transketolase activity but an increase in pyruvate. A magnetic resonance image or computed tomography scan of the brain (*Chapter 18*) of a Wernicke-Korsakoff syndrome patient can show changes in the thalamus or hypothalamus. If the case history is significant for chronic alcohol abuse, then serum or urine alcohol concentrations and liver enzyme activities may be increased above normal.

Wernicke-Korsakoff syndrome is a life-threatening condition with a mortality rate of 10 to 20%. Treatment is aimed at controlling the symptoms as much as possible and preventing any progression of the disorder although some symptoms, particularly the loss of memory and cognitive skills, may be permanent. Hospitalization is required for the initial control of symptoms. In lethargic, unconscious or comatose patients, appropriate monitoring and care is necessary, particularly to prevent obstruction of the airways. Carbohydrate loading, for example by glucose infusion, may precipitate Wernicke's encephalopathy in at-risk patients but can be prevented by supplements of vitamin B_1 prior to the glucose infusion. Injections or oral supplements of vitamin B_1 may improve some symptoms but do not generally lead to the recovery of memory and intellect. Total abstinence from alcohol is required to prevent progressive damage to the brain and peripheral nerves and a well-balanced, nourishing diet is recommended. Joining a support group whose members share common experiences and problems can often help in coping with the stresses of the illness. There may be a need for custodial care if the loss of cognitive skills is severe.

A deficiency of vitamin B_{12} causes megaloblastic anemia (*Chapter 13*) and degeneration of the spinal cord. However, deficiencies are generally uncommon, even in cases of severe malabsorption (*Chapter 11*), since considerable amounts of the vitamin are stored in the liver. Deficiencies can occur in strict vegetarians since sources are animal products. When a deficiency does occur, it is seen most commonly in pernicious anemia, an autoimmune disease characterized by lack of intrinsic factor required for absorption of the vitamin from the GIT (*Chapters 11* and *13*). Bacterial overgrowth in the GIT or diseases affecting the small intestine, such as Crohn's disease (*Chapter 11*), can also lead to a deficiency. Elderly people with atrophic gastritis absorb vitamin B_{12} poorly. However, the first sign of this is neuropathy because the anemia is hidden by folic acid intakes.

Folic acid deficiency is relatively common and occurs because of an inadequate dietary intake in alcoholics and the elderly or because of increased requirements, for example pregnancy and diseases associated with increased cell and nucleic acid turnover, such as leukemia. The deficiency can also arise because of malabsorption and because some drugs affect folic acid metabolism, for example anticonvulsant drugs interfere with its absorption. The major clinical feature of folic acid deficiency is megaloblastic anemia (*Chapter 13*). An increased intake of folic acid before and during pregnancy is associated with a decreased risk of the fetus developing neural tube defects. Taking supplements of folic acid before conception and in the first three months of pregnancy reduces the incidence of spina bifida by more than 20%. In women who have already had a pregnancy affected by a neural tube defect, taking a daily folic acid supplement reduces their chances of a similarly affected pregnancy by approximately 70%. Folic acid supplements are associated with decreased amounts of homocysteine in the plasma with a reduced risk of cardiovascular disease (*Chapter 14*).

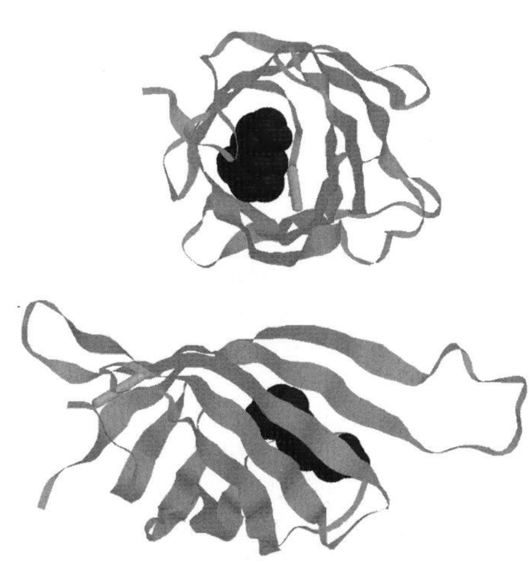

Figure 10.30 Molecular model of vitamin H (biotin), shown in red, bound to avidin. PDB file 1AVD.

Patients suffering a lack of only vitamin H (biotin) are extremely rare. They tend to be individuals who consume extremely large amounts of raw eggs, since the white contains the protein avidin, which binds tightly to biotin (*Figure 10.30*). The vitamin is not released unless the egg is cooked;

Figure 10.31 Molecular model of serum retinol binding protein with a bound retinol form of vitamin A (red). PDB file 1BRP.

however, more than 20 raw eggs must be consumed daily for this to become a serious problem. Biotin deficiency occasionally occurs in hospital patients on parenteral nutrition and leads to anorexia, nausea, dermatitis and depression.

A lack of vitamin C causes scurvy. Inadequate dietary intake occurs in infants aged six to 12 months who receive processed milk without vitamin C supplements and in the elderly who have vitamin C deficient diets. The clinical features of scurvy include skin **papules**, **petechial** and muscle hemorrhages, poor wound healing, gum disease, anemia and osteoporosis. It has been suggested that vitamin C reduces the incidence, duration and severity of the common cold without any side effects, but there is no scientific evidence for these claims.

A deficiency of vitamin A is rare in the developed world; indeed, large amounts are stored in the liver and transported around the body bound to retinol-binding protein in the plasma (*Figure 10.31*). In contrast, deficiencies are major health issues in parts of Africa and South-East Asia. In pregnant women, vitamin A deficiency not only causes night blindness but may increase the risk of maternal mortality. Affected patients may present with anemia and disorders of ectodermal tissues and increased risk of disease and death from severe infections.

A deficiency of vitamin D impairs mineralization of bone matrix causing rickets in children (*Figure 10.32*) and osteomalacia in adults. Rickets results in deformities of the legs that bow due to the weight of the body, skull, rib cage and pelvis. During the 1950s, rickets was virtually eradicated in the UK by fortification of infant foods with vitamin D. Unfortunately, some susceptible children suffered vitamin D toxicity and developed hypercalcemia (*Chapter 8*). As a consequence, the amount of vitamin D added to foods was reduced and rickets reappeared. Overt rickets is now rare in the UK, although subclinical rickets affects about 10% of young children. Osteomalacia causes pain in the bones, especially of the pelvis and legs, and there is an increased susceptibility to fractures of the long bones following minor trauma. It is not uncommon in the UK, especially among women of some ethnic origins because of their lack of exposure to sunlight during purdah and the wearing of traditional clothes.

Vitamin E deficiency is rare. When it does occur, it is most likely in newborn children because transfer of vitamin E across the placenta is poor and these children also have less adipose tissue, where most of the vitamin is stored. Deficiencies of vitamin E can occur with long-term parenteral nutrition and in prolonged and severe steatorrhea (*Chapter 11*). Vitamin E deficiency in children causes irritability, edema, hemolytic anemia and neurological dysfunction. Decreased concentrations of plasma vitamin E have been associated with progression of atherosclerosis and growth of some tumors. Other symptoms such as ataxia, dysarthria, sensory loss and paresthesia have been described in adults.

Margin Note 10.4 Skin papules and petechial hemorrhages

Papules are solid, limited raised areas of skin about 5 mm in diameter but their shape and color may vary. They may be transitional lesions, and can become vesicular or ulcerate. Larger lesions are called nodules. **Petechial hemorrhages** are small round dark red spots caused by bleeding under the skin.

The clinical effects of vitamin K deficiency include a prolonged clotting time and a bleeding tendency as described in *Chapter 13*. A maternal deficiency can lead to serious bone defects in the fetus. Vitamin K deficiency is most common in neonates because it cannot cross the placenta and milk is a poor dietary source so the baby is at risk of bleeding. Newly born babies lack bacteria in their GIT that can synthesize vitamin K, although the significance of intestinal synthesis by bacteria is debatable anyway. For this reason, it has been recommended that all neonates are given a single prophylactic dose of vitamin K. In adults, a deficiency may be seen in fat malabsorption or in people using antibiotics that reduce intestinal bacteria. An inadequate intake may reduce the density of bone and increase the risk of osteoporosis and associated fractures, particularly in postmenopausal women.

Hypervitaminoses

Hypervitaminoses are relatively rare compared with deficiencies. An excess of vitamins B_6 and niacin can be toxic and is usually associated with excessive intake of vitamin supplements. Large doses of niacin are associated with a variety of clinical problems, including abnormalities of liver functions, hyperglycemia, an increase in plasma uric acid and vasodilation. Daily doses of vitamin B_6 greater than 500 mg over an extended period can cause a sensory neuropathy.

Most examples of hypervitaminoses are associated with vitamins A and D. Vitamin A is stored in the liver and excessive dietary intake over prolonged periods can lead to a toxic overload. Typical symptoms are pain in the bones, a scaly dermatitis, nausea and diarrhea with enlargements of the liver and spleen. Most cases of vitamin A toxicity are caused by patients overdosing with vitamin supplements. The only natural food known to contain dangerous levels of the vitamin is polar bear liver; not a common dietary item in most societies! Animal studies have shown that vitamin A can produce teratogenic effects when administered in high doses. The consumption of excessively large amounts of vitamin A during pregnancy may increase the risk of congenital malformations.

Excess of vitamin D is, again, largely associated with the overconsumption of vitamin supplements. Toxicity is due to overstimulation of calcium absorption from the gut and excessive resorption from bone which results in its demineralization. The weakening of the bone and hypercalcemia (*Chapter 8*) promote metastatic calcification and a tendency in the patient to form kidney stones.

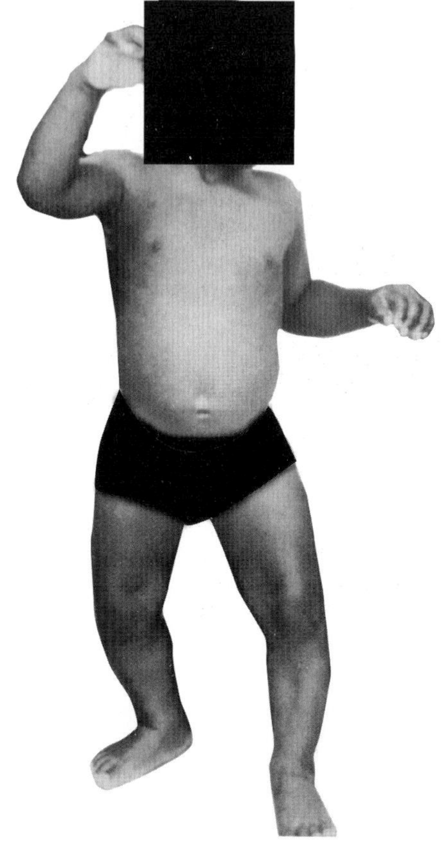

Figure 10.32 A child with the characteristic bowed legs of rickets.

NUTRITIONAL DISORDERS OF MINERALS AND TRACE ELEMENTS

Minerals and trace elements are necessary for numerous and diverse metabolic activities. Clinical disorders arising from deficiencies in the dietary intakes of minerals are not uncommon and a number are described in *Chapter 8*. Conditions caused by excessive mineral ingestion are less common but several are also outlined in the same chapter.

The total quantity of any one trace element in the body is usually less than 5 g and these elements are often required in quantities of less than 20 mg per day, hence dietary deficiencies are uncommon. Chromium deficiency can occur in patients on parenteral nutrition without adequate supplementation and leads to glucose intolerance. An excess of chromium (II) has no known symptoms, although chromium (III) and especially chromium (VI) compounds are toxic. A deficiency of cobalt is rare and causes indigestion, diarrhea, weight loss and a loss of memory. An excess of cobalt is not associated with any known symptoms, although a high intake over a prolonged period may lead to infertility in men. In addition, there are occasional reports of cobalt cardiomyopathy following occupational exposure.

Copper deficiency is uncommon, except in patients on synthetic oral or on long-term parenteral nutrition. It can occur in infants because of malnutrition, malabsorption, chronic diarrhea or prolonged feeding with low copper milk diets. Premature infants are particularly susceptible because of their low copper stores in the liver. Copper deficiency causes neutropenia and hypochromic anemia in the early stages, both of which respond to dietary copper but not iron. This is followed by bone abnormalities such as osteoporosis, decreased pigmentation of the skin, pallor and neurological abnormalities in the later stages. A dietary excess of copper is rare but occasionally happens following food contamination and causes salivation, stomach pain, nausea, vomiting and diarrhea (*Box 10.1*).

Margin Note 10.5 Warfarin and vitamin K 

Warfarin is often used therapeutically as an anticoagulant *(Chapter 13)*. However, because of the similarities in structures, it is a vitamin K antagonist (*Figure 10.33 (A)* and *(B)*). Patients on warfarin therapy require increased amounts of the vitamin.

A)

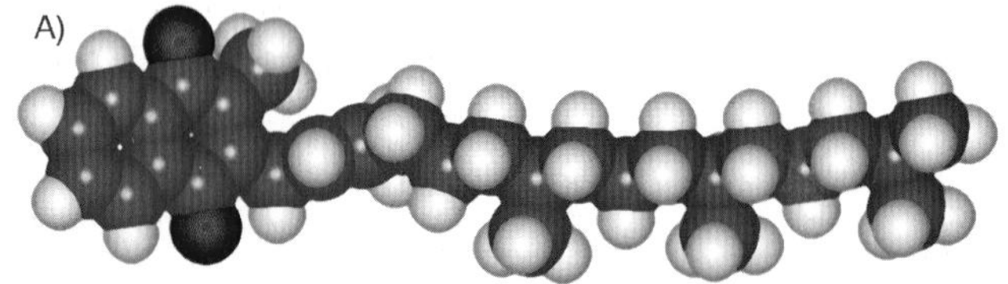

B)

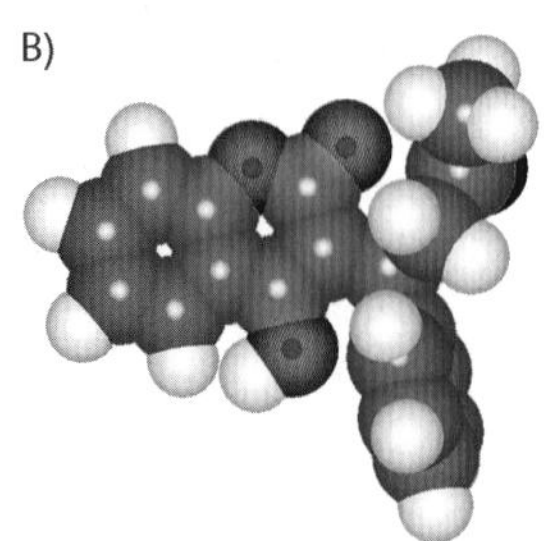

Figure 10.33 Computer-generated molecular models of (A) vitamin K_1 and (B) warfarin. Note the structural similarities in the 'left-hand' sides of the models. Oxygen atoms are shown in red, carbon in black and hydrogen in gray.

Nearly half of dietary fluoride is taken up by bone and can influence its mechanical properties. The incorporation of fluoride in tooth enamel as fluorapatite makes teeth more resistant to dental caries as it resists breakdown by acids. Fluoride deficiency leads to impaired bone formation and dental defects (*Box 10.5*).

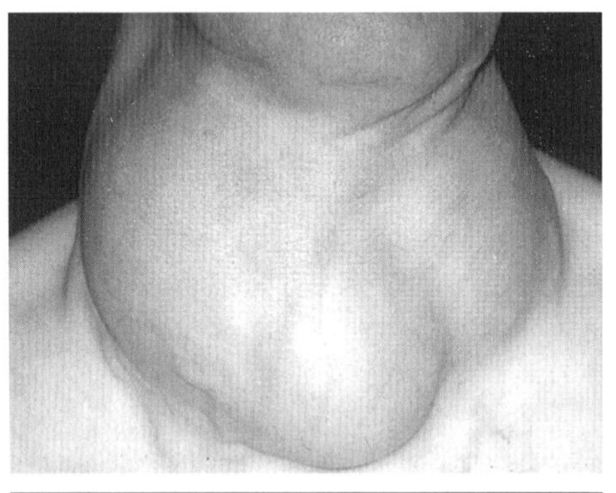

Figure 10.34 A well developed goiter. Courtesy of S.J. Nixon, Royal Infirmary of Edinburgh, Scotland.

Iodine deficiency causes an enlargement of the thyroid gland called goiter (*Figure 10.34*), while a prolonged deficiency in children can lead to cretinism (*Chapter 7*). Iodine deficiency is commonest in upland areas with thin limestone soil because minerals, such as iodine are easily leached from these types of soils and plants in the area are iodine deficient. The situation is exacerbated if commercial iodized salt is not available. In some areas of Africa, Brazil and the Himalayas, more than 90% of the population may develop goiter due to iodine deficiency. Excessive intake of dietary iodide can also lead to goiter. This is most commonly seen in Japanese communities who have a high dietary intake of seaweed.

A lack of iron leads to anemia, which is described in *Chapter 13*. Increased intake of dietary, medicinal or transfused iron can cause hemosiderosis,

BOX 10.5 Dental caries and fluoride

The bulk of the tooth is composed of a hard tissue called dentine. The exposed part of the tooth or crown is covered with enamel. Dental caries or tooth decay is caused by bacteria in the plaque and is the commonest cause of toothache (*Chapter 3*). It is particularly common in children. Bacteria in the plaque catabolize sugars producing organic acids that can demineralize the enamel and the underlying dentine. The development of dental caries depends upon a balance between the plaque and bacteria and the ability of enamel to resist attack. The commonest cause of dental caries is poor dental hygiene and a frequent intake of dietary sugars. Regular brushing of the teeth can remove plaque and reduce acid production since it takes about 24 hours for the bacteria to reestablish themselves. Treatment of caries involves removing the decayed part of the teeth by drilling and replacing the decayed material with a filling.

Epidemiological studies have demonstrated the benefits of adding fluoride to drinking water (fluoridation) because as little as one part per million (ppm) can prevent dental caries. Studies in the UK showed that children living in areas with water fluoride concentrations of about 1 ppm had 50% fewer cases of dental characterized by deposits of iron compounds in organs such as the liver and heart. Acute ingestion of large doses of ferrous salts can be fatal, particularly in children under the age of two. Manganese deficiency is rare but is believed to cause ataxia, hearing loss and dizziness. Inhalation of dust from mining and other industrial sources can lead to manganese toxicity and causes severe neurological dysfunction, similar to Parkinson's disease (*Chapter 18*). A dietary deficiency of molybdenum has not been reported. However, molybdenum deficiency was noted in a patient on parenteral nutrition who suffered from mental disturbances that progressed to coma. Supplementation of molybdenum improved the patient's clinical condition. Levels of manganese in food and water must exceed 100 mg kg^{-1} body weight to produce manganese toxicity. However, few data are available but the major signs include diarrhea and anemia. The high concentrations are thought to stimulate xanthine oxidase, leading to increased serum uric acid and gout (*Chapter 8*).

Selenium deficiency results from a low dietary intake in parts of the world with soils of low selenium content and has been reported in patients on long-term parenteral nutrition. A deficiency of selenium may lead to a cardiomyopathy. In China, this is called Keshan's disease and affects young women and children in selenium deficient regions and although selenium deficiency is a basic factor in Keshan's disease, its occurrence is seasonal and it is associated with viral infections that cause inflammation of the heart. Prophylactic selenium prevents the disease developing but selenium supplements do not reverse heart muscle damage. A large intake of selenium causes selenosis, a condition characterized by loss of hair, skin and nails.

Zinc deficiency is relatively common in populations in rural areas of the Middle East and subtropical and tropical areas where unleavened whole wheat bread can provide up to 75% of the energy intake. The little zinc in wheat is bound by the relatively large amounts of phytic acid and fiber present that inhibit its absorption. This is not a problem in leavened bread, as yeasts produce phytases that inactivate the phytic acid. Zinc deficiency also occurs during prolonged parenteral nutrition if inadequate amounts are provided. A deficiency is associated with a period of severe catabolism, as would occur in PEM. A severe deficiency occurs in the skin condition acrodermatitis enteropathica, where there is an inherited defect in GIT zinc absorption. A deficiency of zinc delays the onset of puberty. High doses of zinc reduce the amount of copper the body can absorb causing anemia and weakness of the bones.

caries than those in areas where the water fluoride was below 0.1 ppm. Indeed, removal of fluoride from the water increased the incidence of caries. These UK studies have been supported by others worldwide, where similar results have been obtained despite ethnic, social, climatic and dietary differences. Fluoride in toothpastes, mouthwashes or when consumed in a liquid or tablet form is also effective in preventing dental caries. The addition of fluoride to drinking water at concentrations greater than 12 ppm promotes dental **fluorosis**, a mottling of the teeth as they form in the jaws. Its effects are generally cosmetic and only cause functional problems when the disorder is severe. Other concerns of fluoridation include its possible association with cancer and arthritis. There is, however, a weak association between fluoridation and an increased susceptibility to bone fractures.

The addition of fluoride to drinking water has also raised ethical issues because it is seen by some as an infringement of personal liberty in that individuals have no choice but to drink water containing fluoride. The use of fluoridation requires careful consideration of the ethical issues and the balance between its beneficial and any potentially harmful effects.

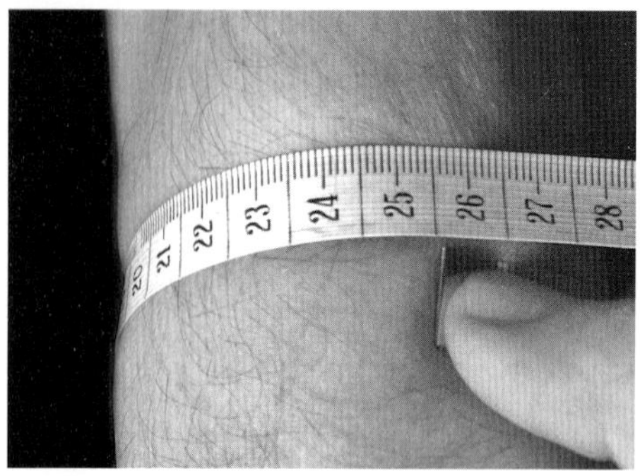

Figure 10.35 Measuring arm circumference.

10.4 INVESTIGATING NUTRITIONAL DISORDERS

The main objectives in investigating nutritional disorders are to detect malnutrition, determine the most appropriate treatment and to monitor the progress of the patient during treatment. A number of fairly general investigations are available for patients suspected of suffering from malnourishment, for example taking a medical history, physical examination and laboratory investigations. A patient's medical history may help to reveal the underlying basis of the complaint. A significant part is determining the dietary history, which usually involves dietary recall of the last one to seven days, or food frequency questionnaires where patients estimate how often a particular food on a list has been eaten. A medical history also includes specific questions concerning recent weight changes, dietary habits, for example alcohol consumption, appetite, GIT function, use of vitamin and mineral supplements and drug intake. A physical examination can include anthropometric measurements, such as weight, height, body mass index (*Margin Note 10.2*), arm circumference and skinfold thickness. Arm circumference is an indicator of muscle mass. The arm circumference is measured at the midarm point in millimeters (*Figure 10.35*) and values are typically 293 mm in males and 285 mm in females. Skinfold thickness indicates the extent of subcutaneous fat stores. It is commonly measured at the biceps, triceps, subscapular and suprailiac sites in both males and females. In, for example, the triceps measurement the arm is allowed to hang freely and a fold of skin above the midpoint of the arm is pulled from the underlying muscle and measured to the nearest millimeter using calipers (*Figure 10.36*). Values are typically 12.5 and 16.5 mm for males and females respectively. Measurements at the four sites should be made in duplicate and should not vary by more than 1 mm. If consecutive measurements become increasingly smaller the fat is being compressed and other sites should be assessed while the tissue recovers. However, the accuracy of the test depends upon selecting the sites correctly, an appropriate technique in taking the measurements and the experience of the tester; hence the trend is more important than the values at any one time. Values for arm circumference and skinfold thickness are compared with standard values for the same sex. Undernutrition is indicated if they are less than 90% of these values.

Figure 10.36 Measuring skin fold thickness.

The appearance of the skin, hair, nails, eyes, bones, teeth and mucous membranes may reveal signs of malnutrition. A number of general laboratory-based investigations on blood and urine samples can aid in diagnosing malnutrition and are listed in *Table 10.8*. These tests are not reliable when

Test	Clinical value in assessment of malnutrition
Serum albumin	reduced values in kwashiorkor, poor protein intake and zinc deficiency
Total iron-binding capacity (*Chapter 13*)	reduced values in kwashiorkor and poor protein intake
Total lymphocyte count (*Chapter 13*)	reduced values in PEM especially kwashiorkor
Serum creatinine (*Chapters 1* and *8*)	reduced values in muscle wasting because of energy deficiency
Creatinine clearance (*Chapter 8*)	reduced values in muscle wasting because of energy deficiency
Hemoglobin (*Chapter 13*)	reduced values in anemia possibly due to nutritional deficiency
Prothrombin time (*Chapter 13*)	increased in vitamin K deficiency

Table 10.8 Laboratory tests for investigation of malnutrition

used in isolation, as their findings are also altered in some disease states. Hair and nail samples are of questionable clinical value and are not routinely taken.

PROTEIN-ENERGY MALNUTRITION

A diagnosis of PEM must identify the type present. Individuals affected by maramus present with decreased anthropometric measurements, loss of muscle mass and body weight. There is also a decline in body temperature, pulse and metabolic rate. Serum protein levels are often low but can be normal. Children suffering from PEM have reduced rates of growth and are shorter, particularly in marasmus, compared with their normal counterparts. This can be easily assessed using charts of growth rates (*Figure 10.37*). They also have an impaired immune system, because of reduced protein synthesis in particular of immunoglobulins, and, as a consequence, infections such as measles that a child would normally be expected to survive are common causes of death in severe cases. A diagnosis of kwashiorkor is made following a thorough physical examination together with a medical and dietary history. Patients affected with kwashiorkor have the characteristic swollen abdomen, show hypoalbuminemia and a reduced lymphocyte count or **lymphopenia**.

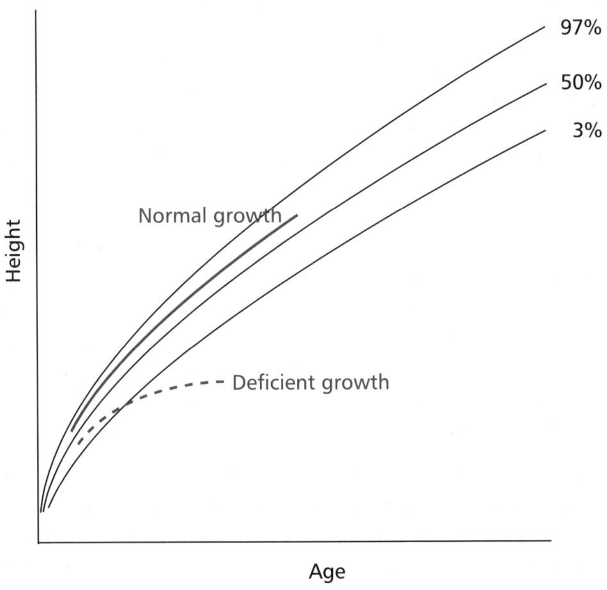

Figure 10.37 A generic growth chart. Charts are available for different sexes and ethnic groups.

OBESITY

In addition to arm circumference and skinfold thickness, the distribution of fat in obese patients can be assessed by determining the waist: hip ratio. The waist circumference is measured at its narrowest point and divided by that of the hip at its widest. Ratios greater than 0.80 for women and 0.95 for men increase the risk of developing clinical problems associated with obesity (*Table 10.7*). Ultrasonography can also be used on soft tissue adipose material. The technique reflects ultrasound waves from internal tissues or organs and the echo patterns are analyzed to form a picture of body tissues called a sonogram. The extent of fat deposition can be accurately determined but is more expensive than other methods and not widely used.

ANOREXIA NERVOSA AND BULIMIA NERVOSA

Anorexia nervosa and BN are difficult to identify, given that the patients often take extreme measures to hide the condition and, indeed, often refuse to admit to any problem. Diagnosis relies on a medical history, clinical investigation supported by anthropometric measurements, coupled with a sympathetic psychological assessment.

VITAMINS

The majority of vitamin disorders encountered in clinical practice are deficiencies. The investigative procedures are varied and depend upon the vitamin in question. Chemical tests can help to confirm the diagnosis of overt vitamin deficiencies and may enable diagnosis to be made at a relatively early stage. The types of tests used include direct measurements of the concentration of the vitamin or one of its metabolites in plasma, serum, erythrocytes, urine or tissue biopsies. The concentration of vitamin in plasma does not necessarily reflect body vitamin status and a measurement of the concentration in blood cells may be a better indicator. Enzyme-based tests are available for some vitamins. Metabolites that accumulate in the blood or urine following the blockage of a metabolic pathway normally catalyzed by an enzyme that requires a vitamin as a cofactor or coenzyme may also be investigated.

Vitamin B_1 (thiamin) deficiency can be assessed by direct measurement of its concentration in plasma or indirectly by determining the increase in erythrocyte transketolase activity in the presence of added TPP. The increase in activity is called the activation coefficient. A coefficient less than 15% is considered normal; an increase of 15–25% indicates a marginal deficiency, while an increase greater than 25% with clinical signs is indicative of severe thiamin deficiency. Thiamin deficiency can also be assessed by the clinical response to administered thiamin, that is, an improvement in the condition after administering thiamin supplements. The nutritional status of vitamin B_2 (riboflavin) is investigated in a similar manner by determining the activation of glutathione reductase activity of erythrocytes in the presence of added FAD.

Assessing the nutritional status of niacin is more difficult. The usual method is to determine the concentrations of metabolites of niacin, for example 1-methylnicotinamide and 1-methyl-3-carboxamido-6-pyridone, in urine samples. Both are reasonably good measures of niacin status, as is the ratio of the concentrations of NAD^+ to $NADP^+$ in erythrocytes. A ratio of less than 1.0 may identify subjects at risk of developing a niacin deficiency. The nutritional status of vitamin B_5 (pantothenic acid) can also be assessed by determining its concentration in plasma or urine samples. In general, plasma pantothenic acid concentrations decrease in patients on a pantothenic acid deficient diet. However, the concentrations of pantothenic acid in blood respond less readily to intake than does the concentration in urine.

The status of vitamin B_6 can be investigated in a manner similar to those for thiamin and riboflavin by determining the activation coefficients of erythrocyte alanine and aspartate transaminase activities (ALT and AST respectively) in the presence of the cofactor pyridoxal phosphate. Alternatively, vitamin B_6 status may be assessed by the tryptophan loading test. Tryptophan is normally catabolized by the pathway shown in *Figure 10.38*. However, the activity of kynureninase decreases markedly in B_6 deficient patients. If such a patient is given an oral dose of 50 mg per kilogram body weight of tryptophan, then there is an increase in the amounts of kynurenic and xanthurenic acids formed and these appear in the urine. Generally less than 30 mg of xanthurenic acid is excreted daily; higher amounts are indicative of vitamin B_6 deficiency. However, some other disorders of tryptophan catabolism can also lead to an increase in xanthurenic acid production so abnormal results must be treated with caution.

A possible deficiency of vitamin H (biotin) can be investigated by measuring its concentration in whole blood, serum or urine. Determining plasma biotin

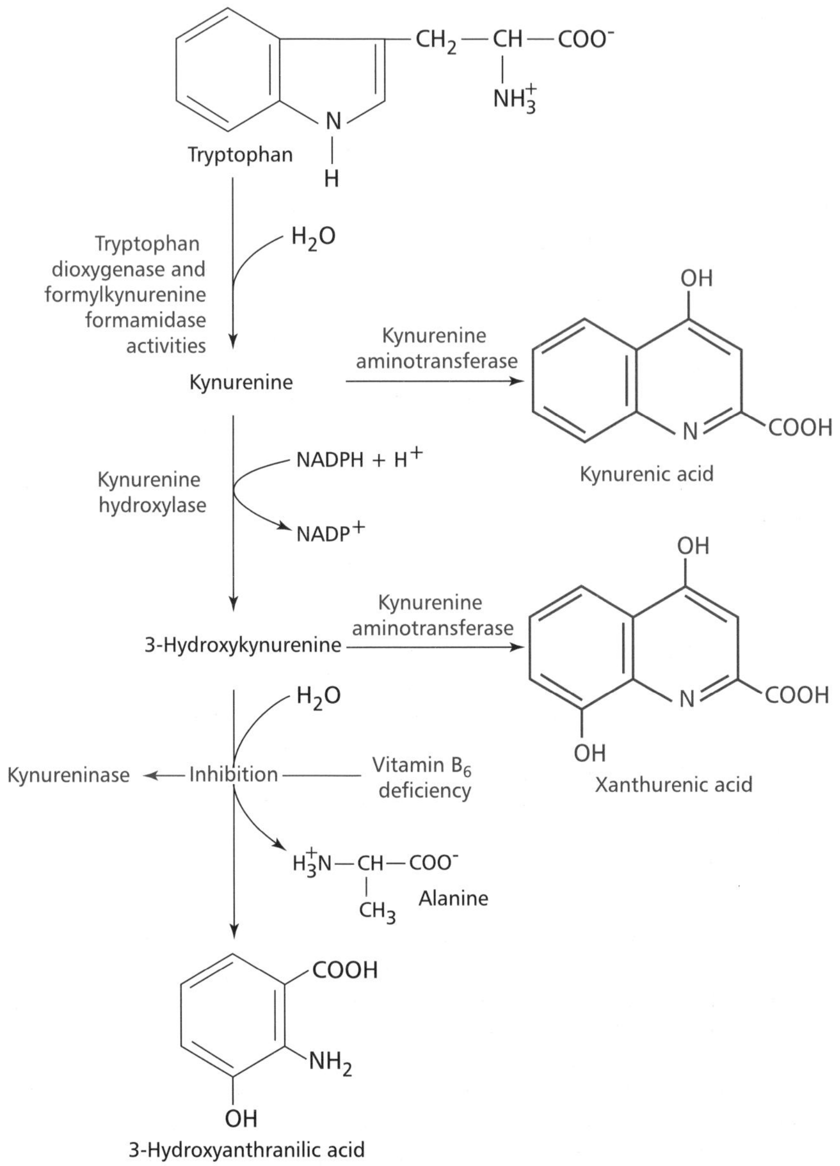

Figure 10.38 The catabolism of xanthurenic acid, which is the basis of the tryptophan load test for vitamin B_6 status. See text for details.

is not a reliable indicator of status. Changes in urinary excretion of biotin or of its metabolites are better indicators of biotin status.

Folic acid status may be investigated by directly measuring its concentration in serum or erythrocytes although these are associated with a number of problems. Serum folic acid tends to reflect dietary intake over the previous few weeks. Patients with both acquired and inherited folic acid deficiency may remain moderately deficient for months or years, taking in just enough folic acid to prevent low erythrocyte folic acid concentrations and frank anemia. Erythrocyte values are not sensitive to short-term variations; depletion occurs only in the later stages of deficiency and is usually accompanied by megaloblastic anemia. Both erythrocyte and serum folic acid studies must be performed. Severe folic acid deficiency is accompanied by a macrocytic anemia (*Chapter 13*) although the size of erythrocytes may be entirely normal in lesser degrees of depletion. Serum vitamin B_{12} concentrations should also be measured when evaluating folic acid deficiency since if either vitamin is deficient it can lead to a failure in absorption by megaloblastic intestinal cells resulting in a secondary deficiency of the other. Formiminoglutamic acid (FIGLU) is a substrate for the folic acid dependent enzyme, formiminoglutamate formiminotransferase required for histidine catabolism. When the vitamin is deficient, FIGLU accumulates and is excreted into urine providing a sensitive test of deficiency. However, FIGLU also increases in vitamin B_{12} deficiency and liver disease, so a high FIGLU excretion is not specific for the diagnosis.

A deficiency of vitamin B_{12} (cobalamin) is investigated by measuring its serum concentration and by hematological examination of blood and bone marrow slides. Serum B_{12} can be measured in isolation or as part of a Schilling test to exclude intrinsic factor deficiency (pernicious anemia, *Chapter 13*). A Schilling test will assess whether vitamin B_{12} is being absorbed correctly by the body. The amount of vitamin B_{12} excreted in urine over a 24 h period is determined after giving the patient a known amount of radioactively labeled vitamin B_{12}. If the GIT is able to absorb vitamin B_{12} normally, then up to 25% of the vitamin will be present in the urine. If there is failure in absorption, then little or no vitamin B_{12} is detected in the urine. In the latter case, the test is repeated following an oral dose of intrinsic factor to determine whether the vitamin deficiency is due to lack of intrinsic factor or a GIT problem.

The concentration of vitamin C in the plasma is a poor indicator of deficiency. Measurements of cellular stores, especially in leukocytes are more useful. However, vitamin C concentrations in leukocytes should always be accompanied by a differential leukocyte count, given that different types of leukocytes vary in their capacity to accumulate the vitamin. A change in the proportion of polymorphonuclear leukocytes, which become saturated with vitamin C at lower concentrations than other leukocytes, would result in a change in total concentration of vitamin C per 10^6 cells, even if the nutritional status of the vitamin is unchanged.

The concentration of vitamin A in the plasma can be measured but this may be misleading as it only declines when tissue stores become severely depleted. Deficiency can also occur in severe protein deficiency which decreases the amount of its carrier protein. In such cases, the concentration of plasma vitamin A would increase once the protein deficiency was corrected. Clinical investigations of possible vitamin D deficiency involve determining serum calcium and phosphate concentrations and measuring serum alkaline phosphatase activity, since the enzyme is lost from cells during bone catabolism. The metabolite, 25-hydroxycholecalciferol in samples of plasma can be measured directly and is a good indicator of vitamin D status in the presence of normal renal function. Vitamin E status can also be assessed by direct measurement of its concentration in the plasma or serum, while vitamin K deficiency is investigated by assessing the prothrombin time of the patient (*Chapter 13*).

MINERALS AND TRACE ELEMENTS

A diagnosis of an overt mineral deficiency or excess can be confirmed by chemical tests that measure its concentration in the serum or, in some cases, urine. These measurements can give an indication of the amounts present in body tissues, although urine values tend to reflect dietary intake rather than the amounts stored in the body. The value of determining the serum concentrations of sodium, potassium, calcium and phosphate have been considered in *Chapter 8*. Investigations of chloride and sulfate are of little clinical relevance.

Measuring the concentrations of trace elements in clinical samples is complex and requires sensitive techniques because of their low values. Samples of plasma are often used but the determined values may not accurately reflect the concentration of the trace element at its site of action, which may be intracellular. However, for many trace elements, a low plasma concentration is indicative of a deficiency and adequate supplementation needs to be provided, while a high value is an indicator of possible toxic levels. Copper deficiency is uncommon, except in patients on synthetic oral or intravenous diets. In these patients, serum copper is reduced to less than 12 μmol dm^{-3}. Low concentrations of plasma copper may indicate depleted stores but are a poor indicator of short-term copper status. Measuring Cu/Zn superoxide dismutase and cytochrome oxidase activities can also indicate copper status as the activities of these enzymes are reduced on a low copper diet. There is no satisfactory test for chromium deficiency and a diagnosis is usually made following improved glucose tolerance after chromium supplementation. Serum fluoride concentration can indicate fluoride exposure and can provide information on endemic fluorosis and allow preventative measures to be taken. Some, but not all studies, have reported a direct relationship between serum fluoride and the degree of fluorosis. Measurements of serum iodine may be a useful assessment of thyroid activity in adults. They may also be of use in investigating cretinism in infants, allowing a diagnosis to be made at an earlier age than is possible by other methods. Serum iodine measurements are also of value in assessing iodine toxicity. Methods to investigate the iron status of a patient are described in *Chapter 13*. Serum iron and total iron binding capacity can be investigated but serum ferritin is a better measure of total body iron stores. The amount of magnesium in serum is less than 1% of the total in the body and is therefore a relatively poor indicator of magnesium status. If hypomagnesemia is present (*Chapter 8*), then magnesium deficiency is likely but a normal serum value does not exclude a significant deficiency. Measurements of molybdenum in biological fluids are rarely required, which is perhaps fortunate since the methods used are inadequate due to the low concentrations involved. Selenium deficiency occurs with poor dietary intake and can be detected by its measurement in plasma or whole blood. However, determining erythrocyte activity of the selenium dependent enzyme, glutathione peroxidase can indirectly assess selenium status. Zinc concentrations less than 8 μmol dm^{-3} in the plasma may indicate zinc deficiency but low values may be associated with hypoalbuminemia, as most zinc is bound to albumin.

Chemical tests for mineral and trace element deficiencies must always be used to complement the medical history and physical examination since many of their findings may reflect underlying disease additional to the nutritional status of a patient. It is therefore necessary to understand these illnesses and how they influence the findings of physical and laboratory investigations.

10.5 GENERAL MANAGEMENT OF NUTRITIONAL DISORDERS

Patients who are malnourished or at risk of developing malnutrition require appropriate therapy, which ranges from simple dietary advice to long-term parenteral nutrition. The dietary needs of the patient must be carefully assessed to provide the correct amounts of energy, protein, vitamins, minerals and trace elements. Patients receive these diets by oral (*Figure 10.39*), tube and parenteral feeding; the last is most commonly administered by intravenous infusion. Oral supplementation should be used wherever possible and the common practice is to encourage consumption of specific foods or supplements that rectify the nutritional disorder in question. In cases where oral feeding is not possible, then liquid food is administered through a nasal tube to the stomach or small intestine. Tube feeding is particularly useful in patients with swallowing difficulties or anorexia. During tube feeding, liquid may be pumped continuously at a constant rate of 75 to 150 cm^3 per hour for 8–24 h. Liquid foods for tube feeding are available commercially as formulae that meet nearly all the patient needs. In some cases, food is administered as a bolus, that is, infusing a discrete volume of formula through the tube under gravity several

Figure 10.39 An example of a liquid food suitable for oral feeding.

times daily. It has the advantages of reducing the cost and allowing stable long-term patients more mobility. Parenteral feeding bypasses the GIT so nutrients are delivered directly into the blood. It is only used when oral or tube feeding have been deemed unsuitable; such as in patients who cannot eat or absorb food from the GIT. Total parenteral nutrition given via the peripheral veins or, in cases of long-term nutrition, through a central venous catheter can provide complete nutrition using preparations containing appropriate amounts of energy, amino acids, vitamins, minerals and trace elements. Like liquid foods for tube feeding, these preparations are available commercially but they are occasionally prepared individually to meet a patient's specific needs. Patients on long-term total parenteral nutrition require careful clinical and laboratory monitoring. Indeed, patients often have an increased risk of infection at the venous catheter site so care is necessary. Biochemical changes usually precede any clinical signs of nutrient deficiency and so regular laboratory monitoring is essential for early detection of any micronutrient deficiency.

Patients with PEM cannot immediately accept normal food because there are digestive enzyme deficiencies and often gastroenteritis. Rehydration is a priority and oral solutions (*Figure 10.40*) achieve this in some cases, while intravenous infusions may be necessary in severe cases. Diluted milk with added sugar may be given initially and as this becomes accepted the proportion of milk can be gradually increased. The cessation of diarrhea indicates that the health of the GIT mucosa is improving and normal foods can be gradually returned to the diet.

A)

CODE 15-611-10

unicef

United Nations Children's Fund

Batch No.

Mfg.Date

Exp. Date

ORAL REHYDRATION SALTS BP

FOR THE TREATMENT OF DEHYDRATION DUE TO DIARRHOEA

SELS DE REHYDRATATION ORALE BP

POUR LE TRAITEMENT D'ETATS DE DESHYDRATATION LORS DE DIARRHEES

SALES DE REHIDRATACIÓN ORAL BP

PARA EL TRATAMIENTO DE LA DESHIDRATACIÓN CAUSADA POR LA DIARREA

أملاح للامهاء الفموي

(أملاح للارواء بالفم)

لمعالجة التجفاف الناجم عن الاسهال

Each sachet contains:	Glucose anhydrous Food Grade	20.0 g
	Sodium chloride BP 98	3.5 g
	Trisodium citrate dihydrate BP 98	2.9 g
Net wt.: 27.9 g	Potassium chloride BP 98	1.5 g

B)

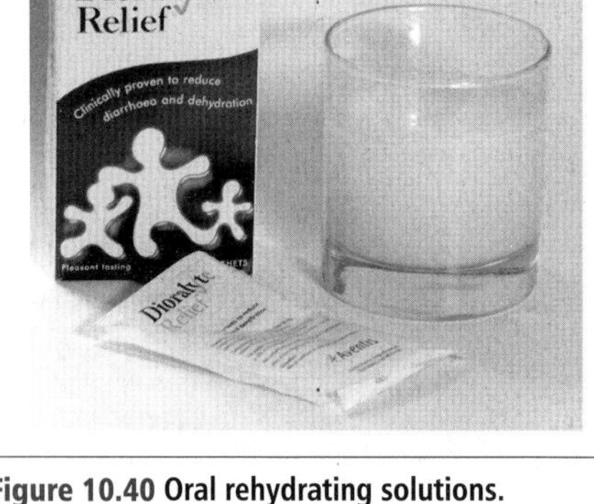

Figure 10.40 Oral rehydrating solutions. (A) The informative cover of a UN approved sachet containing ingredients that form an oral rehydrating solution appropriate for children when dissolved in a suitable volume of water. (B) A commercially available oral rehydrating solution.

The management of obesity aims to reduce food intake, particularly total energy intake, and to encourage regular exercise. This is often achieved by cutting down on high energy foods, such as fats and alcohol. Education and psychological support can be helpful in cases of severe obesity. Orlistat, an inhibitor of pancreatic lipase, has been used to manage obesity since it reduces the digestion and absorption of dietary fats (*Chapter 11*) and sibutramine to suppress the appetite, in conjunction with an energy-controlled diet, has been used to control weight. Surgery is used in some cases of severe obesity. Jejuno-ileal bypass surgery for morbid obesity was first performed in 1952. In this process the end of the jejunum or the beginning of the ileum are removed and the remaining portions joined together. Most of the small intestine is removed leaving only a short length for digestion and absorption. Although jejuno-ileal bypass surgery results in a very good weight loss, severe side effects occur. The technique has been replaced with gastric bypass surgery where the upper part of the stomach is connected to the small intestine about one third of the way along its length. Thus only the lower part of the stomach and approximately two thirds of the small intestine are available for digestion and absorption, reducing energy intake and therefore weight, but dieting is still likely to be required. Again, this operation is not without the risk of developing serious postoperative complications. Liposuction may also be used to remove fat from under the skin though this is typically performed mainly for cosmetic rather than therapeutic reasons.

Eating disorders, such as AN and BN, are difficult to prevent and remain hard to treat. There are seemingly few effective treatments and no universally recognized plan of treatment. However, the goals of any therapy must be to establish normal eating patterns and to restore the patient's nutritional status and weight. Usually a multidisciplinary approach tailored to the individual is used. This will involve specialists in nutrition and mental health in addition to clinicians. Therapy generally involves the family, behavior modification and nutrition counseling, support groups and the use of antidepressants. Most patients with AN or BN are treated as outpatients although in severe cases hospitalization may be necessary. Given that compliance is often problematic, AN and BN are generally considered to be chronic disorders interrupted only by intermittent periods of short-lived remission.

CASE STUDY 10.1

Andrew, a 51-year-old solicitor, had a health check. He was found to weigh 105 kg and his height was 1.82 m. Andrew is a smoker and does not exercise on a regular basis. He had a family history of heart disease. His blood pressure and plasma lipids were analyzed and the results are shown below (reference ranges are shown in parentheses):

Blood pressure	170/100 mmHg (120/80 mmHg)
Total cholesterol	6.3 mmol dm^{-3} (< 6.5 mmol dm^{-3} acceptable, < 5.2 mmol dm^{-3} is desirable)
Triacylglycerols	2.7 mmol dm^{-3} (0.45–1.80 mmol dm^{-3})
HDL cholesterol	0.7 mmol dm^{-3} (1.0–1.5 mmol dm^{-3})

Questions

(a) Calculate Andrew's BMI.

(b) Is Andrew obese?

(c) What risks are associated with obesity?

(d) What advice would you give Andrew?

CASE STUDY 10.2

Emily is a 17-year-old female. Over the past 18 months her periods have become increasingly irregular and intermittent and she consulted her doctor about this problem. During the consultation, Emily explained that she was studying for her preuniversity examinations and wishes to study law. Her parents and teachers expect her to do extremely well. She thought she was 'a bit overweight' and was dieting and about two years ago had started exercising more regularly. She was now jogging about 2 h a day and had recently taken up aerobic classes. A physical examination showed Emily to be 5′ 4″ tall and weigh 110 pounds. Body temperature, pulse rate and blood pressure were all normal. A urine ketostix gave a normal result.

Blood tests give the following results (reference ranges are shown in parentheses):

Glucose	4.3 mmol dm^{-3} (4.5–5.6 mmol dm^{-3})
Ketone bodies	110 μmol dm^{-3} (~70 μmol dm^{-3})

Questions

(a) Is Emily underweight?

(b) If so, by how much?

(c) Account for the clinical results.

(d) What treatment or advice would you give Emily?

CASE STUDY 10.3

John, a 74-year-old man residing in a nursing home, was seen by his doctor because he suffered from mental confusion and had difficulty walking because of paresthesiae and numbness in his legs. There was some general concern that the elderly residents at this nursing home were not being fed appropriately. A blood sample was taken and assessed and an erythrocyte transketolase activity test performed. Results are shown below.

Transketolase activity without added TPP	1.80 mmol h^{-1} per 10^9 erythrocytes
Transketolase activity with added TPP	2.21 mmol h^{-1} per 10^9 erythrocytes

Questions

(a) Does John suffer from deficiency of any vitamin?

(b) If so, how should he be treated?

10.6 SUMMARY

The body requires an adequate supply of proteins, carbohydrates, lipids, vitamins, minerals, trace elements and water for the maintenance of health, which between them provide energy and the raw materials for the synthesis of biological molecules. The amount of each nutrient required to maintain health can be expressed as its dietary reference value (DRV). Nutritional disorders may arise from deficiencies or in some cases an excess of nutrients. Such disorders may be investigated, initially by taking a medical history, by the use of dietary questionnaires and by a physical examination, which may include taking anthropometric measurements. Laboratory tests may be used to confirm suspected diagnoses. Protein-energy malnutrition may lead to marasmus or kwashiorkor, while obesity, an increasing problem in developed countries, predisposes an individual to a variety of disorders including type 2 diabetes and coronary heart disease. Anorexia nervosa and bulimia nervosa are complex eating disorders with underlying psychological bases that require treatment, as well as help for the obvious nutritional deficits. The variety of disorders arising from vitamin deficits reflects the range of vitamins required for health and their diverse roles in the body. Hypervitaminoses are uncommon, although some B vitamins along with vitamin A and D toxicities have been observed when excessive amounts of vitamin supplements have been consumed. Disorders associated with mineral deficiencies are rare, though they do occur, for example is the development of thyroid goiter in diets deficient in iodine.

The management of nutritional disorders may require simple dietary advice, or, where severe, nutritional supplements. Medical interventions may require tube or parenteral feeding to restore a good nutritional status.

QUESTIONS

1. Blood copper binds to which plasma protein?
   a) thyroglobulin;
   b) ferritin;
   c) ceruloplasmin;
   d) hemoglobin;
   e) albumin.

2. Which of the following is **NOT** a complication of obesity?
   a) type 2 diabetes mellitus;
   b) vascular disease;
   c) hypertension;
   d) celiac disease;
   e) cholelithiasis.

3. Arrange the two following lists into their most appropriate pairings.

ATP7A	xanthine oxidase
bulimia nervosa	dental erosion
copper	acetyl CoA
iodide	dental caries
fluoride	Wilson's disease
folic acid	prothrombin time
leptin	Wernicke-Korsakoff syndrome
molybdenum	phytanic acid
edema	osteomalacia

pantothenic acid	heart disease
saturated fats	hemolytic anemia
thiamin	scurvy
vitamin B_1	green leafy vegetables
vitamin B_6	thiaminase
vitamin B_{12}	tryptophan
vitamin C	*OBESE/obese*
vitamin D	kwashiorkor
vitamin E	Crohn's disease
vitamin K	Menkes' disease
zinc	goiter

4. Your adolescent daughter tells you she is going to become a vegan (strict vegetarian). In the light of the accompanying table and your knowledge of protein nutrition, what advice would you give her?

Plant food	**deficient in the following amino acids**
Cereals, for example	
maize	lysine, methionine, tryptophan
rice	histidine, tryptophan
oats	methionine, tryptophan
wheat	lysine, threonine, tryptophan
Legumes	methionine, isoleucine, tryptophan
Nuts	lysine, threonine
Oil seeds	lysine, threonine
Other vegetable	methionine, isoleucine

5. A 72-year-old man who lived by himself was admitted to hospital following an accident. On admission, it was noted that he appeared emaciated, anemic and had widespread petechial hemorrhages. He also presented with gum disease. What types of nutritional deficiencies might be present?

6. *Figures 10.35* and *10.36* show the arm circumference and skinfold thickness being measured on two male patients, A and B respectively. The accompanying table gives some anthropometric data for them.

	A	**B**
Height	1.83 m	1.80 m
	6′ 0″	5′ 10″
Weight	77.5 kg	120 kg
	171 lb	266 lb
Triceps skin fold thickness/mm	12	15
Arm circumference/mm	270	320
Waist : hip ratio	0.90	0.98

What can be concluded from these data about the health risks to A and B?

FURTHER READING

Astrup, A, Meinert Larsen, T and Harper, A (2004) Atkins and other low-carbohydrate diets: hoax or an effective tool for weight loss? *Lancet* **364:** 897–899.

Azzi, A and Zingg, J-M (2005) Vitamin E. *Biochem. Mol. Biol. Edu.* **33:** 184–187.

Bartlett, MG (2003) Biochemistry of the water soluble vitamins: A lecture for first-year pharmacy students. *Am. J. Pharm. Educ.* **67:** 64–67.

Bates, CJ (1999) Diagnosis and detection of vitamin deficiencies. *Br. Med. Bull.* **55:** 643–657.

Bender, DA (2002) *Introduction to Nutrition and Metabolism,* Taylor and Francis, London, UK.

Castiglia, PT (1996) Protein-energy malnutrition (kwashiorkar and marasmus). *J. Pediatr. Health Care* **10:** 28–30.

Fairfield, KM and Fletcher, RH (2002) Vitamins for chronic disease prevention in adults – Scientific review. *J. Am. Med. Assoc.* **287:** 3116–3126.

Gardner, D (2003) The etiology of obesity. *Mo. Med.* **100:** 242–246.

Geissler, C and Powers, HJ (2005) *Human Nutrition.* Churchill Livingstone, London, UK.

Hoek, HW and van Hoeken, D (2003) Review of the prevalence and incidence of eating disorders. *Int. J. Eat. Disord.* **34:** 383–396.

Hollick, MF (2003) Vitamin D: a millennium perspective. *J. Cell. Biochem.* **88:** 296–307.

Jackson, MJ (1999) Diagnosis and detection of micronutrients: minerals. *Br. Med. Bull.* **55:** 634–642.

Klein, DA and Walsh, BT (2004) Eating disorders: clinical features and pathophysiology. *Physiol. Behav.* **81:** 359–374.

Kopelman, PG and Grace, C (2004) New thoughts on managing obesity. *Gut* **53:** 1044–1053.

Murphy, B and Manning, Y (2003) An introduction to anorexia nervosa and bulimia nervosa. *Nurs. Stand.* **18:** 45–52.

Nicholls, D and Viner, R (2005) Eating disorders and weight problems. *BMJ* **330:** 950–953.

Taubes, G (2001) The soft science of dietary fat. *Science* **291:** 2536–2545.

Ten Cate, JM (2004) Fluorides in caries prevention and control: Empiricism or science. *Caries Res.* **38:** 254–257.

Vermeer, G and Schurgers, LJ (2000) A comprehensive review of vitamin K and vitamin K antagonists. *Hematol. Oncol. Clin. North. Am.* **14:** 339–345.

Useful web site:

http://www.foodstandards.gov.uk

chapter 11:

DISORDERS OF THE GASTROINTESTINAL TRACT, PANCREAS, LIVER AND GALL BLADDER

OBJECTIVES

After studying this chapter you should be able to:

- outline the structure of the gastrointestinal tract and its accessory organs;
- describe the processes involved in the digestion of foods;
- outline the process whereby nutrients are absorbed by the gastrointestinal tract;
- review the role of the small intestine in homeostasis;
- explain the causes of some pancreatic, liver and gall bladder disorders;
- describe some common gastrointestinal tract disorders.

11.1 INTRODUCTION

The gastrointestinal tract (GIT) is essentially a long tube extending from the mouth to the anus with a number of specialized regions. Most of the nutrients (*Chapter 10*) in food have to be digested before they can be assimilated. Digestion involves the hydrolysis of polysaccharides, oligosaccharides, proteins and lipids in enzyme-catalyzed reactions. The products of hydrolysis, such as monosaccharides, amino acids, free fatty acids and monoacylglycerols, can then be absorbed. Some nutrients, vitamins and minerals for instance, are absorbed without the need for digestion. The digestion of food and absorption of the released nutrients are the major functions of the GIT together with its accessory organs, the pancreas, liver and gall bladder. The peritoneum lines the abdominal cavity and covers the major organs. Folds of the peritoneum, called the mesentery, and mesocolon hold the intestines in place. The GIT has significant roles in the homeostasis of nutrients and in protecting the body against pathogens (*Chapter 2*) ingested with the food. Digestion and absorption, can, of course, only take place after ingestion of food, unlike the homeostatic and protective functions that operate continuously.

11.2 THE GASTROINTESTINAL TRACT AND ITS ACCESSORY ORGANS

The GIT may be described as a tubular system with distinctive regions that begin with the mouth leading to the pharynx, esophagus, stomach, the small and large intestines and terminating at the anus (*Figure 11.1* (*A*)). It is associated with several accessory digestive organs, such as the pancreas, liver and gall bladder (*Figure 11.1* (*B*)). The walls of the GIT have a fairly common substructure (*Figure 11.2* (*A*) and (*B*)) and its interior is called the lumen. The GIT and accessory organs are bound to each other and to the inner wall of the abdomen by the **peritoneum**. This is a strong, colorless membrane with a smooth surface that consists of two parts; the parietal peritoneum, which lines the abdominal cavity, and the visceral peritoneum that covers most of the organs in the abdomen (*Figure 11.3*). The thin space between the two parts, called the peritoneal cavity, is filled with serous fluid. In males, the peritoneum forms a closed sac but in females it is continuous with the mucous membrane of the uterine tubes. The **mesentery** is a fan shaped portion of peritoneum that extends from the posterior abdominal wall and wraps around the small intestine and returns to its origin forming a double layer (*Figure 11.4*). The mesentery contains the blood and

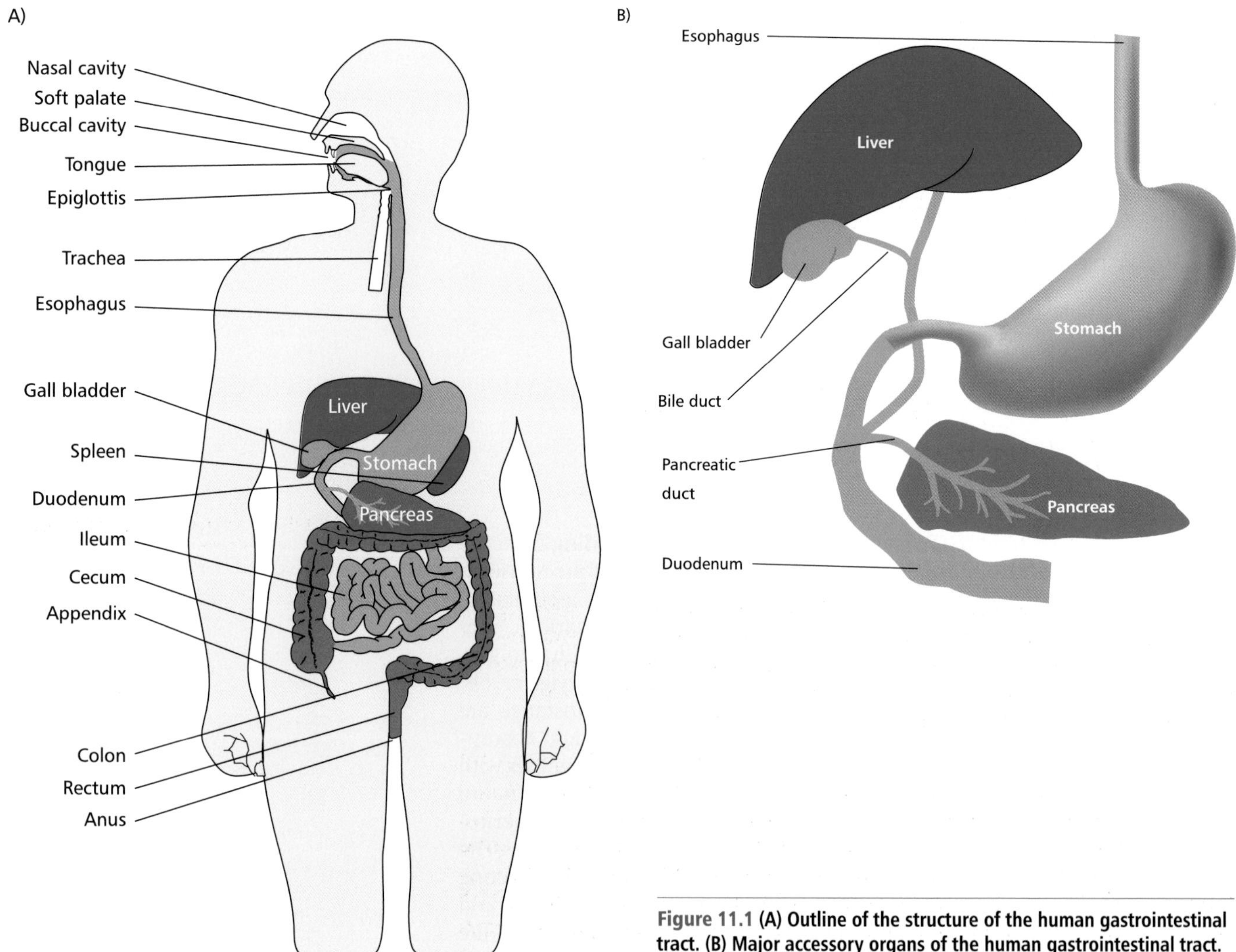

Figure 11.1 (A) Outline of the structure of the human gastrointestinal tract. (B) Major accessory organs of the human gastrointestinal tract.

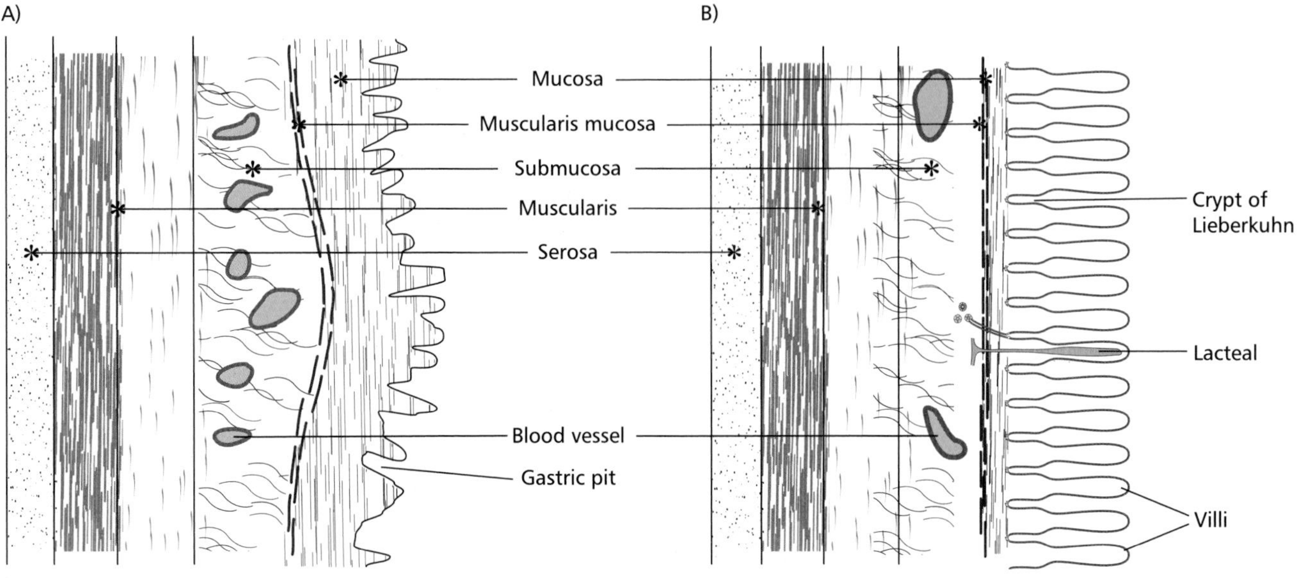

Figure 11.2 Outline of the tissue organization of the walls of (A) the stomach and (B) the small intestine.

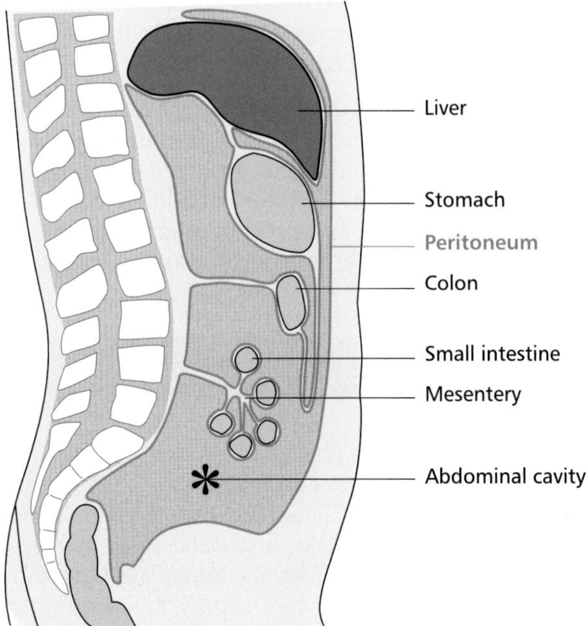

Figure 11.3 Schematic showing the peritoneum.

lymphatic vessels and nerves that supply the small intestine. Another peritoneal fold, the **mesocolon** (*Figure 11.4*), has similar functions with respect to the large intestine. The mesentery and mesocolon anchor the intestines in place but only loosely, allowing them considerable movement as muscular activities mix and move the contents through the GIT.

The pancreas is found near the stomach and small intestine (*Figure 11.1* (*A*) and (*B*)) and functions as both an exocrine and an endocrine gland. In its exocrine role it produces pancreatic juice that is released into the first part of the small intestine, the duodenum, through the pancreatic duct. This secretion contains precursors of digestive enzymes that are activated and function in the GIT, and also hydrogen carbonate (HCO_3^-) that increases the pH of GIT contents. This latter function is vital since the material leaving the stomach is acidic. The pancreas produces about 1200–1500 cm^3 of pancreatic juice per

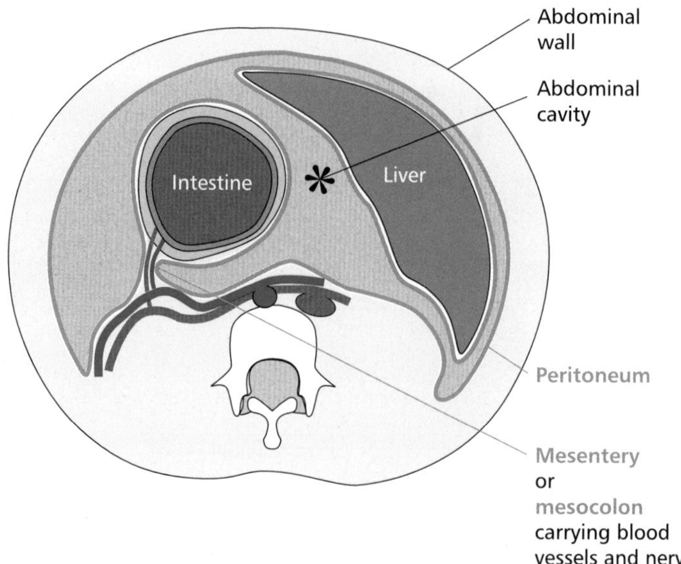

Figure 11.4 Schematic showing a transverse of the mesentery and mesocolon.

day. Within the tissues of the pancreas are groups of cells called the islets of Langerhans. The cells in the islets have endocrine functions and produce and secrete the peptide hormones glucagon and insulin that regulate the concentration of glucose in the blood (*Chapter 7*).

The liver is the largest organ in the body (*Figure 11.1* (*A*) and (*B*)), weighing about 1.5 kg. It has two major lobes, each divided into functional units called lobules. Lobules consist of cuboidal cells arranged around a central vein into which the blood drains. Between the plates of hepatic cells are sinusoids that are supplied by the portal vein and hepatic artery (*Figure 12.2* and *Chapter 12*). The sinusoids are lined with phagocytic Kupffer cells that belong to the

BOX 11.1 Peritonitis

Peritonitis is an inflammation of the peritoneum that lines the abdominal cavity and covers the surfaces of abdominal organs. The condition is marked by exudations into the peritoneum of serum, fibrin, cells and pus. Primary peritonitis is caused by the spread of an infection from the blood and lymph nodes to the peritoneum and accounts for less than 1% of all cases. Secondary peritonitis is the commonest type and occurs when bacteria enter the peritoneum from the GIT or biliary tract. The symptoms of peritonitis may include vomiting, swelling of the abdomen, severe abdominal pain and tenderness, weight loss, constipation and moderate fever. Its major causes are perforations of GIT wall, which allows seepage of the luminal contents into the abdominal cavity. These can arise from a ruptured appendix or from perforations of the stomach, including damage from ulcers, and gall bladder. Pelvic inflammatory disease in sexually active women is also a common cause. Peritonitis can also develop after surgery when bacteria have been allowed to enter the abdomen during an operation.

Peritonitis must be rapidly diagnosed and treated because complications can occur rapidly. Diagnosis relies on taking a medical history and a physical examination, particularly to investigate any abdominal swelling and tenderness. Diagnostic tests include using X-rays or a CT scan (*Chapter 18*) of the abdomen to confirm the presence of fluid, accumulation of pus or infected organs in the abdomen. Microbiological tests on samples of blood or abdominal fluid can identify the causative microorganism. Peritonitis is frequently life-threatening and acute peritonitis is a medical emergency. The outlook for untreated patients is poor. Specific treatments for peritonitis depend on the age, health, medical history and the severity of the condition. Treatment is generally aimed at treating the underlying condition. Antibiotics are given immediately once peritonitis has been diagnosed. A nasal tube may be inserted into the stomach or intestine to drain fluid and gas. Peritoneal lavage, where large amounts of fluid are injected into the peritoneum to wash out the infective microorganisms causing the condition, may be desirable. Intravenous fluids are also given to replace lost body fluids. Morphine may be prescribed in hospitalized patients to reduce pain. Emergency exploratory surgery may be necessary, especially in cases that involve appendicitis, a perforated peptic ulcer or diverticulitis.

reticuloendothelial system and their main function is to engulf bacteria and other foreign particles in blood. Thus there is a dual blood supply to liver, with blood coming from the digestive tract and spleen through the portal vein, and from the aorta in the hepatic artery (*Figure 14.2*). About a third of incoming blood is arterial, and brings oxygen, whereas two thirds is venous from the portal vein.

The large reserve capacity of the liver means that it needs only 10–20% of its tissues to be functioning to sustain life. It also has a remarkable ability to regenerate itself after its tissue has been removed or destroyed by disease. Complete destruction or removal of the liver results in death within 10 h, hence liver disease with loss of function is a serious matter (*Chapter 12*). The liver has numerous functions. It acts as an interface between the GIT and the rest of the body tissues because the hepatic portal vein carries blood directly to it from the GIT. Hence it is able to regulate the post-hepatic blood concentrations of many of the nutrients absorbed by the GIT. Similarly, the liver also regulates the concentrations of many biomolecules produced by the body, for example steroid hormones, and deals with many toxins, such as drugs, pesticides and carcinogens, to render them less harmful and solubilizes them for excretion (*Chapter 12*). The liver also produces many of the plasma proteins including albumin and the clotting factors (*Chapter 13*). Bile produced by the hepatocytes is secreted into bile canaliculi and eventually drains into the bile duct. About 700–1200 cm^3 of bile are produced daily and stored and concentrated in a hollow organ called the gall bladder (*Figure 11.1* (*A*) and (*B*)) prior to its release in the small intestine. Bile does not contain enzymes, but it does contain bile salts, for example sodium glycocholate and taurocholate. These are detergents that aid in the digestion of lipids by emulsifying them to form water-soluble complexes.

Bile pigments are derived from heme which results from the destruction of old erythrocytes (*Chapter 13*). In the spleen (*Figure 11.1* (*A*) and (*B*)), destruction of red blood cells releases hemoglobin, which is then catabolized to free heme and globin. The latter is degraded to amino acids. Bilirubin is derived from heme, the iron-containing protoporphyrin ring of hemoglobin. A typical adult produces around 450 μmol of bilirubin per day. This bilirubin is referred to as unconjugated bilirubin. It is insoluble in water and is transported in the plasma bound to albumin to be taken up by hepatocytes. Here, it is conjugated with glucuronic acid by UDP-glucuronyltransferase to form mono- and diglucuronides in a manner resembling the detoxification reactions described in *Chapter 12*. Conjugated bilirubin is much more water soluble than its unconjugated form and is secreted into the bile duct. In the small intestine, conjugated bilirubin is a substrate for bacteria that convert it to urobilinogen and stercobilin. Most stercobilin is excreted in feces, although some urobilinogen is absorbed and taken to the liver in the hepatic portal vein and re-excreted in bile or by the kidneys (*Figure 11.5*).

11.3 DIGESTION

Digestion is the hydrolytic breakdown of nutrient macromolecules and compound lipids to smaller products that can be absorbed. The hydrolytic reactions are catalyzed by a variety of enzymes: proteases that digest proteins; carbohydrases that digest carbohydrates; lipases that catalyze the hydrolysis of lipids and nucleases that degrade DNA and RNA. Digestion occurs in the mouth, to a small extent, stomach and small intestine, and most absorption of nutrients occurs in the small intestine and that of water in the large intestine.

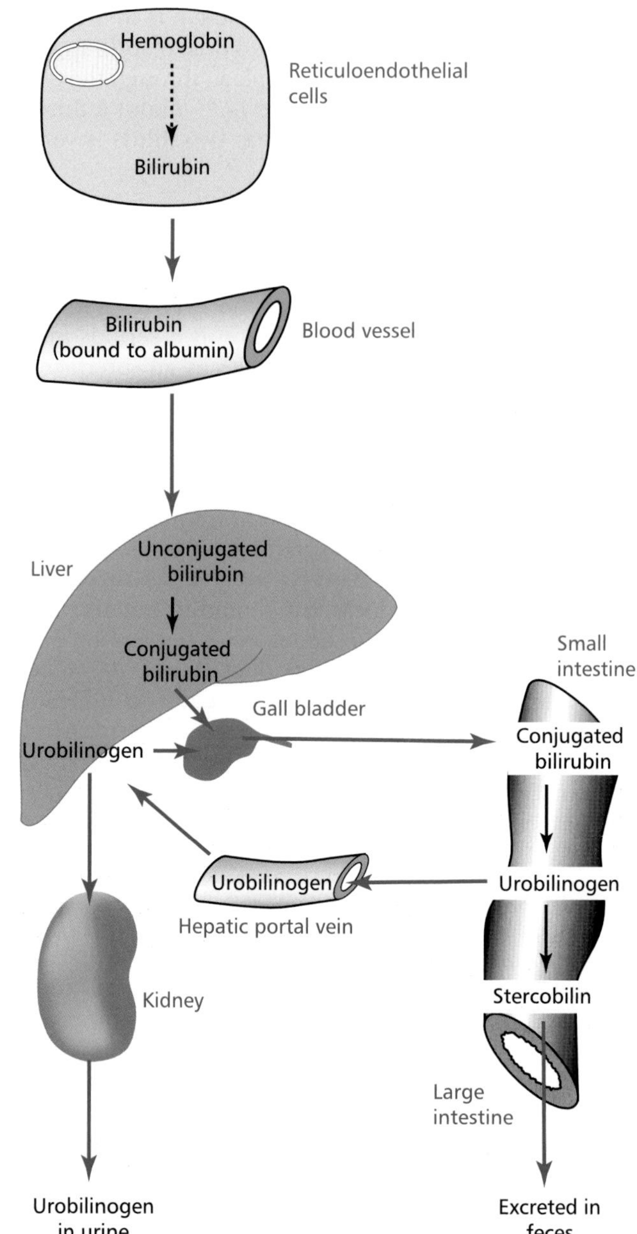

Figure 11.5 Overview of the production and excretion of bilirubin following the catabolism of heme. See text for details.

Margin Note 11.1 Flavor and taste 

Flavor is a complicated amalgam of sensory perceptions, involving the taste, smell and texture of foods, previous exposure to particular foods, especially in childhood, and even the noises associated with chewing some foods. In general, much of the everyday sensation of flavor is derived largely from olfactory stimulation. **Anosmia**, the loss of the sense of smell, for example when infected with a cold, can severely reduce the sense of taste.

Six tastes can be detected by different types of sensory taste buds distributed over the tongue. These are bitter, salty, sour, sweet, fats (fatty acids) and savory or umami. The first four are self-explanatory. The perception of fats is more strictly the detection of small amounts of fatty acids hydrolyzed from lipids by the action of lingual lipase. In 1907, Ikeda working at the Tokyo Imperial University investigated the source of a flavor well recognized in many Asian countries called umami. This most easily translates in English as savory or meaty. He was successful in extracting crystals of the amino acid, glutamate from a type of seaweed called kombu used in traditional Japanese cuisine. Many foods that have a umami flavor are protein rich, for example meat, seafoods, aged cheeses, and contain large amounts of glutamate. Monosodium glutamate (MSG) was developed as a food additive and is probably the most familiar example of an additive that imparts a savory or umami taste.

The evolution of specific taste receptors is most easily explained by the advantages gained in detecting foods rich in essential nutrients (salty, sweet, fats and savory) and in avoiding ones that contain toxins (bitter and sour).

MOUTH

In the mouth, teeth break the food into smaller portions increasing the surface area upon which digestive enzymes can act. Three pairs of salivary glands, the submandibular located under the jaws, the sublingual located under the tongue and the parotid situated in front of the ears (*Figure 11.6*), secrete saliva into the mouth. The saliva contains amylase (*Figure 11.7*), water and mucus. The water helps to dissolve nutrients, while the mucus acts as a lubricant to aid swallowing and lubricates passage of food through the GIT. Chewing and the actions of the tongue mix the food with the saliva. Salivary amylase begins the digestion of carbohydrates (*Figure 11.8*) although the digestion of carbohydrates in the mouth is minimal since the food is not retained there for any length of time. Lingual lipase is secreted by lingual serous glands. Again, little digestion occurs in the mouth but it has been suggested that the fatty acids

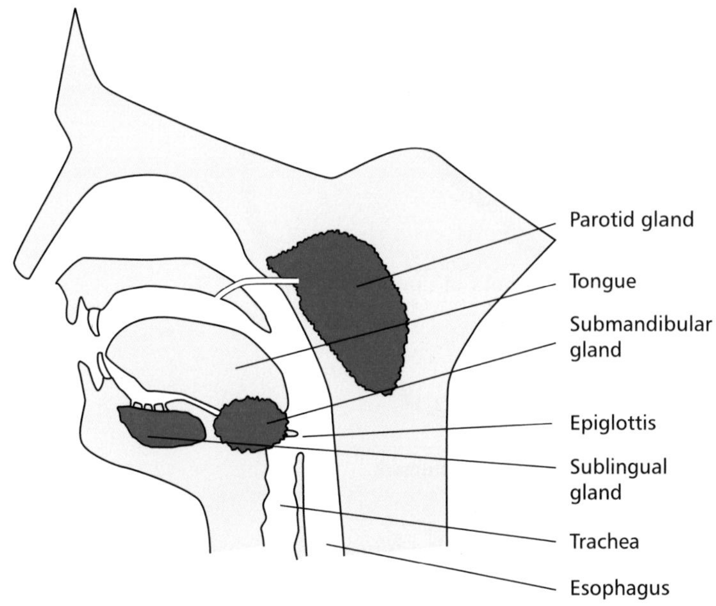

Figure 11.6 The submandibular, parotid and sublingual salivary glands.

Figure 11.7 Molecular model of a salivary amylase molecule. PDB file 1SMD.

released by its action are perceived by specific taste receptors allowing fatty foods, and therefore energy rich nutrients, to be detected.

In swallowing, the soft palate is elevated, which seals off the nasal cavity and pushes against the back wall of the pharynx. This, in turn, triggers an automatic reflex action in muscles that raise the larynx, pushing its opening, the glottis, against a flap of tissue called the epiglottis. These actions prevent food entering the trachea of the respiratory system and ensure food is expelled from the mouth into the esophagus. Rhythmic waves of contraction of smooth muscle tissue in the walls of the GIT, called **peristalsis**, forces food along the GIT. Peristalsis in the esophagus ensures that food is propelled to the stomach even if the person is upside down.

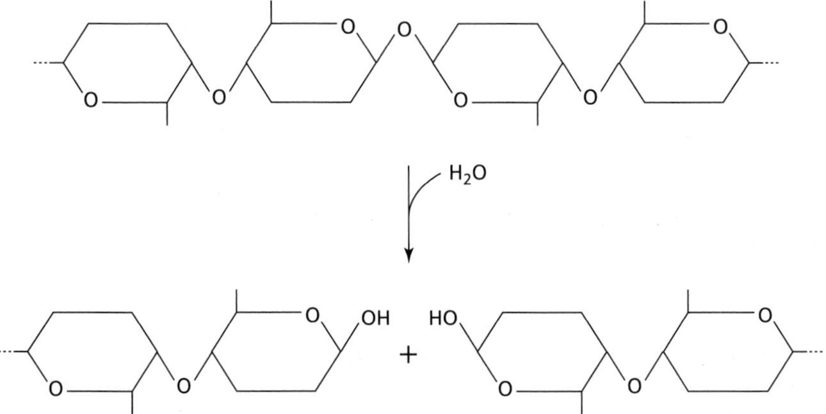

Figure 11.8 Schematic outlining the digestion (hydrolysis) of carbohydrates (polysaccharide).

STOMACH

The stomach is a sac-like region of the GIT (*Figure 11.1* (*A*) and (*B*)). Its inner surface is highly folded (*Figure 11.9*) allowing it to expand up to eightfold from its empty volume of about 50 cm^3 following a meal. Muscular activities of the stomach wall mix the food with gastric juice secreted by gastric glands located in the mucosal lining of the stomach. Gastric glands produce about 1 to 2 dm^3

Margin Note 11.2 Gastric lipase and cystic fibrosis

Patients suffering cystic fibrosis (CF, *Chapter 15*) are defective in pancreatic lipase and other enzyme activities due to exocrine pancreatic insufficiency when the pancreatic duct becomes blocked with a thick secretion. This reduces their ability to digest lipids and leads to lipid malabsorption and steatorrhea (*Section 11.8*). Many CF patients are treated with supplements of porcine pancreatic extracts that contain pancreatic lipase activity. However, these supplements are needed in high doses because the duodenums of many cystic fibrosis patients are more acidic than in healthy individuals and denaturation or inhibition of the enzyme can occur. The treatment is effective in only 30% of cystic fibrosis patients and 15–20% still suffer from steatorrhea. There is considerable interest in using recombinant mammalian gastric lipases since these are resistant to acid damage and inhibition and so should prove a more effective treatment.

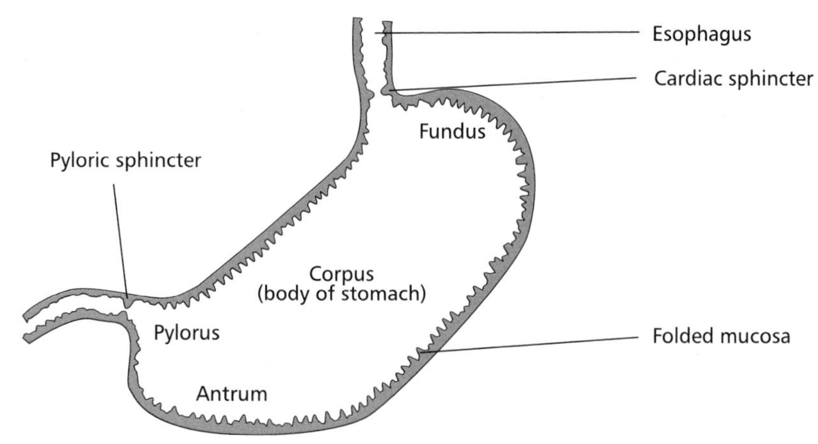

Figure 11.9 The major regions of the stomach.

of gastric juice per day from three types of secretory cells. Parietal or oxyntic cells produce hydrochloric acid (HCl), peptic or chief cells secrete pepsinogen, the precursor of pepsin, and gastric lipase, and mucous cells that secrete mucus that protects the mucosa lining the stomach from the corrosive action of the HCl.

The activity of salivary amylase continues within the bolus and is possibly only inactivated when the bolus contents are completely mixed with the acidic gastric juice. The pH optimum of lingual lipase is between 3.5 and 6.0 and is activated in the upper portion of the stomach. Gastric lipase is secreted by the chief cells of the stomach and also has an acid optimum. About 10 to 30% of dietary fat may be hydrolyzed in the stomach. However, lipase activity in the stomach is highest against triacylglycerols with short or medium length fatty acid residues, which are found in milk, and so may be of most importance in newborn infants.

Pepsinogen is a weakly-active protease. It has an acid optimum pH, hence the need for HCl secretion in the stomach. In these conditions, about pH 1 to 2, protein molecules in the food are denatured making them more susceptible to digestion. Also, at this pH, pepsinogen molecules act on one another at specific sites to produce the fully active protease, pepsin (*Figure 11.10*), that begins the digestion of the denatured proteins to form shorter polypeptides and peptides. The denaturation and digestion of proteins reduces the chance of their absorption and prevents them being immunogenic (*Chapter 4*). The watery mixture of gastric juice and partially digested food is called **chyme**.

Figure 11.10 Molecular model of a pepsin molecule. PDB file 1QRP.

The acid also activates intrinsic factor (IF), a glycoprotein secreted by the parietal cells that is needed for absorption of vitamin B_{12} (*Chapter 13*). Vitamin B_{12} is released from dietary proteins by the action of pepsin and binds to one of two binding proteins present in gastric juice whose affinity for vitamin B_{12} is increased in acid conditions and is greater than that of IF. When chyme enters the small intestine, proteases from pancreatic juice break down the binding proteins and vitamin B_{12} becomes bound to IF. The mucosal lining in the ileum has receptors for IF which bind the IF–vitamin B_{12} complex, so that vitamin B_{12} is absorbed and enters the portal blood. Given that the absorption of vitamin B_{12} is dependent upon IF, any condition that decreases the secretion of IF, for example atrophic gastritis (*Section 11.8*), interferes with digestion of the binding proteins, for example pancreatic exocrine insufficiency, or decreases the binding and internalization of the IF–vitamin B_{12} complex, such as some diseases that affect the ileum, can cause pernicious anemia (*Chapter 13*).

The acid environment of the stomach kills most of the bacteria that are ingested with the food but some survive and enter the small intestine in the chyme. The pyloric sphincter controls the passage of chyme from the stomach to the small intestine allowing only relatively small amounts of chyme through at any one time. This control is necessary because the digestive activity within the small intestine is time-consuming and because its capacity is limited. The activities of the stomach and small intestine are coordinated by the nervous and endocrine systems.

SMALL INTESTINE

The final stages of digestion and the absorption of its products occur in the small intestine (*Figure 11.1 (A)*). Contractions in the small intestine help to break food up, mix it with digestive juices and propel it towards the colon. The small intestine is about 23 feet (7 m) long in an adult. It is divided into three sections: the duodenum constitutes the first 250 mm, followed by the jejunum then the ileum. Digestion occurs primarily in the duodenum and jejunum and absorption in the ileum.

The walls of the duodenum contain goblet cells that secrete mucus that protects it against damage by acidic chyme. Crypts of Lieberkuhn are pits within the wall of the small intestine with secretory cells that release intestinal juice and Paneth cells that secrete lysozyme. Intestinal juice is largely water with mucus and buffered to a pH of about 7.6. Along with pancreatic juice (*see below*), it neutralizes chyme and provides a liquid medium that aids absorption of nutrients. Lysozyme is an antibacterial enzyme. The epithelial lining consists of cells called enterocytes organized into small projections into the lumen called villi (*Figure 11.11*). In turn, each enterocyte

Margin Note 11.3 Disaccharides 

Disaccharides, as their name implies, are composed of two simple sugars (monosaccharide residues) covalently bound together by a glycosidic bond. A variety of disaccharides are known, which can be found naturally or produced by the action of digestive enzymes on polysaccharides. Disaccharides differ from one another in the nature of their constituent monosaccharides and the type of glycosidic bond joining them together (*Table 11.1*).

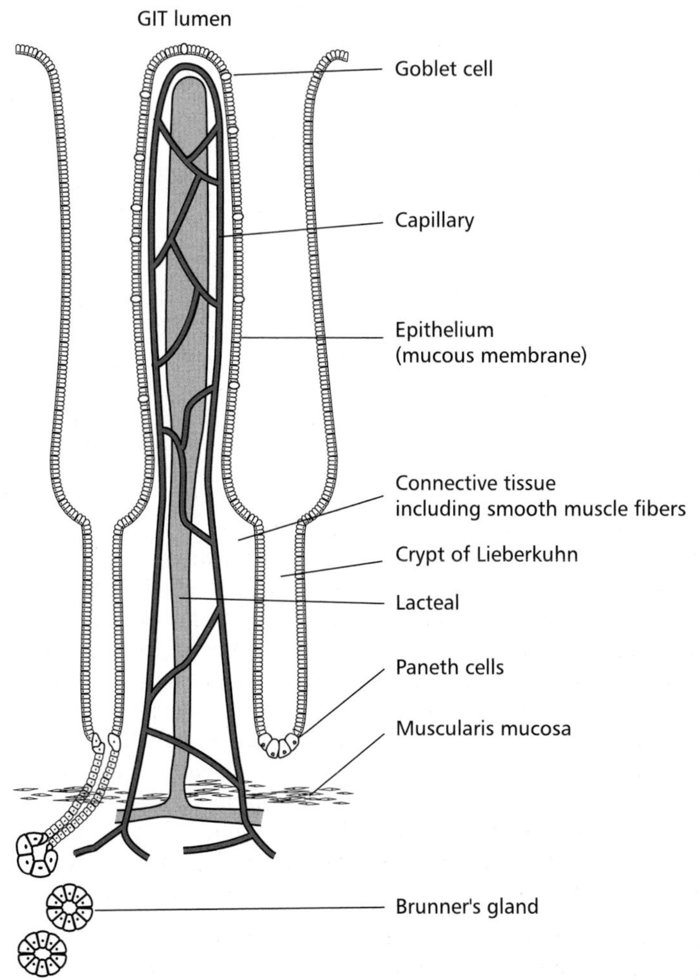

Figure 11.11 A villus with its associated crypts of Lieberkuhn.

BOX 11.2 Cyclic vomiting syndrome

Vomiting or **emesis** is the expulsion of food from the stomach (and sometimes duodenum) through the esophagus and mouth. It is usually experienced as the final of three events, nausea, retching and vomition. Nausea is an unpleasant psychic experience associated with decreased gastric motility and increased muscle tone in the small intestine. Additionally, there is often reverse peristalsis in the proximal small intestine. Retching or dry heaves is the spasmodic respiratory movements conducted with a closed glottis. While this is occurring, the antrum of the stomach contracts and the fundus and cardia relax. Vomition is when the contents of the stomach and sometimes small intestine are propelled up to and out of the mouth.

Vomiting can be initiated by a variety of stimuli, including infections (gastroenteritis), various chemical irritants (emetics) and poisons, distension of the stomach, unpleasant sights and smells, dizziness, anesthetics, a number of drugs, a variety of illnesses, for example brain tumors, and hormonal changes associated with pregnancy. All result in nervous impulses being sent to a vomiting center in the medulla oblongata, which responds by sending motor impulse to muscles of the upper GIT, diaphragm and abdominal muscles (*Figure 11.12*). The resulting contractions squeeze the stomach between the diaphragm and abdominal muscles and expel its contents. A simple episode of vomiting rarely causes problems but may on occasion have more serious consequences, such as aspiration pneumonia. However, severe or repetitive vomiting results in disturbances to acid–base balance, electrolyte depletion and dehydration. In such cases, the underlying cause must be rapidly identified and appropriate therapy initiated. In many cases antiemetic drugs must be administered to suppress vomiting and reduce its sequelae.

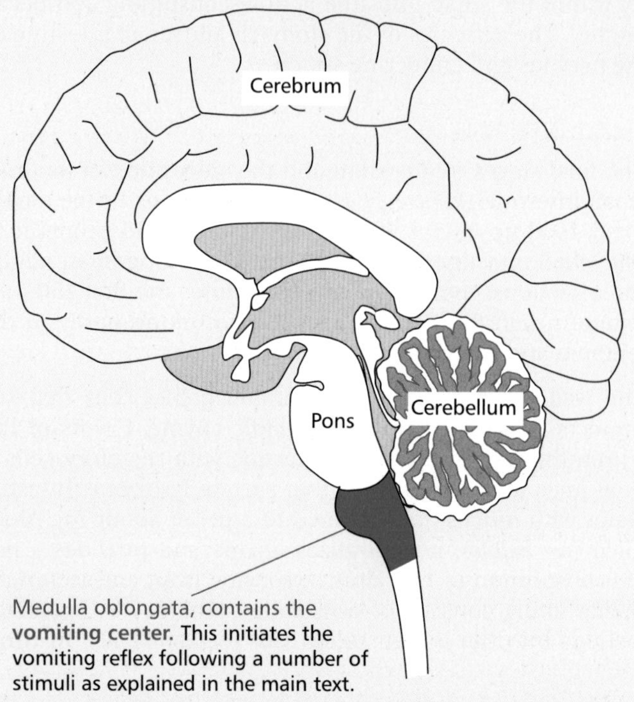
Cerebrum
Pons
Cerebellum
Medulla oblongata, contains the
vomiting center. This initiates the
vomiting reflex following a number of
stimuli as explained in the main text.

Figure 11.12 The site of the vomiting center in the medulla oblongata.

Cyclic vomiting syndrome (CVS) or abdominal migraine is a disorder of the GIT characterized by recurrent, severe and prolonged attacks of nausea, vomiting and abdominal pain that was first described by Gee in 1882. It usually occurs in children of two to 16 years old, most commonly between the ages of three and seven years, but can also occur in adults. The cause of CVS is unknown. Its incidence and prevalence is also uncertain but some evidence suggests that 1 in 50 children in the USA may be affected.

has its apical plasma membrane surface, which is the side facing the lumen of the GIT, arranged into microscopic extensions called microvilli forming a **brush border** that coats the villi (*Figure 11.13*). The opposing face of the enterocyte is called the basolateral membrane. The presence of villi and microvilli means that the interior surface area of the small intestine is enormous, estimated at 300 m^2 in humans. Hydrolytic enzymes that are integral proteins of the plasma membranes forming the microvilli catalyze further digestive reactions. For example: lactose is hydrolyzed to glucose and galactose in a reaction catalyzed by lactase; sucrose to glucose and fructose by sucrase; and the disaccharides maltose and isomaltose, produced by the action of amylase on starch, are hydrolyzed to glucose by maltase and isomaltase activities (*Table 11.1*).

Pancreatic juice is produced by the pancreas and contains water, alkaline salts that give it a pH of 7.8 to 8.0, enzymes and zymogens (*Margin Note 11.4*) and

Margin Note 11.4 Zymogens

Digestive enzymes are produced as inactive precursors called zymogens (*Table 11.2*), which ensures the protection of the cells and tissues that produce them from the catalytic activities of the enzymes. Activation is achieved by a partial and specific hydrolysis of part of the zymogen structure that masks the active site.

Cyclic vomiting syndrome occurs in four distinct phases. First, there is a prodrome phase, which is often accompanied by abdominal pain, and which signals that an episode of nausea and vomiting is about to begin. This can last a few minutes to several hours. However, sometimes a prodrome does not occur and patients enter the second phase, an episode, directly. Episodes consist of nausea followed by severe vomiting, which usually begin at night or first thing in the morning. Abdominal pain, dizziness, headaches, photosensitivity, fever and sometimes diarrhea may also present. Vomiting can be as frequent as six to 12 times an hour during the worst of the episode and can continue for one to five days during which the patient appears pale, listless and exhausted, often to the point of near unconsciousness. The third or recovery phase begins when the nausea and vomiting stop. Generally, appetite and energy return but the time required for this varies considerably. The fourth phase is the symptomless interval between episodes.

Episodes are triggered by specific events or conditions, most commonly infections such as colds and influenza, but also emotional stress, excitement, allergies eating certain foods, for example chocolate or cheese, overeating, excessively hot weather and physical exhaustion. Motion sickness and menstruation can also trigger episodes.

The severity of vomiting in CVS is a risk factor for a number of complications, the most obvious being dehydration and electrolyte imbalance, which are described in *Chapter 8*. Other complications include peptic esophagus caused by stomach acid in the vomit, **hematemesis**, which is blood in the vomit from the damaged esophagus, a Mallory-Weiss tear of the lower end of the esophagus or the stomach may bruise from violent and prolonged vomiting or retching. Finally, tooth corrosion can be caused by stomach acid in a manner similar to that seen in bulimics (*Chapter 10*).

Cyclic vomiting syndrome is difficult to diagnose and many patients are initially misdiagnosed. It is usually identified from the general symptoms and medical history and by excluding more common diseases or disorders that can also cause nausea and vomiting. Diagnosis is time consuming because a repeating pattern of vomiting must be established. There is no cure for CVS. Avoiding known triggers of an episode is an obvious remedy but they cannot always be avoided. Ibuprofen taken in the prodrome phase may help prevent or alleviate an episode. Once an episode begins, treatment is supportive with bed rest and sleep in a dark, quiet room. Other medications that may be helpful are ranitidine or omeprazole which decrease the amount of acid produced. Severe nausea and vomiting may require hospitalization and intravenous fluids to prevent dehydration, since drinking water normally promotes more vomiting although it does dilute the acid in the vomit, making the episode less painful. Sedatives may help if the nausea continues. Drinking water and replacing lost electrolytes is essential in the recovery phase.

Cyclic vomiting syndrome and migraine show a number of similarities and the two may be related. Like CVS, migraine is characterized by headaches that begin and end abruptly, followed by longer periods without pain or other symptoms. They both have a number of similar triggers, including stress and excitement. Many children with CVS have a family history of migraine or suffer migraine attacks as they grow older. Given these similarities, CVS patients subject to frequent and long-lasting episodes are treated with some success using propranolol, cyproheptadine and amitriptyline that are generally used for treating migraine headaches.

Disaccharide	Structure	Products of digestion (hydrolysis)
Lactose (milk sugar)	galactose(β1–4)glucose	galactose + glucose
Isomaltose (product of starch digestion)	glucose(α1–6)glucose	2 glucose
Maltose (product of starch digestion)	glucose(α1–4)glucose	2 glucose
Sucrose (common sugar of plants)	glucose(α1–β2)fructose	fructose + glucose
Trehalose (found in fungi, some insects)	glucose(α1–1)glucose	2 glucose

Table 11.1 Some common disaccharides, the structure (glycosidic bond indicated in parentheses) and products of digestion

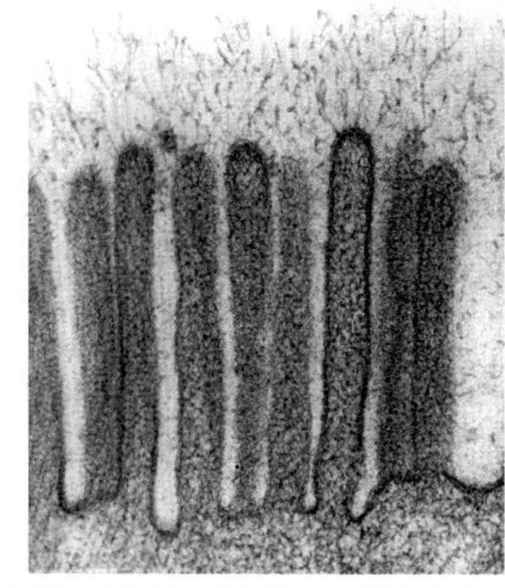

Figure 11.13 Electron micrograph of microvilli of the gastrointestinal tract. Courtesy of Emeritus Professor D.W. Fawcett.

enters the small intestine through the pancreatic duct. The salts neutralize acid from the stomach.

Proteolytic enzymes can be divided into exopeptidases and endopeptidases. The exopeptidases are aminopeptidases and carboboxypeptidases that catalytically remove amino acids from the ends of proteins and peptides (*Figure 11.14 (A)*). The pancreatic proteases, elastase, trypsin and chymotrypsin, are endoproteases (*Figure 11.14 (B)*) and hydrolyze proteins and peptides at different peptide bonds throughout the molecules to produce peptides and amino acids. The actions of endopeptidases increase the number of protein ends, effectively increasing the concentration of substrates for the exopeptidases. Pancreatic amylase continues the digestion of polysaccharides to produce disaccharides. Pancreatic lipase (*Figure 11.15*) hydrolyzes lipids to fatty acids and 2-monoacylglycerols but requires the presence of colipase, another protein secreted by the pancreas, for full activity. Bile released from the gall bladder enters the intestine through the bile duct and emulsifies lipids into droplets of about 1 μm diameter, which greatly increases the surface area available for lipase activity. Pancreatic esterase activity can complete the digestion to glycerol and free fatty acids (*Figure 11.16*).

The activities of many of the enzymes involved in digestion are summarized in *Table 11.3*.

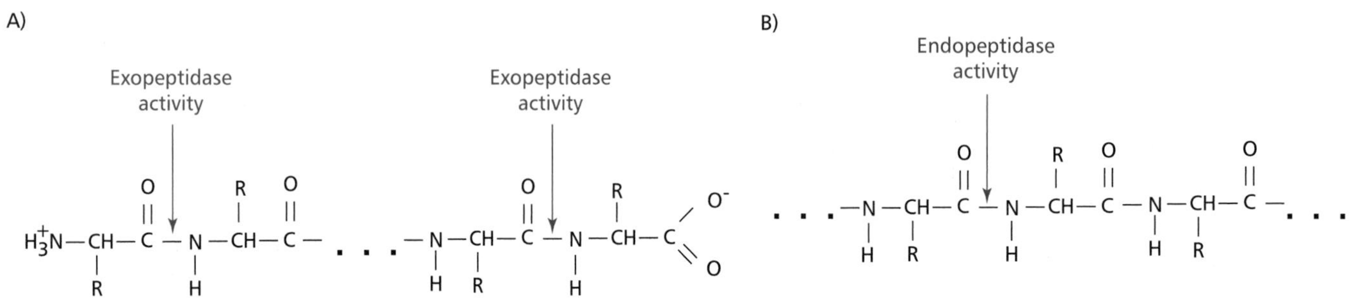

Figure 11.14 The sites of digestion (hydrolysis) of proteins by (A) exopeptidase and (B) endopeptidase activities.

Figure 11.15 Model of the enzymatically active form of a pancreatic lipase molecule. PDB file 1N8S.

Zymogen	Activation mechanism	Active enzyme
Pepsinogen	HCl, pepsin	pepsin
Trypsinogen	enteropeptidase (enterokinase)	trypsin
Proelastase	trypsin	elastase
Chymotrypsinogen	trypsin	chymotrypsin
Procarboxypeptidase	trypsin	carboxypeptidase
Proaminopeptidase	trypsin	aminopeptidase

Table 11.2 Zymogens

Location	Enzyme	Reaction(s) catalyzed
Mouth		
Salivary glands	α amylase	starch → maltose
Lingual glands	lingual lipase	triacylglycerols → diacylglycerols + fatty acids
Stomach		
Chief cells	pepsin	proteins → peptides
	gastric lipase	triacylglycerol → diacylglycerol + fatty acid
Small intestine		
Pancreas	amylase	starch → maltose
	elastase	proteins → peptides
	carboxypeptidase	proteins/peptides$_n$ → proteins/peptides$_{n-1}$ + carboxy terminal amino acid
	trypsin	proteins → peptides
	chymotrypsin	proteins → peptides
	peptidase	peptides → peptides and amino acids
	nuclease	nucleic acids → nucleotides
	lipase	triacylglycerols → monoacylglycerols + fatty acids
	esterase	monoacylglycerols → glycerol + fatty acid
	phospholipase	phosphoacylglycerols → monoacylphospholipids + fatty acids
Intestinal mucosa and brush border	aminopeptidase	proteins/peptides$_n$ → proteins/peptides$_{n-1}$ + amino terminal amino acid
	dipeptidase	dipeptides → 2 amino acids
	maltase	maltose → 2 glucose
	lactase	lactose → glucose + galactose
	trehalase	trehalose → 2 glucose
	sucrase-isomaltase	sucrose → glucose + fructose
		maltose → 2 glucose
	phosphatase	nucleotides → nucleosides + P_i

Table 11.3 Overview of some of the digestive activities of the GIT

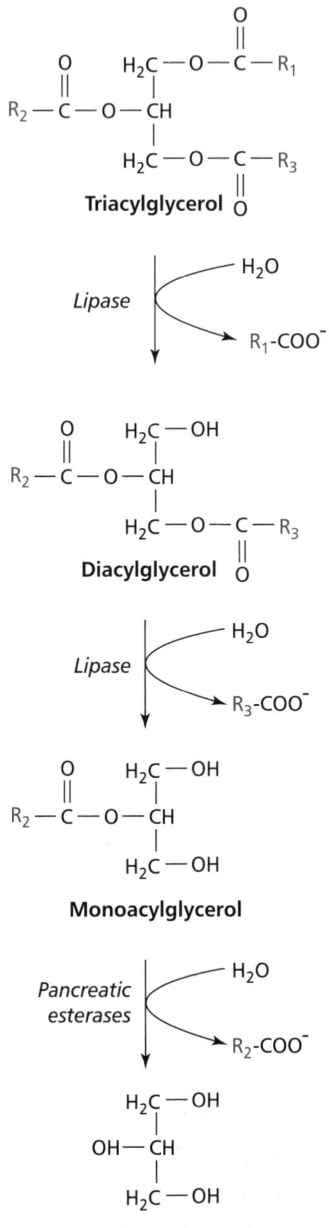

Triacylglycerol: $R_2{-}C(=O){-}O{-}CH(CH_2{-}O{-}C(=O){-}R_1)(CH_2{-}O{-}C(=O){-}R_3)$

↓ Lipase (H_2O in; R_1-COO^- out)

Diacylglycerol: $R_2{-}C(=O){-}O{-}CH(CH_2{-}OH)(CH_2{-}O{-}C(=O){-}R_3)$

↓ Lipase (H_2O in; R_3-COO^- out)

Monoacylglycerol: $R_2{-}C(=O){-}O{-}CH(CH_2{-}OH)(CH_2{-}OH)$

↓ Pancreatic esterases (H_2O in; R_2-COO^- out)

Glycerol: $HO{-}CH(CH_2{-}OH)(CH_2{-}OH)$

Figure 11.16 Overview of the digestion (hydrolysis) of triacylglycerols.

11.4 ABSORPTION OF THE PRODUCTS OF DIGESTION

The large surface area of the small intestine allows the rapid absorption of the products of digestion. The enzymes concerned with the final stages of digestion of a number of nutrients are located in the brush border of the enterocytes as described in *Section 11.3* or even within their cytoplasm. This ensures that the final products of digestion are produced near or within the absorptive surface of the GIT. Enterocytes are joined together by tight junctions that ensure material cannot leak from the lumen. Absorption by enterocytes is largely active and selective and they have a high metabolic rate because the transport of materials across their membranes requires considerable amounts of metabolic energy. A membrane-bound Na^+/K^+-ATPase uses a major proportion of this energy to catalyze the hydrolysis of ATP in the presence of Na^+ and K^+. The free energy from the hydrolysis is used to expel three Na^+ from the cell and to pump two K^+ into the cell. This may be summarized as:

$$3Na^+_{IN} + 2K^+_{OUT} + ATP + H_2O \rightarrow 3Na^+_{OUT} + 2K^+_{IN} + ADP + P_i$$

Since both Na^+ and K^+ are being transported against their electrochemical gradients both movements are an active transport. Also, since more positive charges are being pumped out of the cell than are entering it, the effect contributes to the potential difference across the membrane, called the resting membrane potential, of about –60 mV, the inside of the cell being negative with respect to the outside. The membrane potential of the apical membrane of the enterocyte can be harnessed to facilitate the uptake of a variety of nutrients from the lumen of the GIT. Both amino acids and monosaccharides are transported across the enterocyte luminal membrane in a Na^+-dependent fashion. A variety of different membrane transporter proteins are responsible for the absorption of specific sugars and different groups of amino acids.

Glucose and galactose are transported across the enterocyte luminal membrane in an active, Na^+-dependent fashion by the same transporter. One molecule of these sugars can only move through the transporter into the cell if Na^+ ions move in at the same time (*Figure 11.17*). The concentrations of the sugars can build up within the cytoplasm, such that they are able to leave the cell through the basolateral membrane by facilitated diffusion. Other monosaccharides, for example fructose and trehalose, are absorbed only by facilitated diffusion and are absorbed to a much lesser extent.

The major initial products of protein digestion are small peptides and these are absorbed by enterocytes of the jejunum at their luminal surfaces by peptide transporter protein, called the PepT1 (*Figure 11.18*). This occurs in a H^+-dependent fashion that resembles the uptake of glucose and galactose. Within the cytoplasm, the peptides are hydrolyzed to amino acids, ensuring a continuous sink is present to facilitate peptide uptake by the cells. The exit of the amino acids from the cells on the basolateral side also occurs down their concentration gradients. However, as peptides are moved further along the GIT, they are hydrolyzed by peptidases to free amino acids and their absorption occurs in the ileum using a number of Na^+-dependent transporters (*Figure 11.18*), which have specificities for different amino acid side chains. Peptides

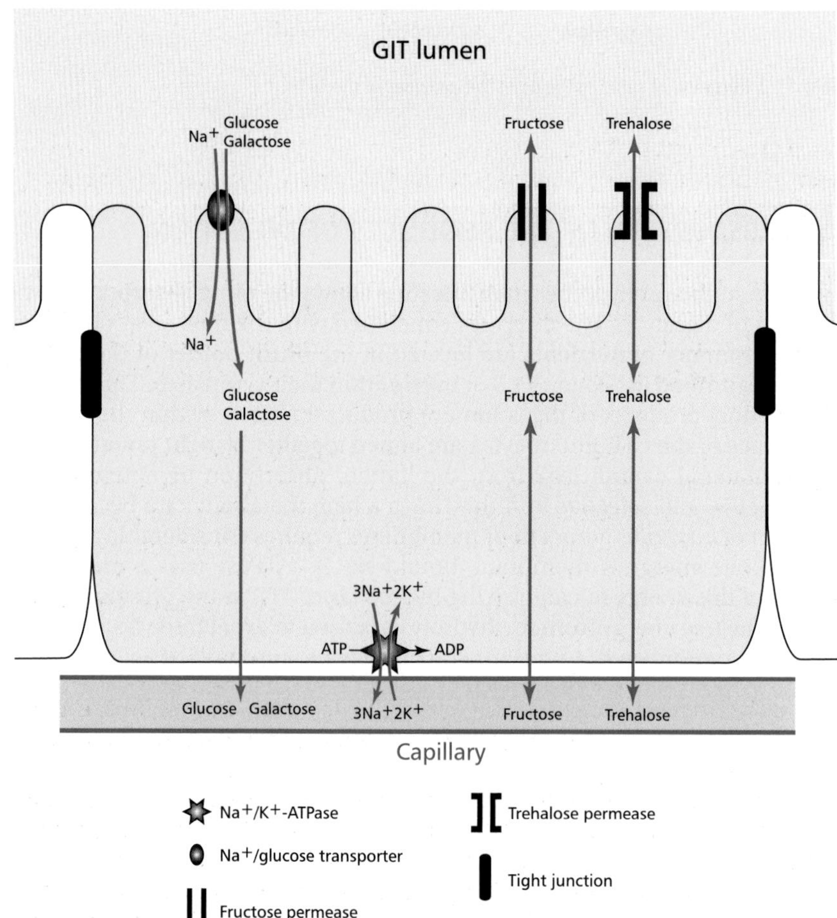

Figure 11.17 Overview of the absorption of monosaccharides by an enterocyte. Transport movements are denoted by colored lines. See text for general details.

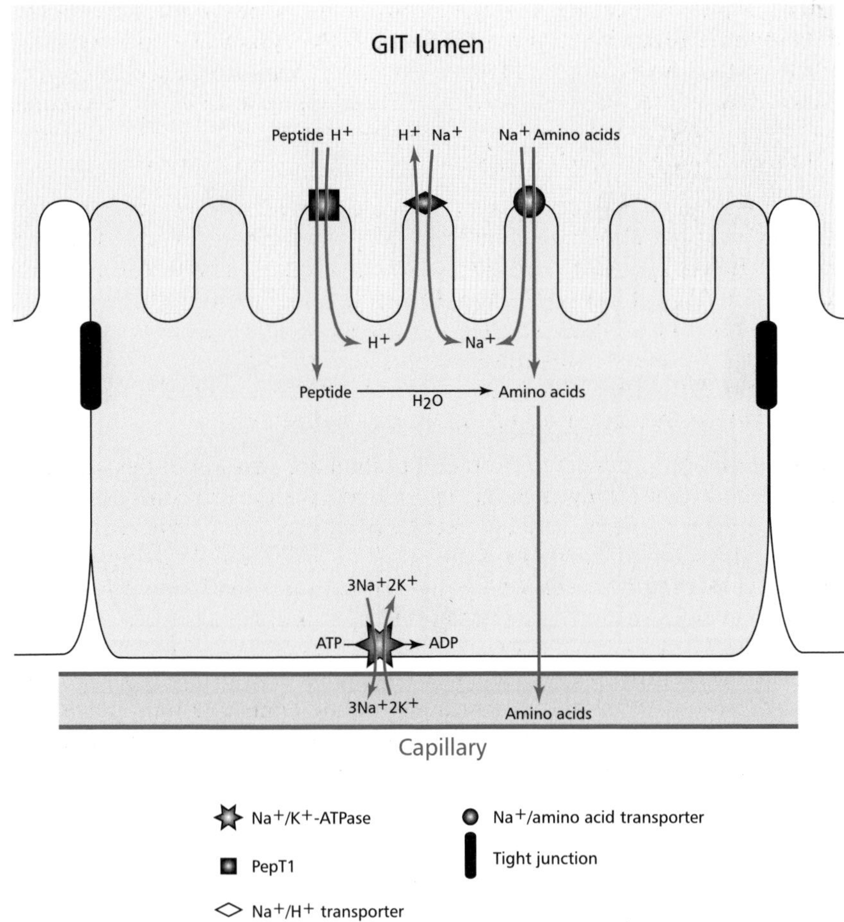

Figure 11.18 Overview of the absorption of amino acids and peptides by an enterocyte. Transport movements are denoted by colored lines, chemical transformations in black. See text for general details.

can also be absorbed by a paracellular route where they pass between enterocytes, rather than being absorbed across the luminal surface. Relatively large peptides can be absorbed by this method and may initiate an allergic reaction leading to food allergies (*Chapter 5* and *10*).

Digestion of RNA produces nucleotides that are further degraded to nucleosides at the brush border and which, again, are absorbed in a Na^+-dependent manner. Catabolism within the cytoplasm converts the nucleotides to ribose phosphate and bases. Eventually the purine bases are converted to urate and the pyrimidines to uracil as shown in *Figure 11.19*.

Fatty acids, monoacylglycerols, monoacylphospholipids and cholesterol are absorbed as mixed micelles by the brush border of the enterocytes (*Figure 11.20*). The triacylglycerols and phospholipids are reformed within the enterocyte cytoplasm and packaged into large lipoprotein complexes called **chylomicrons** (*Chapter 14*) that are transported from the GIT in lacteals of the lymphatic system. This ensures the lipids bypass the liver and are delivered to the blood through the thoracic duct.

Water-soluble vitamins are taken up by enterocytes by a variety of mechanisms. Vitamins B_1 (thiamin) and B_2 (riboflavin) are absorbed in the upper portion of the small intestine. Thiamin is actively transferred to the portal system. Specific transporter proteins actively accumulate niacin (nicotinic acid and nicotinamide), folic acid and biotin (vitamin H) in Na^+-dependent fashions. Pantothenic acid and the vitamers of vitamin B_6 are absorbed by diffusion. Vitamin C is absorbed in the jejunum by a Na^+-dependent mechanism, similar to that described for glucose. The fat-soluble vitamins, A, D, E and K, are absorbed within the mixed micelles of fatty acids, monoacylglyc-

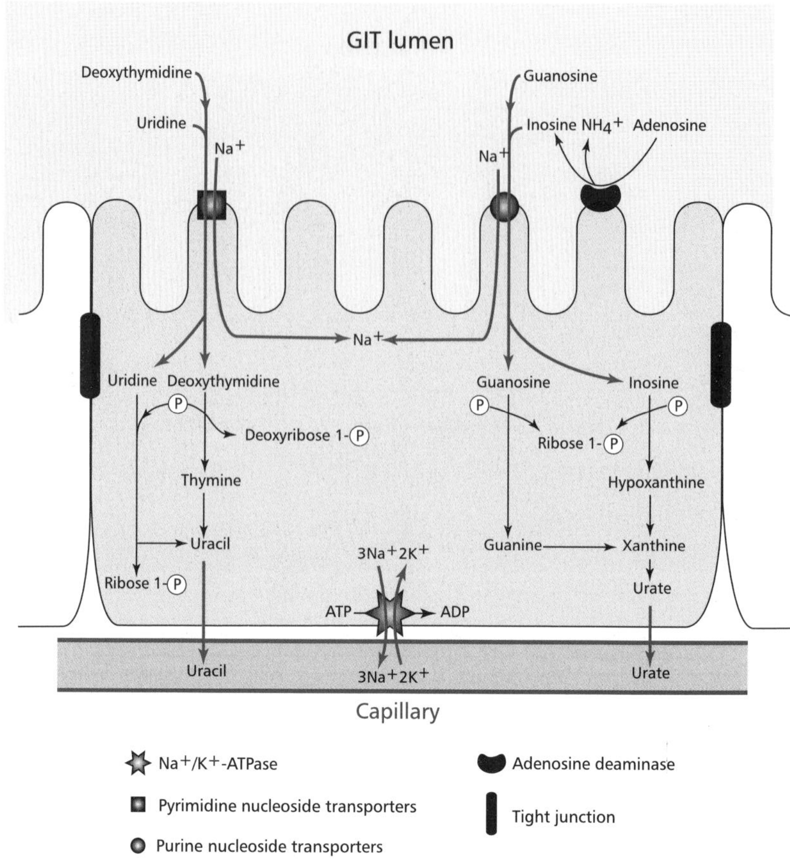

Figure 11.19 Overview of the absorption of the products of nucleic acid digestion by an enterocyte. Transport movements are denoted by colored lines, chemical transformations in black. See text for general details.

erols, monoacylphospholipids and cholesterol described above (*Figure 11.20*) and leave the enterocyte in the chylomicrons. For these reasons, a deficiency in dietary lipids means that the absorption of fat-soluble vitamins is greatly reduced.

Many minerals are absorbed in an energy dependent fashion along the length of the GIT, although Ca^{2+} and iron are mainly absorbed in the duode-

BOX 11.3 Hartnup's disease

Hartnup's disease, also known as Hartnup disorder, Hartnup aminoaciduria or Hartnup syndrome, was first named by Baron and coworkers in 1956 from a disorder that affected the Hartnup family of London. Hartnup disease is inherited as an autosomal recessive trait (*Chapter 15*), particularly where consanguinity is common. It arises from mutations to *SLC6A19*, a gene located on chromosome 5, whose product, a sodium-dependent neutral amino acid transporter, is expressed mainly in the GIT and kidneys. The defective gene product impairs the absorption of tryptophan and other neutral amino acids, such as valine, phenylalanine, leucine, isoleucine, across the brush border membranes of the small intestine and renal tubular epithelium. Two tissue-specific forms have been described; one affects both the GIT and kidneys, the other only the kidneys. The abnormality in amino acid transport can lead to deficiencies in neutral amino acids; the defective absorption of tryptophan may result in a niacin deficiency (*Chapter 10*). The condition clinically resembles pellagra and may be misdiagnosed as a dietary deficiency of niacin. Tryptophan is retained within the GIT lumen and converted by bacteria to toxic indole compounds. Tubular renal transport is also defective and contributes to gross aminoaciduria. Hartnup's disease has an overall prevalence of 1 in 18 000 to 42 000 and, although rare, this makes it among the commonest of amino acid disorders.

Hartnup's disease usually begins at three to nine years of age but it may present as early as 10 days after birth. Most patients are asymptomatic but poor nutrition leads to more frequent and severe attacks. Patients present with pellagra-like light-sensitive rash, aminoaciduria, cerebellar ataxia, emotional instability, neurological and psychiatric symptoms that may considerably diminish their quality of life. Mental retardation and short stature have

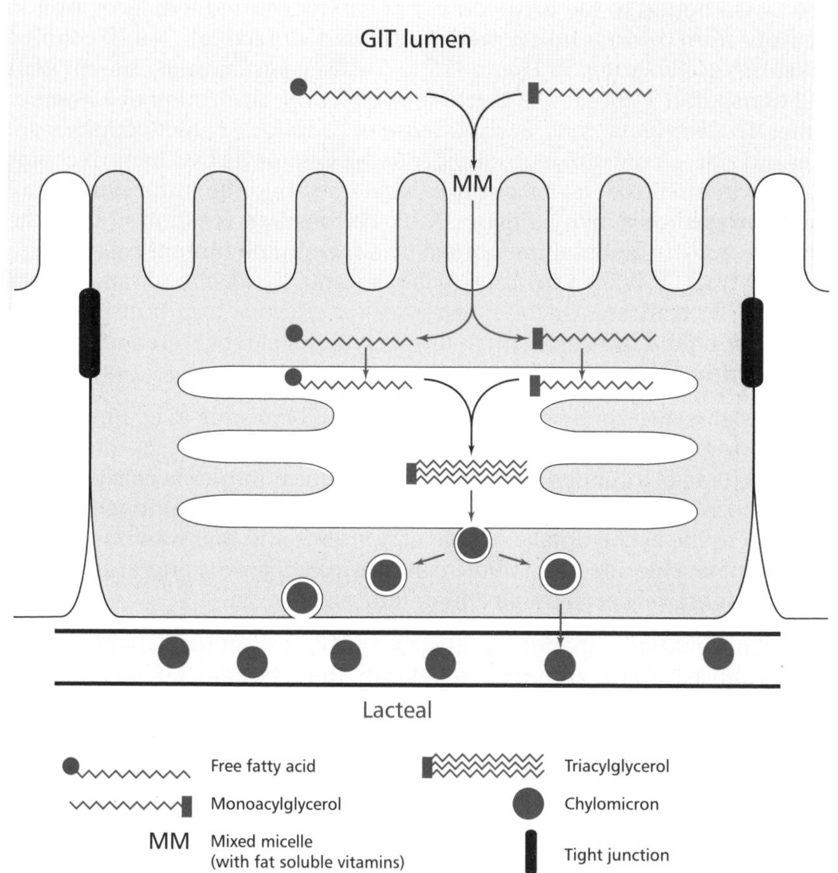

Figure 11.20 The absorption of the products of triacylglycerol digestion and their reformation in an enterocyte. Transport movements are denoted by colored lines, chemical transformations in black. See text for general details.

num. Dietary iron occurs in nonheme and heme forms and their absorption by duodenal enterocytes is by different mechanisms. Dietary nonheme iron occurs in ferrous ($Fe(II)^{2+}$) and ferric ($Fe(III)^{3+}$) forms. Ferrous iron is soluble up to a pH of about 7 but the predominant form, ferric iron, is sparingly soluble above pH 3 and is not available for absorption and must be reduced before it can be transported across the intestinal epithelium. A ferrireductase

been described in some patients. Malnutrition and a low-protein diet are the primary contributing factors to morbidity. In rare cases, severe central nervous system (CNS) damage may lead to death.

Plasma concentrations of amino acids are usually normal in Hartnup's disease. However, it may be diagnosed by demonstrating increased concentrations of neutral amino acids, such as tryptophan, glutamine, valine, phenylalanine, leucine and isoleucine in the urine, with normal levels of proline, hydroxyproline and arginine, which differentiates Hartnup's disease from other causes of gross aminoaciduria. Other tryptophan metabolites, including kynurenine and serotonin, are also found in the urine and neutral amino acids occur in the feces. Urinary indole derivatives, for example 5-hydroxyindoleacetic acid, may be demonstrated following an oral tryptophan load (*Chapter 10*). In selected patients, diagnosis may require the transport defect be identified *in vitro* using a jejunal sample obtained by biopsy.

Most patients can be treated with a high-protein diet which overcomes the deficient transport of neutral amino acids because dipeptides can be actively absorbed as shown in *Figure 11.18* and by paracellular routes. Indeed, in the developed world, cases presenting with extreme clinical symptoms are rare probably because of the protein rich diet available. Patients who are symptomatic are advised to avoid excessive exposure to sunlight and use sunscreens with a skin protection factor of at least 15. Dietary supplements of 40 to 200 mg niacin daily relieve the pellagra-like symptoms. Psychiatric treatment may help patients with severe CNS damage.

called Dcytb bound to the brush border of enterocytes reduces ferric iron to the ferrous form. Ferrous iron is then transported into the cell by a H^+ coupled mechanism as illustrated in *Figure 11.21*. The transporter, called the divalent metal transporter 1 (DMT1), is also able to effect the absorption of a number of other divalent metal ions, such as those of cadmium, cobalt copper, lead, nickel and zinc. Heme iron is absorbed into the enterocyte by a heme receptor and, once internalized, its ferrous iron is released into the intracellular pool by heme oxygenase activity (*Figure 11.21*). Ferrous iron is exported from the enterocyte across the basal membrane by a membrane protein called ferroportin1 or Ireg1. It is then oxidized by hephaestin, a transmembrane copper dependent ferroxidase, which is necessary for effective iron transport. The ferric iron is bound by transferrin in the plasma and can be stored in erythrocytes in ferritin molecules (*Chapter 13*).

Calcium is absorbed in the upper part of the small intestine in an ionic form. This absorption requires the active metabolite of vitamin D, 1,25-dihydroxyvitamin D_3, and is inhibited by substances that form insoluble calcium salts, such as phosphate and oxalate. The uptake of Na^+ has been mentioned already in relation to the active uptake of several nutrients and many anions, hydrogen carbonate, chloride and iodide, can passively follow it into enterocytes. Phosphate is actively accumulated by enterocytes.

Amino acids, monosaccharides, urate and uracil, B vitamins, vitamin C and minerals all leave the enterocytes through their basolateral membranes, enter the hepatic portal vein and are delivered to the liver. Following their absorption, many minerals are bound by intracellular proteins before being expelled through the basolateral membrane into the bloodstream where they

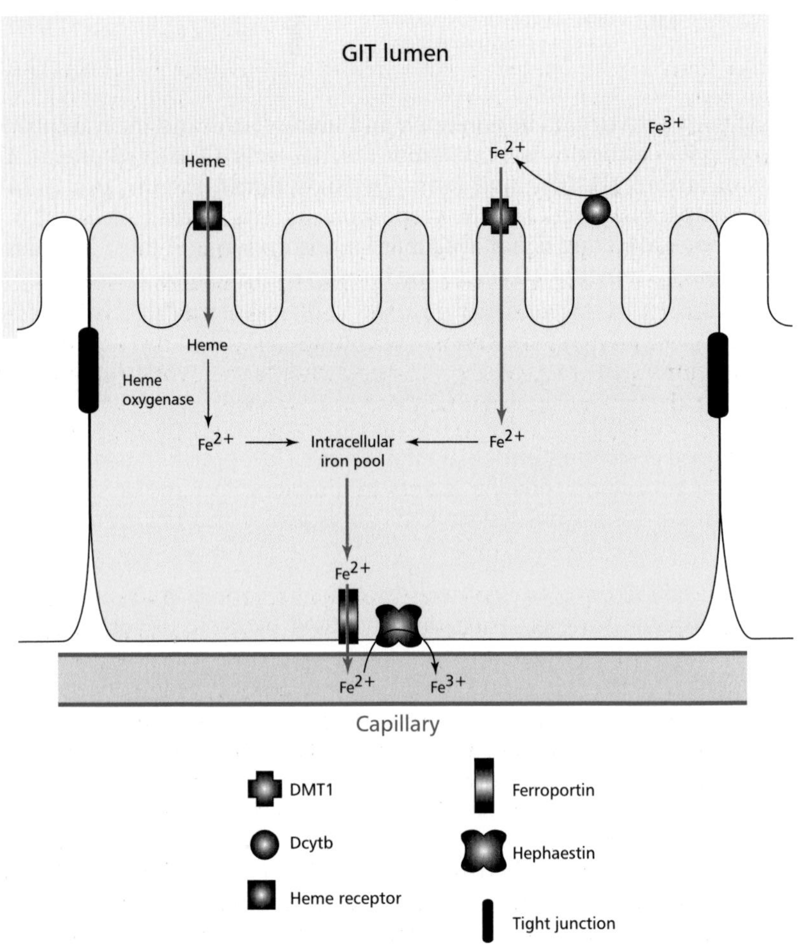

Figure 11.21 Uptake of iron. Transport movements are denoted by colored lines, chemical transformations in black. See text for general details.

are bound by transport proteins, such as transferrin for iron (*Figure 13.4*) and ceruloplasmin for copper. *Table 11.4* summarizes the mechanisms of absorption of the major nutrients.

Approximately 9 dm^3 of fluid pass through the GIT each day. Reabsorption of water from the GIT is essential to prevent dehydration. Most, about 95%, is absorbed by the small intestine, 4% by the large intestine and only 1% is lost from the GIT.

Product of digestion	Uptake is by
Monosaccharides	Na^+-dependent mechanism
Dipeptides	H^+-dependent mechanism
Amino acids	Na^+-dependent mechanism
Monoacylglycerols, monoacylphospholipids, free fatty acids and cholesterol	lipid soluble and absorbed across enterocyte membrane from micelles formed with bile salts
Nucleosides	Na^+-dependent mechanism
Water-soluble vitamins	Na^+-dependent mechanism
Fat-soluble vitamins	lipid soluble and absorbed across enterocyte membrane from micelles formed with bile salts

Table 11.4 Overview of the absorption of nutrients by the small intestine

11.5 ACTIVITIES OF THE LARGE INTESTINE

The large intestine is so named because its diameter is greater than that of the small intestine though it is, in fact, much the shorter of the two. Fluid, containing the unabsorbed products of digestion, directly enters from the small intestine at a junction that is also the site of the vestigial cecum and appendix (*Figure 11.22*). Absorption of Na^+ and water occurs over the surface of the large intestine, which also acts as a reservoir for material resistant to digestion by GIT enzymes. However, bacterial action on this material releases some nutrients from food, for example certain vitamins as well as about 200–2000 cm^3 of gas in 10–14 episodes per day. The final waste together with bacteria forms the feces, which passes to the last section of the GIT, the rectum, and is eliminated through the anus. Two sphincter muscles control elimination: the first of smooth muscle opens involuntarily in response to pressure within the rectum; the second is controlled voluntarily and allows for a conscious decision to defecate.

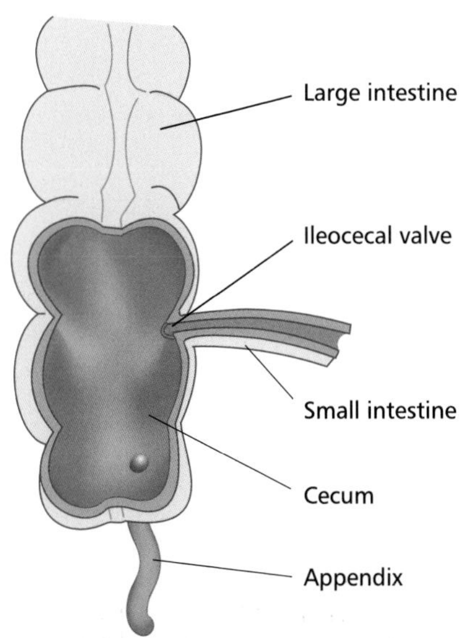

Figure 11.22 The junction of the small and large intestines.

11.6 THE SMALL INTESTINE AND HOMEOSTASIS

Within enterocytes a portion of the monosaccharides absorbed are converted to lactate by glycolysis. Excess nonessential amino acids, especially glutamine, are used to synthesize alanine and ammonia (*Figure 11.23*). These products are then delivered to the liver in the hepatic portal vein. Converting some of the absorbed nutrients to lactate and alanine reduces the metabolic load on the liver because it can easily regenerate pyruvate from them. Pyruvate is a versatile liver metabolite; it is a substrate for the TCA cycle, allowing the formation of ATP during oxidative phosphorylation but it can be used for the biosynthesis of glucose and glycogen, ketone bodies, fatty acids and all but two of the nonessential fatty acids and cholesterol. The GIT is a significant contributor to nutrient homeostasis both during and after nutrient absorption because the formation of lactate and alanine continues even when absorption ceases.

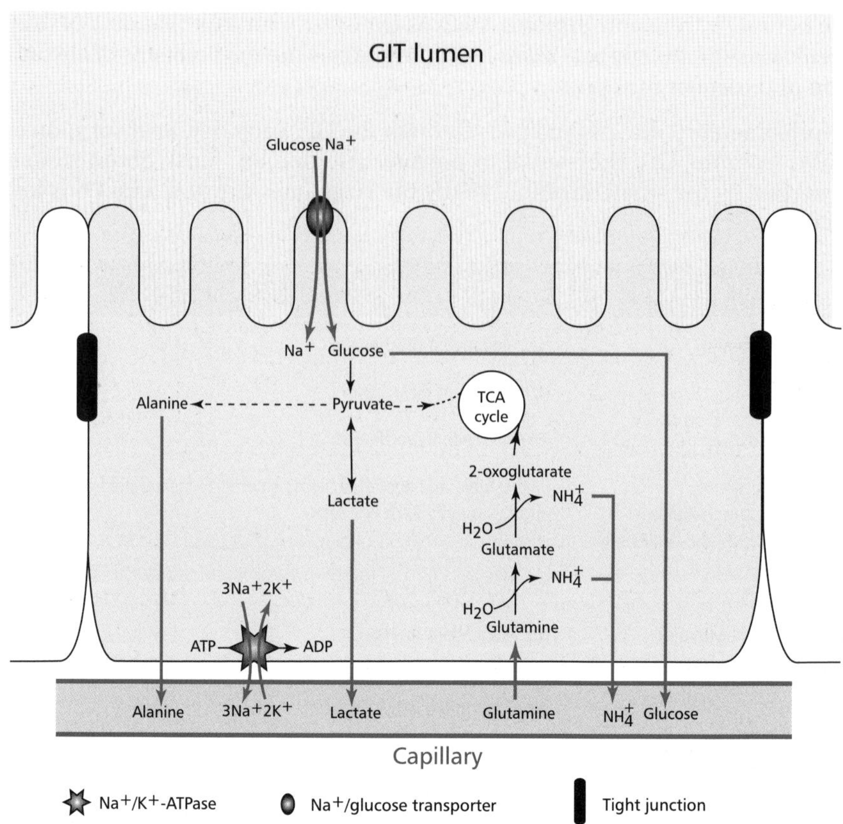

Figure 11.23 The roles of the enterocyte in energy homeostasis. Transport movements are denoted by colored lines, chemical transformations in black. See text for general details.

11.7 HORMONAL CONTROL OF GIT SECRETIONS

The GIT produces a large number of hormones many of whose functions are not well understood, although some of them, together with neuronal activities, are concerned with coordinating the secretions of various digestive juices. Endocrine cells are scattered throughout the entire GIT in clusters forming a diffuse portion of the endocrine system (*Chapter 7*). Over 25 peptides have been extracted and characterized from the GIT. No deficiency states are known for any of these peptides although hormone-secreting tumors have been described.

The G cells in the antral and pyloric regions of the stomach produce gastrin. Gastrin occurs in a number of molecular forms, for example gastrin$_{17}$ and gastrin$_{34}$ are composed of 17 and 34 amino acid residues respectively. Gastrin$_{17}$ is the most active and has a half-life of about 8 min. Its precursor, gastrin$_{34}$, has a half-life of approximately 40 min. The release of gastrin is stimulated by food entering the stomach from the esophagus and its function is, in turn, to stimulate release of gastric juice. Gastric inhibitory peptide (GIP) is a peptide of 43 amino acid residues secreted by the duodenum and upper jejunum. It stimulates insulin release (*Chapter 7*), reduces the secretions of gastrin and pepsin and inhibits gastric movements. The hormone, vasoactive intestinal peptide (VIP) is comprised of 28 amino acid residues. It is released in response to distension of the GIT by food. It stimulates the contraction of smooth muscle tissues of the GIT wall and pancreatic exocrine secretions and it also inhibits gastrin and gastric acid release. Pancreatic polypeptide is formed, as its name implies, by the pancreas and inhibits pancreatic hydrogen carbonate and protein secretions. Secretin is also produced in the duodenum and jejunum. It is a 27-amino acid residue peptide with a half-life of 17 min. Its release is triggered by acid from the stomach and it functions to stimulate the release of pancreatic juice, the hydrogen carbonate of which helps neutralize the acid in chyme. Glucose-dependent insulinotrophic polypeptide (gastric inhibitory polypeptide) is also released in the duodenum and jejunum. It inhibits the secretion of gastric acid and stimulates insulin secretion. Mucosal cells in the upper region of the small intestine secrete cholecystokinin (CCK). Two molecular forms are produced consisting of 33 and 39 amino acid residues respectively. The release of CCK is stimulated by peptides and fatty acids in the food and, in turn, stimulates the release of pancreatic juice and contractions of the gall bladder. Motilin is a 22 amino acid residue peptide that is structurally unrelated to any other GIT hormone produced in the upper small intestine. It controls GIT movements during fasting. The ileum and colon produce peptide YY and neurotensin. The former decreases pancreatic and gastric secretions, while the latter may regulate peristalsis of the ileum. The hormone called substance P is produced along the entire GIT. Its functions include stimulating the secretion of saliva and it is also involved in the vomit reflex.

11.8 DISORDERS OF THE GIT AND ACCESSORY ORGANS

Disorders of the GIT and its accessory organs can affect the mouth, esophagus, stomach, pancreas, liver, bile duct, small and large intestines. Some of the disorders affect the exocrine pancreas, liver, stomach, small and large intestines.

DISORDERS OF THE EXOCRINE PANCREAS

Acute pancreatitis is a severe, rapid inflammation of the pancreas with varying degrees of edema (*Chapter 8*), hemorrhage (*Chapter 13*) and tissue necrosis. It arises because of an inappropriate activation of pancreatic enzymes which then autodigest pancreatic tissue. Normally, these enzymes are inactive until they reach the duodenum. The cause of acute pancreatitis is unclear, although excessive alcohol intake is believed to have a major role, but viral infections, drug reactions and pancreatic cancer have also been implicated. The clinical features of acute pancreatitis include attacks of severe abdominal pain that may extend to the back, vomiting, fever and shock. The leakage of pancreatic enzymes into the bloodstream from damaged tissue and a demonstration of increased plasma amylase activity aids in diagnosis. Typically, there is a five-fold rise in plasma amylase activity over the first two days of an attack that returns to normal within three to five days. Treatment is symptomatic and aimed at maintaining circulation and fluid volume and decreasing pain with analgesics.

Chronic pancreatitis is a slow, progressive destruction of pancreatic tissue accompanied by inflammation and fibrosis. Like acute pancreatitis, the damage is believed to be due to autodigestion of pancreatic tissue by the activation of enzymes *in situ*. Excessive alcohol intake is the leading cause of chronic pancreatitis although many cases are idiopathic. Clinical features of chronic pancreatitis include severe and persistent abdominal pain, weight loss, malabsorption and hyperglycemia. Complications of chronic pancreatitis include diabetes mellitus (*Chapter 7*). Chronic pancreatitis is investigated using X-ray examination of the pancreas to reveal calcification that may result from the release of free fatty acids following the breakdown of fats and by examining the feces to detect steatorrhea. Treatment is directed toward management of pain and rectifying the nutritional disorders that arise from malabsorption (*see below*).

A number of tests are available to assess pancreatic function. The most widely used is the fluorescein dilaurate test which indirectly assesses the activity of pancreatic enzymes. The patient is given an oral dose of the synthetic ester, fluorescein dilaurate, that is hydrolyzed to release fluorescein by pancreatic cholesterol esterase. Fluorescein, but not fluorescein dilaurate, is absorbed by the small intestine and transported to the liver where it is converted to fluorescein glucuronide, which is then excreted in the urine. The latter can be detected by its characteristic fluorescence. The test is controlled for variations in intestinal absorption, hepatic conjugation and renal excretion by repeating the test the next day but using an oral dose of fluorescein. The ratio of fluorescein excreted after administration of fluorescein dilaurate to that excreted after administration of free fluorescein is greater than 0.3 in normal individuals. However, a ratio less than 0.2 is indicative of abnormal pancreatic function. Ratios between 0.2 and 0.3 are inconclusive. Other investigations of pancreatic function include the para-aminobenzoic acid (PABA) test in which the patient is given 0.5 g of the synthetic peptide, benzoyltyrosylaminobenzoic acid (BTPABA). This is hydrolyzed to PABA in the small intestine by pancreatic chymotrypsin (*Figure 11.24*). Following its absorption, the PABA is excreted unchanged in the urine. The patient is also given a known amount of radioactively labeled PABA (^{14}C-PABA) to correct for variations in the absorption, metabolism and excretion of PABA formed from BTPABA. The amount of PABA found in the urine is a measure of the activity of pancreatic chymotrypsin. There is reduced excretion of urinary PABA in individuals with abnormal pancreatic function.

OH

CH$_2$ O

C—NH—CH—C—NH— —COOH

BTPABA

H_2O

OH

CH$_2$ O

C—NH—CH—C—OH + N_2H— —COOH

PABA

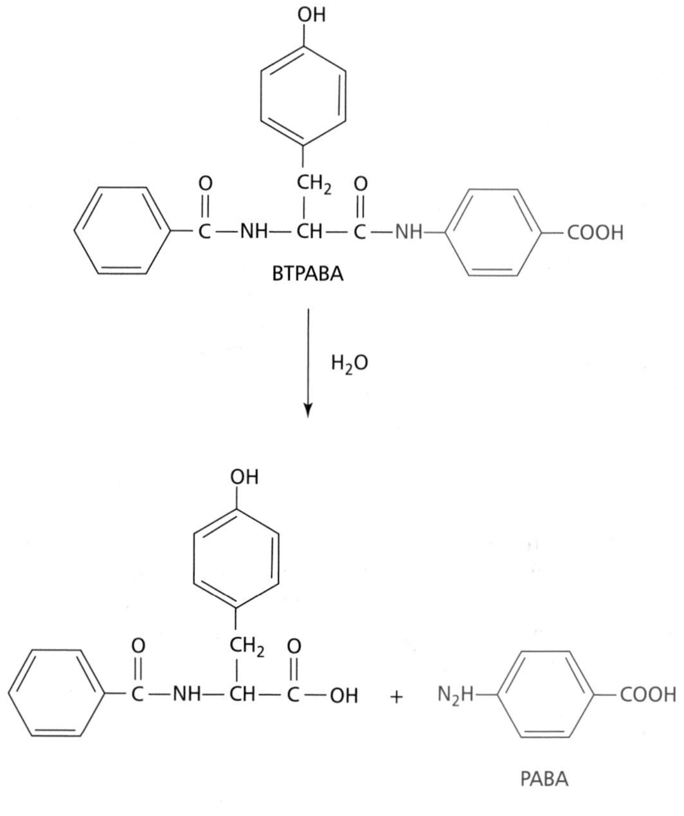

Figure 11.24 The hydrolysis of the synthetic peptide, benzoyltyrosylaminobenzoic acid (BTPABA) by chymotrypsin activity to release para-aminobenzoic acid (PABA).

DISORDERS OF THE LIVER, GALL BLADDER AND BILE DUCT

Jaundice is the yellow discoloration of tissues due to an accumulation of bilirubin (*Figure 11.5*). Many disorders of the liver give rise to jaundice, although clinical jaundice may not be seen until the concentration of bilirubin in the serum is greater than 50 μmol dm^{-3}. The causes of jaundice can be prehepatic, hepatic or posthepatic.

The causes of prehepatic jaundice include hemolysis, where there is an increased breakdown of hemoglobin producing large amounts of bilirubin that overloads the conjugating mechanism. Such bilirubin is mostly unconjugated and commonly occurs in newborn babies. If the concentration of serum bilirubin approaches 200 μmol dm^{-3}, then phototherapy (*Chapter 6* and *Margin Note 11.5*) is used to degrade it, otherwise its high concentration may cause damage to the brain called **kernicterus**. Other causes of prehepatic hyperbilirubinemia include hemolytic disease of the newborn due to Rhesus incompatibility (*Chapter 6*) and ineffective erythropoiesis, which occurs in pernicious anemia (*Chapter 13*). The commonest causes of hepatic hyperbilirubinemia are viral hepatitis and paracetamol (acetaminophen) poisoning (*Chapter 12*). There is also physiological jaundice of the newborn, a mild unconjugated hyperbilirubinemia that develops because of low activity of UDP-glucuronyltransferase, following birth. Activity increases within two weeks and the jaundice disappears. Other causes include Gilbert's and Criggler-Najjar syndromes. In Gilbert's syndrome, the affected individuals have an inherited partial deficiency of hepatic UDP-glucuronyltransferase. Patients present with a mild jaundice and occasionally suffer from abdominal discomfort but otherwise the condition is harmless. Fasting, infection, stress and excessive alcohol intake may aggravate the symptoms. Treatment of Gilbert's syndrome is by administration of phenobarbitone to stimulate glucuronyltransferase activity. Criggler-Najjar syndrome is a rare hereditary disorder characterized by a complete absence of glucuronyltransferase activity from birth. Patients suffer from severe unconjugated hyperbilirubinemia. Treatment using phototherapy in affected newborns may temporarily reduce the unconjugated hyperbilirubinemia but infants generally die within one year of birth.

One of the causes of posthepatic hyperbilirubinemia is **cholestasis** where there is failure of bile to reach the small intestine. Cholesterol is virtually insoluble in water and is maintained in an aqueous environment in vesicles combined with phospholipids and bile salts. In normal conditions, the vesicles maintain the concentration of cholesterol in bile near its saturation point. Cholesterol monohydrate crystals form when the ratio of cholesterol, phospholipids and bile salts exceeds the normal range and results in the formation of gallstones in a process termed **cholelithiasis**. Eighty per cent of gallstones are composed largely of cholesterol; the remaining 20% consist of calcium and bilirubin. They vary in size from that of a grain of sand to the diameter of a golf ball. In many cases, the smaller stones can be excreted in the bile duct without causing harm. Larger gallstones usually cause abdominal pain and are so large that they obstruct the flow of bile into the small intestine. However, in some cases gallstones may exist for years without causing any symptoms. When there is a complete blockage, there is little or no urobilinogen in the feces, which are pale colored due to absence of stercobilinogens. When the blockage is removed, urobilinogen becomes detectable in the urine and the feces regain their normal color. Occasionally intrahepatic obstruction arises where a blockage affects the bile canaliculi in liver cirrhosis (*Chapter 12*) or cancer (*Chapter 17*). This type of blockage causes an increase in the concentration of conjugated bilirubin in the serum.

It is essential to determine whether the cause of the increased amounts of conjugated bilirubin is intra- or extrahepatic because it is of diagnostic significance and determines the subsequent treatment. The degree of obstruction to the flow of bile is usually greater in extrahepatic cholestasis. Extrahepatic cholestasis may benefit from surgery to remove the gall bladder or to remove the gallstone. Nonsurgical treatments are preferred because surgery can be hazardous. Oral dissolution therapy with ursodiol and chenodiol, which are derived from bile salts, is effective in treating small, predominantly cholesterol gallstones. Treatment may be required for months to years before the gallstones are dissolved but is preferred in patients who cannot undergo surgery. In some cases, gallstones may be broken down using ultrasound waves to smaller particles that can easily be excreted.

Margin Note 11.5 Phototherapy for jaundice in newborns 

Phototherapy is the most common way of treating jaundice in newborn children. It involves exposing the baby to blue light (400–500nm). This is absorbed by the bilirubin, degrading it to a form that is more water soluble and can be excreted in the feces and urine. A baby with jaundice may need repeated phototherapy over several days.

BOX 11.4 Pregnancy, cholelithiasis and cholestasis

Cholelithiasis, or the presence of gallstones, in the gall bladder (*Figure 11.25*) has a general incidence one in 1000 but pregnancy induces changes in the composition of bile that increases its frequency up to six in 100. For example, in the second trimester the bile salt pool generally decreases but biliary cholesterol levels may increase producing bile that is more prone to form stones. Additionally, emptying of the gall bladder slows in the second trimester further increasing the risk of cholelithiasis.

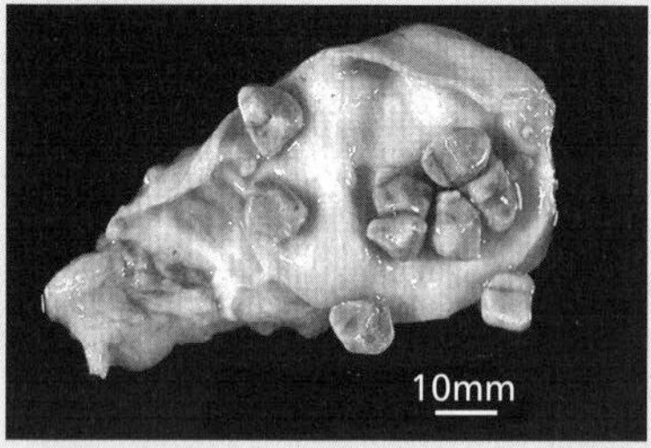

Figure 11.25 Gallstones in an excised gall bladder. Courtesy of Dr H.W. Ray, Southwest Tennessee Community College, USA.

The symptoms of cholelithiasis are similar in pregnant and non-pregnant patients and may present as pain in the middle of the upper abdomen, which can become more severe on eating fatty foods, jaundice and fever. Symptoms usually occur only when the stones block one of the ducts in the biliary system. Cholelithiasis is frequently asymptomatic and often the stones are only discovered by routine X-ray examination, surgery or at autopsy. An ultrasound examination of the liver is helpful in determining the presence of gallstones. Surgical treatment can be safely accomplished in the first or second trimester but should be avoided during the third trimester because of the enlargement of the uterus.

Obstetric cholestasis (OC) or cholestasis of pregnancy is a liver disorder where the flow of bile from the liver in pregnant women is reduced. It has a reported incidence of one in 10 000 pregnancies in the USA. There is some evidence for a genetic link with OC. It also has an ethnic predisposition since it occurs in 0.01 to 0.02% of pregnancies in north America, but 1 to 1.5% in Sweden and 5 to 21% in Chile but is rare in black patients.

Typically OC presents in the third trimester but can occur as early as the thirteenth week. In 80% of patients it presents with severe pruritus (itching), with jaundice developing in the remaining 20% showing typical dark urine and light colored stools. In a normal pregnancy itching is not uncommon and is thought to be related to hormonal changes and stretching of the skin over the stomach as the baby grows. However, in OC the itching generally begins elsewhere, especially on arms, legs, hands and soles of the feet, face, back and breasts. It is usually worse at night, leading to sleeplessness and exhaustion, and can be of such intensity that scratching draws blood. The itching completely disappears within a week or two of the birth and does not cause long-term health problems for the mother but the condition is associated with increased health risks to the fetus. For example, studies in the USA have shown that intrahepatic OC is associated with a 12 to 44% incidence of prematurity, a 16 to 25% incidence of fetal distress and increases the perinatal mortality rate from 1.3 to 3.5%.

Acute hepatitis is caused by infection and subsequent inflammation of the liver, where liver cells are destroyed and the liver becomes necrotic. The commonest cause is viral infections, for example with hepatitis A, B, C, D and E viruses, although drugs, toxins and autoimmune reactions can also lead to acute hepatitis. The initial symptoms of acute viral hepatitis include malaise, anorexia, fever, rashes, abdominal pain, dark urine and jaundice.

Hepatitis A virus causes a mild hepatitis where patients recover usually with no complications. The virus is transmitted by contaminated food or drink, especially where sanitation is poor. Following an incubation period of 15 to 40 days, the patient develops fever, sickness and, shortly afterwards, jaundice. Hepatitis B virus is more serious with a mortality rate of 5–20% although most patients gradually recover. Hepatitis B virus spreads from one person to another via body fluids, such as blood, saliva, semen, vaginal fluids, tears, breast milk and urine. Transmission may occur during sexual activity with an infected person and vertically from an infected mother to the baby. It is commonly present in drug addicts. The symptoms develop suddenly after an incubation period of one to six months and include fever, chills, weakness and jaundice. In contrast to other types of hepatitis, more than 80% of hepatitis C virus (HCV) infections cause chronic liver disease. Approximately 170 million people worldwide may be infected with HCV. This infection is mild in the early stages and is often only diagnosed when it has already caused severe liver damage. For this reason, infection with HCV has been referred to as the 'silent epidemic'. Blood transfusions were the commonest means of

The pathophysiology of intrahepatic cholestasis of pregnancy remains poorly understood and unfortunately no one knows why babies are at risk of being stillborn. One possibility is that the liver cannot cope with the increased amounts of hormones produced during pregnancy, which reduces the flow of bile leading to a build up of bile salts in the blood.

In addition to a complete medical history and physical examination, generalized severe itching without a rash is often the first clue to diagnosis of OC. This can be confirmed by liver function tests (LFTs) and serum bile acid tests; the latter is the most sensitive test. Normally the amounts of bile acids in blood increases before the LFTs can detect any changes. Blood tests to check blood clotting in OC are necessary prior to birth and patients may require extra vitamin K since a lack of the vitamin may decrease the effectiveness of clotting and increase blood loss during the birth (*Chapter 13*). Patients with OC require cardiotocography, that is monitoring the heartbeat of the fetus over a set period of time, ultrasound scans and blood tests. Some pregnant women may be hospitalized to evaluate the progress of the fetus. Close fetal surveillance at delivery is also desirable.

Following a diagnosis of OC, patient care involves giving general support. Specific treatments are determined by the medical history, overall health and tolerance of the patient to specific medications and by the severity of the disease. Resting as much as possible and eating a well-balanced diet that includes large amounts of vegetables, fruit and whole wheat cereals, including bread, may help, as does frequent cold baths, the use of calamine lotion and loose cotton clothing to relieve the itching. Steroids may be used to reduce the levels of bile salts in the blood and relieve the itching. For example, ursodeoxycholic acid at doses of 15 mg kg^{-1} per day helps increase the flow of bile, reduce the level of bile acids in the blood and ameliorate the pruritus and is well tolerated by both mother and fetus. Dexamethasone, another steroid, is sometimes prescribed to increase the maturity of the fetal lungs before delivery and may also help relieve maternal itching. Parenteral vitamin K supplementation is recommended for patients with prolonged cholestasis and when blood clotting factors concentrations are abnormal.

Obstetric cholestasis may also increase the mother's risk of postpartum hemorrhage. If the well being of the mother or the fetus are judged to be at risk, then an early delivery at weeks 37 or 38 may be necessary. There does not appear to be any harmful effects to babies born to mothers with OC. Maternal symptoms usually resolve within two days of delivery and the increases in serum bilirubin and LFTs soon return to normal after delivery. Obstetric cholestasis is not thought to cause any lasting liver damage although it may leave the liver more sensitive to normal changes in the concentration of hormones leading to bouts of mild itching during the menstrual cycle just before ovulation or just prior to the start of a period. A consultant obstetrician familiar with the condition should carefully manage any subsequent pregnancies since the condition recurs in 60 to 70% of cases but it may not follow the same pattern. For example, the itching may be more severe and could begin earlier in the pregnancy.

transmission prior to the testing of blood products for HCV. Infections with hepatitis B and C viruses are associated with liver cancer (*Chapter 17*). The hepatitis D virus occurs only with or after infection with hepatitis B virus and its mode of transmission is identical to that of the B virus. Hepatitis E was initially grouped as a type C virus. It occurs in people who have been to parts of the world where this virus is endemic, such as India. It is transmitted by water contaminated with fecal material.

A clinical history of recent blood transfusions or intravenous drug use may all suggest acute hepatitis. Blood tests based on antigen–antibody reactions are conducted to establish the type of virus causing the hepatitis. Many patients present with proteinuria and bilirubinuria and show increased levels of serum alkaline phosphatase (ALP) activity. A liver biopsy will confirm the initial diagnosis. The HCV is treated with α-interferon (*Chapter 4*), otherwise patients are advised to take plenty of bed rest with adequate food and fluid intakes. A serious complication of many cases of acute hepatitis is the development of chronic hepatitis.

Chronic hepatitis is an inflammation of the liver that persists for more than six months without improvement. Its causes include autoimmune liver damage, chronic infection with hepatitis B virus and excessive drug and alcohol use. Chronic hepatitis can be divided into two histological types, namely, chronic persistent hepatitis, which has a good prognosis, and chronic active hepatitis that may respond to immunosuppressive or antiviral agents but often progresses to cirrhosis, leading to death within five years as a result of liver failure.

Cirrhosis (*Chapter 12*) is a condition where the liver responds to injury or death of some of its cells by producing strands of fibrous tissue between which are nodules of regenerating cells. Patients with cirrhosis may be asymptomatic for a long period of time before vague symptoms such as nausea, vomiting, anorexia, weakness, weight loss and edema of the legs become apparent. Its clinical complications include jaundice, **ascites**, which is an abnormal accumulation of fluid in the abdomen, GIT bleeding and hepatic encephalopathy. Cirrhosis may interfere with intrahepatic circulation causing gradual failure of liver function. Cirrhosis can be divided into three types, namely, alcoholic, postnecrotic and biliary cirrhosis. Alcoholic cirrhosis is discussed in *Chapter 12*.

Postnecrotic cirrhosis accounts for about 25% of all cases of cirrhosis and is associated with viral infections, the use of certain drugs and poisons. About 25% of postnecrotic cirrhosis cases have a prior history of viral hepatitis. Unfortunately 75% of all patients with postnecrotic cirrhosis die within one to five years. Biliary cirrhosis accounts for approximately 15% of all cases of cirrhosis and is characterized by the death of liver cells surrounding bile ducts. It is most commonly caused by an obstruction of the bile duct leading to an accumulation of bile within the liver.

Diagnosis of cirrhosis will involve palpation and X-ray of the abdomen, which often reveal an enlarged liver. A liver biopsy is required to confirm the diagnosis. Other laboratory tests may reveal anemia or hyperbilirubinemia and liver function tests (LFTs) determine increases in the activities of a number of enzymes (*see below*). There are no drugs that can arrest or reverse the fibrotic process in cirrhosis and treatment is aimed at dealing with the underlying cause, for example alcohol abuse or biliary obstruction and by treating any complications.

A number of plasma enzyme activities are used to assess liver function, including those of aspartate transaminase (AST), alanine transaminase (ALT), alkaline phosphatase (ALP) and γ-glutamyltranspeptidase (GGT). Alanine transaminase is present in both the cytosol and mitochondria of hepatocytes whereas ALT is found only in the cytosol. Liver cell damage releases these enzymes increasing their levels in the plasma. Alanine transaminase is specific for the liver whereas AST is also found in pancreatic and skeletal and cardiac muscle tissues. In hepatocellular damage, levels of AST and ALT may increase tenfold but in obstructions of the bile duct or cholestasis, the increases may be relatively slight, usually no more than two to three times their normal levels. Alanine transaminase and AST measurements are useful in monitoring the progress of hepatocellular damage where falling levels suggest an improvement in the disease. Alkaline phosphatase is found on the surface of hepatocytes and in the microvilli of bile ducts but is not specific for liver. Its activity is increased in cholestasis. In hepatocellular disease, ALP levels may be normal or slightly raised. Falling plasma levels of ALP suggest a correction of cholestasis and may be useful for monitoring this defect. Plasma GGT levels are raised in both hepatocellular disease and cholestasis. Although the test for this enzyme is sensitive, it is not specific for liver disease as its activity is increased by some drug therapies and by alcohol. The blood protein albumin is synthesized in the liver and its concentration in plasma reflects the functional capacity of the liver. Plasma albumin concentration is low in chronic liver disease but tends to be normal in the early stages of acute hepatitis.

DISORDERS OF THE STOMACH

Gastritis is the most common disorder affecting the stomach and is characterized by inflammation and erosion of the gastric mucosa. Gastritis is idiopathic in many cases but it can be caused by irritating foods, beverages, ingested poisons, aspirin and staphylococcal exotoxin (*Chapter 2*). Gastritis may present acutely where the patient suffers from GIT bleeding, epigastric pain, that is pain on or over the stomach area, anorexia and hematemesis or vomiting of blood. Patients with chronic gastritis may have no symptoms except for epigastric pain. The possibility of exposure to irritating substances must be determined when assessing the patient's clinical history. Gastroscopy, in which a tube with a camera on its end is passed into the stomach allowing a direct visualization of its wall, can be used to confirm the diagnosis by revealing inflamed portions of the lining of the stomach. Relief from the symptoms of gastritis occurs following removal of the irritant substance or treatment of the underlying cause(s).

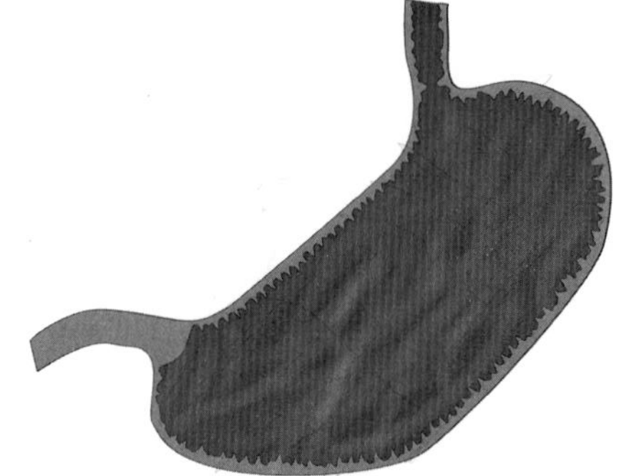

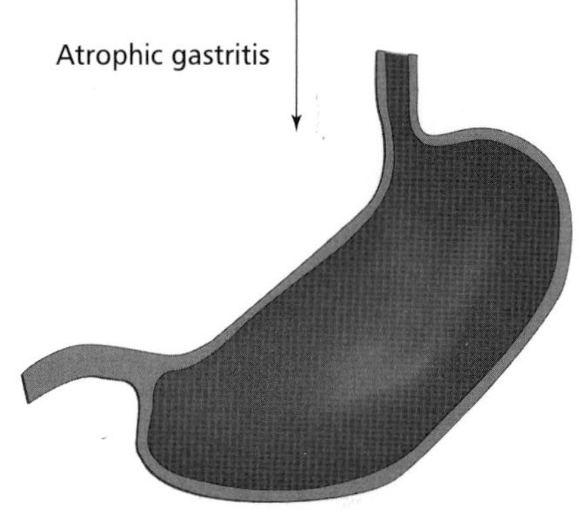

Figure 11.26 Schematic showing the effects of atrophic gastritis on the stomach lining.

Atrophic gastritis is a degenerative stomach disorder characterized by chronic inflammation of the stomach with atrophy of its mucous membrane lining (*Figure 11.26*). This results in loss of gastric glandular cells and their eventual replacement by nonsecretory and fibrous tissues. Secretions of hydrochloric acid, pepsin and intrinsic factor are impaired, leading to digestive problems, vitamin B_{12} deficiency and megaloblastic anemia. Atrophic gastritis is the result of long-term damage to the gastric mucosa and is usually detected late in life. It can be caused by persistent infection with the bacterium *Helicobacter pylori* but it can also have an autoimmune origin.

Helicobacter pylori is able to bind to the stomach lining where the bacteria release urease, which hydrolyzes urea, releasing ammonia that neutralizes the stomach acid. This allows the bacterium to penetrate into the mucosal layer. The release of bacterial and inflammatory toxic products by *Helicobacter pylori* over time results in increasing gastric mucosal atrophy. Some glandular units develop an intestinal-type epithelium; others are simply replaced by fibrous tissue. The loss of gastric mucosa decreases the amount of acid secretion that increases the gastric pH and leads to a reduced ability to kill bacteria. Ingested bacteria can survive and reside in the stomach and the upper part of the small intestine. Infection is usually acquired during childhood and, if left untreated, progresses over the lifespan of the individual in one of two main ways that have different pathological consequences. The first is a gastritis that mainly affects the antrum of the stomach (*Figure 11.9*). This is the most frequently observed pattern in Western countries and individuals with peptic ulcers (*see below*) usually develop this pattern of gastritis. The second pattern is a more widespread atrophic gastritis affecting, for example, the corpus, fundus and antrum with the loss of gastric glands and their partial replacement by an intestinal-type epithelium. This pattern is observed more often in developing countries and Asian individuals who develop gastric carcinoma and gastric ulcers usually present with this pattern of gastritis.

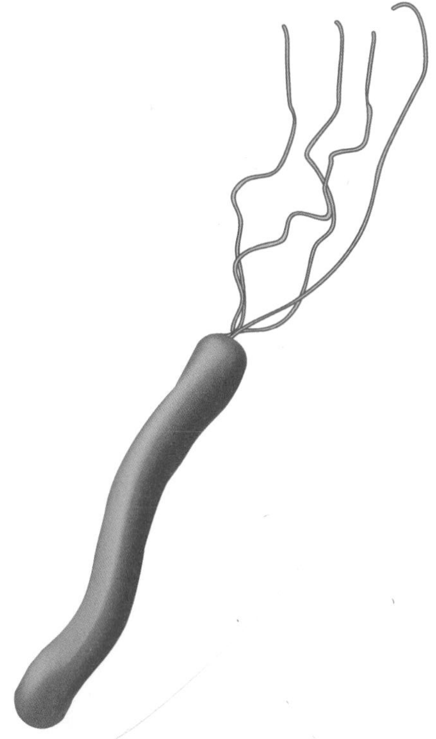

Figure 11.27 Schematic of the bacterium *Helicobacter pylori* (length 2.5–3.5 μm) based on electron micrographs of several specimens.

Autoimmune gastritis is associated with serum anti-intrinsic factor antibodies that reduce the amount of functioning intrinsic factor. This, in turn, decreases the availability of vitamin B_{12} and eventually leads to pernicious anemia (*Chapter 13*) in some patients. Cell-mediated immunity also contributes to the disease because T cell lymphocytes infiltrate the gastric mucosa and contribute to the epithelial cell destruction and resulting gastric atrophy.

Specific data on the incidence of atrophic gastritis are scarse. However, its prevalence mimics that of its two main causes. In both types, atrophic gastritis develops over many years and is detected later in life. *Helicobacter pylori* (*Figure 11.27*) infects approximately 20% of people younger than 40 years and 50% of those older than 60 years in the developed world. Infection is highly prevalent in Asia and in developing countries and it is estimated that 50% of the world's population is infected. Thus chronic gastritis is probably extremely common. In contrast, autoimmune gastritis is a relatively rare condition, which is most frequently observed in patients of northern European descent and in African Americans. The prevalence of pernicious anemia resulting from autoimmune gastritis is estimated to be 127 in 100 000 in the UK.

Chronic gastritis frequently is asymptomatic but can present as nonspecific abdominal pain. Since gastritis often occurs in severely ill, hospitalized people, its symptoms may be eclipsed by other, more severe symptoms.

Atrophic gastritis cannot be reliably diagnosed by gastroscopy but requires a microscopic examination of biopsy specimens. *Helicobacter pylori* infections are normally diagnosed using serological tests, breath tests or antigen tests of the feces. Pernicious anemia resulting from autoimmune atrophic gastritis usually presents in patients approximately 60 years of age.

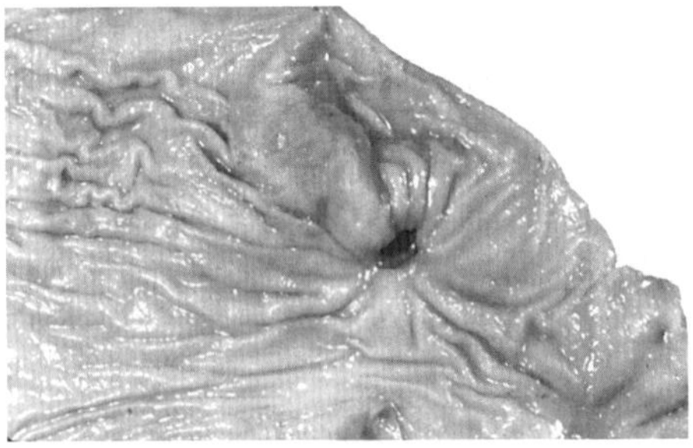

Treatment of atrophic gastritis is directed at eliminating the causative agent, to correct complications of the disease and attempt to revert the atrophic process. When *Helicobacter pylori* is the causative agent, it can be eradicated using a combination of antimicrobial agents and antisecretory agents with a success rate of about 90%. Lack of patient compliance and antimicrobial resistance are the most important factors influencing poor outcome. However, treatment of *Helicobacter pylori* infection may not lead to a reversal of existing damage unless started early but may block further progression of the disease. Some evidence suggests that β-carotene and/or vitamins C and E may reverse or reduce the risk of atrophic gastritis and/or gastric cancer. The major complication in patients with autoimmune atrophic gastritis is the development of pernicious anemia. This requires vitamin B_{12} replacement therapy.

Figure 11.28 Picture of a gastric ulcer. Courtesy of Dr A.S. Mills, Virginia Commonwealth University, USA.

Ulcers are perforations of the GIT wall (*Figure 11.28*), particularly erosions of the mucosal layer related to cancer, that is, malignant ulcers, or to stomach acid, that is, peptic ulcers. Ulcers may also be named from their location, for example esophageal, gastric or stomach and duodenal ulcers. Esophageal ulcers are usually associated with hiatus hernias (*see below*) caused by acid splashing from the stomach into the lower esophagus. Gastric ulcers are relatively rare because the mucosal lining of the stomach is protected from the acid by a layer of alkaline mucus. They generally occur in patients older than 50 years of age. Duodenal ulcers are five times more common than gastric ulcers and generally occur in a younger population. More than 90% of ulcers occur in the duodenal wall, usually after it has been weakened by infection with *Helicobacter pylori*. It used to be thought that ulcers were caused by stress and excessive accumulation of HCl. However, it is now accepted that their commonest cause is infection with *Helicobacter pylori* (*Figure 11.27*) which can colonize and destroy the mucosal layer.

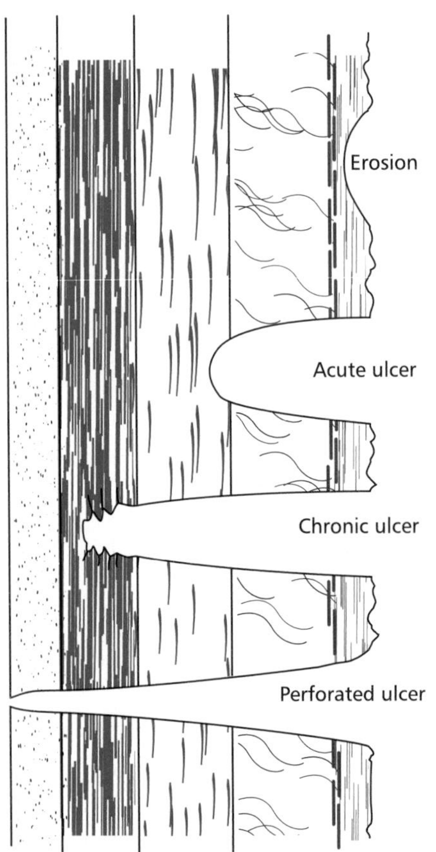

Figure 11.29 Schematic to show the development of an ulcer in the stomach wall.

Peptic ulcers are linked to an increased production of acid and pepsin in gastric juice or to a reduced protection of the mucosa against gastric juice. *Figure 11.29* illustrates diagrammatically the development of a peptic ulcer. Lesions that do not extend through the mucosal lining are referred to as erosions. Acute and chronic ulcers penetrate this layer and, in serious cases, may penetrate the stomach wall. In some patients, blood vessels in the GIT wall ulcerate and lead to heavy, and in some cases fatal, bleeding. Chronic ulcers have an associated basal scarring.

Patients with peptic ulcers present with epigastric pain but their diagnosis is made on clinical grounds, supported by endoscopy, laboratory tests for assessing acid and pepsin secretion and identification of *Helicobacter pylori* infection. Treatment is aimed at eradication of the *Helicobacter pylori* infection and reducing acid output. Antibiotics (*Chapter 3*) that effectively suppress symptoms include amoxycillin, clarithromycin, metronidazole and tetracycline, and they often cure the patient. Bismuth chelate and sucralphate may also be administered to decrease the synthesis of prostaglandins that stimulate inflammation. The resulting decrease in acid production by parietal cells and the increase in hydrogen carbonate production by mucus secreting epithelial cells have cytoprotective effects.

Zollinger-Ellison syndrome is a rare disorder that causes massive, multiple and recurrent peptic ulcers due to the excessive secretion of gastric juice from tumors affecting the pancreas or duodenum. Approximately 60% of the tumors are malignant. They are called gastrinomas because they secrete large amounts of gastrin, hence patients have an increased plasma gastrin concentration and rates of gastric acid secretion greater than 100 compared with normal rates of less than 5 mmol h^{-1}.

A diagnosis of Zollinger-Ellison syndrome usually requires demonstrating an increase in the concentration of gastrin in the patient's serum, combined with an increased release of acid in the stomach. However, in about 30% of cases the plasma gastrin concentration is normal or only slightly above normal. The pentagastrin test is used to assess the acid output of the stomach. Pentagastrin is an analog of gastrin that stimulates the release of stomach acid. Acid output is assessed before and after intramuscular injection of pentagastrin. Patients with Zollinger-Ellison syndrome have a high basal acid output and pentagastrin causes little further increase. Treatment of Zollinger-Ellison syndrome is by surgical removal of the gastrinoma.

A **hernia** is the protrusion of an organ or tissue out of the body cavity in which it is normally found. A hiatus hernia occurs when the upper part of the stomach is dislocated through the hole, called a hiatus, in the diaphragm, into the chest. Sliding hiatus hernias occur when the esophagus and stomach both move upwards so that the top end of the stomach protrudes through the gap in the diaphragm normally occupied by the esophagus (*Figure 11.30* (*A*)) and these constitute 90% of cases. The remaining 10% are rolling hiatus hernias where a portion of the stomach curls upwards adjacent to the esophagus so that both it and an upper part of the stomach protrude through the gap (*Figure 11.30* (*B*)). The causes of hiatus hernias are unknown but they may be due to intra-abdominal pressure or weakening of the gastroesophageal junction caused by trauma or loss of muscle tone. Over 50% of individuals with hiatus hernia are asymptomatic, but when symptoms do occur, they include heartburn, which is aggravated by reclining, chest pain, dysphagia, belching, pain on swallowing hot fluids and a feeling of food sticking in the esophagus. Although hiatus hernia is not usually serious, it can cause inflammation of the lower end of the esophagus leading to a back flow of gastric juices; this is called **reflux esophagitis**, and it may cause bleeding (perhaps anemia) or a stricture. Cancer in a hiatus hernia is very rare, but there is a slight increased risk of it developing in the inflamed area.

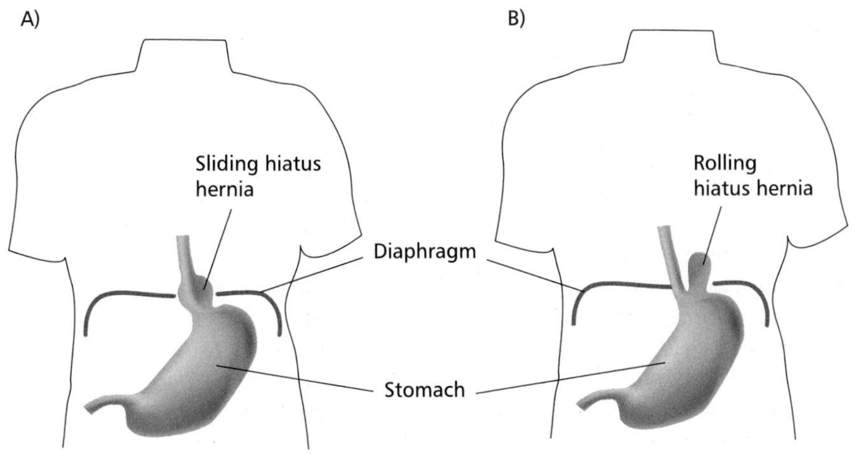

Figure 11.30 Schematic showing (A) sliding and (B) rolling hiatus hernias.

Data on the incidence of hiatus hernia are few but the condition increases with age and is particularly common in overweight middle-aged women and can also occur during pregnancy. The contents of the GIT are often not clearly visible by X-rays and diagnosis requires confirmation with a barium meal. This consists of barium sulfate mixed with liquid and is usually flavored. The barium in the meal lines the inside of the GIT wall and is visible because barium is opaque to X-rays making this a useful method for detecting structural abnormalities of the GIT. The presence of a hiatus hernia can also be investigated by gastroscopy.

The aim of treatment is to alleviate the symptoms. Losing weight, reducing smoking and coffee and alcohol intakes all help to relieve symptoms. The patient may be advised to avoid tight or restrictive clothing. Avoiding food intake before sleep and elevating the head of the bed help in reducing acid reflux. Medication such as antacids may be prescribed. Surgery is only used when there is strangulation of the hernia or the symptoms cannot be controlled.

DISORDERS OF THE SMALL INTESTINE

Lactose intolerance is a condition arising from an inability to express lactase. It is divided into three categories: congenital alactasia, primary acquired and secondary acquired lactose intolerance. Congenital alactasia or hypolactasia is an extremely rare condition and affected babies do not gain weight, are dehydrated and extremely unwell. Human milk is unsuitable for the baby and breastfeeding is precluded, which can also cause emotional distress in some mothers. These babies must be fed dairy-based but lactose-free or lactose-free soya formulae to survive. Primary acquired lactose intolerance usually occurs following weaning and before the age of six years and is the normal condition for approximately 70% of the world's population, the major exception being northern Europeans. It is particularly common in Asian communities and amongst blacks of African origin. Secondary acquired lactose intolerance occurs as a result of damage to the small intestinal mucosa, for example due to gastroenteritis, cows milk protein intolerance or celiac disease (*see below*).

Patients who ingest milk suffer from serious indigestion, nausea and gas, cramps, bloating and diarrhea because of the action of GIT bacteria on ingested lactose, the severity varying with the amount of lactose consumed and the tolerance level of the individual. These rather diffuse symptoms are often associated with other conditions of the GIT, such as infections with parasitic helminths (*Chapter 2*) and the protozoan *Giardia* (*Chapter 3*), inflammatory conditions, for example ulcerative colitis (*see below*), hormonal complaints, such as hypo- and hyperthyroidism (*Chapter 7*) and cancer of the colon and rectum (*Chapter 17*). Lactose intolerant patients are able to ingest a variety of milk products such as cheese, where the lactose has been removed in the whey, and yoghurt, where it has been fermented to lactate.

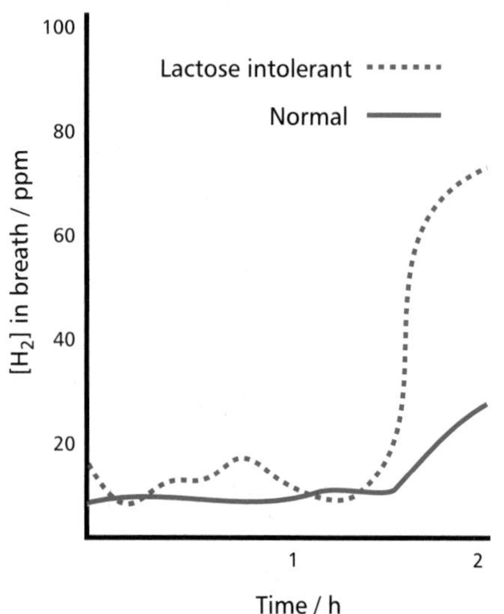

Figure 11.31 Typical results of hydrogen breath test for a control subject and a lactose intolerant patient.

Lactase deficiency can be assessed by the hydrogen breath, stool acidity and lactose tolerance tests. An assay for lactase activity on a tissue sample following a biopsy of the intestinal mucosa would confirm any diagnosis. The hydrogen breath test requires the patient to drink a solution containing 50 g of lactose. If lactase is deficient, the sugar is fermented by colonic bacteria which subsequently produce dihydrogen, some of which will enter the blood and be excreted at the lung surface. Regular analyses of the breath will show the increasing amounts of hydrogen (*Figure 11.31*) in lactose intolerance. Using the appropriate sugars allows both these tests to be used in diagnosing other disaccharide intolerances, although genetic intolerances to these are rather rare. The principle of the stool acidity test is simple: undigested lactose fermented by bacteria in the colon produces lactate and fatty acids, which can then be detected in a stool sample. The lactose tolerance test is still performed but only when other tests are inconclusive since it is more invasive. The patient is required to fast overnight, then drink a solution containing 50 g lactose. Several blood samples are taken over a 2-h period and their glucose concentrations determined. The amount of blood glucose indicates how well the patient is able to digest lactose. The test must be controlled by repeating the procedure using 25 g each of glucose and galactose, the constituent monosaccharides of lactose.

The medical history, age, degree of intolerance and overall health of the patient determine treatment for lactose intolerance. In general, it can be controlled with an appropriate diet. Adding proprietary products containing lactase (*Figure 11.32*) isolated from microorganisms to dairy products prior to ingestion can also help overcome the problem.

Figure 11.32 A proprietary product containing lactase used for the digestion of lactose prior to ingestion.

Celiac disease is a genetically determined chronic inflammatory condition affecting the small intestine and is induced by the ingestion of wheat protein, specifically gluten, and its products. A portion of the gluten molecule forms an autoimmune complex in the GIT mucosa which stimulates T cells to aggregate and release toxins that promote lysis of enterocytes (*Chapters 4* and *5*). This leads to a progressive atrophy and characteristic flattening of the mucosal villi and microvilli in the upper part of the small intestine (*Figure 11.33* (*A*) and (*B*)). Celiac disease is a relatively common disorder with a worldwide prevalence of 1 in 200–300. The mucosa improves morphologically when the patient is placed on a gluten-free diet but relapses occur if gluten is reintroduced in the diet. Celiac disease is accompanied by malabsorption of nutrients due to a decreased surface area over which absorption can take place. The signs and symptoms of celiac disease are surprisingly diverse and include abdominal pain, chronic diarrhea, weight loss, bone pain, fatigue and anemia. Celiac disease is diagnosed by demonstrating villous atrophy in a small intestine biopsy (*Figure 11.34* (*A*) and (*B*)) and by improvements in clinical symptoms or histological tests following restriction to a gluten-free diet. Gluten-free foods are now readily available in supermarkets and health food stores. The management of celiac disease involves adherence to a gluten-free diet, which in most cases helps relieve the symptoms and allows the existing mucosal damage to heal as well as preventing further damage.

Malabsorption is a reduction in the absorption of one or more nutrients by the small intestine. It occurs as a result of a wide range of disorders that affect the GIT, pancreas, liver and gall bladder. Its causes include enzyme deficiencies, such as in lactose intolerance, chronic pancreatitis, bile salt deficiency, as in biliary obstruction or hepatitis, and intestinal diseases such as celiac disease and Crohn's disease (*see below*). The clinical features of malabsorption (*Figure 11.35*) arise because of deficiency of one or more nutrients and include anemia due to iron, folate and vitamin B_{12} deficiencies, osteomalacia due to vitamin D deficiency, edema due to hypoalbuminemia, the tendency to bleeding when vitamin K is suboptimal and generalized weight loss. Malabsorption results in retention of nutrients in the GIT lumen causing diarrhea and abdominal discomfort due to the action of GIT bacteria. Malabsorption of fats leads to their losses in amounts greater than 5 g daily and causes steatorrhea. The feces are greasy, with a pale color and have an offensive smell.

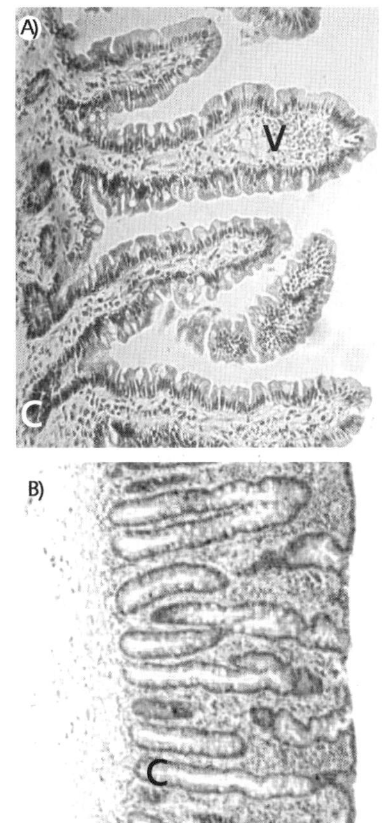

Figure 11.33 Photomicrographs of (A) healthy villi (V) and associated crypts of Lieberkuhn (C) and (B) the mucosal lining from a celiac patient. Note the degeneration of the villi.

A diseased liver may be unable to synthesize bile which is required for digestion of fats and if absent can lead to malabsorption. Liver function tests can indicate the presence of liver disease and a number of blood tests are also useful in assessing liver function. These include determining the concentration of albumin in the plasma and hematological investigations such as full blood count, iron, vitamin B_{12} and folate and can indicate the type of malabsorption. Investigations of malabsorption also include microbiological examination of feces to identify any pathogens present. The pentose sugar xylose is absorbed in the small intestine but is not metabolized and is excreted unchanged in

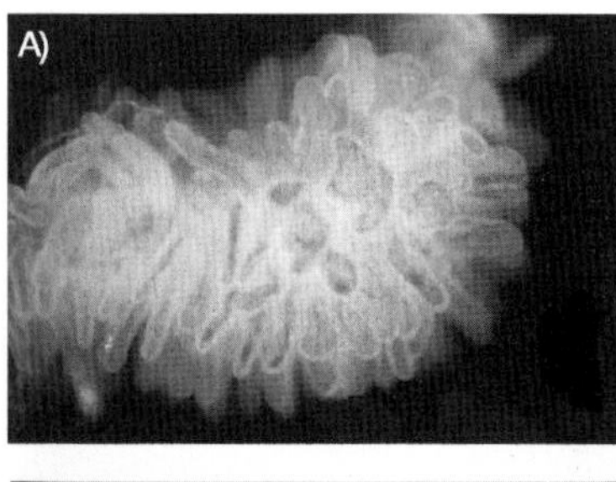

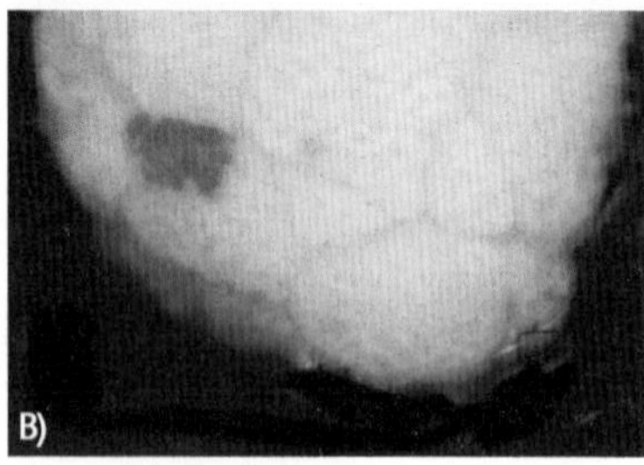

Figure 11.34 Photographs of gastrointestinal tract biopsies from (A) a healthy person and (B) a celiac patient. Note the degeneration of the villi.

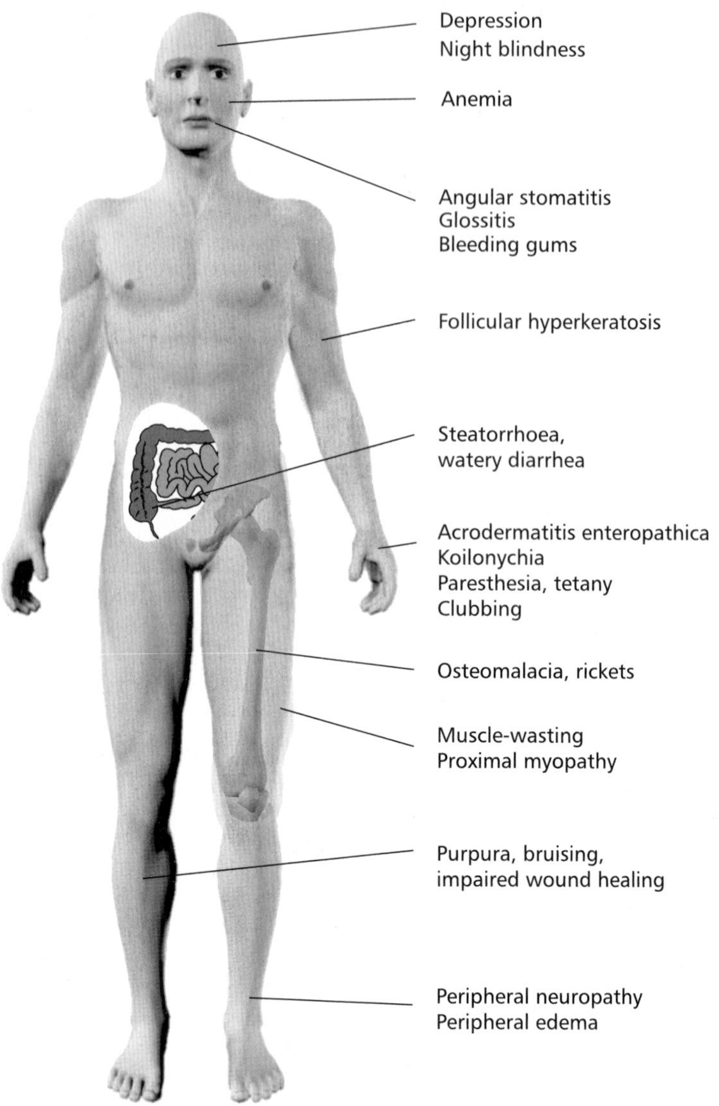

Figure 11.35 Schematic showing the clinical features associated with malabsorption.

the urine. This property is exploited in the xylose absorption test to assess the malabsorption of carbohydrates. The patient fasts overnight, empties the bladder and drinks a 500 cm^3 solution containing 5 g xylose. In normal individuals, the serum xylose concentration increases above 1.3 mmol dm^{-3} one hour after the test. After 5 h, the concentration of xylose in the urine increases to more than 7.0 mmol dm^{-3}. Significantly lower concentrations of xylose occur in the serum of patients with carbohydrate malabsorption. However, care must be exercised since some bacteria colonizing the small intestine are capable of metabolizing xylose and a number of renal diseases can also lead to reduced concentrations.

Malabsorption of fat can occur in a number of pancreatic and intestinal disorders. Bacteria colonizing the small intestine may break down bile acids, reducing their effective concentration and causing malabsorption. A fecal fat test can assess fat malabsorption. The test involves collecting feces over a period of three days after which their fat content is assessed chemically. Normally, up to 5 g of fat is lost in the feces each day but more is lost during malabsorption giving rise to steatorrhea.

BOX 11.5 Food allergies

A food allergy is a reaction of the immune system to food components that are allergens (*Chapter 5*). Foods commonly associated with allergies include milk, eggs, peanuts, tree nuts, soy, wheat, fish and shellfish, which are protein rich, and account for over 90% of all food allergies. The antibodies responsible for food allergies are IgE molecules that react with allergens to trigger the release of histamine (*Figure 11.36*). Histamine is largely responsible for the symptoms but these differ between patients and can differ in the same individual when exposed to different allergens. Symptoms include skin irritations, for example rashes and eczema, itchy nose and eyes, sneezing, excessive nasal mucus, coughing and shortness of breath, nausea, vomiting and diarrhea. Skin rashes, such as nettle rash (also called urticaria or hives), can appear but tend to last only a few days. Occasionally, patients present with long-lasting, chronic skin reactions such as scaly patches. Food allergies are relatively common and affect 1–2% of adults and up to 8% of children in the UK. Some individuals may experience the severe reaction called anaphylaxis described in *Chapter 5*. This is triggered by a small amount of food and is a rare but potentially fatal condition in which several different parts of the body experience allergic reactions including skin rashes, swollen throat and difficulties in breathing, nausea, vomiting, diarrhea, hypotension and unconsciousness. Symptoms usually appear within minutes of exposure to the allergen and can last several hours. An unusual form of anaphylaxis occurs following the consumption of allergens within a few hours of exercising and is referred to as exercise-induced anaphylaxis.

Food intolerance refers to the less well-defined condition occurring in some people when they eat certain foods and is characterized by headaches, muscle and joint aches and pains, and tiredness. The less well-defined symptoms make the condition harder to diagnose compared with classical food allergy. Examples of food intolerance include celiac disease and lactose intolerance.

The diagnosis of food allergy involves taking a clinical history from the patient, focusing on past allergic reactions and considering seasonal or environmental cues. This is followed by a clinical examination to detect characteristic signs and symptoms affecting the eyes, skin and nose. Skin prick testing is often performed. The procedure introduces a small amount of allergen into the skin and is generally safe. A tiny puncture is made with a lancet through a drop of allergen placed on the skin usually of the forearm. A positive reaction to the allergen usually means the patient is allergic and is indicated by itching within a few minutes; the affected site becoming red, swollen and having a raised weal in its center. The weal enlarges to a maximum size within 15 to 20 min and is measured and recorded. A negative response indicates that the patient is insensitive to the allergen under test. Blood tests, such as the radioallergosorbent (RAST) test, to detect the presence of IgE antibodies may be used but can only indicate an allergy if the specific IgE is present in the blood. However, there is no clear relationship between the level of blood IgE and the severity of the allergy. Occasionally, a challenge test may be performed where the suspected food is given to the patient first by touch and then by ingestion in increasing amounts and the response monitored. Challenge tests should always be performed under supervised conditions in a hospital or clinical setting, so that any clinically serious reactions can be detected and managed.

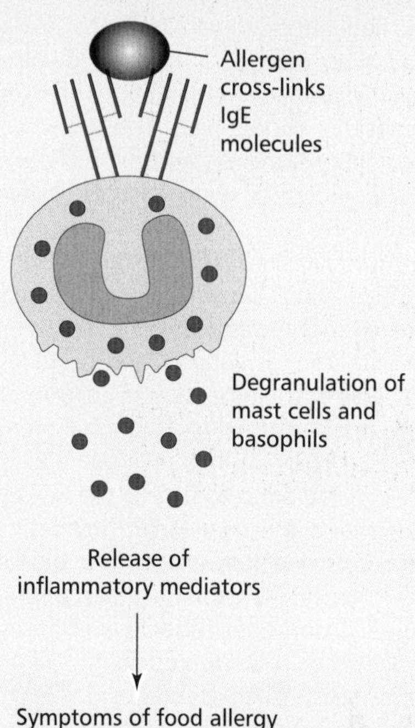

Figure 11.36 Schematic showing the release of inflammatory mediators in response to a food allergy. See text and *Chapter 5* for details.

Food allergy is usually treated by avoiding the offending food. In some cases, complete elimination of the food for one to two years results in a loss of clinical symptoms although certain allergies, for example those associated with the intake of fish or peanuts, may persist for life. It is vital that consumers are provided with clear information on the composition of foods they purchase. Food avoidance may have serious nutritional consequences, particularly when a key nutrient is removed from the diet. Immediate medical attention is necessary when anaphylaxis occurs and usually involves an injection of adrenaline to dilate airways and blood vessels by relaxing smooth muscle tissues.

BOX 11.6 Diarrhea

Diarrhea is the frequent passage of feces that are larger in volume and more fluid than normal. It is not a disease but a symptom of some other underlying conditions that result in abrupt increases in intestinal movements. The accelerated movement of the contents through the GIT leaves insufficient time for fluid reabsorption and produces watery stools. A two-way flow of water and electrolytes occurs between the GIT lumen and the basolateral extracellular fluid (ECF). Water enters the GIT in food and drinks and in various secretions. In the small intestine, the secretion of water and electrolytes normally occurs in the crypts of Lieberkuhn. Sodium chloride is transported from ECF into epithelial cells across their basolateral membranes. Secretory stimuli increase the permeability of luminal membranes of the crypt cells allowing chloride ions to move into the GIT lumen. Sodium ions, however, are returned to the ECF by the action of the Na^+/K^+-ATPase of the basolateral membranes. The movements of these ions generate an osmotic gradient and water flows passively from the ECF into the lumen through intercellular channels. Water reabsorption is also driven by osmotic gradients, which are formed when solutes, particularly sodium ions, actively enter enterocytes. Absorption of sodium ions may occur by direct transport as the ion, or be exchanged for hydrogen ions or linked to the absorption of glucose, amino acids or chloride ions.

Following absorption, sodium is transported out of enterocytes across the basolateral membrane into the ECF by the Na^+/K^+-ATPase. This increases the osmolality of the ECF and water moves passively into it from the GIT lumen. These processes maintain an osmotic balance between GIT contents and ECF in the intestinal tissue. However, since fluid absorption normally exceeds secretion, the net result is fluid absorption. More than 90% of the fluid entering the small intestine is absorbed, with only about one dm^3 of it reaching the large intestine. Here further absorption occurs and only 100 to 200 cm^3 of water is lost in the feces daily.

Diarrhea arises when water and electrolyte transport becomes disordered, for example by increased secretion, decreased absorption or both, and an increased volume of fluid enters the large intestine. When this exceeds the absorptive capacity of the large intestine it results in diarrhea.

Diarrhea may result from one of two principal mechanisms, secretion and osmotic imbalance. These mechanisms are not exclusive; intestinal infections can cause both types and, indeed, both may occur in a single individual. Secretory diarrhea is the more common of the two and is caused by an abnormal secretion of water and salts into the small intestine. This occurs when the reabsorption of sodium ions is impaired but the secretion of chloride ions in the crypts of Lieberkuhn is maintained or increased. This produces a net secretion of fluid resulting in the loss of water and salts from the body in watery stools; this causes dehydration. Infectious diarrhea may result from the actions of bacterial toxins or viruses on the GIT mucosa (*Chapter 2*). Osmotic diarrhea results when a poorly absorbed, osmotically active substance is ingested, causing water and salts to move rapidly across the GIT lining to maintain the osmotic balance. The effects depend upon the osmolarity of the solution. If the substance is taken as an iso-osmotic solution, water and solute will pass through the GIT, with no net absorption causing diarrhea. If taken as a hyperosmotic solution, water and some electrolytes will move from the ECF into the GIT lumen increasing the volume of the feces and causing dehydration. Furthermore, because the loss of body water is proportionally greater than the loss of sodium and chloride ions, hypernatremia also develops (*Chapter 8*).

Diarrhea results in losses of large amounts of sodium, chloride, potassium and hydrogen carbonate ions. Acute effects result from the loss of water and electrolytes, leading to dehydration, metabolic acidosis, because of the loss of hydrogen carbonate and potassium depletion (*Figure 11.37*). The dehydration is the most dangerous in the short term because a decreased blood volume (**hypovolemia**) can result in cardiovascular collapse and death if not treated promptly (*Chapter 8*). The aim of managing diarrhea is to correct dehydration and electrolyte deficits. Fluids can be replaced either orally or intravenously.

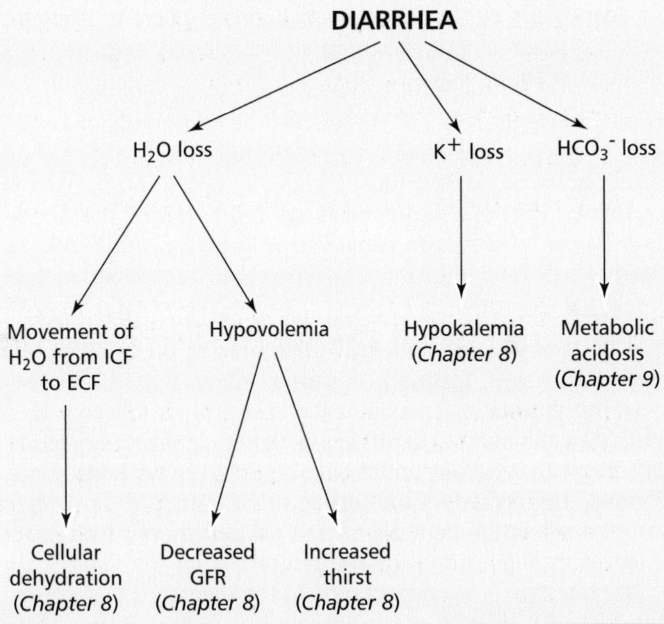

Figure 11.37 The major clinical features that may arise from prolonged diarrhea.

Crohn's disease is a chronic inflammation, usually of the ileum, although it can affect any part of the GIT. The inflammation tends to be patchy but extends throughout the layers of the intestinal wall thickening the wall and narrowing the lumen. The cause of Crohn's disease is unclear although viruses and bacteria have been implicated. Patients with Crohn's disease suffer from lack of appetite, abdominal pain, diarrhea and weight loss. A biopsy of the GIT is used to detect the characteristic changes associated with Crohn's disease. Treatment involves using anti-inflammatory drugs, such as 5-aminosalicylic acid, although surgery may be required in severe cases.

DISORDERS OF THE LARGE INTESTINE

Ulcerative colitis is characterized by chronic inflammation and ulceration of the colon that is distinctive, in that it affects only its mucosal lining. The lining is also affected by numerous hemorrhagic ulcerations. The cause of ulcerative colitis is not entirely clear but is believed to be autoimmune in origin (*Chapter 5*). Patients with ulcerative colitis suffer from abdominal pain, fever, weight loss and bloody diarrhea. Diagnosis of ulcerative colitis is made following colonoscopy where a biopsy from the colon is taken to detect the characteristic histological changes. Ulcerative colitis is treated with anti-inflammatory drugs, such as 5-aminosalicylic acid, although surgical removal of the affected region may be necessary in severe cases.

CASE STUDY 11.1

Alice, a five-year-old child, presented with loss of weight, abdominal distension and obvious signs of anemia. Her stools were loose, bulky, pale colored and had an offensive smell. A variety of clinical tests were ordered with the following results (reference ranges are shown in parentheses):

Serum albumin	29 g dm^{-3} (32–48 g dm^{-3})
Serum iron	4 µmol dm^{-3} (10–30 µmol dm^{-3})
Xylose absorption	0.5 g in 5 h (>1.2 g in 5 h)
Fecal fat	29 g / 3 days (<15 g/3 days)

Jejunal biopsy showed villous atrophy.

Questions

(a) What is the most likely diagnosis?

(b) How should Alice be treated?

CASE STUDY 11.2

It was noticed that Sadaf, a fine art student recovering from an attack of influenza, was slightly jaundiced. Worried that she may have hepatitis, her doctor asked her to provide some blood and urine for laboratory investigation. Test results are shown below (reference ranges are shown in parentheses):

Serum bilirubin	60 µmol dm^{-3} (3–20 µmol dm^{-3})
Serum ALP	74 IU dm^{-3} (30–90 IU dm^{-3})
Serum AST	35 IU dm^{-3} (10–50 IU dm^{-3})
Hemoglobin	140 g dm^{-3} (115–155 g dm^{-3})
Reticulocyte count	1.5 % (1–2%)
Urine bilirubin	Negative (Negative)

Question

What is the most probable diagnosis to account for these signs and symptoms?

CASE STUDY 11.3

Mark, a 52-year-old plumber, was admitted to hospital because of severe abdominal pain. The pain had started suddenly about 15 h previously. On questioning, Mark admitted to being a heavy alcohol drinker over many years. Clinical examination showed that Mark had a tender abdomen and was in mild shock. Radiographic examination did not show any evidence of intestinal obstruction. Biochemical tests gave the following results (reference ranges are shown in parentheses):

Serum	Na^+	140 mmol dm^{-3}	(135–145 mmol dm^{-3})
	K^+	4.2 mmol dm^{-3}	(3.6–5.0 mmol dm^{-3})
	Urea	11 mmol dm^{-3}	(3.3–6.7 mmol dm^{-3})
	Glucose	13 mmol dm^{-3}	(2.8–6.0 mmol dm^{-3})
	Amylase	5000 IU dm^{-3}	(<300 IU dm^{-3})

Questions

(a) What is the most likely diagnosis?

(b) Why does Mark have a high serum urea concentration?

11.9 SUMMARY

The GIT and its associated organs are necessary for the digestion of food and the absorption of nutrients derived from food. Digestion takes place in the mouth, esophagus, stomach and small intestine, with the bulk of ingested water being absorbed in the latter. The pancreas provides enzymes for digestion in the small intestine while the liver secretes bile, containing detergents that aid in the digestion of fats. Most nutrients from digestion are carried to the liver in the hepatic portal vein.

The GIT also contributes to nutrient homeostasis, through metabolic reactions that take place in GIT enterocytes, delivering metabolic products such as lactate to the liver. The GIT also produces a number of hormones that help in coordinating its activities.

Disorders of the pancreas, such as acute and chronic pancreatitis, are serious conditions that may be fatal. Disorders of the liver may be caused by drugs, poisons, viruses, and alcohol or by blockage of the bile duct, as in cholestasis. Liver disorders may be diagnosed by assaying several enzymes, together with an X-ray examination and palpation to detect a swollen liver. Jaundice is a frequent complication of liver disease. Disorders of the GIT include gastritis and ulcers that are associated with infection of *Helicobacter pylori*, intolerance to certain nutrients, such as lactose, celiac disease and malabsorption.

Questions

1. Which of the following is associated with the development of peptic ulcers?

   a) smoking;
   b) *Helicobacter pylori*;
   c) weight loss;
   d) celiac disease;
   e) lactose intolerance.

2. Which one of the following tests is best used to assess pancreatic function?

   a) fecal fat test;
   b) serum bilirubin;

c) oral glucose tolerance test;

d) fluorescein dilaurate test;

e) xylose absorption test.

3. Qiuyu, a 32-year-old from China, entered England to study as a mature research student. Wishing to 'integrate' fully in her new environment she adopted many of the dietary habits typical of the country. After six healthy months, she unfortunately suffered a bout of viral gastroenteritis that required hospitalization. Qiuyu, fortunately made a complete recovery and was able to continue her new lifestyle. However, she again became ill again, but with severe indigestion, abdominal cramps, intestinal bloating and periods of diarrhea. Symptoms were most pronounced about one hour after breakfast.

    a) Suggest the most probable cause of Qiuyu's symptoms.

    b) What test(s) could be used to help your initial diagnosis? Include any precautions that may be necessary in your clinical investigations?

    c) Assuming your initial diagnosis is correct, what treatment do you recommend for Qiuyu?

4. List some possible causes of jaundice.

5 List four causes of malabsorption.

FURTHER READING

Bhatia, M, Wong, FL, Cao, Y, Lau, HY, Huang, J, Puneet, P and Chevali, L (2005) Pathophysiology of acute pancreatitis. *Pancreatology* **5:** 132–144.

DiMagno, MJ and DiMagno, EP (2003) Chronic pancreatitis. *Curr. Opin. Gastroenterol.* **19:** 451–457.

Gibril, F and Jensen, RT (2004) Zollinger-Ellison syndrome revisited: diagnosis, biologic markers, associated inherited disorders, and acid hypersecretion. *Curr. Gastroenterol. Reports* **6:** 454–463.

Higuchi, H and Gores, GJ (2003) Mechanisms of liver injury: an overview. *Curr. Molec. Med.* **3:** 483–490.

Iredale, JP (2003) Cirrhosis: new research provides a basis for rational and targeted treatments. *BMJ* **327:** 143–147.

Johnson, LR (2001) *Gastrointestinal Physiology*, 6th edn, Mosby, St Louis, USA.

Kagnoff, MF (2005) Overview and pathogenesis of celiac disease. *Gastroenterology* **128:** S10–S18.

Karsan, HA, Rojter, SE and Saab, S (2004) Primary prevention of cirrhosis. Public health strategies that make a difference. *J. Postgrad. Med.* **115:** 25–30.

Kashiwagi, H (2005) Ulcers and gastritis. *Endoscopy* **37:** 110–115.

Kastin, DA and Buchman, AL (2002) Malnutrition and gastrointestinal disease. *Curr. Opin. Gastroenterol.* **18:** 221–228.

Kingsnorth, A and O'Reilly, D (2006) Acute pancreatitis. *BMJ* **332:** 1072–1076.

Limdi, JK and Hyde, GM (2003) Evaluation of abnormal liver function tests. *Postgrad. Med. J.* **79:** 307–312.

Louw, JA and Marks, IN (2004) Peptic ulcer disease. *Curr. Opin. Gastroenterol.* **20:** 533–537.

McKevith, B and Theobald, H (2005) Common food allergies. *Nurs. Stand.* **19:** 39–42.

Pearson, C (2004) Inflammatory bowel disease. *Nurs. Times* **100:** 86–90.

Pineiro-Carrero, VM and Pineiro, EO (2004) Liver. *Pediatrics* **113:** 1097–1106.

Roche, SP and Kobos, R (2004) Jaundice in the adult patient. *Am. Fam. Phys.* **69:** 299–304.

Ross, G (2003) Hyperbilirubinemia in the 2000s: what should we do next? *Am. J. Perinatol.* **20:** 415–424.

Sachs, G, Weeks, DL, Melchers, K and Scott, DR (2003) The gastric biology of *Helicobacter pylori. Ann. Rev. Physiol.* **65:** 349–369.

Small, CJ and Bloom, SR (2004) Gut hormones and the control of appetite. *Trends Endocrinol. Metab.* **15:** 259–263.

Tack, J, Bisschops, R and Sarnelli, G (2004) Pathophysiology and treatment of functional dyspepsia. *Gastroenterology* **127:** 1239–1255.

Tomer, G and Ahneider, BL (2003) Disorders of bile formation and biliary transport. *Gastroenterol. Clin. N. Am.* **32:** 839–855.

Various authors (2005) The gut: inner tube of life. *Science* **307:** 1895–1925.

Warren, BF (2004) Classic pathology of ulcerative and Crohn's colitis. *J. Clin. Gastroenterol.* **38:** S33–S35.

Wilson, TR (2005) The ABCs of hepatitis. *Nurse Pract.* **30:** 12–21.

chapter 12:

TOXICOLOGY

OBJECTIVES

After studying this chapter you should be able to:

- define the terms drug, xenobiotic, poison, toxin and toxicology;
- outline the absorption, distribution and excretion of drugs;
- explain the roles of the liver and kidneys in detoxification;
- outline the types and clinical effects of some common poisons;
- describe the general methods for treating and managing poisoning.

12.1 INTRODUCTION

The human body must be prepared for a daily onslaught of thousands of chemicals. Some will be food materials (*Chapter 10*) that are absorbed and metabolized because cells have membrane channels and enzymes to recognize and deal with these compounds. Other compounds may or may not enter the cells of the body. If they do, then the body can recognize such compounds and must deal with them appropriately to prevent deleterious effects. It is not always successful in this aim and many xenobiotics cause diseases and may be fatal if a threshold dose is exceeded.

Toxicology is the study of the adverse effects of chemicals on organisms, that is the study of poisons, where a poison can be defined as any substance that causes injury, illness or death. Toxicology covers the study of the adverse effects of chemicals including drugs, chemicals acquired from the environment and toxins, which are defined here as harmful substances produced by other organisms, often derived from microorganisms, for example the bacterial toxins described in *Chapter 2*.

Xenobiotics, literally meaning 'stranger to life', are substances that do not originate in the body but are pharmacologically, endocrinologically or toxicologically active. Thus, they might be drugs or synthetic chemicals or a substance produced in one organism and introduced into another where they

Margin Note 12.1 A lethal bacterial toxin 

The bacterium *Clostridium botulinum* is an organism that can cause a fatal type of food poisoning. The organism produces several toxins when growing on food in anaerobic conditions. In addition, its spores are resistant to heat, hence food that has not been adequately sterilized and then stored in an anaerobic environment, such as bottled or canned foods, can give rise to food poisoning. These toxins inhibit the release of acetylcholine at neuromuscular junctions, leading to a flaccid paralysis in tissues served by susceptible neuromuscular and peripheral nerves (*Chapter 2*). The lethal dose of botulinum toxin for humans is not known with certainty but estimates from studies with other primates suggest that approximately 0.1 µg given intravenously or intramuscularly and about 1µg if inhaled would kill a 70 kg human.

Surprisingly, given their extreme toxicity, toxins from *Clostridium botulinum* are used cosmetically and therapeutically. Injections of 'Botox', the A toxin, into the skin of the face is used to relax the face muscles reducing the wrinkled appearance of aging (*Chapter 18*). The injections must be repeated every few months, as the effect is temporary. Botulinum toxin is also used in clinical practice to control conditions associated with inappropriate secretions of acetylcholine, for example the muscular spasms associated with the neurological disorder, dystonia.

would not normally occur. An example would be a compound produced by a plant and ingested as food. Equally, it might be a compound that has been completely synthesized chemically and have harmful effects, for example a poison or a carcinogen, or it could be beneficial, such as a medicinal drug.

Drugs are xenobiotics that are used to achieve certain effects. For example, paracetamol alleviates headaches and aspirin controls inflammatory responses. Alternatively, a drug may also be a compound with pharmacological activity used for 'recreational' purposes or taken by an addict. The body has to deal with these drugs and eventually get rid of them. An overdose may exceed the body's capacity to detoxify these compounds, with potentially disastrous effects. However, if the body inactivates the drug too quickly, then its effects will be short-lived. These are all considerations that pharmaceutical companies need to address when developing a new drug.

12.2 DRUG ACTION, METABOLISM, DISTRIBUTION AND EXCRETION

Drugs are xenobiotics and may be defined as any substance, other than food, that affects a living process. **Pharmacology** is the study of the effects of drugs in the prevention, diagnosis, and treatment or cure of disease. Such drugs are often referred to as medicines, which distinguishes them from other drugs that are used for pleasure, such as some narcotics. Pharmacotherapeutics is that branch of pharmacology concerned with the administration of drugs for prevention and treatment of disease.

Drugs can be classified according to their chemical structure but, more often, in terms of their pharmacological effects. For example, they can be divided into three groups: chemotherapeutic drugs, for example the antibiotics described in *Chapter 3*, which are used to treat infectious diseases; pharmacodynamic drugs, such as sedatives that are used in the treatment of noninfectious diseases; and a number of miscellaneous agents including narcotics and analgesics.

Any single drug may have a chemical, brand and a generic name. The chemical name is given according to the rules of chemical nomenclature, whereas the brand name is given by the manufacturer. The generic name is a common, established name given to a drug irrespective of that of its manufacturer.

Most drugs act on cells to alter a biological function. This pharmacological effect occurs as a consequence of the drug reacting with a receptor that controls a particular function, or because the drug alters a physiological mechanism which affects that function. For many drugs, the extent and duration of the pharmacological effect are proportional to the concentration of the drug at the receptor. The site at which the drug acts to produce a pharmacological effect is called its site of action. The mechanism of action of the drug is the biochemical or physiological process occurring at the site of action to produce the pharmacological effect. Drug receptors include enzymes and structural or transport proteins. However, some receptors are nonprotein that bind to the drug to form a complex which alters the permeability of the membranes or the transcription of DNA. Some drugs have a structure similar to endogenous molecules and compete with them for binding sites. Drugs may also act by preventing the formation, release, uptake or transport of key substances in the body or by forming complexes with molecules that can then activate receptors.

The binding of a drug to its receptor usually depends on relatively weak forces, such as van der Waals forces and hydrogen and ionic bonds and thus the formation of the drug–receptor complex that elicits the response is normally freely reversible. Hence the response to any drug is not permanent.

However, the response is dose dependent and, indeed, a dose–response relationship exists between the concentration of drug in the serum and the pharmacological effect. This response eventually reaches a maximum effect because the receptor becomes saturated with the drug (*Figure 12.1*). The therapeutic range is the concentrations of drug in the serum that is appropriate for therapy. The dosage of any drug is planned to give a serum concentration within its therapeutic range. Therapeutic drug monitoring is often necessary to determine which given doses of a drug result in serum concentrations within the therapeutic range. The serum concentration of the drug must not fall below its minimum effective concentration (MEC) otherwise it will be ineffective. However, neither should it rise above its minimum toxic concentration (MTC) because of the danger of metabolic or structural damage. The time required for the concentration of a drug in the blood to decline to half its original value is referred to as its half-life ($t_{1/2}$).

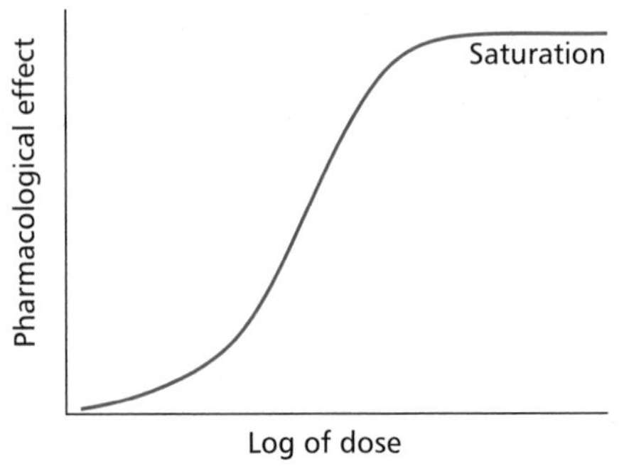

Figure 12.1 A dose–response curve for a typical drug. Saturation occurs when all the receptors are occupied by the drug.

It is essential that a number of properties relating to a medicinal drug, for example its pharmacodynamics and pharmacokinetics, are first ascertained. Pharmacodynamics describes how the drug interacts with its target site and the biochemical and physiological processes that result in any therapeutic or toxic effects. Pharmacokinetics relates to the uptake, distribution, metabolism and excretion from the body.

Most drugs are given orally for convenience, although they can be administered intravenously, intramuscularly or subcutaneously. When given orally, the absorption of the drug depends on its ability to disassociate from its dosing form, dissolve in gastrointestinal fluids and diffuse across the gut wall into the blood. The rate and extent of drug absorption varies with the nature of the drug, the matrix in which it is dissolved and the region of the gastrointestinal tract (GIT) where it is absorbed. The proportion of the drug absorbed into the circulation is referred to as its bioavailability. For an orally delivered drug, this should generally be greater than 70% to be of therapeutic use. However, when the site of action is the GIT lumen itself, for example treating a GIT infection, then a low bioavailability would be advantageous.

A number of drugs undergo what is referred to as first pass metabolism. They are absorbed rapidly and completely by the GIT but, nevertheless, have low bioavailability because they are transported to the liver in the hepatic portal vein and metabolized (*Section 12.3*) and have not entered the systemic circulation. Drugs with delayed absorption are sometimes required and special slow or sustained release formulations have been developed for these cases. Such drugs can be taken orally at less frequent intervals. Certain diseases that affect the GIT and the interaction of some drugs and foods in the GIT can delay their absorption.

Following absorption, drug distribution occurs when the compound enters the vascular system. The physical, chemical and molecular properties of the drug can influence its distribution. Its distribution may also be influenced by its binding to blood components and receptors and its ability to dissolve in lipids and pass through biological membranes. Many drugs bind to plasma proteins and often an equilibrium is established between protein-bound and free drug. Only the free fraction is able to interact with receptors or cross cellular membranes. Any factor that changes drug–protein interactions may alter the distribution, pharmacological effects and excretion of the drug.

Drugs are excreted from the body by the biliary, GIT, pulmonary and/or renal routes. Most drugs are excreted through the renal system, and therefore, alterations in renal function may influence the half-life and serum concentration of the drug. A decline in renal function causes an increase in the serum drug concentration with an associated increased pharmacological effect.

12.3 PHYSIOLOGICAL DETOXIFICATION MECHANISMS

Xenobiotics, whether drugs or toxins, are normally absorbed through the lungs, GIT or the skin. In addition, drugs may be injected through intramuscular, intraperitoneal, subcutaneous or intravenous routes. Following absorption, the xenobiotics are distributed around the body and a proportion will be absorbed by cells, while some may be directly excreted from the kidneys. Humans also have a number of physiological mechanisms devoted to detoxifying ingested xenobiotics, which largely involve enzyme activities in the liver and kidneys. Many of the biochemical detoxification mechanisms convert the xenobiotic to more water-soluble compounds that are more easily excreted. These mechanisms can sometimes prove inadequate. Also, in some unfortunate cases, the detoxification mechanisms form compounds that are even more toxic or carcinogenic than the original compound.

Treatment for poisoning involves administering an antidote to the toxin or drug when one is available. For many common poisons this is not the case and therapy involves general supportive measures, such as decreasing their absorption from the GIT, increasing the rate of elimination from the body or altering the distribution within the body to protect susceptible tissues. An optimal therapeutic concentration of a drug can only be maintained in the plasma if the patient fully complies with the prescribed dose. Unfortunately, the most common reason for emergency admissions to hospitals is because of an excessive intake of a drug prescribed for medication. Other causes of poisoning may be accidental, suicidal or homicidal in intent.

Many poisons are lipophilic and are only sparingly soluble in water and so cannot easily be excreted by the kidneys. Thus a major aim of detoxification in the liver is to convert them to compounds that have increased water solubility and are more readily removed by the kidneys. Detoxification reactions in the liver are favored by its microstructure, which consists of lobules with a central vein and peripheral branches of the hepatic artery and the hepatic portal vein. Blood leaves branches of the hepatic artery and hepatic portal vein and percolates from the periphery of the lobule through sinusoids, bathing the hepatocytes of the lobule as it flows (*Figure 12.2*). While it is unusual for blood to come into such close contact with tissue cells, this arrangement allows poisons to be removed as the blood flows through the lobule to be collected by branches of the hepatic vein. Also, mitotic divisions within the organ replace those liver cells that are irreparably damaged. It goes without saying that a disease of the liver can compromise its ability to deal with toxic substances, which can lead to significant clinical consequences.

Detoxification is achieved in two phases. In Phase I, the xenobiotic is oxidized and/or hydroxylated by mixed function oxidase (MFO) activities of monooxygenase systems. There are two such systems, a flavin-containing monooxygenase (FMO) system used in oxidation reactions of drug metabolism and a cytochrome P-450 monooxygenase system that oxidizes carbon atoms. Both systems require NADPH as a coenzyme and dioxygen and are localized in the smooth endoplasmic reticulum. In Phase II, the oxidized xenobiotic product is linked to a polar compound, such as a glucuronate or sulfate group forming water-soluble conjugates in further enzyme-catalyzed reactions. The enzymes that catalyze both the Phase I and II reactions have broad specificities and are therefore able to detoxify a wide range of organic toxins and drugs. This is essential since the range of xenobiotics to which the body may be exposed is enormous, and individual enzyme systems could not be available to deal with each one separately. At least 50 different members of the cytochrome P-450 enzyme family are found in the smooth endoplasmic reticulum of hepatocytes (*Figure 12.3*). They catalyze the hydroxylation of a wide variety of substances by incorporating one of the oxygen atoms into the xenobiotic to form the hydroxyl group,

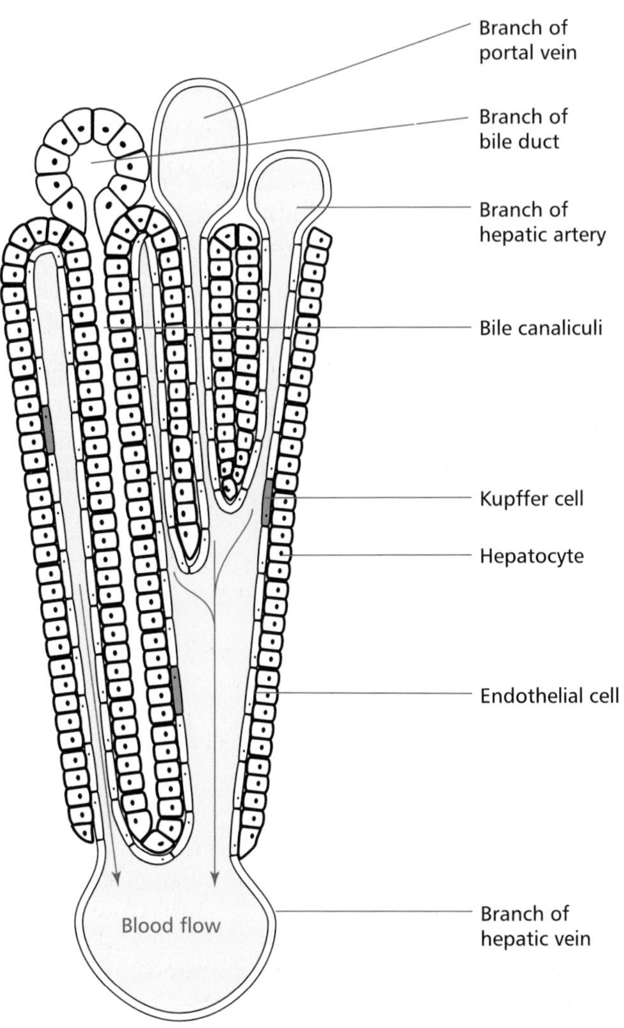

Figure 12.2 Schematic of a portion of a liver lobule. Blood enters at the periphery from a branch of the hepatic portal vein and the hepatic artery and eventually leaves at the center of the lobule via a branch of the hepatic vein. Close contact of the blood with the liver cells means that these cells are the first to contact poisons and toxins after their absorption from the GIT and transport in the portal system. The hepatocytes contain enzymes involved in detoxification.

while the other oxygen atom is reduced to water. The NADPH supplies an electron to complete the Phase I stage.

$$R\text{-}H + O_2 + NADPH + H^+ \longrightarrow R\text{-}OH + H_2O + NADP^+$$

The same enzyme system is able to convert unsaturated compounds to an epoxide, which is a substrate for epoxide hydrolase that catalyzes the conversion of the epoxide to a glycol.

R_1, R_2 —(P-450 oxidase system)→ R_1, R_2, O —(Epoxide hydrolase)→ R_1, R_2, HO, OH —(P-450 oxidase system)→ O, R_1, R_2, HO, OH

The actions of these enzymes form hydroxyl groups that increase the water solubility of the original poison or drug and also form attachment points for the actions of Phase II enzymes. These include glucuronyl transferases of the smooth endoplasmic reticulum membrane and sulfotransferase in the cytosol, which use UDP-glucuronate and 3′-phosphoadenosyl 5′-phosphosulfate (PAPS) as the respective donor substrates:

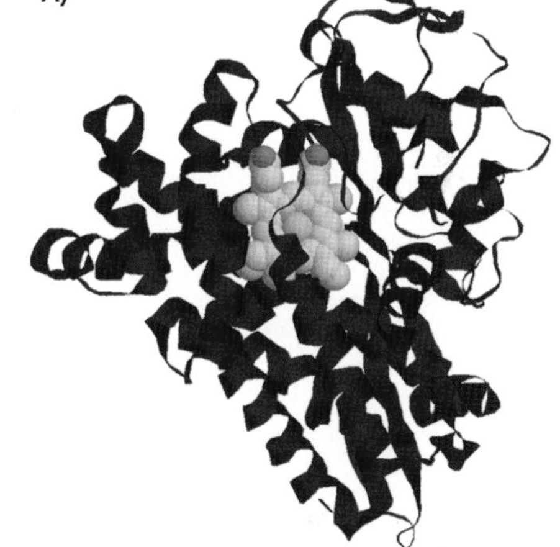

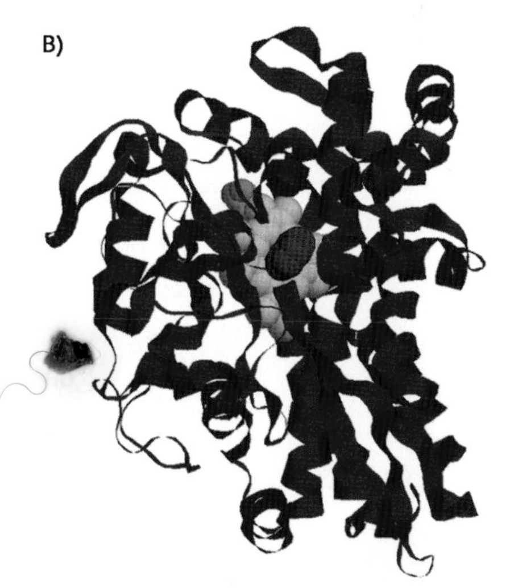

Figure 12.3 Molecular models of (A) a P-450 molecule, with a gray colored heme group containing a central iron atom and (B) with a bound aromatic substrate shown in dark gray. PDB files 1PHC and 1CP4 respectively.

$$\text{R-OH} + \text{UDP-glucuronate} \xrightarrow{\text{glucuronyl transferase}} \text{R-O-glucuronate} + \text{UDP}$$

$$\text{R-OH} + \text{PAPS} \xrightarrow{\text{sulfotransferase}} \text{R-O-SO}_3^- + \text{PAP}$$

The increased water solubility of the products facilitates their excretion by the kidneys, although small amounts are also lost in bile to the feces.

It is unfortunately the case that a number of poisons are rendered more toxic by oxidation by the P-450 system. Indeed, a number of fat-soluble compounds that are relatively harmless because they would normally accumulate in adipose tissue are converted into potentially lethal materials because of the increase in water solubility. Thus, a number of indirectly acting carcinogens, for example, benzo[a]pyrene, found in cigarette smoke and the toxin aflatoxin B1 from *Aspergillus flavus*, are converted to highly carcinogenic products (*Figure 12.4 (A)* and *(B)*).

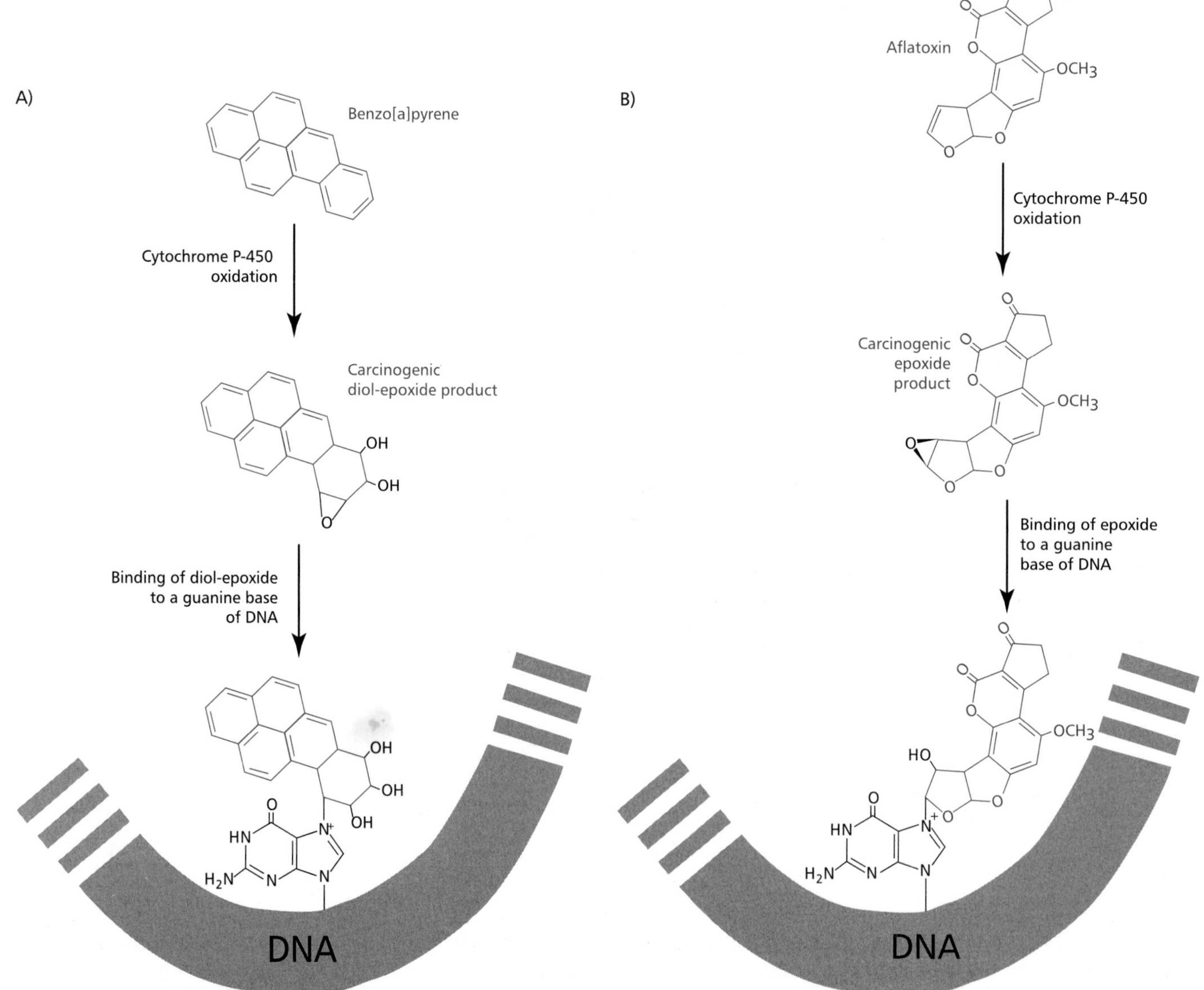

Figure 12.4 The oxidation of (A) benzo [a] pyrene and (B) aflatoxin by the cytochrome P-450 system to reactive chemicals that can bind with guanine bases of DNA and so induce carcinogenesis.

12.4 SYMPTOMATIC POISONING

In the symptomatic patient, the diagnosis of poisoning is usually made on the basis of clinical findings. However, in all cases of suspected poisoning, a number of biochemical investigations should be made as shown in *Table 12.1*. The roles of the clinical laboratory in chemical toxicology in identifying a suspected poison in blood or urine and in monitoring vital functions cannot be overestimated.

Investigation	Purpose
Monitor plasma urea, electrolytes and creatinine levels	assess renal function (*Chapter 8*)
Monitor plasma osmolality	an osmolality gap can indicate the presence of foreign substances in the serum (*Chapter 8*)
Monitor blood gases	assess acid–base status (*Chapter 9*)
Monitor liver functions	assess viability of liver functions (*Chapter 11*)
Monitor blood glucose concentrations	to detect hypo- and hyperglycemia (*Chapter 7*)

Table 12.1 Biochemical investigations of suspected poisoning

The clinical management of patients is directed towards supporting their vital functions and attempting to remove the poison from the body as rapidly as possible. Unfortunately, few of the signs or symptoms that patients present with are specific for any one type of drug or poison. Also, patients may well present in a coma and so be unable to give relevant information. A drug screen on a urine specimen may be carried out but this will only indicate that a drug has been ingested and may give no indication of the severity of the overdose.

12.5 COMMON POISONS

Humans are exposed to numerous xenobiotics, including drugs, pesticides, environmental pollutants, industrial chemicals and food additives. Any of these are potentially capable of perturbing the biochemistry and physiology of the body either directly or after being metabolically transformed. However, the most frequent poisons encountered in emergency toxicology include paracetamol (acetaminophen), aspirin, alcohols (ethanol, methanol, ethylene glycol), barbiturates (though this is now mainly historical), carbon monoxide, paraquat and several metals. The chapter will concentrate only on those most likely to be encountered in clinical practice.

PARACETAMOL

Paracetamol (*Figure 12.5*) is used as an analgesic, that is, to relieve fever and pain. It is safe when taken at recommended doses but is toxic if overdosed. It is the commonest cause of admissions to hospital due to its wide availability. In the UK, overdoses cause approximately 150 deaths annually.

Paracetamol is rapidly absorbed from the stomach and upper GIT. The majority of ingested paracetamol is metabolized by conjugation with sulfate (~30%) or glucuronide (~60%) in the liver as described in *Section 12.3* to form nontoxic metabolites. However, approximately 10% is metabolized by cytochrome P-450 to produce a highly reactive intermediate called *N*-acetyl-*p*-benzoquinoneimine (NABQI). It is possible for NABQI to be

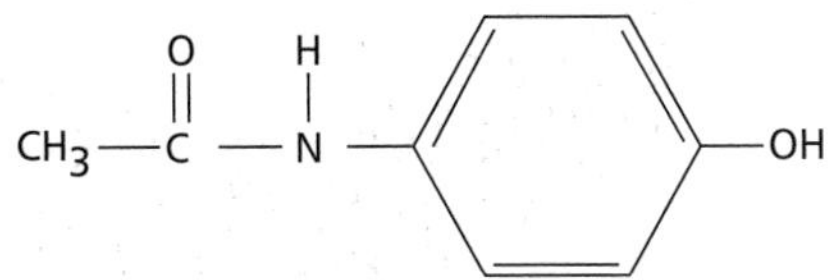

Figure 12.5 Paracetamol (called acetaminophen in the USA).

metabolized by conjugation with glutathione in hepatocytes to produce a nontoxic mercapturic acid. Glutathione is a tripeptide consisting of γ glutamate, cysteine and glycine residues, which will be further mentioned in *Chapters 13* and *18*. It functions as a coenzyme in several oxidation–reduction reactions (*Figure 12.6*). When an overdose of paracetamol is ingested, liver detoxification systems may become saturated and the large amount of NABQI produced exhausts the limited stores of glutathione. As a consequence, NABQI binds to sulfhydryl groups of hepatocyte proteins, forming irreversible complexes that result in acute hepatic necrosis, that is, cell death (*Figure 12.7*). Paracetamol is also metabolized in cells of renal tubules and, in an overdose, renal tubular necrosis may also occur. In the presence of hepatic damage there is usually only a small amount of renal damage but occasionally this may be the major presenting feature of paracetamol poisoning. Alcohol in chronic alcoholics and the drugs phenobarbitone and phenytoin, used to treat epilepsy, may induce the synthesis of cytochrome P-450 enzymes and cause increased production of NABQI. As a consequence, hepatotoxicity may occur following a relatively small overdose of paracetamol in such patients. Severe toxicity is also more likely in people whose intracellular stores of glutathione are depleted as a result of starvation or protein malnutrition (*Chapter 10*).

Substances, such as cimetidine, that inhibit the cytochrome P-450 system without interfering with glucuronidation or sulfation could potentially reduce paracetamol hepatotoxicity.

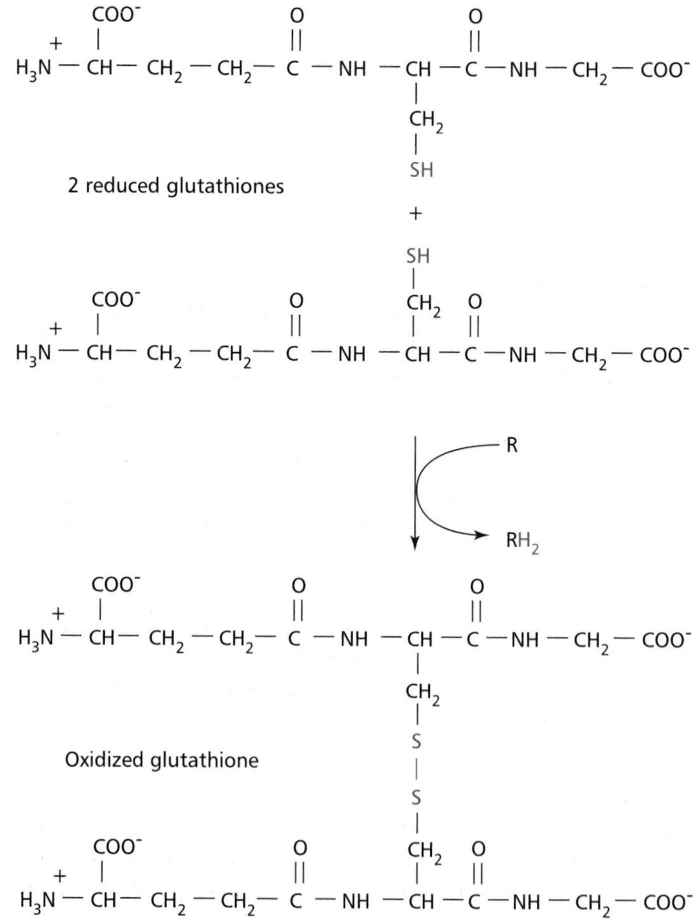

2 reduced glutathiones

Oxidized glutathione

Figure 12.6 The reduction of an organic compound (R) during the oxidation of reduced glutathione. See also *Figures 13.25* and *18.4*.

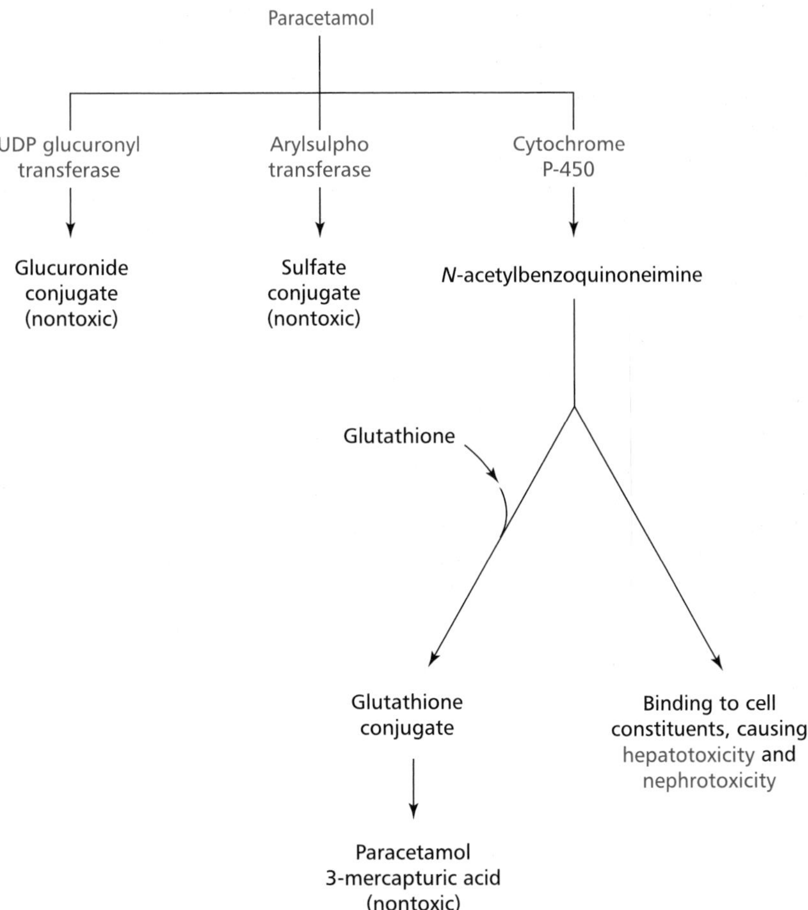

Figure 12.7 An overview of the detoxification and toxic effects of paracetamol. See text for details.

Toxic doses

A dose of 15 g for adults and 4 g for children is normally sufficient to cause hepatotoxicity. It has been suggested that this amount of paracetamol depletes liver glutathione concentrations by 70% in a 70 kg man. However, there is wide variation in the metabolic handling of paracetamol by the body and large overdoses of over 50 g have been known to have little effect in some patients. The incidence of hepatotoxicity in children is significantly lower at concentrations of paracetamol in blood that would be potentially toxic in adults. Animal studies have suggested that turnover of glutathione is age dependent, hence younger animals can tolerate higher doses of paracetamol.

Often estimates of the amount of paracetamol ingested, for example by a potential suicide, are unreliable. Thus, predictions of hepatotoxicity should be made only on the basis of serum concentrations. In general, concentrations of paracetamol greater than 300 mg dm^{-3} cause serious liver damage 4 h after ingestion whereas values below 120 mg dm^{-3} show no toxicity.

Clinical features of paracetamol poisoning

The signs and symptoms of paracetamol overdose are insidious, especially in the earlier stages. The clinical features can be divided into three phases, with a fourth occurring if the person survives toxicity and are described in *Table 12.2*.

Stage	Duration / h	Clinical features
I	0.5–24	loss of appetite and anorexia (*Chapter 10*), nausea and vomiting, general malaise, patient appears normal
II	24–48	less severe symptoms, abnormal blood chemistry with increases in liver enzymes and bilirubin, deterioration in renal functions but blood urea concentration remains low given the decrease in liver function
III	72–96	signs of hepatic necrosis, coagulation defects, jaundice and renal failure, reappearance of nausea and vomiting, death due to hepatic failure
IV	4–14 days	hepatic and renal functions return to normal if patient survives stage III

Table 12.2 Stages of paracetamol poisoning

Laboratory investigations of paracetamol poisoning

Estimating the concentration of paracetamol in plasma is useful for assessing the probability of patients developing hepatotoxicity. A nomogram (*Figure 12.8*) is available for paracetamol poisoning but should only be used when the size of the overdose and the approximate time of ingestion are known. Blood samples for paracetamol determination should be drawn at least 4 h postdose to allow for its complete absorption and the serum concentration to peak. The concentration of paracetamol in plasma can be used as a guide to patient management. Other tests that may be useful are determining the

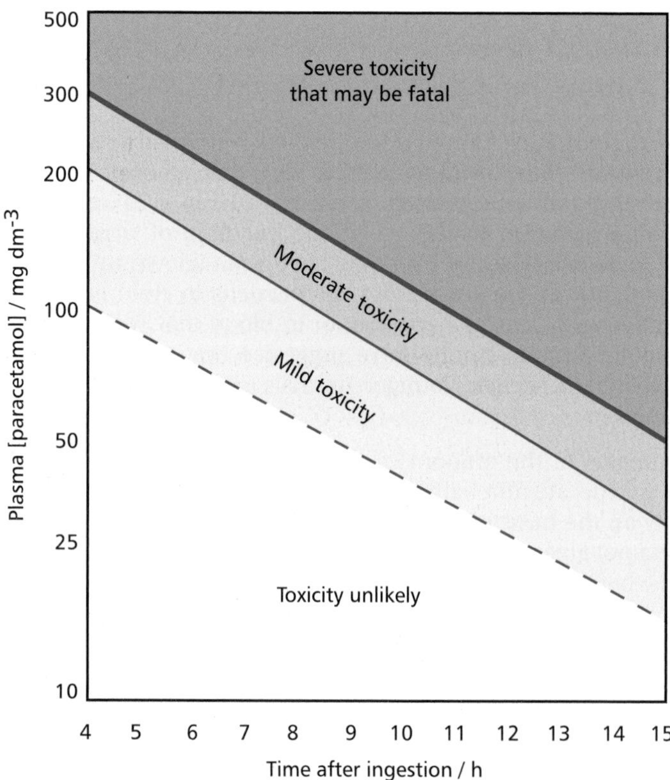

Figure 12.8 A nomogram to assess paracetamol toxicity.

activities of liver transaminases and measuring serum bilirubin to monitor liver functions (*Chapter 11*) and determining serum creatinine concentrations to assess renal function (*Chapter 8*). Paracetamol poisoning causes an increased prothrombin time (*Chapter 13*) which is the time taken for blood clotting to occur in a sample of blood to which calcium and thromboplastin have been added.

Management of paracetamol poisoning

Antidotes to paracetamol poisoning include methionine and *N*-acetylcysteine. Both promote the synthesis of glutathione in the liver, increasing its capacity to detoxify the active metabolite. Methionine may be given orally at 2.5 g every 4 h for 12 h in early or uncomplicated cases, while *N*-acetylcysteine is administered parenterally and is more appropriate for patients who present late, or are comatose or vomiting. *N*-acetylcysteine can act as a glutathione substitute and enhances conjugation with sulfate (*see earlier*). It also limits liver damage by reducing inflammation and improving the microcirculation in the liver. Treatment gives maximal benefit if started within 10 h of ingestion although it may still be beneficial for up to 24–30 h. General procedures, such as administration of activated charcoal, can reduce gastrointestinal absorption of paracetamol if given in the first hour of an overdose. Gastric lavage may be used in patients who have ingested large amounts of paracetamol and present within an hour of ingestion.

ASPIRIN

Aspirin (acetylsalicylic acid) is hydrolyzed in the body to salicylate, the active form of the drug (*Figure 12.9*), which has analgesic, antipyretic and anti-inflammatory properties. Salicylate is eliminated from the body by conjugation with glycine to form salicyluric acid and, to a lesser extent, with glucuronide to form phenol and acylglucuronides. A small amount is hydroxylated to gentisic acid. In an overdose, these pathways may become saturated and a large proportion of salicylate may be excreted unchanged in urine.

Like paracetamol, aspirin is widely and easily available and poisoning by aspirin overdose is therefore relatively common. Salicylate toxicity is due to a number of effects (*Figure 12.10*). An increase in the concentration of salicylate in the brain stimulates the respiratory center leading to hyperventilation, which can result in dehydration. In addition, hyperventilation causes a decline in the PCO_2 and a rise in pH, that is, respiratory alkalosis (*Chapter 9*). As PCO_2 declines, the pH rises and less H_2CO_3 is formed. The kidneys compensate by excreting more hydrogen carbonate (HCO_3^-) and K^+ whilst retaining H^+. These activities, rather than helping correct the respiratory alkalosis, contribute towards a latent metabolic acidosis. Salicylate also uncouples oxidative phosphorylation, decreasing ATP formation and increasing heat production, with sweating and hyperpyrexia, which further contributes to the dehydration and fluid loss. The decline in the amount of ATP stimulates glycolysis, and therefore pyruvate and lactate accumulate. An increase in glycogenolysis (*Margin Note 12.2*) provides the required glucose for this increase. Eventually there may be depletion of glycogen giving rise to

Margin Note 12.2 Glycogenolysis 

Glycogenolysis is the breakdown of glycogen reserves to yield glucose 1-phosphate in a reaction catalyzed by glycogen phosphorylase:

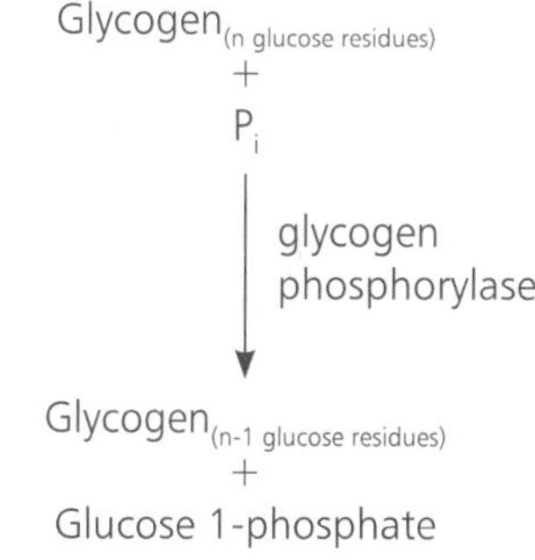

The glucose 1-phosphate can be oxidized to give ATP for use by the cell or, in liver cell particularly, hydrolyzed to form glucose that can be released into the bloodstream to maintain blood glucose concentrations.

Aspirin + H_2O → Salicylic acid + CH_3COOH

Figure 12.9 The conversion of aspirin to salicylic acid.

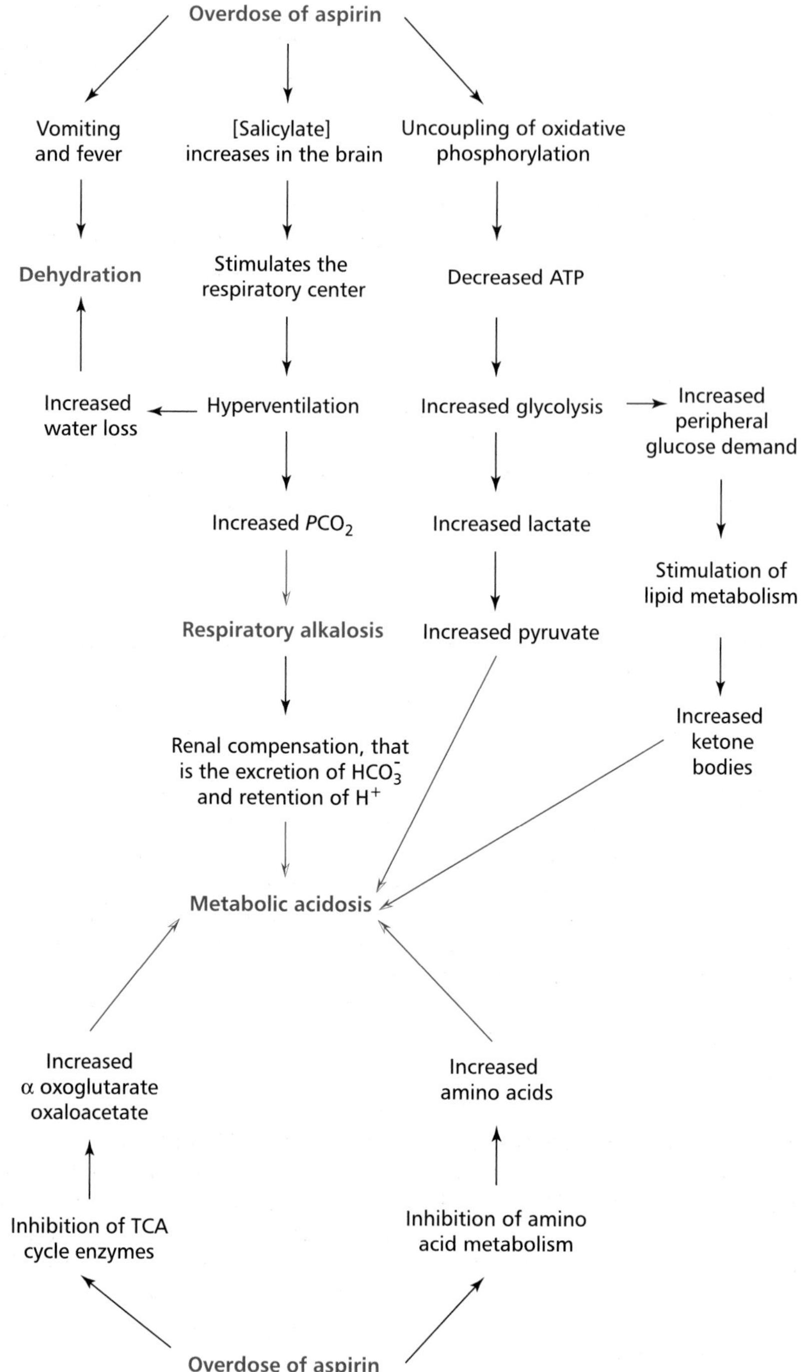

Figure 12.10 An overview of the toxic effects of aspirin. See text for details.

hypoglycemia and the catabolism of lipids producing ketone bodies (*Chapter 7*). Salicylate also inhibits enzymes of the TCA cycle, leading to an accumulation of oxoglutarate and oxaloacetate, and of amino acid metabolism, causing an increase in amino acids that accentuates the metabolic acidosis. Fluid and electrolyte losses are increased by the nausea and vomiting.

Toxic dose

Common therapeutic levels of salicylate are 50 mg dm^{-3} although they can be as high as 250 mg dm^{-3} in the serum of patients with rheumatoid arthritis (*Chapter 5*). The potentially lethal dose of aspirin in adults is 24 to 30 g but death can occur in children under 18 months from as little as 300 mg. Signs of salicylate toxicity occur when concentrations are greater than 300 mg dm^{-3}. In severe cases of aspirin poisoning, the concentrations in serum can be as high as 1000 mg dm^{-3}.

Clinical features of aspirin poisoning

Numerous symptoms are associated with aspirin poisoning, including nausea, vomiting, sweating, hyperventilation, tinnitus (buzzing noise in ear), confusion and/or unconsciousness and a severe loss of fluid.

Laboratory investigations of aspirin poisoning

The concentration of salicylate in plasma should be measured on presentation and every 4 to 6 h until it has fallen below the toxic range. This is necessary as salicylates precipitate in acid conditions and may therefore deposit in large amounts in the stomach. The consequence is delayed absorption and this means that the concentration in plasma may continue to rise for many hours following a severe overdose.

Management of aspirin poisoning

A variety of measures can be taken to alleviate salicylate poisoning, which are aimed at decreasing the absorption of salicylate, increasing its rate of elimination and correcting the acid–base and electrolyte disturbances. These measures include gastric lavage for up to 24 h after ingestion as salicylate may remain unabsorbed in the GIT for long periods. Patients may also be given 50 g of activated charcoal followed by 25 g every 4 h. The charcoal binds salicylate and prevents its absorption. The ionization state of salicylate affects its reabsorption by the kidneys. If the provisional urine is acidic, salicylate is not ionized and is filtered at the glomerulus but reabsorbed from tubules (*Chapter 8*); if the urine is alkaline the salicylate is ionized and its tubular reabsorption is reduced and more salicylate is lost from body. Hence patients are infused with sodium hydrogen carbonate to increase the pH of urine and promote loss of salicylate. Fluid replacement and corrections of acid–base, electrolyte imbalance, especially the hypokalemia, and hypoglycemia are all required.

Hemodialysis may be necessary in severe cases of poisoning, such as when plasma concentrations of salicylate exceed 800 mg dm^{-3}. As well as removing the salicylate, hemodialysis also corrects the acid–base and electrolyte imbalances. The technique requires a dialyzer (*Figure 12.11*). Blood from a patient's artery is circulated through the dialyzer on one side of a semipermeable membrane while a solution of normal electrolytic composition circulates on the other side. Waste products, poisons, including salicylate, and small molecules cross the membrane and the dialyzed blood is returned to the body via a vein.

ETHANOL

Ethanol is an addictive drug and its abuse can lead to dependency and alcoholism. Its abuse is increasingly common in the developed world. The toxic effects of chronic alcohol abuse on the liver, brain and GIT are widely known. Ethanol can also modify the effects of other drugs, for example it inhibits the hydroxylation of barbiturates by the P-450 system preventing their ready excretion by the kidneys.

The metabolism of ethanol occurs mainly in the liver by one of two mechanisms. Normally only small amounts are degraded by the P-450 system (*Figure 12.12*), although this oxidation can become of major importance because the P-450 system is induced by chronic alcohol consumption. The

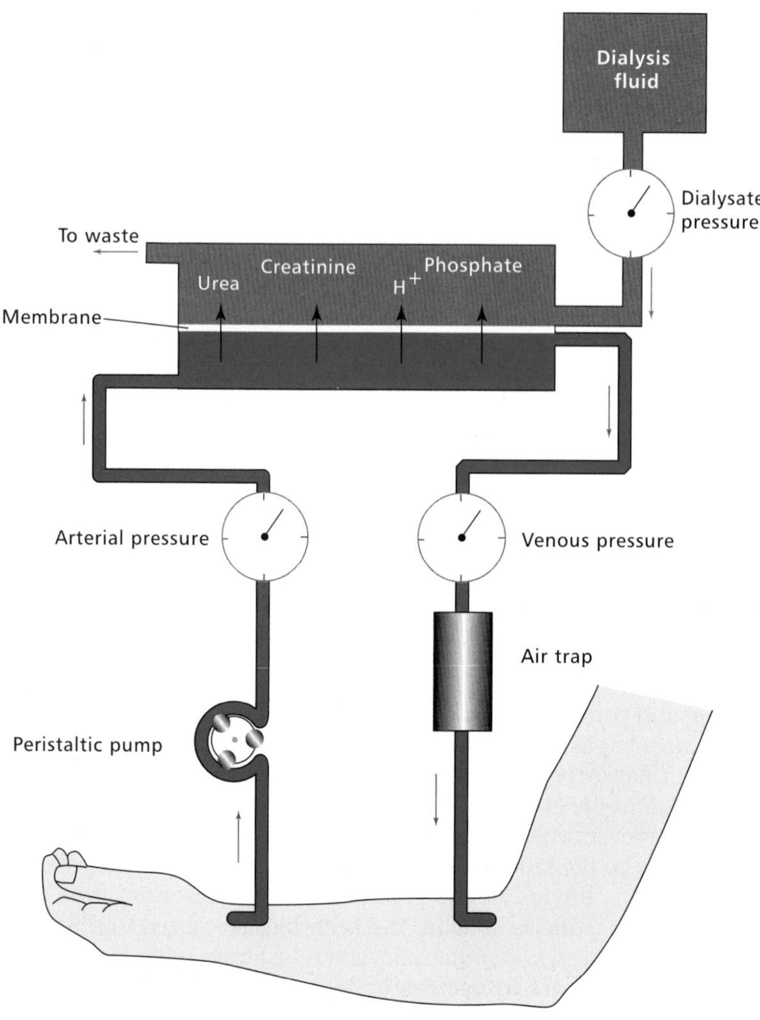

Figure 12.11 A schematic illustrating the procedure of hemodialysis.

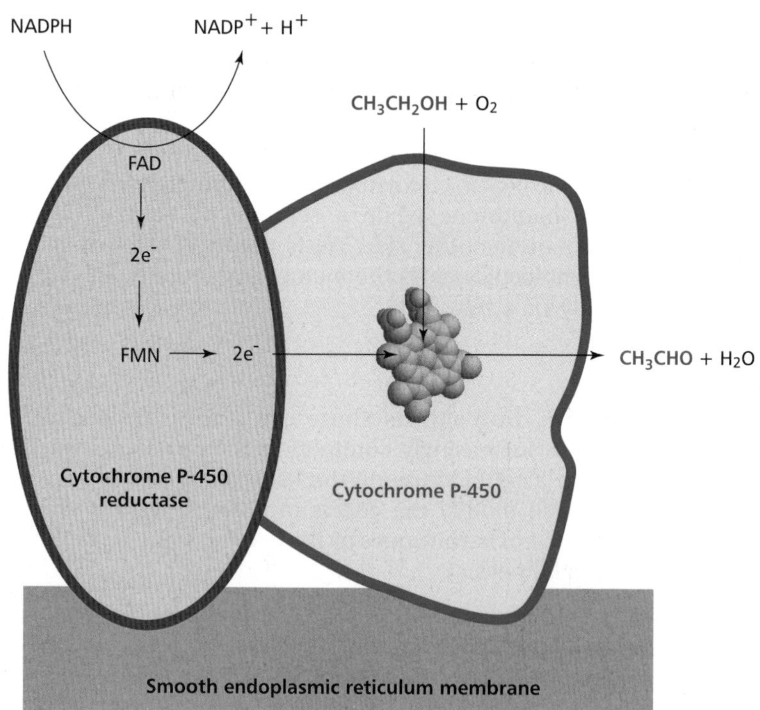

Figure 12.12 The role of P-450 in the detoxification of ethanol.

Figure 12.13

Figure 12.13 Molecular model of alcohol dehydrogenase. The black spheres represent Zn atoms and the bound NADH is shown in gray. PDB file 1HSO.

major catabolic pathway is to oxidize ethanol to the corresponding aldehyde, ethanal, in a reaction catalyzed by alcohol dehydrogenase, ADH (*Figure 12.13*).

$$CH_3CH_2OH + NAD^+ \xrightarrow{\text{Alcohol dehydrogenase}} CH_3CHO + NADH + H^+$$

The effectiveness of ADH varies between different populations and hence the undesirable effects of drinking can appear after widely varied intakes. The ethanal may subsequently be oxidized to ethanoic acid (acetic acid) by aldehyde dehydrogenase, ALDH (*Figure 12.14*).

$$CH_3CHO + NAD^+ \xrightarrow{\text{Aldehyde dehydrogenase}} CH_3COOH + NADH + H^+$$

Much of the acetate made from ethanol escapes into the blood and can result in acidosis (*Chapter 9*). The effect of both enzymes is to increase the NADH/NAD$^+$ ratio, that is, alcohol consumption leads to the accumulation of NADH with consequent severe effects. The increased NADH inhibits fatty acid oxidation and stimulates the synthesis of triacylglycerols in the liver producing a fatty liver. The oxidation of lactate to pyruvate is also inhibited, slowing gluconeogenesis (*Margin Note 12.3*). The increased lactate exacerbates the acidosis and the decreased gluconeogenesis may cause hypoglycemia.

The capacity of liver mitochondria to oxidize acetate to CO_2 is limited because the activation of acetate to acetyl CoA requires ATP:

$$\text{Acetate} + \text{Coenzyme A} + \text{ATP} \longrightarrow \text{acetyl CoA} + \text{AMP} + PP_i$$

Adenosine triphosphate is now in short supply because glycolysis, which requires free NAD, is slowed and because the processing of acetyl CoA by the TCA cycle is blocked since NADH inhibits the regulatory enzymes isocitrate dehydrogenase and 2-oxoglutarate dehydrogenase. Acetyl CoA is converted to ketone bodies (*Chapter 7*) that are released into the blood, intensifying the acidosis caused by acetate and lactate. Acetaldehyde also accumulates and this extremely reactive compound can bind to liver proteins, impairing their functions and causing severe damage to liver cells, leading to their death. Acetaldehyde can also escape from the liver and react with blood proteins to form stable adducts. These can provide useful markers of the past drinking activity of an individual.

Liver damage from excessive alcohol consumption occurs in three stages. The first stage is the formation of the fatty liver described above. This condition, in the absence of other complications, is readily reversible within four to six weeks if alcohol is avoided. The second stage is the occurrence of alcoholic hepatitis when groups of liver cells die. This leads to inflammation and can be fatal. In the third stage, the patient may develop cirrhosis, a condition seen in 10% to 15% of alcoholics (*Figure 12.15 (A)* and *(B)*). Approximately half of all cases of cirrhosis are due to alcoholic liver disease. Cirrhosis occurs when fibrous structures and scar tissue are produced around the dead cells. This impairs many of the biochemical functions of the liver, for example, cirrhotic liver cannot convert ammonia to urea and the concentration of ammonia in the blood rises. Ammonia is toxic to the nervous system and can cause coma

Figure 12.14

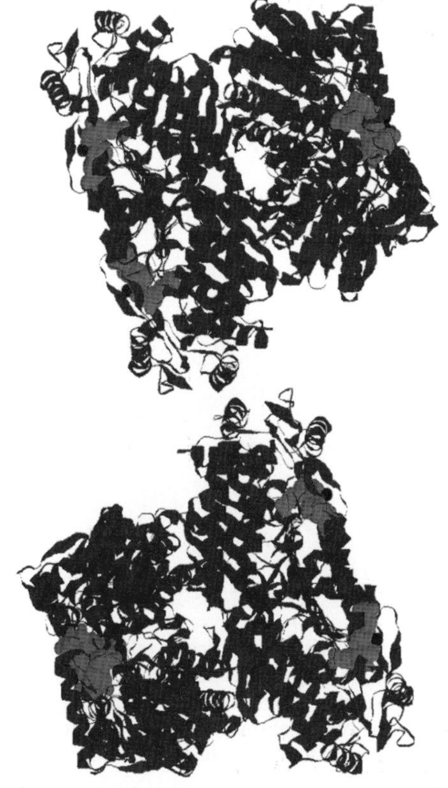

Figure 12.14 Molecular model of aldehyde dehydrogenase. The black spheres represent Mg atoms and the bound NADH is shown in gray. PDB file 1O02.

Margin Note 12.3 Gluconeogenesis

Gluconeogenesis is the synthesis of glucose using materials that are not carbohydrates as precursor molecules. These include pyruvate, the end product of glycolysis, oxaloacetate, an intermediate of the TCA cycle, and dihydroxyacetone phosphate, which can be made from glycerol obtained from the hydrolysis of triacylglycerols.

and death. Ethanol also directly affects the central nervous system (CNS). For example, it enhances the inhibitory affects of γ aminobutyric acid (GABA) at the $GABA_A$ receptor and the functions of some 5-hydroxytryptamine receptors, in addition to numerous other membrane proteins. The resulting depressive effects of these activities are well known.

Chronic liver damage invariably results in malnutrition, partly due to malabsorption and partly because the metabolism of nutrients by the liver is defective. However, many alcoholics are malnourished because of dietary inadequacies. While ethanol supplies most of their energy needs they may be ingesting insufficient amounts of other nutrients, particularly proteins and vitamins (*Chapter 10*). In addition, chronic alcohol ingestion can damage the mucosal lining of the GIT and pancreas as well as the liver (*Chapter 11*). Alcoholics require increased amounts of vitamins and some trace elements because of the metabolic load experienced and their increased excretion. For example, niacin, although not strictly a vitamin since it can be formed, albeit very inefficiently, from tryptophan (*Chapter 10*), is necessary to form the coenzymes, NAD^+ and $NADP^+$. A borderline deficiency of niacin leads to **glossitis** (redness) of the tongue while a pronounced deficiency leads to pellagra, with dermatitis, diarrhea and dementia. In developed countries, pellagra is rarely encountered other than in alcoholics, given their severe malabsorption problems, hence dietary supplies of niacin and tryptophan are required. A severe zinc deficiency (*Chapter 10*) also occurs primarily in alcoholics, especially those suffering from cirrhosis. Heavy drinkers frequently suffer from GIT varicose veins (*Chapter 14*) and diarrhea caused by a variety of factors, including ethanol-exacerbated lactase deficiency and interference with normal peristalsis. Steatorrhea is also common, due to deficiencies in folic acid and bile salts in the GIT (*Chapter 11*).

The obvious treatment for alcoholic liver disease and cirrhosis is abstinence. Given the addictive nature of ethanol, this may be difficult to maintain. Unfortunately, patients presenting with severe alcoholic hepatitis have a high mortality even when abstinence is successful. Although a number of pharmacological treatments have been attempted as therapies, none has been successful. However, supportive nutritional management in cases of acute and chronic liver disease is essential to achieve electrolyte, vitamin and protein replenishments. The only treatment for advanced cirrhosis is a liver transplant (*Chapter 6*).

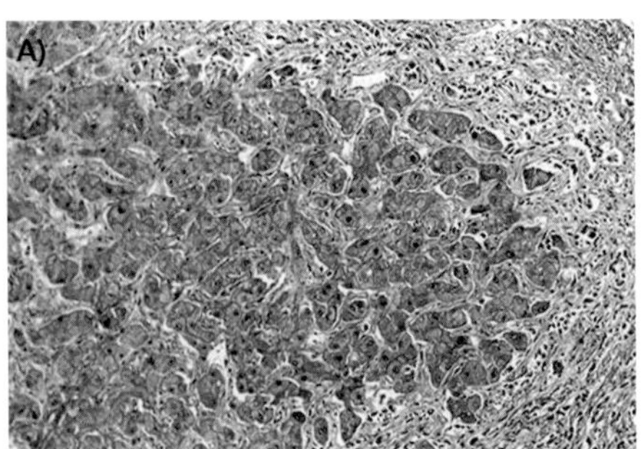

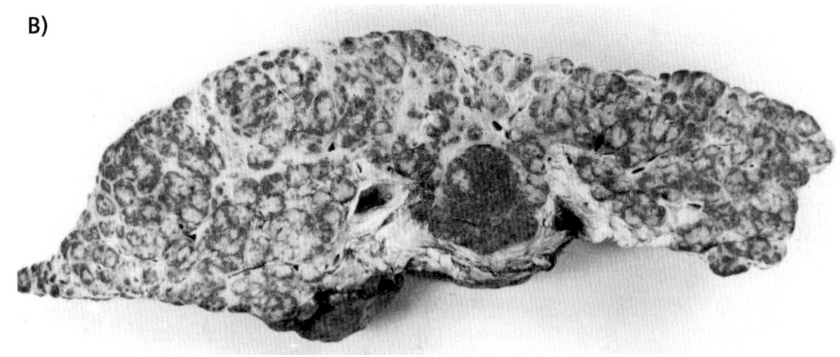

Figure 12.15 Cirrhosis in alcohol related liver disease. (A) A slide showing how alcohol abuse has led to diffuse scarring of the liver which has regenerated at numerous points to form nodules. The sinusoidal fibrosis is well established in this small regenerative nodule where the regenerating cells are darkly stained compared with the lighter fibrous tissue. As the nodules get larger, the fibrous tissue round the outside becomes compressed giving the characteristic hob-nail appearance to the liver as shown in (B), which is a transected cirrhotic liver from an alcoholic. The extensive damage in the form of nodules of varying sizes separated by bands of fibers is apparent. Courtesy of Professor F.A. Mitros, The University of Iowa, USA and Dr N.P. Kennedy, Trinity College Dublin, Republic of Ireland respectively.

The major enzyme involved in ethanol metabolism, alcohol dehydrogenase, has a broad specificity and can catalyze the oxidation of a range of alcohols. This allows ethanol to be used as a competitive inhibitor of the enzyme in the treatment of poisoning by other alcohols, such as methanol and ethylene glycol.

METHANOL

Methanol is widely used as a solvent and as antifreeze. It is a potent poison with as little as 10 cm^3 causing blindness and 30 cm^3 can cause death. Toxicity can come from ingestion, inhalation and skin exposure. Mass poisonings have occurred from drinking alcoholic beverages made with contaminated ethanol and from accidental exposure. The major cause of methanol toxicity is its initial oxidation to formaldehyde, which is then converted to formate. The first step in the pathway is catalyzed by alcohol dehydrogenase and the production of formate is the result of several enzyme activities.

While both formaldehyde and formate are toxic, formaldehyde has a short metabolic half-life, whereas formate accumulates since its metabolism to CO_2 is slow in humans. This leads to metabolic acidosis. The symptoms of methanol poisoning are an initial mild inebriation and drowsiness. Visual disturbances, such as blurred vision, diminished visual acuity, dilated pupils occur after about 6 h. Within 8–36 h nausea, vomiting, abdominal pain, headaches and possibly coma occur. Treatment for methanol poisoning is, first, to administer ethanol. This blocks metabolism since alcohol dehydrogenase has a greater affinity for ethanol than methanol. Second, sodium hydrogen carbonate is given intravenously to correct the metabolic acidosis.

ETHYLENE GLYCOL

The alcohol ethylene glycol is a commonly used component of antifreeze, paints, polishes and cosmetics. It is a poison with a minimum lethal dose of approximately 100 cm^3. The toxicity of ethylene glycol is due to its oxidation to oxalate, which humans cannot rapidly excrete. The oxidative steps produce excessive NADH resulting in lactate production (*Figure 12.16*). The oxalate and lactate can result in metabolic acidosis, while the organic acids produced by the breakdown of ethylene glycol inhibit a number of metabolic processes

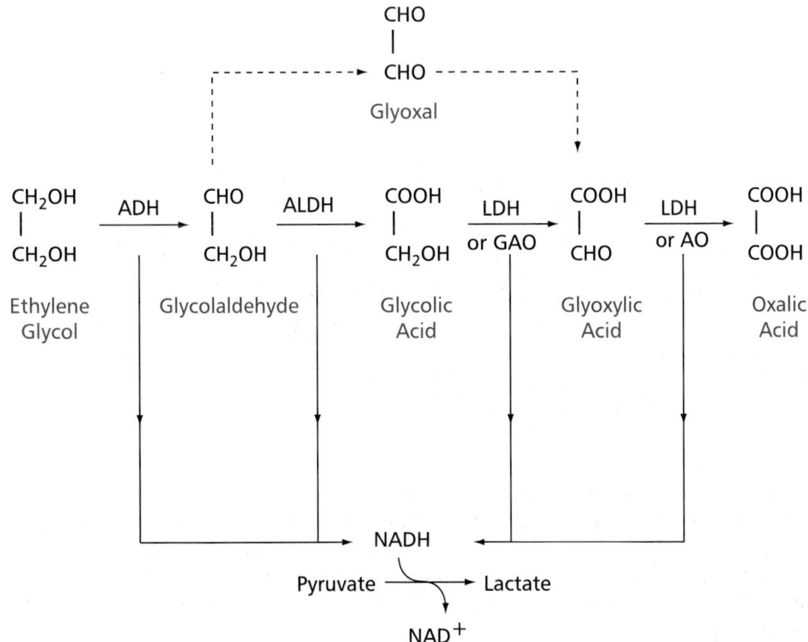

Figure 12.16 The formation of lactate during the catabolism of ethylene glycol.

including oxidative phosphorylation. The production of oxalate also results in its deposition as insoluble calcium oxalate in renal tubules and the brain. These biochemical events reduce the concentrations of hydrogen carbonate and Ca^{2+}, but increase that of K^+ in plasma. Crystals, blood and protein may all leak into the urine, giving rise to three distinct clinical phases. Within 30 min to 12 h, intoxication, nausea, vomiting, coma, convulsions, nystagmus, papilloedema, depressed reflexes, myclonic jerks, tetanic contractions and permanent optic atrophy may occur. In 12 to 23 h tachypnea, tachycardia, hypertension, pulmonary edema and congestive cardiac failure may all present. In the following 24 to 72 h kidney damage with pain and acute renal tubular necrosis may feature. Death may occur within 24 h due to damage to the CNS, or between eight to 12 days due to renal failure.

Treatment of ethylene glycol poisoning is to apply gastric lavage to reduce its absorption, combined with supportive therapies for the shock and respiratory distress. Administration of ethanol is standard since it competes effectively for the active site on alcohol dehydrogenase, inhibiting the metabolism of the absorbed ethylene glycol. Sodium hydrogen carbonate, administered intravenously, and calcium gluconate are used to correct the acidosis and hypocalcemia respectively. Dialysis is also used to remove ethylene glycol.

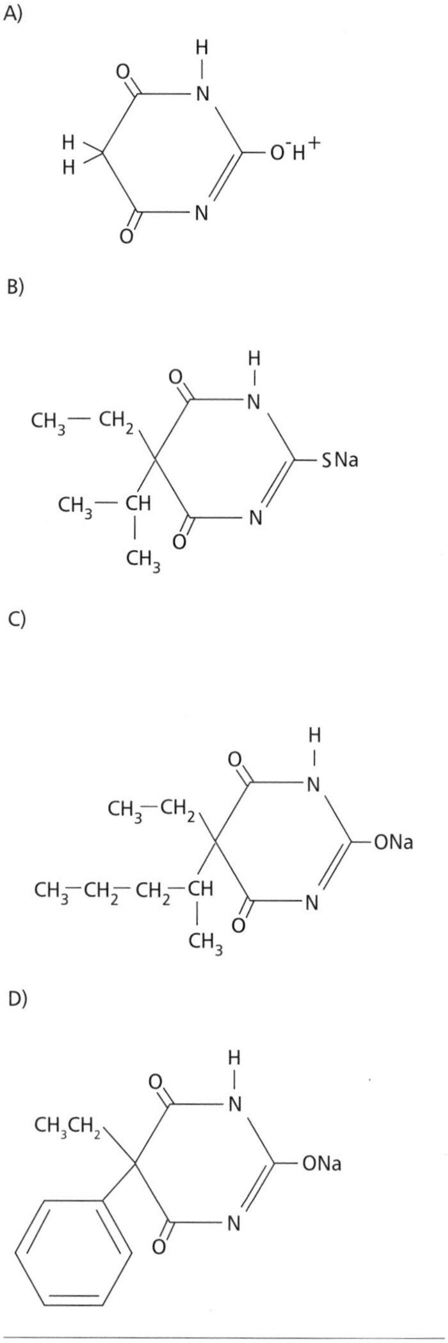

A)

B)

C)

D)

Figure 12.17 The barbiturates (A) barbituric acid, (B) thiobarbitone, (C) pentobarbitone and (D) phenobarbitone.

BARBITURATES

Barbiturates are a group of drugs based on the parent compound, barbituric acid (*Figure 12.17 (A)*). All are sedatives, that is, they depress certain activities of the CNS. Although they have well-established therapeutic uses, barbiturates can be toxic if taken as an overdose, indeed, they are one of the commonest methods for attempting suicide. They were once used as recreational drugs, for example in purple hearts, and this too led to accidental overdoses. Barbiturates have now largely been replaced by benzodiazepines and barbiturate overdose is less likely to be encountered today.

Barbiturates induce drug dependence and this involves three distinct and independent components: tolerance, physical dependence and compulsive abuse or psychic craving. Barbiturate dependence stimulates all three components to such an extent that they produce major problems for the individual user as well as for society at large. Once barbiturate dependence has developed, abrupt withdrawal from the drug induces a particularly unpleasant withdrawal syndrome. This is characterized by weakness, tremors, anxiety, increased respiratory and pulse rates with a corresponding increased blood pressure, vomiting, insomnia, loss of weight, convulsions of the grand mal type and a psychosis resembling alcoholic delirium tremens. Thus during withdrawal, it is appropriate to reduce the dosage gradually over an extended period.

The length of time barbiturates act *in vivo* varies. Some, for example thiobarbitone (*Figure 12.17 (B)*), are short acting, others, such as pentobarbitone (*Figure 12.17 (C)*), act in the medium term, while the group including the best known barbiturate, phenobarbitone (*Figure 12.17 (D)*), induce long-lasting effects. Barbiturates act by enhancing the effects of GABA by binding to a different site on the receptors on target neurons of the CNS to those for GABA itself. The blood–brain barrier (*Figure 3.4*) prevents the easy entry of many substances into the brain compared with their uptake into other tissues. However, lipid-soluble substances, such as the barbiturate thiopentone, enter the brain quickly by passive diffusion. This rapid uptake allows thiopentone to exert its anesthetic effects extremely quickly. In contrast, other barbiturates, such as phenobarbitone, are weak acids and so may be ionized. This slows their entry into the CNS but produces a longer lasting effect. The life-threatening effects of barbiturate poisoning include depression of the centers in the CNS that control respiration and blood circulation. Some barbiturates, for example phenobarbitone, can also induce the synthesis of monooxygenase enzymes and so, by altering the rate or route of metabolism of other drugs, can alter their toxicities.

There is no antidote for barbiturate poisoning. Primary care is to maintain a free airway, administer artificial respiration if required and forced alkaline diuresis. The pH dependence of ionization of many common barbiturates is exploited by infusions of large volumes of sodium hydrogen carbonate containing the osmotic diuretics urea or mannitol. This increases the pH of plasma relative to the cytoplasm of cells and increases the proportion of ionized barbiturate in the plasma causing more of the un-ionized drug to diffuse out of the tissues, including the brain, into the plasma. This can promote a diuresis as large as 12 dm^3 in 24 h. The ionized form is also excreted more rapidly since the alkalinity of the provisional urine ensures it remains ionized and cannot cross the tubule wall back into the plasma.

CARBON MONOXIDE

Carbon monoxide (CO) is a poisonous gas. However it is not an irritant and is odorless. Hence it is insidious and concentrations can build up with the victim being unaware of any danger. Carbon monoxide is no longer present in domestic gas in the UK but is still a major cause of poisoning, both accidental and intentional and results in the deaths of several hundred people in the UK each year. Sources of CO include domestic fires, ovens and boilers, coal gas, furnace gas, cigarette smoke, burning plastic and car exhausts, although catalytic converters have reduced the CO output from car petrol engines. Thus traffic policemen, firemen, those trapped in fires and some factory workers are all potentially at a greater risk. In the UK, the major cause of poisoning results from exposure to inefficient oxidation in engines and poorly maintained gas fires, ovens and boilers, especially where ventilation is inadequate.

The mechanism of CO poisoning is well understood at the biochemical level. The gas is absorbed rapidly through the lungs and binds to the iron atom of hemoglobin at same site as dioxygen, but about 240 times more strongly. This prevents the efficient distribution of oxygen to the tissues. The product of CO binding is carboxyhemoglobin. Carbon monoxide is potentially extremely poisonous at low concentrations. Given that air contains 21% O_2, approximately 0.1% CO will saturate 50% of the hemoglobin. Concentrations of 60% carboxyhemoglobin in the blood are usually fatal, even if maintained for only a few minutes. A concentration of 20% carboxyhemoglobin may not present obvious symptoms but the ability to perform tasks can be impaired. At 20–30%, the victim may have a headache, with raised pulse, dulling of the senses and feelings of weariness. Concentrations of 30–40% accentuate these symptoms and decrease the blood pressure so that exertions may lead to faintness. At 40–60% and above, the victim becomes unconscious and will suffer convulsions. Other clinical features include pink skin, nausea, vomiting, loss of hearing, hyperpyrexia, hyperventilation, a decrease in light sensitivity, renal failure and acidosis.

The main target organs of CO poisoning are the heart and brain. These organs extensively utilize aerobic metabolic pathways and so their abilities to sustain an oxygen debt are relatively poor. Death is due to brain tissue hypoxia although cardiac arrythmias and heart and respiratory failures may also occur. The duration of exposure is also a factor since hypoxic cell death is not instantaneous. Also, some individuals, for example those with anemia (*Chapter 13*), are more sensitive to CO poisoning than healthy people.

The treatment of CO poisoning involves removing its source and supplying the victim with fresh air or oxygen. The use of 100% oxygen at 2.5×10^5 Pa pressure, that is hyperbaric oxygen, increases the rate of dissociation of carboxyhemoglobin and reduces its plasma half-life from 250 min in a patient breathing air to 23 min. Adding carbon dioxide can be useful as this reduces the half-life to 12 min at normal pressure.

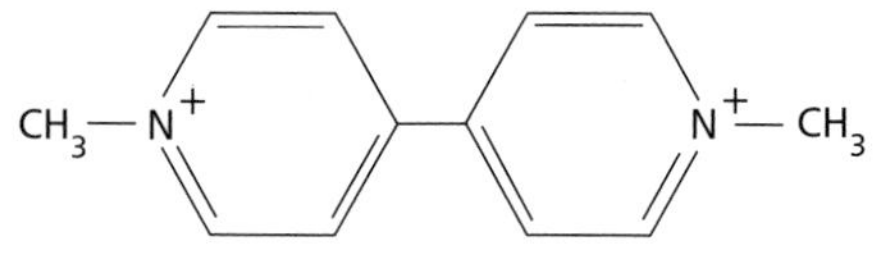

Figure 12.18 **Paraquat.**

PARAQUAT

Paraquat (*Figure 12.18*) is used widely as a weed killer, hence is often found in the home. Unfortunately paraquat has caused many hundreds of deaths both by accidental and deliberate poisoning, the latter including both suicide and homicide.

Paraquat is a local skin irritant, causing inflammation, although poisoning usually follows oral ingestion. The toxic effects are dose related and with small amounts there may be minimal damage that is reversible. Fatal doses cause a painful death within several days or weeks, with extreme abdominal pain, vomiting and diarrhea. The major target organs are the lungs with larger doses causing alveolar edema, resulting in destruction of lung tissues and fibrosis, if the patient survives beyond a few days. Pulmonary fibrosis and respiratory failure can, however, develop up to six weeks after ingestion. The kidneys, heart and liver may also be damaged but the lungs are particularly susceptible in paraquat poisoning because alveolar epithelial cells actively accumulate paraquat to toxic concentrations. Furthermore, the presence of large concentrations of oxygen in the organs exacerbates the morbid effects. Paraquat in alveolar epithelial cells is reduced by electron donors, such as NADH, to a stable reduced form (*Figure 12.19*). Given the

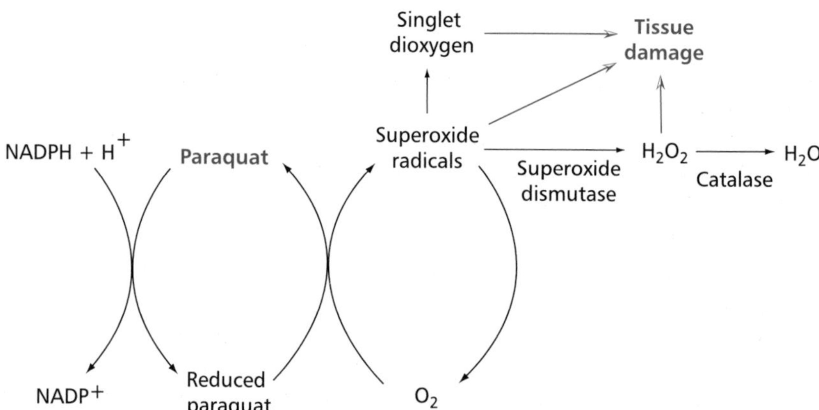

Figure 12.19 **An overview of the detoxification and toxic effects of paraquat.** See text for details.

prevailing aerobic conditions, this transfers an electron to dioxygen to form a superoxide radical:

$$O_2 + e^- \longrightarrow O_2^{\bar{\bullet}}$$

(superoxide radical)

Superoxide dismutases (SOD) can remove superoxide radicals by converting them to hydrogen peroxide:

$$2H^+ + O_2^{\bar{\bullet}} + O_2^{\bar{\bullet}} \xrightarrow{\text{SOD}} H_2O_2 + O_2$$

(hydrogen peroxide)

which can be degraded to water and dioxygen by the action of catalase:

$$2H_2O_2 \xrightarrow{\text{Catalase}} 2H_2O + O_2$$

However, the actions of SOD and catalase are likely to be overwhelmed in aerobic conditions and the superoxide radicals accumulate and react with hydrogen peroxide to give highly toxic hydroxyl radicals, particularly in the presence of ions of transition metals, such as iron and copper:

$$O_2^{\bar{\bullet}} + H_2O_2 \xrightarrow{Fe^{2+}/Cu^{2+}} OH^{\bar{\bullet}} + OH^- + O_2$$

(hydroxyl radical)

These lead to a variety of toxic effects, such as lipid peroxidation (*Figure 12.20*). The resulting lipid peroxides may give rise to lipid radicals and membrane damage leading to tissue damage and fibrosis. Lipid peroxides will also oxidize glutathione (*Figure 12.6*) and its reoxidation further reduces the depleted amount of NADPH. This reduces the ability of the alveolar cells to carry out essential functions, such as biosynthetic repairs.

There is no antidote for paraquat poisoning and once it has accumulated in the lungs little can be done to prevent its toxic effects. Treatment largely consists of trying to prevent absorption by the gut by gastric lavage and by using Fullers Earth as an adsorbent. Hemoperfusion may also be used to reduce the concentration of any paraquat already absorbed.

METALS

The presence of certain metals in high concentrations may be toxic to humans. To diagnose metal toxicity, three features have to be identified, namely, a source of the toxic metal, the presence of signs and symptoms typical of toxicity by that metal and increased concentrations of the metal in the body tissues. Metals that are commonly screened for toxicity include lithium, aluminum and the heavy metals lead, arsenic, cadmium and mercury. Lithium is widely used therapeutically to treat patients with certain psychiatric disorders. However, plasma concentrations of lithium in excess of 1.5 mmol dm^{-3} should be avoided and regular measurements of serum lithium concentrations are important in monitoring therapy. Lithium toxicity is associated with tremors, drowsiness, tinnitus, blurred vision, polyuria, hypothyroidism and, in severe cases, renal failure and coma.

Acute poisoning with aluminum is extremely rare. Indeed, aluminum compounds are used for their antacid properties. The dietary intake of aluminum is 5 to 10 mg day^{-1} and this amount is removed completely by the kidneys. Unfortunately, patients with renal failure are susceptible to aluminum toxicity. They cannot remove the aluminum and, as the water used in dialysis may contain aluminum that can enter the body through the dialysis membrane, the metal can build up to toxic concentrations and cause osteodystrophy and encephalopathy. Aluminum toxicity is diagnosed by determining its concentration in plasma. Chronic toxicity occurs at concentrations above only 3 μmol dm^{-3} whereas 10 μmol dm^{-3} can cause acute poisoning. The treatment is aimed mainly at prevention. When aluminum poisoning does occur, then its excretion may be enhanced using chelating agents such as desferrioxamine.

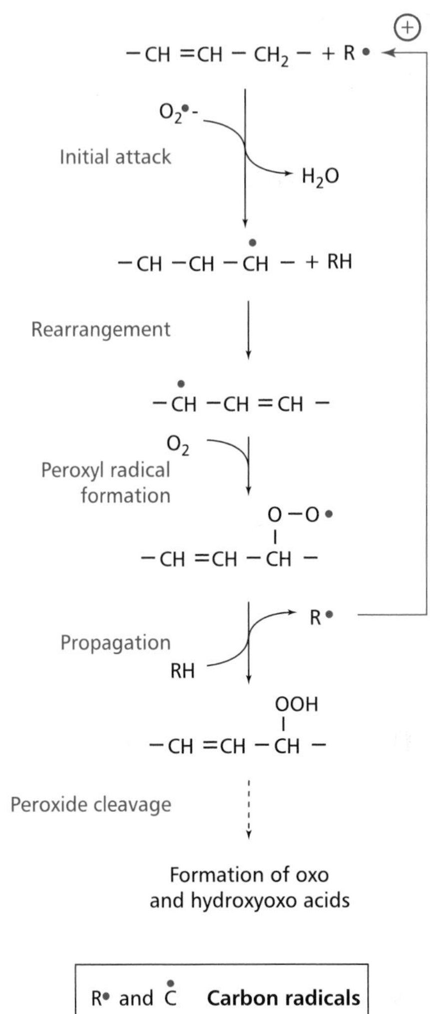

Figure 12.20 An overview of lipid peroxidation reactions. The peroxidation is initially started by a reactive radical. However, as carbon radicals are formed, the peroxidation becomes self sustaining.

BOX 12.1 Aluminum poisoning and Camelford, Cornwall

If aluminum sulfate, $Al_2(SO_4)_3.18H_2O$ (*Figure 12.21 (A)*) is dissolved in water it hydrolyzes to form sulfuric acid and a gelatinous precipitate of aluminum hydroxide:

$$Al_2(SO_4)_3 + 6H_2O \rightarrow 3H_2SO_4 + 2Al(OH)_3$$

An 8% solution of aluminum sulfate is commonly used in sewage treatment and water purification (*Figure 12.21 (B)*) because it is a highly effective coagulant that reacts with suspended organic matter to form flocs of large size. These rapidly settle, reducing the load of suspended solids and the turbidity of the water or sewage so improving their quality and lengthening the effective life of water and sewage filters.

The discharge of aluminum sulfate into aquatic environments should be minimized as it can clog the gills of marine organisms. With regard to handling, it is regarded as a weak acid and should be treated accordingly: contact with eyes, skin and clothing and repeated and prolonged exposure should be avoided. However, tests have shown it to be a minimal irritant and a nonirritant to rabbit eyes and skin respectively. Its LD_{50} dose in rabbits is greater than 5 g kg^{-1} so, in general, its ingestion is thought to be relatively harmless.

Unfortunately, in 1988 a contractor accidentally dumped 20 tons of aluminum sulfate into the wrong tank at a water

Figure 12.21 (A) Hydrated aluminum sulfate as seen using polarized light. (B) A modern water purification plant. Courtesy of United Utilities, Warrington, UK.

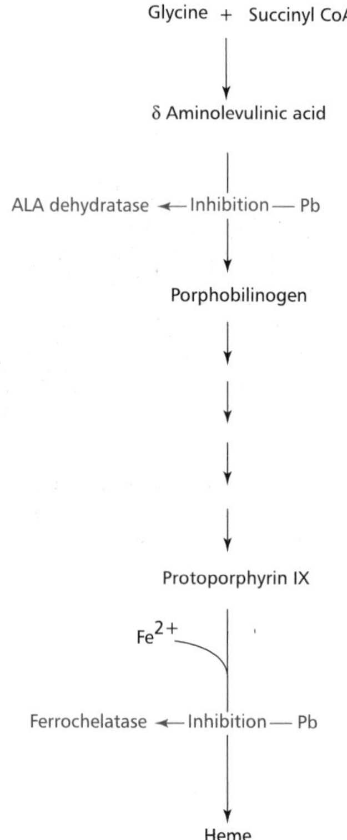

Lead is a heavy metal found in the environment. Common sources include paints, old water pipes, petrol fumes, industrial pollution and medication and cosmetics from the Indian subcontinent. Poisoning by lead can be acute, which is rare, or, more commonly, chronic. It results from inhalation, dermal absorption or ingestion, where it is actively absorbed by the GIT through the calcium transport system. All tissues are affected by lead poisoning although organ pathology is associated primarily with the nervous system and the blood. In the blood, lead is concentrated in the erythrocytes where it inhibits the activities of aminolevulinic acid (ALA) dehydratase and ferrochelatase, enzymes involved in the synthesis of heme (*Figure 12.22*). Inhibition of ALA dehydratase is the most sensitive measure of lead poisoning. Concentrations of lead in the blood in excess of 0.49 μmol dm^{-3} are associated with illness in children. If the concentration of lead increases above 3.4 μmol dm^{-3}, individuals

Figure 12.22 A schematic showing the enzymes of heme biosynthesis affected by lead poisoning. ALA is aminolevulinic acid. See text for details.

treatment plant that supplied the 2500 residents of the small town of Camelford in Cornwall, UK, with domestic water. The mistake was later attributed to the contractor being a relief driver who was not familiar with the layout of the plant, itself an unmanned installation and so on site advice was not available. The contaminated water entered the town's supplies. Its pH was as low as 3.9 and contained concentrations of aluminum and sulfate of 620 and 4500 mg dm^{-3} respectively. Within two days, many residents of Camelford complained about the poor taste of the water, skin irritation and corrosive effects of the water on plumbing and fixtures. Although the cause of the problem with the water was not solved for two days, the responsible water authority assured the towns people that the water, while tasting slightly acidic, was safe to drink. Once the cause of the problem was determined, a program of flushing the water supplies rapidly reduced the concentration of aluminum to 1 mg dm^{-3}. After several months, over 400 complaints had been made relating to skin rashes, sore throats, painful joints, memory losses and exhaustion. Following the incident, the water authority responsible was prosecuted for causing a public nuisance and fined £10 000 with additional legal costs of £25 000.

Reports were published in 1989 and 1991 following investigations by a UK government appointed Lowermoor Incident Health Advisory Group into complaints of long-standing ill health in residents of Camelford. Both studies concluded that there was no evidence of any increase in ill health in the community related to any toxic effects of the aluminum contamination of the water. Rather, the symptoms suffered by affected individuals were largely attributed to anxiety. However, these findings are inconsistent with those of a scientific study conducted in 1991, but not published until 1999 for legal reasons, which showed that there was an association between the contamination and reduced cerebral functions in affected residents. Criticisms have been aimed at this study. The incident is still surrounded with considerable controversy, with multiple claims for damages, and a more in-depth third inquiry into possible delayed or persistent health effects was established by the UK government in 2001. This inquiry reported in 2005 and concluded that although it was unlikely that any of the chemicals involved in the water contamination were responsible for any long-term effects, it did recommend further studies on the effects of the contaminants on neurological health, on diseased joints in the area and on the development of children below the age of 12 months at the time of the aluminum sulfate discharge.

whose occupation exposes them to lead must be removed from its source. Clinical features of lead poisoning include lethargy, abdominal discomfort, anemia, constipation, encephalopathy and motor peripheral neuropathy.

The management of patients with lead poisoning involves identifying the lead source and removing the patients from it. Patients are placed on chelating agents, such as EDTA (*Figure 12.23*) that is effective in both acute and chronic lead poisoning, but has to be given intravenously. Another chelating agent, dimercaptosuccinic acid is less efficient as a chelator but can be given orally.

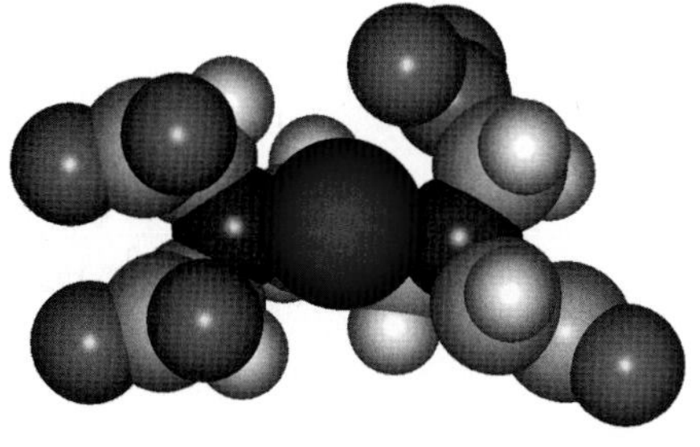

Figure 12.23 Computer generated structure of lead EDTA. The large central red sphere represents the bound lead atom.

Chronic arsenic poisoning is associated with well water contaminated with arsenical pesticides or, classically, with murder! Arsenic occurs in a number of different forms, most of which are toxic although arsenite (AsO_2^-) is much more

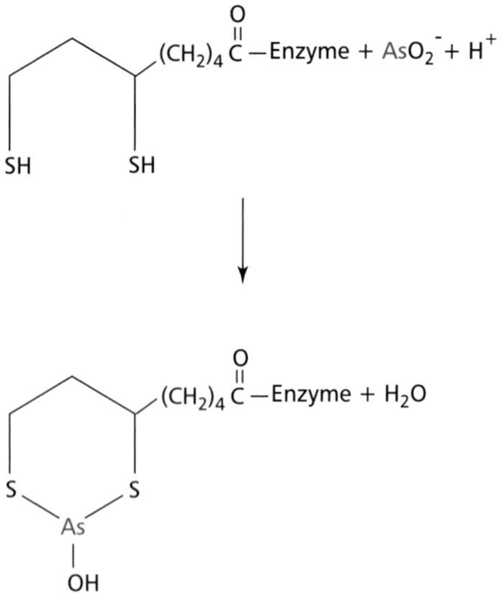

$(CH_2)_4C(=O)$—Enzyme + AsO_2^- + H^+

SH SH

↓

$(CH_2)_4C(=O)$—Enzyme + H_2O

S S As OH

Figure 12.24 The reaction of arsenite with enzyme-bound lipoic acid.

so than arsenate ($HAsO_4^-$). Arsenate substitutes for P_i in biological systems, hence less ATP is produced. Arsenite toxicity is associated with the formation of a stable complex with enzyme-bound lipoic acids (*Figure 12.24*). Indeed, arsenic poisoning can be explained by its ability to inhibit enzymes, such as pyruvate dehydrogenase, 2-oxoglutarate dehydrogenase and branched chain α-oxoacid dehydrogenase, that require lipoic acid as a coenzyme. Chronic poisoning with arsenic is usually associated with diarrhea, polyneuropathy and dermatitis whereas acute poisoning with arsenic gives rise to severe gastrointestinal pain, vomiting and shock. Poisoning can be diagnosed by determining the concentration of arsenic in the hair or fingernails of the victim. Values larger than 0.5 $\mu g\ g^{-1}$ of hair indicate a significant exposure to arsenic. The hair of a person chronically exposed to arsenic could have 1000 times as much as this. Treatment of arsenic poisoning is aimed at enhancing its excretion using chelating agents.

Chronic cadmium toxicity may occur in workers exposed to fumes in cadmium-related industries, although concentrations of cadmium are twice as high in tobacco smokers compared with nonsmokers. Concentrations in the serum greater than 90 nmol dm^{-3} are associated with toxicity. Cadmium poisoning causes influenza-like symptoms, such as chills, fever, muscular aches, nausea, vomiting, abdominal pain, and diarrhea. However, these symptoms may resolve after a week provided there is no respiratory damage. More severe exposures can cause bronchitis and pulmonary edema and occasionally cardiovascular collapse. Long-term exposure may lead to nephrotoxicity with proteinuria, bone disease and hepatotoxicity. The treatment of cadmium

BOX 12.2 The death of Napoleon: arsenic and wallpaper

Following his defeat at the Battle of Waterloo in 1815, Napoleon Bonaparte was exiled on St Helena. This is a volcanic island only about 6 by 8 miles in size in a remote region of the south Atlantic. During most of his exile, Napoleon lived in Longwood House with a retinue of about 20. Napoleon never left St Helena and he died in 1821 (*Figure 12.25*). A postmortem of the body showed an enlarged liver and stomach lesions. It was concluded he had died of a perforated stomach ulcer that had turned cancerous. Napoleon was initially buried on St Helena but his body was removed 20 years later and reburied at Les Invalides on the banks of the Seine in Paris, as had been his wish.

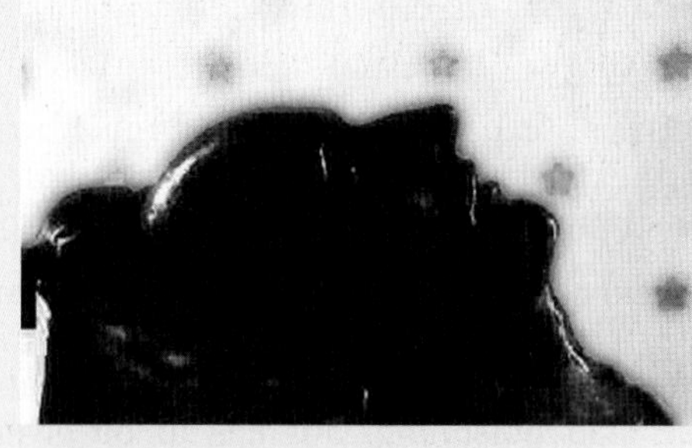

Figure 12.25 Napoleon's deathmask. Courtesy of H. Ball, www.grand-illusions.com.

In 1952, Forshufvud read an account of Napoleon's death and, given his symptoms, concluded that Napoleon may have been murdered by arsenic poisoning. White arsenic or arsenic oxide is extremely poisonous and its symptoms, stomach pains, diarrhea, shivering and swollen limbs, can be confused with other illnesses. Indeed, a few days prior to his death, Napoleon had requested that his doctor make a full examination, particularly of his stomach and some of his symptoms did correspond to those of arsenic poisoning.

It is possible to poison a person by slow exposure to small quantities of arsenic over an extended period. A number of

Napoleon's staff had kept locks of his hair, which were subsequently passed down the generations. Hair is largely composed of keratin, a protein that contains sulfur. If arsenic is ingested, some of it will bind to the sulfur atoms and since hair is constantly growing it can even show how the concentrations of arsenic in the body change with time. Hair is resistant to degradation and some of the samples of Napoleon's hair could still be analyzed in the 1960s using neutron activation poisoning usually involves removal from exposure. Treatment with chelating agents is not effective because the soluble form of cadmium damages the kidney.

Mercury poisoning can be acute or chronic. It occurs following exposure to mercury vapor, inorganic salts or organic compounds of mercury. Mercury poisoning is primarily occupational but can be caused by contaminated food. The clinical features of acute mercury poisoning include coughing, bronchiolitis, pulmonary edema, pneumonitis, peripheral neuropathy and neuropsychiatric problems. Chronic mercury poisoning causes anorexia, sweating, insomnia, impaired memory, paresthesiae of the lips and extremities and renal tubular damage. Mercury poisoning is usually diagnosed by determining the concentration of mercury in serum and urine. Urine mercury/creatinine ratios are often used to assess exposure. Ratios of 40 to 100 nmol mmol^{-1} require monitoring and further investigation, while values of greater than 100 nmol mmol^{-1} require the patient to be removed from the source of mercury.

Acute mercury poisoning is treated with chelating agents such as dimercaprol, to increase its excretion into bile and urine. Methylmercury is an organic form of mercury that has been used to preserve seed grain. Methylmercury poisoning can be caused by eating meat from animals which have been fed treated seedgrain. An outbreak of mercury poisoning began in the 1950s in Minamata bay in Japan and affected over 3000 villagers who ate fish contaminated with methylmercury.

techniques. A number of hairs were found to contain abnormal amounts of arsenic. Naturally this led to speculation that Napoleon had been murdered. However, other possible sources of arsenic poisoning are paints and wallpapers. The pigment, Scheele's Green contains copper arsenite and has been used in fabrics and wallpapers since about 1770. Unfortunately throughout the nineteenth century many people were made ill or killed by this wallpaper. In 1893, Gosio determined the pathological mechanism: if wallpapers containing Scheele's Green become damp and contaminated with molds, such as *Scopulariopsis brevicaulis*, the molds can metabolize the arsenic salts to volatile poisonous compounds, for example trimethyl arsine, which are released in a vaporous form into the atmosphere. Following a public appeal in 1980 for help in making a radio program, a small piece of wallpaper from the wall of the drawing room of Longwood House was located (*Figure 12.26*). The sample of paper showed a single star and its main colors are green and brown. Gold and green were the Imperial colors but it is possible that the brown was originally gold in color. More significantly, X-ray fluorescence spectroscopy analysis showed the wallpaper to contain appreciable quantities of arsenic. Longwood House was notoriously damp, and the paper would have degraded to release volatile arsenic compounds. Further, many of its inhabitants other than Napoleon had apparently also become ill and complained of the 'bad air', so it is possible that he might have been a victim of British wallpaper makers rather than a deliberate murder plot. However, the amount of arsenic released could not have been large, and apparently was insufficient to have killed Napoleon although once he did become ill with a stomach ulcer, the arsenic could have exacerbated his condition.

Figure 12.26 A sample of Napoleon's wallpaper.
Courtesy of H. Ball, www.grand-illusions.com.

CASE STUDY 12.1

Helen, an 18-year-old, was admitted to the Accident and Emergency Department of her local hospital. She was found in her bedroom with an empty bottle of aspirin tablets. On admission, Helen was in a confused state, sweating and breathing heavily. There was clinical evidence for mild dehydration. She had a pulse rate of 112 per min and blood pressure of 110/70 mmHg. Her body temperature was 39°C. Laboratory investigations give the following profile (reference ranges are shown in parentheses).

Na^+	132 mmol dm^{-3} (135–145 mmol dm^{-3})
K^+	3.4 mmol dm^{-3} (3.6–5.0 mmol dm^{-3})
HCO_3^-	11 mmol dm^{-3} (22–30 mmol dm^{-3})
Urea	11 mmol dm^{-3} (3.3–6.7 mmol dm^{-3})
Glucose	3.5 mmol dm^{-3} (2.8–6.0 mmol dm^{-3})
Salicylate	4.6 mmol dm^{-3} (up to 2.5 mmol dm^{-3} is therapeutic)
pH	7.20 (7.36–7.44)
*P*CO2	3.6 kPa (4.5–6.0 kPa)
Prothrombin time	18 s (14 s)

Questions

(a) What can be concluded from the clinical data?

(b) What treatment would be suitable?

CASE STUDY 12.2

Olga, a 25-year-old, was admitted to hospital after being found unconscious by her flatmate. It was reported that empty vodka and paracetamol bottles had also been discovered in her flat. On admission to an Accident and Emergency Department, Olga's breath smelt of alcohol and her serum paracetamol concentration was 105 mg cm^{-3}. Her liver function tests were normal as was the prothrombin time. Her serum alanine aminotransferase levels showed a transient increase over the next few days but gradually fell back within the reference range.

Questions

(a) Comment on these clinical findings.

(b) What treatment should Olga be given?

CASE STUDY 12.3

Ed is a building site worker aged 58 who has been a heavy drinker for over 30 years. He was admitted to hospital after collapsing outside a public house. He smelt strongly of alcohol, was confused and unsteady on standing. He complained of fatigue, nausea, particularly early in the morning, and a loss of appetite. A physical examination detected abdominal tenderness and enlargement of the liver and gastrointestinal varicose veins. His blood alcohol concentration was 79 mmol dm^{-3}.

Maureen aged 52 was admitted into hospital at a similar time. She presented with general GIT problems and frequent diarrhea. She was subjected to considerable stress at work and admitted she had been drinking 'rather heavily'.

Ed's Ca^{2+} and Mg^{2+} blood concentrations were low and his clotting time extended. His urinary urea excretion was also subnormal. Liver function tests were performed on sera from both patients. The results are shown below (reference ranges are shown in parentheses).

	Ed	Maureen	
Total protein / g dm^{-3}	67	78	(60–84)
Albumin / g dm^{-3}	31	39	(35–50)
Total bilirubin	57	14	(3–15)
Alkaline phosphatase / U dm^{-3}	720	335	(100–300)
Alanine transaminase / U dm^{-3}	34	91	(5–35)
Aspartate transaminase / U dm^{-3}	41	162	(10–40)
γ-glutamyltransferase / U dm^{-3}	780	455	(7–45)

Questions

(a) Suggest why the tests above are particularly indicative of the state of the health of the liver.

(b) Does Ed or Maureen show the greater liver degeneration?

CASE STUDY 12.4

John's girlfriend found him in his closed garage unconscious in his car with the engine running. He was taken to the hospital immediately. The following blood results were obtained (reference ranges are shown in parentheses).

H^+	50 nmol dm^{-3}	(36–46 nmol dm^{-3})
HCO_3^-	15 mmol dm^{-3}	(22–30 mmol dm^{-3})
PCO_2	3.5 kPa	(4.5–6.0 kPa)
Hemoglobin	150 g dm^{-3}	(130–180 g dm^{-3})
Lactate	10 mmol dm^{-3}	(0.65–2.0 mmol dm^{-3})

Questions

(a) Suggest a plausible explanation for these data?

(b) How should John be treated?

12.6 SUMMARY

The body is exposed daily to many chemicals, some of which have the potential for harm. These chemicals may be ingested with food and drink, they may be therapeutic or recreational drugs or they may be toxins from microorganisms or the abiotic environment. Xenobiotics may be detoxified in the liver, by conversion to more water-soluble compounds, which are then excreted through the kidneys. Sometimes this process results in more harmful compounds being produced, as is the case with paracetamol poisoning. Clinical investigations of suspected poisoning include liver function tests, analysis of blood gases and determination of acid–base status.

The commonest cases of poisoning involve accidental or deliberate overdose with common analgesics such as paracetamol or aspirin but abuse of alcohol is an increasing problem in developed countries, leading to liver damage in the patient. In addition, carbon monoxide poisoning may be insidious and fatal, in badly ventilated houses warmed by faulty gas heaters. Treatment of poisoning first involves identification and removal of the poison; thereafter treatment depends on the nature of the poison involved. Unfortunately, some poisons, such as paraquat, are invariably fatal.

QUESTIONS

1. Which of the following produces a combination of respiratory alkalosis and metabolic acidosis?

   (a) paracetamol poisoning;
   (b) ethylene glycol poisoning;
   (c) aspirin poisoning;
   (d) lead poisoning;
   (e) carbon monoxide.

2. Which of the following is the toxin produced during paracetamol poisoning?

   (a) paracetamol-3-mercapturic acid;
   (b) glutathione;
   (c) *N*-acetylbenzoquinoneimine;
   (d) UDP-glucuronyl transferase;
   (e) trimethylarsine.

3. What are the possible prognoses for a patient who has taken an overdose of paracetamol if treated within hours or if treatment is delayed for several days?

4. Disulfiram is an inhibitor of aldehyde dehydrogenase. It was used as the drug antabuse to deter alcohol consumption in chronic alcoholics. Suggest a plausible reason why treatment with it should deter alcoholics from consuming alcoholic drinks.

5. Suggest why alcohol should be consumed with caution or avoided when taking other drugs.

6. A 10-year-old Asian girl was admitted to hospital where she presented with vomiting, nausea, and neurological symptoms. She had been using a traditional cosmetic called surma that was applied around the eyes. What laboratory investigations should be applied to this patient?

7. A 50-year-old scientific instrument technician repaired old scientific gauges that often contain mercury. One winter, he complained of weight loss, weakness, symptoms that resembled influenza and depression. On examination he had muscle weakness and wasting. A urine sample was examined and was found to contain 180 nmol Hg per mmol creatinine. How should this patient be treated?

FURTHER READING

Bartlett, D (2004) Acetaminophen toxicity. *J. Emerg. Nurs.* **30:** 281–283.

Bellinger, DC (2004) Lead. *Pediatrics* **113:** 1016–1022.

Furge, LL and Guengerich, FP (2006) Cytochrome P450 enzymes in drug metabolism and chemical toxicology. *Biochem. Mol. Biol. Educ.* **34:** 66–74.

Geffreys, D (2005) *Aspirin: the remarkable story of a wonder drug.* Bloomsbury, London.

Greene, SL, Dargan, PI and Jones, AL (2005) Acute poisoning: understanding 90% of cases in a nutshell. *Postgrad. Med. J.* **81:** 204–216.

Jarup, L (2003) Hazards of heavy metal contamination. *Br. Med. Bull.* **68:** 167–182.

Jones, DEH and Ledingham, KWL (1982) Arsenic in Napoleon's Wallpaper. *Nature* **299:** 626–627.

Kao, LW and Nanagas, KA (2004) Carbon monoxide poisoning. *Emerg. Med. Clin. North. Am.* **22:** 985–1018.

Kennedy, NP and Tipton, KF (1990) Ethanol metabolism and alcoholic liver disease. *Essays in Biochem.* **25:** 137–195.

Levesque, H and Lafont, O (2000) Aspirin throughout the ages: an historical review. *Review de Medicine Interne,* **21** Suppl 1:8s–17s.

Lieber, CS (2000) Alcoholism: its metabolism and interaction with nutrients. *Annu. Rev. Nutr.* **20:** 395–430.

Mehta, P (2002) Aspirin in the prophylaxis of coronary artery disease. *Curr. Opin. Cardiol.* **17:** 552–558.

Nadarajah, P and Hayden, MJ (2004) Poisoning. *Hosp. Med.* **65:** 174–177.

National Poisons Information Service and Association of Clinical Biochemists (2002) Laboratory analyses for poisoned patients: joint position paper. *Ann. Clin. Biochem.* **39:** 328–339.

Needleman, H (2004) Lead poisoning. *Annu. Rev. Med.* **55:** 209–222.

Nelson, DR (1999) Cytochrome and the individuality of species. *Arch. Biochem. Biophys.* **369:** 1–10.

Oremaland, RS and Stolz, JF (2003) The ecology of arsenic. *Science* **300:** 939–944.

Ryter, SW and Otterbein, LE (2004) Carbon monoxide in biology and medicine. *BioEssays* **26:** 270–280.

Scally, RD, Ferguson, DR, Piccaro, JC, Smart, ML and Archie, TE (2002) Treatment of ethylene glycol poisoning. *Am. Fam. Physician* **66:** 807–812.

Song, Z, Joshi-Barve, S, Barve, S and McClain, CJ (2004) Advances in alcoholic liver disease. *Curr. Gastroenterol. Rep.* **6:** 71–76.

Stachulski, AV and Lennard, MS (2000) Drug metabolism: the body's defence against chemical attack. *J. Chem. Educ.* **77:** 349–353.

Stewart, S, Jones, D and Day, CP (2001) Alcoholic liver disease: new insights into mechanisms and preventative strategies. *Trends Mol. Med.* **7:** 408–413.

Tchounwou, PB, Centeno, JA and Patlolla, AK (2004) Arsenic toxicity, mutagenesis, and carcinogenesis – a health risk assessment and management approach. *Mol. Cell Biochem.* **255:** 47–55.

Underwood, M (2006) Diagnosis and management of gout. *BMJ* **332:** 1315–1319.

Watson, I and Proudfoot, A (2002) *Poisoning and Laboratory Medicine,* ACB Venture Publications, London.

Yip, L, Dart, RC and Gabow, PA (1994) Concepts and controversies in salicylate toxicity. *Emerg. Med. Clin. North. Am.* **12:** 351–364.

chapter 13:

DISORDERS OF THE BLOOD

OBJECTIVES

After studying this chapter you should be able to:

- list the components of the blood;
- describe the structures of blood cells and platelets;
- explain how hemoglobin functions;
- explain how blood clotting occurs;
- give examples of some types of anemias;
- describe the consequences of the genetic defects in sickle cell anemia and the thalassemias;
- explain the diagnoses and treatments of some types of anemias;
- describe some clinical disorders associated with clotting factors.

13.1 INTRODUCTION

Blood is a protein-rich fluid called **plasma** in which erythrocytes and leukocytes, sometimes called red and white blood cells respectively, and platelets are suspended (*Figure 13.1*). The cells constitute about 40–45% of the volume of the blood. The blood is pumped around the body by the heart through the arteries that supply the capillaries and is returned to the heart in the veins (*Chapter 14*). The main functions of the blood are to distribute oxygen, nutrients and hormones and other signaling molecules between tissues and to remove carbon dioxide and other waste products. Plasma contains the proteins of the clotting system and of the immune systems (*Chapters 4* and *5*).

Plasma is blood from which the cells have been removed. It contains a range of plasma proteins in addition to the clotting and immune system proteins mentioned above, nutrients, such as glucose, waste materials, for example urea, and a range of electrolytes in solution. If it is allowed to clot, the clear straw-colored liquid remaining after removal of the clot is called **serum**. The composition of the blood and the plasma is given in *Table 13.1*.

Plasma	Concentration
[Total protein] / g dm^{-3}	66
[Fibrinogen] / g dm^{-3}	3.1
[Albumins] / g dm^{-3}	32.6
[Globulins] (excluding fibrinogen)] / g dm^{-3}	30.1
Cells and platelets	**Number**
Erythrocytes male / dm^{-3} female / dm^{-3}	 $4.4–5.9 \times 10^{12}$ $3.8–5.2 \times 10^{12}$
Leukocytes / dm^{-3}	$4–11 \times 10^{9}$
Platelets / dm^{-3}	$2.5–5.0 \times 10^{9}$

*Blood volume of 78 and 66 cm^3 kg^{-1} body weight in males and females respectively

Table 13.1 Composition of the blood*

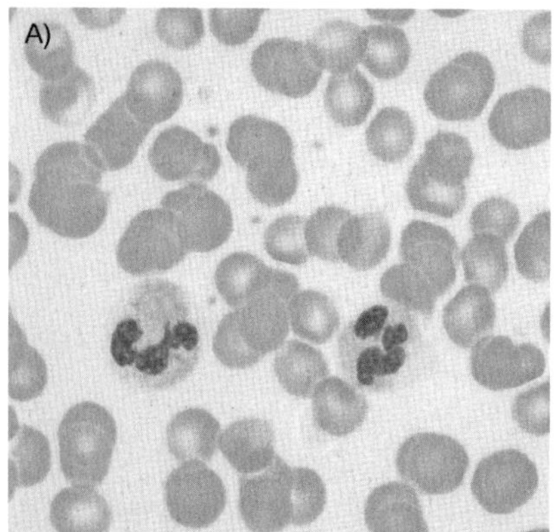

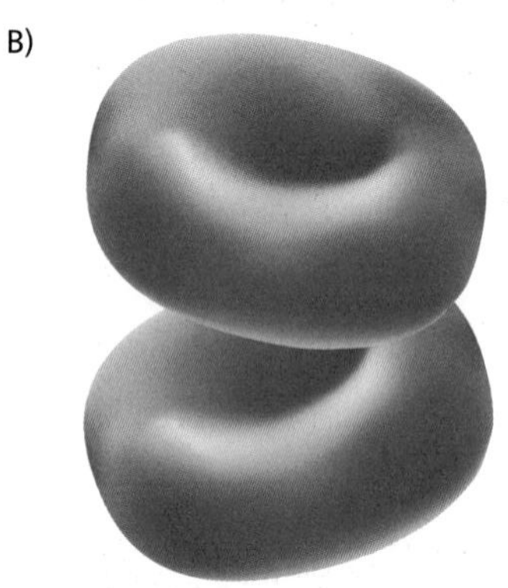

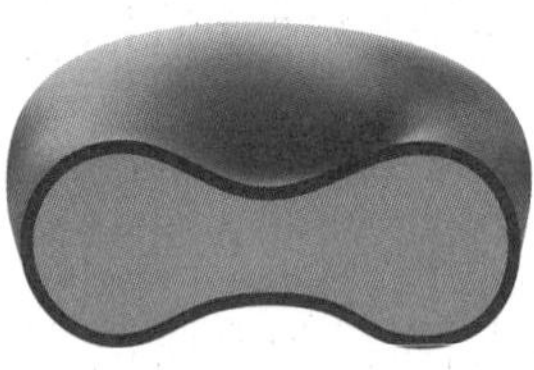

Figure 13.1 (A) Shows the appearance of a normal blood smear when examined with a microscope. The erythrocytes predominate with the occasional leukocyte being visible. (B) The erythrocytes are biconcave in shape and their centers appear lighter in color.

In a text of this size it is not, of course, possible to discuss each type of blood disorder and attention will focus only on the major types of diseases likely to be normally encountered.

13.2 BLOOD CELLS AND PLATELETS

All of the cells of the blood originate from pluripotent stem cells in the bone marrow (*Figure 13.2*). Chemical signals, such as cytokines (*Chapter 4*), direct primordial stem cells to develop in different ways to produce **erythrocytes, leukocytes** of various types, and **megakaryocytes**, which are the precursors of **platelets**.

Normoblasts are erythroid cells that arise from divisions of pluripotent stem cells. Eventually these lose their nuclei giving rise to **reticulocytes,** which contain mRNAs for globins and are still able to synthesize hemoglobin (Hb), and which are the precursors of the erythrocytes. The reticulocytes circulate in the blood for 1–2 days before maturing to erythrocytes, and normally constitute 1–2% of the circulating red cells. Erythrocytes are the most numerous cells in the blood. Adult males and females have erythrocyte counts of about 5.5 and 4.8×10^{12} dm^{-3}, respectively. The number of cells in a given volume of blood can be determined using a hemocytometer (*Figure 13.3*). About 2×10^{11} mature erythrocytes are formed daily. They have no nuclei or other organelles and are biconcave in shape (*Figure 13.1*). Erythrocytes circulate for about 120 days and are then removed from circulation and destroyed by macrophages in the liver and spleen. Hemoglobin is the red protein found in the erythrocytes that carries dioxygen (O_2) and which also plays an important role in buffering, maintaining the pH at 7.4 ± 0.1 (*Chapter 9*). The iron-containing heme is removed from the Hb of defunct erythrocytes and its porphyrin ring is converted to bilirubin, which is excreted in the bile. The iron is conserved and recycled (*Box 13.1*). Iron circulates in the blood attached to a transport protein called transferrin and stored in the liver and spleen cells in a storage protein called ferritin (*Figure 13.4 (A)* and *(B)*). Excessive breakdown of Hb produces more bilirubin than can be excreted and this accumulates in the tissues causing jaundice (*Chapter 11*).

Pluripotent stem cell
Lymphoid stem cell
Pre-NK cell
NK cell
Thymus
T cell
Pre-B cell
B cell
Myeloid stem cell
Granulocyte-monocyte progenitor
Neutrophil
Monocyte
Eosinophil progenitor
Eosinophil
Basophil progenitor
Basophil
Mast cell
Megakaryocyte
Platelets
Nucleated reticulocyte
Erythrocyte

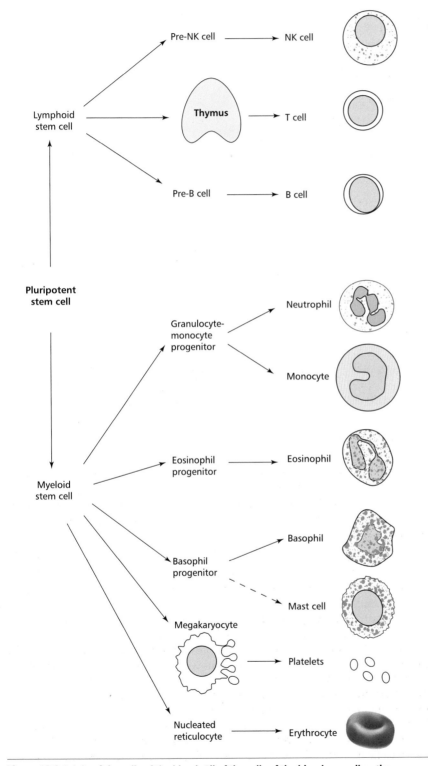

Figure 13.2 Origin of the cells of the blood. All of the cells of the blood, as well as the platelets, originate in the bone marrow where progenitor stem cells divide and differentiate to produce the different cell types. The process of differentiation is controlled by growth factors. See *Chapter 4* for more information on lymphocytes (natural killer (NK), T and B cells).

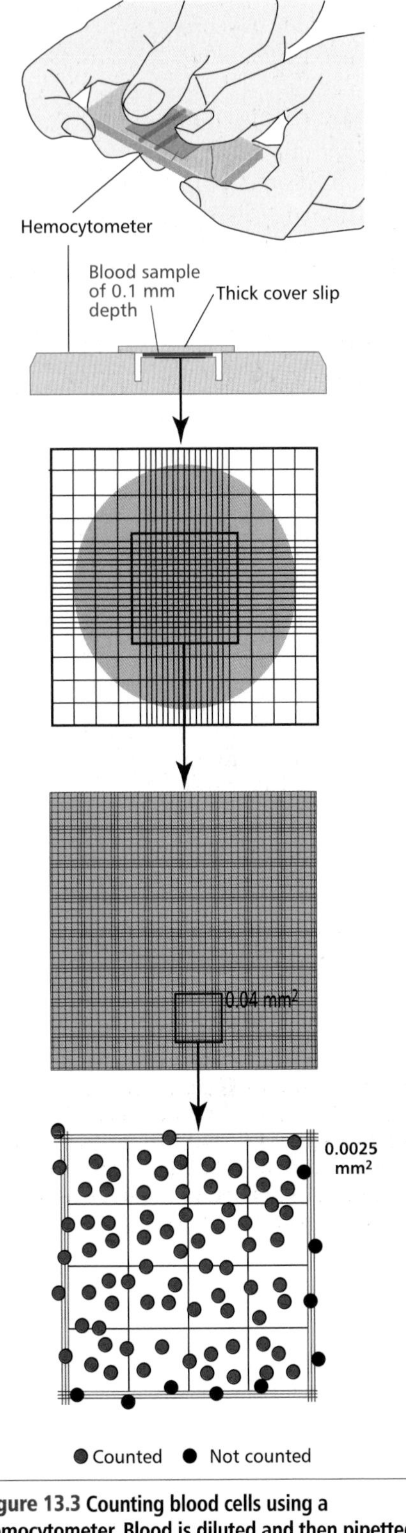

Figure 13.3 Counting blood cells using a hemocytometer. Blood is diluted and then pipetted into a glass chamber of known volume. The number of cells within the indicated grid is counted under the microscope and this number, multiplied by the dilution, gives the number of cells in the blood sample. Different dilutions need to be done for erythrocytes and leukocytes. In hospital laboratories cells counts are done automatically using instruments called cell counters.

BOX 13.1 Iron metabolism

Only about 10% of the average daily intake of iron, approximately 20 mg in the UK, is absorbed, mostly by the duodenum and jejunum (*Chapter 11*), although more is absorbed in pregnancy and in iron deficiency anemia (*Section 13.5*). There are obviously increased demands for iron in growth periods and menstruation, when about 0.7 mg iron is lost daily, and in pregnancy (*Chapter 7*). Heme iron from Hb and myoglobin in red meats is better absorbed than nonheme iron. Absorption is controlled by the mucosal cells of the small intestine and the iron is to some extent stored in these cells before being passed to the hepatic portal blood. Stored iron may be lost when these cells are shed. The iron is bound to the iron binding protein transferrin (*Figure 13.4 (A)*) during its transport in the blood. Transferrin can carry two atoms of iron per molecule but is only about one-third saturated on average. The iron is detached from the transferrin in the bone marrow when the protein interacts with specific receptors on erythroblasts and reticulocytes supplying the iron needed for Hb synthesis. Although 60–70% of the body's iron is found in circulation as Hb, with more in the cytochromes and other iron proteins; some is stored in the protein, ferritin (*Figure 13.4 (B)*). This protein is found in most cells, but particularly in those of the liver and spleen.

The body does not have an excretory route for iron, but this is normally a problem only when repeated transfusions are given, such as in the cases of patients with sickle cell anemia and thalassemia. As the transfused erythrocytes are broken down the iron accumulates and may form deposits of hemosiderin in the liver and spleen.

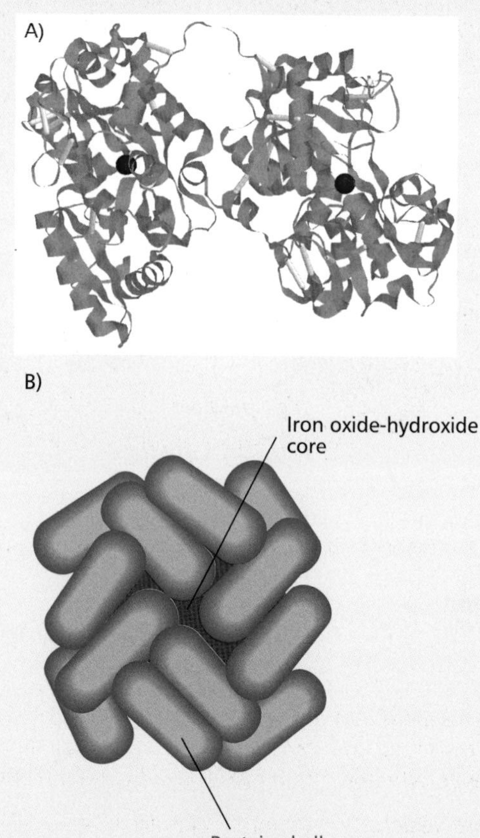

Figure 13.4 (A) Molecular model of transferrin. PDB code 1JNF. (B) Schematic of ferritin that consists of a protein coat of 24 subunits with a central cavity. When completely filled one multisubunit of ferritin can hold more than 2000 iron atoms as hydrated ferric hydroxide.

There are several types of leukocytes or white cells in the blood, each with its own function. The total white cell count in an adult is between 4 and 11 x 10^9 dm^{-3} but there is considerable variation. White cells were originally classified on the basis of which microscopic stain they took up, whether they had a granular cytoplasm, and whether the nucleus was lobed. The three main types of leukocyte are called **polymorphonucleocytes** (PMN, but sometimes referred to as *polymorphs*), **lymphocytes** and **monocytes** (*Chapter 4*). Polymorphonucleocytes are further subdivided into **neutrophils** (57% of the total white cell population) that contain neutral staining granules, **eosinophils** (3.5%) which contain acid-staining granules, and **basophils** (0.5%) which contain basic staining granules (*Figure 13.2*). Polymorphonucleocytes release chemokines some of which are mediators of inflammation as described in *Chapter 4*. Neutrophils can migrate to areas of infection and phagocytose bacteria. Eosinophils seem to be more concerned with dealing with larger parasites and their number also increases in allergic diseases. Basophils and the similar **mast cells** (*Chapter 4*), which are found mostly in the skin, can release histamine from their granules and this also contributes to some types of allergic responses.

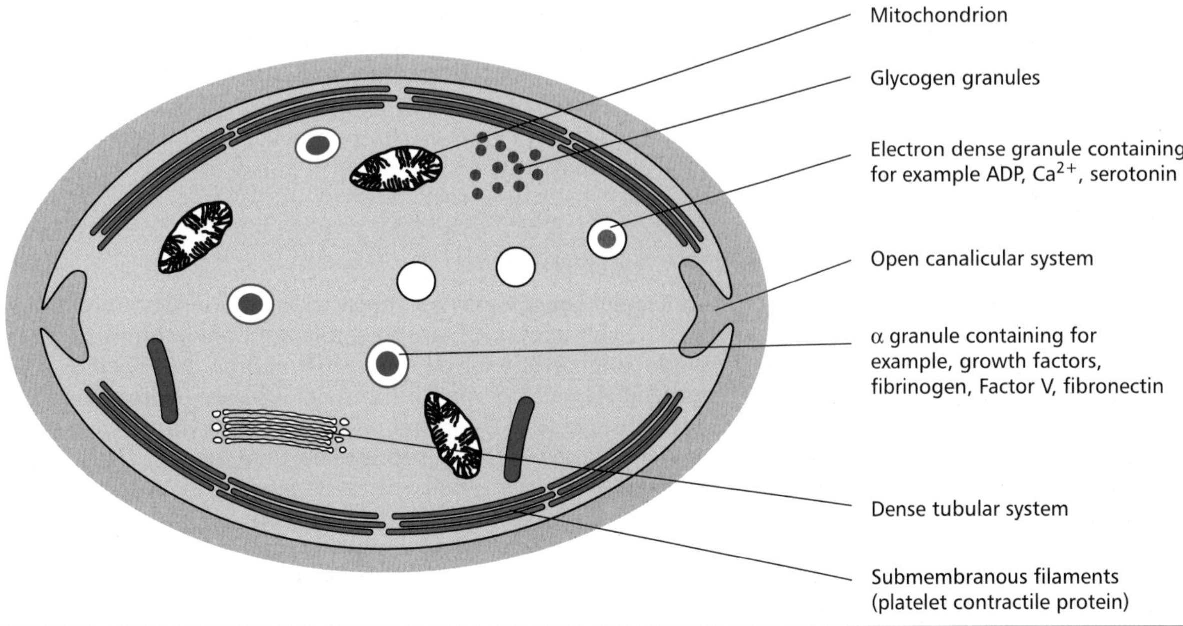

Figure 13.5 Schematic of a blood platelet. The platelet is packed with granules that have a major role in blood clotting.

Lymphocytes are key components of the immune system. Monocytes circulate in the blood for about 72 h and then enter the tissues and transform into macrophages, which play a key role in inflammation and defense (*Chapter 4*).

Platelets are vesicle-like structures about 3 μm in diameter with a volume of 7 fdm^3 (femtodecimeters^3 or femtoliters). They are formed by the fragmentation of large precursor cells called megakaryocytes and, like erythrocytes, are not complete cells although they contain numerous granules, some organelles and a tubular system (*Figure 13.5*). Their lifespan is 10–12 days and they function in blood clotting or hemostasis (*Section 13.4*).

Margin Note 13.1 Platelet concentrates 

Platelet concentrates can be prepared by centrifugation and may be stored for up to five days (*Chapter 6*). Such concentrates are used to treat patients suffering from thrombocytopenia (*Section 13.9*) and who have insufficient platelets and to prevent bleeding in patients with bone marrow failure.

13.3 HEMOGLOBINS

Hemoglobin is the red-colored, oxygen-transporting protein in erythrocytes. Its M_r is about 64 000 and it is made up of four subunits, each containing an iron-containing heme group (*Figure 13.6*). Each molecule can carry up to four O_2 molecules. Oxygen is taken up as the blood passes through the lungs and is transported to all parts of the body allowing respiration, the oxidation of fuels, to occur in the mitochondria. The iron in the heme group of Hb remains in the ferrous (Fe(II)) state throughout. Should the iron become oxidized to Fe(III), **methemoglobin** is formed, which is incapable of carrying oxygen. This oxidation happens to a small extent continuously, so that normal blood always contains a few percent of methemoglobin. However, methemoglobin reductase, present in erythrocytes, constantly catalyzes its reduction back to Hb. The rare individuals with a genetic deficiency of this enzyme have severe problems and tend to be cyanosed unless they are treated with a reducing agent.

Margin Note 13.2 Cyanosis – going blue in the face 

Cyanosis (from Greek *cyan*, blue) is the bluish complexion resulting from lack of oxygen in the circulating blood. It is most frequently observed under the nails, lips as well as the skin. Cyanosis occurs following an inadequate oxygen intake in the lungs or from many other reasons, for example the stagnation of blood in the circulation during heart failure.

Hemoglobin was one of the first proteins to have its complete structure determined. Indeed, because Hb is so important medically, it is fair to say that more is probably known about Hb than any other protein. It is vital to life because the low solubility of oxygen in water means that insufficient amounts can be carried by blood in solution. The binding of O_2 to Hb in erythrocytes

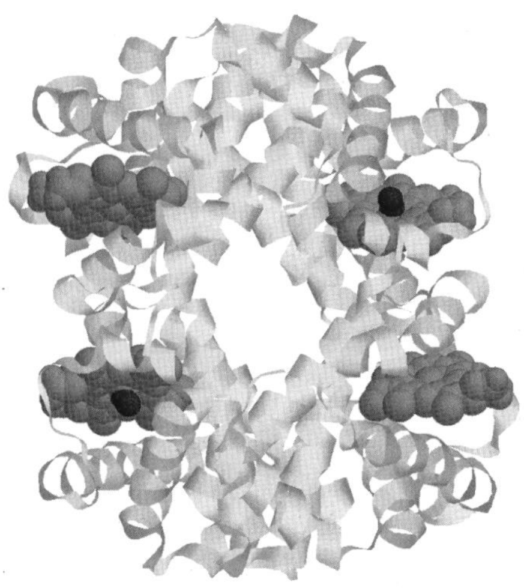

Figure 13.6 Molecular model of adult hemoglobin. PDB code 1GZX. Note how the four subunits interact closely with each other. The bound O_2 are shown in red.

increases the oxygen carrying capacity several thousand-fold. Furthermore, the body has a limited capacity to work anaerobically and so O_2 is vital. Even a small reduction in the amount of Hb in the blood leads to **anemia**, which causes serious clinical problems (*Section 13.5*). Anemia can result from there being too few erythrocytes or from each erythrocyte having too little or a defective Hb and this may be a consequence of a lack of iron. In all cases, medical treatment is usually necessary.

KEY PROPERTIES OF HEMOGLOBIN

Hemoglobin molecules are roughly spherical with the four subunits fitting tightly together. The subunits of Hb are all similar and are in identical pairs. Thus, for example, adult Hb, referred to as HbA, can be described as $\alpha_2\beta_2$. Embryonic and fetal Hbs also occur. The heme groups are hydrophobic and sit inside hydrophobic clefts in the protein. The iron atom within each heme binds one O_2 molecule and so Hb can successively bind four O_2. The strength of this binding, that is the oxygen affinity, can be measured. In the lungs, where there is abundant oxygen, Hb should bind its maximum of four molecules and become saturated. In the tissues, the oxygen must 'unbind' and be released. The strength of binding is critical; too weak and Hb would be ineffective as a carrier, too tight and the tissues would not be supplied with oxygen because the oxyhemoglobin would not release its oxygen.

When an O_2 binds to one subunit, it induces a small change in the shape of the protein making the binding of the next O_2 slightly easier, that is, the strength of binding changes with each successive addition. The consequence is that the graph of oxygen bound against oxygen concentration, the oxygen binding curve (*Figure 13.7*), is S-shaped or sigmoidal. This means that in the lungs the HbA molecule can become nearly 100% saturated with oxygen but in the tissues can release almost all of it.

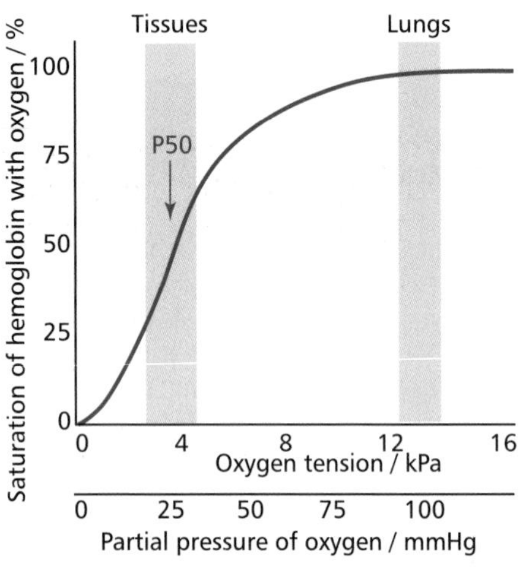

Figure 13.7 Oxygen binding curve for hemoglobin. Note that it is sigmoidal, indicating that the affinity of O_2 changes as each successive O_2 binds. Thus in the lungs, where the oxygen tension is high, the hemoglobin becomes almost saturated with oxygen. In the tissues, where the oxygen tension is low, the hemoglobin is able to give up almost all of its oxygen.

BOX 13.2 Nomenclature of mutant hemoglobins

Two nomenclature systems are in use, which may be a little confusing. Originally, normal adult Hb was called hemoglobin A (HbA, A for adult, $\alpha_2\beta_2$). Fetal Hb was hemoglobin F (HbF, $\alpha_2\gamma_2$), sickle cell Hb was HbS and so on. However, as more Hb mutations were identified, and hundreds are known, it was realized that the number of mutations known exceeded the number of letters in the alphabet. Subsequently, Hbs began to be named after the geographic location where they were first discovered, for example Hb Dakar, Hb Lepore, Hb Sydney. An additional complication occurs when newly discovered Hbs have the same characteristics as a 'letter' Hb and both nomenclatures may be combined, as in, for instance, HbJ-Capetown. It is appreciated that this nomenclature is not perfect but it is too difficult to change now.

The precise mutation can of course be described in terms of the base change(s) when as is usual the gene sequence is known, but this is a little cumbersome for everyday use. Thus Hb Sydney is caused by a GTG to GGG mutation at position 67 in the gene for β–globin, causing an amino acid residue change from valine to alanine. This results in unstable Hb with poor heme-binding properties, resulting in mild hemolysis. *Table 13.2* gives some examples of mutant Hbs.

Hemoglobin	Codon change	Amino acid changed	Comments
Torino	TTC to GTC	Phe to Val	α-chain (43); decreased O_2 affinity, unstable
Ann Arbor	CTG to CGG	Leu to Arg	α-chain (80); unstable
Bibba	CTG to CCG	Leu to Pro	α-chain (136); dissociates
M-Iwate	CAC to TAC	His to Tyr	α-chain (87); forms met-Hb, benign cyanosis in heterozygotes
Constant Spring	UAA to CAA	STOP to Gln	α-chain mutation of the chain termination codon (142) gives extended α-chain
S	GAG to GTG	Glu to Val	β-chain (6); cells sickle, forms fibrils
C	GAG to AAG	Glu to Lys	β-chain (6); enhances sickling when with HbS
St Louis	CTG to CAG	Leu to Gln	β-chain (28); Fe readily oxidized, polar group in heme pocket, increased O_2 affinity, unstable
Seattle	GCC to GAC	Ala to Asp	β-chain (70); decreased O_2 affinity, unstable
Gun Hill	–	deletion	β-chain (91-95); increased O_2 affinity, unstable
M-Saskatoon	CAT to TAT	His to Tyr	β-chain (63); forms met-Hb, benign cyanosis in heterozygotes

Table 13.2 Some of the many know mutations in hemoglobin genes

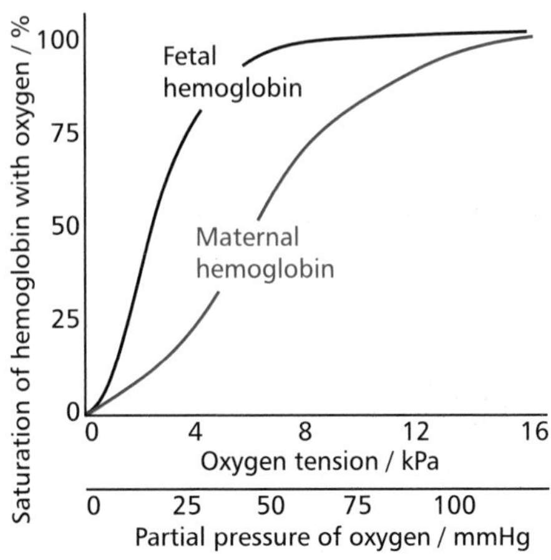

Figure 13.8 Graph showing that fetal hemoglobin has a higher affinity for oxygen than adult, maternal hemoglobin. This allows the fetus to obtain oxygen from the maternal blood.

EMBRYONIC AND FETAL HEMOGLOBINS

The embryonic and the fetal forms of Hb differ slightly from HbA. There are several types of embryonic Hb present early in embryonic life but at about 6 weeks there is a switch to fetal Hb (HbF). Fetal Hb has two α subunits, as in the adult, but two γ subunits, ($\alpha_2\gamma_2$, and there are actually two types of γ subunit). The embryo and fetus obtain their oxygen from the mother's blood in the placenta. Thus their Hbs need to become saturated with oxygen at lower oxygen tensions than maternal HbA so they can obtain it from the mother (*Figure 13.8*). This is possible because embryonic and fetal Hbs have a greater affinity for O_2 than HbA. Adult Hb production starts shortly before birth and by 30 weeks of age it should have replaced all the HbF (*Figure 13.9*).

Margin Note 13.3 Hereditary persistence of fetal hemoglobin

In *hereditary persistence of fetal hemoglobin* (HPFH), the HbF is not replaced. This is presumably due to a failure of the switching mechanism that normally occurs at around the time of birth. Although individuals with HPFH have high concentrations of HbF, the condition does not cause any major hematological abnormalities and does not prevent those affected from having a normal life.

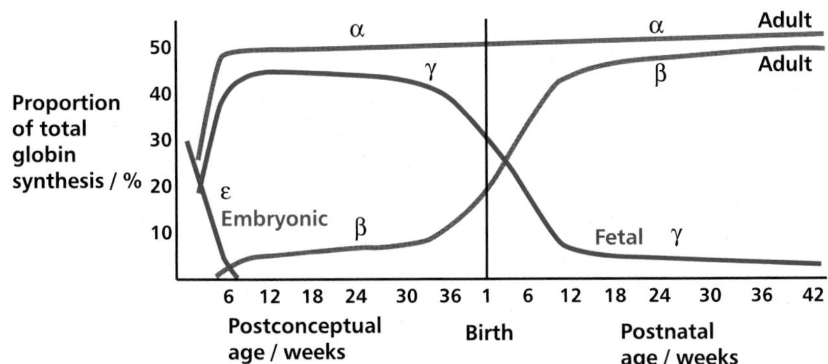

Figure 13.9 The production of human globins during development. There are several types of β-globins in the embryo and the fetus. Any given erythrocyte contains only one type of α- and one type of β-globin.

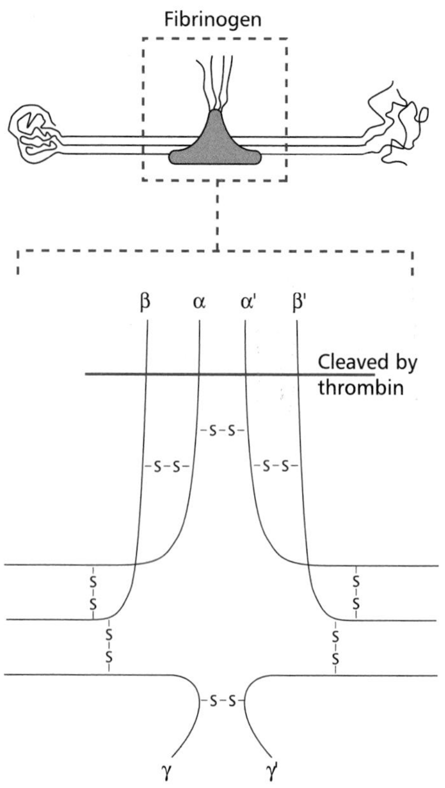

Figure 13.10 The structure of fibrinogen. Fibrinogen has a M_r about 340 000 and consists of two units, each containing three polypeptides (α, β, γ) joined together at their *N*-terminal ends by a number of disulfide bonds. Short lengths of the amino terminal regions of the two α and two β polypeptides project outwards and cleavage of these by thrombin, as indicated, allows the resulting fibrin molecules to aggregate to form a 'soft clot'. This is subsequently strengthened by the cross-linking action of Factor XIIIa which is a transglutaminase.

13.4 HEMOSTASIS AND BLOOD CLOTTING

The circulatory system is self-sealing. Hemostasis rapidly stops all but the most catastrophic bleeding in normal individuals. If the lining of a blood vessel is damaged, eventually a platelet plug is formed that prevents further blood loss. Blood clotting is then initiated leading to the deposition of fibrin and the formation of a clot to seal the wound. Wound healing can then begin.

Blood clotting occurs in one of two pathways, the so-called intrinsic and extrinsic pathways. These pathways each have a number of unique reactions but, in the end, both pathways activate the final clotting stage, which is the formation of fibrin. The clotting pathways involve a group of plasma proteins that act in sequence, each activating the next in line. The end result is the conversion of soluble fibrinogen to the insoluble fibrin, which polymerizes to form a clot at the site of the damage and critically not elsewhere (*Figure 13.10*).

About 20 plasma proteins or plasma clotting factors are produced by the liver and circulate in the blood as inactive proteins. They were called factors long before anything was known of their chemical nature. They are identified by Roman numerals although some of them also have common names (*Table 13.3*). However, their numbering is unfortunately not in a logical order with

Factor number	Name	Active form/ function	Associated diseases
I	fibrinogen	fibrin subunit	afibrinogenemia (uncommon)
II	prothrombin	serine protease	defective function (extremely rare)
III*	tissue factor	receptor	–
IV	Ca^{2+}	–	
V	labile factor	cofactor	–
VI	initially wrongly identified, now known to be Va		
VII	proconvertin	serine protease	deficiency known (but rare)
VIII	antihemophilia	cofactor	hemophilia A (classical hemophilia,
	factor		see main text)
IX	Christmas factor	serine protease	Christmas disease (hemophilia B)
X	Stuart-Prower factor	serine protease	deficiency known (but extremely rare)
XI	plasma thromboplastin antecedent	serine protease	rare, primarily in Ashkenazi Jews
XII	Hageman factor	serine protease	–
XIII	fibrin stabilizing factor	transglutaminase	rare, poor wound healing, scarring
	prekallikrein	precursor of kallikrein, a serine endopeptidase	–
	von Willebrand factor (vWF)	–	vWB disease (up to 125 in 10^6)
	high M_r kininogen	–	–

*factor III is used as a synonym for thromboplastin, a mixture of tissue factor and phospholipid.

Table 13.3 Clotting factors

reference to the clotting process. Their activated forms are distinguished by a lower case 'a'; thus the activation of Factor VIII produces Factor VIIIa. Many of them become proteases when activated but, as each factor catalytically activates the next one in sequence, a small amount of the first factor produces a very large reaction at the end of the pathway. Thus the cascade of clotting reactions is an amplification of an initial small stimulus and leads to the production of large amounts of fibrin forming a clot.

It is convenient to divide clotting factors into four groups based on what they do in the clotting process. The first group consists of **zymogens,** which become active proteases when subjected to specific proteolytic cleavage. The second group contains **cofactor proteins** that bind to zymogens and their protease products, increasing the specificity and speed of the activation. The third group of factors contains protease inhibitors, which inactivate the proteases after their roles in the clotting process are complete. The fourth group is a miscellaneous group. It includes fibrinogen, which is cleaved to form fibrin, Factor XIII/XIIIa, a transglutaminase that catalyzes the formation of covalent cross-links between fibrin molecules, which stabilizes the clot, and von Willebrand factor which can anchor platelets to the endothelium of blood vessels and which also carries Factor VIII in the plasma.

THE INTRINSIC PATHWAY

The intrinsic pathway is a sequence of reactions that are catalyzed by enzymes, which become activated when tissue is injured (*Figure 13.11*). This pathway can take several minutes for completion. Damage to the wall of a blood vessel or to the endothelium that lines it (*Chapter 14*) results in the exposure of collagen to which platelets stick and become activated. The platelets degranulate, releasing a range of highly active substances, including serotonin (5-hydroxytryptamine or 5-HT) and ADP, as well as certain growth factors, such as platelet derived growth factor (PDGF), which has a role in subsequent wound healing. Serotonin (*Figure 7.5*) is a powerful vasoconstrictor and causes the blood vessel to constrict, temporarily limiting the blood supply to the damaged area until clotting takes place. This is a temporary effect, and a clot must form soon. The period before blood flow stops is called the **bleeding time** and is normally two to six min. **Platelet aggregation** occurs as the released ADP activates other platelets and these in turn activate others to extend pseudopodia and become sticky and clump with those already adhering to the damaged area. This forms a plug that prevents further blood loss. Fibrin is then deposited in the loose platelet plug, which traps more platelets as well as blood cells, strengthening the clot. The **clotting time** is usually six to 12 min.

A complex cascade of reactions is required to produce fibrin. These are initiated by the binding of Factor XII (Hageman factor) to tissue collagen and anionic surfaces that somehow changes its conformation to Factor VIIa allowing it to convert prekallikrein to kallikrein. Kallikreins are serine endopeptidases that are widely distributed in mammalian tissues and body fluids. Plasma kallikrein, also known as kininogenin or Fletcher Factor, cleaves bonds in kininogen to produce a varied group of polypeptides, which include angiotensin, bradykinin, substance P and secretin. Kallikrein also activates the bound Factor XII to form Factor XIIa, in a reaction that also requires high M_r kininogen as a cofactor. Factor XIIa and high M_r kininogen activate Factor XI (plasma thromboplastin antecedent) to Factor XIa. Factor XIa also binds to the exposed surfaces of the tissue where injury has occurred. Here it catalyzes the conversion of Factor IX (Christmas factor) to Factor IXa in a reaction that requires Ca^{2+}. The final reaction, unique to the intrinsic pathway, involves the conversion of Factor X (Stuart factor) to Factor Xa, the final product, and is catalyzed by Factor IXa. This reaction also requires Factor VIII, the antihemolytic factor found on the surface of aggregated platelets and Ca^{2+}.

THE EXTRINSIC PATHWAY

The extrinsic pathway is so called because it requires the nonplasma protein, thromboplastin to initiate the cascade (*Figure 13.11*). Thromboplastin is an integral membrane protein found in many tissues but especially the walls of blood vessels, brain, lung and placenta. The initial reaction is the conversion of Factor VII (proconvertin) to Factor VIIa in a reaction that requires Ca^{2+} and phospholipids released from the injured tissue or from the surface of aggregated platelets. Damaged tissues also release thromboplastin which, in combination with Factor VIIa and Factor IXa from the intrinsic pathway, directly activates Factor X to Xa. The final result of this part of the pathway is the production of Factor Xa. Thus the final product of the extrinsic pathway, like the intrinsic, is the production of Factor Xa. Despite this, deficiencies in either pathway result in prolonged bleeding times.

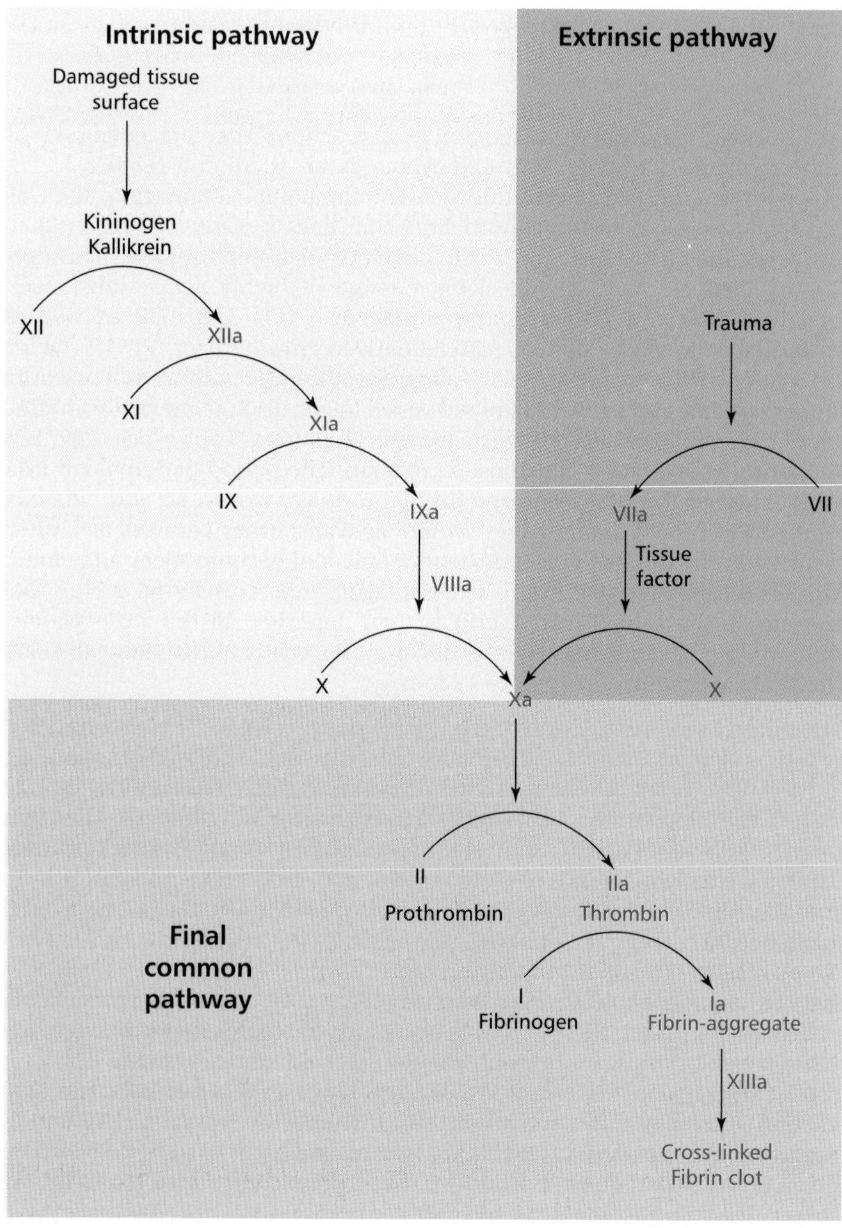

Figure 13.11 A simplified form of the blood clotting cascade consisting of the intrinsic, extrinsic and final common pathways. See text for details.

COMMON FINAL PATHWAY

Factor Xa from the intrinsic and extrinsic pathways catalyzes the hydrolysis of Factor II, or prothrombin to thrombin (*Figure 13.11*). There is an intermediate stage as prothrombin is first converted into prethrombin, which is slowly converted to thrombin. Factor Xa by itself is a relatively slow prothrombin activator, but its activity is enhanced about 20 000-fold by Ca^{2+}, Factor V (proaccelerin) and negatively charged phospholipids, for example phosphatidylserine, from damaged cell membranes. These types of phospholipids occur almost exclusively on the cytosolic sides of cell membranes, which, of course, are not usually in contact with the blood. Hence clotting reactions take place on the surface of the platelets so that the clotting action is confined to the sites of injuries.

Thrombin is a serine protease that catalyzes the hydrolysis of fibrinogen to fibrin, the final reaction of the clotting cascade. Fibrinogen forms 2–3% of the plasma proteins. It consists of three pairs of polypeptides and two pairs of oligosaccharides (*Figure 13.12*).

Thrombin cleaves four peptide bonds in the fibrinogen molecule, releasing two A and two B fibrin peptides from the amino terminal ends of the fibrinogen polypeptides changing the net charge from positive to negative. In the presence of Factor XIIIa and Ca^{2+}, the fibrin monomers polymerize producing a stable clot. The clots formed are strengthened by cross-links formed by the transaminase, Factor XIIIa, which is also activated by thrombin. Within the clot, the platelets contract, reducing the clot to less than half its original size and one that is tougher and more elastic. The process also draws the edges of the wound together.

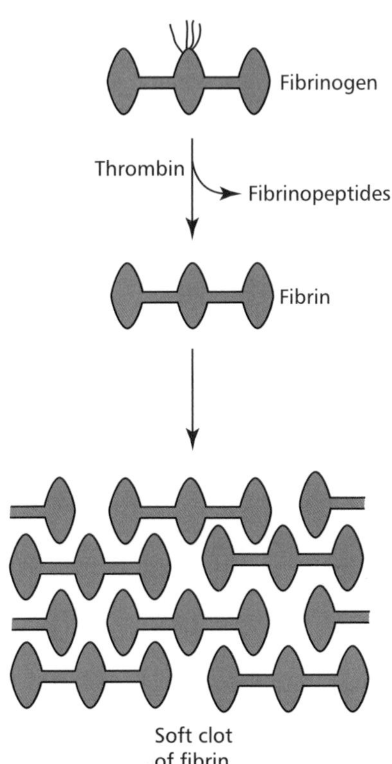

Figure 13.12 Schematic showing the conversion of fibrinogen to fibrin. The cleavage of small peptides from fibrinogen catalyzed by thrombin forms fibrin monomers that aggregate to give a soft clot (see *Figure 13.10*).

Margin Note 13.4 Fibrinogen 

Fibrinogen is an **acute phase protein** whose concentration is substantially increased in certain clinical situations such as acute inflammation caused by surgery, infections and myocardial infarction.

PREVENTION OF CLOTTING

It is clearly important that clots do not form in the absence of injury. There are several systems to prevent this occurring. The protein factors constantly circulate in the blood but, of course, require specific activation before clotting can take place; in addition, the liver removes activated factors. However, this is a slow preventative measure. A more rapid mechanism is effected by antithrombin III, which is a natural clotting inhibitor that binds to all of the serine proteases of the clotting cascade, especially thrombin, and inhibits their proteolytic activities. Heparin (*Figure 13.13*), a sulfated polysaccharide found in the circulation, activates antithrombin III. In addition, the plasma protein, α_2-macroglobulin also inhibits the clotting cascade. Patients with antithrombin III deficiency have an increased risk of thrombosis and resistance to the action of heparin.

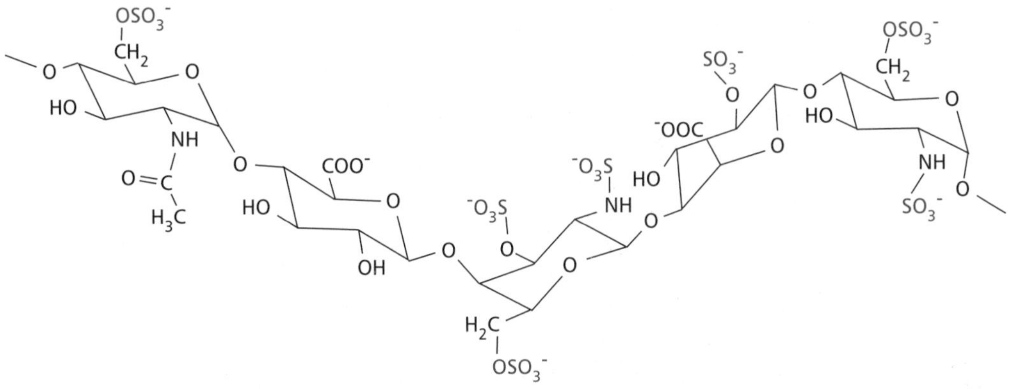

Figure 13.13 A pentasaccharide portion of the anticoagulant polysaccharide, heparin. The M_r of a complete molecule is approximately 17 000.

DISSOLUTION OF CLOTS

Clots on the skin surface eventually scab and are largely removed by abrasion. However, internal clots are eventually destroyed by a process called **fibrinolysis**. Again, a cascade of reactions is involved. Fibrin activates plasminogen activator, a protein that as its name implies, converts plasminogen to plasmin. Plasmin is a hydrolytic enzyme that catalyzes the digestion of fibrin and dissolves the clot. Plasminogen activator has a high affinity for fibrin clots and it and a tissue-type plasminogen activator (t-PA), bind to the clot and activate plasminogen.

In some instances, such as prolonged bleeding, shock and some types of cancer, plasminogen activator can be activated in the absence of fibrin. In these cases, plasminogen activators, such as streptokinase or t-PA are given to patients to dissolve blood clots to try to reduce the damage caused by myocardial infarction (*Chapter 14*).

13.5 ANEMIAS

Anemia develops when the amount of Hb in the blood falls below the reference levels for an individual's age and sex (*Table 13.4*) and there is insufficient iron for Hb synthesis. Anemia may be caused by major blood loss, or as a consequence of defects with Hb, the hemoglobinopathies, and by deficiencies of, for example, iron or some vitamins (*Section 13.6*). The characteristic signs of anemia are pallor, tachycardia, a fast heart rate, cardiac failure (*Chapter 14*) and epithelial changes including brittle nails, spoon-shaped nails, atrophy of the tongue papillae, angular stomatitis and brittle hair. Other signs specific to the type of anemia may also be present. However, anemic patients may be asymptomatic, even when the anemia is quite severe, or may present with various nonspecific symptoms, such as fatigue, headache, breathlessness, angina on effort or palpitations (*Chapter 14*). Rapid onset of anemia tends to cause more symptoms than slow onset, and the elderly tolerate anemia less well than the young when the normal compensation of increased cardiac output is impaired. The body responds to anemia with a variety of physiological responses. For example, the heart responds with an increased stroke volume and tachycardia to increase its output (*Chapter 14*). The oxygen binding curve of Hb can be modified by the production of 2,3 bisphosphoglycerate (BPG), which increases the release of O_2 from oxyhemoglobin to the tissues. In iron deficiency anemia, the concentration of BPG can increase by 40–75%.

Margin Note 13.5 Hematological indices 

The measured erythrocyte values for Hb concentration, packed cell volume (PCV) and erythrocyte count (RBC) allows four hematological indices to be calculated. These are the packed cell volume (PCV), mean cell volume (MCV), mean cell Hb (MCH) and mean cell Hb concentration (MCHC). The PCV can be measured using a hematocrit (*Figure 13.14*) or an automated cell counter (*Figure 13.15*). It can also be derived as the product of the MCV and RBC. The MCV is obtained by dividing the PCV by the RBC. The MCH can be calculated by dividing the Hb concentration of whole blood by RBC. The MCHC can be obtained by dividing the Hb concentration by the PCV. The normal values for these indices are given in *Table 13.4*.

	Males	Females
[Hemoglobin] / g dm^{-3}	135–175	115–155
PCV* (hematocrit) / %	40–52	36–48
MCV* / fdm^3	80–95	80–95
MCH* / pg	27–34	27–34
MCHC * / g dm^{-3}	200–350	200–350
[Serum iron] / μmol dm^{-3}	10–30	10–30
Total iron-binding capacity / μmol dm^{-3}	40–75	40–75

*PCV, packed cell volume; MCV, mean cell volume; MCH, mean cell Hb; MCHC, mean cell Hb concentration (*Margin Note 13.5*).

Table 13.4 Normal blood values

The reduced amount of Hb in anemia is usually accompanied by a reduction in the erythrocyte count and the packed cell volume that are usually determined in hospital laboratories using an automated blood analysis system (*Figure 13.15*). These types of systems can determine all hematological indices, for example the Hb concentration, packed cell volume (PCV) and the erythrocyte count (RBC), for numerous samples rapidly and efficiently. The mean cell volume (MCV), mean cell Hb (MCH) and mean cell Hb concentration (MCHC) can be derived from these measured values (*Margin Note 13.5*). The investigation of anemia should also consider the reticulocyte, white cell and platelet counts and any abnormal morphology as seen in a blood and/or marrow when examined with a microscope.

The three major types of anemias are the normocytic, microcytic and macrocytic, which are classified largely in terms of erythrocyte indices, especially their MCVs. Normocytic anemias, that is with normal sized erythrocytes, are associated with acute blood loss and a variety of disease states. However, the microcytic and macrocytic anemias are associated with observable changes to the sizes of erythrocytes in the blood sample and distinctive changes to the bone marrow appearance (*Table 13.5*). Thus the classification of anemias starts from routine hematological investigations. Their diagnosis involves taking a medical history and a clinical investigation of the patient, especially, of course, blood and marrow examinations to determine any changes in erythrocyte size.

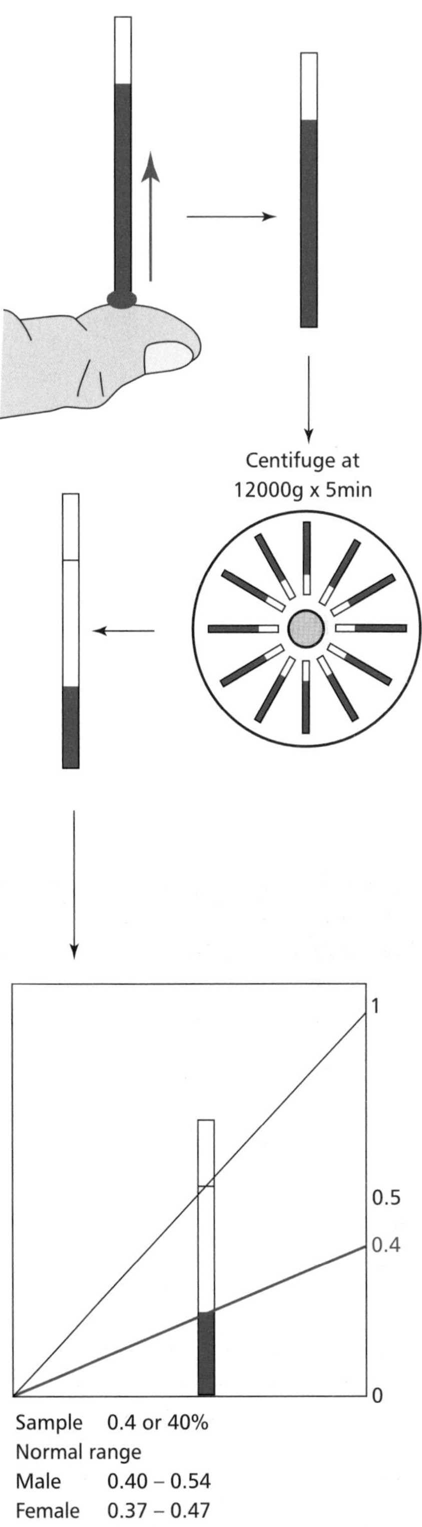

Figure 13.14 Determining PCV with a hematocrit following the collection of blood in a capillary from a thumb prick.

Erythrocyte appearance	Diameter / μm; MCV / fdm^3	Causes
Microcytic and hypochromic (small cells, pale due to reduced Hb content)	< 7; < 80	iron deficiency, thalassemias, sideroblastic anemia, chronic disease
Normocytic and normochromic (normal size and color)	7; 76 to 96	acute blood loss, infection, collagen disease, malignancy, endocrine disease, chronic disease
Macrocytic (large cells; oval or round in shape)	> 9; 96	deficiencies of vitamins B_{12} or folate (oval), alcoholism (round), liver disease (round)

Table 13.5 A classification of anemias

MICROCYTIC ANEMIAS

The major causes of microcytic anemias are iron deficiency, thalassemias (*Section 13.6*), sideroblastic anemia and the anemia of chronic disease. Iron is difficult to absorb because of problems connected with the low solubility of its salts, its oxidation state, and interaction with other components of the diet. Loss of iron also occurs in hemorrhage and menstruation (*Box 13.1*).

Iron deficiency anemia, the commonest cause of anemia worldwide, shows a number of characteristic features. The erythrocytes (*Figure 13.16*) are microcytic, with an MCV of less than 80 fdm^3, and hypochromic, the MCH being less than 27 pg. There is variation in, cell sizes, anisocytosis and poikilocytosis, that is abnormal shapes, and a reduced reticulocyte count. There are also changes in the bone marrow, for example erythroid hypoplasia and decreased iron deposits. The serum iron decreases while the serum iron binding capacity increases compared with their normal concentration

Figure 13.15 An automated blood analysis system that can determine all hematological indices automatically. Courtesy of Department of Clinical Biochemistry, Manchester Royal Infirmary, UK.

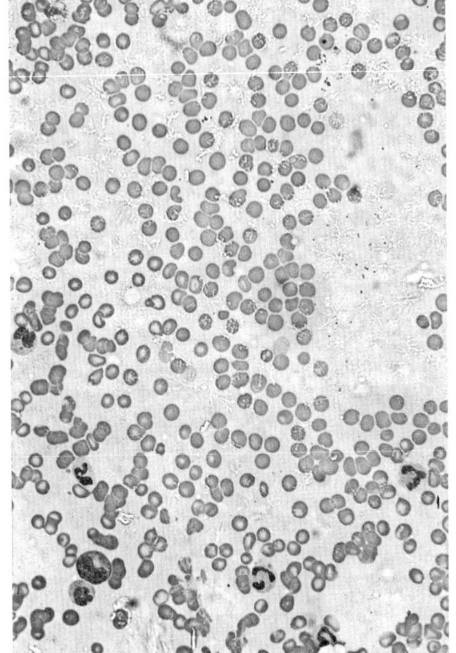

Figure 13.16 A photomicrograph of a peripheral blood smear from a patient suffering from severe iron deficiency. The cells are microcytic and hypochromic.

(*Table 13.4*). The transferrin saturation, that is the proportion of serum iron to its total iron binding capacity, falls below 19%, compared with a more usual 30% or so. The concentration of serum ferritin resulting from cellular degradation is regarded as the most reliable measurement of anemia. Patients may also show an impaired ability to maintain body temperature, depressed muscle function and abnormal thyroid hormone metabolism.

The underlying cause of the iron deficiency should be identified by the appropriate tests and taking a careful history of the patient. To counteract the deficiency, 600 mg of ferrous sulfate is given orally each day. If there are side effects, such as nausea, diarrhea or constipation, then ferrous gluconate may be substituted. Failure to respond to the treatment may be due to lack of patient compliance, continuing hemorrhage, severe malabsorption, or another cause for the anemia. It may be necessary to give iron parentally if absorption is defective, as for example in patients with ulcerative colitis or Crohn's disease (*Chapter 11*).

MACROCYTIC ANEMIA

Macrocytic anemias are characterized by the presence of anemia and erythrocytes of variable shapes but with diameters in excess of 9 μm and MCVs characteristically greater than 96 fdm^3 (*Table 13.4*). The condition may be caused by certain liver diseases, including alcoholism, that produces large rounded cells or by megaloblastic anemia, which is associated with enlarged oval cells. The latter is also indicated by the presence of the erythrocyte precursors, erythroblasts (megaloblasts) in blood. The increased proportion of immature forms of all cell lines reflects the premature death of cells in the process of development (*Figure 13.17*). The cells are large, although there is a substantial variation in size, and they have large, immature nuclei. The basis of the problem is the inability to synthesize deoxythymidine monophosphate from methylated deoxyuridine monophosphate. The methyl group is supplied by the folate coenzyme, methylene tetrahydrofolate polyglutamate and deficiency of folate reduces its supply. A deficiency of vitamin B_{12} (*Figure 13.18(A)*) also reduces its supply by slowing the demethylation of methyl tetrahydrofolate. Thus deficiencies of vitamins B_{12} or folate (*Chapter 10*) or other defects, for example genetic ones, that affect DNA biosynthesis in the bone marrow, produce an asynchrony between nuclear and cytoplasmic development and a delayed maturation of blood cells and result in megaloblastic anemia.

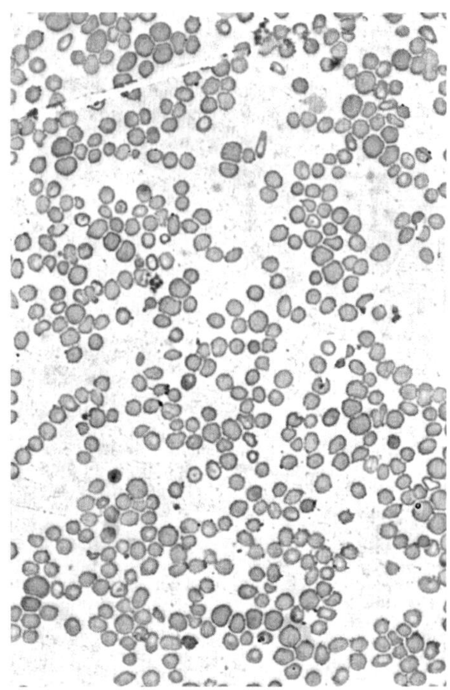

Figure 13.17 A photomicrograph of a marrow smear from a patient with megaloblastic anemia. The MCV is over 95 fdm^3 and the macrocytes are typically oval in shape. The bone marrow is usually hypercellular and the normoblasts (erythroblasts) are large and show failure of nuclear maturation with an open, fine, primitive (stippled) chromatin pattern. Giant and abnormally shaped cells called myelocytes are present.

Vitamin B_{12}, also called cobalamin, is found in animal products and is produced by certain microorganisms but not by plants. It is liberated from protein complexes by gastric enzymes and binds to a glycoprotein called **intrinsic factor** (*Chapters 10* and *11*). This is secreted by the gastric parietal cells along with H^+ and carries vitamin B_{12} to specific receptors on the mucosal surface of the ileum. Although the vitamin enters the ileal enterocytes, the intrinsic factor remains in the lumen of the gut. Transport in the blood is by another protein, transcobalamin. Atrophy of the gastric mucosa and consequent failure to produce intrinsic factor leads to the malabsorption of the vitamin, whose deficiency results in pernicious anemia. Cytotoxic IgG antibodies directed against gastric parietal cells and/or against intrinsic factor are found in the serum in about 90% of individuals with pernicious anemia. In a majority of these individuals the antibodies are also present in the gastric juice and either prevent the binding of vitamin B_{12} to intrinsic factor or inhibit the absorption of the vitamin B_{12}: intrinsic factor complex.

The onset of the disease is insidious with progressively increasing symptoms of anemia. Patients show **achlorhydria**, a low or absence of gastric acid secretion, and lack of secreted intrinsic factor. There may be jaundice because of excessive breakdown of Hb and because erythropoiesis in the bone marrow is deficient. The serum bilirubin may be increased and the serum vitamin B_{12} concentration is usually considerably below its physiological value of approximately 160 ng dm^{-3}. However, the polyneuropathogical symptoms make it important that treatment is not delayed as they can become irreversible; patients present with symmetrical paraesthesia in the fingers and toes, an early loss of vibration sense and ataxia. Paraplegia may be the result. Pernicious anemia is predominantly a disease of the elderly with one in 8000 of the over 60 population being affected in the UK. It also seems to be associated with certain autoimmune diseases, such as thyroid and Addison's diseases (*Chapter 7*).

The causes of folate (*Figure 13.18 (B)*) deficiency are nutritional, for example a poor intake of green vegetables, such as broccoli and spinach and offal, alcohol excess, cancer, or excessive utilization in pregnancy and lactation and the use of antifolate drugs, such as methotrexate, phenytoin and pyrimethamine. The clinical manifestations of folate deficiency are megaloblastic anemia with a serum folate concentration that is lower than the reference value of 4 to 18 μg dm^{-3}.

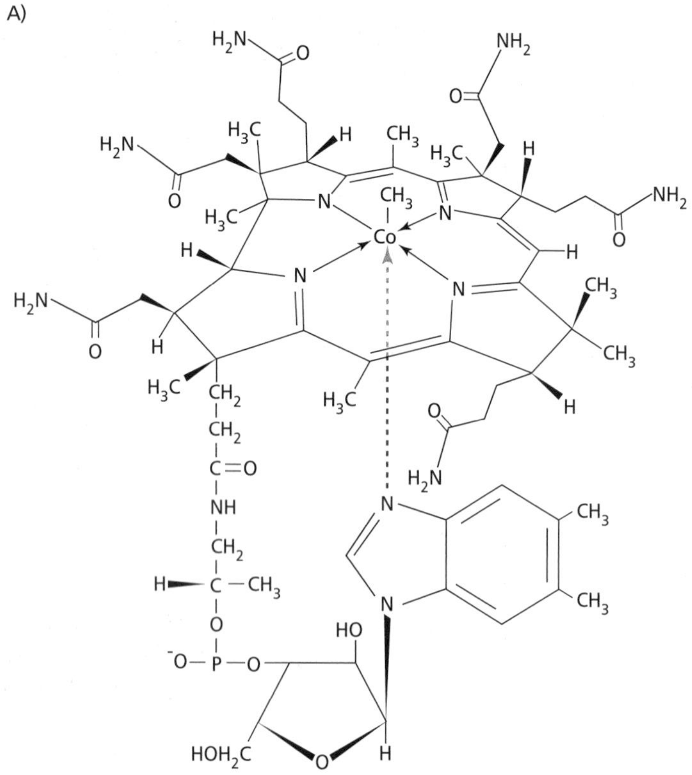

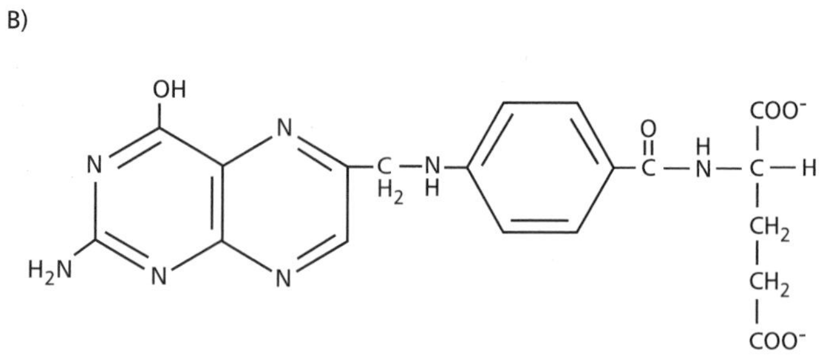

Figure 13.18 Structures of (A) vitamin B_{12} (cobalamin) and (B) folic acid.

The deoxyuridine suppression test for megaloblastic anemia is performed by adding tritiated thymidine (^{3}H-thymidine) to a sample of bone marrow. Bone marrow samples may be obtained by aspiration or by trephine. Aspiration, using a specialized needle, is usually carried out at the iliac crest with a local anesthetic. In normal marrow less than 5% of the ^{3}H-thymidine is usually taken up, but in megaloblastic marrow up to 50% of it may be used. The microscopic picture of the bone marrow can be investigated by using the aspirate to make a smear on a microscopy slide. If a larger sample of bone marrow is required, the posterior iliac crest is used but a longer and wider needle is used to obtain a 'core' of bone. This core is fixed and decalcified over several days and then stained for microscopy.

It is important to distinguish pernicious anemia from other causes of megaloblastic anemia, such as folate deficiency, because this will affect treatment. However, this is usually clear from the blood concentrations of these two vitamins. Also, the ability to absorb vitamin B_{12} can be measured using the Schilling test in which patients are given vitamin B_{12} radioactively-labeled with ^{58}Co. The urine is collected over a period of 24 h and the amount of radioactivity measured.

Vitamin B_{12} deficiency is treated by intramuscular injection of 1 mg of the pure vitamin, to a total of 6 mg over a period of three weeks. Oral administration is obviously unsuccessful in pernicious anemia because of the lack of intrinsic factor. A maintenance dose of 1 mg every three months is then given for the rest of the patient's life. Clinical improvement may occur within a few days (provided that the neuropathy has not been long-standing) and a reticulocytosis is observed a few days later. Folate deficiency can be corrected by giving 5 mg of folic acid daily, and this usually produces a rapid hematological response. Prophylactic folate is recommended for all women during pregnancy and especially for women who have had a previous child with a neural tube defect.

13.6 HEMOGLOBINOPATHIES

Hemoglobinopathies are clinical conditions that result from mutations that change the sequences of bases in DNA of the genes for globins (*Chapter 15*). If the bases in the DNA are changed even by a single one, then a modified protein may be produced (or no protein at all). The consequences can be negligible, severe or fatal. Mutations are inherited and, if the disease is not fatal, then the disease symptoms will be inherited too. The severity of the disease may depend on whether one or both copies of the gene in question carry the mutation, in other words, whether the individual is homozygous or a heterozygote. The mutations involved in hemoglobinopathies include point mutations, the largest group, that substitute one amino acid residue for another, insertions or deletion of one or more residues, drastic changes caused by frameshift mutations (*Margin Note 13.6*) and alterations in the lengths of the polypeptide chains by mutations that produce or destroy stop codons.

In normal adult humans, there are two α- and one β-globin genes, coding for polypeptides of 141 and 146 amino acid residues respectively, which go to form HbA, $\alpha_2\beta_2$. In a diploid cell there are actually four α and two β genes. Each of these genes has two introns (*Margin Note 13.7*). The α genes are located on chromosome 16 and the β genes on chromosome 11. If there is a mutation, it may have been inherited from one or both parents giving a heterozygous or homozygous condition respectively. A mutation in an α gene tends to have less serious consequences than one in a β gene because there may still be nonmutated copies of the α gene present. Nevertheless, even small changes in the structure of the Hb protein can sometimes result in disastrous clinical effects. Over 750 Hb mutations are known. They usually only affect one type of subunit because there are separate genes for the α- and the β-globins (*Table 13.2*).

Originally, many of the different mutant Hbs were identified by their mobilities in electrophoresis (*Figure 13.19*) and peptide mapping (*Box 13.3*) but now, of course, the DNA can be analyzed directly. A major technical advance has been the ability to make DNA probes that are specific for α- or β-chains. This means it is possible to identify which mRNAs are being produced and identify any mutations present. Thus the different *clinical* variants can be understood at the *molecular* level. For example, in so-called 'hemoglobin H disease' it has been shown that there is only one of the four possible α-globin genes present and functioning, so that only 25% of the normal amount of α-chain mRNA is produced. The mutation causing this situation is a deletion not a point mutation.

The majority of mutations are harmless and therefore do not produce a hemoglobinopathy because they do not cause disease. For example, mutations

Margin Note 13.6 Frameshift mutations

In the genetic code, sequences of three bases or **codons** code for each amino acid residue. A change in one base may cause the incorporation of a different (or 'wrong') residue, such as occurs in HbS. However, if one or more bases is lost or added then the reading frame of the code is shifted. Instead of a single amino acid being changed, a totally new sequence may be produced, which may result in the production of a different protein or often no protein at all, depending upon where in the sequence the frameshift occurs.

Margin Note 13.7 Introns and exons

The majority of genes in eukaryotic cells are not continuous but are arranged in sections along the DNA of the chromosome. The coding sequences are called **exons**, and the noncoding portions in between are called intervening sequences or introns. In order to make the messenger RNA that can be translated to produce a polypeptide, an RNA transcript of the DNA is made, that is one with introns and exons transcribed, and then the intron coded sections are cut out and the exposed ends joined together in a process called splicing. This is a normal part of the processes that produce a mRNA that can be translated to produce a polypeptide in eukaryotic cells.

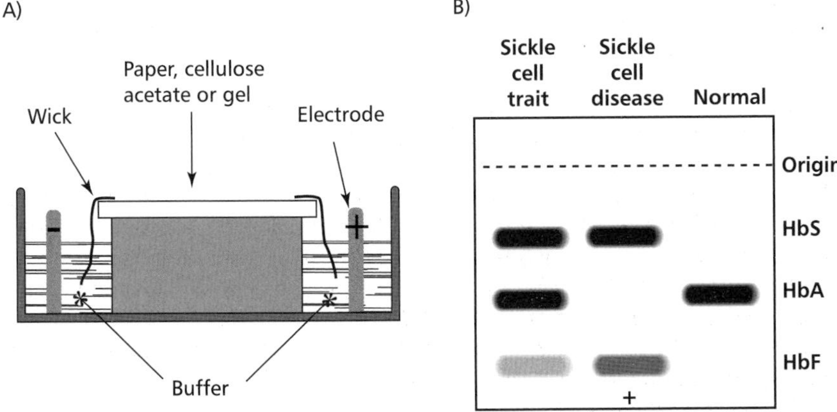

Figure 13.19 Electrophoresis to identify mutant hemoglobins that have different charges from normal adult hemoglobin. A hemolysate of erythrocytes is subjected to electrophoresis, for example sickle cell hemoglobin (HbS) moves more slowly towards the positive electrode because a glutamic acid residue in the β-chain (negatively charged) is replaced by a valine residue (zero charge) so that the whole molecule of HbS has two fewer negative charges than HbA.

Margin Note 13.8 Heinz bodies

Heinz bodies or Heinz-Ehrlich bodies were first reported in 1890 by the German physician Heinz (1865–1924) as inclusions in the erythrocytes of some patients with hemolytic anemia. They are known to be aggregates of denatured, precipitated Hb, which associate with the erythrocyte membrane (*Figure 13.20*). This causes the erythrocyte to become misshaped and leads to anemia. The bodies are best seen when the blood is stained with crystal violet. Heinz bodies are associated with certain type of hereditary hemolytic anemia, for example hemolytic anemia of infancy. Oxidative damage to Hb by a number of toxic chemicals, including nitrobenzene, diphenylamine, naphthalene and hydroxylamine and a number of its derivatives can also result in Heinz body formation. They also have iatrogenic causes and may result from sensitivity to some drugs, for example primaquine especially in glucose 6-phosphate dehydrogenase deficiency (*Section 13.7*) and sulfonamides and can appear after a splenectomy.

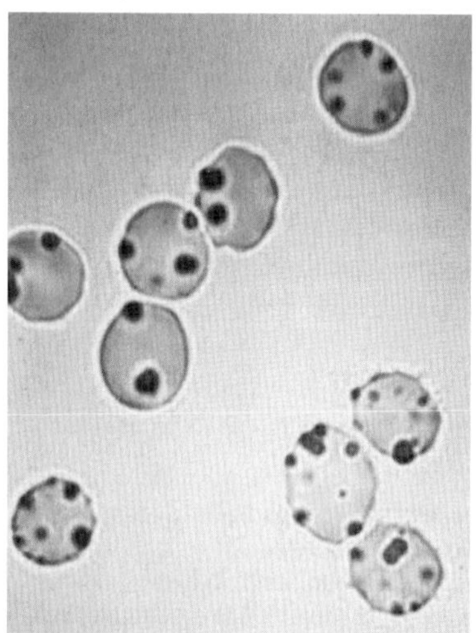

Figure 13.20 Heinz bodies in erythrocytes are formed from denatured, precipitated hemoglobin. Courtesy of Dr Ian Quirt, Department of Medical Oncology and Hematology, Princess Margaret Hospital, Toronto, Canada.

distant from the heme binding cleft, or from the regions of subunit contact may have little effect on the properties of the Hb. However, mutations may change the shape of the globin subunit(s), the binding of the heme groups or even prevent globin synthesis, all with severe clinical consequences. To function properly, the four subunits in the Hb molecule must fit together tightly but still produce a molecule that is flexible. The regions of contact have been conserved in evolution and are essential for normal functions, such as the cooperative binding of O_2 (*Figure 13.7*). Thus mutations can upset the delicate balance of interactions between the amino acid side chains with several consequences. The molecule may dissociate upon deoxygenation and, in some cases, the monomers may precipitate in the erythrocytes reducing O_2 affinity. Microscopically, the denatured and precipitated Hb can be seen as Heinz bodies (*Margin Note 13.8* and *Figure 13.20*). A deletion of one or more amino acid residues or substitution mutations can produce this effect, as in Hb Leiden and Hb Philly respectively. There may also be cell membrane damage, with intravascular hemolysis, anemia, reticulocytosis and splenomegaly as consequences. In other cases, a small change in the regions that bind the heme groups may make the pockets slightly less hydrophobic so that it does not bind appropriately, and again, the denatured Hb can precipitate to form Heinz bodies. Thus only two of the four subunits may have heme groups. In other cases, the change in the pocket allows the iron to become oxidized to the Fe^{3+}(III) state (methemoglobin), which will not bind O_2. The resulting condition is referred to as methemoglobinemia and patients become cyanosed because they lack oxygen (*Margin Note 13.2*).

This chapter will concentrate on two hemoglobinopathies, **sickle cell anemia**, which arises from a point mutation, and a group of diseases called the **thalassemias** that seem to originate from point mutations or very small deletions.

SICKLE CELL ANEMIA

Sickle cell anemia was first described in a black patient in the USA in 1904, a time when little was known of the structures of proteins. The patient presented with severe pain and a microscopic examination of a blood sample showed sickle shaped erythrocytes (*Figure 13.21*). The gene for sickle cell Hb (HbS) differs from that for HbA by a single point mutation in the β-globin gene at the codon responsible for the amino acid residue at position 6. This substitutes a thymine for an adenine base. Given that one β-globin gene is inherited from each parent, the condition may be homozygous (Hb^SHb^S) or heterozygous (Hb^AHb^S).

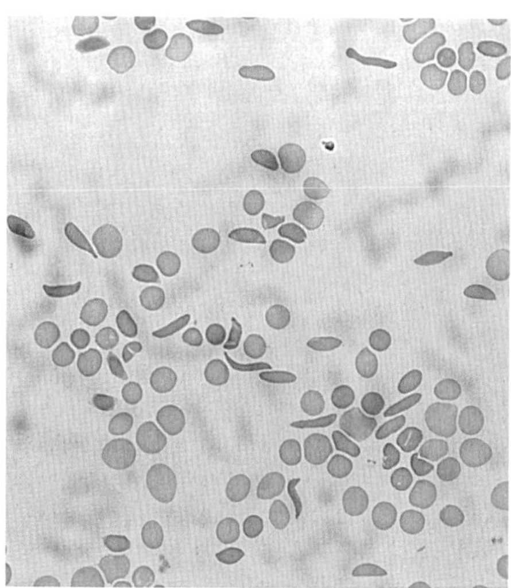

Figure 13.21 A photomicrograph of a blood smear from a patient with sickle cell anemia. Note the sickled erythrocytes.

BOX 13.3 Sickle cell anemia: a molecular disease

The identification of the precise mutation in sickle cell disease was a significant step in the understanding of molecular diseases. The use of peptide mapping in two dimensions on large sheets of chromatography paper enabled the differences between the HbA and HbS to be identified. The change from a glutamate to a valine residue means the β-globin molecule has lost one negative charge and become more hydrophobic leading to sickle cell anemia, as described in the main text:

Hemoglobin A ...Pro-**Glu**...

Hemoglobin S ...Pro-**Val**...

PEPTIDE MAPPING

The sequence of amino acids in a protein may be determined by hydrolyzing the polypeptide into small fragments, for example by digestion with trypsin, and then separating the fragments and determining the sequence of each. It is relatively easy to determine the amino acid sequence of short peptides. The sequences of these then have to be aligned to give the sequence for the complete polypeptide.

The traditional way to separate many short peptides was two-dimensional separation on a large sheet of filter paper. The separation was usually by electrophoresis in one dimension, followed by chromatography in the second to produce a peptide map, sometimes called a fingerprint. The colorless peptides were located on the paper by staining them with ninhydrin. When a point mutation has occurred, it is often possible to observe that just one peptide has changed its position provided that the mutation has produced a change in charge, as is the case with HbS (*Figure 13.22*), or that the substitute amino acid residue is substantially different in M_r to the original. The stained 'spots' can then be cut out and the peptide eluted from the paper and its sequence determined. This method is no longer used. Typically HPLC is now used to separate and purify the small peptides and the use of mass spectrometry can give the sequence quickly.

The identification of the precise defect in the mutant Hb in sickle cell patients as an adenine to thymine mutation in codon 6 of the β-globin gene was a major step in the understanding of genetic diseases. The condition was the first one to be referred to as a *molecular disease*.

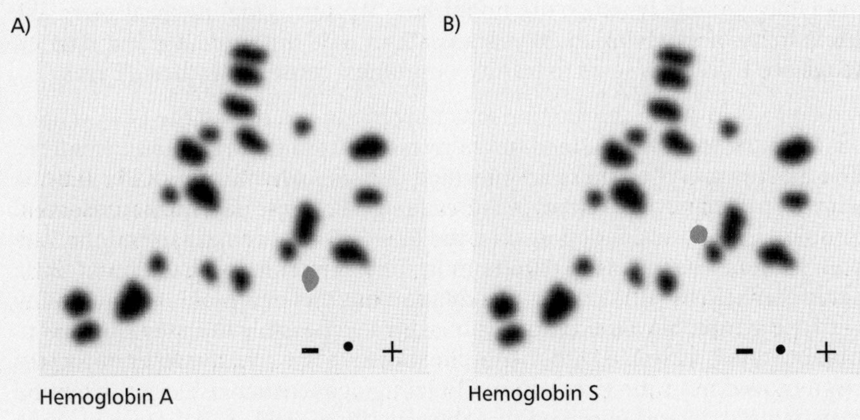

Figure 13.22 Peptide mapping to show the difference, highlighted in red, between (A) normal and (B) sickle cell hemoglobins. See text for details.

The mutation means that an acidic, hydrophilic glutamate residue is replaced by a hydrophobic valine. The presence of the valine residue means that the Hb molecule is a little more hydrophobic or 'sticky' in two places on its surface because there are two β-chains present. The sticky patches are more exposed in the deoxygenated state when the conformation of HbS changes as the molecule releases its oxygen in the tissues. The HbS molecules therefore aggregate forming stiff fibrils that cause the sickling of the erythrocytes although, even after years of study, it is still not completely understood how these changes occur. The deformed erythrocytes are less flexible than normal ones and cannot squeeze through the capillaries in the tissues and block them. This leads to hemostasis, anoxia and severe pain and, because the sickling occurs as a result of changes occurring in deoxyhemoglobin, the effects are exacerbated and more cells become sickle-shaped. The life of a typical erythrocyte is reduced from 120 to about 10–12 days in sickle cell patients: the abnormal cells are destroyed in the spleen and consequently anemia ensues.

At low concentrations, HbS shows a normal oxygen binding curve, but at high concentrations, as would occur in the erythrocytes of a homozygote for sickle cell anemia, the oxygen affinity is decreased. Again, the reasons for this are not fully understood, although the resulting shift in the oxygen dissociation curve to the right means a greater proportion of the oxygen is released and ameliorates the effects of the anemia. Indeed, the amino acid substitution that causes the condition does not affect the structure of the oxygen binding site or the ability of the molecule to bind and carry oxygen.

Patients who are homozygous for sickle cell anemia present with crises of intense pain that can occur anywhere in the body caused by blockage of capillaries. Crises tend to occur when the circulation is slow or when there is hypoxia; about 15 s of low oxygen tension are required to produce sickling so when the circulation is reasonably rapid there is insufficient time for this to happen. Clinical complications of sickle cell disease are highly variable, and the clinical consequences may include megaloblastic erythropoiesis, aplastic crises, stroke, bone pain crises, proneness to infection, especially by *Pneumococcus*, *Salmonella* and *Haemophilus* due to hyposplenism, and acute chest syndrome. Acute chest syndrome is a common form of crisis in children with sickle cell disease and is sometimes fatal. It occurs in about 40% of all people with sickle cell disease. It is characterized by severe chest pain and difficulty in breathing. It is probably caused either by a chest infection or by blocked pulmonary capillaries resulting from a blood clot. In developed countries the mortality in sickle cell disease is relatively low but this is not the case in developing countries. In general, there seems to be an approximate 10% mortality in the first few years of life but, again, this depends on the treatment available, namely whether the infant has 'Western-style' medical care. The probability of surviving to 29 years is about 84% but there are few data on longevity. Infections seem to be the commonest cause of death at all ages.

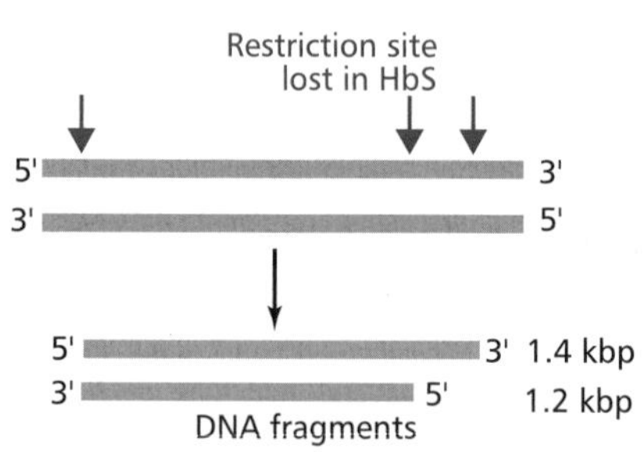

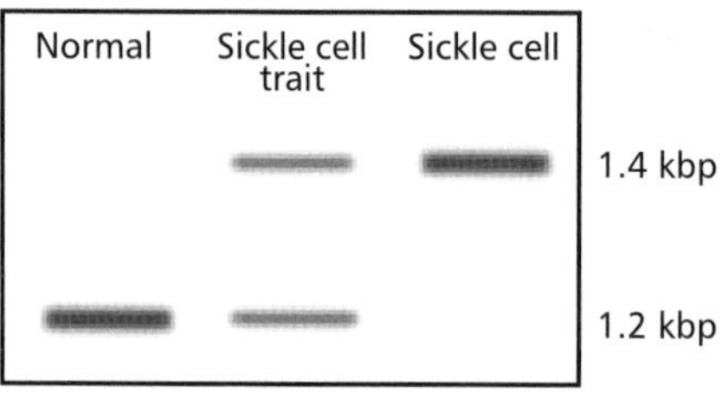

Figure 13.23 The mutation in sickle cell anemia changes one nucleotide base in the gene for the β-globin producing a new site, which the restriction enzyme, *Mst*II cannot attack. Consequently, different sized fragments of DNA are produced from normal compared with sickle cell hemoglobin genes. Gel electrophoresis separates the fragments, which may be detected with a probe for the β-globin gene to reveal the fragments of different sizes. This technique can also be applied to the prenatal diagnosis of sickle cell disease.

A diagnosis of sickle cell anemia may now be carried out on the DNA (*Figure 13.23*) of the embryo obtained by chorionic villus sampling or amniocentesis. The parents may then make an informed decision whether or not to continue with the pregnancy. Treatment includes analgesia for the pain during crises and antibiotics and vaccination against the likely life-threatening infections. The pain is often so acute as to require morphine. Sometimes inhalation of nitric oxide can help by producing a vasodilation but this treatment is only dealing with the symptoms. Blood transfusions are also possible but they can lead to iron overload, as well as other complications, as the transfused erythrocytes are removed from the circulation. Chelating agents such as desferrioxamine may be used. In was observed that the severity of sickle cell disease in some populations was reduced by the presence of high concentrations of HbF. Fetal Hb is almost as good as HbA in transporting oxygen and, of course, does not sickle. Hydroxyurea and butyrate are used as therapeutic agents to try to induce higher levels of HbF in sickle cell patients. Hydroxyurea is thought to kill selectively precursor cells in the bone marrow whilst sparing the erythroblasts that produce HbF. However, this compound is an antineoplastic agent and its long-term effects are unknown. Butyrate seems to activate transcription of the γ-globin gene so that HbF is produced in the adult. Both agents have met with reasonable success in treating sickle cell patients and, in some cases, may be used synergistically to increase HbF up to 20% with a marked clinical improvement.

The mutation that causes HbS production is not the only one that leads to sickling of erythrocytes but the many other variants are rather rare. The second commonest of these in black Americans occurs in HbC, in which a lysine residue replaces glutamate at position 6 in the β chain. Hemoglobin C is rather insoluble and crystals of it can sometimes be seen in peripheral blood smears. Heterozygotes for HbC are asymptomatic but homozygotes have a mild hemolytic anemia.

BOX 13.4 Sickle cell trait and malaria

Heterozygotes, those with genes for HbA and HbS, have sickle cell trait. Individuals typically have about 30% HbS and life expectancy is about the same as for normal persons. The condition causes relatively few problems except at high altitudes and when flying in nonpressurized aircraft. Sickle cell disease, the homozygous condition, is present in approximately 8% of American blacks and may be as high as 45% in some African populations. It is thought to cause 60 000–80 000 deaths in African children annually. It may be asked why such a deleterious gene should have persisted. The answer is probably that the possession of a single HbS gene, that is sickle cell trait, increases resistance to malaria caused by *Plasmodium* spp (*Chapter 2*). Malaria is typically endemic in areas where the sickle cell trait reaches high levels (*Figure 13.24*). The same may also be true of thalassemia (*see below*). The reason for resistance seems to be that as *Plasmodium* parasites grow in the erythrocytes, they lower the intracellular pH and generate hydrogen peroxide. The lower pH promotes sickling of the erythrocytes and the hydrogen peroxide damages cell membranes, which become more permeable to K^+. The resulting intracellular decrease in K^+ kills the parasites.

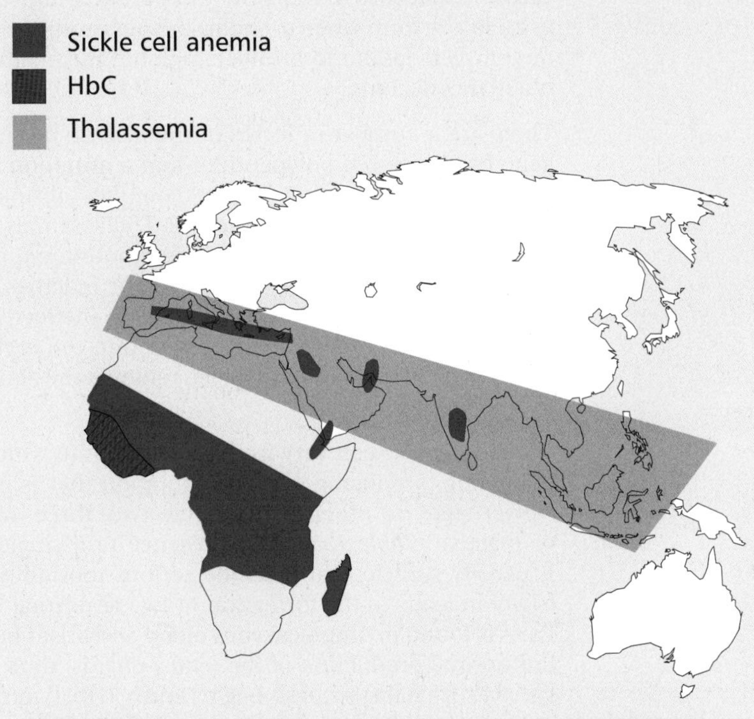

Figure 13.24 Areas of the world where sickle cell disease and thalassemia are common.

As a result of the coincident distribution of the genes for HbS and HbC, heterozygotes for both Hbs are not uncommon producing HbSC disease. This is milder than true sickle cell disease but patients can show practically all the same complications. Furthermore, it is symptomatic in the heterozygous state. There can also be co-inheritance of the sickle cell and thalassemia genes, which generates a wide spectrum of clinical symptoms whose severity depends on the type of thalassemia mutation (*see below*).

THE THALASSEMIAS

Thalassemia, from 'thalassa', which is Greek for 'the sea', are a group of hemoglobinopathies originally discovered in people living near the

Mediterranean sea. However, the thalassemias are also relatively common in southeastern Asia, the Philippines, China and worldwide and is perhaps the most common group of hereditary diseases. Cooley first accurately described them clinically in 1925 and the disease used to be called Cooley's anemia, a term now reserved for β-thalassemia.

In the late 1930s, thalassemia was shown to be an inherited disorder. However, it was not until protein analytical techniques improved in the 1960s that the disease was shown to be the result of an imbalance in the amounts of α- and β-globins synthesized. The severe anemias associated with some thalassemias stimulate the production of erythrocyte precursors. As a result, bone marrow is able to expand to all areas of the skeleton leading to skeletal deformities. If this occurs within the spine and compresses the spinal cord it can cause intense pain. Some forms of the disease are fatal causing death *in utero*, while others require copious blood transfusions for the anemia. α-Thalassemia, for example, is caused by a complete or partial failure to produce α-globin. It is fatal in its severe form when α-chains are not produced. In β-thalassemia, a partial or complete failure to produce β-globin means that patients require lifelong blood transfusions.

There are a number of forms of the disease because there is more than one gene for the globin polypeptides and a mutation may not affect all of them. Consequently there may not be complete absence of a given globin chain and the disease will be less severe. Thalassemias are classified according to which globin chains are reduced in amount, are mutated or are absent. The more copies of the gene are missing or inactive, the greater the severity of the disease. The anemia is caused by an ineffective erythropoiesis and from precipitation of excess free globin within the erythrocytes. The cells have a shortened life span and the spleen removes the abnormal erythrocytes leading to splenomegaly.

α-Thalassemias can vary from a condition in which only one α-globin chain is missing, producing a 'silent' mutation that is practically a symptom-free, carrier state, to other forms where two, three or all four genes are absent or inactive (*Table 13.6*). The presence of a single functional α-globin gene is usually sufficient to preclude serious morbidity. A decreased synthesis of α-globin leads to the formation of two abnormal Hb tetramers. Hemoglobin Barts is found in umbilical cord blood and arises because of a lack of α-chains but normal production of the fetal γ-chains; thus Hb Barts is γ_4. If the infant survives, β-chain synthesis begins and β_4 (HbH) tetramers form. Unfortunately, neither γ_4 nor β_4 can take up cooperatively O_2 but bind it so tightly it cannot be delivered to the tissues. Hemoglobin H precipitates in older erythrocytes

Syndrome	Genotype	Number of α-genes	Clinical severity	Hemoglobin
Hydrops fetalis	--/--	0	lethal *in utero*	mostly γ_4, little β_4
HbH disease	--/-α	1	severe microcytosis	about 25% HbA, mostly β_4
α-thalassemia trait	--/αα or -α/α-	2	asymptomatic	variable
Silent carrier	-α/αα	3	none	functional

Table 13.6 Characteristics of α-thalassemias

and under oxidant stress, such as when the patient is treated for malaria with primaquine. A total absence of a α-globin gene results in the lethal *in utero* condition, hydrops fetalis.

Some mutations that produce thalassemia are referred to as nondeletion mutations. The mutation occurs in the STOP codon for the α-chain, consequently translation continues beyond the normal end point to the next STOP codon, extending the polypeptide by 31 residues. An example of this produces Hb Constant Spring, the incidence of which is fairly common in southeastern Asian populations where the usual STOP codon UAA is mutated to CAA. It appears that the extended mRNA for the α-chain is unstable and leads to a reduced rate of Hb synthesis. If this mutation is present together with a lack of one of the α-globin genes, then HbH disease results. However, in heterozygotes only about 1% of the α-globin produced is the Constant Spring mutant type.

β-Thalassemias result from mutations in the β-globin genes. In some ways, it would be more appropriate to call them *'thalassemias* of the β-globin gene family' because the genes for δ-, and γ-globins and the single β-globin are all grouped together on chromosome 11. In β-thalassemia there is reduced synthesis of β-globin with or without a reduced synthesis of δ- or γ-globins. β-Thalassemia is not a single disease. The molecular defects in this and related disorders are highly heterogeneous, with about 200 mutations having been identified to date. These include many single nucleotide substitutions that affect the expression of β-globin genes. Examples include nonsense and frameshift mutations in the exons, point mutations in the intron–exon splice junctions and mutations in the 5′ region and the 3′-polyadenylation sites. The latter are extended sequences of adenine nucleotides at the 3′ end of mRNA molecules that stabilize the mRNA molecules and help in their transport from the nucleus to the cytoplasm. A number of heterozygous states are also known but these usually only give rise to mild clinical symptoms. In some forms of β-thalassemia, abnormal δ-β fusion polypeptides may be formed. For example, in Hb Lepore there is a fusion gene caused by nonhomologous crossing over between the δ-gene on one chromosome and the β-gene on the other. Thus normal β- and δ-genes are absent.

β-Globin chains are not required until after birth (*Figure 13.9*) and β-thalassemia infants are usually born normally at term. Clinical problems begin two to six months later, when γ-globin synthesis, and therefore the amount of HbF, has declined. Consequently β-thalassemia is a crippling disease of childhood, characterized by the precipitation of excess α-globin chains, destruction of erythrocytes in the bone marrow and circulation, and deficiency of functional Hb tetramers. In β-thalassemia major, formerly Cooley's anemia, β-globin synthesis is strongly depressed or absent causing massive erythroid proliferation and skeletal deformities.

Thalassemias can be diagnosed from their general clinical symptoms and the anemia, including the precipitation of excess free globin chains. Hypochromic erythrocytes with a clear center and a darker rim containing the Hb are visible in blood smears as are poikilocytes, which are abnormally shaped cells produced by the spleen when it removes target cells. There is also splenomegaly. Electrophoresis of hemolyzed erythrocytes will provide information about the relative proportions of A, A_2 and F globins. In HbH disease, the Hb (tetramer of β-globin) is detected as a rapidly moving band at pH 8.4.

The treatment for thalassemias, as with sickle cell disease, is to give repeated blood transfusions combined with chelation therapy to remove the iron with, for example, desferrioxamine. The latter is necessary because the body has no real excretion route for iron, and an iron overload may be fatal due to deposition cardiomyopathy by the second decade of life. Transfusions may also put the patient at risk from hepatitis and AIDS, especially in developing countries. Other treatments, such as giving hydroxyurea to augment HbF synthesis, are also used (*see above*) and local irradiation may bring immediate relief when expansion of bone marrow has happened.

In general, β-thalassemias are more severe, cause more patient suffering and are more expensive to treat because more medical intervention is required, than is the case with α-thalassemias.

13.7 GLUCOSE 6-PHOSPHATE DEHYDROGENASE DEFICIENCY (G6PD)

Glucose 6-phosphate dehydrogenase is the first enzyme of the pentose phosphate pathway (PPP), sometimes called the hexose monophosphate shunt, of glucose metabolism. It catalyzes the reaction:

$$\text{Glucose 6-phosphate} + \text{NADP}^+ \rightarrow \text{6-phosphogluconate} + \text{NADPH} + \text{H}^+$$

The PPP offers an alternative route from glycolysis and the TCA cycle for the complete oxidation of glucose. Erythrocytes can carry out glycolysis and generate ATP but have no mitochondria and so cannot use the TCA cycle. Their pentose phosphate pathway is also necessary to produce the NADPH, a form of reducing power essential for a number of metabolic activities. The gene for glucose 6-phosphate dehydrogenase is on the X chromosome and G6PD is therefore a sex-linked condition affecting males (*Chapter 15*). It is, however, carried by females who have half the normal level of the enzyme. Female carriers, like sickle cell patients, are more resistant to the malarial parasite, presumably because the host cells provide a less suitable habitat for the malarial parasite and/or because the cells lyse before the parasite can mature. However, the presence of G6PD in males, who only have one X chromosome, or in homozygous females, has little antimalarial effect for reasons that are not clear.

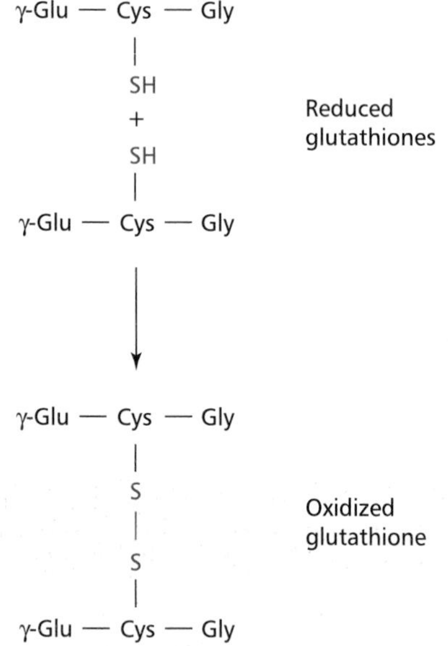

Figure 13.25 Structure of glutathione (GSH), a γ-linked tripeptide of glutamate, cysteine and glycine residues. Oxidation of the SH groups between two GSH molecules produces one molecule of the oxidized form (GSSG). See also *Figures 12.6* and *18.4*.

Genetic deficiency of glucose 6-phosphate dehydrogenase (G6PD) is common in Africa, the Middle East, South East Asia and the Mediterranean region. It is estimated that about 400 million people are affected making it the most common inherited disease. Many hundreds of different mutations are known but the commonest is the African or A type present in about 11% of blacks. The degree of deficiency is mild with enzyme activity being about 10% of usual levels and erythrocytes can manufacture sufficient NADPH under normal circumstances.

Several hundred different mutant variants of the enzyme are known that are unstable or have abnormal kinetics resulting in a reduced enzyme activity. Erythrocytes are most severely affected in G6PD because they have a long life in circulation and cannot carry out protein synthesis to replace the defective enzyme. However, most patients can make enough NADPH under normal conditions and the defect may only become apparent when the person takes a drug, such as the antimalarial, primaquine (*Table 13.7*), that greatly increases the demand for NADPH. Many different drugs besides antimalarials, that require NADPH for their detoxification, can bring on a crisis. In individuals with a severe form of the disease, oxidative stress may lead to severe hemolytic anemia with a loss of 30–50% of the erythrocytes. Heinz bodies (*Margin Note 13.8* and *Figure 13.20*) may be present. The urine may turn black because of the high concentrations of Hb and its degradation products, and a high urine flow must be maintained to prevent renal damage.

The requirement for NADPH relates to the need for glutathione (GSH) a sulfur containing tripeptide that was met in *Chapter 12*, also *18*. Glutathione contains a thiol (–SH) group that is readily oxidized (*Figure 13.25*). A major function of glutathione in erythrocytes is to eliminate hydrogen peroxide, H_2O_2, which is

Group	Examples
Antimalarials	pamaquine, primaquine, pentaquine, atabrine, quinine*
Analgesics	aspirin (high dose), phenacetin
Antibacterials	chloramphenicol*, nitrofurantoin, furazolidine
Sulfonamides	sulfacetamide, sulfanilamide, sulfapyridine
Sulfones	dapsone, thiazolesulfone, diphenylsulfone
Arsenicals	neoarsphenamine
Chemicals	methylene blue, naphthalene, phenylhydrazine, toluidine blue, nitrite, ascorbic acid (large doses), nalidixic acid

*only in the Mediterranean type.

Table 13.7 Examples of drugs and chemicals that can cause acute hemolytic anemia in individuals with G6PD deficiency

a toxic product of a number of reactions. Hydrogen peroxide can react with unsaturated fatty acids in the erythrocyte membrane forming hydroperoxides, damaging the membrane (*Chapter 12*) and leading to premature cell lysis. The peroxides are eliminated by the action of glutathione peroxidase, which requires glutathione as a reducing agent:

$$\underset{\text{reduced glutathione}}{2GSH} + \underset{\text{peroxide}}{R\text{-}O\text{-}O\text{-}H} \xrightarrow{\text{Glutathione peroxidase}} \underset{\text{oxidized glutathione}}{GSSG} + ROH + H_2O$$

There is only a limited amount of glutathione in the cell and the reduced form must be regenerated. This takes place in an NADPH-requiring reaction, catalyzed by glutathione reductase:

$$GSSG + NADPH + H^+ \xrightarrow{\text{Glutathione reductase}} 2GSH + NADP^+$$

Consequently a steady supply of NADPH is required for erythrocyte integrity.

Glucose 6-phosphate dehydrogenase activity is highest in the immature erythrocytes or reticulocytes and declines as the mature erythrocytes age. In a hemolytic crisis, the older erythrocytes are destroyed first, leaving behind reticulocytes and so giving a higher reticulocyte count. Consequently, measuring the G6PDH activity in the erythrocytes following a crisis can lead to spuriously high values. If the patient survives the crisis, with or without transfusions, recovery will usually occur quite rapidly as the reticulocyte count increases even if the individual continues to take the antimalarial, because the reticulocytes can synthesize glucose 6-phosphate dehydrogenase.

FAVISM

Favism is a hemolytic crisis brought on by the consumption of fava or broad beans (*Vicia fava*) by some, but not all, individuals with G6PD. Infants are especially susceptible. It is not understood why only some patients are susceptible but it has been suggested that another mutation must also be present for favism to be shown. Broad beans contain small quantities of toxic glycosides, which, like the antimalarials, increase the demand for NADPH.

Margin Note 13.9 Favism 

Broad beans were the only edible bean known in Europe until new species were brought from the New World. However, the philosopher Pythagoras forbade his followers to eat broad beans on the grounds that the beans contained the souls of the dead.

They are a popular item in the diet in the Mediterranean, an area where G6PD is endemic. Favism is frequently fatal unless a large volume of blood is transfused promptly.

13.8 CLINICAL ASPECTS OF CLOTTING

Clinical problems occur if the blood fails to clot as, for example, in hemophilia (*Box 13.5*), or if it clots too easily or inappropriately, as in a coronary thrombosis (*Chapter 14*). Thus analytical methods to clinically investigate the clotting abilities of blood samples are necessary. These blood samples require the addition of Ca^{2+} chelating agents, such as oxalate, citrate, EDTA or heparin to prevent them clotting and centrifugation to remove blood cells. The addition of excess Ca^{2+} or the removal of the chelating agents to the remaining plasma allows it to clot. Thus, the clotting of blood from patients can be studied in the pathology laboratory and any defects identified to help with the diagnosis and monitoring of treatment. For example, the prothrombin time, a measure of the activity of the extrinsic pathway, can be estimated. Thromboplastin is added to blood from the patient into a tube containing citrate to chelate the Ca^{2+} and the time taken for a fibrin clot to form noted. The thrombin time is a measure of the activity of thrombin. It is determined by adding thrombin to plasma and waiting for the clot to form. It is also possible to measure the activity of the intrinsic pathway by estimating the partial thromboplastin time. Calcium ions and phospholipids are added to plasma which is then exposed to a surface to activate Factor XII and the time taken for a clot to form is noted.

VITAMIN K AND ANTICOAGULANTS

Like most of the other blood clotting proteins, prothrombin is synthesized in the liver in a process that requires vitamin K, one of the fat-soluble vitamins (*Chapter 10*). A lack of vitamin K leads to the production of abnormal prothrombin that is activated by Factor X at only about 1% of the normal rate. Normal prothrombin contains 10 γ-carboxyglutamate residues. Vitamin K (*Figure 13.26 (A)*) is a cofactor required for the enzymic conversion of the relevant glutamates in the protein to the γ-carboxyglutamate residues (*Figure 13.26 (B)*). Dicoumarol and warfarin (*Figure 13.27 (A)* and *(B)*) are competitive

A)

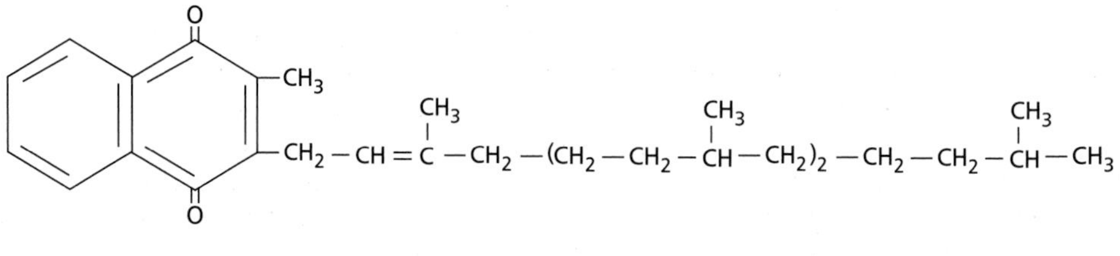

B)

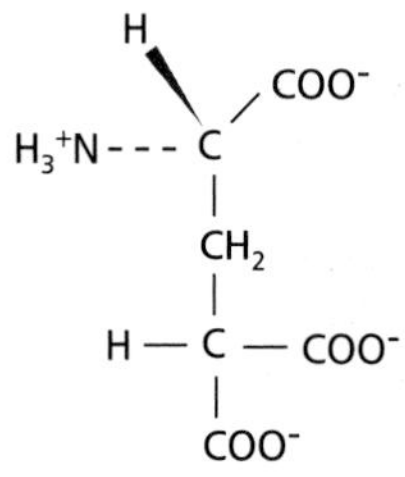

Figure 13.26 Structures of (A) vitamin K_1 and (B) γ-carboxyglutamate.

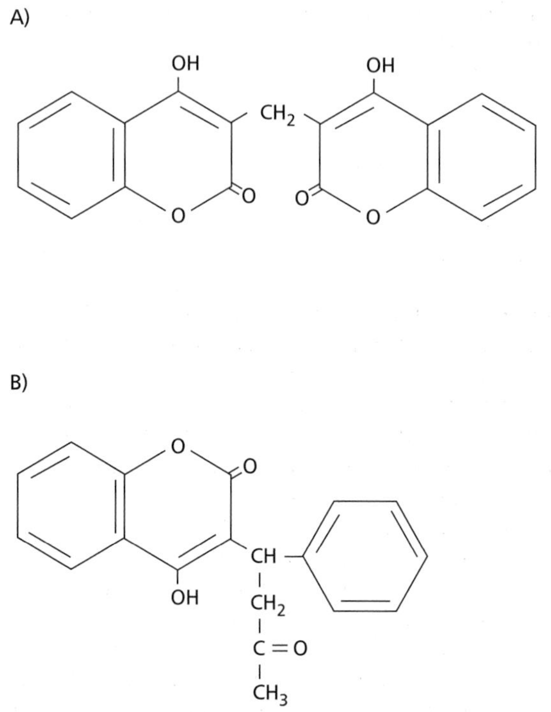

A)

B)

Figure 13.27 Structures of the clotting inhibitors (A) dicoumarol and (B) warfarin.

inhibitors of this process. Dicoumarol was first discovered in spoiled sweet clover because it causes fatal hemorrhages in cattle, and warfarin was originally developed as rat poison. Agents such as warfarin are used for anticoagulant therapy. They are slow to produce effects but they have long plasma half-lives and so can be active for days (*Chapter 14*).

HEMOPHILIA

The bleeding disease, hemophilia, is caused by a genetic lack of clotting factors. In these disorders, the bleeding time is normal but the clotting (or coagulation) time prolonged. The symptoms are a tendency for hemorrhages, that is blood loss, either spontaneous or from even small injuries. Some patients with hemophilia A may have a normal prothrombin time because their concentration of tissue factor is high. The commonest of the hemophilias is hemophilia A, in which Factor VIII is deficient. In north America, about 80% of the cases are hemophilia A. This deficiency is sex-linked to males, with a frequency of about one in 5000 to 10 000. The plasma levels of Factor VIII in patients with severe hemophilia A are less than 5% of normal. Hemophilia B, the second commonest form of hemophilia, is due to a lack of Factor IX. Factor IX is also called Christmas Factor because it was first found to be missing in a patient named Stephen Christmas.

In the past, hemophilias were treated by blood transfusion. However, repeated transfusion brings the possibility of infection with HIV or hepatitis, as well as the possibility of immune reactions and iron overload. Initial treatments involved purifying the factor from human plasma obtained from pooled blood. Modern treatment involves injections of Factor VIII prepared by recombinant DNA technology. This eliminates any chance of infection but is expensive. The same treatments are applicable to hemophilia B.

BOX 13.5 Hemophilia and Queen Victoria

Hemophilia, the inability of blood to clot properly, results from deficiency in one of the clotting factor proteins of the clotting cascade. If untreated it is accompanied by internal bleeding into the joints and muscles as well as there being a risk of uncontrolled bleeding in the case of injury or surgical operation. Treatment involves injections of the purified clotting factor that is absent. The commonest form of the disease, hemophilia A, is caused by a recessive gene carried on the X chromosome leading to a deficiency of Factor VIII. Factor VIII has a half-life of 12 h, so it needs to be administered twice daily to maintain the required therapeutic concentrations in the plasma, such as after an injury or before an operation.

The incidence of type A hemophilia varies from one in 5000 to 10 000 of the male population. In males, the presence of the defective allele on their sole X chromosome means that they will show the disease, because the Y chromosome does not carry an allele of the gene. Females can have the recessive gene on one X chromosome and a normal gene on their other X chromosome and in this case their blood will clot normally. Such females are carriers. Given hemophilia affects only about 0.02%, or less, of males in the population, so the frequency of X chromosomes carrying the allele for hemophilia is about 0.0002 among human males. Therefore among human females the occurrence cannot exceed $(0.0002)^2$; so female hemophiliacs are very uncommon. Thus hemophilia affects mainly the males and is normally transmitted by a female carrier who shows no symptoms. The following genotypes are seen in:

females	X^HX^H (normal)	X^HX^h (carrier)	X^hX^h (hemophiliac)
males	X^HY (normal)	X^hY (hemophiliac)	

The human gene for Factor VIII was cloned in 1984. It is an enormous gene of 186 kilobases forming about 0.1% of the DNA in the X chromosome. It is subjected to various genetic defects, including deletions, point mutations and insertions. Spontaneous mutations in the Factor VIII gene are fairly common.

Hemophilia is associated with members of the royal families of Europe. Queen Victoria (*Figure 13.28*) appears to have received a mutant allele from one of her parents and, as shown in *Figure 13.29*, this was passed on. Prince Albert could not have been responsible because male-to-male inheritance is impossible. One of Victoria's sons, Leopold, Duke of Albany, died of hemophilia at the age of 31 and at least two of Victoria's daughters were carriers. Through various intermarriages the disease spread from throne to throne across Europe, including the son of the last Tsar of Russia.

Figure 13.28 Queen Victoria.

13.9 THROMBOCYTOPENIA AND THROMBOCYTOSIS

In thrombocytopenia there is reduced platelet production. The clotting time may be only slightly prolonged but the clot formed is soft and does not retract. The condition can result from many diseases where bone marrow does not produce enough platelets, or where they become entrapped in an enlarged spleen and destroyed, or is caused by some drugs. The symptoms of thrombocytopenia include bleeding in the skin, pinpoint bruises, bleeding gums and it can be life threatening. Suspected thrombocytopenia can be investigated by platelet counts, assessing the size of the spleen and by bone marrow biopsy. Treatments would be cessation of drugs, if that is the cause, and by transfusions of platelets.

In thrombocytosis, in contrast, there is a high platelet count and an increased chance of thrombosis. The condition can be primary, myeloproliferative disease, or secondary, for example following splenectomy or some other operations, bleeding following extreme exercise or by inflammatory diseases. Clinically patients present with bleeding disorders. The usual treatments are to give antiplatelet agents, such as aspirin or dipyridamole.

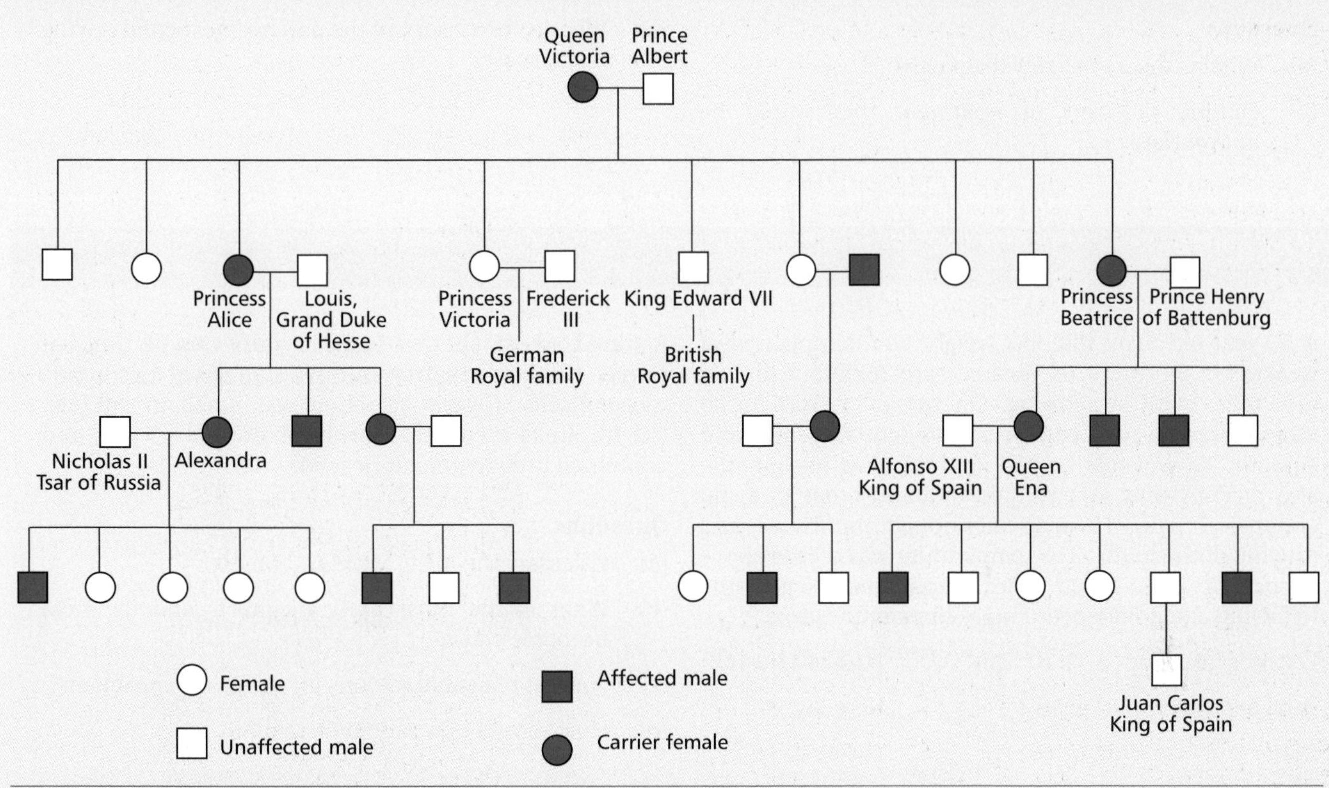

Figure 13.29 Family tree to show the inheritance of the hemophila A gene in descendants of Queen Victoria.

CASE STUDY 13.1

Daniel, a four-year-old child of Jamaican origin, was brought into the hospital because of frequent severe headaches and abdominal pain. He had also been tired and sleepy for several months. His sclerae were yellow and abdomen distended. Palpation revealed an enlarged spleen. His blood samples give the following data (references ranges in parentheses):

Erythrocyte count	2.2×10^{12} dm^{-3}	(4.4 – 5.9 × 10^{12} dm^{-3} (men))
Hemoglobin	47 g dm^{-3}	(140–180 g dm^{-3})
Serum bilirubin (unconjugated)	+++	(±)

A fresh smear of blood showed a few crescent-shaped cells and so an electrophoresis strip was run of hemolyzed erythrocytes from Daniel and both his parents, with the following result:

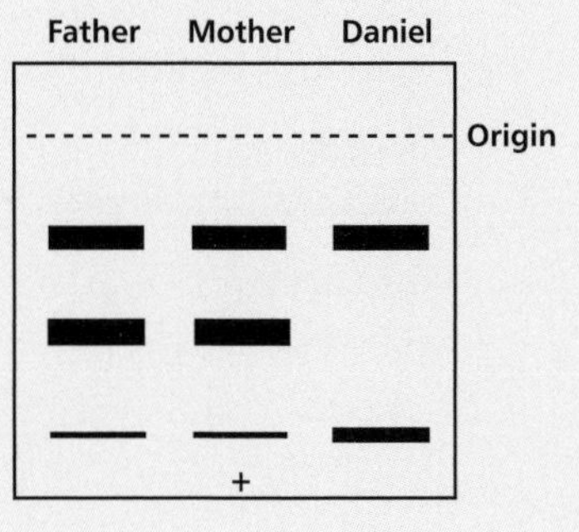

CASE STUDY 13.1 *continued*

Questions

(a) What is the most likely diagnosis?

(b) Outline the sort of treatment that would be appropriate.

(c) What is the chance of the parents' next child having this disease?

CASE STUDY 13.2

A 57-year-old man, Bill, lost weight and complained of weakness, shortness of breath, sore tongue and had difficulties with swallowing. On referral it was found that his skin was yellowish and his tongue was shiny and smooth. He said that he had gradually lost his appetite and recently only taken liquid foods in order to avoid abdominal pain. He also mentioned numbness and tingling of the hands. His temperature was a little above normal at 39°C. Blood index measurements give the following data (references ranges in parentheses):

Erythrocyte count	3.7×10^{12} dm^{-3}	(4.4 to 5.9×10^{12})
Total hemoglobin	82 g dm^{-3}	(135 g dm^{-3})
PCV	37%	(40–52%)

A blood smear showed large erythrocytes of unusual shape. His bone marrow sample contained basophilic megaloblasts. Gastric secretion was small in volume, 0.3 dm^3 in 24 h compared with the normal 2.5 dm^3 and contained little hydrochloric acid.

Questions

(a) Calculate the MCV, MCH and MCHC.

(b) What is the most likely diagnosis and its most probable cause?

(c) Suggest plausible reasons for the gastric problems.

(d) How should this patient be treated?

CASE STUDY 13.3

Alan has sickle cell anemia and has frequent sickling crises. His brother, Michael, is known to have sickle cell trait. However, their sister Liza seems to be normal and does not show any symptoms. However, she is about to marry and wants to start a family. She is therefore concerned to know her sickle cell status. A Hb electrophoresis on her blood was carried out and showed that she had 53% HbA, 46% HbS and 1% HbF .

Questions

(a) What molecular biology test could be carried out to confirm Liza's Hb distribution?

(b) If Liza's proposed marriage partner was heterozygous for sickle cell anemia, what are the chances that their first child would have sickle cell anemia?

13.10 SUMMARY

The blood system has many components, erythrocytes, leukocytes and platelets suspended in the complex protein rich plasma. Hemoglobin transports oxygen from the lungs to the tissues. Anemia results when the Hb level decreases and this is usually accompanied with a reduction in the erythrocyte count. Anemias may be associated with normal sized erythrocytes or ones that are microcytic or macrocytic, depending upon the cause.

Many hundreds of mutations are known that affect Hb giving rise to hemoglobinopathies. These range from relatively mild conditions to ones that are severe and life-threatening. In sickle cell anemia, the molecules of deoxyhemoglobin aggregate to form fibrils, which distort the erythrocytes, making them less flexible when passing through the capillaries. This leads to a lack of oxygen in the tissues, severe pain and hemolysis. It is due to a point mutation in the gene for β-globin. Thalassemias are a group of relatively common hemoglobinopathies, caused by the defective synthesis of the α- or β-globins. Thalassemias, again, vary from the relatively mild to very severe, depending upon the precise defect.

The commonest blood disease is glucose 6-phosphate dehydrogenase deficiency. Typically this is asymptomatic until the patient takes a particular drug that produces an excessive demand for glutathione in the erythrocytes and this leads to a severe hemolytic anemia.

Many defects of the blood clotting system are known. The commonest is hemophilia A, which is caused by lack of Factor VIII, one of the protein factors active in the clotting mechanism.

QUESTIONS

1. Which of the following statements about iron-deficiency anemia is/are true?
   - (a) The erythrocytes are small and hypochromic.
   - (b) The serum iron-binding capacity falls.
   - (c) There is angular stomatitis and brittle hair.
   - (d) Deposits of hemosiderin will be found in the liver and spleen.
   - (e) The erythrocytes are microcytic in appearance.

2. Identify the blood cells, A and B, in the accompanying figure.

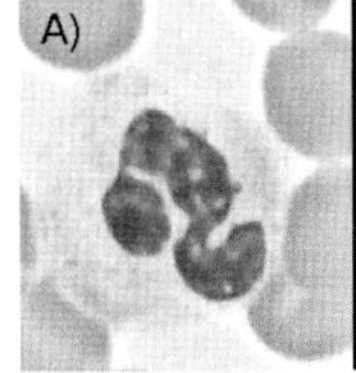

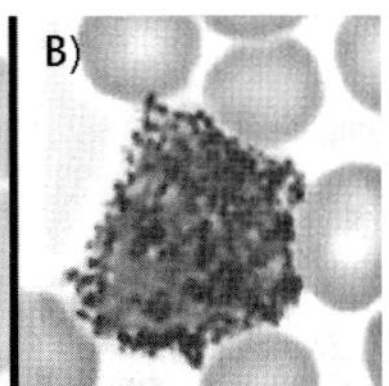

3. The accompanying figure shows an electrophoresis of Hb from four individuals. In sickle cell disease the change at position 6 in the β-globin chain is from Glu to Val, and in HbC from Glu to Lys. Individual C is normal, with HbA present. What can be said about individuals A, B and D?

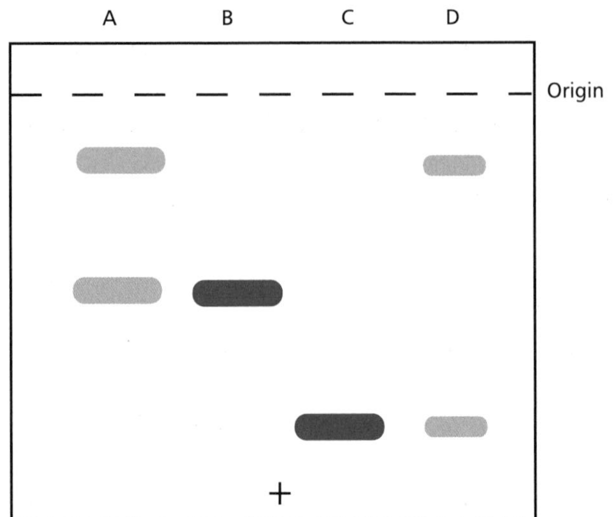

4. Explain why hereditary persistence of fetal hemoglobin (HbF) does not have serious consequences.

5. Indicate in the following table whether the listed hematological indices are not affected (NA), decreased (D) or increased (I) in iron deficiency and pernicious anemias.

Index	iron deficiency anemia	pernicious anemia
PCV		
MCV		
MCH		
MCHC		

6. Elsa went to her doctor complaining of pain in her lower spine. The doctor noticed some deformity of her skull and facial bones. A CT scan of the bones of her spine showed an area of spinal cord compression in the upper lumbar region. Blood tests indicated severe anemia and thalassemia was diagnosed. Explain these findings and comment on possible treatment.

7. Explain the role of platelets in blood clotting.

8. Simon, a nine-month-old male infant, was brought to the Accident & Emergency Department with a painful, expanding mass in his left thigh. His mother has noticed this a few hours after he had fallen down on a hard floor, and the child was in considerable pain. An X-ray revealed that there were no fractures, but the soft swelling was shown to be a hematoma caused by bleeding into the tissues. On questioning, Simon's mother said that soon after he began to crawl Simon's knees became painful and swollen. The pediatrician suspected a coagulation disorder. Tests showed that Simon had only about 5% of the normal level of plasma Factor VIII. What is the most likely diagnosis and treatment?

FURTHER READING

Bhagavan, NV (2002) *Medical Biochemistry*, 4th edn, Harcourt Academic Press, San Diego.

Campbell, K (2004) Pathophysiology of anaemia. *Nurs. Times* **100:** 40–43.

Catlin, AJ (2003) Thalassemia: the facts and the controversies. *Pediatr. Nurs.* **29:** 447–449.

Clark, BE and Thein, SL (2004) Molecular diagnosis of haemoglobin disorders. *Clin. Lab. Haematol.* **26:** 159–176.

Claster, S and Vichinsky, EP (2003) Managing sickle cell disease. *BMJ* **327:** 1151–1155.

Dame, C and Sutor, AH (2005) Primary and secondary thrombocytosis in childhood. *Br. J. Haematol.* **129:** 165–177.

Dhaliwal, G, Cornett, PA and Tierney, LM (2004) Hemolytic anemia. *Am. Fam. Physician* **69:** 2599–2606.

Gottlieb, S (2001) Gene therapy shows promise for haemophilia. *BMJ* **322:** 1442.

Hoffbrand, AV and Pettit, JE (2002) *Essential Haematology*, 4th edn, Blackwell Science, Oxford.

Howard, MR and Hamilton, P (2002) *Haematology*. Churchill Livingstone, London.

Hughes-Jones, NC, Wickramsinghe, SN and Hatton, C (2004) *Lecture Notes on Haematology*, Blackwell Publishing, Oxford, UK.

Mannucci, PM and Tuddenham, EGD (2001) The hemophilias – from royal genes to gene therapy. *New Engl. J. Med.* **344:** 1773–1779.

Mehta, A, Mason, PJ and Vulliamy, TJ (2000) Glucose-6-phosphate dehydrogenase deficiency. *Baillière's Best Pract. Res. Clinical Haematol.* **13:** 21–38.

Potts, DM and Potts, WT (1999) *Queen Victoria's Gene: Haemophilia and the Royal Family*, 2nd edn. Sutton Publishing, NY.

Serjeant, GR (2005) Mortality from Sickle cell disease in Africa. *BMJ* **330:** 432–433.

Stuart, MJ and Nagel, RL (2004) Sickle-cell disease. *Lancet* **364:** 1343–1360.

Vidler, V (2004) Haemophilia. *Br. J. Perioper. Nurs.* **14:** 110–113.

Wajeman, H and Galacteros, F (2005) Hemoglobins with high oxygen affinity leading to erythrocytosis. New variants and new concepts. *Hemoglobin* **29:** 91–106.

Weatherall, DJ (2000) Single gene disorders or complex traits: lessons from the thalassaemias and other monogenic disorders. *BMJ* **321:** 1117–1120.

Working Party of the General Haematology Task Force of the British Committee for Standards in Haematology (1998) The laboratory diagnosis of haemoglobinopathies. *Br. J. Haematol.* **101:** 783–792.

Useful web sites:

http://academic.sun.ac.za/haema/dept/rbc.htm

http://www.medicine.ox.ac.uk/ohc/

osulibrary.oregonstate.edu/specialcollections/coll/pauling/blood/index.html

chapter 14:

DISORDERS OF THE CARDIOVASCULAR SYSTEM

OBJECTIVES

After studying this chapter you should be able to:

- outline the structure of the heart and circulatory system;
- describe the general functioning of the heart;
- list the major types of heart diseases and their causes;
- review some of the methods for detecting and investigating heart disease;
- discuss the role of cholesterol and apolipoproteins in atherosclerosis;
- explain the use of antithrombolytic therapy;
- describe the major peripheral vascular diseases.

14.1 INTRODUCTION

The cardiovascular system comprises the heart, with its covering pericardium, and blood vessels, that is, arteries, arterioles, capillaries, venules and veins, which enclose and distribute blood to the tissues.

The heart is a muscular pump weighing about 500 g in an adult and is the strongest muscle in the body. It can contract for over a hundred years nonstop, beating about 100 000 times a day at a rate of about 70 beats min^{-1}, to pump blood (*Chapter 13*)) around the body. This supplies, for example, oxygen and nutrients to the tissues and removes their waste products. However, in such a complicated system many things can go wrong. In the developed countries, heart disease is responsible for about half of the annual deaths. Much of this is due to lifestyle, for example smoking, poor diets and excess weight. In developing countries rheumatic heart disease is more of a problem but this has decreased in the developed countries.

14.2 THE HEART AND CIRCULATORY SYSTEMS

The heart (*Figure 14.1 (A)* and *(B)*) is a hollow, four-chambered, muscular organ situated approximately in the center of the chest. A smooth layer called the **endocardium** lines the inside of the chambers. The wall of the heart or **myocardium** is rich in cardiac muscle tissue arranged into three layers. The inner layer is circular and thicker in the wall of the left ventricle than in the right. The outer layers of muscle spiral around the ventricles and extend to the fibrous attachments of the four valve rings. On contraction they tend to pull the chamber of the ventricle towards the valve rings. The exterior of the heart is a tough, fibrous layer that is partly covered by fat. The heart is enclosed by a double membrane system called the pericardium.

Internally, the heart is centrally divided by a septum that prevents oxygenated and deoxygenated blood from mixing. Each side is subdivided into an upper chamber or **atrium,** which collects blood and passes it to a lower chamber or **ventricle** that ejects blood. Each of the heart's ventricles has a one-way inlet valve and a one-way outlet valve (*Figure 14.1 (B)*). The tricuspid valve opens from the right atrium to the right ventricle and the pulmonary valve opens from this ventricle into the pulmonary arteries. The mitral valve separates the left atrium and ventricle and the aortic valve opens from the ventricle into the aorta. All these valves ensure that blood flows only in one direction.

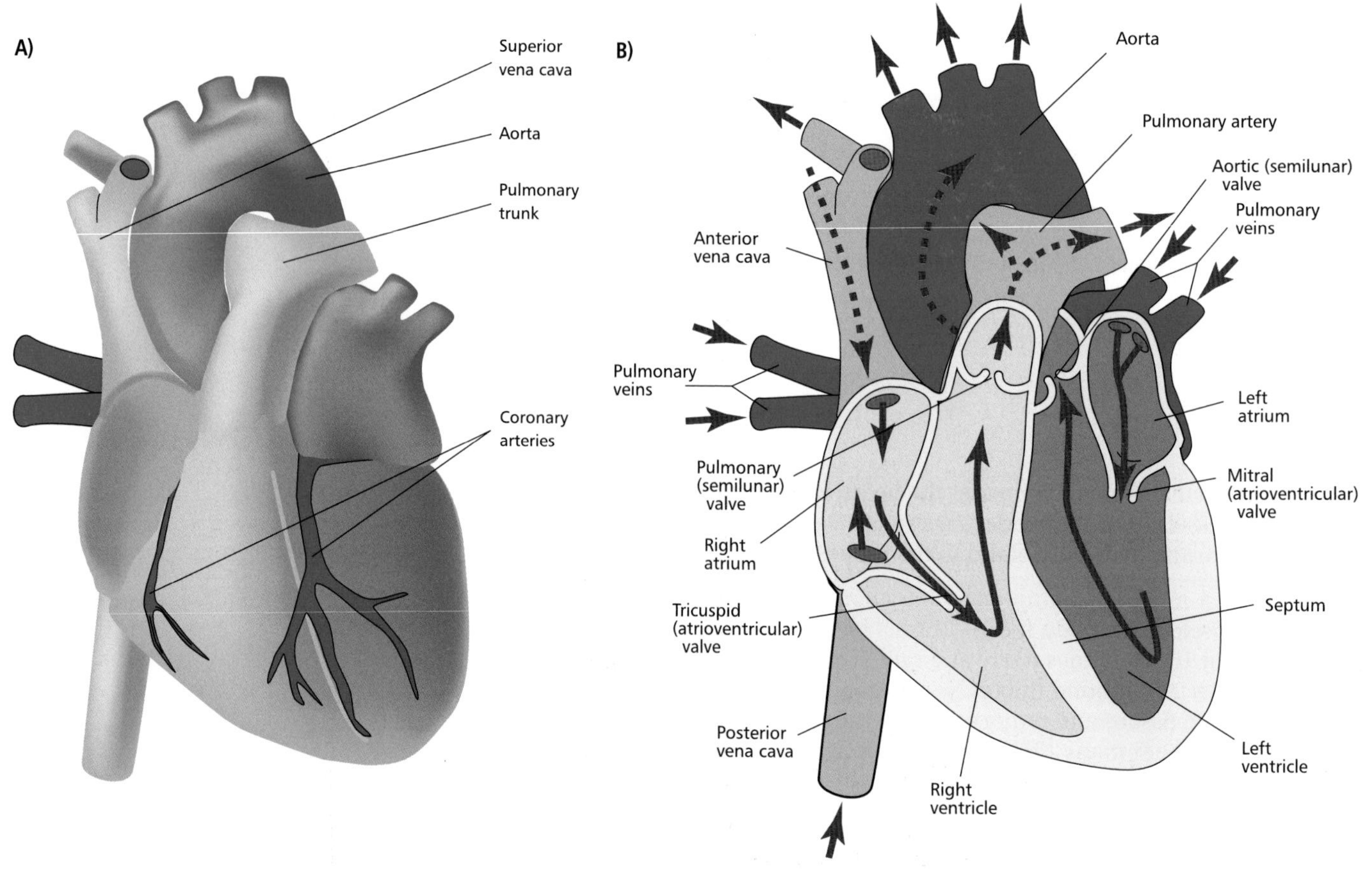

Figure 14.1 Structure of the heart. (A) Overall view of the heart showing the major blood vessels. (B) A cutaway view from the same perspective showing the heart valves and the direction of blood flow.

The flow of blood through the pulmonary and systemic circulations is illustrated in *Figure 14.2*. Oxygen-depleted blood from the body flows through the two largest veins, the venae cavae, into the right atrium. When this chamber is full it propels the blood into the right ventricle and when this is full it pumps blood through the pulmonary valve into the pulmonary artery, which takes it to the lungs. As the blood flows through the network of capillaries that surround the lung alveoli it absorbs oxygen and gives up its carbon dioxide (*Chapter 9*). The resulting oxygen-enriched blood flows through the pulmonary veins into the left atrium of the heart. The complete circuit between the right side of the heart, the lungs and the left atrium is called the **pulmonary circulation**. When the left atrium is full, it sends the oxygen-rich blood into the left ventricle, which, in turn, propels the blood into the **aorta**, which is the largest artery in the body. From there it flows to all the tissues of the body, with the exception of the lungs, in the **systemic circulation**. The major arteries to the head, neck and arms branch off from the aorta (*Section 14.3*) and there are also smaller vessels, the coronary arteries that supply the heart itself with blood. The aorta arches over the

Margin Note 14.1 Heart sounds 

Listening to the heart with a stethoscope ('ausculation') reveals the distinctive sounds caused by the opening and closing of the heart's valves and the flow of blood. If there are abnormalities of the valves and heart structures then these may cause turbulent flow of the blood that creates characteristic sounds called **murmurs**. Typically the murmur is caused by blood flowing through narrowed or leaking valves. However, not all heart diseases cause murmurs and not all murmurs indicate heart disease. For example, pregnant women usually have heart murmurs because of the increase in blood flow. In infants and young children, harmless murmurs are commonly caused by the rapid flow of blood through the smaller sized heart structures.

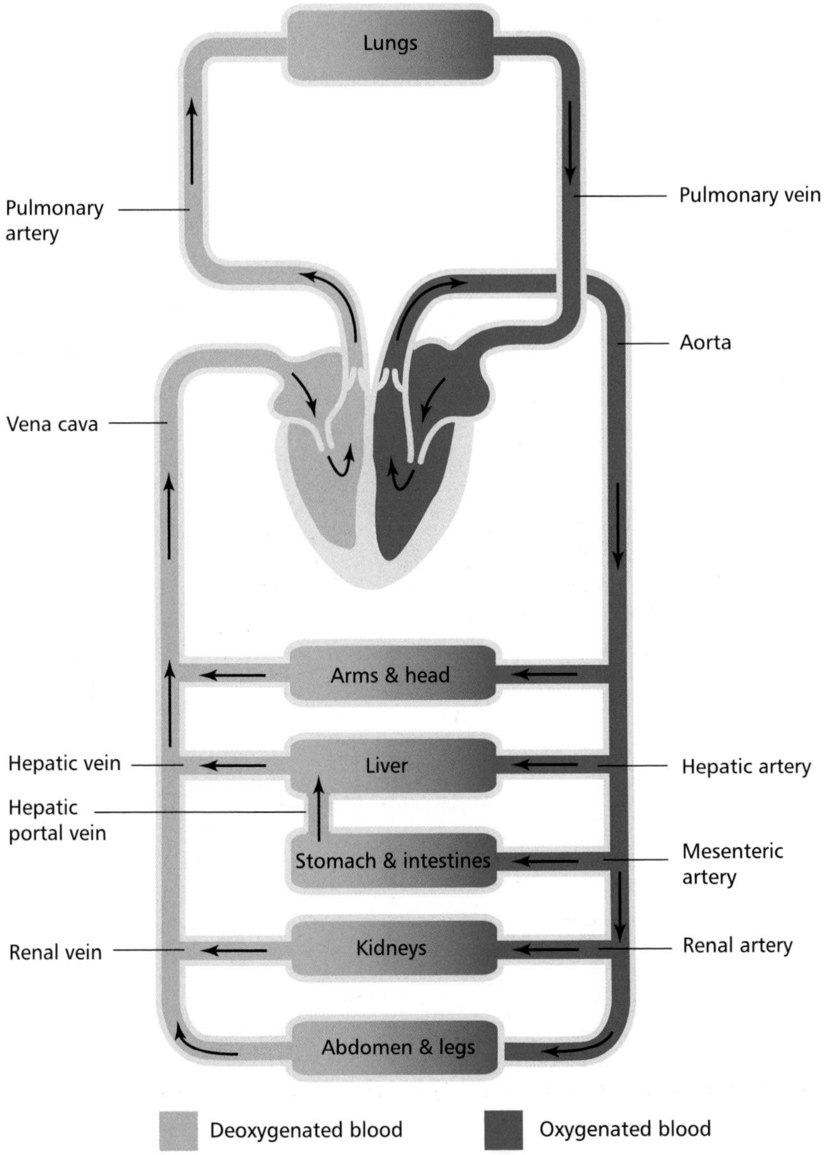

Figure 14.2 Schematic illustrating the pulmonary and systemic circulations. See text for details.

heart and then goes downwards carrying blood to the abdomen and legs. In the capillaries of the tissues, oxygen is exchanged for carbon dioxide and nutrients and hormones, for example, supplied to the tissues. The blood then returns to the heart through the veins (*Figure 14.2*).

To work efficiently, the four chambers of the heart must beat in a coordinated way, with the atria and then the ventricles contracting simultaneously (*Figure 14.3*). A given chamber of the heart contracts when an electrical impulse moves across it. Each signal originates in a small bundle of specialized cells in the right atrium called the **sinoatrial** or **SA node** (*Figure 14.4*). This is the natural **pacemaker** that ensures the heart beats regularly by generating impulses at a given rate. Although it produces a natural rate, this can be modified by emotional and physical reactions and by hormones, especially adrenaline and noradrenaline (*Chapter 7*), which speed up the heart rate, enabling it to respond to varying demands. The electrical impulses generated by the SA node travel throughout the right and left atria causing the heart muscle to contract (*Figure 14.4*) and arrive at the **atrioventricular** or **AV** node situated between the atria and the ventricles. This node delays the transmission of the impulse to allow the atria to contract completely and the ventricles to fill with as much blood as possible. The phase of relaxation of the ventricles is called **diastole**. After passing through the AV node, the impulse travels through the **bundle of His**. This is a group of modified cardiac muscle fibers, called **Purkinje fibers**, that divides into two branches that serves the left and right ventricles respectively (*Figure 14.4*). The fibers spread over the surface of the ventricles in an orderly arrangement and thus initiate **systole** or ventricular

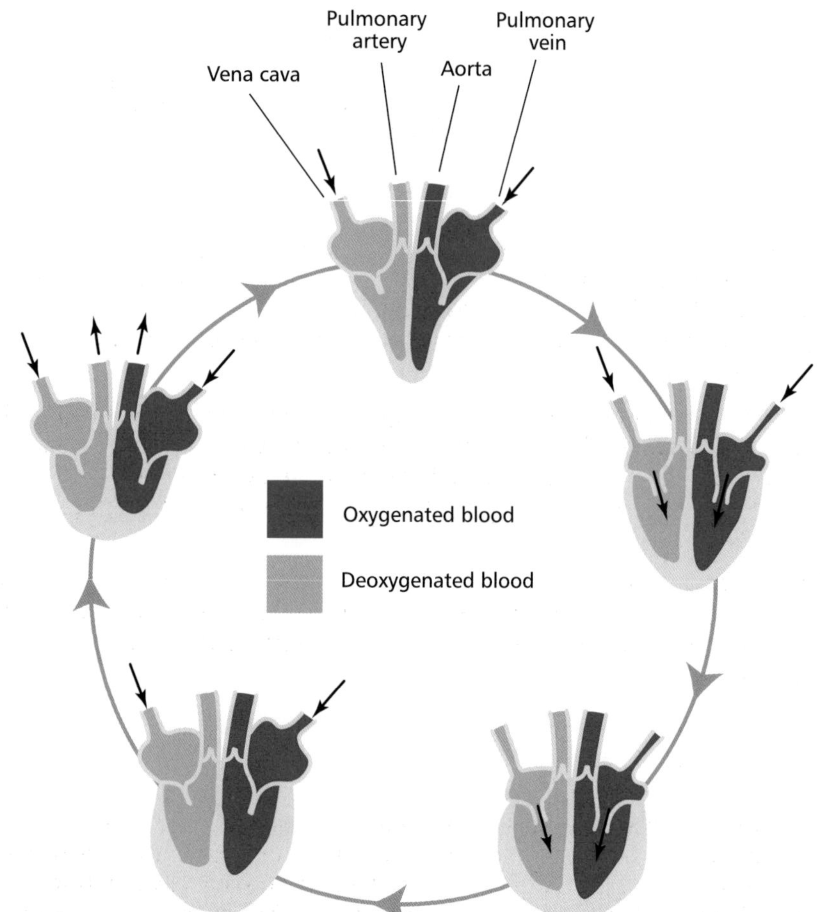

Figure 14.3 Diagrammatic representation of the contractions and relaxation of the heart chambers during one cardiac cycle. Start at the top and observe oxygenated and deoxygenated blood entering the heart, its passage from the atria to the ventricles and finally its expulsion from the ventricles as the cycle starts again.

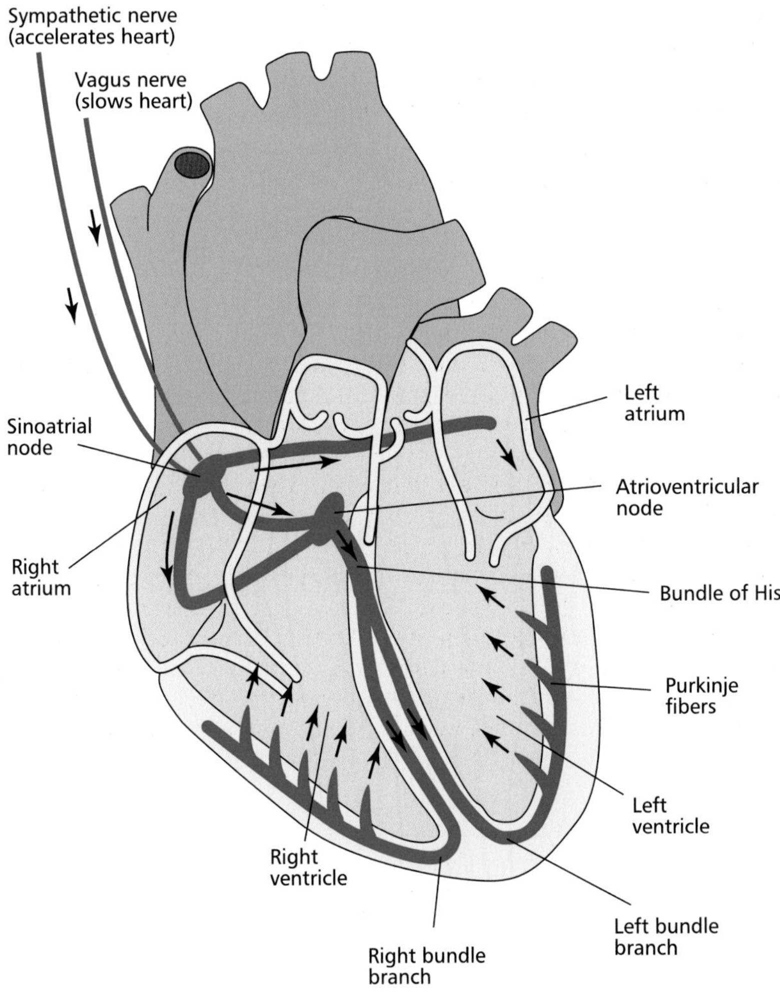

Figure 14.4 The SA and AV nodes and the control of the heart rate. See text for details.

contraction. Both ventricles contract together in a wringing fashion so that blood is squeezed out from their bases, pumping blood from the heart. These coordinated actions keep the four chambers of the heart working in the appropriate sequence. When things go wrong people may have to have artificial pacemakers fitted.

The part of the nervous system that regulates the heart rate automatically is, not surprisingly, the autonomic nervous system and consists of the sympathetic and the parasympathetic fibers of the vagus nerve (*Figure 14.4*). The sympathetic system increases the heart rate and the parasympathetic system slows it down.

The pericardium is a thin, flexible, two-layered bag that surrounds the heart. It contains just sufficient lubricating fluid between the two layers so that they slide easily over each other as the heart beats. It keeps the heart in position and prevents it overfilling with blood.

14.3 BLOOD VESSELS

The blood vessels comprise arteries, arterioles, capillaries, venules and veins (*Figures 14.5 and 14.6*). Arteries are strong and flexible and carry blood away from the heart. They are subjected to the highest pressure (*Section 14.4*), and their resilience helps to maintain blood pressure while the heart is in between

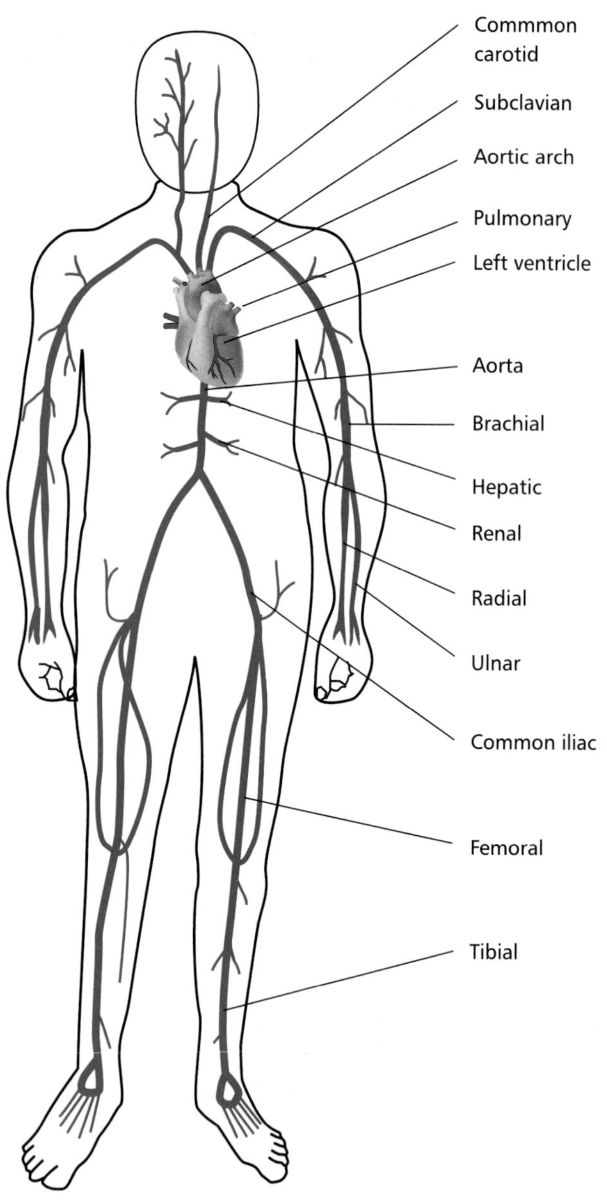

Figure 14.5 Schematic showing the main arteries of the body.

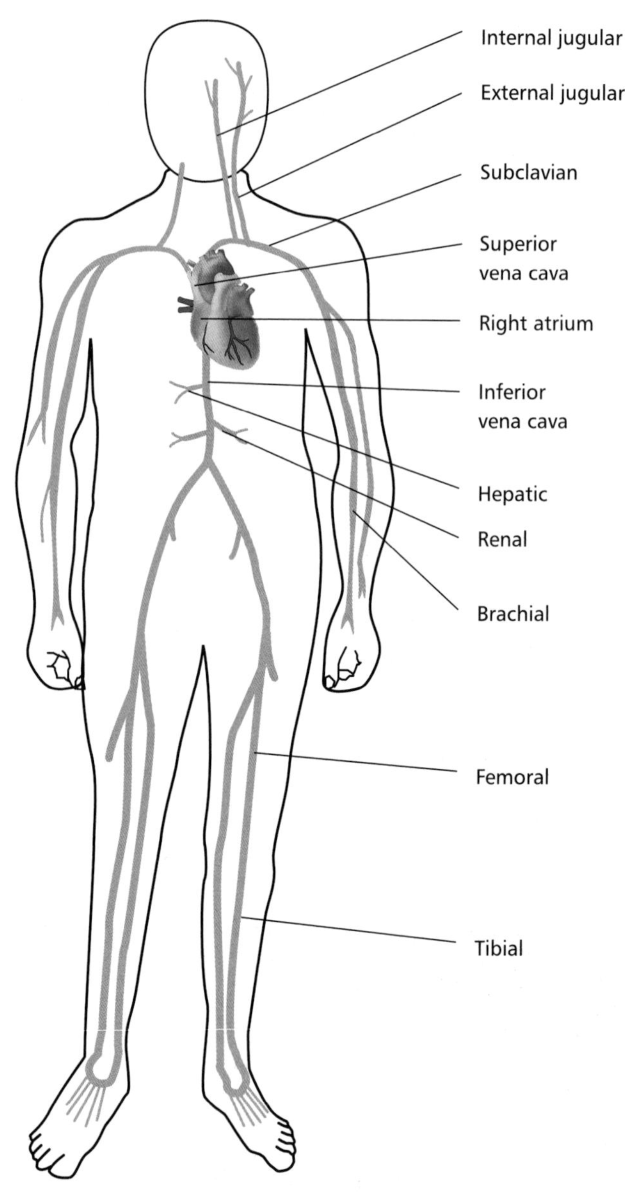

Figure 14.6 Schematic showing the main veins of the body.

beats. The smaller arteries and arterioles have muscular walls and their diameters can be adjusted to regulate blood flow to a particular region of the body. Capillaries are thin-walled vessels that allow oxygen, nutrients and other materials, for example hormones, to diffuse from the blood to the tissues and waste products to pass from the tissues into the blood. The capillaries are links between the arteries and arterioles and the venules and veins. Veins are, in general, larger in diameter than arteries but have much thinner walls. They transport blood back to the heart but at much lower pressures and speeds than is found in arteries.

ARTERIES

Arteries are elastic tubes with circular cross-sections (*Figures 14.7 (A)* and *(B)*). They are built up from three layers called tunics. The tunica intima is

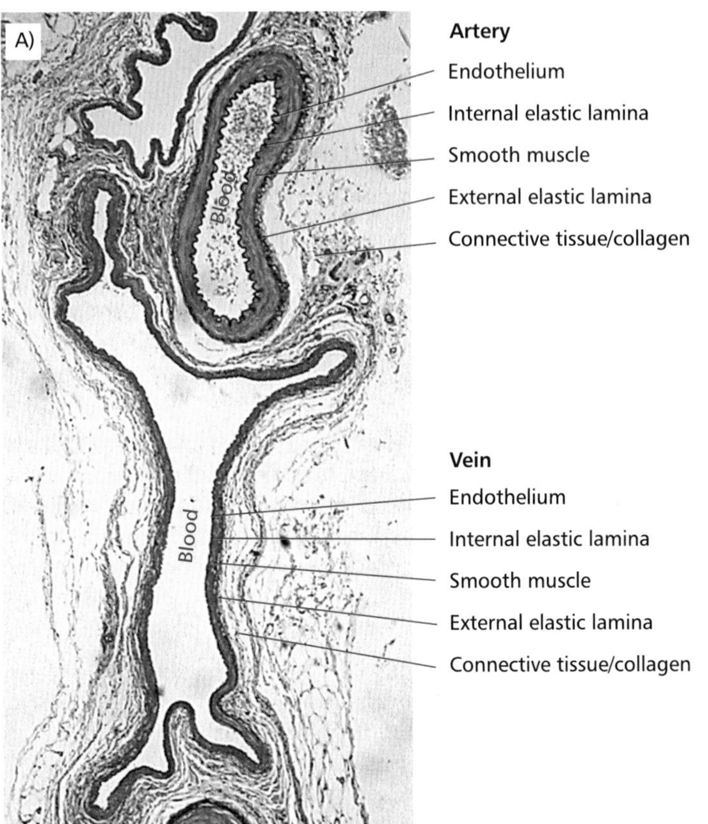

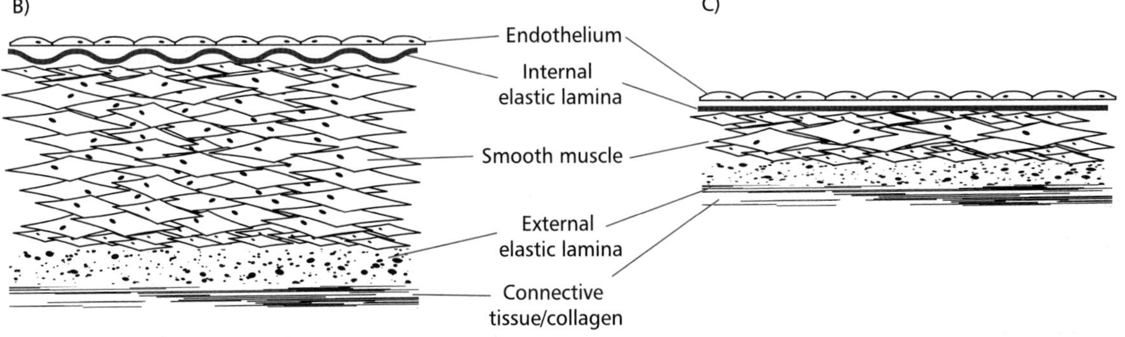

Figure 14.7 (A) Photomicrograph showing the cross sections of an artery and a vein. Schematics illustrating the layers in the walls of (B) an artery and (C) a vein.

the layer of endothelial cells that provides a smooth, low friction innermost surface. This layer is in direct contact with the blood, and damage to it leads to serious clinical problems. The tunica media is a relatively thick middle layer composed of smooth muscle and elastic fibers, which are arranged both lengthways and circumferentially. The amount of elastic material varies with the vessel. Closer to the heart, the aorta and large branches contain larger amounts because they have to accommodate the stroke volume. In contrast, the pulmonary artery has rather thin walls. The smooth muscle is under autonomic nervous control that regulates the diameter of the vessel. The fibrous components limit the amount of stretch under a surge of pressure. The tunica adventitia is the outer covering fibrous layer composed of tough collagenous material.

Arteries branch repeatedly in order to supply all parts of the body. The diameter of the aorta is about 25 mm, and that of a medium-sized artery about 4 mm.

When their diameter is less than 0.1 mm, they are called arterioles. Junctions between the branches are called anastomoses and these may occur between arteries and arterioles.

The pulmonary arteries distribute the output of the right ventricle to the lungs. They are shorter, have thinner walls and contain blood at a lower pressure than systemic arteries. Arteries also have a storage function since the output of the ventricles is discontinuous as the heart beats. For example, the aorta stores up blood during systole and its elastic recoil propels it on during diastole. This means that the discontinuous blood flow from the heart is converted to a more continuous flow through the peripheral circulation, conserving energy and reducing the pressure that the smaller arteries have to withstand.

VEINS

The veins form a blood collecting system that returns blood from the periphery to the heart. In general, veins have the same pattern of distribution as the arteries and frequently run alongside them (*Figure 14.6*). The blood collected from the systemic circuit is eventually delivered to the right atrium of the heart. However, the venous drainage from the stomach, spleen and intestine is carried by the hepatic portal vein to the liver (*Chapter 11*). Blood from the liver is then returned to the heart in the hepatic vein.

Veins, like arteries, have walls composed of three layers but they contain considerably less muscle and elastic tissues and their walls are much thinner (*Figure 14.7 (A)* and *(C)*). The internal lining is the same endothelium as in the arteries. Veins are much more easily distended than arteries and are also more easily collapsed because they are less able to withstand high pressures. At intervals, especially along the lengths of long veins, the endothelial lining forms cup-shaped valves, rather similar to the semilunar valves in the heart. These allow blood to flow in only one direction, which is back towards the heart, and also prevent its back flow under gravity. In addition, during muscular activity especially in the legs, the compression and relaxation of the muscular tissue surrounding the veins forms a 'muscle pump' that expedites the transport of blood back to the heart. In general, the blood flow through veins is less discontinuous than it is in arteries.

CAPILLARIES

The arterioles supply beds of capillaries found deep in the tissues and these eventually deliver their blood to the venules, and hence to the veins for return to the heart. Capillaries have internal diameters about the same as that of erythrocytes, about 7 μm.Their walls are composed of a single layer of endothelial cells on a basement membrane (*Figure 14.8*) and so they lack muscle and elastic tissues. The capillary network is the site where gases are exchanged, nutrients and other biomolecules are delivered and waste products removed. The flow of blood through capillaries is relatively slow compared with that in arteries and veins and this allows adequate exchanges to take place.

14.4 BLOOD PRESSURE

The blood pressure is the hydrostatic force that the blood exerts against the wall of a blood vessel and that propels blood around the body. It is determined partly by cardiac output and partly by the peripheral resistance. As already stated, it is greater in arteries than in veins and is highest in the arteries when the heart contracts (systole). This is called the systolic pressure. When the heart contracts blood enters the arteries faster than it can leave through the capillaries so the vessels stretch under the pressure. This bulging of the arteries is the **pulse** that can be felt at a number of sites in the body. During diastole,

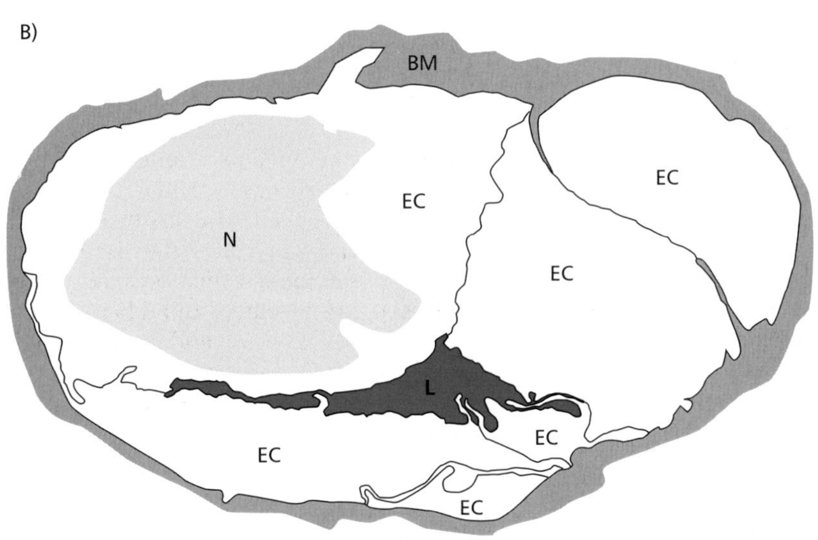

Figure 14.8 An electron micrograph of (A) a capillary and (B) its tracing. BM, basement membrane; EC, endothelial cell; L, lumen; N, nucleus of endothelial cell. Courtesy of Dr P. Kumar, School of Biology, Chemistry and Health Science, Manchester Metropolitan University, UK.

the artery walls snap back, but the heart contracts again before enough blood has flowed into the capillary bed to relieve the pressure. This is the peripheral resistance. As a result, there is still a substantial blood pressure even during diastole, the diastolic pressure, and so blood flows into the arterioles more or less continuously.

Typically, at rest the systolic pressure is about 120 mmHg and the diastolic pressure approximately 70 mmHg. Blood pressure can be measured using a manual sphygmomanometer combined with a stethoscope (*Figure 14.9 (A)* and *Box 14.1*) or with an automatic electronic one (*Figure 14.9 (B)*). Blood pressure is not constant because as demands on the circulatory system change, regulatory mechanisms ensure that adequate blood flow is maintained. Thus, during heavy exercise the arterioles in the working muscles dilate to allow increased delivery of oxygen to the muscles. This of course decreases the peripheral resistance, which, by itself, would cause a fall in blood pressure. However, the cardiac output is increased to counteract this effect.

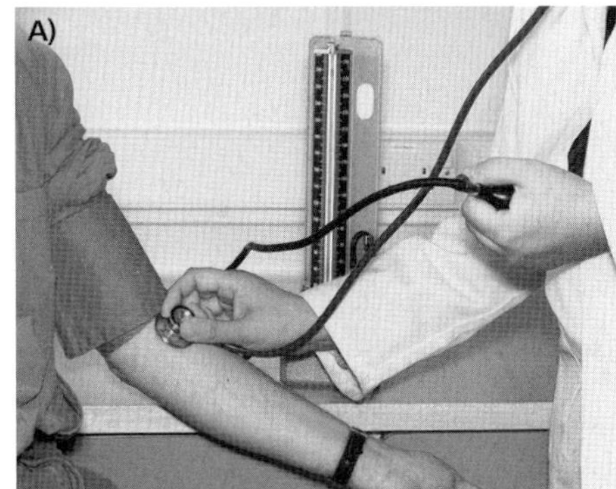

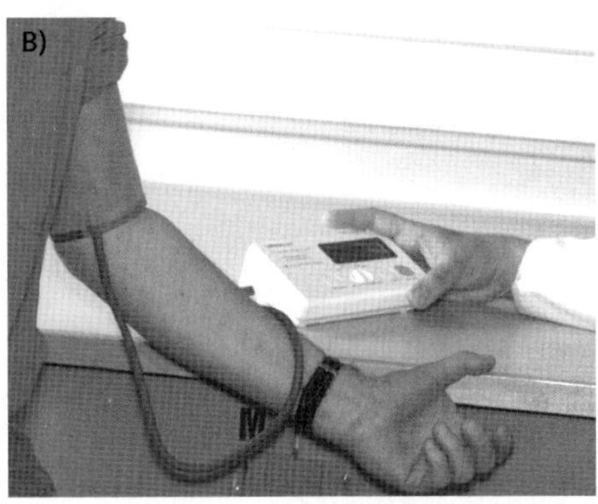

Figure 14.9 Measuring blood pressure using (A) a manual sphygmomanometer and a stethoscope (*Box 14.1*) and (B) electronic sphygmomanometer.

BOX 14.1 Measuring blood pressure

Traditionally, blood pressure is measured using a manual sphygmomanometer and stethoscope (Greek, *sphygmos*, pulse), as illustrated in *Figure 14.10 (A)–(D)*. There are also electronic instruments that can be used (*Figure 14.9 (B)*). With a sphygmomanometer, an inflatable cuff connected to a pressure gauge, in this case a mercury manometer, is placed around the upper arm over the brachial artery and inflated (*Figure 14.10 (B)*). A stethoscope is placed on the skin above the brachial artery so that blood sounds may be heard. When the pressure in the cuff is greater than the systolic pressure the arteries are compressed and no brachial pulse can be heard. The cuff pressure is then slowly reduced and when it falls to that of the systolic pressure, clear sounds (**Korotkoff sounds**) can be heard due to the turbulence generated as the blood flows through the partially occluded artery (*Figure 14.10 (B)–(C)*). As the pressure further reduces the sounds become suddenly muffled and then disappear completely. The pressure at this point is generally taken as the diastolic pressure (*Figure 14.10 (D)*).

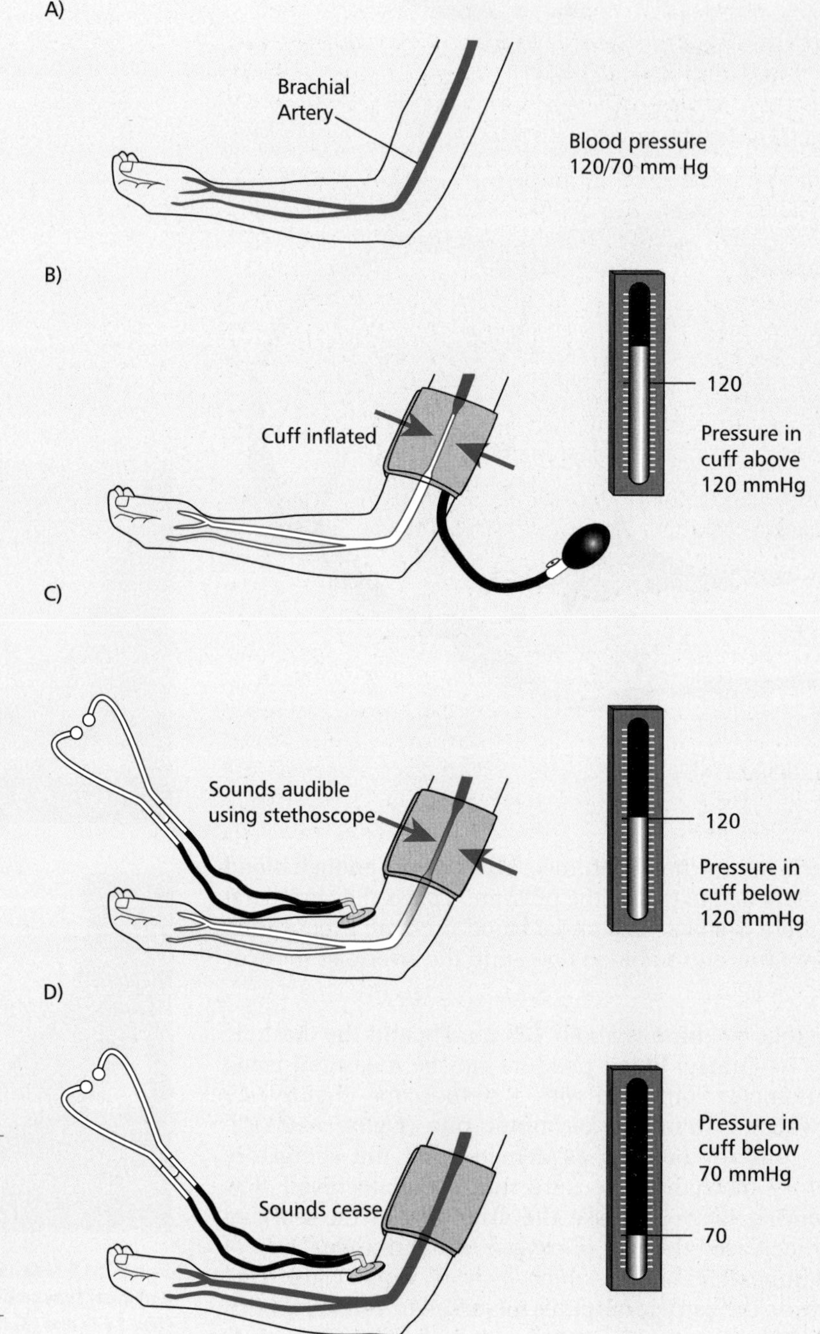

Figure 14.10 Panels (A) to (D) schematically show the use of a manual sphygmomanometer and stethoscope to measure blood pressure. See text for details.

14.5 INVESTIGATING CARDIAC FUNCTION IN HEALTH AND DISEASE

Clinicians need to be able to diagnose cardiovascular problems and to monitor their treatment. Obviously, noninvasive methods are to be preferred as far as possible for the comfort of the patient.

THE ELECTROCARDIOGRAM (ECG)

The spread of depolarization and repolarization through the muscle mass of the heart is accompanied by measurable electrical potentials, which may be recorded through electrodes placed on the skin. The record of these potentials is called an **electrocardiogram** (**ECG**) and represents the aggregate or resultant electrical activity associated with the action potentials of millions of individual cells, each of which has an amplitude and direction. Recording a patient's ECG is a way of investigating the electrical activity of the heart and when the pattern is abnormal it is invaluable in diagnosing a variety of heart complaints. The test is quick, simple and painless, and provides information on the heart rhythm and underlying cardiac morphology.

An ECG is recorded by placing electrodes, these days usually disposable, self-adhesive ones, on the chest and limbs. In practice the ECG is recorded simultaneously from six electrodes connected to the limbs and six to the chest that measure the direction and flow of electric currents during each heartbeat. The results are recorded (*Figure 14.11 (A)*), with each trace representing a particular 'view' of the heart's electrical activity; these views are referred to as leads. The spikes and troughs on the graph correspond to specific events in the cycle of a heartbeat and these are lettered alphabetically (*Figure 14.11 (B)*). Modern ECG instruments have a built-in computer that analyzes the recordings and produces a printout of the analysis.

In the normal ECG waveform, the first deflection is caused by atrial depolarization and is a low-amplitude slow deflection called a P wave. The following QRS complex results from ventricular depolarization, and as can be seen in *Figure 14.11 (B)*, it is sharper and larger in amplitude than the P wave. The T wave is another slow, low-amplitude wave resulting from ventricular repolarization. Atrial repolarization is not usually seen because it is low voltage and is hidden by the QRS complex. The PR interval is the time from the start of the P wave to the start of the QRS complex and represents the time taken for activation to pass from the SA node, through the atrium to the AV node. The QRS complex is a measure of the time associated with impulses passing through the His-Purkinje system and the subsequent contractions of the ventricles. The QT interval begins at the start of the QRS complex and finishes at the end of the T wave and this represents the time taken to depolarize and repolarize the ventricular myocardium. The ST segment is the time between the end of the QRS complex and the start of the T wave. At this point all the cells of the normal heart are depolarized.

The print out from an ECG examination can help the cardiologist to identify a number of heart problems including abnormal rhythms, inadequate blood supply to the heart and excessive thickening of the heart muscle (hypertrophy). For example, a heightened P wave indicates an enlarged atrium; a deeper than normal Q wave may indicate a myocardial infarction (*Section 14.14*) and a heightened R wave usually indicates a thickening of the ventricular wall. If the ST segment is raised above the horizontal it may indicate acute myocardial infarction or, if it is below the horizontal, it can imply high blood K^+ concentrations (*Chapter 8*) or cardiac ischemia (*Section 14.9*). Any deviation from the normal rate or sequence of the ECG is referred to as a cardiac arrhythmia. If the sinoatrial (SA) node is damaged the heart rate may slow to 40–50 beats per min and if damage occurs to both the SA and AV nodes it may fall to 20–40 beats per min and the patient will require a pacemaker.

A)

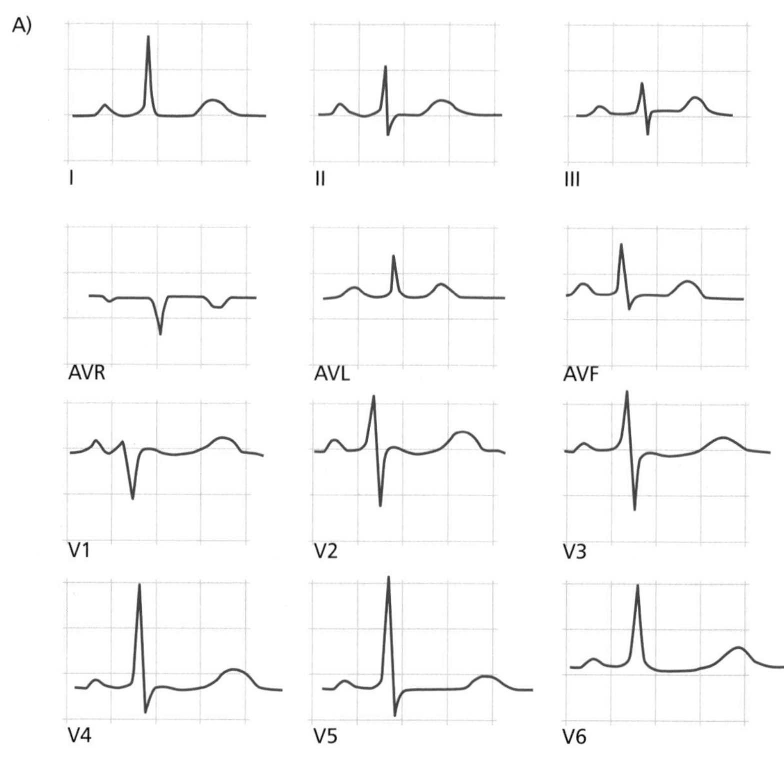

B)

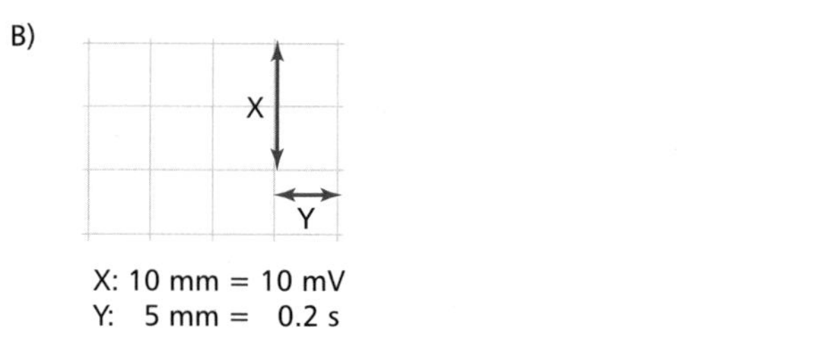

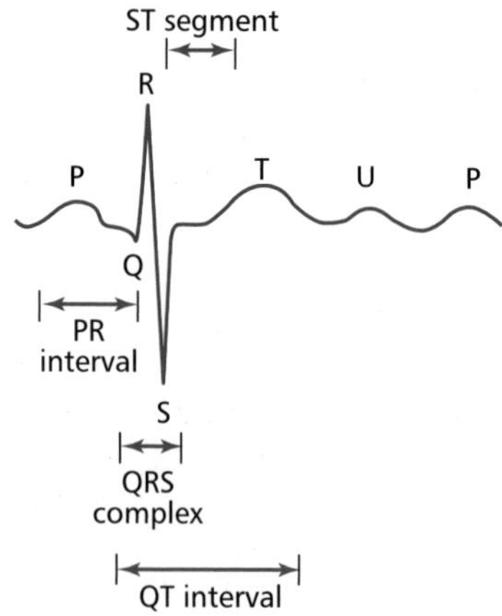

Figure 14.11 Electrocardiogram arising from (A) the 12 leads and (B) explanation of the parts of the wave observable in the graphs. See text for details.

In addition to the basic technique, the cardiac response may be recorded during exercise as the patient walks or runs on a treadmill and this can give further clinical information. In other situations, the ECG may be recorded over 24 h using a portable ECG machine to monitor intermittent arrhythmias (*Section 14.7*).

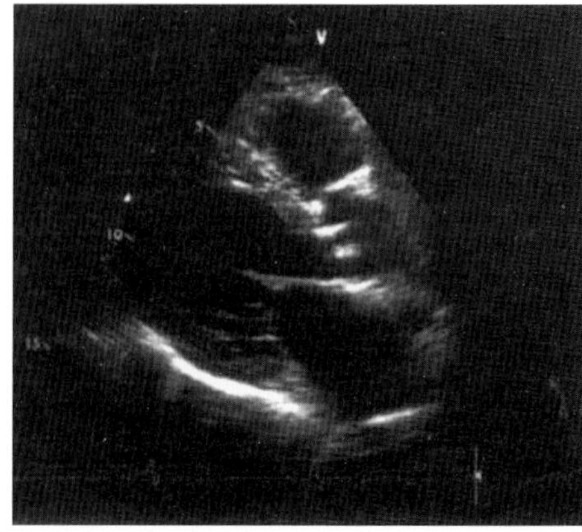

Figure 14.12 Echocardiography uses ultrasound waves that are reflected from the interfaces between the various tissues and analyzed to produce the type of image shown.

ECHOCARDIOGRAPHY

Echocardiography uses high-frequency ultrasound to map the heart and study its various functions. It is one of the most widely used diagnostic techniques, is painless and does not use X-rays and can indicate the extent of cardiac damage. Ultrasound waves from a probe at wavelengths of 1 mm or less (corresponding to frequencies of around 2 MHz) are generated in short bursts of a few microseconds. When the probe is pressed against the body the emitted pulses encounter the interfaces between the various body tissues. In crossing each interface some sound energy is reflected and is detected by a transducer in the probe and recorded as an echo. The time delays for the echoes to return are analyzed by a computer to produce a video picture on a screen with a moving image of the heart and blood vessels (*Figure 14.12*). This allows the cardiologist to see if the heart valves are functioning properly, for example whether they leak when closed and if blood is flowing normally. Abnormal connections between blood vessels of heart chambers are also revealed, as well as contractibility of muscle walls.

CARDIAC CATHETERIZATION (ANGIOGRAPHY)

Cardiac catheterization is a further method for investigating heart function and abnormalities. However, it is an invasive technique and involves a certain amount of discomfort for the patient. Under local anesthetic, a thin plastic tube is inserted into a vein or artery in an arm or leg (groin) and advanced into the major vessels and heart chambers. The catheter can be used to measure pressure, take a view of the inside of the heart or to take blood samples to measure oxygen and lactate concentrations. Dyes which are X-ray opaque can be injected into the catheter allowing moving X-ray pictures to be made that can show up anatomical abnormalities or abnormal blood flow. The coronary arteries can also be investigated by catheterization to check for coronary artery disease.

ANGIOGRAPHY AND RADIOGRAPHIC VISUALIZATION

A chest X-ray taken from the back also enables the size of the heart in relation to the lungs and the major blood vessels to be assessed. It may also reveal areas of calcium deposition, which are a sign of tissue damage and death.

EXERCISE STRESS TEST

The exercise stress test measures oxygen uptake, CO_2 production, heart rate and lung ventilation during progressively more strenuous treadmill or cycle ergometer exercises. It can detect lung and heart diseases in their early stage and also be used to assess fitness. Patients with lung disease stop exercising before achieving their maximal predicted heart rate. Also, their levels of ventilation are disproportionately high for a given oxygen uptake. Thus the more expensive and complicated equipment can differentiate breathlessness due to lung or to heart disease. For patients who cannot physically exercise, stress test measurements can be obtained using the dobutamine stress test. Dobutamine is an inotropic drug that increases the heart rate, hence giving increasing doses makes the heart work harder.

14.6 ENDOCARDITIS

Endocarditis is an inflammation of the endocardium, the interior lining of the heart and its valves. It most often results from bacterial infection that may originate from bacteria in the blood or as a result of heart surgery. The bacteria in the blood may enter from a skin wound or even from small injuries occurring when chewing food or brushing the teeth. Injecting drug users and patients with prolonged catheter use are also at risk. Abnormal or damaged valves are more susceptible than normal ones, and people with artificial valves are at risk. Bacteria and blood clots can accumulate on the valves (called **vegetation**) and can then break loose and block vessels elsewhere in the body causing strokes, heart attacks (*Section 14.14*) pulmonary embolisms or infecting the area where they lodge.

Acute infective endocarditis has a rapid onset and can be life threatening, unlike subacute infective endocarditis which develops slowly over weeks and months. In the acute form, the symptoms are usually the sudden onset of a high fever, a fast heart rate and tiredness. There can be extensive valve damage as well as the blood clot damage elsewhere in the body. Individuals may go into shock and be subject to renal failure. Prompt diagnosis and hospitalization are vital. The subacute form is associated with mild fever, tiredness, weight loss, sweating and a low erythrocyte count. However, because the symptoms of the subacute form are more vague, damage may occur before the condition is recognized; it is just as life threatening as the acute form.

Patients with heart valve abnormalities or artificial valves are more susceptible to endocarditis, as mentioned above. If they are about to undergo medical or dental procedures, they must inform their surgeon or dentist and be given antibiotics prior to invasive treatments. If endocarditis occurs and is identified, the treatment usually consists of at least two weeks of high-dose, intravenous antibiotics. However, heart surgery may also be necessary to repair or replace damaged valves.

14.7 ABNORMAL HEART RHYTHMS

The normal heart rate is between 60 and 100 beats per min but much lower rates may be encountered and are quite normal in young adults who are physically fit. As has been mentioned, the rate also responds to exercise or inactivity and also to pain and anger. An inappropriately fast heartbeat is called **tachycardia** and an abnormally low rate **bradycardia**. Abnormal rhythms are frequently encountered and can be regular or irregular. The contractions of the heart muscle fibers and therefore heartbeat is controlled by electrical discharges that flow through the heart along distinct pathways at controlled speeds (*Section 14.2* and *Figure 14.4*). If disturbances occur with the flow of the electrical discharge then **arrhythmias** in these contractions can occur. These range from the harmless to the life threatening. For example, minor arrhythmias can be caused by excessive alcohol consumption, smoking, stress or exercise. Thyroid hormones also affect the heart rate and an over- or under-active thyroid gland (*Chapter 7*) may affect the rate and rhythm of the heart. Some of the drugs used to treat lung disease or high blood pressure can have similar effects.

The commonest causes of arrhythmia are heart disease, especially coronary heart disease, heart failure or abnormal valve function. In many cases patients are aware of an abnormal heartbeat and this is referred to as **palpitations**. This awareness may be disturbing but there are many possible causes of arrhythmias and they are often not the result of an underlying disease.

However, when they are, it is the nature and severity of the disease that is more important than the arrhythmia itself. If brought to the attention of a clinician, he or she will want to know if they are fast or slow, regular or irregular, whether they make the person feel dizzy, light-headed or even lose consciousness, whether they occur at rest or during exercise and are they accompanied by a shortness of breath or chest pain.

An ECG is, of course, helpful in diagnosis but the arrhythmias may only occur over a short period. Consequently a portable monitor may be placed on the patient to record them over 24 h. The prognosis and treatment will depend on whether the arrhythmias start in the pacemaker, the atria or the ventricles. Most arrhythmias are harmless and do not interfere with the heart's pumping action. However, antiarrhythmic drugs are available if the patient is anxious or if the arrhythmias cause intolerable symptoms or pose a risk. Sometimes it is necessary to fit the patient with an artificial pacemaker that is programmed to replace the heart's own pacemaker. These are usually implanted surgically under the skin of the chest and are wired to the heart. Most commonly they are used to correct abnormally slow heart rates. Sometimes an externally applied electric shock to the heart, called cardioversion, electroversion, or defibrillation, can correct an abnormal arrhythmia. Arrhythmias caused by coronary artery disease may be controlled by drugs, pacemakers or by surgery (*see below*). After a coronary infarction, some people have life-threatening episodes of ventricular tachycardia that may be triggered in an injured area of heart muscle that may have to be removed during open heart surgery.

ATRIAL FIBRILLATION AND FLUTTER

Atrial fibrillation and flutter are rapid electrical discharges which make the atria contract rapidly but each contraction may not conduct to the ventricles. This causes the ventricles to contract less efficiently and irregularly, producing a condition that may be sporadic or persistent. During the fibrillation or flutter, the atrial walls simply quiver and blood is not pumped into the ventricles properly. The consequence is that inadequate amounts of blood are pumped from the heart, blood pressure falls and heart failure may occur. The diminished pumping ability may make the patient feel weak, faint and short of breath. Sometimes, especially with elderly patients, there is chest pain and heart failure. If the atria do not contract completely, blood may stagnate in the atria and clot. If pieces of clot or emboli break off they may move and block an artery elsewhere in the body. If this is in the brain it may cause an embolism or stroke (*Section 14.16*).

The treatment for atrial fibrillation is to correct the disorder that causes the abnormal rhythm, restoring it to normal, and to slow the rate at which the ventricles contract so as to improve the pumping efficiency. The latter can often be achieved with digoxin but a β-blocker, such as propanolol, or other drugs may also be required. Frequently a normal rhythm has to be restored by antiarrhythmic drugs or cardioversion. In these conditions blood can pool in the ventricles and clot, hence anticoagulant drugs may be used when atrial fibrillation is present.

VENTRICULAR TACHYCARDIA

Sustained ventricular tachycardia, with a ventricular rate of at least 120 beats per min, occurs in various heart diseases that damage the ventricles. Most often it occurs over weeks or months after a heart attack (*Margin Note 14.2*). It is characterized by palpitations and will usually require emergency treatment because the blood pressure falls and heart failure may follow. Cardioversion is needed immediately.

Margin Note 14.2 Heart attack or myocardial infarction 

A heart attack is a **myocardial infarction**. It is a severe medical emergency during which some or all of the blood supply to the heart muscle through the coronary arteries is cut off. If this continues for more than a few minutes the heart tissue dies with serious clinical consequences or death of the individual. The term heart attack is sometimes used loosely for other heart conditions: strictly it should only be applied to myocardial infarctions.

VENTRICULAR FIBRILLATION

Ventricular fibrillation is a form of cardiac arrest. It is similar to atrial fibrillation, but the prognosis is very serious and potentially fatal if not treated immediately. It is the product of an uncoordinated series of rapid but ineffective contractions throughout the ventricles. These, in turn, arise from multiple chaotic electrical impulses. Its commonest cause is an insufficient flow of blood to the heart muscle because of coronary heart disease or a heart attack. Given that blood is not pumped from the heart, it can lead to unconsciousness in seconds and, if untreated, the patient usually has convulsions and develops irreversible brain damage because of oxygen starvation. Ventricular fibrillation needs to be treated as a medical emergency. Cardiopulmonary resuscitation must be started within the minimum time possible, usually three minutes. This should be followed by cardioversion also as soon as possible. Subsequently drugs are needed to restore and maintain the normal heart rhythm.

HEART BLOCK

Heart block describes a delay in electrical conduction through the AV node. There are various degrees of seriousness; the least may not require treatment but the most serious may require the fitting of an artificial pacemaker. First-degree heart block is common in well-trained athletes, teenagers, and young adults, but it may also be caused by rheumatic fever (*Box 14.2*) or by certain drugs. At the other extreme in third-degree heart block, electrical impulses from the atria to the ventricles are completely blocked. The ventricles beat very slowly and the pumping ability of the heart is compromized. Fainting, dizziness and sudden heart failure are common.

BOX 14.2 Rheumatic fever

Rheumatic fever occurs mostly in children and young adults and is caused by infection with group A Streptococci. It is now much less common in the developed countries than was previously the case: for example 10% of children in the 1920s compared with about 0.01% now. This is mainly due to the use of antibiotics (*Chapter 3*). Rheumatic fever is thought to result from an autoimmune reaction triggered by the bacteria rather than any bacterial toxin. The skin, joints and the central nervous system and all the layers of the heart may be affected. The disease presents with fever, joint pains, malaise, loss of appetite and a characteristic fleeting polyarthritis affecting the larger joints, such as knee, elbows, ankles, which become swollen, red and tender. Effects on the heart include new or changed murmurs, cardiac enlargement or failure and pericardial effusion.

INVESTIGATION

There will usually be nonspecific indicators of inflammation, such as the erythrocyte sedimentation rate and C-reactive protein (*Chapter 13*), both of which may be elevated. Throat swabs should be cultured for group A *Streptococcus* and there may be serological changes indicative of a recent streptococcal infection.

TREATMENT

If patients have fever, active arthritis or active carditis, they should be completely bed rested. Residual streptococcal infection should be eradicated with a single intramuscular injection of benzathine penicillin or four daily oral doses of phenoxymethyl penicillin for a week. Salicylate and steroids may also be given if carditis is present. Recurrence is common. More than half of those with acute rheumatic fever with **carditis**, a general inflammation of the heart, will develop conditions after 10–20 years that affect the mitral and aortic valves.

14.8 CARDIAC FAILURE

Cardiac failure is the inability of the heart to maintain an adequate cardiac output, that is pump a volume of blood per minute sufficient to meet the demands of the body. The heart does not stop beating as is often thought, but its diminished ability imposes severe demands. It is a serious condition but commoner in older people. The incidence is about one in a hundred for individuals over 65 years, and irrespective of the cause, the prognosis is poor. Approximately 50% of patients will die within two years, although new drug treatments are improving mortality and morbidity.

There are many possible causes of heart failure and, indeed, any disease that affects the heart and circulatory system can lead to heart failure. The commonest of these is coronary artery disease that limits the flow of blood, and hence oxygen and nutrients to the heart muscle and can lead to heart attack. Bacterial and viral infections (*Chapter 2*) can also damage the heart muscle, as can diabetes, an overactive thyroid (*Chapter 7*) and obesity (*Chapter 10*). Obstruction of the heart valves or heart valves that leak increases its workload and this eventually weakens the contractions. Similarly, a narrowed aortic valve means that the heart has to work harder because it has to force blood through a smaller exit, again imposing an extra metabolic burden. High blood pressure (*Section 14.17*) also means that the heart has to work too hard. Diseases that affect electrical conduction in the heart can result in an abnormal heartbeat that reduces the pumping efficiency. Other causes are also known. Although the increased workload initially results in enlargement, or hypertrophy, of the heart muscle so that it can contract with greater force, eventually the heart malfunctions making the heart failure worse.

Heart failure results in tiredness and weakness during physical activities because the skeletal muscles are starved of blood. The disease may be on one side of the heart or the other, but the condition usually affects the whole heart. Nevertheless, there are characteristic symptoms depending on which side is affected. Thus right-sided disease tends to cause a build-up of blood flowing into the right side of the heart, which leads to swelling of the feet, ankles, legs and liver. In contrast, left-sided disease increases fluid in the lungs (pulmonary edema) causing, in turn, shortness of breath. At first this is only experienced during exertion but it gradually increases in severity so that the breathlessness occurs even at rest. If this happens at night, the patient may wake up gasping for breath and may find it better to sleep in a sitting position. Cardiac failure gradually worsens with time if the underlying disease is not treated, although patients may continue to live for many years.

INVESTIGATION

The symptoms described above are usually sufficient for an initial diagnosis of heart failure, which would be confirmed by a weak and rapid pulse, lowered blood pressure and abnormal heart sounds. However, its underlying cause must also be identified. In many cases taking a clinical history and examining the patients will be sufficient. General diagnostic tests include chest X-ray to demonstrate an enlarged heart and fluid accumulation in the lungs, ECG, echocardiography, blood tests, for example full blood count, liver function, urea and electrolytes (*Chapters 8, 11* and *13*), and analysis of cardiac enzymes in acute heart failure to diagnose myocardial infarction (*Box 14.4*) will then usually be carried out. Functional tests may also be performed, including exercise testing, ECG monitoring and angiography at rest and under stress.

TREATMENT

The treatment for heart failure is focused on relieving the symptoms, retarding the progression of the disease and aiming to improve the chances

Margin Note 14.3 ACE inhibitors 

Angiotensin converting enzyme (ACE) inhibitors are used to treat chronic heart failure and high blood pressure. The drugs block the conversion of angiotensin I to angiotensin II which is both a vasoconstrictor and the most important stimulus for the release of aldosterone (*Chapter 8*) from the adrenal cortex. Angiotensin converting enzyme inhibitors are therefore effective antihypertensives. Their effect is to lower the systemic vascular resistance, venous pressure and reduce levels of circulating catecholamines, thus improving myocardial performance. The first ACE inhibitor used was captopril but a number of other drugs are now available. The way in which they work is well understood. If the flow of blood through the kidneys is low they release renin which converts angiotensinogen to angiotensin I. Subsequently ACE converts angiotensin I into angiotensin II. Among its other effects on blood vessels, angiotensin II causes the division of heart muscle cells and fibrosis that can make heart failure more serious in the long run. Giving ACE inhibitors reduces or eliminates these effects. Angiotensin converting enzyme inhibitors are used routinely postmyocardial infarction, to maintain good cardiac action and prevent heart failure.

of survival. This means that any factor aggravating the failure should be identified and treated. The precise cause of failure should be identified and if possible corrected. Patients should be nursed in a comfortable, upright position.

CHRONIC HEART FAILURE

In chronic heart failure, the circulation at rest is adequate but there is an inadequate reserve to pursue daily activities. Its treatment depends upon the underlying disease to be dealt with. For example, heart surgery can correct narrowed or leaking heart valves, and bypass surgery can correct blocked coronary arteries. If the disease is caused by an infection the condition may be improved by antibiotics without surgery. Additionally, there are many things a patient can be advised to do to help the condition, including giving up smoking, eating less salt, reducing excessive weight and controlling alcohol consumption.

The best treatment for heart failure is to prevent it happening in the first place or by reversing its underlying cause as soon as possible. Nevertheless, there is still much that can be done. For example, if reducing the salt intake does not lower fluid retention then diuretics may be prescribed. A reduction in the amount of body fluid reduces the volume of blood to be pumped and so alleviates the strain on the heart. Digoxin may be given to increase the power of each heart contraction and will slow a rate that is too rapid. Vasodilatory drugs may be prescribed to expand blood vessels, lowering the blood pressure. Some of the older drugs dilate arteries more than veins or *vice versa*. However, the ACE inhibitors (angiotensin converting enzyme inhibitors (*Margin Note 14.3*)) dilate both arteries and veins are, perhaps, the most commonly used. These improve the symptoms and prolong life. A heart transplant is perhaps the ultimate possibility but there are never enough good hearts to go round!

ACUTE HEART FAILURE

In acute heart failure, the hemodynamic derangement is so severe that it results in symptoms even at rest. If fluid suddenly accumulates in the lungs the condition is known as acute pulmonary edema and the person has to gasp for breath and emergency treatment is required. Oxygen is given by a facemask and together with intravenous diuretics may result in a rapid and dramatic improvement. Glyceryltrinitrate may be given intravenously or placed under the tongue and this leads to dilation of the veins, reducing the amount of blood flowing through the lungs. It may be necessary to insert a tube into the patient's airway to help breathing. The treatment for acute heart failure is essentially as for the chronic condition described above.

14.9 ISCHEMIC HEART DISEASE

Myocardial ischemia, meaning a lack of oxygen to the myocardium, is the result of an imbalance between the demand of the myocardium for oxygen and the amount supplied. There are several possible causes of this condition. Firstly, the blood flow through the coronary arteries may be reduced because of mechanical blockages, such as atheroma (plaque), thrombosis (a clot), spasm, ostial stenosis, arteritis or any sort of blockage (an embolism) due to, for example, tumor cells or an air bubble. Secondly, a decreased flow of oxygenated blood to the myocardium because of anemia (*Chapter 13*), **hypotension** (low blood pressure) or carbon monoxide poisoning (*Chapter 12*). Thirdly, an increased demand for oxygen caused by exercise or myocardial hypertrophy that requires an increase in cardiac output. Note that in the last two scenarios, the coronary arteries may be healthy. The commonest cause of ischemic heart disease is coronary atheroma, which obstructs the flow of blood through the coronary arteries.

14.10 CARDIOMYOPATHIES

Cardiomyopathy is a progressive disorder that impairs the function of the ventricular muscle walls. It may come about as a result of a number of diseases or may have no identifiable cause.

DILATED CONGESTIVE CARDIOMYOPATHY

Dilated congestive cardiomyopathy is not a single condition but a group of heart disorders in which the ventricles have enlarged but are still not able to pump enough blood to meet the needs of the body. Heart failure may result. The commonest cause of the defect in developed countries is widespread coronary artery disease, which leads to an inadequate blood supply to the heart muscle. This causes damage and the undamaged muscle then stretches in compensation. If this is inadequate to meet body needs, then dilated congestive cardiomyopathy develops. Its symptoms are shortness of breath on exertion and a rapid onset of tiredness due to the weakening of the heart's pumping action. The heart rate speeds up so blood pressure is normal or low, but fluid is retained in the legs, abdomen and lungs. Enlargement of the heart can mean that the valves do not close properly, leading to leakage, and this improper closing may be heard as murmurs using a stethoscope. The stretching may also increase the potential for arrhythmias. Electrocardiography or magnetic resonance imaging (MRI, *see Chapter 18*) may be used to confirm the initial diagnosis.

About 70% of people with the condition die within five years from the onset of symptoms and the prognosis declines as the heart walls become thinner with reduced contractibility. Men tend to survive only half as long as women and blacks half as long as whites. In about half of the cases there is sudden death. Treating the underlying cause, for example reducing alcohol abuse can prolong life. When there is coronary artery disease there may be angina (*Section 14.13*), which is treated with glyceryltrinitrate, β-blockers or calcium channel blockers. There may also be pooling of blood in the swollen heart that can cause clots to form and therefore the patient is given anticoagulants (*Chapter 13* and *Box 14.3*).

BOX 14.3 Anticoagulant therapy

Venous thromboembolism is a common problem after surgery, especially in patients who are elderly, have malignancies or have a previous history of thrombosis. There is also a high incidence in patients confined to bed after trauma or a variety of heart diseases. Its prevention and treatment includes the use of anticoagulants, such as heparin, warfarin and aspirin.

HEPARIN

Heparin is a mixture of polysaccharides prepared from pig gastric mucosa. It potentiates the formation of irreversible complexes between antithrombin III (AT-III), a potent inhibitor of coagulation and several coagulation factors, for example thrombin, and factors XIIa, XIa, Xa, IXa and VIIa (*Chapter 13*). Its injection has an immediate effect on blood coagulation, but its action is quite short-lived. Bleeding complications of heparin treatment occasionally occur and are treated by giving the positively charged protein, protamine. Low M_r heparins, produced by chemical degradation of standard heparin, have somewhat different properties. They have a longer half-life than heparin and so can be given as a once or twice daily subcutaneous injection (preferred) or as a continuous infusion or six-hourly injection, as is required with regular heparin. They also seem to cause less inhibition of platelet function and thus there is a reduced risk of causing bleeding.

Myocarditis, which is an acute inflammation of the heart muscle, occurs as a result of a viral infection, most often coxsackie virus B, hence it is sometimes called viral cardiomyopathy. It may weaken the heart muscle, producing a condition similar to dilated congestive cardiomyopathy. A number of chronic hormonal disorders (*Chapter 7*) including diabetes and thyroid disease can also produce cardiomyopathy, as can prolonged alcohol abuse (*Chapter 12*).

HYPERTROPHIC CARDIOMYOPATHY

Hypertrophic cardiomyopathy is a group of conditions in which the ventricular walls thicken. It may occur as a birth defect, in adults with acromegaly or in people with pheochromocytoma, a tumor of the adrenal gland (*Chapter 7*). The heart becomes thicker and stiffer than normal and more resistant to filling with blood from the lungs, leading to a backpressure on the lung veins causing a pulmonary edema. The patient therefore becomes chronically short of breath, with symptoms that include faintness, chest pains and palpitations brought on by irregular heartbeats; the heart sounds through a stethoscope are usually characteristic. Younger patients tended to die suddenly of hypertrophic cardiomyopathy but with better and earlier diagnosis and drug therapies this is now less of a problem.

The main treatments are administering β-blockers and calcium channel blockers. Surgery to remove some of the heart muscle may improve the outflow of blood but this is only undertaken when the patient is incapacitated despite drug therapy.

Margin Note 14.4 Amyloidosis

There are several forms of this disease in which amyloid, an unusual form of protein that is not normally present in the body, accumulates in various tissues. One type of amyloidosis affects the heart and is associated with normal aging, but what causes the build up of the amyloid protein is not known with any certainty. The condition may not need treatment and, in any case, treatments are not usually very successful. A patient with heart problems may be given a heart transplant, but the transplanted organ itself may be affected later.

RESTRICTIVE CARDIOMYOPATHY

Restrictive cardiomyopathy is the least common of the cardiomyopathies. It has many features in common with hypertrophic cardiomyopathy. The replacement of heart muscle tissue with scar tissue or its infiltration with abnormal material, such as white blood cells, amyloidosis and sarcoidosis (*Margin Notes 14.4* and *14.5*) can all lead to the condition, although its cause is frequently unknown. The major symptoms are shortness of breath and an edematous swelling of the tissues. About 70% of individuals with the disease die within about five years of the symptoms commencing. In most cases therapy is unsatisfactory. The use of diuretics, which are normally given to treat heart failure and reduce the amount of fluid accumulating in the tissues, may actually reduce the amount of blood entering the heart and worsen the condition.

ORAL ANTICOAGULANTS

Oral anticoagulants work by interfering with vitamin K metabolism (*Chapters 10* and *13*). Vitamin K is a cofactor required for the formation of γ-carboxyglutamate residues in prothrombin. The coumarin anticoagulant, warfarin, is most frequently used because of its low incidence of side effects other than promoting bleeding. However, many drugs interact with warfarin. Tricyclic antidepressants, clofibrate, aspirin, and alcohol increase the anticoagulant effect of warfarin, while other drugs such as rifampicin and barbiturates decrease this effect. Contraindications include pregnancy, peptic ulcers, severe liver and renal disease and preexisting hemostatic disease.

ASPIRIN

Low doses of aspirin inhibits platelet aggregation by blocking cyclooxygenase activity irreversibly thus altering the balance between the complex fatty acids, prostacyclin and thromboxane A_2 which is required for platelet activation. Clinical trials have confirmed that long-term treatment with aspirin greatly reduces the risk of myocardial infarction (and death) in patients with angina.

14.11 HEART VALVE DISORDERS

Major problems encountered with heart valves include an improper closing of the valves leading to leakage (**regurgitation**) or failure to open fully (**stenosis**). These conditions interfere with the heart's capacity to pump blood. If the mitral valve leaks, then regurgitation occurs each time the ventricle contracts. As blood is pumped into the aorta some leaks back into the left atrium increasing the volume and the pressure in that compartment. This, in turn, increases the blood pressure in the vessels leading from the lungs to the heart resulting in a pulmonary edema. Rheumatic fever was once the commonest cause of mitral valve regurgitation (*Box 14.2*) and heart attacks that damage the structures supporting the valve is its commonest cause. However, in countries where there is poor preventive medicine, rheumatic fever is still common. Repair or replacement of the valve is required if the regurgitation is severe.

The aortic valve may also become leaky, and the most common causes were rheumatic fever and syphilis but this is now rare because of the use of antibiotics. In contrast, aortic valve stenosis is mainly a disease of the elderly (over 60 years) and is the result of the valve becoming calcified. The left ventricle wall thickens as the heart strains to pump sufficient blood through the narrow opening and the enlarged heart muscle requires extra oxygen and nutrients from the coronary arteries. Eventually the output of blood from the heart becomes insufficient for the body's needs and the resulting heart failure causes shortness of breath and fatigue. The treatment is to replace the aortic valve, preferably before the left ventricle becomes irreparably damaged.

Problems can also occur with the tricuspid valve. However, regurgitation usually requires little treatment. Stenosis of the tricuspid valve is rare, again, because the damage was mainly associated with rheumatic fever.

Margin Note 14.5 Sarcoidosis 

In sarcoidosis, abnormal collections of inflammatory cells called granulomas, commonly appear in lymph nodes, lungs, liver, eyes and skin, but can also occur in the spleen, bones, joints, skeletal muscles, heart and nervous system. The cause of the disease is unknown but seems to result from an abnormal response of the immune system. In some instances there are no symptoms, while in others there can be fever, weight loss and aching joints. Often the condition clears up spontaneously and the granulomas may eventually disappear or become scar tissue. In other cases, there may be permanent damage, such as lung scarring. Most people do not need treatment but corticosteroids may be used to suppress severe symptoms, such as shortness of breath or severe skin lesions.

14.12 HEART TUMORS

Heart tumors (*Chapter 17*) may be asymptomatic or precipitate life-threatening crises, such as sudden heart failure, sudden onset of irregular heartbeat or a sudden drop in blood pressure caused by bleeding into the pericardium. They are relatively uncommon and are difficult to diagnose.

14.13 ATHEROSCLEROSIS OR ARTERIOSCLEROSIS

Atherosclerosis or **arteriosclerosis** refers to the simultaneous development of an **atheroma** in an artery and the sclerosis of its wall. An atheroma (from the Greek word for porridge) is a hard yellow plaque that gradually builds up on the inside of medium-sized arteries. The plaque consists of a necrotic (dead) core rich in cholesterol, surrounded by fibrous tissue. Sclerosis (from the Greek word for hard) means an abnormal hardening or fibrosis, which is the formation of excess fibrous material within a tissue. Sufferers may experience a sudden heart attack or stroke, but this belies the fact that in most cases the arteries of the victims have gradually become blocked by atherosclerosis.

Initially atheromatous plaques start at the site in an artery where the smooth muscle layer has thickened and been infiltrated with fibrous connective tissue (fibrosis) when cholesterol and other lipids have been deposited, and may become calcified. This condition is commonly referred to as hardening of the arteries. During the course of the disease, the affected artery expands as the plaque becomes larger so as to allow a more or less normal flow of blood. However, as the plaque increases in size this becomes less possible and the lumen of the artery becomes narrower and a ballooning of the arterial wall causes it to weaken. Also, there is more likelihood of an embolus becoming trapped in the narrowed artery making the blockage worse. Healthy arteries are lined with endothelial cells, but the rough lining of a plaque-damaged artery seems to encourage the adhesion of platelets which means they are common sites for the formation of a clot (thrombus).

The progresses of atheromatous disease means the arteries become increasingly occluded and the threat of a heart attack or stroke increases. Some patients may receive warning in the form of chest pains if, for example, a coronary artery is partially blocked. The condition known as angina pectoris (*see below*) is a signal that the heart is not receiving sufficient oxygen. This is most likely to occur when the heart is working hard because of physical or emotional stress. However, for many people there are no symptoms and they are completely unaware of their condition until the catastrophic event occurs. Some individuals have an inherited tendency to develop hypertension (*Section 14.17*), which promotes atherosclerosis and increases the risk of heart attack or stroke and can cause chronic damage to the endothelium lining the arteries promoting atherosclerosis.

ANGINA

Angina pectoris is caused by myocardial ischemia. It presents as a crushing or squeezing pain in the chest and the discomfort may radiate into the neck, jaw, arms (especially the left) and sometimes into the back. There may also be shortness of breath, abdominal pain, nausea and dizziness. Myocardial oxygen demand relates to the heart rate, left ventricular contractility and systolic wall stress. The demand for oxygen is increased by exercise, hypertension (*Section 14.17*) and left ventricular dilation, which may happen in chronic heart failure.

Several types of angina are recognized. Stable angina occurs when atherosclerotic plaques block one or more of the coronary arteries. Under resting conditions, cardiac oxygen demand is quite low and is satisfied even by the diminished blood flow. However, when exertion or emotional stresses increase this oxygen demand, ischemia develops on the inner part of the myocardial wall. However, the response to exercise is variable: some patients may have excellent exercise tolerance one day and then develop angina with minimal exertion the next. In addition to causing pain, the ischemia causes a decline in the output of ATP and creatine phosphate and hence contractility is impaired. Stable angina is normally relieved by a short rest or by administering glyceryltrinitrate. The latter dilates the arteries, increasing blood, and therefore oxygen, supplies to the muscle leading to less pain.

Variant angina is an intensely painful, transient spasm caused by a blockage of one of the coronary arteries. It is relatively uncommon, but can occur at rest. It is exacerbated by smoking and by cocaine use. About one-third of patients show no evidence of atherosclerotic lesions.

Ischemic heart disease and stable angina can be distinguished from other conditions that cause chest pain on the basis of their characteristic symptoms and by a number of types of diagnostic tests, for example the ECG, exercise stress test and by using coronary angiography to obtain a direct radiographic visualization, as described earlier.

The management of angina is designed to control the symptoms and reduce any underlying risk factors. The drugs used include the nitrovasodilators, for example glyceryltrinitrate, β-adrenoceptor blockers, calcium channel antagonists, as well as drugs that inhibit platelet aggregation and thrombosis (*Margin Note 14.6* and *Chapter 13*). In the case of stable angina, mortality is 2 to 4% a year if only one coronary artery is diseased but increases with the number of diseased arteries.

The other main variant of angina is the so-called unstable angina, which is a dangerous condition, often heralding an impending myocardial infarction (*Section 14.14*). In general, the symptoms resemble those of stable angina but are more intense and persistent, often lasting 30 min, and the pain is often resistant to glyceryltrinitrate treatment. The attacks may be frequent, becoming progressively more severe and prolonged, may be brought on by minimal exertion (or even during sleep) or may occur several days after a myocardial infarction. The episodes are preceded by a fall in coronary blood flow, which is thought to be the result of the periodic development of coronary thrombosis and vasoconstriction. These are triggered by coronary arterial disease. The thrombosis may be promoted by the turbulent blood flow associated with atherosclerotic plaques: there may also be damage to the endothelial lining of the blood vessels.

An ECG is taken to help in the diagnosis but, in addition, serum levels of C-reactive protein and amyloid-A protein may be increased; these are classic markers of inflammation. Unstable angina is a medical emergency and treatment usually begins with aggressive drug therapy to control the symptoms and prevent further episodes, and to try to reverse coronary vasospasm. Platelet glycoproteins IIIA and IIb (Tirofiban and ReoPro) are now used in unstable angina to stabilize the clot in the narrowed coronary artery, which is causing the pain, prior to angiography and possibly angioplasty and stenting. Angioplasty is a procedure similar to angiography (*Section 14.5*) but the catheter delivers a small inflatable balloon to the narrowed portion of the coronary artery. When the balloon is inflated it opens the restricted section of the artery. Stents are very small, coated spring-like structures that are deployed, by cardiac catheters into the narrowed section following angioplasty. They act like miniature struts to maintain the opening of the vessel.

Urgent revascularization needs to be considered for patients at high risk, or unsuitable for angioplasty/stenting due to significant coronary arterial disease. In a coronary artery bypass grafting a length of healthy 'surplus' blood vessel, such as the saphenous vein from the leg (*Figure 14.13*), is obtained and pieces of it are inserted between the aorta and the coronary arteries distal to any stenosis (narrowing). The left internal mammary artery may also be used. A bypass improves survival in patients with severe atherosclerotic disease in all the major coronary arteries.

Margin Note 14.6 Thrombolysis 

Thrombolysis is the dissolution of a blood clot blocking an artery. Fibrinolysis occurs when the inactive zymogen, plasminogen, is converted to the fibrin-dissolving active enzyme plasmin by endothelium-derived **tissue** plasminogen activator (t-PA). Plasmin also inactivates fibrinogen and coagulation factors V and VIII (*Chapter 13*).

Streptokinase is a bacterial protein that binds to a molecule of plasminogen. The resulting complex cleaves other molecules of plasminogen to produce more plasmin, which dissolves the fibrin. It can cause hemorrhage and also it can only be used once since it may cause the patient to produce antibodies with the danger of an allergic reaction. Tissue plasminogen activator is now produced commercially by recombinant DNA methods and marketed as tenecteplase and alteplase. Tenecteplase is now the thrombolytic of choice. It binds to fibrin and this has a greater effect on clot-associated plasminogen than on plasma plasminogen. It has the advantage that it is cleared from the plasma in a few minutes and is nonantigenic. Urokinase (u-PA) is another endogenous plasminogen activator with properties similar to those of t-PA.

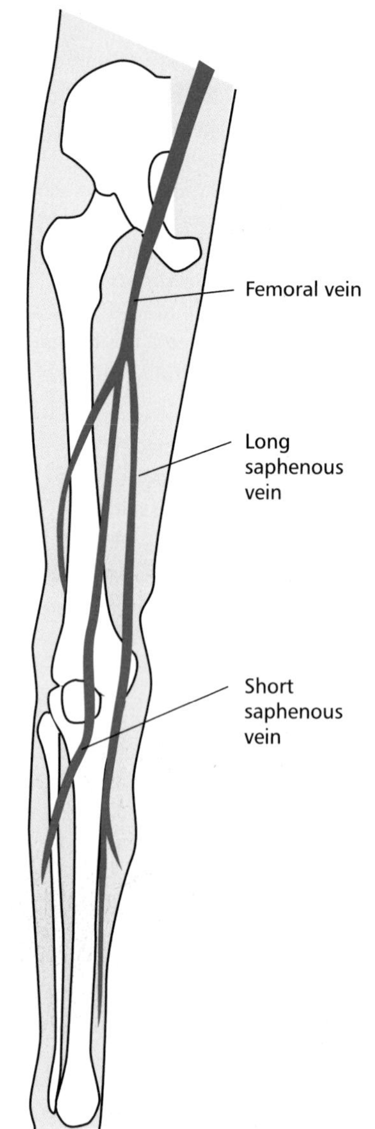

Figure 14.13 Schematic to show the saphenous veins of the leg.

CHOLESTEROL AND LIPOPROTEINS

Aside from any inherited tendency to develop hypertension (*Section 14.17*), there are a number of nongenetic factors that correlate with an increased risk of cardiovascular disease. These include smoking, lack of exercise, a diet containing too much and inappropriate fat types (*Chapter 10*); all lead to an abnormal concentration of cholesterol in the blood. Homocysteine is known to be raised in certain genetic conditions and coronary heart disease and is suspected to have a role in increasing cholesterol levels. Homocysteine concentrations can be lowered by treatments with folate and vitamin B_{12} with a subsequent reduction in blood cholesterol.

Cholesterol is essential because it forms part of the plasma membrane of cells and is used in the biosynthesis of bile salts and steroid hormones. However, atherosclerosis is largely due to problems with cholesterol. The body can synthesize it but it is also obtained from the diet. Cholesterol is practically insoluble in plasma and, like triacylglycerols, is transported in the blood in **lipoprotein particles**. It is an imbalance between the different types of lipoprotein particles that leads to clinical problems.

Plasma lipoprotein particles consist of a core of triacylglycerols and cholesterol esters surrounded by phospholipids, proteins and free cholesterol (*Figure 14.14*). They are classified by their densities, the greater the proportion of triacylglycerol the lower the density (*Table 14.1*). There are several different types of lipoproteins in lipoprotein particles, called apolipoprotein-A and –B, -C, -D and -E, usually abbreviated to apoA, apoB, apoC, apoD and apoE.

The transport of cholesterol round the body is a complicated process (*Figure 14.15*). There is some cholesterol in the chylomicrons derived from the diet, but the liver exports the cholesterol obtained from the diet or synthesized, together with triacylglycerols it synthesized from dietary carbohydrate,

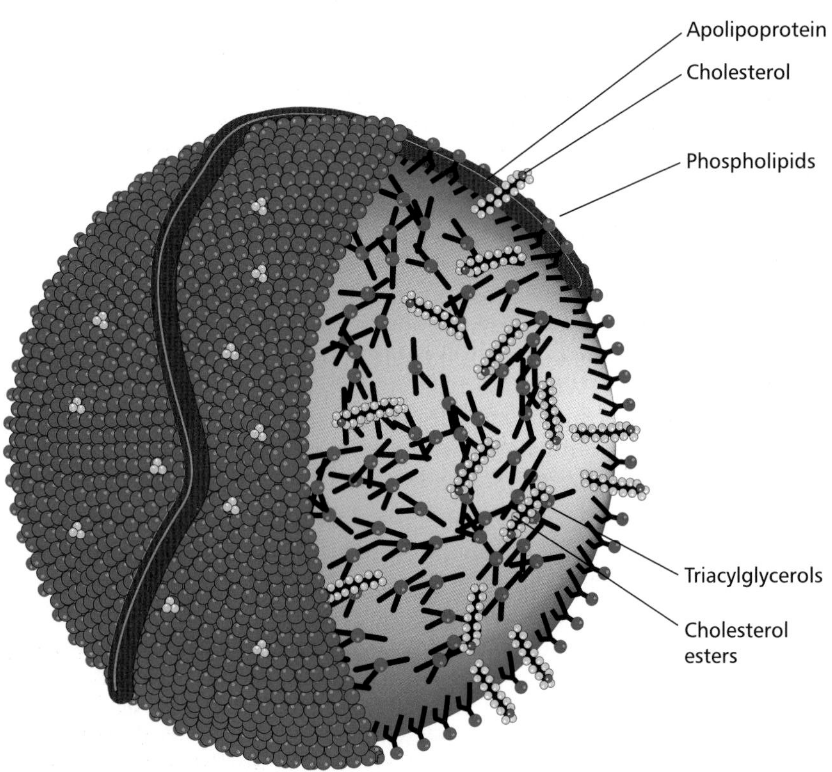

Figure 14.14 Schematic of a generalized lipoprotein particle.

	Chylomicrons	VLDL	LDL	HDL
Density / g cm^{-3}	0.93	<1.006	1.019-1.063	1.063-1.21
Protein / %	2	9	20	45
Phospholipid / %	8	20	20	25
Free cholesterol / %	1	7	10	5
Triacylglycerols / %	85	55	10	8
Cholesteryl esters / %	2	10	35	15
Apolipoproteins	B-48, C, E	A, B-100, C, E	B-100	A, C, D, E

Table 14.1 Composition of lipoprotein particles

in the form of very low density lipoproteins (VLDL). These contain the apolipoprotein, B-100, but the major lipids are triacylglycerols, with less than 20% free cholesterol and cholesteryl esters (*Table 14.1*). A primary determinant of the amount of VLDL secreted is the amount of free fatty acids entering the liver. Further, if saturated fatty acids predominate, the VLDL particles are smaller but more numerous than if polyunsaturated fatty acids are in excess. A high carbohydrate diet also substantially increases the concentration of plasma VLDL particles. Newly secreted VLDL particles undergo a series of changes in the plasma. They acquire apoC and apoE proteins from high density lipoproteins (HDL, *see below*). The catalyzed hydrolysis of triacylglycerols by lipoprotein lipase of the endothelial cells of capillaries allows the fatty acids and glycerol to be taken up by tissues. The reduced triacylglycerol content increases the density of the VLDL so that they become low density lipoproteins (LDL), and these particles are the principal carriers of cholesterol in the plasma. Low density lipoprotein particles serve as the major source of cholesterol for most of the tissues. Although most cells can synthesize cholesterol, the bulk of synthesis occurs in the liver and intestinal enterocytes.

The concentration of LDL in the plasma correlates positively with the incidence of coronary heart disease. Hence LDL is often referred to as 'bad' cholesterol. However, it is only 'bad' in excess, and when combined with other risk factors. High density lipoproteins are secreted by the liver, but they are also formed by modifications to chylomicrons and VLDL. High density lipoproteins (*Table 14.1*) can pick up cholesterol from tissues, essentially the opposite function to that of LDL. The particles probably acquire cholesterol from the cell surface membranes and convert it enzymatically to cholesteryl esters. Consequently HDL may be considered 'good' cholesterol.

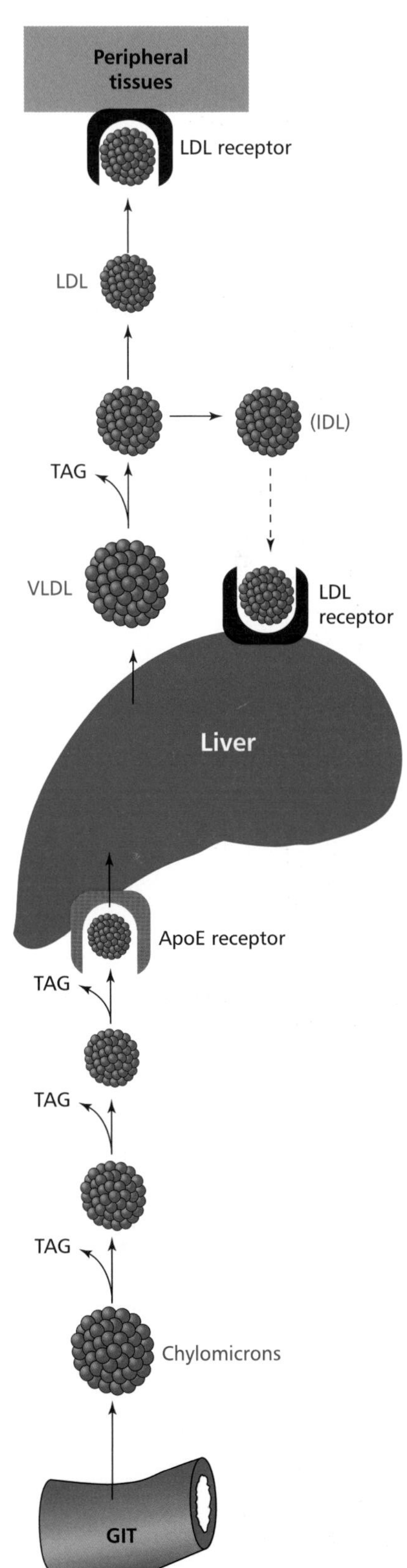

Figure 14.15 Schematic illustrating the transport of cholesterol. IDL, intermediate density lipoprotein particle; LDL, low density lipoprotein particle; VLDL, very low density lipoprotein particle; TAG, triacylglycerols. See text for details.

Margin Note 14.7 Tangier disease

The removal of cholesterol from cells uses a transporter protein that hydrolyzes ATP in an active transport process. The rare autosomal recessive disease called Tangier disease is associated with mutations in the gene encoding this transporter, and leads to an accumulation of cholesterol esters in the tissues and an almost complete absence of cholesterol in HDL particles. The name 'Tangier' comes from an island off the coast of Virginia, USA, where inbreeding amongst the families that lived there in isolation made it relatively common. The outlook for sufferers of this mutation is premature coronary heart disease.

Measuring serum cholesterol

It is necessary to know the concentration of LDL cholesterol in serum (*Table 14.2*) to assess risks to a person, diagnose an ailment and monitor the progress of a patient. However, until recently this could only be determined by a rather complicated laboratory technique not available for routine clinical use. An empirical formula (the Friedewald Formula) was used to derive the LDL cholesterol concentration from the HDL cholesterol and the triacylglycerol levels. However, this formula is inaccurate and sometime fails completely, such as when serum triacylglycerol concentrations are high. A newer test is the LDL Direct method that isolates LDL cholesterol using an immunological assay. It is more accurate, does not require the patient to fast and is not affected by high concentrations of triacylglycerols in the serum.

Lipid component	Concentration
Total cholesterol / mmol dm^{-3}	<5.2
LDL cholesterol / mmol dm^{-3}	<3.6
HDL cholesterol / mmol dm^{-3}	
Female	>1.2
Male	>0.9
Triacylglycerols / mmol dm^{-3}	0.4 – 1.8

Table 14.2 Desirable concentrations of cholesterol, lipoproteins and triacylglycerols in adults

Familial hypercholesterolemia

The circulating LDL particles are recognized by protein receptors on the surfaces of liver cells and removed from the plasma by endocytosis into the cells. Here their various components are metabolized, stored or recycled. The disease, **familial hypercholesterolemia**, is an autosomal dominant condition that affects heterozygotes, with a frequency of one in 500. It is associated with a defective receptor on the liver cells. This disease has been intensively investigated and a great deal is known about it. For the sufferers, the problems are whitish-yellow deposits largely of cholesterol on the tendons called **xanthomas**, and opaque fatty deposits around the periphery of the cornea called **corneal arcuses** and an onset of coronary heart disease before the age of 10 years in homozygotes. The blood cholesterol concentrations in homozygotes are between 15 and 30 mmol dm^{-3} compared with normal values of about 5.0.

14.14 MYOCARDIAL INFARCTION

An **infarction** is the death of a section of tissue because its blood supply has been cut off; an **infarct** is the segment of tissue affected. In general, if one artery is blocked neighboring arteries with communicating branches can compensate and tissue death is limited. Infarction occurs in places where small arteries do not communicate with one another, such as in the kidney, or where all the arteries together supply only enough blood for the whole organ, such as in the brain; or where alternative arteries are also blocked and cannot take over. The latter is what happens to the coronary vessels in many middle-aged hearts, particularly in the developed world. Thus myocardial infarction almost always occurs in patients with atheroma in the coronary arteries resulting from sudden coronary thrombosis, usually at the site of a fissure or rupture of the surface of an atheromatous plaque. There may be hemorrhage into the plaque with local coronary spasms. Irreparable damage can begin after only 20 min of occlusion. After about six h, the site of infarction of the myocardium is pale and swollen and after 24 h necrotic tissue appears deep red owing to the hemorrhage. Subsequently an inflammatory reaction develops and the infarcted tissue turns gray in color.

Myocardial infarction is the commonest cause of death in the UK but surprisingly was hardly known before 1910. Patients present with severe intermittent chest pain that is similar in character to the angina that can occur on exertion, but usually occurring at rest and lasting several hours. Sometimes, however, the pain is less severe and may be mistaken for indigestion. The episodes of pain may become more frequent, but about 20% of patients have no pain. If there is pain, the onset is usually, but not always, sudden. The patient may feel restless and there is often sweating, nausea and vomiting. The most recognizable pain is in the middle of the chest that may spread to the back, jaw or left arm. The condition, once recognized, is a medical emergency. Half of the deaths occur in the first three to four h after the symptoms begin, so the sooner treatment begins the better the chances of survival.

PLASMA ENZYMES IN MYOCARDIAL INFARCTION

The diagnosis of myocardial infarction is usually made on the basis of the clinical symptoms and ECG findings, and is confirmed by the characteristic changes in plasma enzyme activities (*Box 14.4*). The enzyme activities that are of the greatest value are creatine kinase (CK), lactate dehydrogenase (LDH) and aspartate transaminase (AST, previously known as GOT, glutamate oxaloacetate transaminase). Plasma enzyme activities are increased in about 95% of cases of myocardial infarction and sometimes increase to high levels. The degree of increase gives a rough estimation of the size of the infarct but is of little prognostic value. A second and subsequent rise after their return to normal may indicate extension of the damage. All tend to show normal serum activities until at least four h after the onset of chest pain due to the infarction and so blood samples should not be taken until after this time. If the initial serum CK activity is approximately normal, a second blood sample should be taken four to six h later. An increase in plasma CK activity supports the diagnosis of an infarction. The sequence of changes in plasma AST activity after a myocardial infarction are similar to those for CK but the increases are significantly less.

TREATMENT

Usually the patient is given an aspirin to chew, which should improve the chances of survival by reducing the clot in a coronary artery. A β-blocker may also be given to slow the heart rate and reduce its workload. Oxygen may be given through a facemask to deliver more oxygen to the heart. Blood clots in an artery can often be cleared by intravenous thrombolytic therapy (*Box 14.3* and *Margin Note 14.6*). The indication for thrombolytic treatment is usually based on the clinical presentation and the ECG picture rather than on the activities of plasma enzymes. Treatment must be given within 6 h of the start of the heart attack to be effective. After 6 h it is likely that some of the damage will be permanent and the patient could be compromized and some may die. Most patients who survive for a few days after the attack can expect a full recovery but about 10% will die within a year. The majority of deaths occur in the next three to four months in patients who continue to have angina, arrhythmias and subsequent heart failure.

CORONARY BYPASS SURGERY

In individuals who have angina and coronary arterial disease that is not too widespread, coronary bypass surgery is a possible treatment that improves exercise tolerance, reduces symptoms and decreases the number of drugs that are needed. Bypass surgery involves grafting arteries or veins taken from the leg to take blood from the aorta past the obstructed region, replacing the role of the coronary arteries in supplying blood to the heart muscle. Such a graft often works well for up to 10 years or more.

BOX 14.4 Diagnostic value of various plasma markers in heart disease

The activities of a number of enzymes in blood samples are used in diagnosing and monitoring some types of heart damage and some other diseases (*Figure 14.16*). The activities of these enzymes in a patient depend on the rate of their release into the plasma from damaged cells and on the extent of cell damage. Other factors that need to be considered include the rate of cell proliferation and the rate of clearance of enzymes from the circulation. The rate at which damage is occurring is also important. Thus acute cell damage in viral hepatitis may lead to high activities in plasma but these will fall as the condition resolves. In contrast, in advanced cirrhosis of the liver, the rate of cell damage may be low and consequently the plasma enzyme levels may only be a little above normal.

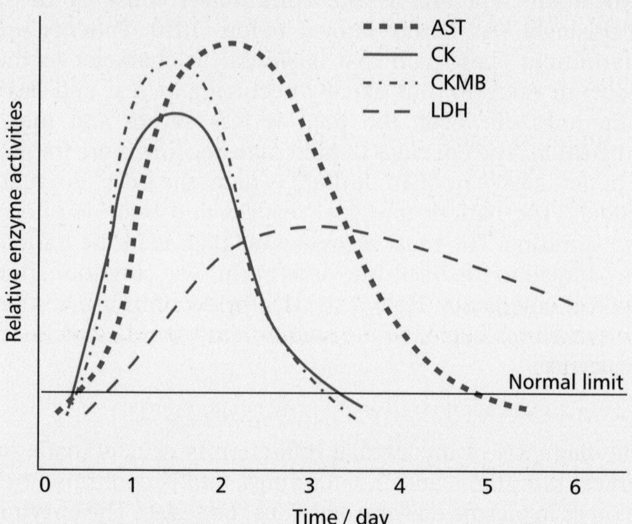

Figure 14.16 Increases in the activities of blood enzyme markers for a myocardial infarction. See text for details.

Aspartate transaminase (AST, formerly glutamate oxaloacetate transaminase, GOT) is present in high concentrations in cardiac and skeletal muscle tissues, liver, kidney and erythrocytes. Damage to any of these tissues will increase the plasma level. In myocardial infarction there may be a 10- to 100-fold increase on the upper reference limit. The level will also increase after cardiac surgery.

Lactate dehydrogenase (LDH) is widely distributed in the tissues of the body and is a relatively nonspecific marker of tissue damage. The plasma activity may increase some five-fold above the upper reference limit in myocardial infarction. There are five isoenzymes of LDH (LDH_1–LDH_5) and estimation of the relative levels of their activities may help to identify which tissue is damaged. Thus increase in the activities of LDH_1 and LDH_2 occurs predominantly after myocardial infarction, although the levels of all the isoenzymes may be increased. In contrast the level of LDH_5 is characteristically elevated after damage to liver or muscle tissue.

Creatine kinase (CK) is abundant in the cells of cardiac and skeletal muscle and in brain. Consequently, a marked rise in its plasma activity occurs after myocardial infarction but, since the enzyme is present in so many tissues, this by itself may not be all that helpful. However, the enzyme consists of combinations of two distinct subunits called M and B respectively, which combine to form dimers characteristic of the tissue in which they are found. Thus the isoenzyme, MM is predominant in cardiac and skeletal muscle, whereas the BB isoenzyme is characteristic of brain and smooth muscle. The third isoenzyme, MB, accounts for about 35% of the cardiac muscle activity but less than 5% of that in skeletal muscle. The concentration of this isoenzyme in plasma is always high after myocardial infarction.

The triple marker blood test more easily distinguishes between cardiac and skeletal or other muscle damage. The triple marker consists of three components, which are myoglobin, troponin I and CKMB. All three are cardiac specific and therefore more reliable. Myoglobin is released from damaged cardiac muscle and peaks by six h. Troponin I is evident after six h and peaks at 12 to 16 h and remains in the system for two to four weeks. Creatine kinase MB is evident after 12 h and is the cardiac isoform of CK and therefore more accurate than total CK activity. The triple marker blood test is especially valuable when there are no or nonspecific ECG changes and can assist in clinical decision making.

14.15 PERICARDIAL DISEASES

The pericardium can become inflamed producing acute or chronic **pericarditis**. However, the pericardium is not absolutely essential to life and can be removed without significantly affecting the functioning of the heart.

Acute pericarditis has a sudden and often painful onset. There are also characteristic heart sounds. Symptoms include fever and chest pain that typically extends to the left shoulder and down the left arm. The inflammation causes fluid and blood components, such as fibrin, erythrocytes and leukocytes, to pour into the pericardial space. The inflammation may be caused by a viral infection, in which case the condition may be painful but short-lived and have no lasting effects, or result from a number of other causes, for example cancer, heart attack, AIDS, kidney failure, heart surgery and the side effects of certain drugs, some of which are life-threatening. The treatment for acute pericarditis is to hospitalize the patient and treat with antiinflammatory drugs, such as aspirin or ibuprofen that also reduce the pain. Further treatment depends on the underlying cause. Individuals with cancer that has invaded the pericardium rarely survive longer than 12–18 months.

The chronic form of the disease develops gradually and is long-lasting. Usually the cause is unknown, but cancer and a reduced thyroid function have been implicated.

14.16 DISORDERS OF ARTERIES AND VEINS

A number of clinical disorders are associated with arteries and veins. These include peripheral arterial disease, arterial aneurysms and dissections, strokes, varicose veins and deep vein thromboses.

PERIPHERAL ARTERIAL DISEASE

Peripheral arterial disease can affect the abdominal aorta and its major branches including the arteries to the legs. Obstruction can be sudden or gradual. Most patients with peripheral arterial disease have atherosclerosis that gradually narrows the arteries. Partial occlusion can also result from a blood clot resulting in a sudden decrease in the oxygen supply. A sudden, complete obstruction normally results from a clot lodging in a narrowed artery. Emergency surgery or the use of thrombolytic drugs may be needed to remove the obstruction.

ANEURYSMS

An **aneurysm** is a round or tube-like bulge that usually develops in weak areas of an arterial wall. There are many reasons for the development of aneurysms, but high blood pressure and cigarette smoking increase the risks. Also, a blood clot may form in the aneurysm. If an aneurysm occurs in the aorta, an aortic aneurysm, rupture, hemorrhage and separation of the layers of the wall (called **dissection**, *see below*) can occur with disastrous results. Such conditions can be immediately fatal although most take years to develop. Aortic aneurysm can develop anywhere along its length but over 70% of them occur in the segment that run through the abdomen. An individual with abdominal aortic aneurysm usually becomes aware of a pulsing sensation in the abdomen, with a deep penetrating pain mainly in the back. The aneurysm may rupture, with severe internal bleeding, and the patient will typically go into severe shock (*Section 14.17*). Aortic rupture is frequently fatal. The treatment of ruptured aneurysms involves surgical repair, which is extremely risky. However, many patients are diagnosed early during a routine examination, and then surgery to insert a synthetic graft can be used to repair the aneurysm with a good chance of success.

AORTIC DISSECTION

In an aortic dissection the inner lining of the vessel wall tears allowing blood to surge through the tear, splitting the middle layer and creating a new channel. The condition usually results from a deterioration of the arterial wall caused, in most cases, by high blood pressure, but it can also be a consequence of certain hereditary conditions. The clinical picture includes stroke, heart attack, sudden abdominal pain, nerve damage and an inability to move a limb. Treatment in intensive care is required. Drugs are given to reduce the heart rate and lower the blood pressure. Thereafter a decision needs to be taken as to whether to carry out surgery to replace the portion of damaged blood vessel with a synthetic graft. Untreated, about 75% of patients die within two weeks, but following treatment several years of life are possible in the majority of cases even though the death rate from surgery is 15% or more.

STROKE

If the blood supply to the brain is disrupted for any length of time the brain cells can be permanently damaged or die due to the lack of oxygen and this is called a **cerebrovascular accident** (CVA) or **stroke** (*Figure 14.17*). Brain cells are also damaged if bleeding into the brain occurs. Therefore strokes can be either ischemic or hemorrhagic. In ischemic stroke the blood supply to part of the brain is cut off either because of atherosclerosis or a clot blocking a blood vessel. In hemorrhagic stroke, a blood vessel bursts, preventing normal flow of blood and allowing it to leak into an area of the brain and destroy it.

In the developed world, strokes are the most common cause of disabling neurological damage resulting, typically, in a loss of speech and/or loss of motor function on one side of the body. High blood pressure and atherosclerosis are the major risk factors. The incidence is falling because the importance of controlling high blood pressure and dealing with inappropriate blood cholesterol levels has been recognized. Clinicians can usually diagnose a stroke from the clinical history of events and a physical examination. Computerized tomography scans and MRI (*Chapter 18*) are used for differential diagnoses.

Many people who have had a stroke recover some or all of their normal functions but others may be mentally and physically devastated, unable to speak or move normally. About 20% of people who have had a stroke die in hospital and the older the patient the greater the risk of this happening. Since each area of the brain is served by specific blood vessels, the area where the cerebrovascular accident occurs decides which part of the body becomes disabled. The loss of function is greatest just after the stroke and some function may return subsequently. This is because although some brain cells die, others may recover and take over a given function at least partially. The immediate treatment is to give oxygen to try to prevent further damage. Anticoagulants may be given if the stroke is ischemic but these are of little use

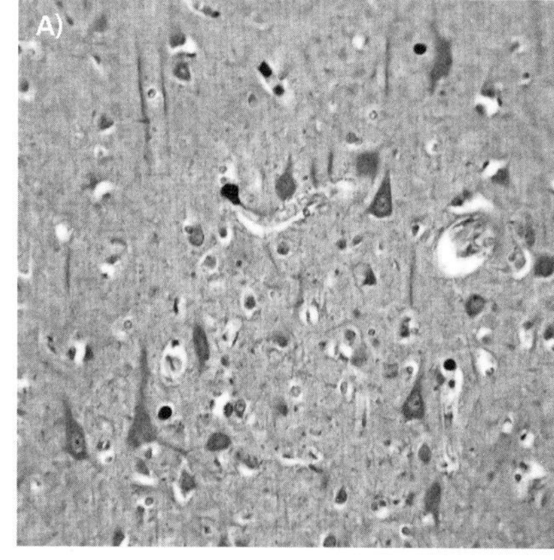

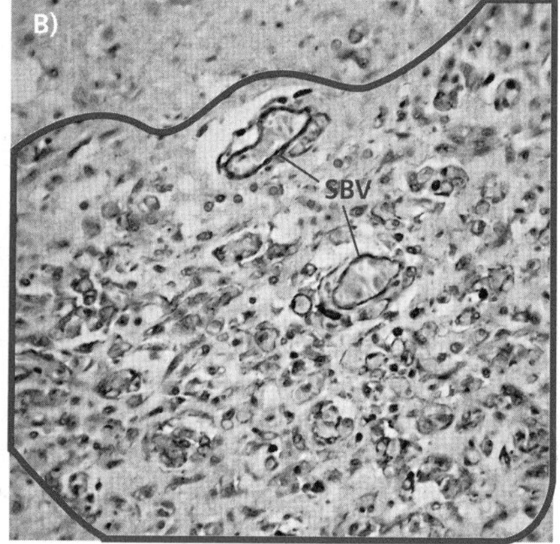

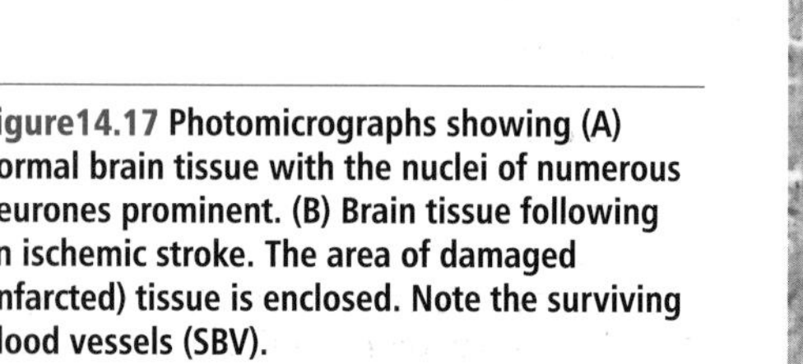

Figure 14.17 Photomicrographs showing (A) normal brain tissue with the nuclei of numerous neurones prominent. (B) Brain tissue following an ischemic stroke. The area of damaged (infarcted) tissue is enclosed. Note the surviving blood vessels (SBV).

once the stroke is completed. Surgery is not normally of much use since the brain cells are already dead. For the survivors, a long period of rehabilitation may be required.

VARICOSE VEINS

Varicose veins are enlarged superficial veins, which occur particularly in the legs (*Figure 14.18*). The cause(s) of this condition is(are) not precisely known, but is probably due to a weakness in the walls of the veins which may be inherited. Over time, the veins lose their elasticity and they stretch and become wider. They may take on a tortuous, snake-like appearance, and cause bulges in the skin over them. The widening causes the valve cusps to separate (*Figure 14.19*) and as a result the veins fill with blood when the person stands and the veins bulge even more. As well as being unsightly, the legs feel tired and the veins ache, and after removing socks or stockings the legs may itch. There may be other complications and minor injuries may cause an ulcer that fails to heal. Varicose veins are common during pregnancy but these usually improve during the two to three weeks following delivery. Hypertension and obesity may have parts to play in the propensity for varicose veins.

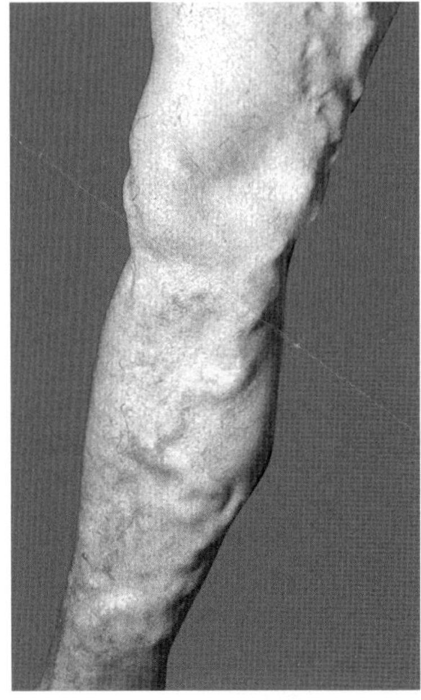

Figure 14.18 Picture showing external appearance of varicose veins of the long saphenous system of the thigh and calf.
Courtesy of J. Guy, The Royal College of Surgeons of Edinburgh, Scotland.

Treatment for varicose veins

Varicose veins cannot be cured, but the symptoms may be relieved in various ways. Elevating the legs on a stool when sitting down and wearing elasticated stockings compress the veins and prevent them hurting, but these are not cures. They may be treated surgically by either stripping or by injection therapy. Stripping involves removing as many of the varicose veins as possible. The superficial veins play a less important role than the deep veins in returning blood to the heart and their removal does not impair the circulation significantly. Two incisions are made; one in the groin and one at the ankle while the patient is under general anesthetic. The saphenous vein (*Figure 14.13*) is then removed by threading a flexible wire through the vein, which is pulled to remove the entire vein. However, surgery does not remove the tendency to develop new varicose veins. Injection therapy involves sealing the veins by injecting an irritant solution that causes a thrombus to form so that no blood flows. Healing of the thrombus causes scar tissue, which can block the vein. However, the thrombus may dissolve allowing the vein to re-open. Injection therapy was popular but has fallen into disrepute, probably because of poor techniques with resulting complications. The more modern techniques, if carried out carefully, seem to be successful.

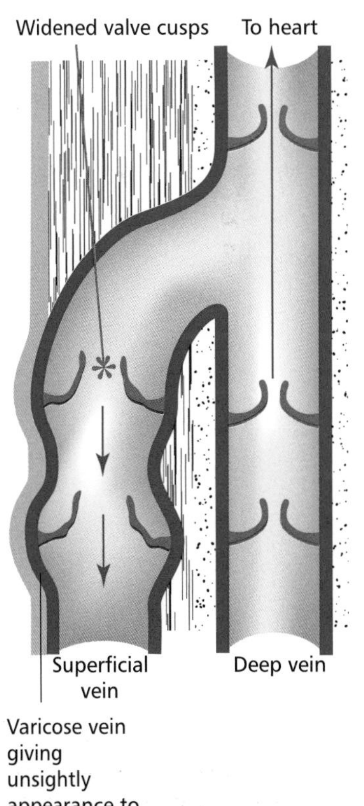

Figure 14.19 Schematic to show the widening of the valve cusps in varicose veins.

DEEP VEIN THROMBOSIS

Deep vein thrombosis is blood clotting in the deep veins. Like varicose veins, they primarily affect the legs. They are potentially dangerous. All or part of the clot or thrombus can break loose and lodge in a narrow artery in the lung obstructing blood flow, causing a pulmonary embolism. In serious cases this can result in blockage of all or nearly all of the blood travelling from the right side of the heart to the lungs rapidly causing death. Such serious consequences are not common but it is impossible to predict what will happen once a thrombus has formed. It is believed that the possible causes of this condition are an increased tendency of the blood to clot, which can happen with some cancers (*Chapter 17*) and very occasionally with oral contraceptives; slowing of the blood movement in the veins, as may also occur in prolonged bed rest or, sitting on a long flight and some types of injury or major surgery. The condition is difficult to diagnose until the thrombus moves. Deep vein thrombosis may be prevented to some extent by flexing and extending the ankles from time to time and by wearing elastic stockings. Anticoagulant therapy may also be appropriate in some cases.

14.17 CLINICAL PROBLEMS ASSOCIATED WITH BLOOD PRESSURE

The blood pressure of any individual varies with their age, sex, physical activity and emotional state. For example, the normal upper limits at age 20 are 150/90 mmHg, which can increase to 160/95 and 170/95 at 40 and 60 years of age. If these pressures are found consistently then they are abnormally high. Blood pressure also varies with ethnic origin. American blacks, for example, are much more likely to have high blood pressure than whites.

The three major factors that determine the blood pressure are the amount of blood pumped by the heart, the volume of blood in the blood vessels and the elasticity of the blood vessels. Thus the more blood pumped by the heart, the higher the pressure. If the heart beats more slowly or its contractions are weakened, such as may happen after myocardial infarction (*Section 14.14*), then less blood will be pumped. Equally, a rapid heartbeat can result in inefficient pumping. The greater the volume of blood in circulation, the more likely the blood pressure is to be high. Conversely, a loss of blood by bleeding or as a result of dehydration will have the opposite effect. The smaller the capacity of the blood vessels, the higher will be the resulting blood pressure. Consequently, a dilation of the blood vessels will lower the blood pressure.

Sensors in the neck and chest continually monitor the blood pressure and trigger physiological changes if the blood pressure changes (*Chapter 8*). The actions taken might be to modify and strengthen the heartbeat, to regulate the kidneys to alter the amount of water excreted which, in turn, changes the volume of blood circulating or to constrict or dilate the blood vessels. However, these compensatory mechanisms have limits, for example if too much blood is lost as a result of bleeding there is little that can be done. Action must be taken to stop the bleeding and to transfuse blood or fluid to make up the volume. Furthermore, these compensatory mechanisms may themselves fail in certain conditions.

LOW BLOOD PRESSURE (HYPOTENSION)

Clearly the heart must pump hard enough so that the blood pressure is maintained. Pressures below 100/60 mmHg at any age are abnormally low. If an individual's blood pressure is too low it can lead to dizziness and fainting. Fainting or **syncope** is the result of a temporary inadequacy of oxygen and nutrients and is usually associated with a temporary decrease in blood flow. This can happen in people with an abnormal heart rhythm when they suddenly begin to exercise or the heart rate is too slow. However, there are many other possible causes of fainting including anemia, hypoglycemia, **hypocapnia** (a lower than usual concentration of CO_2 in the blood) or hyperventilation. The latter may be caused by anxiety. Usually lying flat is all that is needed for the individual to regain consciousness but checks should be performed to eliminate more serious conditions and this may mean, for example, an ECG examination.

SHOCK

If the blood pressure falls too low to sustain life, the body is said to have gone into **shock**. This is more severe and prolonged than in fainting, since if body cells are deprived of oxygen and nutrients for any length of time then they quickly become irreversibly damaged and die. Shock may result from low blood volume, inadequate pumping by the heart or excessive vasodilation as can occur in extreme allergic reactions (*Chapter 5*). Low blood volume may be the result of bleeding following serious trauma. Inadequate pumping by the heart may be the result of a heart attack, pulmonary embolism, the failure of a heart valve or an irregular heartbeat or drug toxicity. Head injuries, liver failure, poisoning, severe bacterial infections or drug reactions may all lead to excessive blood vessel dilation.

Unless treated promptly, shock is usually fatal. When shock results from a sudden loss of blood for example in an accident or a hemorrhage the first person on the scene should aim to stop the bleeding, keep the victim warm and raise the legs slightly to improve the return of blood to the heart. Emergency personnel may provide mechanically assisted breathing, if it has stopped, and fluid or blood to increase the blood volume. Other treatments will depend on the cause of the shock.

HIGH BLOOD PRESSURE (HYPERTENSION)

Hypertension refers simply to the condition where a person has a blood pressure that is higher than that which is regarded as normal, regardless of its cause. High blood pressure is defined as a systolic pressure at rest of 140 mmHg or greater and a diastolic pressure of 90 mmHg or greater, or both of these. In fact both are usually elevated in hypertension. It does not usually cause symptoms, at least for many years, and often tends to go undetected unless the person's blood pressure happens to be measured for some other reason. Nevertheless, it is sometimes referred to as the 'silent killer' because there tend to be no symptoms until some vital organ is damaged. Mortality and morbidity rise continuously with increasing blood pressure. However, the risk is not linear and rises more steeply at higher pressures.

Initially an abnormal heart sound indicating hypertension may be detected using a stethoscope. A diagnosis of hypertension can be made on the basis of an elevated blood pressure reading of 140/90 mmHg or more when measured several times. A single reading on a given day is unreliable. Obviously the higher the values, the more serious the condition must be considered. It is possible to judge the seriousness of the condition by examining the arterioles at the back of the eye to determine the degree of damage to the retina, as hypertension is known to cause retinopathy. In addition, ECG and echocardiography can detect an enlargement of the heart brought on by the increased workload. Kidney damage may be detected by urine analyses.

The cause of hypertension can be identified in less than 10% of patients. These are usually kidney disease, a hormonal disorder or the use of oral contraceptives. Thus in most cases the primary cause cannot be identified and this form is referred to as essential hypertension. Many factors are probably responsible. For example, in older people the larger arteries lose their flexibility and become stiffer. Consequently when the heart pumps blood they cannot expand and the pressure increases. If the kidneys malfunction such that the urinary output is decreased, more fluid will be added to the system. Obesity, stress, a sedentary lifestyle, excessive amounts of alcohol and too much salt in the diet can also contribute.

Untreated hypertension increases the chances of a person developing heart diseases, such as cardiac failure or myocardial infarction (*Section 14.14*), kidney failure (*Chapter 8*), or a stroke (*see above*). Stopping smoking, reducing weight, salt intake and cholesterol levels reduces the risk. In general, patients do not have to restrict their activities as long as their blood pressure is controlled. Various drugs are available as part of a treatment program. These include diuretics to help the kidneys eliminate water and salt, adrenergic blockers to block the effects of the sympathetic nervous system and ACE inhibitors (*Margin Note 14.3*) which lower the blood pressure by stimulating arterial dilation.

CASE STUDY 14.1

Jim, a 52- year-old man, has suffered from angina over the past 18 months. He has also been suffering from lethargy, constipation and loss of concentration. His serum specimen was analyzed and yielded the following results (reference ranges are given in parentheses).

TSH	>100 mU dm^{-3}	(0.35 – 4.1 mU dm^{-3})
Cholesterol	12.8 mmol dm^{-3}	(<5.2 mmol dm^{-3})
Triacylglycerols	1.4 mmol dm^{-3}	(0.5 – 0.9 mmol dm^{-3})

Question

Explain these results.

CASE STUDY 14.2

Ted is a 60-year-old accountant with a past history of myocardial infarction. He was admitted to the hospital about 3 h after developing acute chest pain, fainting with a heavy fall. The results for his ECG were equivocal. His serum enzymes were measured on admission and after 24 and 48 h and give the following activities:

Enzyme	admission	24 h	48 h	reference range
CK	1160	570	190	<90 U dm^{-3}
AST	90	45	25	<37 U dm^{-3}

Question

Has Ted suffered a myocardial infarction?

CASE STUDY 14.3

Roger, a 37-year-old bank manager, has corneal arcus and xanthomata. He had normal blood pressure and weight and is a nonsmoker. However, he had family history of myocardial infarction, as his father died following one at the age of 40. A blood specimen was taken and analyzed for its lipid content (reference ranges are given in parentheses).

Cholesterol	16.8 mmol dm^{-3}	(<5.2 mmol dm^{-3})
Triacylglycerols	2.1 mmol dm^{-3}	(0.4–1.8 mmol dm^{-3})
LDL Cholesterol	14.2 mmol dm^{-3}	(<3.6 mmol dm^{-3})
HDL Cholesterol	1.3 mmol dm^{-3}	(>1.2 mmol dm^{-3})

Questions

(a) How do you account for this abnormal lipid profile?

(b) How should Roger be treated?

14.18 SUMMARY

The cardiovascular system comprises the heart and the blood vessels that supply the organs and tissues of the body with, for example oxygen and nutrients, and remove CO_2 and other wastes. The heart pumps blood through the pulmonary circulation that oxygenates the blood, and the systemic circulation around the rest of the body. The heart pumps nonstop for millions of beats during a person's lifetime. However, there are a number of conditions that can cause serious clinical problems. These can be investigated by listening to the heart sounds using a stethoscope, by measuring its electrical activities by an ECG, as well as studying its function by echocardiography. The rhythm of the heart beat may be faulty and this may require correction, including the insertion of an artificial pacemaker. The valves may function inappropriately due to calcification or other diseases and may need to be replaced. Clinical problems are also associated with blood vessels. These may become partially or completely blocked because of atherosclerosis. When this happens to the coronary arteries, bypass surgery with a grafted blood vessel may be required. Atherosclerosis, caused by high levels of cholesterol in the blood, leads to inappropriate changes in the plasma lipoproteins. A change of diet and drugs may help this condition. The blood pressure may be excessively high in some individuals. This can be dangerous and lead to heart failure and stroke. Here again, drugs may be used to alleviate the condition. Heart failure needs to be treated promptly otherwise permanent disability or death may occur. The peripheral blood vessels can also become diseased. For example, the veins returning blood to the heart may develop faulty valves leading to varicose veins.

QUESTIONS

1. Which of the following is the odd one out?
   (a) endocardium;
   (b) bundle of His;
   (c) mitral valve;
   (d) pericardium;
   (e) semilunar valve.

2. Hypertension is a risk factor for which of the following?
   (a) asthma;
   (b) cerebral hemorrhage;
   (c) narrowing of the aorta;
   (d) a high blood cholesterol level;
   (e) venous thrombosis.

3. The lesions found in atherosclerosis
   (a) are initiated as a response to damage to the venous endothelium;
   (b) can develop at sites of intact endothelium;
   (c) do not contain smooth muscle cells;
   (d) do not contain macrophages;
   (e) do not contain thrombi.

4. Arrange the two following lists into their most appropriate pairings:

Blood pressure	tricuspid valve
Bundle of His	pacemaker
Coronary arteries	ventricular tachycardia
Pericardial diseases	sphygmomanometer
Palpitations	Purkinje fibers
Right ventricle	tenecteplase
SA node	lowered blood pressure
Thrombolysis	myocardium infarction
Vasodilation	familial hypercholesterolemia
Xanthoma	aspirin, ibuprofen

5. Account for the development of hypertrophic cardiomyopathy in some patients with pheochromocytoma. You may find it helpful to consult *Chapter 7.*

6. Barry, a 55-year-old, was admitted to hospital with severe chest pain, which had been present for the past 30 min. He had a previous history of angina. Which serum markers should be measured for this patient on admission?

7. If a patient is found to have high VLDL levels in the blood, what change in diet could help to lower the concentration?

8. Suggest why chylomicrons are not taken up by LDL receptors.

FURTHER READING

Chalela, JA, Merino, JG and Warach, S (2004) Update on stroke. *Curr. Opin. Neurol.* **17:** 447–451.

Chilton, RJ (2004) Pathophysiology of coronary heart disease: a brief review. *J. Am. Osteopath Assoc.* **104:** S5–S8.

Choy, PC, Siow, YL and Mymin, DOK (2004) Lipids and atherosclerosis. *Biochem. Cell Biol.* **82:** 212–214.

Fox, KF (2005) investigation and management of chest pain. *Heart* **91:** 105–110.

Nicholas, M (2004) Heart failure: pathophysiology, treatment and nursing care. *Nursing Stand.* **19:** 46–51.

Page, RL (2004) Clinical practice. Newly diagnosed atrial fibrillation. *N. Engl. J. Med.* **351:** 2408–2416.

Rullan, F and Sigal, LH (2001) Rheumatic fever. *Curr. Rheumatol. Rep.* **3:** 445–452.

Scott, J (2004) Pathophysiology and biochemistry of cardiovascular disease. *Curr. Opin. Genet. Dev.* **14:** 271–279.

Torney, WT (1996) The diagnosis and management of acute myocardial infarction – a role for biochemical markers? *Ann. Clin. Biochem.* **33:** 477–481.

Yee, CA (2005) Endocarditis: the infected heart. *Nursing Manage.* **36:** 25–30.

Useful web sites:

'Target Heart Diseases' is a free booklet produced by the Association of the British Pharmaceutical Industry (ABPI) copies of which are available from the ABPI, 12 Whitehall, London SW1A 2DY or from their website: www.abpi.org.uk

Cox, D and Dougall, H (2001) What's so difficult about ECGs – a bundle of what? – a series of articles for medical students about understanding ECGs, available on www.studentbmj.com

chapter 15:

GENETIC DISEASES

OBJECTIVES

After studying this chapter you should be able to:

- outline the major features of the cell cycle;
- explain how genetic traits, including genetic diseases, are inherited;
- describe the general effects of single gene mutations;
- list the ways in which single gene disorders are treated;
- list the types of chromosomal mutations or aberrations;
- outline how genetic diseases are screened *in utero* and in neonates.

15.1 INTRODUCTION

Genes are the fundamental units of heredity and encode specific functional products, such as RNA molecules and polypeptides. They are encoded by sequences of bases in DNA molecules and are found at particular positions in chromosomes in the nucleus and also in the relatively small circular DNA molecules in the mitochondria. Only 37 of the approximately 22 000 human genes occur in mitochondria although mutations of these may become clinically significant as described in *Chapter 16*. The genes constitute the blueprint or the set of instructions which affects hereditary characteristics, for example hair and eye color, height and the susceptibility to certain diseases.

When a cell divides the genetic information needs to be replicated accurately so that these instructions pass on to the daughter cells. When changes occur in the base sequence of DNA, either as a result of incorrect replication or from random changes caused by physical or chemical agents, then the instructions become corrupted. This is a **mutation**, and may eventually lead to disease because the cell is unable to make, for example, a particular enzyme, hormone, transporter or structural protein.

Chromosomes have a complex structure (*Figure 15.1*). Each is comprised of a single double-stranded DNA molecule associated with numerous proteins.

Figure 15.1 Scanning electronmicrograph of a chromosome. Courtesy of Dr C.J. Harrison, Christie Hospital, Manchester, UK.

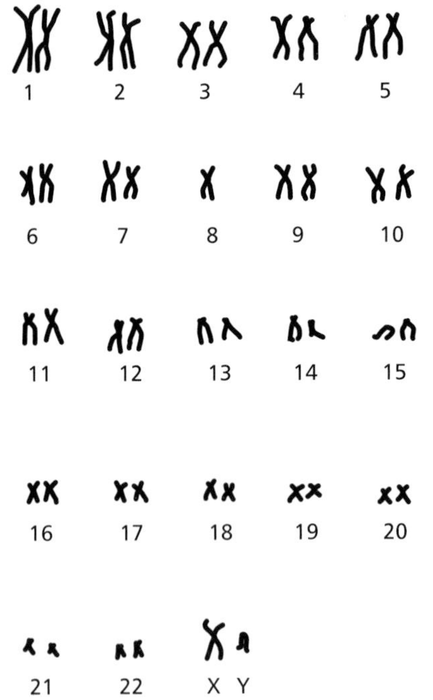

Figure 15.2 **The chromosome complement of a normal human male. That of a normal female is shown in *Figure 1.13*.**

In general, chromosomes occur in matching or homologous pairs, with each member of a pair containing **alleles** or different forms of the same gene, which are found at the same loci (singular locus) in each member of the pairs. Normal human somatic (body) cells contain 23 pairs of chromosomes and are said to be **diploid** (2N). The 46 chromosomes in diploid cells comprise 22 homologous pairs of **autosomes** (nonsex chromosomes) and one pair of sex chromosomes; XX in females and XY in males (*Figure 15.2*; see also *Figure 1.13*). Oocytes and spermatozoa have half the diploid number and are said to be **haploid** (N); oocytes can only contain an X chromosome but sperm can have an X or a Y chromosome.

15.2 GENETICS AND DNA

Genetics, from the Greek *genno* meaning 'to give birth', is that branch of biology concerned with heredity, genes and DNA (*Figure 15.3*), the genetic material. It is also the scientific study of the variations in inherited characteristics, often called traits, and how these are transmitted from one generation to the next. Inherited characteristics include a number of clinical conditions and diseases that are described in other chapters, for example sickle-cell anemia and hemophilia (*Chapter 13*), the muscular dystrophies and cystic fibrosis (*Chapter 16*). Others, such as phenylketonuria and Down syndrome, will be described in this chapter. Genomics is the study of the full complement of bases in the DNA of an organism.

Genes are the stretches of bases in DNA that carry the code for making RNA or proteins. The code is contained in sequences of the four nucleotide bases, adenine, cytosine, guanine and thymine (A, C, G and T, respectively). DNA normally occurs as the famous double helical molecule (*Figure 15.3 (A)*) that consists of two long polymers of alternating phosphates and deoxyribose sugars, linked together by hydrogen bonds between pairs of complementary bases across the center of the double helix, rather like the steps of a spiral staircase. Adenine always pairs with T and C always pairs with G (*Figure 15.3 (B)*).

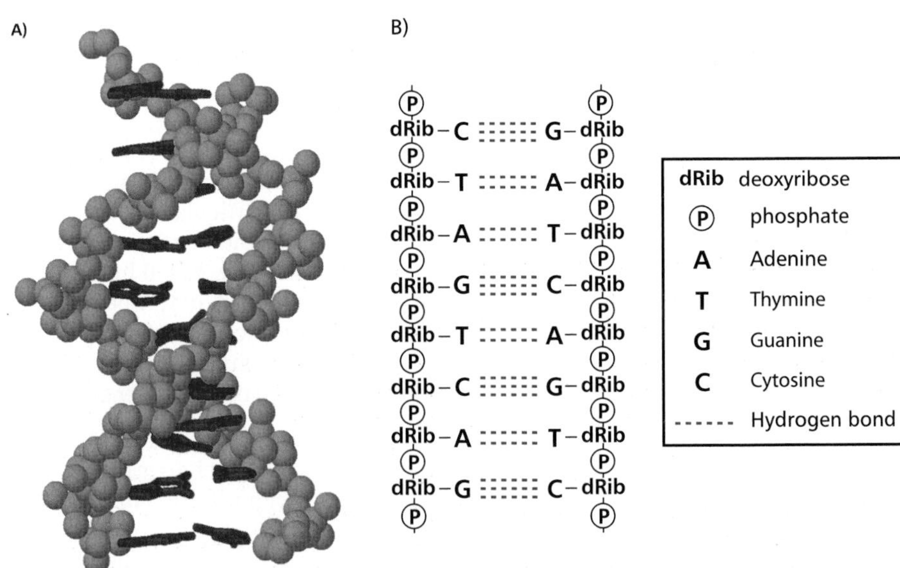

Figure 15.3 **(A) Molecular model of DNA, showing the base pairs in red and the sugar-phosphate backbones in gray. PDB file 1ZFM. (B) A schematic to illustrate the structure of DNA.**

For a gene to be expressed, the nucleotide base code in a gene in the DNA must be copied or transcribed to form an RNA molecule. The strands of the double helical DNA separate and one of them, 'the gene', acts as a template for the synthesis of a complementary new strand (*Figure 15.4 (A)*). However, the new strand is an RNA molecule not DNA. The base pairing rules are similar to the complementary base pairing in DNA but uridine (U) is used in RNA, not T and the sugar ribose, not deoxyribose. A number of RNA molecules may be transcribed from one gene. In some cases, the formation of an RNA molecule is the major event in gene expression. However, in the vast majority of cases, the expression of a gene results in the formation of a protein (*Figure 15.4 (B)*). In these cases, the RNA formed by transcription is a messenger RNA or mRNA molecule. However, DNA is transcribed in the nucleus but the synthesis of proteins takes place in the cytosol. Thus the mRNA molecules are transported out of the nucleus (*Chapter 16*) and go to the ribosomes in the cytosol where their message is translated into a linear sequence of amino acids to make a protein. Each sequence of three bases in the mRNA codes for the addition of one specific amino acid to the growing polypeptide chain of the protein; for example AUG codes for the amino acid methionine, UUU codes for phenylalanine and so on. Each mRNA molecule may be translated numerous times so that many molecules of protein are produced.

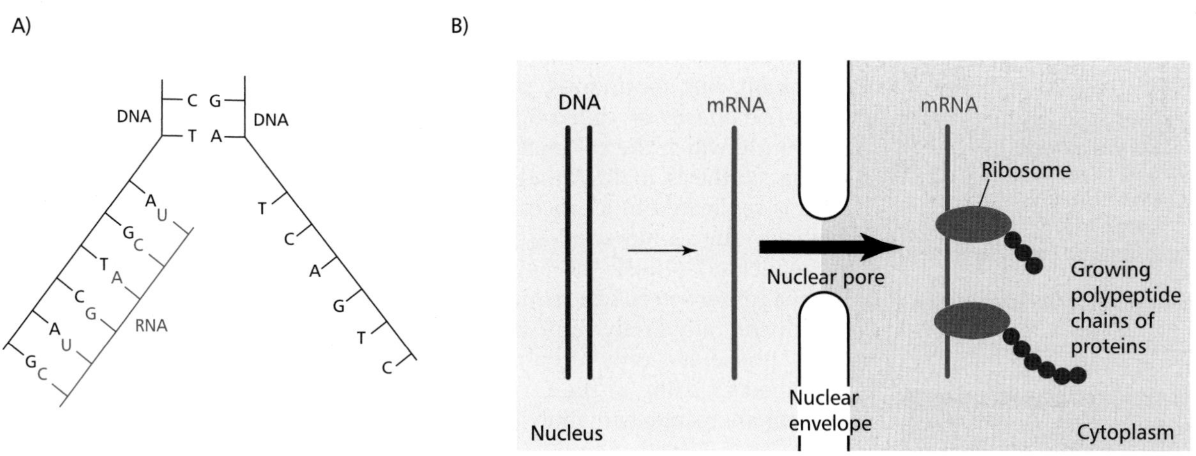

Figure 15.4 (A) A schematic to illustrate the formation of RNA by transcribing a strand of DNA. (B) An overview of protein synthesis as summarized in the main text.

15.3 DNA REPLICATION AND THE CELL CYCLE

When a cell divides the genetic information must be passed on to the two daughter cells. The series of biochemical and morphological events that occur in a population of reproducing cells is called the **cell cycle**. This results in the replication of the genetic material (DNA molecules) and division of the cell into two daughter cells. The replication of DNA involves separating the two DNA strands of the double helix and aligning new bases according to the usual pairing rules; A with T and G with C. The new nucleotides are linked together to form two new strands, each of which is complementary to one of the original (parental) strands. This action forms two new double-stranded DNA molecules, each of which consists of one parental strand and one daughter strand (*Figure 15.5*) and, for this reason, is often called semiconservative replication. This is a very simplified account of an extremely complex process, which is catalyzed by a range of enzymes. This replication is very accurate; it needs to be because the genetic instructions must be retained from generation to generation. Most of the few errors that inevitably occur are corrected by error-detecting enzyme systems in the cell.

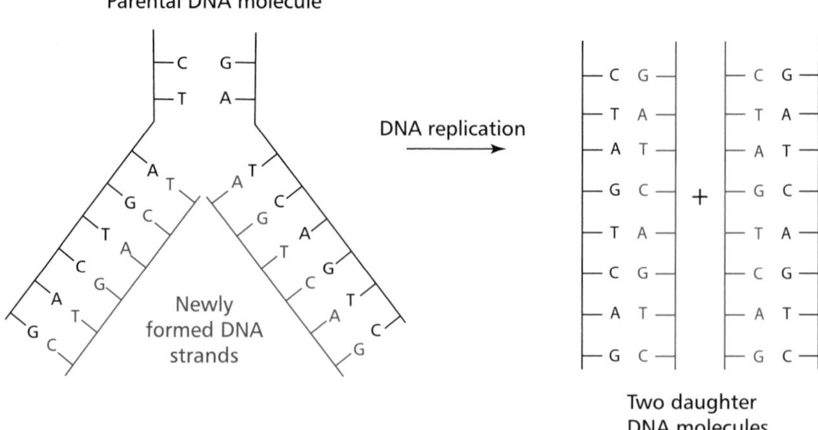

Figure 15.5 A schematic illustration of DNA replication, which emphasizes its semiconservative nature.

The cell cycle is an orderly sequence of events consisting of **interphase**, the period between cell divisions, and mitosis (or meiosis if gamete producing cells are concerned) when the cell divides. Thus interphase, which lasts about 24 h, prepares the cell for division by building up large energy stores and synthesizing new organelles for the daughter cells; a high metabolic rate is typical of cells about to undergo division. The cell cycle can be divided into four main, but continuous, phases that are often drawn as a circle (*Figure 15.6*). The phases are called G_1, S, G_2 and M. The G_1 (for gap) phase lasts about 8 h during which the cell makes a commitment to divide. It is characterized by the synthesis of RNA and protein. In the S phase, S for synthesis, the DNA is replicated in a process lasting approximately 6 h. The G_2 phase is a relatively quiescent period, which typically lasts 4 h, during which organelles are replicated. Mitosis (or meiosis) occurs in the M phase. Following mitosis, the cells enter interphase, which lasts until the S phase of the next cycle. Cells that divide only rarely, for example neurons, are said to be in a stage called the G_0 phase. It is only when they become committed to divide that they are described as being in the G_1 phase of the cycle. Two different types of cell division are recognized, mitosis and meiosis.

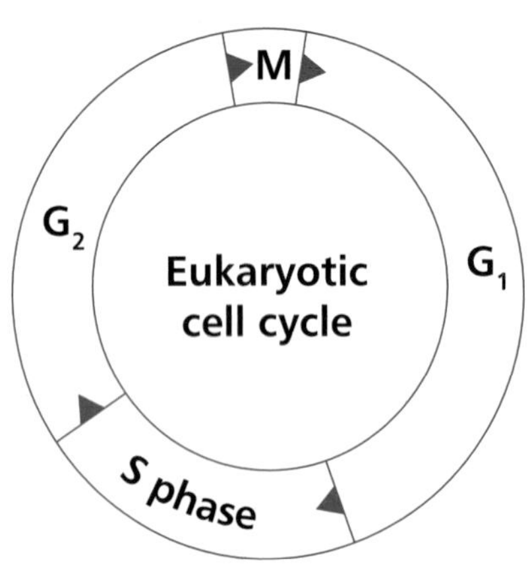

M	Mitosis or meiosis
G_1	Gap 1
S phase	DNA sysnthesis
G_2	Gap 2

Figure 15.6 Overview showing the stages of the cell cycle.

Margin Note 15.1 The cell cycle and malignancy 

The timed sequence of events in the cell cycle is a highly regulated process controlled by numerous different proteins. Mutations in the genes encoding these proteins can perturb the regulation of the cell cycle and cells can die or their growth can become uncontrolled depending on the nature of mutation. The latter may lead to malignancies (*Chapter 17*). A number of chemotherapeutic drugs act by interfering with the cell cycle. For example, some immunosuppressive drugs, such as aminopterin and cyclophosphamide, disrupt the synthesis of DNA during the S phase. Taxol (*Figure 15.8 (A) and (B)*) is a drug that interferes with mitosis by preventing depolymerization of the microtubules that make up the spindle.

MITOSIS AND MEIOSIS

Mitosis is the type of division that occurs during growth and the renewal of tissues. The daughter cells produced have the same diploid complement of chromosomes as the parental cell. Mitosis is a continuous process that lasts about 1 h. For convenience it is divided into four stages; prophase, metaphase, anaphase and telophase (*Figure 15.7*). In interphase the chromosomes occur as dispersed, thread-like material called chromatin, which cannot be seen with a light microscope. In prophase the chromosomes begin to condense to form distinct chromosomes that are visible with microscopy. Since their DNA has been replicated, each chromosome is present as an identical pair of chromosomes, although at this stage each member of the pair is referred to as a sister **chromatid**, which are joined together by centromeres (*Section 15.7*). The centrioles, normally located just outside the nuclear envelope, undergo replication and migrate to opposite poles of the cell. This leads to the microtubules of the cytoskeleton rearranging to form the spindle, which spans the cell from one end to the other. The ends of the spindle are known as poles whereas its middle region is called an equator. During prophase, the cell's nucleolus disappears and prophase concludes with dissipation of the nuclear envelope. In metaphase, the chromosomes migrate to the center of the cell and are arranged around the equator of the spindle, where the centromere of each chromosome (paired chromatids) becomes attached to spindle fibers. The chromatids are drawn apart at the centromere region towards opposite poles of the spindle. During anaphase, the chromatids of each chromosome are pulled apart and each chromatid moves towards the poles of the spindles reaching the ends by telophase. In telophase, the cell undergoes division to two daughter cells by the plasma membrane constricting and cutting across the spindle equator. The spindle breaks down and nuclear envelopes form around each separated grouped of chromatids, now called chromosomes. The nucleoli also become apparent in the new nuclei and the chromosomes return to the nonvisible forms typical of cells in interphase. Hence the parental cell has divided to form two daughter cells that are genetically identical to the parent.

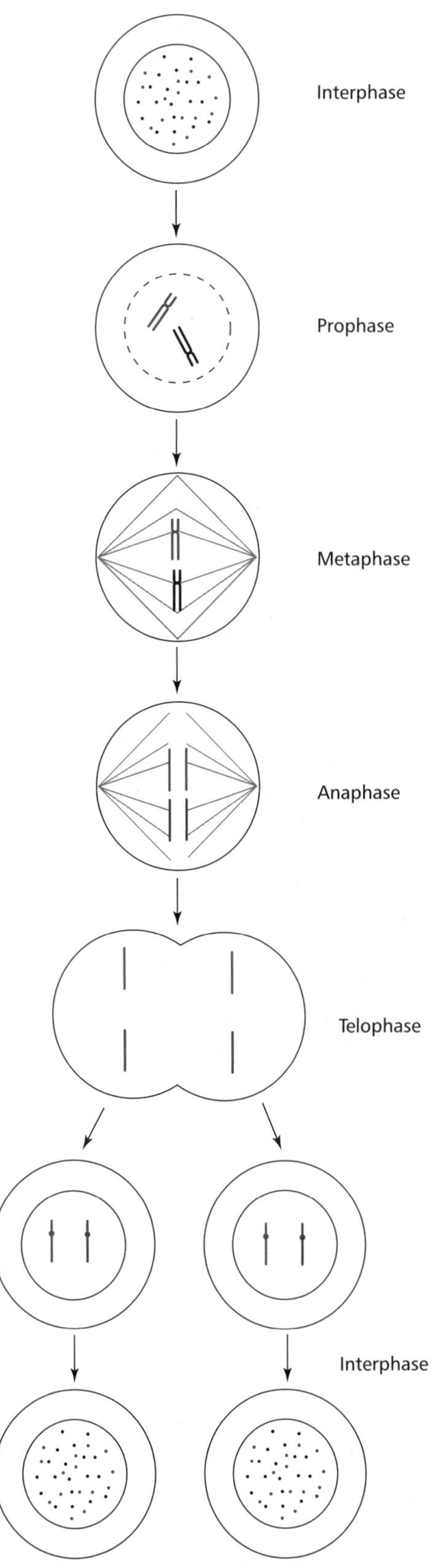

Figure 15.7 An outline of mitosis as described in the main text. For simplicity only two chromosomes, each consisting of two chromatids, are shown.

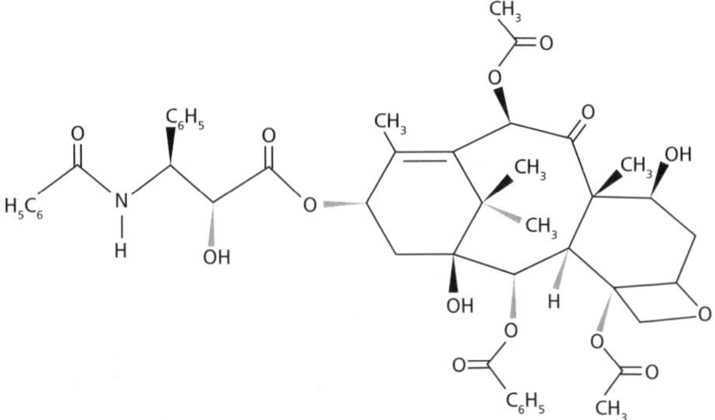

Figure 15.8 (A) The structure of taxol, which is obtained from (B) the leaves of the pacific yew tree (*Taxus brevifolia*).

Meiosis occurs prior to reproduction during the formation of gametes (*Chapter 7*). The parental cell has a diploid number of chromosomes, whereas the daughter cells are now gametes with the corresponding haploid number of chromosomes. During meiosis, the number of chromosomes is halved and the daughter cells receive only one of each type of chromosome and, for this reason, meiosis is sometimes called reduction division.

Meiosis may be thought of as consisting of two separate divisions (*Figure 15.9*). In the first meiotic division (prophase I through to telophase I) the parent cell divides into two cells each of which receives one of each pair of homologous chromosomes. Each of these chromosomes consists of two chromatids. The second meiotic division (prophase II through to telophase II) results in each chromosome being separated into chromatids, with the result that four daughter cells each with a haploid chromosome complement are formed. In prophase I, the chromosomes contract and the nucleolus shrinks in size. Homologous chromosomes lie side by side in pairs, a situation called synapsis. Each member of the pair is bivalent. It is at this stage that genetic recombination or crossing over occurs. While they are paired, the nonsister chromatids, that is one maternal and one paternal chromatid, of a homologous pair are broken at equivalent positions and exchange homologous pieces of material (*Figure 15.9*). The crossed strands of the chromatids formed during recombination are called chiasmata (singular chiasma). Recombination

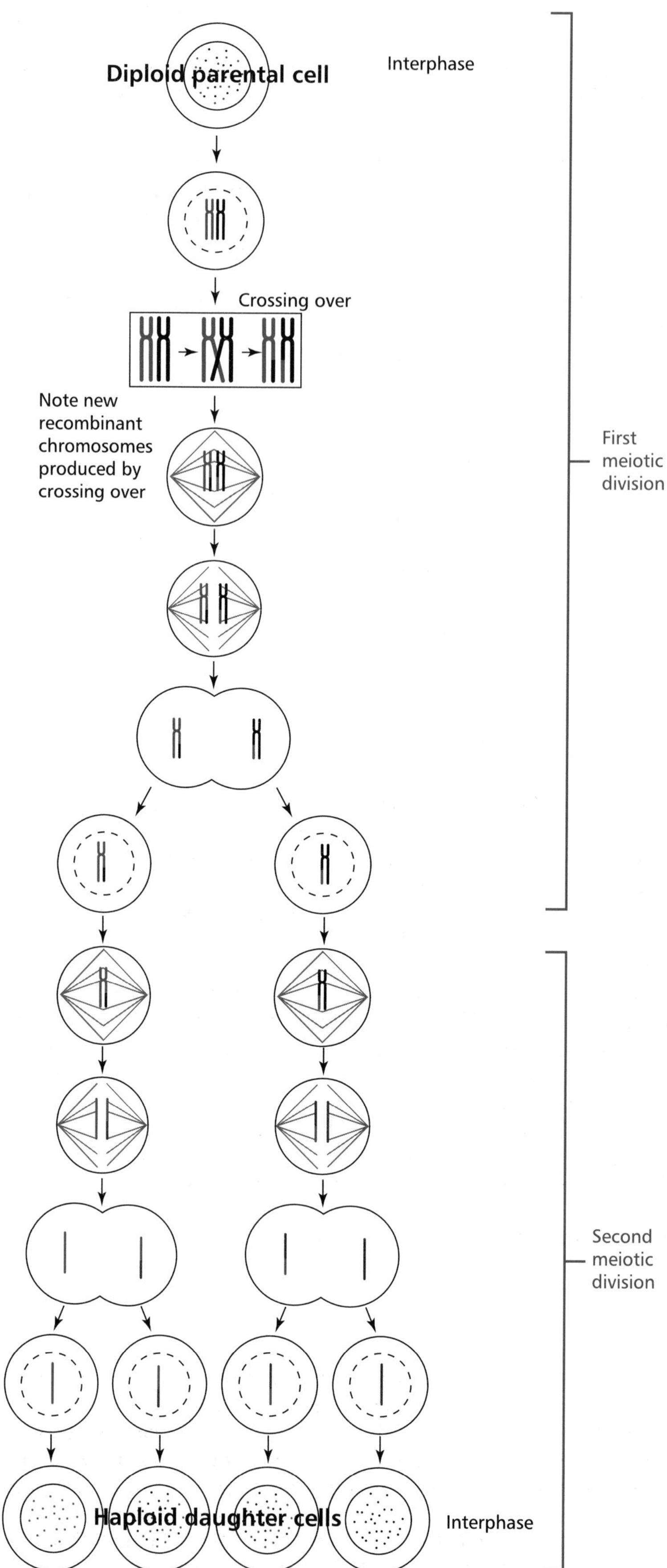

Figure 15.9 An outline of meiosis as described in the main text. For simplicity only a single pair of homologous chromosomes is shown. Each member of the pair consists of two chromatids. The insert illustrates recombination between the two nonsister chromatids.

results in each chromatid acquiring genes or parts of genes from the other and the process leads to the formation of new combinations of genes or parts of genes in chromosomes. Eventually recombinant gametes are formed that differ from their paternal cells in their gene content. Thus crossing over promotes genetic variation.

During metaphase I, the homologous chromosomes or bivalents move to the equator of the spindle. The sister chromatids orientate towards the same pole whereas the homologous chromosomes orientate themselves towards opposite poles. During anaphase I, the homologous chromosomes, each of course consisting of a pair of chromatids, migrate towards opposite poles of the cell. During telophase I, the cell divides as in mitosis to give rise to two daughter cells whose chromosomes each consist of paired chromatids. Following a brief interphase, these cells enter the second meiotic division. In prophase II, the two daughter cells essentially prepare for the second division with formation of a new spindle. In metaphase II, the chromosomes move to the equator of the spindle and the chromatids arrange themselves towards opposite poles and in anaphase II, the chromatids separate from each other and move to opposite poles of the cell. Finally, in telophase II, each cell divides into two daughter cells but these have only the haploid number of chromosomes. Thus the diploid parental cell has produced four haploid daughter cells (sperm or ova (*Chapter 7*)).

Failure of chromosomes to separate at metaphase in mitosis or either of the metaphases in meiosis is called nondisjunction. Nondisjunction can have serious clinical consequences as explained in *Section 15.9*.

15.4 GENOTYPE AND PHENOTYPE

The genetic or hereditary constitution of an individual, which is the whole complement of genes present, forms the **genotype**. The term can also be applied to any particular pair of alleles that an individual possesses at a specific locus on a chromosome. In contrast, the visible or measurable characteristics of an individual constitute the **phenotype**. A phenotype includes biochemical, physiological, morphological and behavioral characteristics or, indeed, any observable biological trait that is apparent throughout life, such as the total physical appearance and constitution of an individual or any specific trait, such as size, weight or eye color and, of course, includes characteristics of clinical importance and the presence of a disease. Some phenotypic traits, for example eye color, are directly observable but others, such as the blood group of a patient (*Chapter 6*), may only become apparent following specific tests. Phenotypic traits do not necessarily occur merely following the expression of the genotype of an individual; some, such as the blood groups, are completely determined by heredity but many others, for example weight and height, result from interactions between the genotype and the environment.

15.5 INHERITANCE AND MUTATIONS

Genes occur as paired alleles. Each corresponding allele is carried by one of a pair of homologous chromosomes. If the two alleles are identical, the individual is **homozygous** for that gene and, if they differ, the individual is said to be **heterozygous**. In the heterozygous state, one allele may be **dominant** over the other which is therefore **recessive**. In this situation, only the characteristic encoded by the dominant trait will be expressed, as would also be the case if the individual was homozygous for both dominant alleles. The recessive trait will only become apparent in a homozygous recessive individual.

Dominant genes are conventionally written as an upper case italic letter, for example *G*, while its recessive counterpart is given the lower case form, *g*. *Figure15.10 (A)* illustrates the normal inheritance pattern first established by Mendel (1822–1884). If one parent is homozygous for an autosomal dominant gene (*GG*) and the other parent is homozygous for the recessive form (*gg*), then all the offspring will be genetically heterozygous (*Gg*) and phenotypically will express the dominant trait. If both parents are heterozygotes (*Figure 15.10 (B)*), then 25% of offspring will be homozygous for the dominant gene (*GG*), 25% homozygous for the recessive gene (*gg*) and the remaining 50% of offspring will be heterozygous (*Gg*).

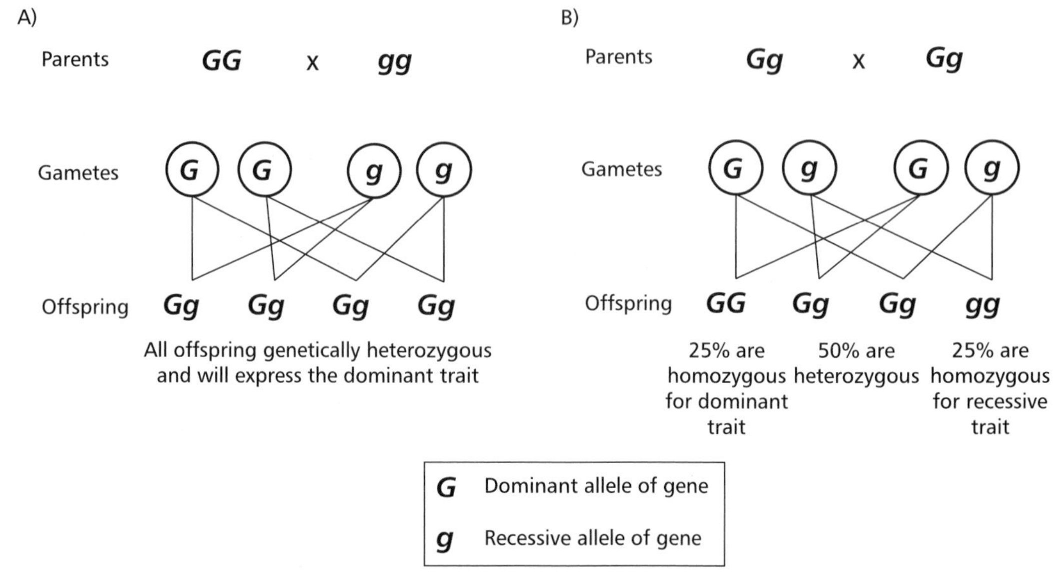

Figure 15.10 The inheritance patterns shown for a single pair of genes between (A) two contrasting homozygous parents and (B) two heterozygous parents.

Margin Note 15.2 The human genome

The genome is the complete sequence of bases in the DNA molecules, which is all of the hereditary material possessed by an individual. The human haploid genome contains about 3 000 000 000 pairs of nucleotides. The total length of this DNA is about a meter and is divided into 23 individual molecules; 22 of which are found in the autosomes and one in the sex chromosome. Mitochondrial DNA contains 37 genes.

Mutations are changes that occur in the genome and can give rise to clinical disorders (*Margin Note 15.2*). Mutations include changes within single genes and changes to whole chromosomes. They may be simple substitutions of one nucleotide for another (point mutations), involve the insertion or deletion of one or more nucleotides within the normal sequence of DNA within a chromosome or even alter the structures of individual chromosomes or the number of chromosomes present. When considering the effects of mutations, it is important to distinguish between a genetic change which occurs in somatic cells and one occurring in gametes. Mutations arising in somatic cells will not be transmitted to future generations although they may represent the first step in the development of cancer (*Chapter 17*). In somatic cells, mutations that produce a recessive autosomal allele are unlikely to have clinical consequences because their expression is masked by the corresponding dominant allele. However, somatic mutations that are dominant or X-linked (*see below*) can have a greater impact because they are likely to be expressed. Similarly, their impact is greater if they arise early in development before undifferentiated cells give rise to differentiated tissues or organs. In adult tissues, the activities of many nonmutant cells often mask mutations in a few other cells. Mutations in gametes or gamete forming tissues are part of the germ line (*Chapter 17*) and are of greater clinical concern because as well as affecting that individual, they will also be transmitted to offspring.

Dominant autosomal mutations *M* will be expressed phenotypically in both the homozygous and heterozygous condition. However, if the mutation is recessive (*m*) then it is not likely to affect an individual unless both chromosomes carry the same mutation. The inheritance patterns shown by dominant/recessive alleles that are associated with a clinical condition follow the Mendelian rules explained above. If only a single member of a homologous pair of chromosomes carries the mutation, it can, however, be passed on to the next generation and the parent is described as a carrier (*Figure 15.11 (A)* and *(B)*). If one parent is homozygous for the normal gene (*MM*) and the other parent heterozygous for the mutated allele (*Mm*) then 50% of the offspring will be homozygous for the normal gene (*MM*) and the other 50% heterozygous (*Mm*) and carriers. None of the children will be affected. If both parents are heterozygotes (*Mm*) then 25% of offspring will be dominant homozygotes (*MM*), 50% of offspring will be heterozygotes and carriers (*Mm*) but the remaining 25% will be recessive homozygotes (*mm*) and express the condition (*Figure 15.11 (B)*).

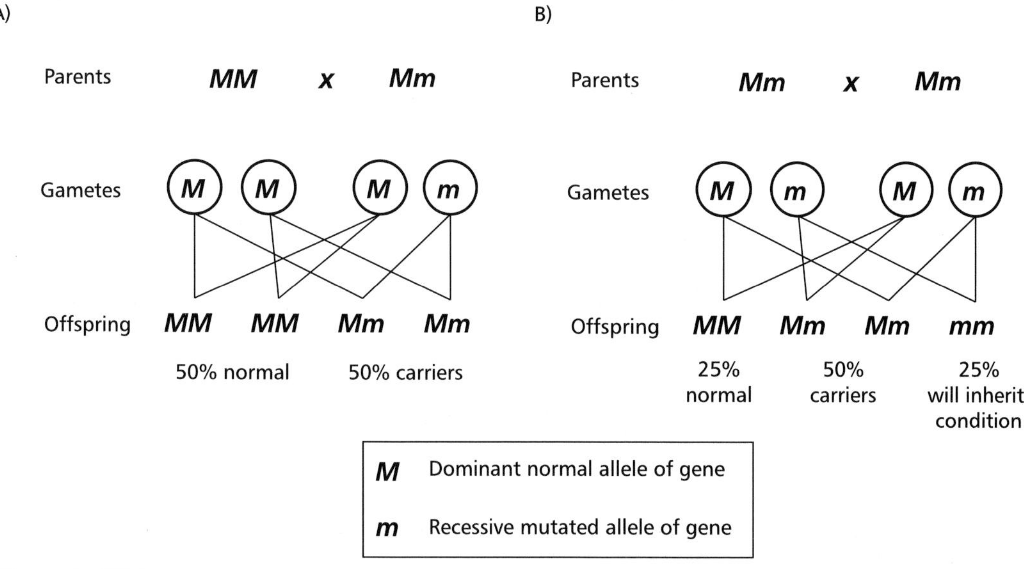

Figure 15.11 The inheritance patterns shown for a recessive allele of a gene between (A) dominant homozygous and heterozygous parents and (B) two heterozygous parents.

Heterozygosity means that autosomal recessive mutations, even one resulting in a lethal allele, may go unnoticed and be maintained in the population for many generations, until the resultant allele has become widespread in the population. The new allele will become evident only when a chance mating brings two copies of it together in the homozygous condition.

SEX-LINKED GENETIC DISEASES

A number of genetic diseases are caused by defective alleles of genes of the sex (X and Y) chromosomes. The X chromosome contains many more genes than the Y, although they do have some genes in common, thus any defective (mutated) gene on the X chromosome is likely to be expressed in males (XY) but be masked in females (XX). Genetic diseases associated with the X chromosome are commonest and they are often referred to as X- or sex-linked genetic conditions. X-linked diseases can be recessive or dominant, although the former, for example hemophilia described in *Chapter 13*, are the better known (*Table 15.1*). Given that females have two X chromosomes but males only one together with a Y chromosome, then the expression of sex-linked genes differs between females and males because many genes on the Y chromosome lack a corresponding allele on the X chromosome. Thus X-linked recessive genes are only expressed in females if there are two copies of the gene; one on each of the X chromosomes. However, for males,

Recessive	Dominant
Duchenne and Becker muscular dystrophies (*Chapter 16*)	Coffin-Lowry syndrome
Hemophilia (*Chapter 13*)	incontinentia pigmenti
Red–green color blindness	
Wiskott-Aldrich syndrome (*Chapter 5*)	
X-linked agammaglobulinemia (*Chapter 5*)	

Table 15.1 Examples of X-linked diseases

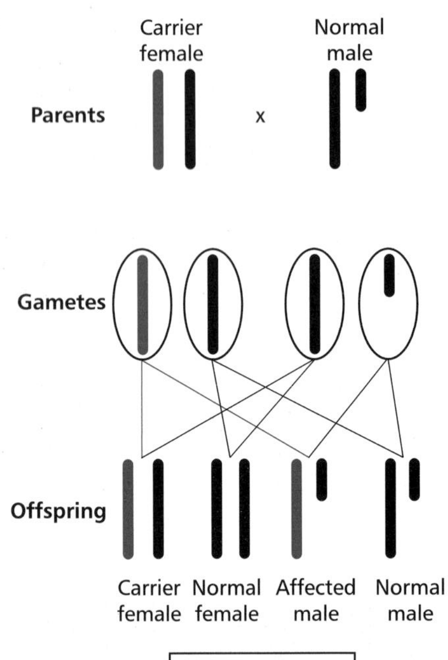

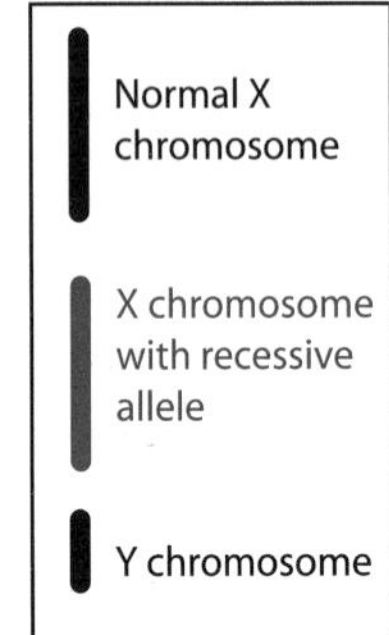

Figure 15.12 The inheritance pattern shown for a recessive allele carried on one of the X chromosomes of the mother with a normal male father.

there only needs to be one copy of a defective (mutated) X-linked recessive gene for the disorder to be expressed. For example, if a woman carried a mutated recessive gene on one of the X chromosomes (*Figure 15.12*) then statistically, 50% of her sons would inherit the defective gene and show the disorder; however, 50% of her sons would not receive the gene and would not express the condition. Similarly, half of her daughters would not receive the gene and therefore would be unable to pass it to future generations. The other half would receive the defective gene and be able to transmit it to the next generation. Like their mother, they are asymptomatic carriers of the disorder.

A common recessive X-linked condition only expressed in males is red–green color blindness. This is an inability to distinguish between red and green colors although visual acuity (keenness of vision) is normal. It is not associated with any serious complications but affected individuals may not be considered for some occupations that involve transport or the armed forces, where the ability to distinguish colors is essential. The defective gene is located on the X chromosome and males are 16 times more likely to be affected than females. Its prevalence in males is about 10%.

15.6 INHERITED GENE DISORDERS

Most inherited diseases are due to mutations in genes in the nuclear chromosomes although they can also occur as a result of mutations in mitochondrial genes as described in *Chapter 16*. The mutation of a single gene may lead to the absence or modification of a specific protein, for example, the abnormal hemoglobin in sickle cell anemia (*Chapter 13*). In some cases, the inherited disorder may result in defective receptor synthesis, such as in familial hypercholesterolemia (*Chapter 14*) where there is a defect in low density lipoprotein (LDL) receptors, or in defective carrier proteins, such as in cystinuria where renal reabsorption of cystine (formed by the oxidation of two cysteines) is impaired. If the defective or absent protein is an enzyme the result is a metabolic disorder. Most inherited metabolic disorders are autosomal recessive diseases, that is, symptoms are only seen in the homozygous condition and heterozygotes are phenotypically normal because sufficient amounts of the protein are produced. Nevertheless a number of these conditions have an autosomal dominant mode of inheritance and consequently heterozygotes are affected. Examples of these include porphyrias (*Chapter 13*) and familial hypercholesterolemia (*Chapter 14*).

CONSEQUENCES OF AN ENZYME DEFICIENCY

In inherited metabolic disorders caused by a complete or partial deficiency of an enzyme that controls a particular reaction in a metabolic pathway, the

BOX 15.1 Spongiform encephalopathies or prion diseases

The spongiform encephalopathies (SEs) or prion diseases (*Table 15.2*) are a rather peculiar group of diseases that can be inherited although this is not their usual method of transmission. It was noted in *Chapter 2* that these are infectious diseases but also have a low sporadic occurrence of about one in a million for CJD. However, they are unusual in that they can also be familial, that is they are also inherited (genetic) diseases. Irrespective of cause, these diseases generally develop slowly over 10 to 20 years in older individuals and are characterized by the presence of holes or plaques in brain tissue that can only be observed postmortem, giving it a spongy appearance (*Figure 15.13*), hence the name spongiform. There are no cures for these diseases and all are fatal. Variant CJD, which first appeared in the UK in the 1980s, differs from 'conventional' CJD in that it occurs in younger people and death occurs relatively rapidly within about 2 years following the first appearance of symptoms.

Prions are proteins that are normally found in a predominantly α helical conformation, the native form. However, these molecules can change shape to a form with an increased β sheet content that is a pathological conformation (*Figure 15.14*). In a poorly understood manner, the β sheet-rich prion protein somehow induces conformational changes in native α helical-rich molecules to change them to the β type conformation. These newly misfolded molecules can, in turn, stimulate conformational changes in other molecules in a chain reaction that deposits aggregates of prions in the brain leading to the destruction of neurons and finally the lethal spongiform condition. Thus SEs are a subdivision of a group of diseases called protein conformational diseases.

The sporadic forms of these diseases occur in individuals with point mutations in the gene that encodes the prion protein. These mutations alter the sequence of amino acid residues in the prion protein molecules and predispose them to misfold to the β sheet-rich, pathological form. Different mutations in the gene are associated with different SEs (*Table 15.3*). The mutations are, of course, heritable but since the symptoms of the diseases usually only become apparent after reproductive life is over, they can run in families producing the familial forms of the disease.

Name of disease
Atypical dementias
Creutzfeldt-Jacob disease (CJD)
Variant Creutzfeldt-Jacob disease (vCJD)
Fatal familial insomnia (FFI)
Gerstmann-Sträussler-Scheinker disease (GSS)
Kuru

Table 15.2 Examples of human spongiform encephalopathies or prion diseases

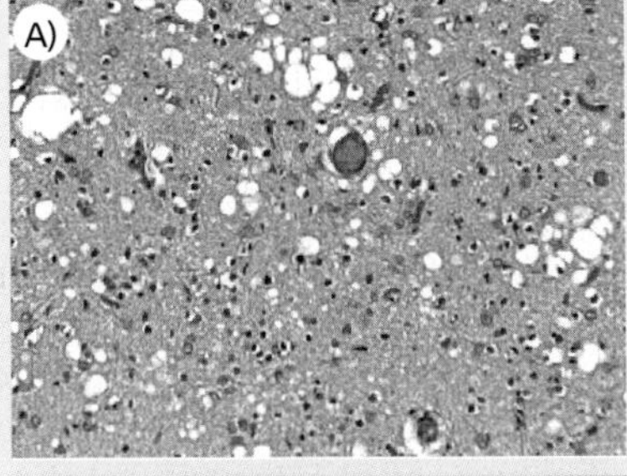

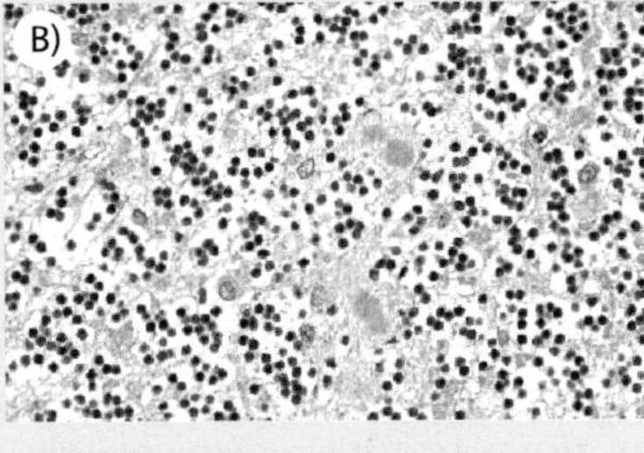

Figure 15.13 (A) The distinctive spongiform appearance of the cortex of the brain associated with CJD and (B) Aggregates of the pathological form of the protein deposited in the cerebellum. Courtesy of National CJD Surveillance Unit, UK.

Mutation	Disease
Pro102Leu	GSS
Pro105Leu	GSS
Ala117Val	GSS
Tyr145Stop	GSS
Asp178Asn	familial CJD, FFI
Val180Ile	GSS
Phe198Ser	GSS
Glu200Lys	familial CJD
Arg208His	CJD
Val210Ile	familial CJD
Gln217Arg	GSS
Met232Arg	GSS(?)
Octarepeat insert	familial CJD

Table 15.3 Some mutations of the human prion protein associated with spongiform encephalopathies

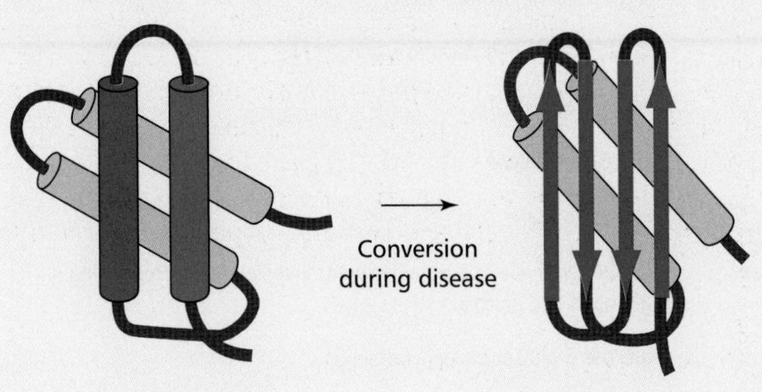

Figure 15.14 Schematic to show the change in conformation of the normal (α helical-rich) prion protein to the pathological (β sheet-rich) conformation.

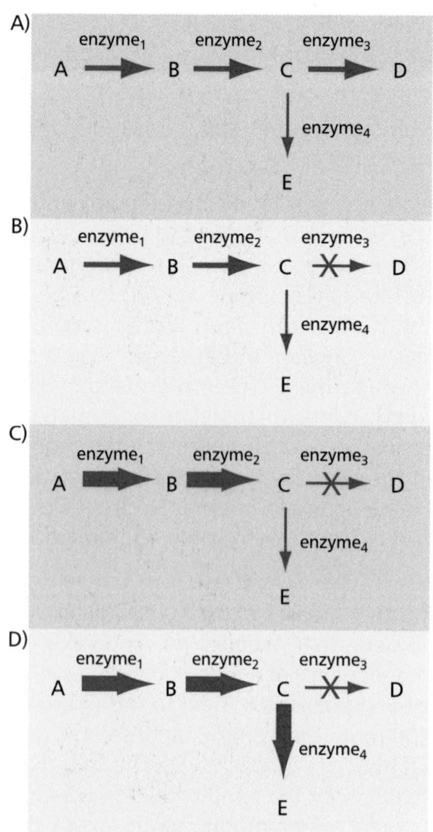

Figure 15.15 Idealized metabolic pathways to show effects of an enzyme deficiency. Metabolites are shown in upper case letters, while the relative rates of the reactions are indicated by the size of the arrows. (A) The metabolic pathway illustrates synthesis of product D from substrate A by a series of reactions catalyzed by enzymes 1, 2 and 3. Product E is derived from a minor pathway by the action of enzyme 4 on substrate C. A deficiency of enzyme 3 has a number of consequences including (B) where conversion of C to D is blocked resulting in a decline in the production of D. In (C) the concentration of B and C increases due to the increased activities of enzymes 1 and 2 because the lack of D means there is no negative feedback on enzyme 1 and possibly 2 and (D) shows the increase in the conversion of C to E by enzyme 4.

clinical features are a consequence of the enzyme deficiency. *Figure 15.15* shows how this may occur. Symptoms arise from a lack of product D if this is an essential substance and alternative pathways for its synthesis do not exist. Moreover, an accumulation of precursor C will occur if the enzyme is absent and may produce clinical features if the substrate is toxic when it accumulates. Intermediates of the pathway may also accumulate especially since there will be no negative feedback effect as the final product of the pathway is absent and cannot inhibit the first enzyme of the pathway. Finally, accumulation of a product of a minor pathway, E, may occur and if this is toxic in excess it may produce clinical features.

The treatment of inherited metabolic disorders aims at trying to prevent the accumulation of precursor(s) and provide the necessary product of the pathway. The removal of toxic products of any minor pathways may also be necessary. Future strategies are aimed at replacing the deficient enzyme or correcting the defective gene by gene therapy. Many inherited metabolic disorders caused by a deficiency of an enzyme are known (*Table 15.4*); one of the most thoroughly documented is phenylketonuria.

Phenylketonuria

Phenylketonuria (PKU) is the commonest disorder of amino acid metabolism. It has an autosomal recessive mode of inheritance, which leads to a deficiency, mostly in the liver, of phenylalanine hydroxylase (*Figure 15.16*), which catalyzes the hydroxylation of phenylalanine to tyrosine (*Figure 15.17*). Tyrosine is required for the synthesis of proteins, the pigment melanin, thyroxine and the catecholamine hormones (*Chapter 7*). However, if the enzyme is absent, then phenylalanine and its metabolites accumulate and are toxic to the developing brain. The manner in which damage occurs is not completely understood but it is believed that hyperphenylalaninemia interferes with brain amino acid metabolism and inhibits the release of neurotransmitters. There is also an increase in the level of phenylpyruvic acid, a phenylketone, which is normally a minor metabolite of phenylalanine (*Figure 15.17*). Excess phenylpyruvic acid is excreted in urine, hence the name phenylketonuria. The incidence of PKU is one in 10 000 in the UK; in other countries it varies from one in 5000 to one in 20 000 births.

The clinical features of PKU are absent at birth but develop within a few days if the newborn is untreated. These signs include a characteristic mousy odor, irritability, poor feeding, vomiting, eczema, mental retardation, as well as a pale skin, fair hair and blue eyes due to decreased melanin synthesis. The most serious of these features is irreversible mental retardation which develops within three to six months following birth. A diagnosis of PKU is made on the

Disorder	Enzyme deficiency
Phenylketonuria	phenylalanine hydroxylase
Porphyria cutanea tarda	uroporphyrinogen decarboxylase
Gaucher's disease	glucocerebrosidase
Forbe's disease	α,1,6-glucosidase
Fabry disease	α galactosidase
Glycogen storage diseases	deficiency of one of several enzymes involved in interconverting glycogen and glucose
G6PD deficiency	glucose 6-phosphate dehydrogenase

Table 15.4 Inherited enzyme deficiencies

demonstration of a concentration of phenylalanine in the serum in excess of 0.7 mmol dm^{-3}, compared with the reference range value of less than 0.1 mmol dm^{-3}. Clinical determinations are performed a few days following birth since it is vital to begin any treatments as soon as possible. The management of PKU involves restricting the dietary intake of phenylalanine so that the serum phenylalanine concentration does not exceed the limits shown in *Table 15.5*.

In the early stages of life, when the brain is developing rapidly, strict control of phenylalanine concentrations must be imposed to prevent brain damage. Commercially prepared diets with a low phenylalanine content are available. The concentration of this amino acid is low but cannot be zero since phenylalanine is an essential amino acid (*Chapter 10*) and some must be provided in the diet to support protein synthesis. Tyrosine is not an essential amino acid unless phenylalanine intake is limited. Therefore adequate quantities of tyrosine must also be provided in the diet of PKU patients. Regular monitoring of treatment is advisable. Blood is collected as a dried spot or as liquid plasma for laboratory analysis. The diet can become somewhat less rigorous after the age of 10, although many clinicians now believe that dietary restriction should be continued throughout life.

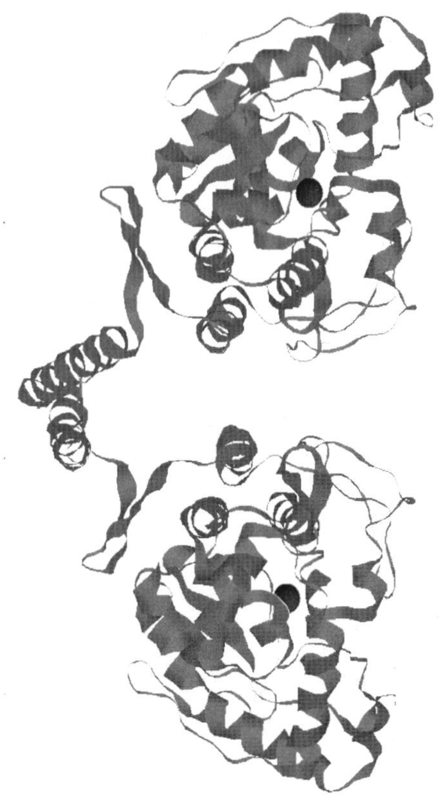

Figure 15.16 Molecular model of a phenylalanine hydroxylase molecule. The colored spheres represent Fe atoms. PDB file 2PAH.

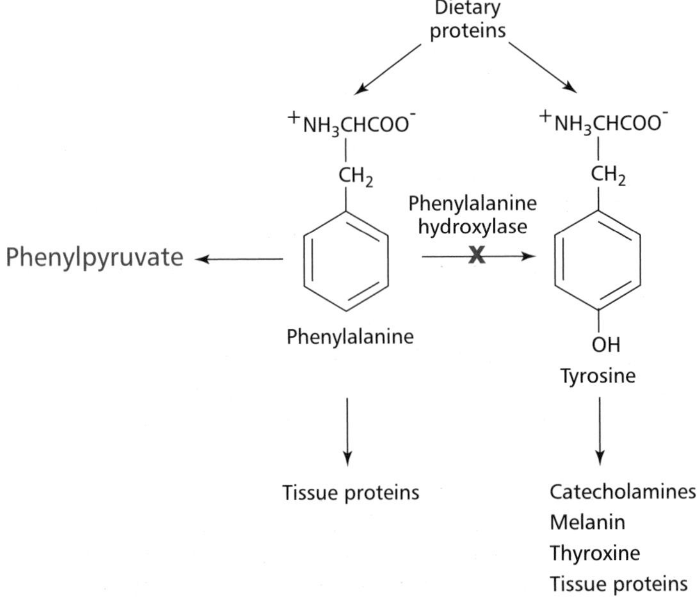

Figure 15.17 The metabolism of phenylalanine. The red cross indicates the step blocked in PKU.

Age/years	[Phenylalanine]/mmol dm^{-3}
0–5	<0.36
5–10	<0.48
10+	<0.70

Patients should be tested regularly to check that they are adhering to their diets.

Table 15.5 Recommended limits of serum phenylalanine in PKU patients

Children who have been diagnosed shortly after birth and properly treated by dietary management develop normally. Early treatment is crucial because the IQ of an affected individual rarely exceeds 70 and brain damage caused by untreated PKU is irreversible. Strict dietary control is also necessary in pregnant women who have PKU, since maternal hyperphenylalaninemia can affect the fetus *in utero*, even if the fetus itself does not have PKU. Mental retardation and congenital abnormalities can occur in a large proportion of these infants.

NATIONAL SCREENING PROGRAMS FOR INHERITED DISEASES

A number of factors need to be considered before a screening program (*Chapter 1*) for any inherited disease is instituted. These include:

- does the disease have a relatively high incidence;
- can the disease be detected within days of birth;
- can the disease be identified by a biochemical marker that is easily measured;
- will there be a failure in diagnosing the disease early and would this cause irreversible damage to the baby;
- can the disease be treated and will the result of any screening test be available before irreversible damage to the baby occurs?

Thus, for example, neonatal screening programs for PKU are well established in practically all the countries of the developed world, including the UK. Any screening program has to be cost effective. Screening for PKU involves collecting a specimen of capillary blood from the baby at 6 to 10 days after birth, which allows for sufficient time for feeding and protein intake to become established. The test involves determining the concentration of phenylalanine in the plasma. If the result is indicative of PKU, further definitive tests are performed. The plasma phenylalanine concentration used to be determined by the Guthrie test, which involves determining the ability of plasma to support the growth of the bacterium *Bacillus subtilis*, which can only grow if phenylalanine is present in the medium. Nowadays, however, most laboratories use chromatographic, fluorimetric or mass spectrometric methods for the estimation of phenylalanine.

15.7 CHROMOSOMES AND THE HUMAN KARYOTYPE

Interphase chromosomes are present as extended structures and cannot be seen with the light microscope. At the onset of cell division, both mitotic and meiotic, the chromosomes condense to form compact structures, referred to as mitotic figures. All chromosomes have a narrowed region called the centromere that divides the chromosome into two portions and allows them to be classified as metacentric, submetacentric, acrocentric or telocentric as shown in *Figure 15.18*. When the chromosome is divided into two unequal lengths, the shorter is called the *p* arm and the longer the *q* arm. When stained with Giemsa stain (*Figure 15.19*), most arms are divided into two or more regions by prominent bands and each region is further subdivided into subbands that can be numbered unambiguously. For example, band Xp21.2 is to be found on the *p* arm of the X chromosome in region 2, band 1, subband 2.

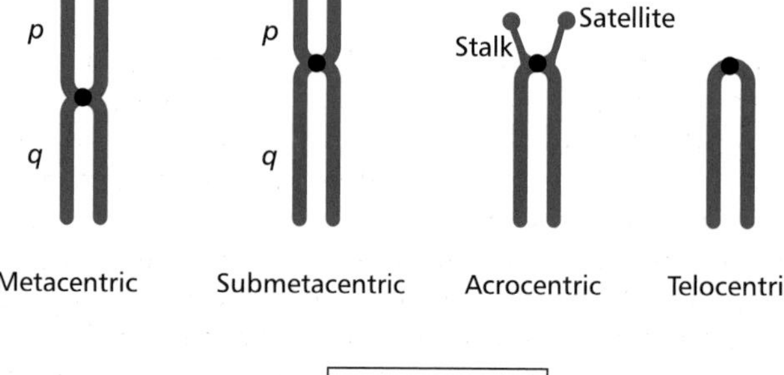

Figure 15.18 A classification of chromosomes into metacentric, submetacentric, acrocentric and telocentric types based on the position of the centromere.

A **karyotype** is the characteristic number, size and shape of chromosomes of a species. A **karyogram** is a photographic representation of these chromosomes stained and arranged in order, as for example in *Figures 15.2* and *15.20*. An **idiogram** is a diagrammatic representation or interpretive drawing of the chromosomes based on the physical features seen in the karyogram. The karyotype of normal humans is 46. Human autosomal chromosomes are divided into seven groups (*Table 15.6*) on the basis of their sizes and the positions of their centromeres.

Figure 15.19 A single human A2 (*Table 15.6*) chromosome showing G banding. Courtesy of J.S. Haslam and K.P. O'Craft, Tameside General Hospital, Ashton under Lyne, UK.

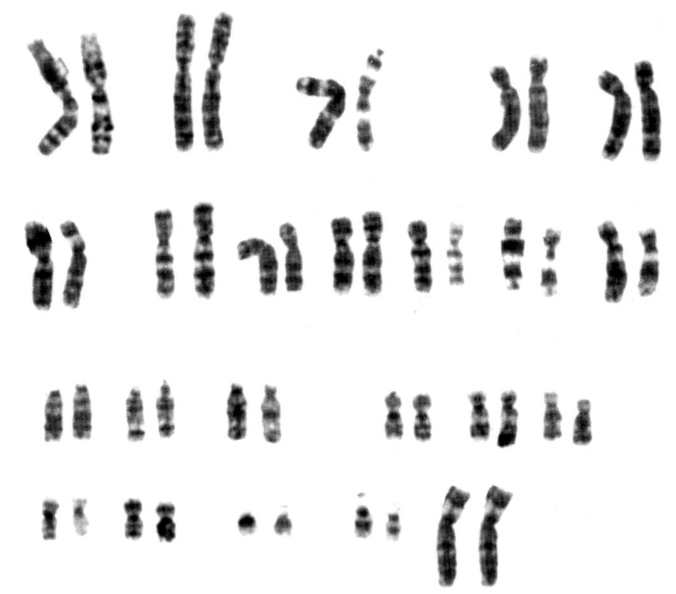

Figure 15.20 Karyogram of a normal human female showing G banding. Courtesy of J.S. Haslam and K.P. O'Craft, Tameside General Hospital, Ashton under Lyne, UK.

Chromosome group	Chromosome numbers	Structures
A	1 to 3	
B	4 to 5	
C	6 to 12, X	
D	13 to 15	
E	16 to 18	
F	19 to 20	
G	21 to 22, Y	

Table 15.6 The seven groups of human chromosomes

Cytogenetics is the microscopic study of chromosomes. Small lymphocytes isolated from a blood sample or cells obtained by amniocentesis or chorionic villus sampling (*Section 15.10*) are stimulated to divide by treatment with the plant lectin, phytohemagglutinin (PHA) and mitosis is then arrested at metaphase by an inhibitor such as colchicine. Metaphase chromosomes are visible by microscopy when stained in one of several different ways to allow their accurate identification when examined by microscopy. For routine karyotyping, Giemsa (G) staining is usually the preferred procedure since this produces a pattern of alternating dark and light bands characteristic for each pair of chromosomes. These patterns reflect differences in the detailed structure of each chromosome. Karyotyping can be performed on white cells from whole blood as described above or from amniotic fluid, which contains cells from the developing fetus. When a patient's chromosomes are examined using a microscope, it is possible to identify aberrations in chromosome number and structure.

The phenotype of a patient with a chromosomal disorder depends on the nature of their chromosomal defect. The first human chromosomal disorder was discovered in 1959 when three copies of chromosome 21 were found to be associated with Down syndrome (*Section 15.9*). The development of chromosomal banding in 1970 has markedly increased the ability to resolve small chromosomal aberrations.

FRAGILE SITES

When human cells are grown in culture, some of the chromosomes in cells derived from certain individuals fail to stain in particular regions, giving the appearance of a gap. These sites are known as **fragile sites**, since they are susceptible to breakage when the cells are cultured in the absence of certain chemicals such as folic acid, which is normally present in the culture medium. More than 80 such sites have been identified since they were first discovered in 1965. The cause of the fragility at these sites is not known with certainty but they may represent regions where the DNA has been incompletely replicated.

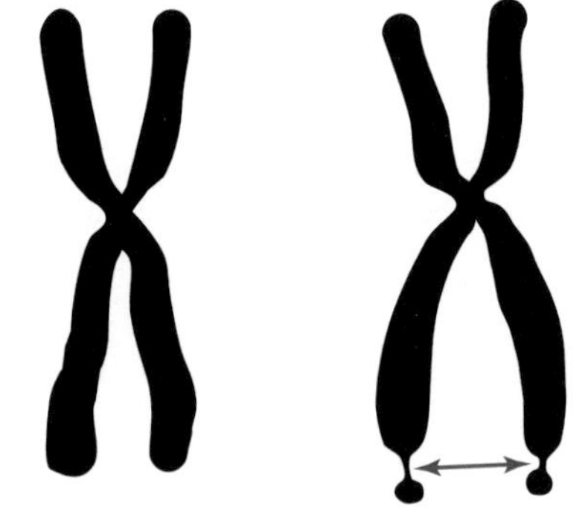

Figure 15.21 Schematic of a human X chromosome with a fragile site (arrowed).

Almost all studies on fragile sites have been carried out *in vitro* on cells halted in mitosis. Initially they were not considered to be clinically relevant and, indeed, most fragile sites do not appear to be associated with any clinical syndrome. However, a strong association has been shown to exist between a form of mental retardation called fragile X or Martin-Bell syndrome and a fragile site on the X chromosome at position Xq27.3 (*Figure 15.21*) associated with the *FMR1* gene. It is a dominant trait but fortunately fails to be fully expressed (incomplete penetrance) in many individuals. However, it is the commonest cause of mental retardation and has been estimated to affect one in about 4100 males in the UK. Most suffer mental retardation to the point that they are unable to live an independent life, and have a distinct physical appearance, including long, narrow faces with protruding chins, enlarged ears, and increased testicular size particularly after puberty. The syndrome also affects about one in 8000 females, who tend to suffer milder forms of retardation. Most humans carry a stable version of *FMR1* which has about 30 CGG repeats (*Box 15.2*). Individuals who have genes with about 45 to about 55 CGG repeats are in a gray zone; they do not have fragile X syndrome and, while they are likely to pass on a stable gene to their children, they have an increased chance of having children with a larger number of CGG repeats. People with about 55 to about 200 are said to have a *premutation* since although they generally have few or lack symptoms of fragile X syndrome, they can have children with more than 200 CGG repeats. This is the full mutation that initiates inappropriate methylation of cytosine bases and unfortunately can lead to the full blown syndrome.

The *FMR1* gene is carried on the X chromosome and shows Mendelian patterns of inheritance for X-linked disorders (*Figure 15.12*), although if it reaches the premutation stage it has a high probability of mutating (by repeat amplification) from one generation to the next. However, amplification of the CGG repeats can only occur in females, not males. Thus although there may be no family history of fragile X syndrome, it can suddenly appear in a number of offspring. The patterns of inheritance of fragile X syndrome are also somewhat complicated compared with a disorder like phenylketonuria, described in *Section 15.6*. Given that *FMR1* is on the X chromosome, a father cannot pass on any form of it to male offspring. However, daughters generally

BOX 15.2 Dynamic mutations and DNA methylation

In the early 1990s types of mutation called dynamic or expansion mutations were identified that were associated with a number of genetic disorders that increased in severity, or had an earlier onset over several generations. These disorders are referred to as trinucleotide repeat disorders. Repeating combinations of three nucleotides occur commonly in DNA molecules. All combinations of triplets, such as CGG, CAG, AGG and ACC, are found but the first two are commonest. Such sequences seem to be a normal part of the DNA and are thought to have regulatory roles in gene expression. However, if the number of repeats becomes too large, clinical problems that result in an identifiable disease are triggered. Fragile X syndrome is one such disease and is associated with an increase in the number of CGG repeats that are normally found in the *FMR1* (fragile X mental retardation 1) gene at the Xq27.3 fragile site. One consequence of the expansion in the number of CGG repeats is that methylation of the regulatory region of the gene occurs and prevents the cell from expressing *FMR1* and synthesizing fragile X mental retardation protein (fmrp), an RNA binding protein that is expressed in the brain. The lack of fmrp leads to fragile X syndrome. Methylation of DNA is the addition of methyl ($-CH_3$) groups (*Chapter 10*) to some of its bases from the donor molecule, *S*-adenosylmethionine (SAM). Only a few percent of A and C nucleotides are methylated and in vertebrate eukaryotic cells only the formation of 5-methylcytosine (Me^5C) occurs (*Figure 15.22*). Methylation at particular CpG sequences may inhibit transcription and have a role in gene regulation by switching off the expression of that gene, although the mechanism by which this occurs is unclear. However, the methyl group is known to project into the DNA molecule and interfere with the attachment of DNA binding proteins. The patterns of methylation are inherited, that is they are repeated from generation to generation. In general, *FMR1* is copied by the DNA polymerase during DNA replication with high fidelity and this stable version is almost always inherited. However, in some circumstances amplification occurs that increases the number of CGG repeats in the daughter chromosome. This is possible because repeats containing G and C nucleotides can base pair with themselves to form hairpin structures (*Figure 15.23*), which increases the risk of slipped mispairing occurring during DNA replication leading to an increase in the number of triplet repeats. The number of CGG repeats in the *FMR1* is the major factor that determines the presence or absence of fragile X syndrome. In fragile X syndrome, the mutant *FMR1* gene is nonmethylated in asymptomatic males, methylated in the inactive X chromosome of females and totally methylated in most fragile X males, which prevents its expression.

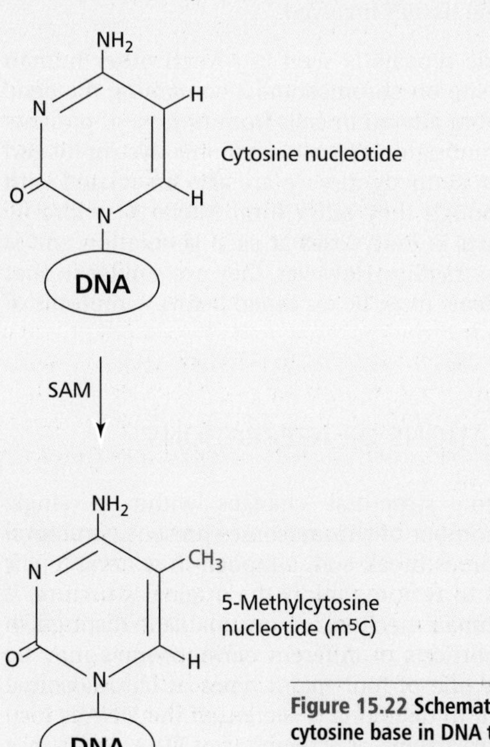

Figure 15.22 Schematic illustrating the methylation of a cytosine base in DNA to 5-methylcytosine. SAM, *S*-adenosylmethionine.

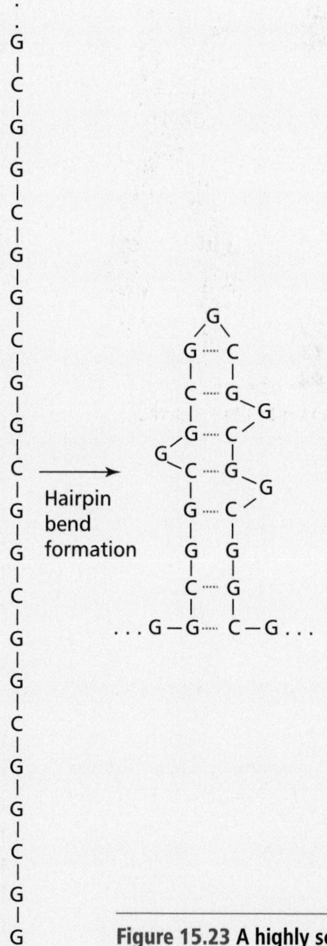

Figure 15.23 A highly schematic representation of a hairpin bend formed by a series of repeated CGG trinucleotides.

receive the paternal type. For example, almost all males with the stable version generally have daughters with the stable version. However, males with premutations, who are generally phenotypically normal and called normal transmitting males, have premutation type daughters. This inheritance can have severe consequences for any male grandchild as explained below. Most full mutation males do not have children. Those few who do would give the full mutated version to their daughters but surprisingly the daughters only express the premutation. Hence the father is passing on a reduced number of CGG repeats presumably because there are protected cells in the testes that never expand to the full mutation or a reduction in the repeat number occurs in some male reproductive cells. This means that all females who do have the full mutation must have received it from their mothers since they cannot receive it from their fathers.

Females have two X chromosomes and every child, male or female, has an equal, random chance of receiving one or the other of them. A female who has a copy of the premutation from her normal transmitting father can pass it on to her children. Most daughters who receive the premutation will show an increase in repeat number compared with their mother and, while most will show only the premutation, others will express the full mutation. Sons who inherit an X chromosome from a mother carrying a premutation are the principal group affected by fragile X syndrome since they are much more likely than females to have amplification to the full mutation. The probability of the full mutation is dependent upon the mother; those at the lower end of the premutation range, about 56–70 repeats, are less likely to have a son with the full mutation than those at the higher end with more than 100 repeats.

All males with the full mutation will experience significant symptoms. Some females with the full mutation will have symptoms of fragile X but, in general, the severity is less. Finally, there are individuals who cannot be assigned to these categories but have cells that vary regarding repeat size or the extent of methylation (*Box 15.2*). The severity of their symptoms depends on the proportion of cells affected and the tissues involved.

The phenomenon of trinucleotide repeats is seen in several other human disorders. For example, a fragile site on chromosome 3 containing the gene *FHIT* (fragile histidine triad) is often altered in cells from tumors of patients with lung cancer (*Chapter 17*). Huntington disease, myotonic dystrophy and spinobulbar muscular atrophy or Kennedy disease are also associated with trinucleotide amplifications, although they differ from fragile X syndrome in that the amplification can occur in both sexes at each generation and is not associated with chromosome fragility. However, they are similar in that a threshold number of triplet repeats must be exceeded before symptoms of the disease appear.

15.8 CHROMOSOMAL MUTATIONS OR ABERRATIONS

Chromosomal mutations include structural changes within a single chromosome or changes in the number of chromosomes present. Structural mutations occur when chromosomes break and, although in general repair mechanisms rejoin the two ends to restore rapidly the original structure, if more than one break occurs, the repair mechanisms are unable to distinguish between the broken ends and portions of different chromosomes may be joined together. This can lead to one of four major types of chromosomal structural aberration or mutation. In deletions, a section of the DNA is lost; in duplications, one or more extra copies of a segment of DNA occur in a chromosome; in inversions, there is a reversal of direction of a portion of the DNA in the chromosome and in **translocations**, segments of the DNA are moved to a different chromosome. Deletions and duplications change the amount of DNA in a chromosome. Inversions and translocations change the arrangement of bases in a length of DNA but do not change the amount of DNA present in the chromosome.

If the exchange of chromosomal material during a translocation and inversion does not involve breaks within a gene or alter the amount of DNA, then the individual will be clinically normal and is said to have a balanced translocation. However, if the structural alteration occurs in the gonads, even a balanced translocation has clinical significance for future generations since it may lead to offspring who are chromosomally unbalanced, that is, who have lost DNA (*Figure 15.24*). Since most fetuses with unbalanced translocations tend to be spontaneously aborted, people with balanced translocations may only attract clinical attention if they and their partner are investigated because of a history of miscarriages. Infants with unbalanced translocations that do survive are mentally retarded and show multiple **dysmorphic** features, that is alterations (abnormalities) to the accepted appearance. Specific disorders can also occur when discrete genes are damaged at the translocation fractures, the resulting disorder being dependent on which genes are damaged.

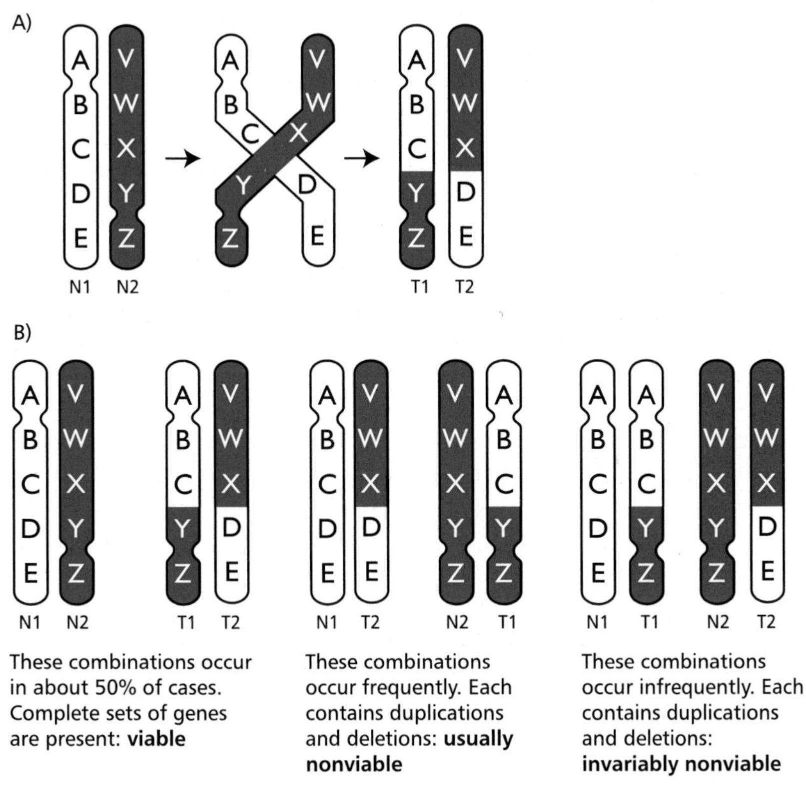
A)
N1 N2
T1 T2
B)
N1 N2
T1 T2
N1 T2
N2 T1
N1 T1
N2 T2
These combinations occur in about 50% of cases. Complete sets of genes are present: **viable**
These combinations occur frequently. Each contains duplications and deletions: **usually nonviable**
These combinations occur infrequently. Each contains duplications and deletions: **invariably nonviable**

Figure 15.24 (A) Schematic showing a possible balanced translocation between two chromosomes. (B) Illustrates the types of gametes, both normal and genetically deficient that could result from translocations shown in (A). For simplicity, it has been assumed that recombination did not occur during meiosis. See text for details.

DELETIONS

Deletion of part of a chromosome can arise between two breakpoints as a result of unequal crossing over during meiosis (*Section 15.3*) or as a result of a parental translocation. Several clinical disorders are caused by deletions. In many cases, the abnormalities only occur in individuals who are heterozygous since the homozygous condition is lethal, especially if the deleted portion of the chromosome is large. However, any chromosomal deletion that can

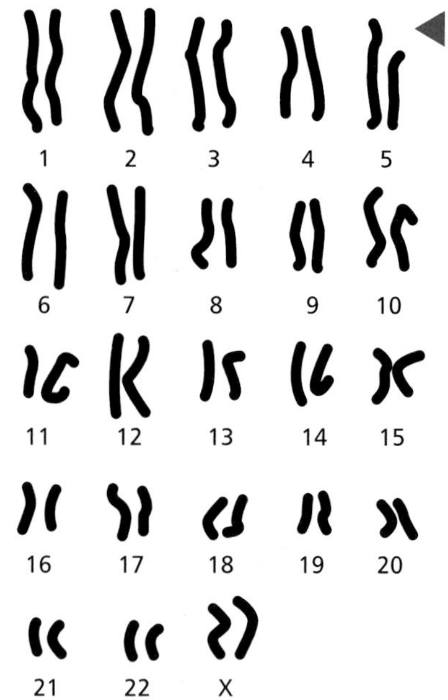

Figure 15.25 Deletion of part of one chromosome 5 (arrowed) in a karyogram of a child showing cri-du-chat syndrome.

be observed microscopically almost invariably produces a phenotype with multiple abnormal features and mental retardation because of the absence of expression of the deleted genes. For example, cri-du-chat syndrome is a heterozygous condition that occurs in about one in 50 000 births. Infants with this syndrome show anatomical malformations including gastrointestinal and cardiac complications and are often mentally retarded. The glottis and larynx also develop abnormally giving the characteristic cry, similar to the meowing of a cat, which names the syndrome. Cri-du-chat syndrome is caused by a deletion of part of the short arm of chromosome 5 (46,5p-, *Figure 15.25*). The length of the deleted portion varies; the longer the deletion the more defective the syndrome in surviving children. The effects of the syndrome are severe, although individuals who receive good home care and early special schooling can walk and communicate verbally and develop self-care skills.

Prader-Willi syndrome (frequency one in 10 000 to 25 000) and Angelman syndrome (exact frequency unknown) both result from the 'microdeletion' of the band 15q11–13 although they have different phenotypes depending upon the parental origin of the deletion. In Prader-Willi syndrome the deletion occurs invariably on the chromosome 15 inherited from the father whereas in Angelman syndrome the deletion occurs almost exclusively on the chromosome 15 from the mother. Hence inheritance of paternal and maternal copies of this region of chromosome 15 is important for normal development, a phenomenon known as **genetic imprinting**. People with Prader-Willi syndrome are mentally retarded, have small external genitalia and characteristic facial features. Babies with Prader-Willi syndrome have a poor sucking reflex hence feeding is difficult and results in weakness and stunted growth. Strangely, children with Prader-Willi syndrome become compulsive eaters at five to six years of age and suffer obesity and its related health problems (*Chapter 10*). If untreated, afflicted individuals may feed themselves to death.

Angelman syndrome is characterized by developmental delay, absence of speech, jerky movements, paroxysms of inappropriate laughter and characteristic facial features which differ from those of Prader-Willi syndrome.

Many other portions of the genome are also subject to genetic imprinting although the precise mechanisms controlling whether the paternal or maternal copies of a gene are expressed are not fully understood.

DUPLICATIONS

Chromosomal mutations that result in the doubling of a part of a chromosome are called duplications. The size of the duplicated segment varies enormously and duplicated segments may occur in a tandem configuration, that is, adjacent to each other or in different locations in the genome (*Figure 15.26*). Duplications can result in gene redundancy and may produce phenotypic variation, since there are now two copies of the gene and one copy may mutate independently of the other. This is thought to be a significant source of genetic variability during evolution. For example, gene duplications have been essential in the evolution of multigene families. These are groups of several genes whose products are similar in structure or functions. The genes for globins are a particularly well studied multigene family. Hemoglobins are tetrameric proteins consisting of two pairs of differing polypeptides each with an attached heme group able to bind and release a dioxygen molecule. Human individuals produce different hemoglobins at different stages during their lives as described in *Chapter 13*. Each of the globin polypeptides has slightly different primary structures, but consist of the α types that are found as a cluster on chromosome 16 and the β types clustered on chromosome 11. It is thought that each group of genes evolved from one original ancestral gene that underwent duplications and subsequent sequence divergence.

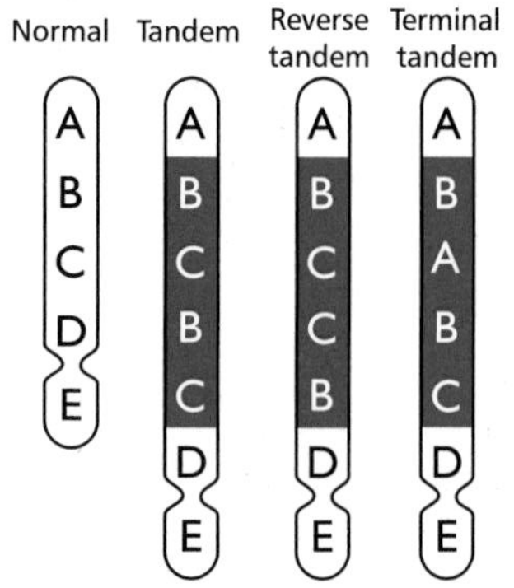

Figure 15.26 Schematic to illustrate some possible types of chromosomal duplications.

INVERSIONS

Figure 15.27 illustrates how inversions might arise. A chromosomal loop forms before fractures occur at two places on the chromosome. The insertion of the inverted segment at the newly created sticky ends and their subsequent joining within the chromosome completes the inversion. There are two types of inversions: paracentric inversions do not include the centromeres whereas pericentric inversions do. Genetic material is not lost during inversions although there can be clinical problems when fractures occur within genes or within regions that control gene expression. The meiotic consequences of a chromosomal inversion depend on the type of inversion encountered and the resulting gametes may be nonviable leading to reduced fertility.

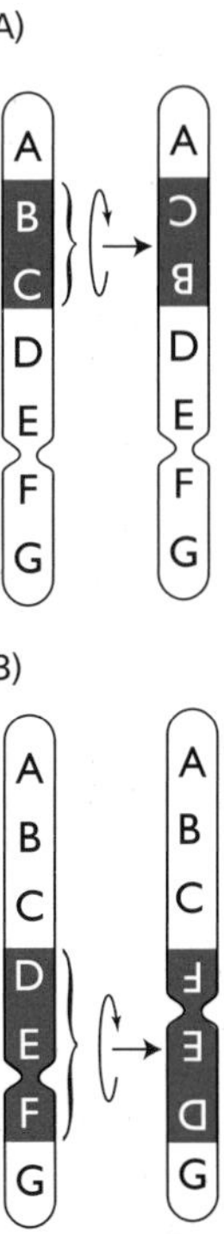

Figure 15.27 Schematics to show (A) a paracentric and (B) a pericentric inversion.

TRANSLOCATIONS

Numerous translocations occur in the human population. The simplest kinds are intrachromosomal translocations that move part of a chromosome to a different position within the same chromosome. Interchromosomal translocations transfer part of a chromosome to a nonhomologous chromosome (*Figure 15.28* and *Box 15.3*). Reciprocal translocation involves an interchromosomal translocation between two nonhomologous chromosomes. The least complex way for this event to occur is for two nonhomologous chromosome arms to come close to each other so that an exchange is facilitated.

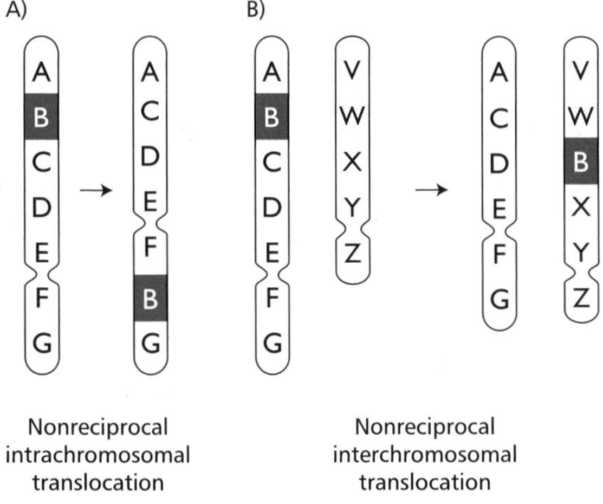

Figure 15.28 (A) Intrachromosomal and (B) interchromosomal translocations.

Homologues that are heterozygous for a reciprocal translocation undergo unorthodox synapsis during meiosis and pairing results in mitotic figures with a cross-like configuration. These chromosomes produce genetically unbalanced gametes and often result in reduced fertility. As few as 50% of the progeny of parents that are heterozygous for a reciprocal translocation survive, a condition known as semisterility. In humans, translocation can result in variations from the normal diploid number of chromosomes leading to a variety of birth defects. Translocations may transfer a gene to a region of a chromosome that is more transcriptionally active. This can lead to the development of some forms of cancer (*Chapter 17*).

A common type of translocation involves breaks at the extreme ends of the short arms of two nonhomologous acrocentric chromosomes (*Figures 15.29* and *15.34*). The small fragments produced are lost but the larger ones fuse together at their centromeric regions. This type of translocation produces a new, large submetacentric or metacentric chromosome and is often called a Robertsonian translocation.

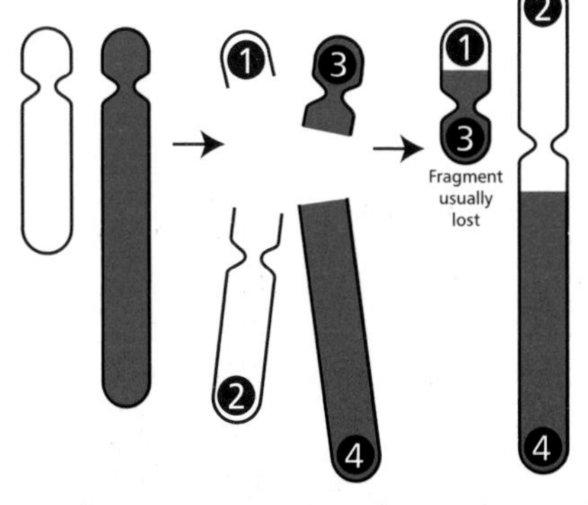

Figure 15.29 The formation of a Robertsonian translocation following breaks in two acrocentric chromosomes.

BOX 15.3 Acute promyelocytic leukemia

The biological functions of vitamin A are varied. In its aldehyde form, retinal, it participates in vision and, as the acid, retinoic acid, it controls embryonic development and the development of skin and other organs, by regulating cell proliferation and differentiation. The naturally occurring and many synthetic forms of vitamin A produced by pharmaceutical companies are called **retinoids**. The main natural retinoids are all-*trans*-retinoic acid (ATRA) and its isomer, 9-*cis*-retinoic acid. They act in similar ways to steroid hormones (*Chapter 7*). Retinoids penetrate the plasma membrane of target cells and interact with intracellular receptor proteins. The retinoid–receptor complexes are translocated to the nucleus where they interact with specific sections of DNA, leading to changes in gene expression, in other words, genes are turned on or off (*Figure 15.30 (A) and (B)*). There are two types of intracellular receptors for retinoids. One is called RAR, for retinoic acid receptor, and interacts with ATRA. The other receptor is called RXR because originally the retinoid which it recognized was unknown. However, it was subsequently found to interact with the 9-*cis* isomer. These receptors must be present in the cytosol of the cell if retinoids are to exert their influence on gene transcription.

Leukemia is a tumor that originates in the bone marrow and results in the overproduction of immature leukocytes (*Chapter 17*). Several types of leukemia have been linked with chromosomal disorders, including acute promyelocytic leukemia (APL). In APL there are abnormal hypergranular promyelocytes or immature granulocytes (*Chapter 13*) and the bleeding disorder called disseminated intravascular coagulation (DIC), which is thought to be linked to procoagulant phospholipids present in the leukemic cells. Patients may present with severe bleeding and this tends to worsen when treatment is started as the leukemic cells break down and consume large amounts of clotting factors and platelets. Eventually it was discovered that the defect in APL was a chromosomal translocation between parts of the long arms of chromosomes 15 and 17. The translocation is balanced and reciprocal and results is one abnormally long chromosome 15 (15q+) and one abnormally short chromosome 17 (17q–). This is clinically significant because of effects on the *PML* gene located on chromosome 15 that encodes the so-called PML protein and the *RAR* gene located on chromosome 17. The net result of the translocation is that a PML-RAR fusion protein is formed. This protein interferes with the normal function of PML as a growth suppressor and with that of RAR which is involved in myeloid differentiation to produce different types of blood cells from progenitor cells in the bone marrow.

In the early 1990s it was found that ATRA treatment was beneficial to patients with APL, although at the time the reason for this was not understood. It is now known that ATRA influences the genes affected by the chromosomal translocation involving chromosomes 15 and 17 as described above. Treatment with ATRA produces remission of the leukemia by promoting the conversion of leukemic blast cells into mature leukocytes. Unfortunately the remission does not usually last and normally needs to be consolidated with conventional chemotherapy.

A)

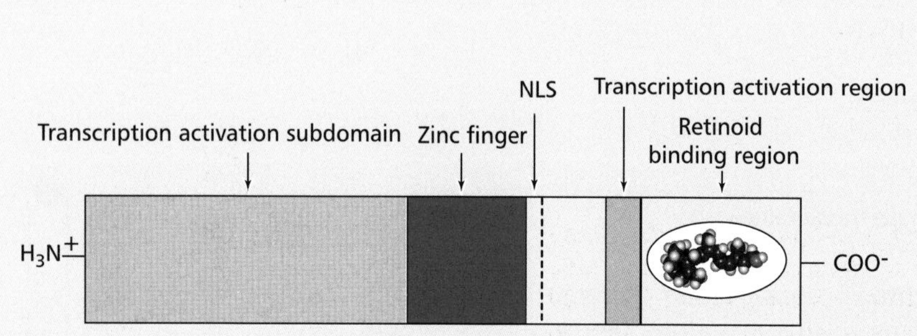

B)

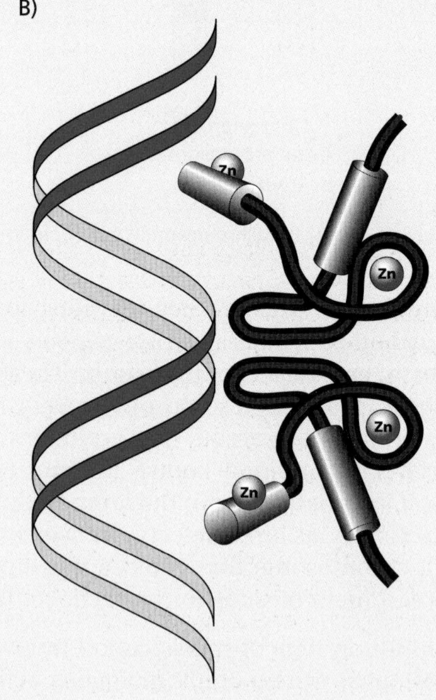

Figure 15.30 Retinoids (vitamin A derivatives) like steroid and thyroid hormones (*Chapter 7*) and vitamin D all function in the same general way. The vitamin binds to a specific receptor protein (a zinc finger protein) in the cytosol. This complex then dimerizes before being translocated to the nucleus where it binds with a specific sequence of DNA bases. Other proteins then also bind and the transcription of specific genes is either turned on or turned off. (A) The cytosolic binding protein has several characteristic regions. The one nearest to the carboxy terminus complexes with the retinoid while the middle portion, labeled 'zinc finger', binds to the DNA. NLS is a nuclear locating signal, a sequence of basic amino acids that ensures the complex enters the nucleus. (B) Schematic showing the DNA double helix binding to the dimeric form of the zinc finger protein.

15.9 VARIATIONS IN CHROMOSOME NUMBERS

Eukaryotic organisms are normally diploid and produce haploid gametes (*Section 15.3*). However, chromosomal mutations with numerical aberrations in the number of chromosomes present occur. These can be divided into two major types. **Aneuploidy** occurs when the number of chromosomes differs in having more or fewer than an exact multiple of the haploid number of chromosomes. *Table 15.7* lists a number of human aneuploid abnormalities of autosomes and sex chromosomes. In contrast, **euploidy** is the presence of an exact multiple of the haploid number of chromosomes.

Chromosomes	Syndrome
Autosomes	
Trisomy 13	Patau syndrome
Trisomy 18	Edward syndrome
Trisomy 21	Down syndrome
Sex chromosomes, female	
X0	Turner syndrome
XXX	triple X syndrome (trisomy X)
XXXX	tetrasomy
Sex chromosomes, male	
Y0	nonviable
XYY	XYY syndrome
XXY	Klinefelter syndrome
XXXY	Klinefelter syndrome
XXYY	Klinefelter syndrome

Table 15.7 Aneuploid abnormalities of human chromosomes

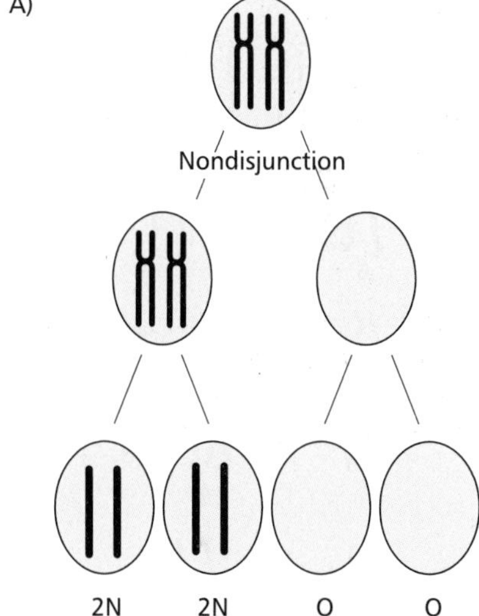

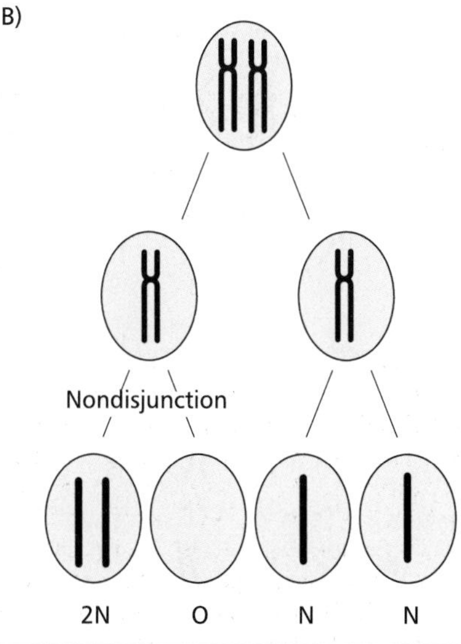

Figure 15.31 The consequences of nondisjunction of a single chromosome at (A) meiosis I and (B) meiosis II.

ANEUPLOIDY

Aneuploidy is usually caused by the nondisjunction of paired chromosomes at meiosis I or of sister chromatids at meiosis II or by delayed movement of a chromosome at anaphase. Nondisjunction is caused by the failure of pairs of homologues to separate or disjoin during segregation. *Figure 15.31 (A)* and *(B)* illustrates the consequences of nondisjunction during first meiosis and second meiosis for a single chromosome. Thus gametes are formed that either lack the chromosome or contain two copies of it. If these are fertilized by a normal haploid gamete, then zygotes are produced with one or three chromosomes. Thus nondisjunction can lead to a variety of aneuploid conditions.

The loss of a single chromosome from an otherwise diploid genome is called monosomy (2N – 1). Nullisomy results from the loss of one pair of homologous chromosomes (2N – 2). The gain of one chromosome results in trisomy (2N + 1). Tetrasomy describes the presence of four copies of a specific chromosome rather than the normal two (2N + 2). Aneuploidy can also involve the loss or the addition of more than one particular chromosome or pair of chromosomes. Thus a double monosomy involves the loss of two separate nonhomologous chromosomes (2N – 1 – 1), while a double tetrasomy would describe the presence of four copies of two chromosomes (2N + 2 + 2). Both

A)

1 2 3 4 5
6 7 8 9 10
11 12 13 14 15
16 17 18 19 20
21 22 X Y

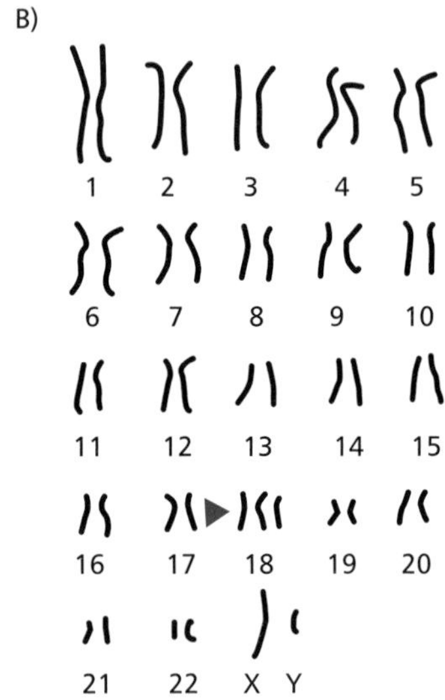

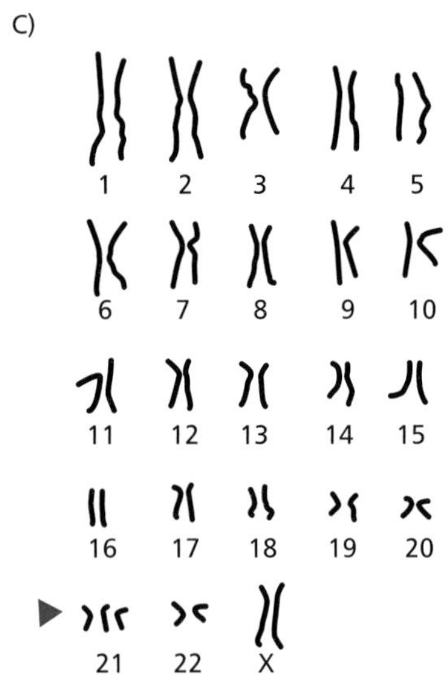

Figure 15.32 Karyotypes showing (A) trisomy-13, Patau syndrome, (B) trisomy-18, Edward syndrome and (C) trisomy-21, Down syndrome.

these cases involved meiotic nondisjunction in two different chromosomes prior to gamete formation.

Monosomy results in two types of haploid gametes, N and N – 1. Also, the single unpaired chromosome in the 2N – 1 cell is easily lost during meiosis resulting in the formation of two gametes with N – 1 chromosomes. Other types of aneuploidy also have serious, often lethal consequences in humans. Approximately 90% of all chromosomal aberrations lead to a termination of pregnancy in a spontaneous abortion.

Autosomal monosomy is rare since monosomic embryos do not develop significantly and are usually lost early in pregnancy although the abnormalities can be detected in aborted fetuses. The extra chromosome in trisomy produces individuals who are likely to have more chance of being viable provided that the chromosome involved is relatively small. The addition of a large autosomal chromosome has severe effects and is usually lethal during development. Autosomal trisomies are found in about half of the chromosomal abnormalities that lead to fetal death.

Trisomy-13 (47,13+) produces Patau syndrome (*Figure 15.32 (A)*) and affects about one in 5000 live births. The syndrome is characterized by many abnormalities including mental and physical retardation, cardiac anomalies, **polydactyly**, that is extra fingers or toes, cleft lip and palate and small eyes. Most such babies die before they are three months old.

Trisomy-18 (47,18+) is associated with Edwards syndrome (*Figure 15.32 (B)*) and occurs in about one in 4000 live births. About 80% of cases of Edwards syndrome are female. Sufferers are small at birth and have multiple congenital malformations that affect almost all body systems. Among the many abnormalities associated with the syndrome are mental and developmental retardation, elongated skull, low-set malformed ears and clenched fists. Ninety percent of infants with trisomy-18 die within six months, often from cardiac problems.

Trisomy-21 (*Figure 15.32 (C)*) leads to Down syndrome, the only human autosomal trisomy in which significant numbers of individuals survive more than a year following birth. It was named after its 'discoverer', a doctor called (Langdon) Down (1828–1896), in 1866. Affected individuals have common physical features and affectionate, loving natures. They generally have flat faces with epicanthic folds over the eyes, round heads with a protruding furrowed tongue that causes the mouth to remain partially open. They are below average height and have short, broad hands. Their physical and mental development is retarded and muscle tone and motor skills are poor. Down patients are prone to respiratory disease, 50% of them have heart problems and their incidence of leukemia is approximately 15 times higher than that of the normal population. Not surprisingly, life expectancy is reduced and few survive to 50 years of age. Many die of Alzheimer's disease (*Chapter 18*).

Down syndrome affects, on average, one in every 700 live births. However, the incidence increases with the age of the mother but not that of the father. *Figure 15.33* illustrates the relationship between the incidence of Down syndrome and maternal age, although in terms of gross numbers, most affected children are born to women under 35 years old since the majority of pregnancies occur below this age.

The relationship between the age of the mother and the incidence of Down syndrome is explainable in terms of the production of oocytes (*Chapter 7*). Females have a full complement of primary oocytes that developed in the ovary of the developing female fetus. These oocytes have commenced meiosis but are arrested at the prophase I before birth. In an adult fertile female, the nucleus of a secondary oocyte begins the second meiotic division at each monthly ovulation but progresses only to metaphase II, when division again stops. The second meiotic division is not completed unless a sperm penetrates the secondary oocyte. Thus the succeeding ovum has been arrested in meiosis for about a month longer than the preceding one. Thus older ovulating women produce ova that are significantly older and have been arrested in meiosis longer than those of younger women. It is possible that the probability of nondisjunction increases with the length of time the primary oocyte spends in the ovary but whether ovum age is the specific cause of the increased nondisjunction leading to Down syndrome is not yet known.

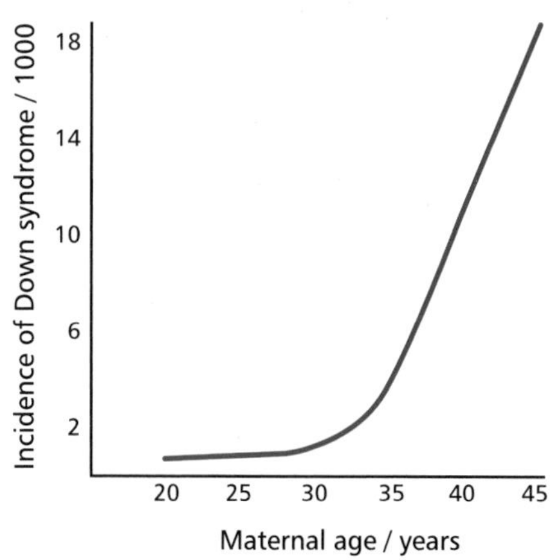

Figure 15.33 The effect of increasing maternal age on the incidence of Down syndrome babies.

Over 95% of all cases of Down syndrome are caused by trisomy 21 (*Figure 15.32 (C)*) following a nondisjunction event during meiosis in one parent, nearly always the mother. Thus, most people with Down syndrome have 47 chromosomes in all their cells. Approximately 3–4 % have the normal number (46) but have a type of Down syndrome that runs in families called familial Down syndrome, which is the result of a Robertsonian translocation (*Section 15.8*) that produces three copies of the long arm of chromosome 21 by joining the long arm of chromosome 21 with the long arm of chromosome 14 or sometimes 15 (*Figure 15.34*). The heterozygous carrier is normal because there are two copies of all major chromosome arms and hence two copies of all essential genes (*Figure 15.34*). However, meiosis will result in a 25% of the gametes formed having two copies of chromosome 21; one normal chromosome 21 and a copy attached to chromosome 14. When this gamete is fertilized by a normal haploid gamete, it forms a zygote with the normal 46 chromosomes but with three copies of chromosome 21 (*Figure 15.34*). These individuals exhibit Down syndrome.

Some individuals with Down syndrome must be institutionalized but most can be cared for at home and benefit greatly from special education programs. Advances in several areas of medical treatment have resulted in a greater life expectancy for modern day Down syndrome children.

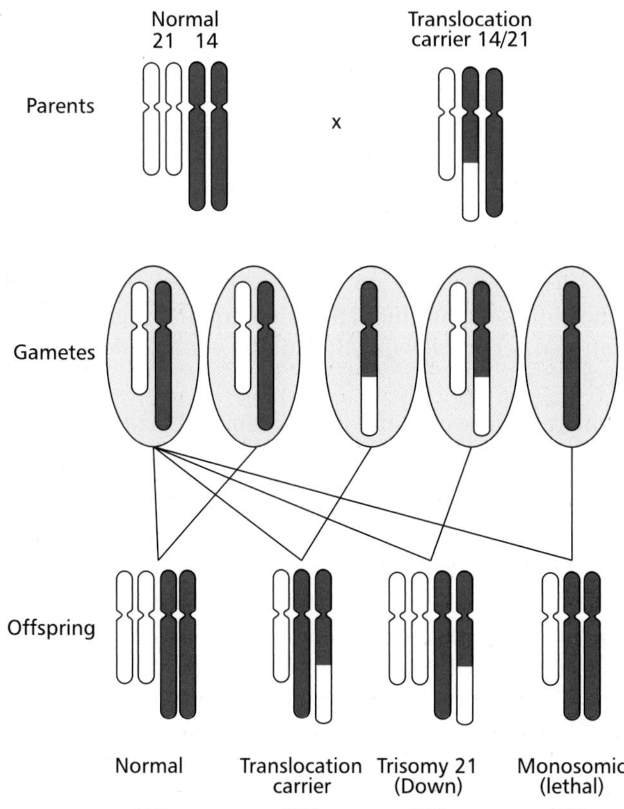

Figure 15.34 A Robertsonian translocation involving chromosomes 14 and 21 resulting in trisomy-21, Down syndrome. See text for details.

Aneuploidy involving sex chromosomes

Aneuploidy involving sex chromosomes gives rise to a number of well defined syndromes. These include Turner, Klinefelter and XYY syndromes. A female born with a single X chromosome (45X0) shows Turner syndrome (*Table 15.7*). The incidence is one per 5000 female births although the vast majority of 45X0 aneuploids are spontaneously aborted. The condition may arise from nondisjunction in either parent but in 75% of cases of Turner syndrome only the maternal X chromosome is present, implying that the problem originates in spermatogenesis in the father. Females with Turner syndrome are short in stature and rarely undergo secondary sexual development and so are mostly infertile, although their intelligence and life span are normal. A male born with one or more extra X chromosomes (*Table 15.7*) exhibits Klinefelter syndrome. The incidence of 47XXY is one for every 1000 boys born although the risk increases with an increase in the age of the mother. The extra chromosome is donated from the mother in 60% and from the father in 40% of cases and arises by nondisjunction during both maternal and paternal meiotic division. Affected males have underdeveloped testes and are infertile, are often above average height and have mild mental retardation. Approximately one in 1000 male children exhibit XYY syndrome, which arises from a nondisjunction of the Y chromosome. Males with XYY syndrome are above average in height and may be less fertile.

EUPLOIDY

In **monoploidy** only a single set of chromosomes (23 in humans) is present, rather than the normal diploid two (46). Monoploid fetuses do not reach full term presumably because the recessive lethal mutations, which are usually counteracted by dominant alleles in heterozygous individuals, are expressed.

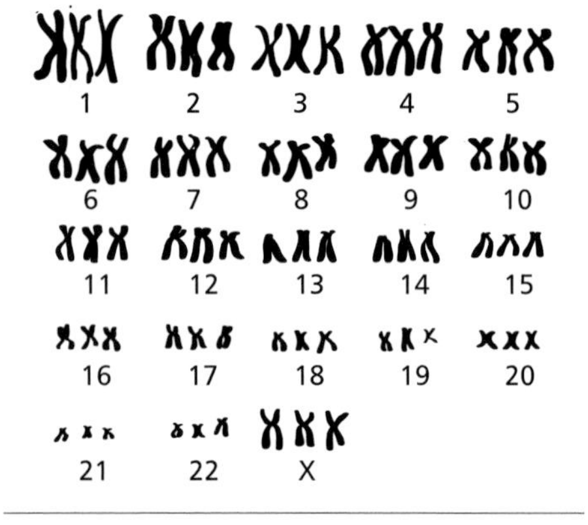

In **polyploidy**, the chromosome number is an exact multiple of the haploid number but exceeds the diploid number. Organisms with three sets of chromosomes are triploid (*Figure 15.35*), those with four sets are tetraploid and so on. Polyploidy usually arises from fertilization of the egg by two spermatozoa, which increases the total number of chromosomes to 69, or from a failure at one of the maturation divisions of the egg or spermatozoon so that a diploid gamete is produced. Thus if nondisjunction occurs at meiosis I, 50% of the gametes lack chromosomes and the other 50% have two sets of chromosomes (*Figure 15.36*). If nondisjunction occurs at meiosis II, 50% of the gametes possess the normal single set of chromosomes, 25% have two sets of chromosomes and 25% lack chromosomes (*Figure 15.36*). Fusion of a gamete with two chromosome sets with a normal gamete produces a zygote with a triploid set of chromosomes (3N). Similarly, fusion of two gametes, each with two chromosome sets, produces a tetraploid (4N) zygote. Polyploidy of somatic cells can also occur following the mitotic nondisjunction of complete chromosome sets.

Figure 15.35 A triploid karyotype. Courtesy of J.S. Haslam and K.P. O'Craft, Tameside General Hospital, Ashton under Lyne, UK.

Polyploidy normally causes an early spontaneous abortion and survival of the fetus to full term is rare. The commonest type of polyploidy in humans is triploidy. However, polyploids with an odd number of chromosome sets always possess an unpaired chromosome for each type, hence the probability of producing a balanced gamete is low. Indeed, triploidy in humans is always lethal and is seen in 15–20% of spontaneous abortions and in only about one in 10 000 births, where death invariably occurs within a month. Triploid babies show many abnormalities, including a characteristically enlarged head. Polyploids with even numbers of sets of chromosomes generally have a better chance of being at least partially fertile because there is the potential for homologous chromosomes to be segregated equally during meiosis. However, tetraploidy is also always lethal in humans and is seen in approximately 5% of spontaneous abortions. Very rarely, tetraploid humans are born although they only survive for a short time.

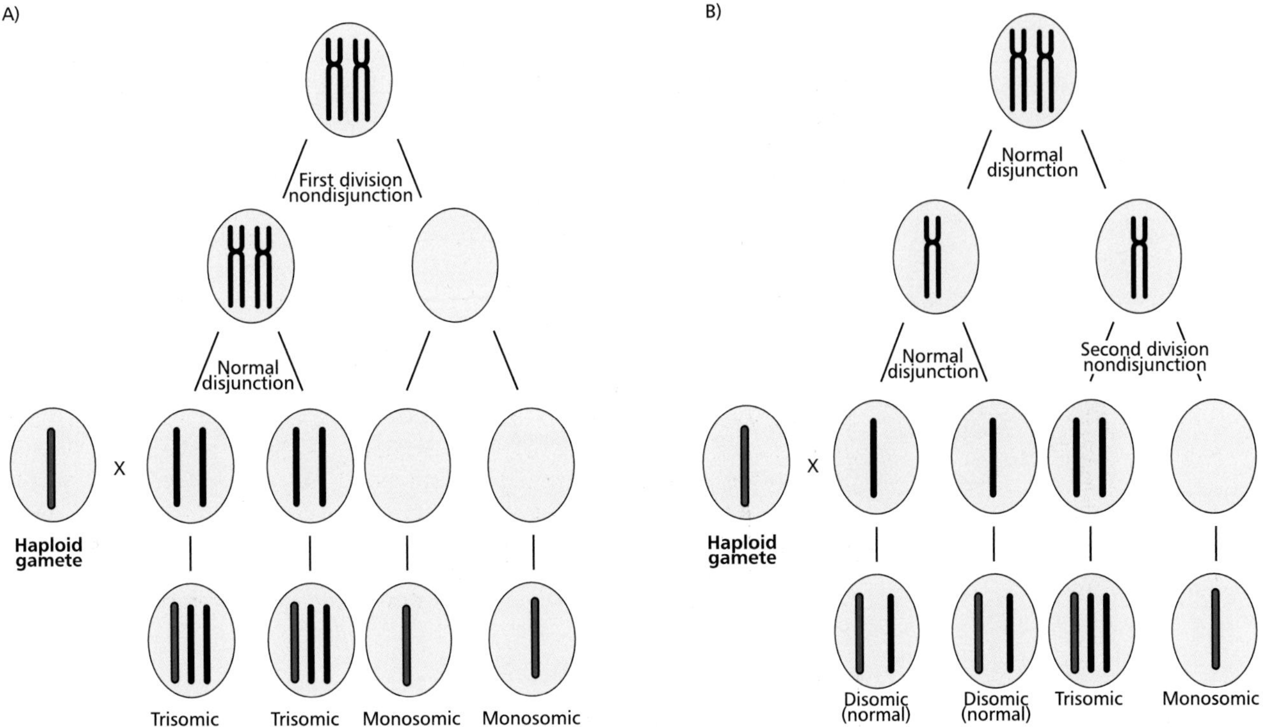

Figure 15.36 Schematic to show the production of abnormal gametes following nondisjunction at (A) at meiosis I and (B) meiosis II resulting in trisomic zygotes following fertilization with a normal haploid gamete.

15.10 DETECTING, DIAGNOSING AND SCREENING HUMAN GENETIC DISEASES

Identifying the genetic basis for a human disorder usually requires an analysis of the family history as far back as possible. If this shows that the trait is inherited, it is possible to predict whether the mutant allele is dominant or recessive and whether it is X- or autosomal-linked. Obviously, dominant mutations are the simplest to detect. If they are X-linked, then affected fathers will pass the trait to all their daughters. However, if autosomal, then dominant mutations would be expected to occur in approximately 50% of the children of an affected heterozygous individual.

In many cases, the diagnosis of a genetic disease can be made prenatally. The most widely used methods for this are amniocentesis and chorionic villus sampling (CVS). In the former, a needle is used to withdraw amniotic fluid (*Figure 15.37 (A)*) containing cells genetically identical to those of the fetus. In CVS, a catheter is inserted into the uterus and a small tissue sample of the fetal chorion removed (*Figure 15.37 (B)*). In both cases, a variety of cytogenetic and biochemical tests can then be performed on the cells and tissue to identify any single gene disorders or chromosomal abnormalities.

People in whom a genetic defect has been detected will be advised to seek genetic counseling, especially if they wish to have children. The advice they are given is based on analyses of the risks that they may produce a child with a genetic abnormality or that they themselves may develop a late onset genetic disease. A broad range of information is required and genetic counselors require training to provide this information with appropriate consideration.

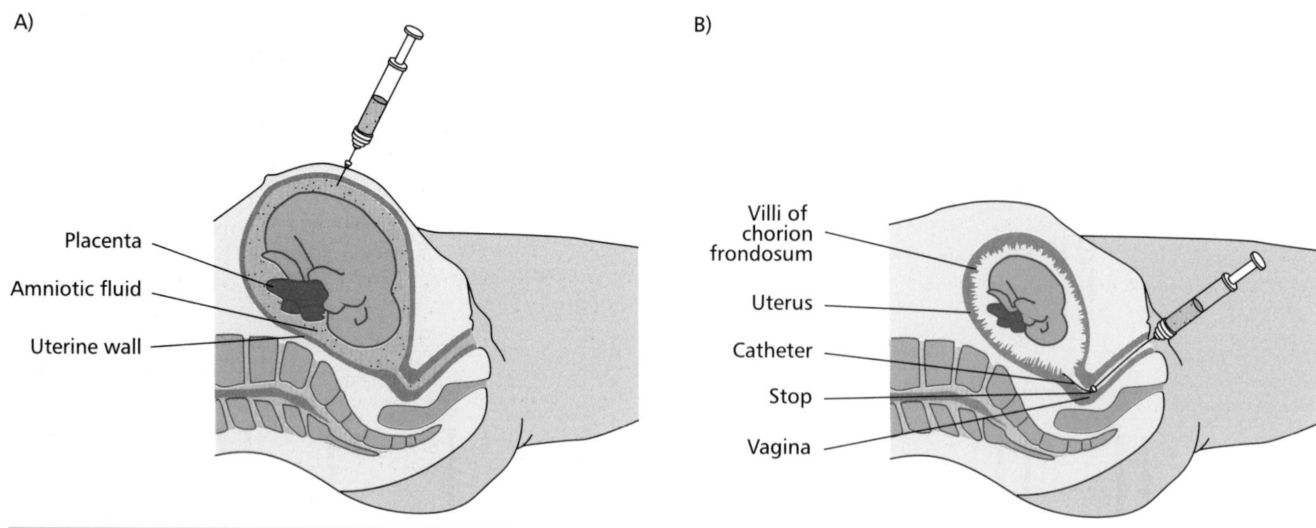

Figure 15.37 Outlines of (A) amniocentesis and (B) chorionic villus sampling to obtain specimens for genetic analysis of the fetus.

DETECTION OF HEREDITARY DISEASES

Tests based on recombinant DNA technology have increased in recent years because of their high sensitivity and accuracy in detecting genetic disorders at the prenatal stage. The detection of hereditary diseases in a given genome was traditionally a long and laborious task. However, modern techniques of chromosome and DNA analyses have resulted in a dramatic increase in the number of tests for genetic diseases with a considerable improvement in the time required for analysis. For example, using PCR, polymerase chain reaction (*Chapter 3*), virtually any portion of a gene or even whole genes can be amplified for analysis by electrophoretic techniques or sequenced to detect mutations. This not only allows for diagnosing patients with inherited disorders but also the detection of mutations in carriers, even though they do not express any symptom of the disorder. It is also a direct way of distinguishing different mutations within a single gene, each of which can lead to disorders, for example muscular dystrophies which are described in *Chapter 16*. It is hoped that eventually predictive tests for disorders that have only some genetic component, such as heart disease and cancers, will eventually be developed. Indeed, PCR analysis of cells lost in feces has demonstrated premalignant changes in the gastrointestinal tract and allowed patients at risk of developing colon cancer to be identified. This is of clinical importance since the earlier treatments of malignant conditions are started, the more favorable the prognosis.

Cloned DNA sequences have expanded the range of prenatal testing because they allow the fetal genotype to be examined directly, rather than relying on secondary tests for the products of the normal or mutant genes. Thus, mutations in DNA can be detected in those cases where an aberrant product cannot be detected prior to birth, even though a test is available. A test for sickle cell anemia using a restriction enzyme has been described in *Chapter 13*. However, RFLP analysis requires that the mutation in the gene alters the restriction sites recognized by restriction endonucleases and this is not always the case. In contrast, allele-specific oligonucleotides (ASOs) are synthetic nucleotide probes, which bind only to their complementary DNA. They will not hybridize to other sequences and, in appropriate highly stringent conditions, differentially bind to and distinguish between alleles that differ by as little as one nucleotide. They can thus distinguish between the native and mutant forms of a gene with excellent resolution and potential versatility.

BOX 15.4 Prenatal diagnosis for Down syndrome

The Quadruple Test for Down syndrome calculates the risk of having a Down fetus taking into account the maternal age at term and four markers, α-fetoprotein (AFP), unconjugated estriol, total human chorionic gonadotropin, hCG (or more usually its free β subunit) and inhibin-A at 14–22 weeks, in the maternal serum. There have been several variations on this test. In Down syndrome, the α-fetoprotein tends to be low compared with the normal (while in neural tube defects, for example spina bifida, it is high). However, concentrations of hCG are increased. A screening protocol was proposed in 1998 that combined maternal serum AFP, the urinary β-core fragment of hCG (a breakdown product of hCG) and total urine estradiol. This test is said to be superior to the 75% sensitivity at the 5% false-positive rate compared with when β-core/estradiol and maternal age alone are used.

Maternal age is important since birth prevalence of Down syndrome increases 100-fold between the maternal ages of 15 and 50 years. Statistics that relate the risk of bearing a Down syndrome child to the age of the mother (*Figure 15.33*) raise serious issues for women who become pregnant later in life. Genetic counseling may also be useful in informing the parents about the probability that their child will be affected and in educating them about Down syndrome. It is important that older pregnant women consider tests to determine whether the fetus has a normal complement of chromosomes. Amniocentesis or chorionic villus sampling and the culture of fetal cells will allow its karyotype to be determined.

The definitive test for Down syndrome is prenatal cytogenetic screening, that is karyotyping (*Section 15.7*), and in the developed countries this is routinely offered to pregnant women. It requires either amniocentesis or chorionic villus sampling. Karyotyping detects a range of numerical and structural abnormalities in the chromosomes in addition to Down and other common autosomal trisomies. However, the cells obtained from amniocentesis or chorionic villus sampling must be cultured to provided sufficient biological material and so full cytogenetic analysis involves a delay of 14 days or longer before a result can be given. Obviously the earlier a definitive positive result is obtained, the sooner the prospective parents can make clinically relevant decisions. Therefore 'molecular methods' have been investigated since these have the potential to give a result much more rapidly. Fluorescence *in situ* hybridization (FISH) is a technique in which fluorescently-labeled DNA probes that will hybridize to relevant regions of the chromosomes are applied to cell preparations (*Chapter 6*). When the preparations are examined with a fluorescent microscope the presence or absence of the target regions are revealed. Alternatively, PCR (*Chapter 3*) can be used to amplify these regions and the results can be seen on electrophoresis gels. Both methods can potentially give a result in 24–48 h. These molecular methods have been shown to be accurate since they would have failed to detect an abnormal karyotype in only about one in 100 amniocentesis samples. Obviously such tests will only give information about regions of the chromosomes for which DNA probes are applied. Many laboratories in the UK now offer FISH or PCR along with karyotyping, mostly to target pregnancies which are considered to be at high risk after screening by the Quadruple test. If the fetus is diagnosed as having Down syndrome, a therapeutic abortion is an option the parents may consider.

Tests based on ASOs are now available to screen for mutations associated with cystic fibrosis (*Chapter 16*) and glaucoma.

Allele-specific oligonucleotides are being increasingly combined with DNA microarrays, or chips. These are small glass slides or nylon membranes divided into squares, called fields, each of which contains a specific probe about 20 oligonucleotides long bound to the slide. DNA microarrays can be produced containing thousands of probes and can thus be used to analyze many different genes simultaneously. DNA is extracted from the cells and digested with one or more restriction endonucleases to produce small sized fragments that are labeled with a fluorescent dye. These are heated to separate the DNA into single strands which are then added to the microarray. Fragments whose sequence is complementary to a probe will hybridize with it and bind to the microarray; others will be washed off even if their complementary sequence differs from the probe by as little as one base. A laser based device is used to scan the array and determine where the binding has occurred to identify specific mutations (*Figure 15.38*). DNA chips are in development that will contain probes for all of the approximately 22 000 human genes to allow the simultaneous analysis of the DNA from a single person for hundreds of genetic diseases or genetic predispositions. The generation of such data has profound social, ethical and legal implications.

Potential parents in whom a genetic defect is detected must be advised to have genetic counseling. The advice they are given is based on analyses of the risks of them having a child with a genetic abnormality or that they may develop a late onset genetic disease. Information drawn from a wide range of sources is required in such circumstances and genetic counselors must provide such information with tact and sensitivity.

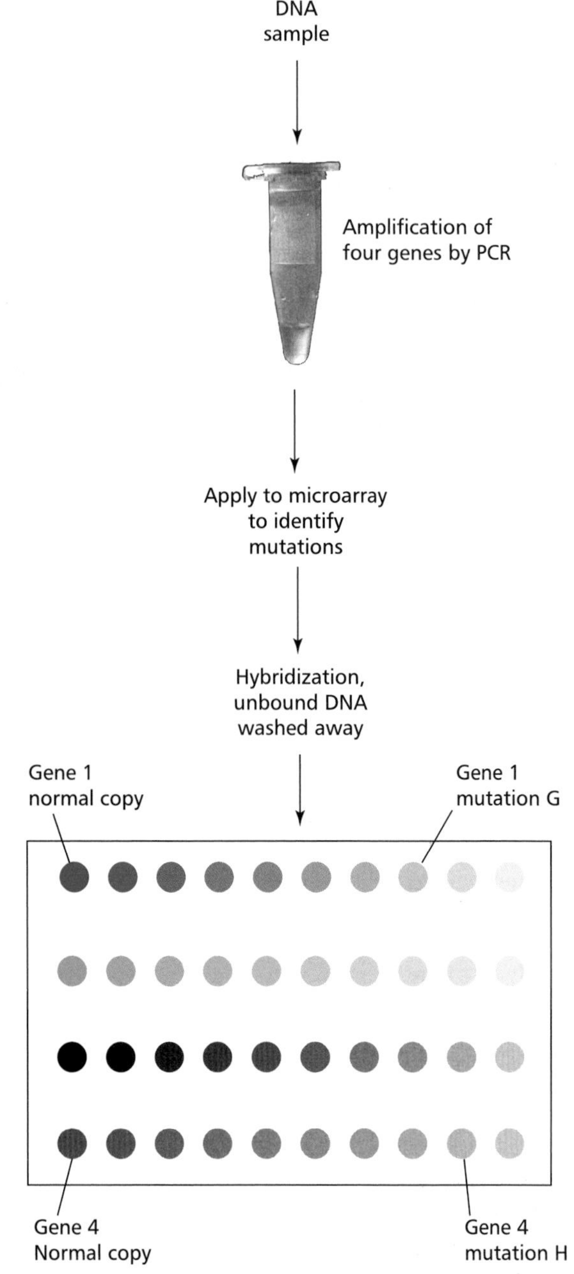

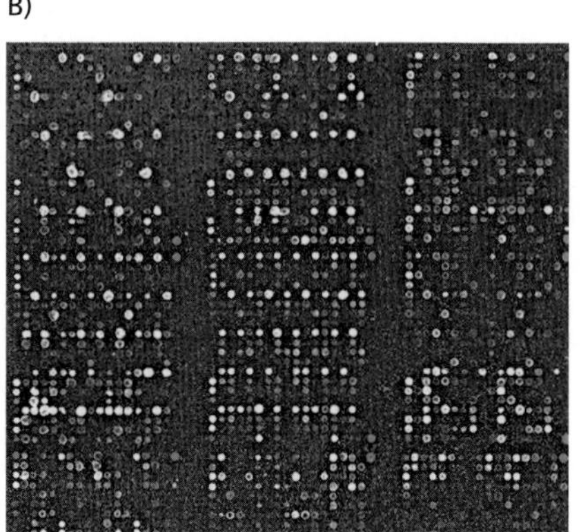

Figure 15.38 (A) Schematic to show the use of the polymerase chain reaction and microarray to amplify genetic material and detect specific mutations. For clarity, only four genes each with nine (A–I) mutations have been shown. See texts for details. Redrawn from Klug, WS and Cummings, MR (2002) *Essentials of Genetics*, 4th edn, Prentice Hall, NJ, USA. **(B) Detail from a developed microarray. The hybridization of nucleic acid samples to the probes of the microarray determines the pattern of fluorescence spots that can be analyzed by a laser.** Courtesy of Dr Q. Wang, School of Biology, Chemistry and Health Science, Manchester Metropolitan University, UK.

CASE STUDY 15.1

Christine was born after a normal pregnancy. For the first few days she was healthy but then started to vomit frequently. Her mother noted that her urine had a peculiar mousy smell. About a month after birth, Christine was admitted to hospital for the frequent vomiting where it was realized that the genetic screening test after her birth had been deficient. Her blood was analysed for phenylalanine as shown (reference value in parentheses).

Plasma [phenylalanine] 1.6 mmol dm^{-3} (<0.1 mmol dm^{-3})

Questions

(a) What is the diagnosis for Christine?

(b) How should Christine be treated?

(c) Why does Christine's urine have a characteristic mousy odor?

(d) What would happen to Christine if she were not treated?

CASE STUDY 15.2

Jane is a 39-year-old healthy woman who has just given birth to her first child, a son, Peter. She had wanted the pregnancy in her words 'to be completely natural' and had refused an amniocentesis. Unfortunately, Peter shows the very obvious physical characteristics of Down syndrome. Below is Peter's karyogram.

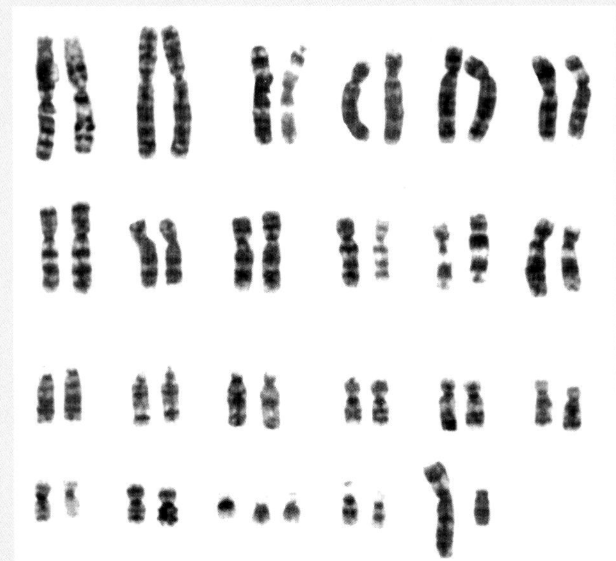

Questions

(a) Examine the karyogram. What is the cause of Down syndrome in this case?

(b) Given that a clinical diagnosis of Down syndrome is nonproblematical, why is it necessary to obtain a karyotype?

(c) What advice would you give Jane?

15.11 SUMMARY

The sequence of nucleotides of the genes in the DNA forms the genetic blueprint of cells, controlling all the cell's activities. This DNA sequence is transmitted to offspring in the gametes. Alterations in the sequence of nucleotides or to the chromosomal content of cells, that is, mutations, can lead to disease and/or death. Mutations may result in absence or abnormality of a protein, perhaps an enzyme. Chromosomal aberrations involve structural changes to individual chromosomes, such as deletions,

A)

1 2 3 4 5
6 7 8 9 10
11 12 13 14 15
16 17 18 19 20
21 22 X

B)

1 2 3 4 5
6 7 8 9 10
11 12 13 14 15
16 17 18 19 20
21 22 X Y

C)

1 2 3 4 5
6 7 8 9 10
11 12 13 14 15
16 17 18 19 20
21 22 X Y

duplications, inversions and translocations. Changes to the chromosomal complement include aneuploidy and euploidy. Down syndrome is the commonest aneuploid condition and is characterized by trisomy of the whole, or part of, chromosome 21. Genetic disease can be diagnosed using molecular techniques, including the use of the polymerase chain reaction and DNA probes in microarrays. Once a genetic defect has been diagnosed, genetic counseling can be offered to affected individuals as part of the treatment process.

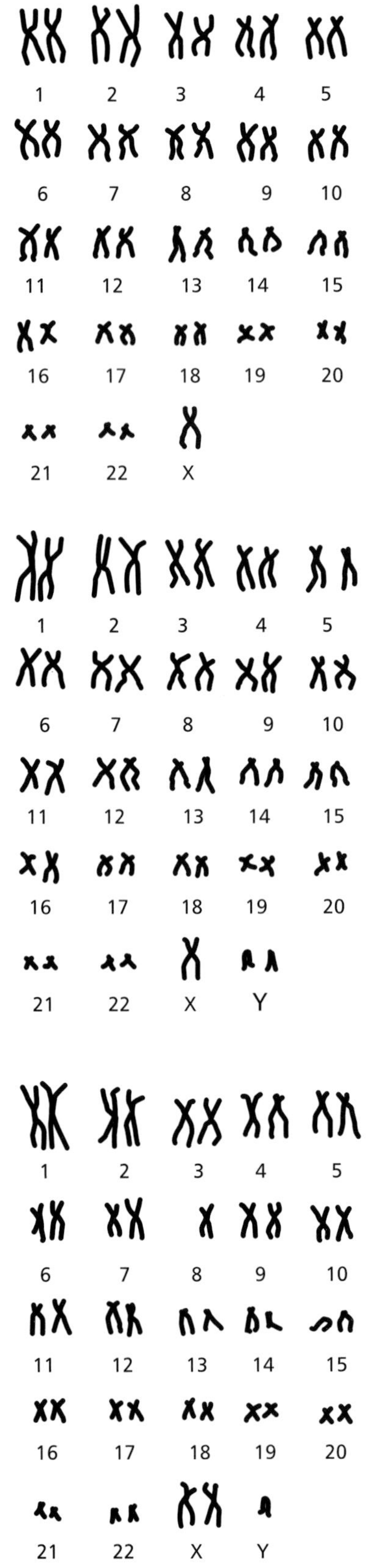

QUESTIONS

1. Which **ONE** of the following statements about phenylketonuria is **INCORRECT**?
   a) PKU sufferers should be placed on a diet low in phenylalanine.
   b) PKU has an autosomal dominant mode of inheritance.
   c) Individuals with untreated PKU may suffer from mental retardation.
   d) PKU has an incidence of around 1:10 000 in the UK.
   e) PKU sufferers have insufficient melanin in their skin and hair.

2. Which of the following has an aneuploid karyotype?
   a) A male with a balanced Robertsonian translocation involving chromosomes 13 and 21.
   b) A male sufferer of phenylketonuria.
   c) A female sufferer of fatal familial insomnia.
   d) A female with a balanced translocation involving chromosome arms 11q and 22q.
   e) A male with Klinefelter's syndrome.

3. Duchenne muscular dystrophy is an X-linked recessive trait. On average, what proportion of the children of a normal father and carrier mother would be affected?

4. A female is heterozygous for fragile X syndrome. Describe the inheritance pattern you would expect to see in her children, given the father is a normal male.

5. Autosomal dominant and X-linked dominant diseases can both be passed on to children. Suggest characteristics that would help identify whether a particular syndrome was inherited in an autosomal or X-linked fashion.

6. Examine the three accompanying karyotypes (A)–(C). In each case, name the syndrome caused by these karyotypes.

FURTHER READING

Bahado-Singh, RO, Oz, U, Kovanci, E, Cermic, D, Flores, D, Copel, J, Mahoney, M and Cole, L (1998) New triple screen test for Down syndrome: combined urine analytes and serum AFP. *J. Maternal-Fetal Med.* **7:** 111–114.

Caine, A, Maltby, AE, Parkin, CA, Waters, JJ and Crolla, JA (2005) Prenatal detection of Down's syndrome by rapid aneuploidy testing for chromosomes 13, 18, and 21 by FISH or PCR without a full karyotype: a cytogenetic risk assessment. *Lancet* **366:** 123–128.

Cederbaum, S (2002) Phenylketonuria: an update. *Curr. Opin. Pediatrics* **14:** 702–706.

Chakraborty, C, Nandi, S and Jana, S (2005) Prion disease: a deadly disease for protein misfolding. *Curr. Pharm. Biotechnol.* **6:** 167–177.

Clague, A and Thomas, A (2002) Neonatal biochemical screening for disease. *Clin. Chim. Acta* **315:** 99–110.

Cockle, H (1999) Maternal age-standardisation of prevalence of Down's syndrome. *Lancet* **354:** 529–530.

DeArmond, SJ and Bouzamondo, E (2002) Fundamentals of prion biology and diseases. *Toxicology* **181–182:** 9–16.

Kooy, RF, Willemsen, R and Oostra, BA (2000) Fragile X syndrome at the turn of the century. *Molec. Med. Today* **6:** 193–198.

Lanfranco, F, Kamischke, A, Zitzmann, M and Nieschlag, E (2004) Klinefelter's syndrome. *Lancet* **364:** 273–283.

Migeon, CJ and Wisniewski, AB (2003) Human sex differentiation and its abnormalities. *Best Pract. Res. Clin. Obst. Gynaecol.* **17:** 1–18.

Nativio, DG (2002) The genetics, diagnosis and management of Prader-Willi syndrome. *J. Ped. Health Care* **16:** 298–303.

Parmar, S and Tallman, MS (2003) Acute promyelocytic leukaemia: a review. *Expert Opin. Pharmacotherapy* **4:** 1379–1392.

Patterson, GD (1987) The causes of Down syndrome. *Sci. Am.* **257 (1):** 52–61.

Roizen, NJ and Patterson, D (2003) Down's syndrome. *Lancet* **361:** 99–110.

Schafer, AJ (1996) Sex determination in humans. *BioEssays* **18:** 955–963.

Scrivener, CR, Beaudet, AL, Sly, WS, Valle, D, Kinzler, KW and Vogelstein, B (eds) (2001) *The Metabolic and Molecular Bases of Inherited Disease.* McGraw-Hill, NY.

Stankiewicz, P and Lupski, JR (2002) Genomic architecture, rearrangements and genetic disorders. *Trends Gen.* **18:** 74–82.

Sybert, VP and McCauley, E (2004) Turner's syndrome. *New Engl. J. Med.* **351:** 1227–1238.

Taylor, AI (1968) Autosomal trisomy syndromes: A detailed study of 27 cases of Edwards syndrome and 27 cases of Patau syndrome. *J Med. Gen.* **5:** 227–252.

Terracciano, A, Chiurazzi, P and Neri, G (2005) Fragile X syndrome. *Am. J. Med. Gen.* **137C:** 32–37.

Wattendorf, DJ and Muenke, M (2005) Diagnosis and management of fragile X syndrome. *Am. Fam. Phys.* **72:** 111–113.

Interesting web sites:

http://www.cjd.ed.ac.uk/

www.doh.gov.uk/cjd

www.orpha.net/consor/cgi-bin/home.php?Lng=GB

The Online Mendelian Inheritance in Man (OMIM™) website. Address: http://www.ncbi.nlm.nih.gov/omim/

chapter 16:

MEMBRANE, ORGANELLE AND CYTOSKELETAL DISORDERS

OBJECTIVES

After studying this chapter you should be able to:

- list some clinical conditions that arise from specific defects in membranes, organelles and the cytoskeleton;
- relate the symptoms of some of these disorders of membranes, organelles and cytoskeletons to specific defects in their structures and activities;
- discuss the management and treatment of some membrane, organelle and cytoskeletal disorders.

16.1 INTRODUCTION

Membranes are essential for biological activities. A plasma membrane surrounds all eukaryotic cells. Individual organelles are also surrounded by a single and, in some cases, a double membrane or envelope. All these membranes share a common basic structure but differ in their individual compositions that are characteristically adapted to the functions of the cell or organelles in question. Defects in their compositions or structures lead to clinical problems that in many cases are extremely severe or fatal. The recent successes of genome sequencing projects have indicated that as many as 25% of all protein-coding genes may specify the structures of membrane proteins. The shapes and locomotion of cells depend on a highly organized arrangement of fibrous proteins called the cytoskeleton. This is also responsible for the active transport of some materials around the cytoplasm.

This chapter will describe a selected number of diseases or disorders associated with membranes, organelles and the cytoskeleton. These will include diseases associated with defects in nucleocytoplasmic transport; the plasma membrane disorder, cystic fibrosis, and diseases linked to mitochondria, lysosomes, peroxisomes and some cytoskeletal disorders. Many other membrane and organelle associated diseases are not included owing to a lack of space.

16.2 NUCLEUS AND NUCLEOCYTOPLASMIC TRANSPORT

The nucleus contains the genome of the cell stored as sequences of nucleotide bases in DNA molecules most of which are contained in chromosomes (*Chapter 15*). Other than a few mitochondrial genes (*Section 10.4*), all the human genes are found here. Surprisingly, it seems that less than 5% of the three billion pairs of nucleotides constitute the less than 30 000 genes of humans. The rest of the DNA is found interspersed as introns (*Chapter 13*) in between individual genes and as repetitive sequences. The function of these short sequences is unknown, although telomeres (*Chapter 18*) stabilize the ends of chromosomes and centromeres (*Chapter 15*) allow the spindle to attach to the chromosomes during cell division.

All cells, with the exception of mature mammalian erythrocytes, possess a nucleus that contains the chromosomes. The nucleus is separated from the cytoplasm by a **nuclear envelope** consisting of outer and inner nuclear membranes (*Figure 16.1*). Nuclear pores in the envelope (*Figure 16.1 (B)* and *(C)*) allow the transport of proteins into the nucleus and the export of ribosomal subunits, transfer RNA and messenger RNA molecules to the cytoplasm. This movement between the nucleus and cytoplasm is called **nucleocytoplasmic transport**. Nuclear pores have elaborate structures called nuclear pore complexes (NPCs) that regulate nucleocytoplasmic transport. The best studied NPCs are those from amphibians, such as the toad *Xenopus laevis*. Each NPC consists of around 100 proteins and has a M_r of approximately 125 x 10^6. These complexes are cylindrical with a ring-like structure of eightfold symmetry containing a central core. Fibrils, 50 to 100 nm long, extend from the cylinder into the cytoplasm while the inner surface has a basket-like attachment extending into the nucleoplasm. The 5600 pores of a typical mammalian nucleus are held in place by attachments to the nuclear envelope and to a scaffold of fibrous proteins called the **lamina** that lines and supports the inner face of the nucleus (*Figure 16.1(A)*).

Nucleocytoplasmic transport is a rather complicated process and only a simplified view of protein import is given here (*Figure 16.2*). Generally, proteins with a M_r larger than 30 000 cannot enter the nucleus by free diffusion but they can enter by active transport if they have a nuclear locating signal (NLS). This also requires the participation of a number of soluble import factors. The best known NLSs are rich in basic amino acid residues, such as that of the large T antigen of the simian virus 40 which has a sequence: Pro-Lys-Lys-Lys-Arg-Lys-Val. A protein with a NLS is called a cargo. An adapter protein recognizes the NLS signal of the cargo and binds to it. This complex then binds to an import receptor (*Figure 16.2*). Importin α and importin β are the best characterized adapter and import receptors respectively. This tripartite complex then passes through a NPC into the nucleus where it binds to a small protein called Ran. Ran is a GTPase, which can bind and then hydrolyze GTP to GDP. The conformation of Ran depends upon whether it has a bound GTP or GDP. The binding of Ran-GTP to the complex releases the cargo-adaptor in the nucleus and allows the import receptor-Ran-GTP to be shuttled to the cytoplasm. The adapter now releases its cargo within the nucleoplasm. Within the cytoplasm, yet another protein, GTPase-accelerating protein (Ran-Gap, not shown in *Figure 16.2* for the sake of clarity) stimulates Ran to hydrolyze its bound GTP. The conformation of Ran-GDP has only a low affinity for the import receptor, which is then released and can participate in a new cycle of nuclear import. The Ran-GDP is returned to the nucleus where it releases its GDP and binds a fresh GTP under the influence of a specific guanine nucleotide-exchange factor (Ran-GEF). The adapter in the nucleus binds to a nuclear export receptor that allows it to be returned to the cytoplasm.

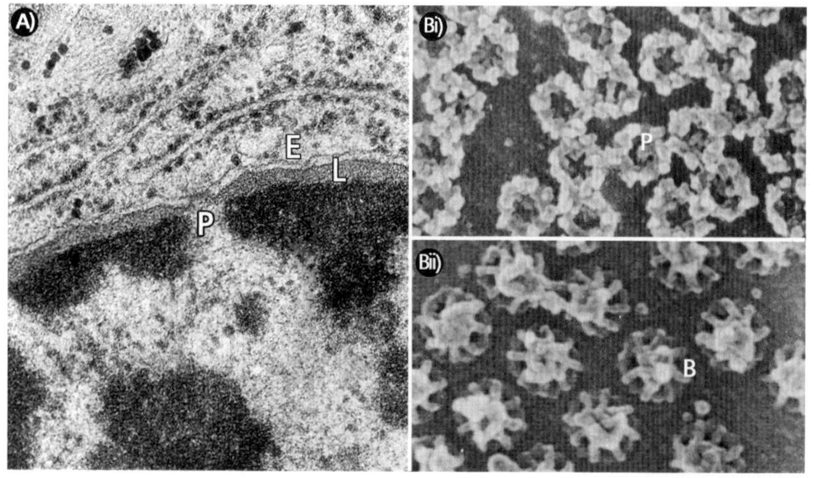

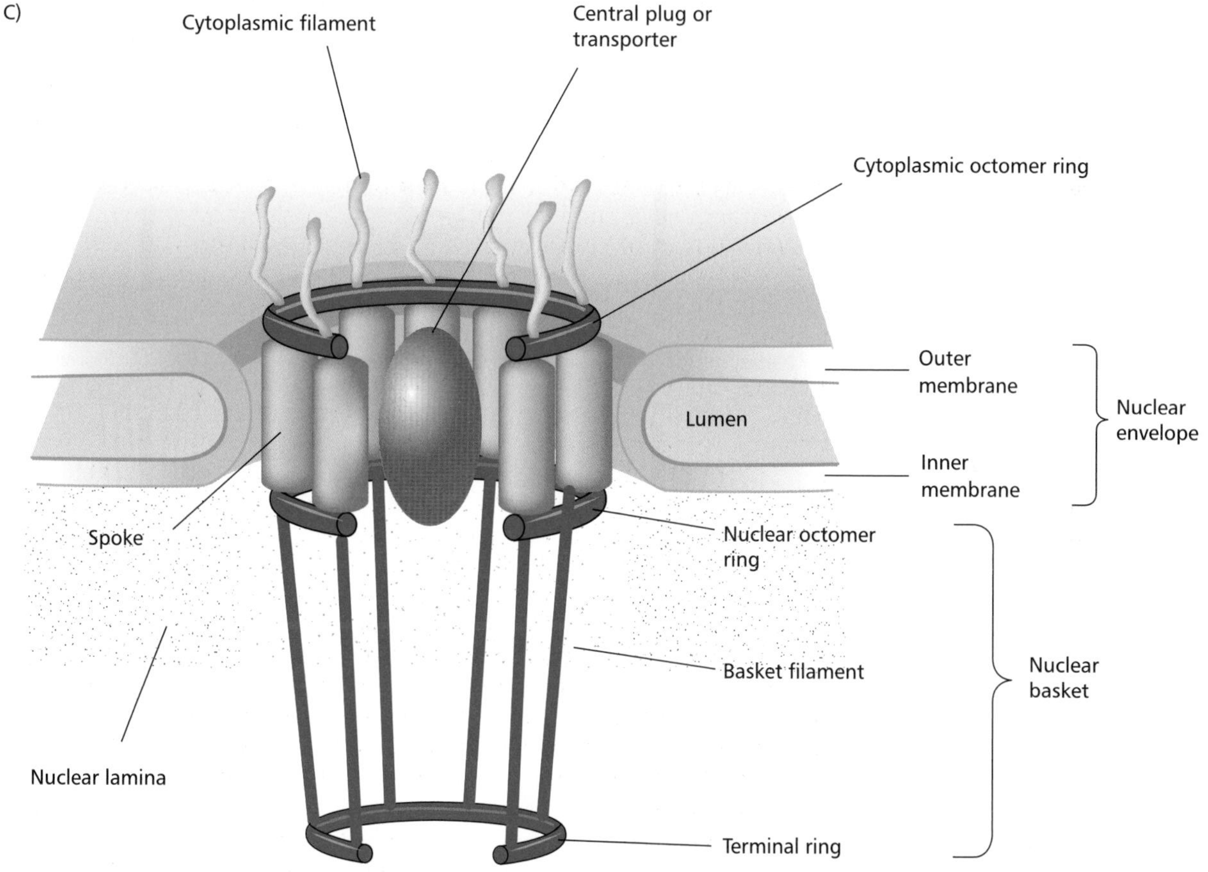

Figure 16.1 (A) Electron micrograph showing the nuclear envelope (E), internal lamina (L), nuclear basket (B) and nuclear pore (P). Picture by G. Szabo from Fawcett, D.W. (1981) *The cell*, 2nd edn, W.B. Saunders, with permission. Courtesy of Professor D.W. Fawcett. **(B) Electron micrographs of the (i) outer and (ii) inner faces of nuclear pore complexes approximately 90 nm in diameter.** Courtesy of Drs T.A. Allen and M.W. Goldberg, Department of Structural Cell Biology, Paterson Institute, Manchester, UK. **(C) Schematic of a nuclear pore complex.** See text for details.

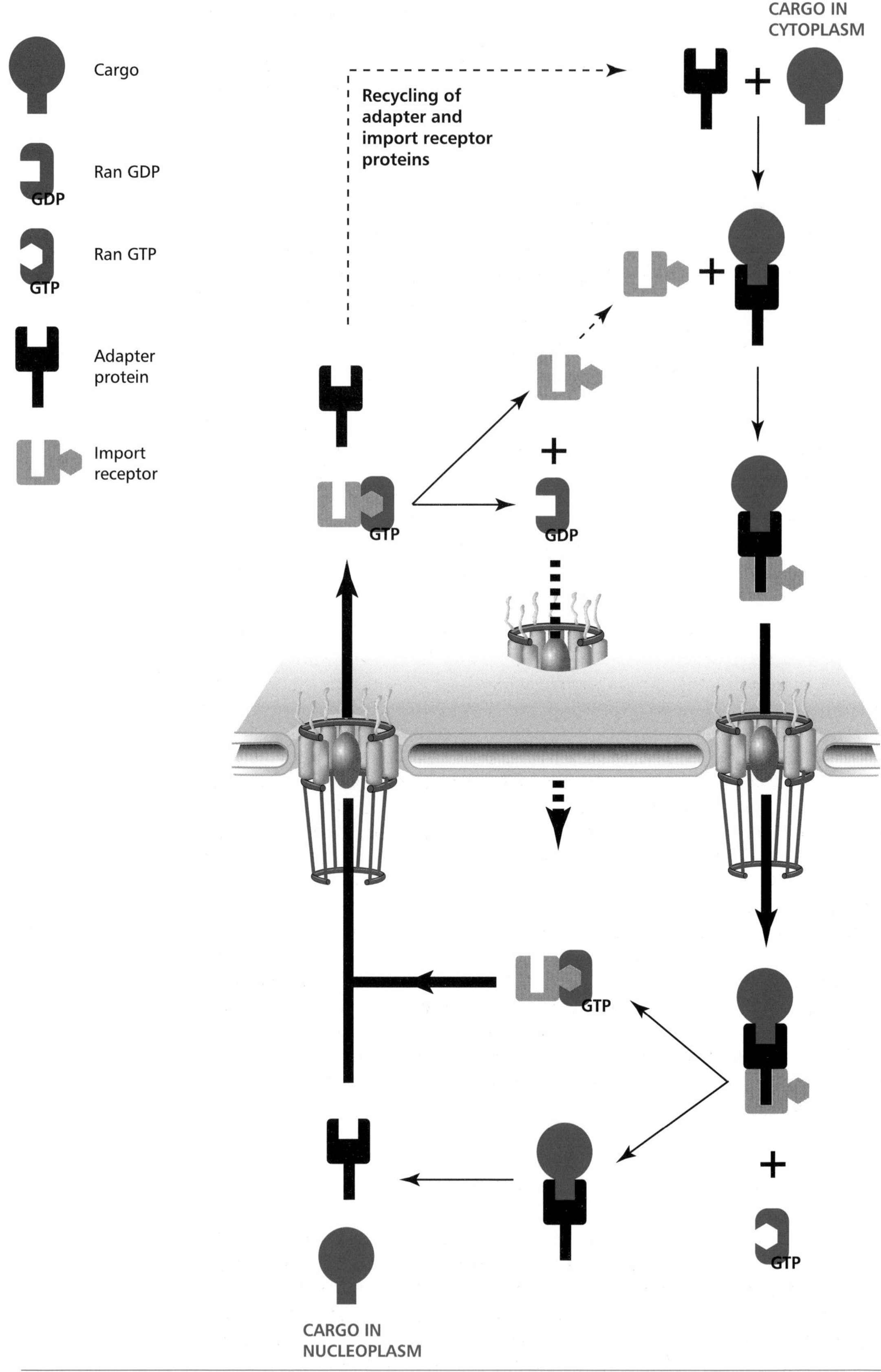

Figure 16.2 A highly simplified outline of the nucleocytoplasmic transport of a protein. Thin continuous arrows represent the formation and dissociation of protein complexes and the thick arrows their transport. The thin broken arrows indicate the recycling of the adapter and import receptor proteins. See text for details.

The export of proteins, transfer RNA molecules and ribosomal subunits from the nucleus to the cytoplasm appears to operate by similar mechanisms to that described above for nuclear import. The export of mRNA molecules, however, seems to involve an mRNA exporter protein that directs their passage through the NPCs.

CLINICAL ASPECTS OF NUCLEOCYTOPLASMIC TRANSPORT

Given that NPCs are the pivotal junctions between the cytoplasm and the nucleus, they are closely involved in the pathology of viral infections (*Chapter 2*). Virus particles fuse with the plasma membrane or are taken up into the cells by endocytosis. However, they must leave the cytoplasm and enter the nucleus. For example, the hepatitis B virus (HBV) has a capsid of diameter 32–36 nm and is small enough to cross the NPC directly. *In vivo*, phosphorylation of HBV capsids makes their NLSs accessible to the nuclear transport machinery. Following their transport through the central pore of the NPC, capsid protein and DNA from mature viruses are released into the nucleoplasm. In contrast, adenovirus particles and herpes simplex virus (HSV) have capsids with diameters of 90 nm and 125 nm respectively that are too large to traverse the NPC. These virus particles dock at the cytoplasmic side of the NPC in an importin dependent manner. The HSV then releases its DNA through the NPC into the nucleoplasm. Adenovirus particles, however, first trap a variety of transport factors that are necessary for the subsequent disassembly of the capsid before the translocation of the viral DNA into the nucleus is possible.

Malfunctions of nucleocytoplasmic transport are associated with a number of diseases including Huntington's disease (*Chapter 15*). In Huntington's disease, a single gene is mutated and the corresponding protein, **huntingtin**, accumulates in the nucleus rather than in the cytoplasm. Nuclear pore complexes have also been implicated in several autoimmune diseases (*Chapter 5*) that are caused by autoantibodies recognizing proteins of the NPCs. This contributes to a number of diverse illnesses, such as systemic lupus erythematosus, rheumatoid arthritis and primary biliary cirrhosis. Nucleoporins have been implicated in several types of cancers (*Chapter 17*). In some cases, these are associated with the overexpression of NPC proteins, but in most other cases tumorigenesis stems from chromosomal rearrangements that result in oncogenic fusion proteins. Examples include tumors such as acute myeloid leukemia, adenocarcinoma, chronic myeloid leukemia, inflammatory myofibroblastic tumor, myeloblastic leukemia, osteosarcoma and papillary thyroid carcinomas, which affect a variety of tissues. At least 11 chromosomal rearrangements in acute leukemia involve genes that encode nuclear pore proteins; nine of these involve a single protein called nucleoporin 98. Some studies have suggested that the density of NPCs in aneuploid bladder tumors that have cells with abnormal numbers of chromosomes is significantly higher than in normal cells.

16.3 PLASMA MEMBRANE DISORDERS

Numerous clinical conditions are associated with defective receptors of the plasma membrane and/or subsequent defective signal transduction or defective transport of materials across the membrane. Defects of receptor proteins and signal transduction can result in the development of cancerous states (*Chapter 17*), while cases of type 2 diabetes are associated with an ineffective insulin receptor (*Chapter 7*). Familial hypercholesterolemia, which was described in *Chapter 14*, arises from defective receptors for LDL particles. The rare Tangier disease mentioned in *Chapter 14* is caused by mutations in the gene for cholesterol transport protein of the plasma membrane. This protein is one of a group of transporters called ATP-binding cassette (ABC) proteins. These are multidomain structures: all have two cytosolic ATP-binding domains or cassettes and two transmembrane domains each consisting of several helices. However, the most studied of the ABC transporters is the transmembrane conductance regulator protein, which regulates a chloride channel. Unlike Tangier disease, which is extremely rare, defects with this transporter lead to the much more common condition, cystic fibrosis.

CYSTIC FIBROSIS

Cystic fibrosis (CF) is a multisystem, inherited disorder characterized by the secretion of very viscous mucus secretions in the lungs, digestive tract and associated organs, and epididymis. This results in chronic respiratory disease, malabsorption, cirrhosis and electrolyte disturbances. The clinical features of CF are shown in *Table 16.1*.

Feature	Consequences
Recurrent respiratory infections	irreversible lung disease
Pancreatic insufficiency	malabsorption; poor growth; steatorrhea (excessive fat in feces)
Intestinal obstruction in neonates (meconium ileus)	failure to thrive
Diabetes mellitus (late onset)	failure to control blood sugar levels
Epididymal epithelial malfunction	male sterility

Table 16.1 Clinical features of cystic fibrosis

Cystic fibrosis is the commonest fatal, homozygous recessive disorder of the Caucasian population affecting about one in 2000 people in the UK, with one in 20 Caucasians carrying one copy of the mutated gene. The prevalence of CF varies throughout the world with certain populations reporting higher incidences. The onset may be at birth or later in childhood. Cystic fibrosis is characterized by decreased permeability of the apical membrane of epithelial tissues lining the lungs and other organs to chloride ions (Cl^-) that results from mutations (*Chapter 15*) in the *CFTR* gene that encodes the cystic fibrosis transmembrane conductance regulator (*Figure 16.3*). This protein is a

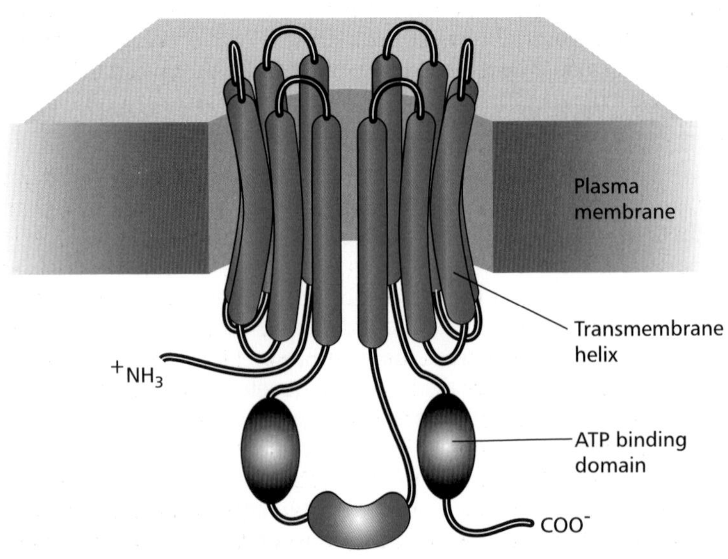

Figure 16.3 Schematic of the cystic fibrosis transmembrane conductance regulator protein (CFTR). Redrawn from Akabas, MH (2000) Cystic fibrosis transmembrane conductance regulator. *J. Biol. Chem.* **275:** 3729–3732.

cyclic AMP-regulated chloride channel and has been identified, cloned and sequenced and most of its mutations are known. The decreased permeability results in many exocrine secretions with altered ion and reduced water contents and high viscosities. The appropriate viscosity is essential for normal functions of lubrication, digestion and protection by proteins secreted at epithelial surfaces. The highly viscous secretions can cause obstructions in ducts and lead to an accumulation of mucus and colonization by bacteria. Considering that CF is essentially a fatal disease, one may wonder why the mutated genes persist in the population. One theory is that mutations in *CFTR* increase resistance to cholera (*Chapters 2* and *3*). The toxin of *Vibrio cholerae* (*Figure 16.4*) binds to surface receptors on the plasma membranes of gut enterocytes, leading to a voluminous secretion of fluid and Cl^- resulting in a secretory diarrhea that can be fatal if untreated. Cystic fibrosis mice that do not express *cftr* do not secrete fluid in response to cholera toxin. Heterozygous CF mice that express half the normal amount of cftr protein secrete about half the normal fluid and Cl^- in response to cholera toxin. Thus CF heterozygotes might have a selective advantage in being resistant to cholera.

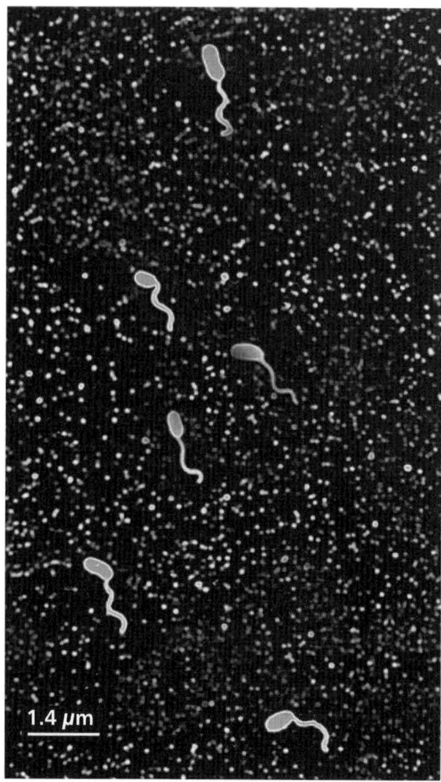

Figure 16.4 Light micrograph of *Vibrio cholerae*. Courtesy of Public Health Image Library, Centers for Disease Control and Prevention, USA.

Diagnosis and treatment of cystic fibrosis

A diagnosis of CF can be confirmed by demonstrating a concentration of Cl^- in sweat higher than 70 mmol dm^{-3}. Pilocarpine is first applied topically to patients to stimulate the flow of sweat, samples of which are then collected and analyzed for Cl^-.

A number of screening tests for CF are now available based on the effects of CF on pancreatic function. The most popular of these relies on measurement of immunoreactive trypsin (IRT) in dried blood spots, similar to those collected for PKU (*Chapter 15*). Immunoreactive trypsin is greatly increased in infants with CF in the first month of life compared with that of healthy infants. However, the test cannot be used after the first few weeks of life since IRT levels fall as pancreatic insufficiency develops.

The identification of mutations in the *CFTR* of a cystic fibrosis patient will confirm the diagnosis and it can also assist with carrier testing of the relatives of the patient. It is also used to give an accurate and rapid prenatal diagnosis of CF once the parental genotypes have been confirmed. Currently there are over 1300 mutations listed in the *CFTR* mutation database with the largest number (500) being attributed to missense mutations. When it is not possible to detect mutant *CFTR* alleles, then the sweat test remains the definitive test for diagnosis of CF but, unfortunately, the sweat test is not always reliable in the first six weeks of life or in adulthood. Furthermore, some patients with CF, including a number with severe lung disease, give a normal sweat test.

The techniques for genetic analysis of CF are based on the detection of an alteration in the DNA sequence of the *CTFR* gene. In principle, any method capable of detecting a change in the DNA sequence can be used in mutation detection. Current techniques rely in one way or another on the hybridization of two single strands of DNA and on the ability of changes in the sequences of DNA to alter the kinetics of hybridization. These techniques depend on the polymerase chain reaction or PCR (*Chapter 3*) and this, in turn, requires that the normal sequence of the gene, including its intron–exon boundaries (described in *Chapter 13*) be known, as is the case for CF. Polymerase chain reaction primers are designed that will bind to adjacent regions of the gene of interest and allow it to be amplified. In most cases, whole exons are amplified together with a short region of their neighboring intron sequences to give a PCR product 300 to 500 base pairs long. These products are then analyzed for specific point mutations or for generalized, nonspecific evidence of a sequence change. The testing strategy may use both approaches depending on the purpose of the analysis.

The commonest CF mutation is called ΔF508. This is the deletion of a codon, in exon 10 of the *CFTR* gene, that codes for a phenylalanine (F) residue at position 508 in the amino acid sequence of the encoded protein. The simplest way to detect this mutation (*Figure 16.5*) is by amplifying that part of the gene

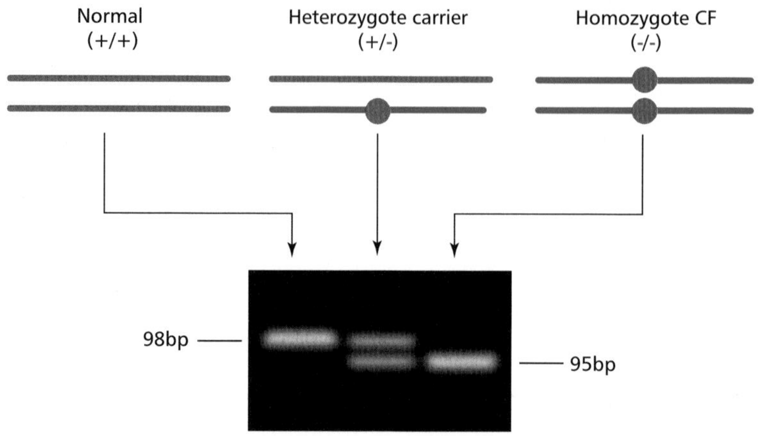

Figure 16.5 Detection of the ΔF508 mutation. Shows the differences in sizes of the relevant portions of normal and mutated genes and a simplified schematic of the separation of these by polyacrylamide gel electrophoresis (PAGE).

and looking for differences in the sizes of the PCR products. Polyacrylamide gel electrophoresis (PAGE) can be used to separate the products of PCR of the small fragment of exon 10 of the *CFTR* gene and these may be visualized with ethidium bromide or silver staining. The primers most often used are called C16B and C16D and generate a 98bp product of exon 10. However, if the ΔF508 mutation is present, the PCR product is only 95bp long. This is a relatively small difference but careful PAGE can easily separate them. This test is not absolutely specific for ΔF508 given that other 3bp deletions in this region would produce the same difference in size. Thus the mutation ΔI507, the second identified CF mutation, also gives a 95bp PCR product. Mutations other than ΔF508, however, can be identified in many cases because they produce differing banding patterns following PAGE.

The treatment for cystic fibrosis is aimed at preventing respiratory infections by regular physiotherapy to try to remove the mucus, along with the use of antibiotics to counter bacterial infections (*Chapter 3*). Pancreatic enzymes, in tablet form, are given with food to counter the effects of pancreatic insufficiency.

Mice with deletions of *cftr*, that is gene knockout mice, have been bred to use as experimental models to test possible therapies. However, such mice do not develop lung disease in the same manner as humans and therefore do not provide an effective animal model. This has hampered experimental gene therapy studies that have tried to insert the normal gene into the lung epithelial cells of CF patients by using DNA-containing liposomes. To date, none of these therapies have been successful. Cystic fibrosis used to be fatal by the age of about 20 years but improvements in treatment mean that many sufferers now live to be 30 years and over. However, although the prognosis for CF has improved, many patients still die in early adult life.

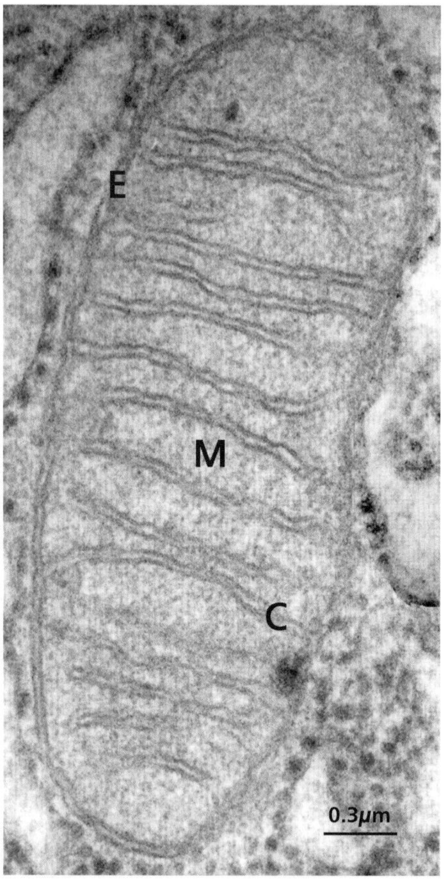

Figure 16.6 Electron micrograph of a mitochondrion. C, cristae, E, envelope and M, matrix.

16.4 MITOCHONDRIAL DISORDERS

Mitochondria are organelles found in almost all eukaryotic cells although numbers vary from one to several hundred per cell. They are bounded by a mitochondrial envelope that consists of outer and inner mitochondrial membranes with infoldings called cristae, and encloses a central region called the mitochondrial matrix (*Figure 16.6*). The inner mitochondrial membrane contains five protein complexes (*Table 16.2*) that together with cytochrome *c* perform the energy transduction reactions with the formation of ATP by oxidative phosphorylation. The matrix contains enzymes that catabolize fuel molecules to yield the reduced coenzymes NADH and $FADH_2$ necessary for oxidative phosphorylation, and produces small organic molecules that are the precursors in biosynthetic metabolism. The mitochondria also help in maintaining the intracellular homeostasis of many metabolites and ions, including Ca^{2+} and H^+.

Complex	Name	Comment
Complex I Complex II	reduced nicotinamide adenine dinucleotide (NADH): ubiquinone reductase succinate: ubiquinone oxidoreductase	complexes I and II transfer electrons from NADH and succinate respectively to the mobile carrier, **ubiquinone** which then transfers them to complex III.
Compex III	ubiquinone: ferricytochrome *c* oxidoreductase	complex III then transfers electrons to the second mobile carrier, cytochrome *c*, which transfers them to complex IV.
Complex IV	cytochrome *c* oxidase	complex IV donates electrons, in turn, to dioxygen.
Complex V	ATP synthase	synthesizes ATP from ADP and P_i when stimulated by H^+.

Table 16.2 The protein complexes of the inner mitochondrial membrane involved in oxidative phosphorylation

Mitochondria are also involved in cellular responses to environmental toxins that can stimulate them to release a potent mixture of enzymes and other proteins that eventually bring about the death of the cell. This is one form of **apoptosis** or cell suicide. These apoptotic mechanisms are, of course, under strict regulation, but defects in these controls cause some neurodegenerative and autoimmune diseases and some cancers.

Mitochondria are unusual organelles, in that the matrix also contains molecules of mitochondrial DNA (mtDNA) containing mitochondrial genes, in addition to mitochondrial ribosomes and mitochondrial transfer RNA molecules. Mitochondrial ribosomes differ in size and structure from those present in the cytosol.

Human mitochondrial DNA is a circular double-stranded molecule of 16 569 base pairs present in multiple copies in each mitochondrion. Human mtDNA has been completely sequenced and contains only 37 genes (*Figure 16.7*). Twenty-eight genes are encoded on one of the strands of the DNA and nine genes on the other. These genes encode 22 transfer RNAs, the 12S and 16S mitochondrial ribosomal RNAs, and 13 mitochondrial polypeptides that are involved in oxidative phosphorylation. Thus, mitochondria have the genetic information and machinery to synthesize proteins. Mitochondria contain approximately 1000 different proteins, that is about 3% of all cellular proteins. It follows that the majority of mitochondrial proteins are encoded by nuclear genes and are synthesized on ribosomes in the cytosol. These proteins require transportation across one or both of the mitochondrial membranes to reach their sites of activity.

CAUSES OF MITOCHONDRIAL DISORDERS

Mitochondrial disorders are a heterogeneous group of disorders resulting from impairment of the mitochondrial oxidative phosphorylation system.

Figure 16.7 Schematic of a human mtDNA molecule showing its major regions and the sites of mutations associated with some mitochondrial disorders (see *Table 16.3*). This diagram was compiled from several sources.

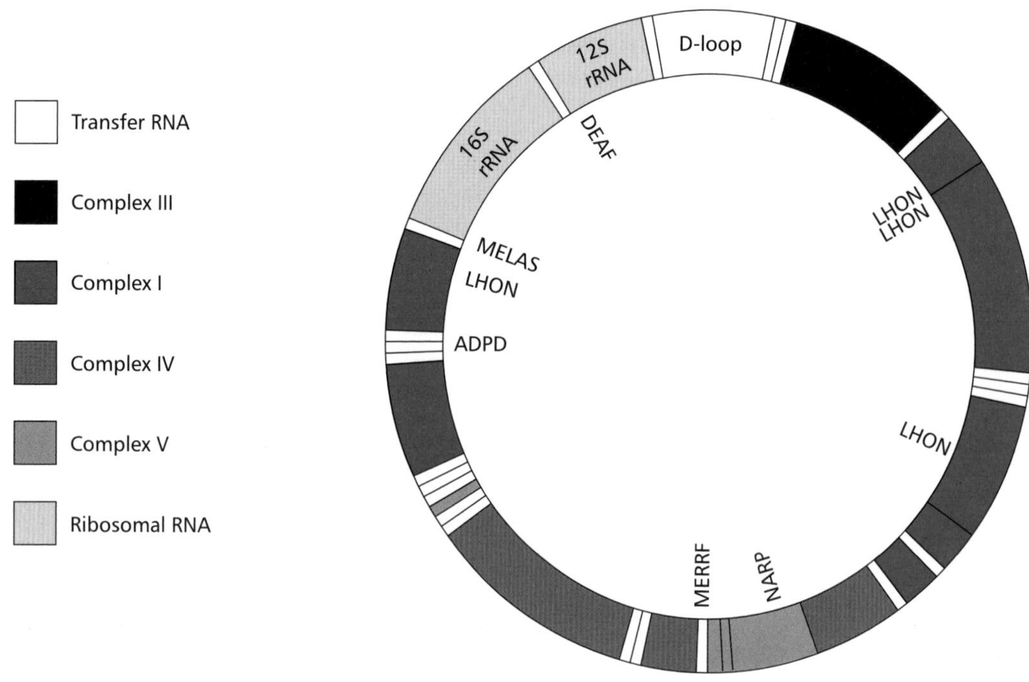

They are associated not only with numerous mutations of the mtDNA but also mutations in nuclear genes that affect processes such as the assembly of protein subunits and the import of proteins from the cytosol into the mitochondrion. Such mutations can affect the morphology of the mitochondrion (*Figure 16.8*).

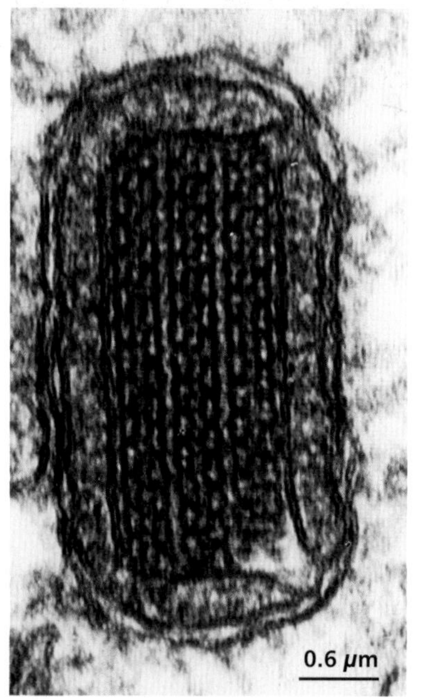

Figure 16.8 Electron micrograph showing abnormal morphology of a mitochondrion associated with a mitochondrial disorder. Compare with *Figure 16.6*. Courtesy of M.J. Cullen, Muscular Dystrophy Research Laboratories, Newcastle General Hospital, UK.

Mitochondria replicate by simple divisions that are independent of mitosis and meiosis. The mitochondrial chromosomes are copied prior to replication. However, the replicating enzyme, mtDNA-dependent DNA polymerase or DNA polymerase γ, replicates DNA with much poorer fidelity than the nuclear DNA polymerases. Thus the mutation rates of mitochondrial genes are estimated to be about 10 times greater than those of nuclear genes. This may contribute to the aging process (*Chapter 18*). Furthermore, the mitochondrial genome resembles that of bacteria in that the genes lack introns and repeated DNA sequences so that about 93% of the DNA is transcribed as opposed to the 5% or less for the nuclear genome. Thus, despite its relatively small size, a relatively large number of mutations are associated with mtDNA and these mutations are significant contributors to human disease.

Mitochondrial disorders are caused by similar types of mutations to those that cause diseases associated with the nuclear genome, that is, point mutations, deletions, and duplications (*Chapter 15*). However, the inheritance of these diseases differs strikingly to that of the nuclear genetic diseases. In zygote formation, a spermatozoon contributes its nuclear genome but not its mitochondrial genome to the egg cell. The resulting fertilized zygote contains only the mitochondria that were present in the unfertilized egg and are therefore entirely maternal in origin. Thus all children, male and female, inherit their mitochondria only from their mother, and males cannot transmit their mitochondria to subsequent generations (*Figure 16.9*). Thus a typical disease resulting from a mutation in mtDNA is an inherited condition that can affect both sexes but can only be passed on by affected mothers.

The first disease to be directly linked to mutations in mtDNA was Leber hereditary optic neuropathy (LHON), an inherited neuropathology. This

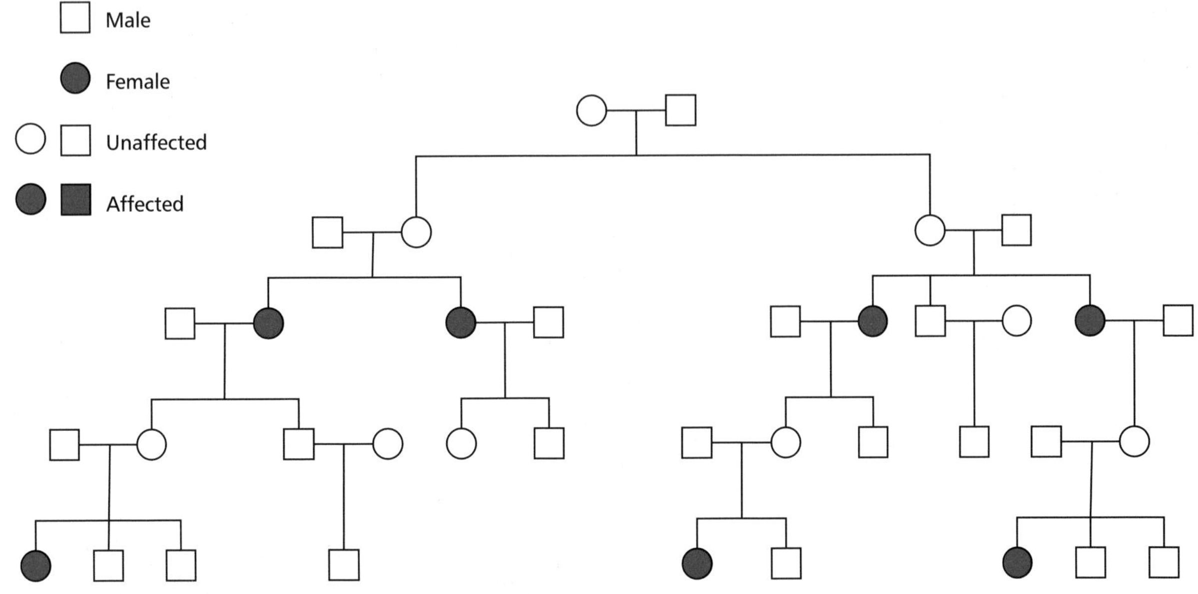

Figure 16.9 An outline of the mode of inheritance associated with many mitochondrial diseases. Redrawn from Wallis, C. (1999) *The Genetic Basis of Human Disease,* Portland Press, UK.

disease presents in midlife with a rapid bilateral loss of central vision due to atrophy of the optic nerve leading to blindness. Early studies of LHON showed a puzzling maternal pattern of inheritance, with more males than females affected. In 1988 it was demonstrated that LHON was caused by mutations in the mtDNA. Three main mutations are present in at least 90% of families with the disease and all cause substitutions of highly conserved amino acid residues in the affected mitochondrial proteins. Thus the mtDNA mutations explain the maternal transmission of LHON although the reason why a higher proportion of males than females are affected is not known. It has been suggested that the expression of the disease could also require the coinheritance of an X-linked recessive mutation; the development of LHON could be hormonally influenced by androgens or by environmental factors, such as heavy tobacco smoking, may also contribute to the progression of the condition.

More than 100 pathogenic point mutations and innumerable rearrangements of mtDNA are known that lead to general, multisystem disorders (*Table 16.3*), which, like LHON, are often neuromuscular in nature. Perhaps not surprisingly, most mitochondrial disorders tend to affect the most energy demanding tissues, such as the central nervous system, heart and skeletal muscles, kidneys and secretory tissues, although this is not always the case since mutant proteins may interfere with mitochondrial functions in addition to ATP formation. The diseases also vary in severity in patients with the same mtDNA mutation. The term **homoplasmy** refers to the condition in which every mtDNA molecule has the same causative mutation. In contrast, **heteroplasmy** occurs when the cell has a mixed population of normal and mutant mitochondria. If heteroplasmy occurs, the severity and symptoms of the disease may depend on the proportion of abnormal mtDNA in some critical tissue. The proportion can also differ between mother and child because of the random segregation of mtDNA molecules at cell division.

Mitochondrial disorders are far more common than was previously thought and a substantial increase in the number of genetic defects have been

Margin Note 16.1 Acquired mitochondrial DNA defects 

The most common cause of acquired mitochondrial DNA defects occurs in HIV-1 sufferers treated with highly active antiretroviral therapy (HAART). Highly active antiretroviral therapy for HIV-1 is a combination therapy (*Chapter 3*) which uses inhibitors of the viral reverse transcriptase and the HIV protease. The first inhibitor of reverse transcriptase to be used clinically was the nucleoside analog Zidovudine or AZT (*Figure 3.32*). An unwanted effect of this drug is to inhibit the replication of mtDNA and this depletes the amount of mtDNA and leads to side effects that mimic the symptoms of inherited mitochondrial disorders. Unfortunately such effects are common to most nucleoside analog inhibitors of reverse transcriptase. It is now routine practice to monitor mitochondrial functions in patients treated with HAART.

Disease	Mutation	General features
Chronic progressive external ophthalmoplegia (CPEO)	deletion with loss of several genes	paralysis of eye muscle, myopathy
Kearns-Sayre syndrome (KSS)	deletion with loss of several genes	CPEO and retinal deterioration, ataxia, heart disease, loss of hearing, kidney failure and diabetes
Leber hereditary optic neuropathy (LHON)	mutations in genes for subunits of NADH dehydrogenase	damage to optic nerve leading to blindness
Leigh syndrome	mutations in genes for subunits of ATP synthase and inner membrane oxidation complexes	loss of verbal and motor skills following degeneration of basal ganglia
Mitochondrial encephalomyopathy, lactic acidosis and stroke-like episodes (MELAS)	mutations in genes for $tRNA^{Leu}$, $tRNA^{Glu}$ and NADH dehydrogenase	myopathy, dementia, seizures, lactic acidosis
Myoneurogastrointestinal encephalopathy (MNGIE)	multiple mtDNA deletions	progressive external ophthalmoplegia, gastrointestinal dysmotility (often pseudo obstruction), diffuse leukoencephalopathy, myopathy
Myoclonic epilepsy and ragged-red fiber (MERRF)	mutations in genes for $tRNA^{Lys}$, $tRNA^{Leu}$	myoclonic seizures, ataxia, lactic acidosis, 'ragged-red fibers' – an abnormality of the tissue when seen in a microscope
Neurogenic muscle weakness, ataxia and retinitis pigmentosa (NARP)	mutations in genes for subunits of ATP synthase	muscle wasting, ataxia, blindness
Pearson syndrome	deletion with loss of several genes	childhood bone marrow dysfunction, multiple blood disorders and pancreatic failure

Table 16.3 Examples of mitochondrial diseases

reported in patients. In the UK, it is estimated that mitochondrial disorders occur with a frequency of about seven in 100 000.

DIAGNOSIS AND TREATMENT OF MITOCHONDRIAL DISORDERS

Considerable advances have been made in our understanding of the pathology of mitochondrial disorders. However, the diagnosis and detection of mtDNA mutations is problematic because of the varied etiology and clinical features of these diseases (*Table 16.3*). Similarly, predictors of disease progression are also highly unsatisfactory. The concentrations of lactate in plasma, cerebrospinal fluid and urine may, individually or collectively, be increased relative to normal, although these changes are also seen in numerous other clinical conditions. However, a plasma lactate : pyruvate ratio of greater than 40 is usually considered a significant indicator of mitochondrial dysfunction in adults. Other indicators may be the presence of organic acids and myoglobin in the urine, ketoacidosis (*Chapter 7*), and impaired renal, liver and glandular functions.

It would appear that several different approaches to treating and managing mitochondrial disorders will be necessary. Most current treatments of mitochondrial disorders are unsatisfactory and concentrate merely on relieving symptoms. For example, administration of hydrogen carbonate and dialysis may be used to relieve lactate acidosis. Exercise may decrease the proportion of mutant mtDNA in muscle. Treatment with standard vitamins and ubiquinone supplements has been tried but any benefits have yet to be established.

Gene therapy using **allotopic expression** may offer a permanent cure in the future. The strategy behind this technique is to insert a mitochondrial gene into the nucleus where it is transcribed. The resulting messenger RNA is translated in the cytosol and the protein product is then imported into the mitochondria where it alleviates the effects of the mtDNA mutation.

Families affected by mitochondrial disorders should be offered counseling on prognosis, risk of recurrence and prevention but even this is complex, because the degree of mutant mtDNA can change dramatically during each transmission of mitochondria from a mother to a child. The advice to affected families on potential reproductive options may therefore be unsatisfactory. Consequently, the clinical management of patients with mitochondrial disorders is largely supportive.

16.5 PEROXISOMAL DISORDERS

Peroxisomes (*Figure 16.10*) are organelles surrounded by a single membrane, that carry out certain oxidation reactions, particularly those involved in the partial degradation of long chain fatty acids. Unlike mitochondria, these oxidations are not coupled to the formation of ATP. Peroxisomal proteins are synthesized in the cytosol and imported into the organelles by a complex process that relies upon the relevant protein possessing a peroxisomal targeting signal (PTS). The major signal (PTS1) consists of the carboxyl terminal three amino acid residues being the consensus sequence –Ser-Lys-Leu-COO^-.

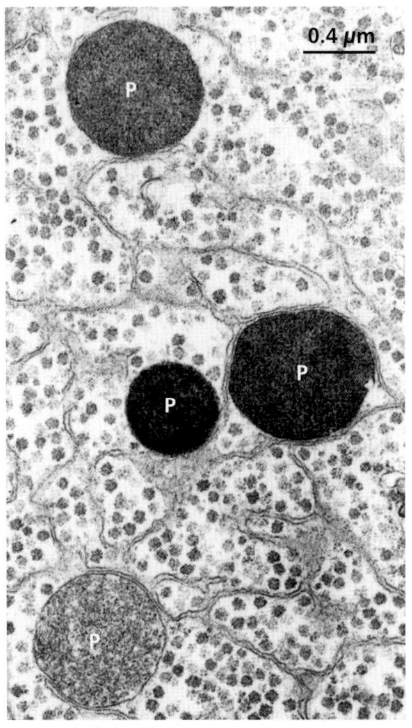

Figure 16.10 Electron micrograph of liver peroxisomes (P) identified by catalase activity. Note the intensity of staining varies between individual organelles. Courtesy of Drs M. Espeel and F. Roels, Department of Anatomy, Embryology and Histology, University of Ghent, Belgium.

Defects in peroxisomes produce a variety of lethal human diseases. The most common of these are adrenoleukodystrophy (ALD), Zellweger syndrome and Refsum's disease. These diseases are caused by defects in individual peroxisomal enzymes or faults in the transport system necessary to convey proteins from their site of synthesis in the cytosol to the peroxisome. All are rare, occurring in approximately one in 50 000 live births.

Adrenoleukodystrophy is an X-linked disease (*Chapter 15*) caused by mutations in a gene encoding an integral peroxisome membrane protein that acts as an active carrier across the peroxisomal membrane. The disease is characterized by increased concentrations of very long chain fatty acids (VLCFAs) in the plasma and general tissues and progressive adrenal cortex dysfunction. Several types of ALD occur, the most severe affecting males of four to eight years of age. The major symptoms are learning disabilities, seizures, deafness, dysarthria, dysphagia, poor muscular coordination, vomiting, increased pigmentation of the skin, progressive dementia and poor behavior, such as aggression and hyperactivity. The adult form normally presents at 20 to 36 years. Symptoms include stiffness and a progressive **paraparesis** or weakness of the legs and **ataxia**, an inability to control muscular movements. Progress of the adult form is slower than childhood ALD, but may also result in loss of brain function and may be confused with multiple sclerosis. Occasionally, women carriers can show mild symptoms of ALD.

Zellweger syndrome is caused by defects in the general ability to import proteins into peroxisomes. The peroxisomal enzymes remain in the cytosol and

Figure 16.11 Outline of the catabolism of phytanic acid.

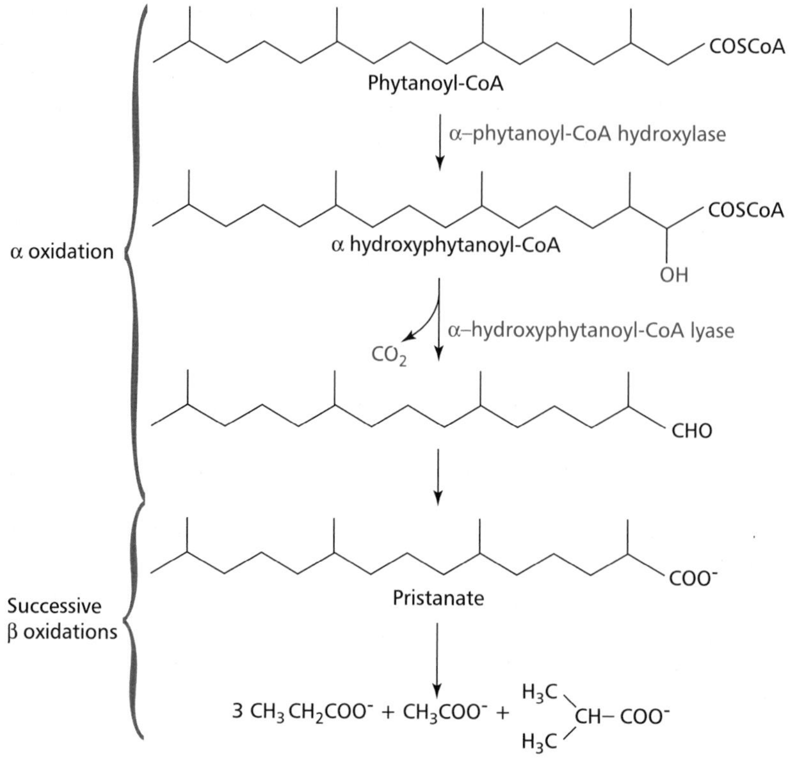

α oxidation

Phytanoyl-CoA (COSCoA)

α–phytanoyl-CoA hydroxylase

α hydroxyphytanoyl-CoA (COSCoA, OH)

α–hydroxyphytanoyl-CoA lyase (CO_2)

CHO

Successive β oxidations

Pristanate (COO^-)

$3\ CH_3CH_2COO^- + CH_3COO^- + (H_3C)_2CH{-}COO^-$

are rapidly degraded leading to a severe peroxisomal deficiency. Individuals have severe abnormalities in their brain, liver and kidneys and die soon after birth. Refsum's disease or hereditary motor sensory neuropathy type IV, heredopathia atactica polyneuritiformis, was first recognized in the 1940s by Refsum, who observed markedly increased concentrations of phytanic acid, a major constituent of dairy foods, in certain patients. The clinical features include the accumulation of phytanic acid in plasma and lipid-containing tissues, retinitis pigmentosa, blindness, **anosmia** (loss of sense of smell), deafness, sensory neuropathy and ataxia.

Refsum's disease is an autosomal recessive disorder that affects the α-oxidation of phytanic acid. In contrast to Zellweger syndrome, Refsum's disease is due to a reduced enzyme activity. Two enzymes, α-phytanoyl-CoA hydroxylase (PAHX) and α-hydroxyphytanoyl-CoA lyase, found in peroxisomes, are necessary for the oxidation of phytanic acid (*Figure 16.11*). Mutant forms of PAHX have been shown to be responsible for some cases of Refsum's disease. An infantile form of Refsum's disease also leads to an accumulation of phytanic acid but resembles Zellweger syndrome in that it is due to general defects in the transport of appropriate peroxisomal enzymes into the organelle. It is relatively less severe than Zellweger syndrome because the enzymes in the cytosol are thought to have a longer half-life.

DIAGNOSIS AND TREATMENT OF PEROXISOMAL DISORDERS

The diagnosis of peroxisomal disorders usually involves biochemical tests, such as determining the concentrations of VLCFAs in the plasma. Observations of deficient secretions of aldosterone and cortisol (*Chapter 7*) by the adrenal cortex can imply ALD. Prenatal diagnoses are available using cultured amniocytes and chorionic villus cells (*Chapter 15*).

The treatment of patients with peroxisomal defects is problematic since many, especially those with Zellweger syndrome, have significant brain damage at birth so a full recovery is not possible even with postnatal therapy. Patients with some of the relatively milder conditions can live into their second decade.

Margin Note 16.2 Lorenzo's oil

Lorenzo's oil gained considerable publicity due to the 1992 film of the same name. The film is based on the story of Augusto and Michaela Odone's efforts to save the life of their son, Lorenzo, who was diagnosed with incurable adrenoleukodystrophy (ALD) in 1984. His parents were, however, unwilling to give up the struggle to save him and after participating in several failed therapies, began to investigate the disease themselves. This resulted in the treatment of Lorenzo with a mixture glyceryltrioleate ($C_{18:1}$) and glyceryltrierucate ($C_{22:1}$) hence 'Lorenzo's oil', in the hope it would reduce the concentration of longer chain fatty acids in the body and also reduce demyelination and clinical progression.

It would appear that Lorenzo's oil rapidly reduces very long chain fatty acids in plasma to normal or near normal levels. Unfortunately, in patients with neurological symptoms, its use does not alleviate symptoms nor delay the progression of the disease. Postmortem studies show that very long chain fatty acids in brain are apparently unaffected.

Treatment largely involves supportive care and symptomatic therapy. Patients with ALD may be given Lorenzo's oil (glyceroyl trioleate and trierucate oils) and a strict dietary regimen that returns VLCFAs to normal concentrations but does not prevent neurological degeneration. A bone marrow transplant usually has a successful outcome but only if diagnosis is early and the donor is a perfect HLA match (*Chapter 6*). The treatment of Zellweger syndrome is limited. The most beneficial is the administration of the essential long chain fatty acid, docosahexaenoic acid (*Chapter 10*). Sufferers of Refsum's disease are given a dietary regimen free of phytanic acid.

16.6 LYSOSOMAL DISORDERS

Lysosomes are organelles bounded by a single, impermeable membrane. They contain about 60 different hydrolytic enzymes (hydrolases) that are active at acid pH and are able to degrade all biological materials (*Figure 16.12*). Other membrane-bound vesicles, endosomes and phagosomes, fuse with lysosomes and deliver materials taken up by the cell by endocytosis or phagocytosis for degradation by these enzymes.

Lysosomal enzymes are synthesized by the ribosomes of the rough endoplasmic reticulum and enzymes within the Golgi apparatus. In the Golgi apparatus, synthetic enzymes recognize sites on the surface of the prolysosomal enzymes and tag them with mannose 6-phosphate residues that act as molecular addresses, ensuring that they are integrated into vesicles that eventually become lysosomes.

Lysosomal storage diseases (LSDs) are caused by mutations in genes encoding lysosomal hydrolases leading to an accumulation of the substrate for the inactive enzyme within the lysosome (*Figure 16.13*). Eventually the cell dies, impairing the functions of major organs such as the brain and liver. More than 40 lysosomal storage diseases are known (*Table 16.4*). All are relatively uncommon in the general population although several are more prevalent among Ashkenazi Jews.

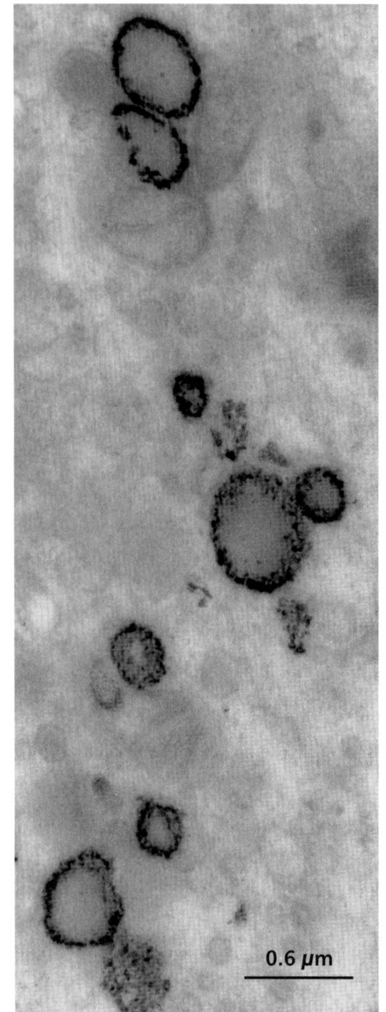

Figure 16.12 Electron micrograph of lysosomes in a synovial cell identified by aryl sulfatase activity. Courtesy of Dr C.J.P. Jones, Department of Pathology, University of Manchester, UK.

Disease	Enzyme deficiency	Material (substrate) accumulated
Sphingolipidosis GM_1 gangliosidosis	β-galactosidase	GM_1 gangliosides
Tay-Sachs disease (GM_2)	hexoseaminidase A	GM_2 gangliosides
Gaucher's disease	β-glucocerebrosidase	glucosylceramide
Niemann-Pick disease	sphingomyelinase	sphingomyelin
Metachromatic leukodystrophy	arylsulfatase A	sulfatides
Inclusion cell disease (I-cell disease or mucolipidosis II)	numerous lysosomal enzymes absent	glycolipids, glycoproteins, sialyloligosaccharides

Table 16.4 Examples of lysosomal storage diseases

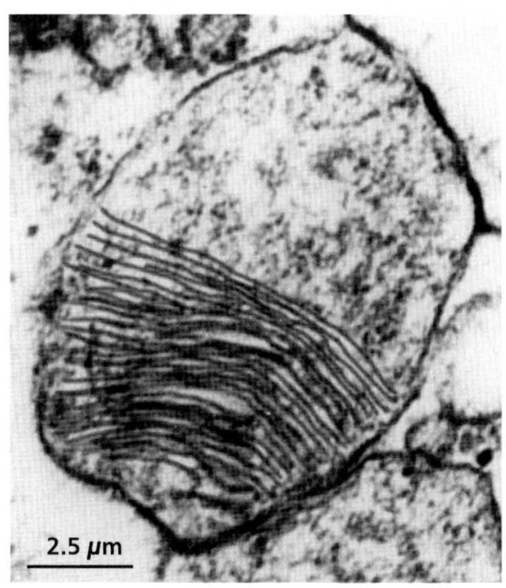

Figure 16.13 Electron micrograph of enlarged lysosomes containing a zebra body of accumulated lipid (GM_2 ganglioside, see *Table 16.4*) from the cerebral cortex of a 19-week-old fetus. A diagnosis of Tay-Sachs disease was made by showing a deficiency of hexoseaminidase A activity in cultured amniotic fluid cells obtained by amniocentesis, following which the fetus was aborted. Courtesy of Dr A. Cooper, Willink Biochemical Genetics Unit, Royal Manchester Children's Hospital, UK.

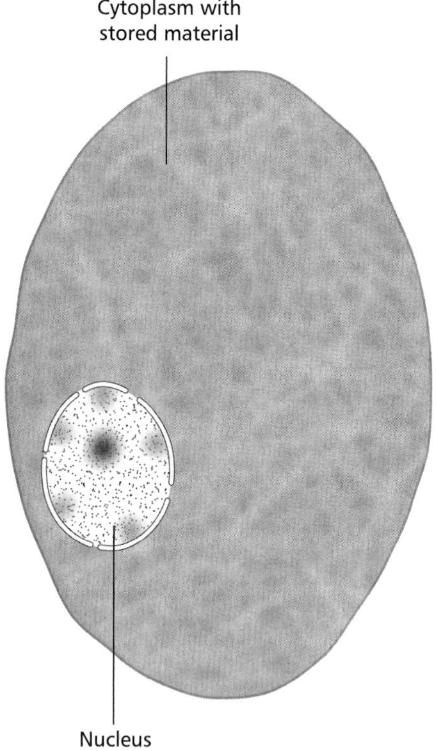

Figure 16.14 Schematic of a Gaucher's cell based on several light micrographs. Note how the stored material (glucosylceramide, see *Table 16.4*) gives the cytoplasm a characteristic 'wrinkled' appearance.

Lysosomal storage diseases are congenital disorders that seriously reduce the quality of life and life expectancy. For example, Tay-Sachs disease leads to blindness, paralysis, dementia and death, usually by two years of age. In most cases LSDs are caused by the deficiency of a single enzyme. The diseases can be subdivided according to which pathway is affected. For example, enzyme deficiencies in the pathway for degrading glycosaminoglycans cause mucopolysaccharidoses, deficiencies affecting glycopeptides cause glycoproteinosis and glycolipid storage diseases result from deficiencies in enzymes involved in the degradation of sphingolipids.

The commonest lysosomal storage disease is Gaucher's disease with an incidence of one in 30 000 to one in 50 000 although it is about 30 times more common in Ashkenazi Jews. The disease is characterized by the presence of histiocytes, cells belonging either to the macrophage or Langerhans cell lineage (*Chapter 4*), enlarged with accumulated lipid. These cells are known as Gaucher's cells (*Figure 16.14*) and are found in the bone marrow, reticuloendothelial system, parts of the circulation and other organs. The disease is an autosomal inherited recessive condition leading to a deficiency of lysosomal β-glucocerebrosidase and the accumulation of glucosylceramide within lysosomes. The pathogenic mechanisms are poorly understood but three forms are recognized: type 1 or adult, type 2 or infantile and type 3 or juvenile; the differences presumably arise from different mutations in the gene encoding the enzyme.

Adult, type 1, Gaucher's disease may present from infancy to adulthood. Although some sufferers may die at a young age due to pulmonary infections, most survive to late adult life. Type 1 disease is characterized by hepatosplenomegaly, the enlargement of the liver typically occurring after that of the spleen. The abdomen can also enlarge due to distention of the colon. Most of these visceral features arise from macrophage dysfunction. The skin bruises very easily and becomes pale yellow in color. Febrile episodes with generalized pain in the body and limbs occur. Painful bones, joints, degeneration of vertebral bodies and arthritic hip joints are common. Pathological fractures may also occur. Mild anemia and recurrent pulmonary infections are characteristic. Some patients show neurological abnormalities, decreased intelligence, with ataxia and fits.

Infantile, type 2, Gaucher's disease is much more severe with death occurring by the age of two years following progressive psychomotor regression. The earliest symptoms include a protuberant abdomen usually due to hepatosplenomegaly and a general failure to thrive. Indeed, after six months developmental arrest and neurological regression are obvious characteristics. Fits may occur. The patient becomes spastic, with laryngeal spasms causing difficulties in swallowing, and **strabismus** (a squint or abnormal alignment of the eyes). Deterioration is rapid, leading to recurrent pulmonary infections and death.

Juvenile, type 3, Gaucher's disease is the rarest of the three. Although it presents later than type 2 and has a slower progression it also has a characteristic neurological deterioration. The major features are hepatosplenomegaly, convulsions and psychomotor retardation. Ataxia, spasticity and strabismus may also occur.

Inclusion cell (I-cell) disease or mucolipidosis II is rare, with an incidence of about one in 640 000. All lysosomal hydrolases are actually produced but a lack of *N*-acetylglucosylaminyl-1-phosphotransferase activity means they fail to be tagged with mannose 6-phosphate residues, with the result that they are not delivered to lysosomes but are secreted from the cell. The disease is named from the densely staining intracytoplasmic regions present in the fibroblasts of sufferers (*Figure 16.15*). Overt symptoms are present at birth or appear within the first few years of life. There is psychomotor developmental delay and generally failure to thrive. Hernias, hepatomegaly, joint limitation and hip dislocations are common. Patients experience frequent upper respiratory tract infections with recurrent attacks of pneumonia, bronchitis and otitis media (*Chapter 3*). Cardiac murmurs from pathologies of heart valves are common and survival beyond the age of five years is unusual.

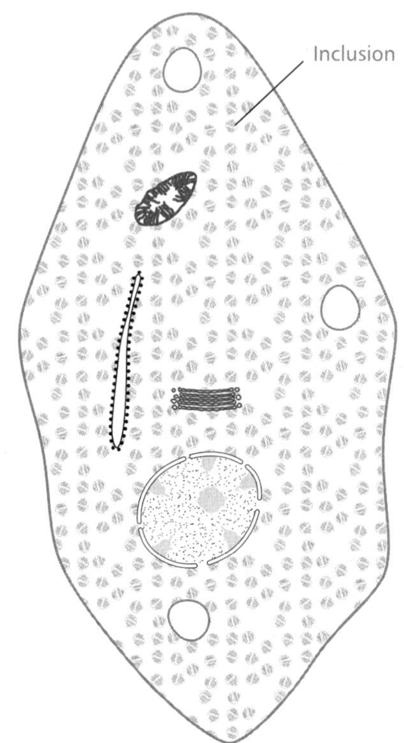

Figure 16.15 Schematic of an I cell based on several light and electron micrographs. Note the numerous vesicles (inclusion bodies) containing undigested material that has accumulated because of deficiencies in lysosomal enzyme activities.

DIAGNOSIS AND TREATMENT OF LYSOSOMAL STORAGE DISORDERS

Experienced laboratory practitioners using relatively simple methods can accurately diagnose patients exhibiting the typical traits of LSDs. However, atypical patients usually require more detailed investigations. An early diagnosis is important to prevent serious damage to the nervous and skeletal systems. Almost all lysosomal storage diseases can be diagnosed using samples of leukocytes or plasma. Diagnosis is based upon assays that determine the reduced activities of lysosomal enzymes or an increase in substrate in skin fibroblasts, plasma or urine. For example, Gaucher's and I-cell diseases can be confirmed by demonstrating deficiencies in β-glucocerebrosidase and *N*-acetylglucosylaminyl-1-phosphotransferase activities in leukocytes or fibroblasts respectively.

The prognosis for adult Gaucher's disease patients is relatively good. The hypersplenism is usually relieved by splenectomy although this may hasten bone deterioration. Symptomatic treatment includes analgesics for the bone pains, blood transfusions to relieve the anemia and orthopedic relief for fractures and degeneration of the hip joints. Enzyme replacement therapy, involving intravenous administration of purified β-glucocerebrosidase, is now common. Receptors on the surfaces of macrophages selectively bind mannose residues on the enzyme and allow it to be absorbed by the cells. Within the macrophages, the enzyme is delivered to the lysosomes where it catalyzes the breakdown of the accumulated glucosylceramide. In the majority of cases this is an effective and safe treatment, reducing the sizes of the liver and spleen and allowing them and the bone marrow to function effectively. Again, bone marrow transplants (*Chapter 6*) may be useful. Gene therapy, in which a functional gene for β-glucocerebrosidase is inserted into stem cells in the bone marrow, may in the future provide a complete cure.

Treatment for I-cell disease is limited. Bone marrow transplants can potentially replace the defective hemopoietic system with stem cells from a healthy donor, providing a replacement for the defective enzymes. In a limited number of cases, bone marrow transplants have normalized lysosomal enzyme activities, but the preexisting clinical damage from the disease is usually so extensive by this stage that this option is of questionable long-term benefit. In addition, bone marrow transplants can fail and the problems of finding compatible donors are immense.

Genetic counseling and prenatal diagnosis (*Chapter 15*) have roles in relation to LSDs. In cases where the likelihood of a lysosomal storage disease being present in a fetus is high, prenatal diagnosis is possible and a therapeutic abortion can be offered. Methods for screening all newborns without a family history of such conditions have been proposed but such screening programs would require a large amount of effort for comparatively little gain.

16.7 CYTOSKELETAL DISORDERS

The cytoplasm of nucleated cells is supported by a cytoskeleton consisting of three types of fibers and a number of associated proteins. The functions of the fibers are to resist forces that would deform the cell, to allow the cell to change shape and move and some types of intracellular transport. The three types of fibers are microfilaments (MF), intermediate filaments (IF) and microtubules (MT). Although the lengths of the fibers are indeterminate as they are actively extended and shortened during cellular activities, their diameters are fairly uniform between cell types. Microfilaments and MTs have diameters of approximately 7 and 25 nm respectively. As their name implies, IFs have diameters between these values of 8–11 nm.

Microfilaments are made of the protein actin. They are relatively flexible filaments but, cross-linked into bundles, they can withstand compression. Microtubules are composed of tubulin proteins arranged into hollow rods that are rigid and can resist both compression and tension. Intermediate filaments are built up from a number of types of proteins that are tissue specific, keratins in epidermal cells, desmin in muscles, for example. They form flexible cables whose high tensile strength allows the cell to resist excessive stretching.

Microfilaments and MTs form defined tracks within the cell for the transport of macromolecules and membranous structures. The two most common methods for this involves the movements of motor proteins along the filaments that are driven by the hydrolysis of ATP. The motor proteins of the MTs are dyneins and kinesins; those of the MFs are the myosins. Actin–myosin complexes are probably best known as the contractile apparatus of skeletal muscle tissues. Skeletal muscle tissue shows a multinuclear organization or syncitium arranged into fibers, which are surrounded by a basal lamina of extracellular matrix proteins, which forms a supporting sheath. Each fiber contains sarcoplasm (cytoplasm) that houses the contractile fibers of actin and myosin and is surrounded by a sarcolemma (plasma membrane). A network of elongated protein molecules about 150 nm long of the protein dystrophin is found within the sarcoplasm The dystrophin links actin filaments to a transmembrane complex of proteins that, in turn, is linked to components of the basal lamina (*Figure 16.16*). This complex arrangement

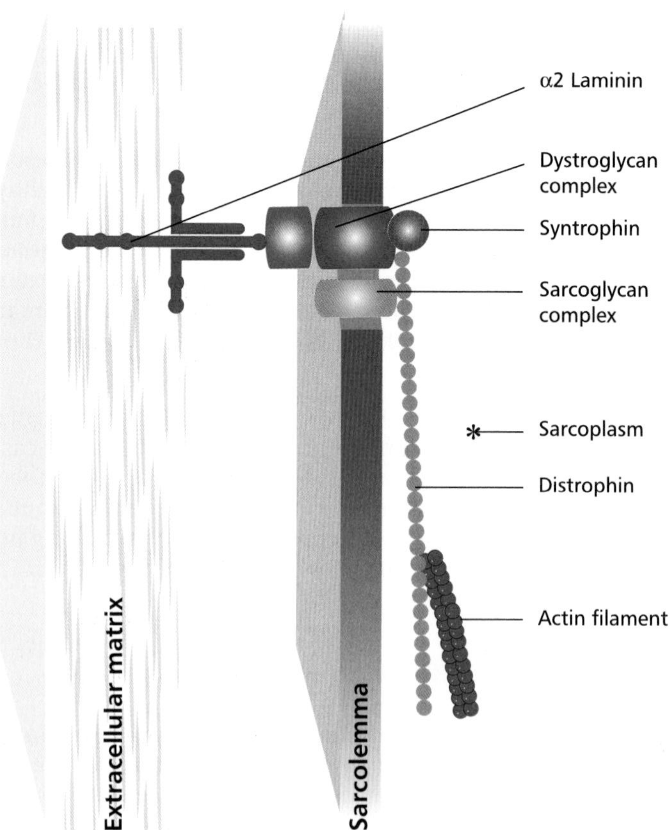

Figure 16.16 Schematic indicating how dystrophin, a transmembrane protein complex and the basal lamina may all interact to stabilize the sarcolemma of muscle fibers.

of proteins provides mechanical support to the sarcolemma during muscle contraction.

The blood contains about 5×10^{12} erythrocytes per dm^3 (*Chapter 13*). Their major function is to carry dioxygen from the lungs to the general body tissues. The unique biconcave shape of erythrocytes is maintained by a cytoskeleton composed of five major proteins that form a network lining the inner surface of their plasma membranes (*Figure 16.17*). The spectrin–actin complex is thought to act in a manner that resembles that of the dystrophin–actin complex of skeletal muscle and provides mechanical support to the plasma membrane preventing its lysis during circulation. The network of proteins also allows erythrocytes to deform and spring back into shape as they are pumped through the narrow capillaries of the vascular system. The numbers of erythrocytes are maintained by a constant production in the bone marrow and the destruction of worn out or misshapen erythrocytes by the spleen. This destruction releases bilirubin, which is converted to bile salts in the liver and released into the gastrointestinal tract in bile (*Chapter 11*). The iron from the hemoglobin is largely retained and reused by the body.

MUSCULAR DYSTROPHIES

Mutations involving the genes for dystrophin, its associated transmembrane complex proteins or the α2 laminin of the basal lamina lead to genetic

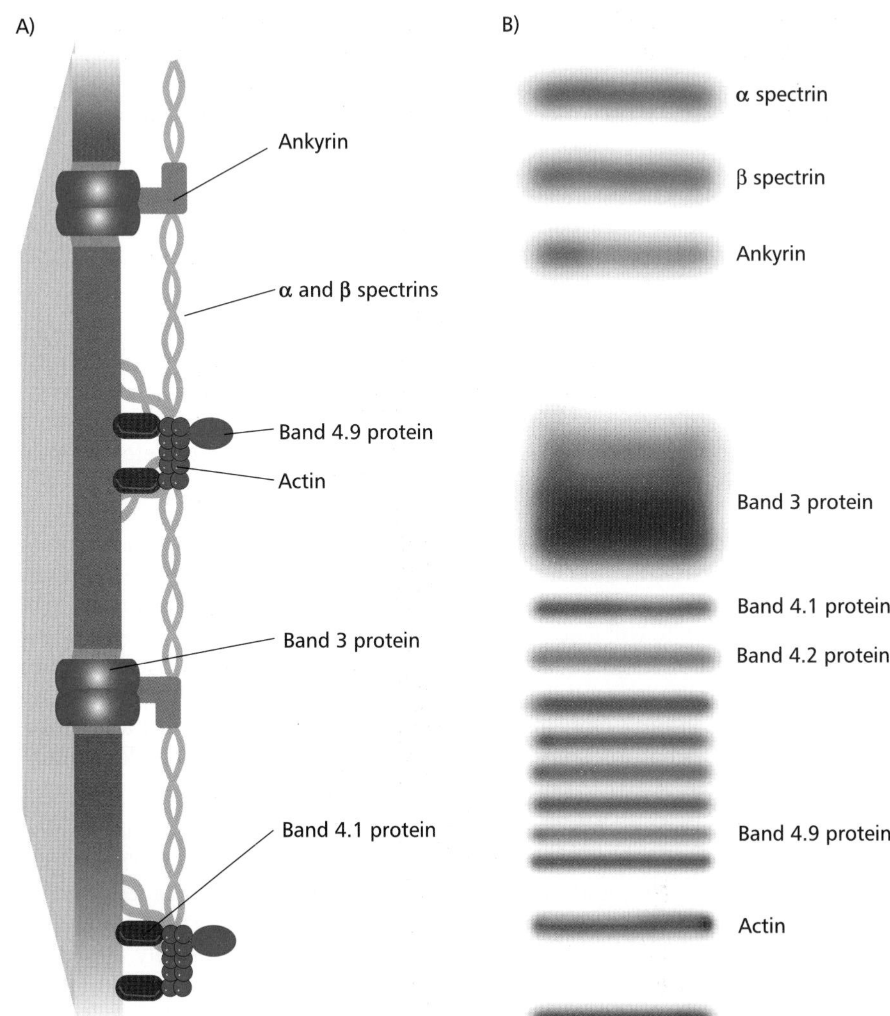

Figure 16.17 (A) Simplified schematic of the cytoskeleton of an erythrocyte. (B) Schematic of the separation of proteins of the cytoskeleton of a red blood cell by SDS polyacrylamide gel electrophoresis.

disorders called muscular dystrophies (MDs). Defects in these proteins compromise the mechanical strength of the muscle fibers leading to their rupture and death during years of contraction, hence MDs are characterized by a progressive wasting of the muscle tissues (*Table 16.5*). The two best described MDs are Duchenne muscular dystrophy (DMD also called pseudohypertrophic muscular dystrophy), which was first described by the French neurologist Duchenne (1806–1875) in the 1860s, and Becker's muscular dystrophy (BMD, or benign pseudohypertrophic muscular dystrophy), named after the German doctor, Becker (1908–2000), who described this variant of DMD in the 1950s.

Muscular dystrophy	Proteins involved (*Figure 16.16*)
Autosomal recessive muscular dystrophy (ARMD)	α sarcoglycan β sarcoglycan
Becker's muscular dystrophy (BMD)	dystrophin
Congenital muscular dystrophy (CMD)	α2 laminin
Duchenne muscular dystrophy (DMD)	dystrophin
Limb girdle muscular dystrophy (LGMD)	α sarcoglycans β sarcoglycans γ sarcoglycans
Severe childhood autosomal recessive muscular dystrophy (SCARMD)	γ sarcoglycan

Table 16.5 Muscular dystrophies

The incidences of MDs vary depending on the specific type. Duchenne MD is the commonest with an incidence of three in 10 000 live-born males. Becker's MD is the second most common form, at one per 30 000 live male births. Other types of MD are rare. For example, limb-girdle dystrophy occurs in only about 1% of patients with MDs.

Duchenne and Becker's MDs are caused by different mutations in the gene for dystrophin. Mutations that produce STOP codons or deletions of even relatively small portions that change the reading frame lead to DMD. Surprisingly, BMD is the result of much larger deletion mutations but the reading frame of the rest of the gene is unaltered, leading to the production of defective dystrophin. This, apparently, is sufficient to protect the muscles from degenerating as badly or as quickly as is the case with DMD. The gene for dystrophin is located at Xp21 (*Chapter 15*). Thus both DMD and BMD are inherited in an X-linked recessive manner although up to 30% of cases of DMD are the result of a *de novo* mutation. Like all X-linked recessive characters, the presence of the gene in a male will lead to the development of symptoms because of the absence of a homologous, normal gene on the Y chromosome. Heterozygous females will be carriers and generally symptomless, although a number do show mild symptoms. Only in the unlikely event of a female being a homozygote, with two X chromosomes carrying a mutated gene, will females present with the disease.

Duchenne MD is a rapidly progressive disease characterized by loss of muscle functions that are associated with the muscle tissue wasting. This begins in the legs and pelvis and progresses to the shoulders and neck followed by the respiratory muscles. The course of DMD is fairly predictable, in that it follows an aggressive and progressive course. Symptoms usually appear in male children aged one to six years. Boys with the disorder are often late in learning to walk. In toddlers, it may present with enlarged calf muscles and a clumsy, unsteady gait. The initial enlarged muscle mass of the calf muscles is a compensation for loss of muscle strength. However, the enlarged muscle tissue is eventually replaced by fat and connective tissue (pseudohypertrophy). Muscle contractures occur in the legs, rendering the muscles unusable as muscle fibers shorten and fibrosis occurs in connective tissue. Intellectual impairment may also occur although it is not inevitable and does not worsen as the disorder progresses. At school age, there is trouble climbing stairs and the child may walk on his toes or the balls of his feet and can easily fall over. Characteristically, the belly is stuck out and shoulders held back to maintain balance. Nearly all children with DMD lose the ability to walk between seven and 12 years and, by 10, braces may be required for walking. Bones develop abnormally leading to skeletal deformities of the spine and other areas. In the early teens, or earlier, the heart and respiratory muscles may also be affected. Cardiomyopathy is commonly present, but signs of congestive heart failure or arrhythmias (*Chapter 14*) are rare. The muscular weakness and skeletal deformities frequently lead to breathing disorders, including pneumonia. The aspiration of food or fluid into the lungs is common during the later stages. Death usually occurs by age 25, typically from these respiratory disorders.

The symptoms of BMD are considerably milder than those of DMD. They usually begin about the age of 12 or in early adulthood. The pattern of development resembles that of DMD although the course is slower and far less predictable. Generally, muscle wasting begins in the legs and pelvis and then progresses to the muscles of the shoulders and neck, followed by loss of arm muscles and respiratory muscles. Calf muscle enlargement (pseudohypertrophy) is again apparent. The rate of muscle tissue degeneration in BMD varies greatly. Some men require wheelchairs by their thirties, others manage for many years with minor aids, such as walking sticks, and, indeed, the ability to walk may continue to the age of 40 or over. Cardiomyopathy is less common than with DMD. If it does occur, then, like DMD, congestive heart failure or arrhythmias are rare. Death usually occurs in the fifth decade but some patients live to an advanced age.

Diagnosis and treatment of muscular dystrophies

A diagnosis of MDs begins with a family history and physical examination. The pattern of weakness can be particularly helpful and usually the source of weakness, whether in motor neurons or muscle tissue, can be identified by a physical examination. Nerve conduction studies can indicate whether the underlying problem is in the muscles or the nerves. However, an electromyography (EMG) measures the electrical activity of the muscles and can show if the weakness is caused by destruction of muscle tissue. Another early diagnostic test would be to measure creatine kinase (CK) activity of the blood to assess any leakage of the enzyme from damaged muscle. An increase in CK levels is seen in MDs but also in other conditions, such as an inflammation. A magnetic resonance (MR) scan can be useful in visualizing intramuscular activity. Genetic testing and a muscle biopsy can confirm a diagnosis and both are essential since other diseases, such as limb-girdle MD and spinal muscular atrophy, present with some of the same symptoms. Biopsy material can also distinguish MDs from inflammatory and other disorders as well as identify the type of dystrophy present. Diagnostic tests using DNA of blood or muscle cells can give precise clinical information, and their use is expanding. A diagnosis of BMD is often not made until adolescence or even adulthood, when physical activities become difficult.

There are no known cures for DMD or BMD. In both diseases, the goal of treatment is to control the symptoms to promote the quality of life. Inactivity, such as prolonged bed rest, generally worsens muscle disease and so physical activity is encouraged. Physiotherapy may be helpful in maintaining muscle strength and coordination. Orthopedic appliances, for example leg braces, walking sticks and wheelchairs, may improve mobility and the ability for self-care. However, associated stress means that a support group where members pool common experiences and problems can be of use. Genetic counseling is advisable, especially if there is a family history of the disorders. Note that sons of a man with BMD (or DMD) will not develop the disease but daughters can be carriers. An amniocentesis and appropriate genetic studies means that DMD is detectable with an approximately 95% accuracy during pregnancy.

HEREDITARY ELLIPTOCYTOSIS AND SPHEROCYTOSIS

Two relatively common disorders associated with the erythrocyte cytoskeleton are hereditary elliptocytosis (HE) and spherocytosis (HS). Their incidences are variable. For example, in the USA, HS is the commonest of the hereditary hemolytic anemias (*Chapter 13*) among people of Northern European descent, with an incidence of approximately one in 5000, although many regard this as an underestimation. The prevalence of HE is thought to be one in 2000 to 4000. Like sickle cell anemia, which was described in *Chapter 13*, the condition shows a much higher incidence in areas endemic for malaria, in equatorial Africa, for instance the incidence is approximately six per 1000.

Patients with HE or HS synthesize lower amounts of cytoskeletal proteins than normal, or the proteins themselves are defective. Most cases of hereditary spherocytosis are caused by mutations in the *ANK1*, *SPTB*, *SLC4A1*, *EPB42* and *SPTA1* genes that encode ankyrin, spectrin P-chain, band 3, protein 4.2 and α-spectrin respectively (*Figure 16.17*). The mutations responsible for hereditary elliptocytosis occur in the *SPTB* and *SPTA1* and the *EPB41* gene encoding protein 4.1. These mutations mean that the erythrocytes are misshapen (*Figure 16.18*) and less deformable and so become trapped in the spleen where they are prematurely degraded. The consequent excessive release of bilirubin leads to jaundice and gallstones (*Chapter 11*). The anemia caused by reduced numbers of circulating erythrocytes results in the release of immature erythrocytes from the bone marrow.

Diagnosis and treatment of hereditary elliptocytosis and spherocytosis

The simplest way to test for HE and HS is the microscopic examination of peripheral blood smears to observe the misshapen erythrocytes (*Figure 16.18*). Other tests include demonstration of the increased resistance of the erythrocytes to osmotic shock using the osmotic fragility test, observation of reduced fluorescence following staining with eosin-5-maliemide and determination of the amounts of cytoskeletal proteins present following their separation by electrophoresis.

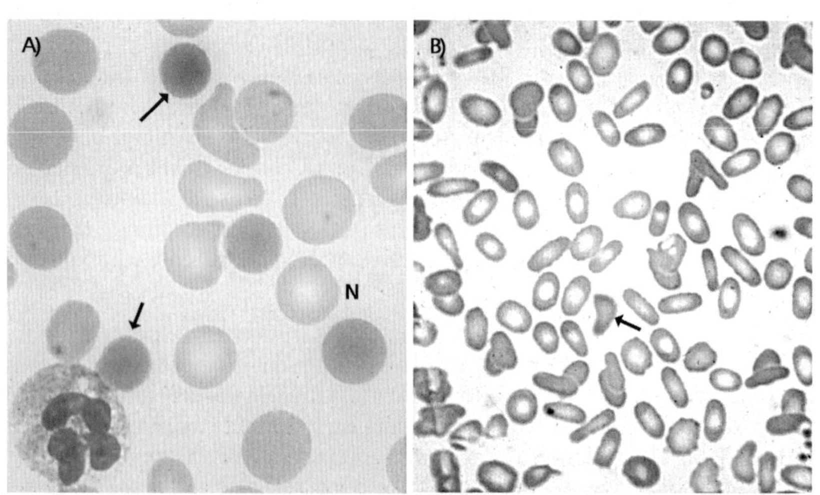

Figure 16.18 Light micrographs of misshapen erythrocytes in (A) hereditary spherocytosis and (B) elliptocytosis. In (A) the abnormal spherocytes (examples highlighted) lack the usual biconcave shape (N) and appear a uniform dark color. Note the presence in (B) of abnormal elliptical erythrocytes (example highlighted). Courtesy of A. Will, Royal Manchester Children's Hospital, UK.

General treatments include daily ingestion of folate when hemolysis is significant, and blood transfusions. The symptoms of HE and HS are alleviated by surgical removal of the spleen (*Table 16.6*). This increases the time the defective erythrocytes stay in the circulation to about 80% of that of a normal erythrocyte.

	Before splenectomy	After splenectomy
Erythrocyte count x 10^{-12} dm^3	3.1	5.2
Reticulocytes / %	16.0	0.95
[Hemoglobin] / g dm^{-3}	8.5	15.1
Serum [bilirubin] / mmol dm^{-3}	83.0	8.2

Table 16.6 Typical hematological data before and after splenectomy in a patient suffering from hereditary spherocytosis (*see Chapter 13*)

CASE STUDY 16.1

Bethany, a female infant, suffers from recurrent respiratory infections and presented with clinical evidence of dietary malabsorption. Her brother and sister are asymptomatic. She was sent to hospital for a variety of clinical investigations. After a number of negative tests, she was subjected to a sweat chloride test. The result is given below:

Sweat [Cl^-] 76.0 mmol dm^{-3} (Normally less than 30 mmol dm^{-3})

Questions

(a) Suggest a plausible diagnosis.

(b) Explain how the diagnosis for Bethany was reached.

(c) How should Bethany's siblings be investigated?

CASE STUDY 16.2

Janice is an 18-year-old. She was admitted to hospital for clinical tests following a series of frequent heavy nose bleeds and extensive bruising following comparatively minor accident bumps. She also complained of stiff aching joints and some pains in her long bones. Her erythrocyte and platelet counts were found to be depressed. There was some extension of her abdomen. Routine clinical tests eliminated leukemia. Enzyme assays showed β glucocerebrosidase activity in leukocytes to be greatly reduced.

Questions

(a) What is Janice suffering from?

(b) What would be the likely progression of this disease if untreated?

(c) What treatment would you recommend for Janice?

16.8 SUMMARY

Defects in the plasma membrane or those that surround organelles and in the cytoskeleton cause a variety of disorders, some of which have been outlined. Disorders of nucleocytoplasmic transport have been implicated in a number of diseases, including some cancers. Cystic fibrosis, which is characterized by alterations in the permeability of epithelial cell membranes to CL^-, is a very common inherited disorder, with a prevalence of one in 2000 in the UK. Mitochondrial diseases are associated largely with defects in the oxidative phosphorylation system, leading to a variety of neurological and neuromuscular disorders. Disorders of peroxisomes result in number of fatal diseases that are characterized by the accumulation of very long chain fatty acids in tissues. Lysosomal storage disorders, of which more than 40 are known, are rare and often fatal, though the adult form of Gaucher's disease is usually associated with a longer life span. Diseases associated with defects of the cytoskeleton of muscle fibers are Duchenne and Becker's muscular dystrophies. Mutation in genes for proteins of the cytoskeleton of erythrocytes can lead to them being deformed and to their premature removal by the spleen, with accompanying anemia.

QUESTIONS

1. Which of the following modes of inheritance is shown by cystic fibrosis?
   (a) autosomal dominant;
   (b) autosomal recessive;
   (c) sex-linked;
   (d) spontaneous mutation;
   (e) none of the above.

2. Which of the following genes is associated with cystic fibrosis?
   (a) *EPB41*;
   (b) *ANK1*;
   (c) *CFTR*;
   (d) *SPTB*;
   (e) Mitochondrial genome.

3. Arrange the two following lists to show the most appropriate pairings:

α-spectrin	LGMD
EPB41	hereditary elliptocytosis
Hexoseaminidase A	sphingomyelin
Sarcoglycans	chloride channel
Niemann-Pick disease	MELAS
Refsum's disease	BMD, DMD
Adrenoleukodystrophy	Tay-Sachs disease
CFTR	hereditary spherocytosis
Dystrophin	VLCFAs
Mitochondrial gene for $tRNA^{Leu}$	phytanic acid

4. Suggest why some mitochondrial diseases lead to myoglobinuria (myoglobin in the urine).

5. The accompanying graph shows the extent of hemolysis of two blood specimens, A and B, in response to decreasing concentrations of NaCl. One specimen was from a patient with hereditary spherocytosis and the other was a control.

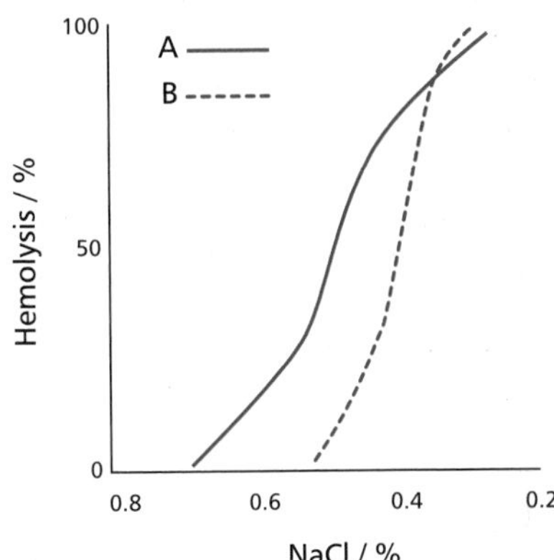

(a) Which of the bloods has the greater osmotic fragility?

(b) Which of the two specimens was from the patient?

6. Gallstones are hard lumps of cholesterol, bile pigments and calcium salts. Explain why patients with hereditary spherocytosis and elliptocytosis show an increase in their production of gallstones. You may also find it useful to refer to *Chapter 11*.

FURTHER READING

Chinnery, PF and Turnbull, DM (2001) Epidemiology and treatment of mitochondrial disorders. *Am. J. Med. Genet.* **106:** 94–101.

Cronshaw, JM and Matunis, MJ (2004) The nuclear pore complex: disease associations and functional correlations. *Trends Endocrinol. Metab.* **15:** 34–39.

Delaunay, J (2002) Molecular basis of red cell membrane disorders. *Acta Haematol.* **108:** 210–218.

DiMauro, S (2004) Mitochondrial diseases. *Biochim. Biophys. Acta* **1658:** 80–88.

DiMauro, S and Schon, EA (2001) Mitochondrial DNA mutations in human disease. *Am. J. Med. Genet.* **106:** 18–26.

Emery, AEH (2002) Muscular dystrophy into the new millennium. *Neuromuscul. Disord.* **12:** 343–349.

Fahrenkrog, B, Köser, J and Aebi, U (2004) The nuclear pore complex: a jack of all trades? *Trends Biochem. Sci.* **29:** 175–182.

Futerman, AH, Sussman, JL, Horowitz, M, Silman, I and Zimran, A (2004) New directions in the treatment of Gaucher disease. *Trends Pharmacol. Sci.* **25:** 147–151.

Gabriel, SE, Brigman, KN, Koller, BH, Boucher, RC and Stutts, MJ (1994) Cystic fibrosis heterozygote resistance to cholera toxin in the cystic fibrosis mouse model. *Science* **266:** 107–109.

Gieselmann, V, Matzner, U, Klein, D, Mansson, JE, D'Hooge, R, DeDeyn, PD, Rauch, RL, Hartmann, D and Harzer, K (2003) Gene therapy: Prospects for glycolipid storage diseases. *Philos. Trans. R. Soc. Lond. B. Biol. Sci.* **358:** 921–925.

Goldfarb, DS, Corbett, AH, Mason, DA, Harreman, MT and Adam, SA (2004) Importin α: a multipurpose nuclear-transport receptor. *Trends Cell Biol.* **14:** 505–514.

Hancock, MR (2002) Mitochondrial dysfunction and the role of the non-specialist laboratory. *Ann. Clin. Biochem.* **39:** 456–463.

Hyde, SC, Gill, DR, Higgins, CF and Trezise, EO (1993) Correction of the ion transport defect in cystic fibrosis transgenic mice by gene therapy. *Nature* **362:** 250–255.

McGuinness, MC, Wei, HM and Smith, KD (2000) Therapeutic developments in peroxisome biogenesis disorders. *Expert Opin. Investig. Drugs* **9:** 1985–1992.

Ratjen, F and Doring, G (2003) Cystic fibrosis. *Lancet* **361:** 681–689.

Schwarz, MJ (1998) DNA diagnosis of cystic fibrosis. *Ann. Clin. Biochem.* **35:** 584–610.

Towbin, JA (1998) The role of cytoskeletal proteins in cardiomyopathies. *Curr. Opin. Cell Biol.* **10:** 131–139.

Wanders, RJA (2004) Metabolic and molecular basis of peroxisomal disorders: a review. *Am. J. Med. Genet.* **126A:** 355–375.

Wenger, DA, Coppola, S and Liu, SL (2002) Lysosomal storage disorders: diagnostic dilemmas and prospects for therapy. *Gen. Med.* **4:** 412–419.

Yashiroda, Y and Yoshida, M (2003) Nucleo-cytoplasmic transport of proteins as a target for therapeutic drugs. *Curr. Med. Chem.* **10:** 741–748.

Yew, NS and Cheng, SH (2001) Gene therapy for lysosomal storage disorders. *Curr. Opin. Mol. Ther.* **3:** 399–406.

Useful web sites:

The Online Mendelian Inheritance in Man (OMIM™) website. Address: http://www.ncbi.nlm.nih.gov/omim/

http://www.genetsickkids.on.ca/cftr/

chapter 17:

CANCER

OBJECTIVES

After studying this chapter you should be able to:

- explain what is meant by the following terms: cancer, malignant, neoplasm, tumor;
- discuss the incidence and epidemiology of cancer;
- describe and explain some of the environmental causes of cancer;
- outline the relationship between gene mutations and cancer;
- describe how cancer may be clinically detected;
- describe the clinical features of some cancers;
- explain how cancers may be treated.

17.1 INTRODUCTION

In multicellular organisms, the growth of cells by division is under tight control (*Chapter 15*). Cells divide when stimulated with the requisite internal and external signals (*Figure 17.1*). This strict control of cell division ensures that division occurs at a rate appropriate to the structure of the tissue or organ. In some tissues, for example bone marrow, skin and gastrointestinal (GIT) endothelium, cells divide constantly, replacing ones that have died. However, in other tissues, such as nervous tissue, mitosis is rare and lost cells are not replaced. In yet other tissues, cells can commence dividing when required as, for example, in the regeneration of liver or during the healing of wounds in the skin. The term **cancer** covers a number of diseases in which the growth of cells becomes uncontrolled. Cancer cells fail to respond to the usual controling signals and their growth becomes unregulated. Indeed, the name *cancer* comes from a Latin word meaning 'a crab', and describes the manner in which the pattern of penetration into normal tissues by the abnormal growth bears a superficial resemblance to a crab's claw. These abnormal cells may

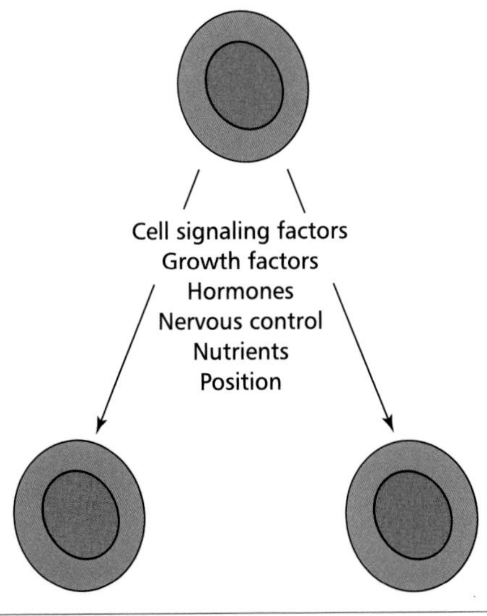

Figure 17.1 Some intrinsic and extrinsic factors that influence cell division.

invade nearby tissues and may enter the blood and lymph systems and spread to remoter areas.

Today, the term cancer is used popularly to describe what is known as a **malignant tumor**. **Oncology**, derived from the Greek word *oncos*, a lump, is the branch of medicine involved with the study of the development of tumors, their epidemiology, diagnosis and treatment. A tumor is an abnormal mass of cells that may be benign or malignant. Benign tumors generally grow slowly and do not spread to other tissues, though they may continue to grow *in situ*. Such tumors are only harmful if they interfere with the normal function of a tissue, or if they cause pressure by growing within a confined space, such as in the brain. A malignant tumor is one that spreads from its initial site, where it is known as the **primary** tumor, through the blood and lymph to establish **secondary** tumors in other organs. Such movement from the primary tumor and the formation of secondary tumors is known as **metastasis**. A term that is often used in the context of cancer is **neoplasm**. This means, literally, a new tumor or new mass, but is generally used to describe a cancer.

The causes of cancer are complex and varied. Some arise from environmental agents called **carcinogens**, others are brought about by oncogenic, that is cancer-inducing, viruses. Most cancers arise, ultimately, from mutations in DNA. These mutations may be caused by environmental agents, or may be inherited in the germ line, making individuals more susceptible to cancer.

Margin Note 17.1 The germ line 

Germline DNA refers to the DNA which is present in the cells that give rise to the gametes, that is, the sperm and eggs (*Chapter 7*). The egg and sperm fuse to form a zygote, and, as further divisions occur, that DNA is passed to all the cells in the developing embryo. Mutations which occur in germline DNA are present in the gametes and in all the cells of the individuals to which they give rise.

Cancers can arise from any tissue in the body; indeed, they have been detected in over 200 different sites. Some sites are more susceptible than others, the commonest being the lungs, breasts, prostate, GIT and skin. While most cancers occur more frequently in the old than in the young, with cancer generally being regarded as a disease of aging (*Chapter 18*), certain cancers occur typically in children. In the UK it has been estimated that one in three individuals will develop a cancer at some time in their life and that cancer causes one in four deaths. Treatment of cancer represents 6% of all NHS hospital expenditure. Similar incidence rates are seen in the USA.

This chapter will review the biology of cancer and its consequences to the individual. The causes of cancer, including genetic aspects, environmental insults and viruses will be examined. The involvement of the pathology laboratory in the screening, diagnosis and treatment of cancers will also be discussed.

17.2 CLASSIFICATION OF CANCER TYPES

Cancers are often classified according to their tissue of origin. Thus, a carcinoma is derived from epithelial tissue, whereas a sarcoma is derived from tissue of mesodermal origin such as muscle, bone or cartilage. The term leukemia refers to tumors of the bone marrow that result in excess cells of a single type appearing in the blood. Lymphoma refers to a tumor which arises from lymphoid tissue, such as a lymph node. There is also a group of highly malignant tumors that occur in childhood, which all have the suffix 'blastoma'. These include neuroblastomas, which arise in the neuroblasts of the adrenal medulla, retinoblastoma, which originates in the retina of the eye, and nephroblastoma, derived from the embryonic cells of the kidney.

17.3 EPIDEMIOLOGY OF CANCER

In England in 2002 there were 276700 new cases of malignant cancer registered, approximately evenly divided amongst men and women. However, individual cancers are often unevenly distributed between the sexes and, clearly, some cancers occur only in men or women. Obvious examples of the latter are cancer of the *cervix uteri* or 'cervical cancer' in women and testicular and prostate cancer in men. Contrary to common belief, men can develop breast cancer, although the incidence is much lower than in women. Moreover, the incidence of breast cancer in both sexes appears to be increasing. A study in the USA showed increases in male and female breast cancer of 26% and 52% respectively between two separate studies in 1973–1978 and 1994–1998. The commonest forms of cancer in men, women and children in the USA are shown in *Table 17.1*.

Men	Women	Children
Prostate cancer	breast cancer	leukemias
Lung cancer	lung cancer	brain tumors
Colorectal cancer	colorectal cancer	lymphomas

Table 17.1 Commonest forms of cancer in the US population

The annual incidence of different forms of cancer is shown in *Figure 17.2.* With the exception of the childhood cancers, the incidence of cancer increases with age, as shown in *Figure 17.3*. This increasing incidence is due to a number of factors including increased length of exposure to environmental agents associated with cancer, and an accumulation within cells of mutations in the DNA (*Chapter 18*), coupled with decreased efficiency of the cellular DNA repair mechanisms. In addition the immune system, which may play a role in eliminating early cancerous cells, also decreases in efficiency with age and this may lead to a failure to eliminate malignant cells as they arise.

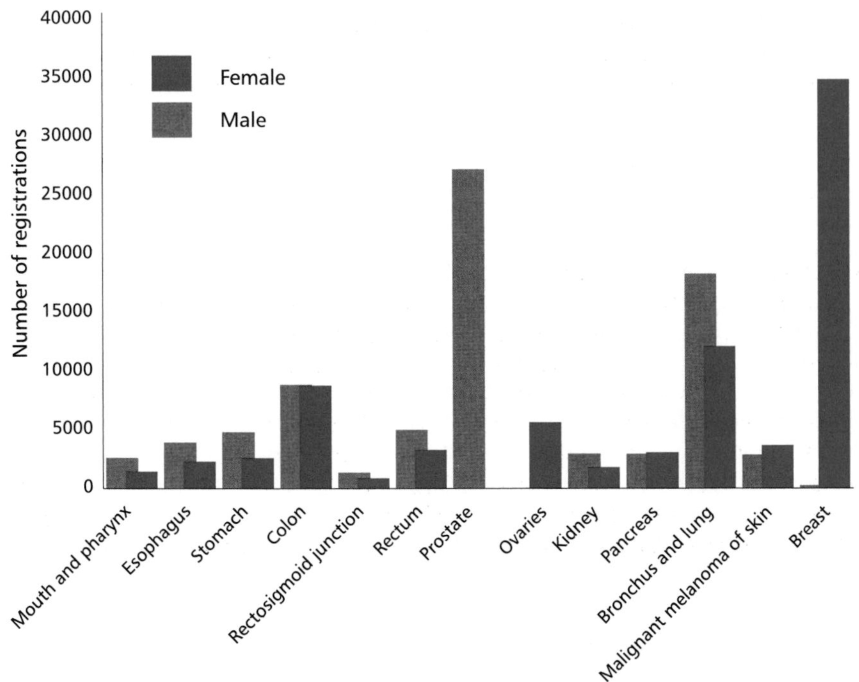

Figure 17.2 Annual incidence of different forms of cancer in the UK. Statistics obtained from the Office for National Statistics, UK (2005).

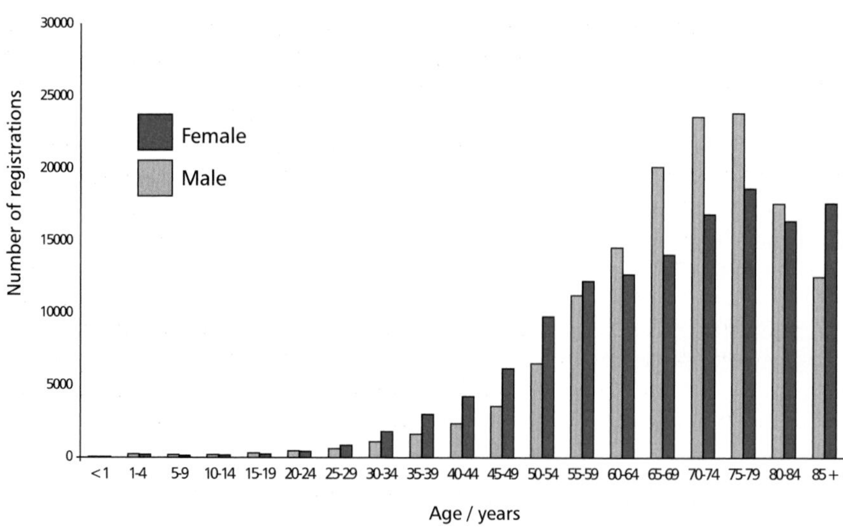

Figure 17.3 The incidence of cancer with age. Based on statistics published in 2005 by the Office for National Statistics, UK.

17.4 MOLECULAR BIOLOGY OF CANCER

In 1914, Boveri (1862–1915) suggested that a malignant tumor arose from a cell that had acquired chromosomal abnormalities. In other words, cancer was caused by mutations in the DNA of cells. The multistep theory of cancer suggests that between five and seven successive mutations may have to occur before a cancer can develop (*Figure 17.4*). Each successive mutation gives the cell a selective growth advantage over the normal cells which surround it. For example, one mutation may confer the ability to grow at a faster rate than normal cells. The cell would then proliferate to form a tissue which is said to be hyperplastic. One of the daughter cells may then undergo another mutation, which allows uncontrolled division of the progeny, to form a mass of cells which have abnormal morphology. At this stage the tumor is known as a carcinoma *in situ*. Accumulation of further mutations within the clone may mean that the cells can detach from the tumor and invade surrounding tissue, that is, they metastasize.

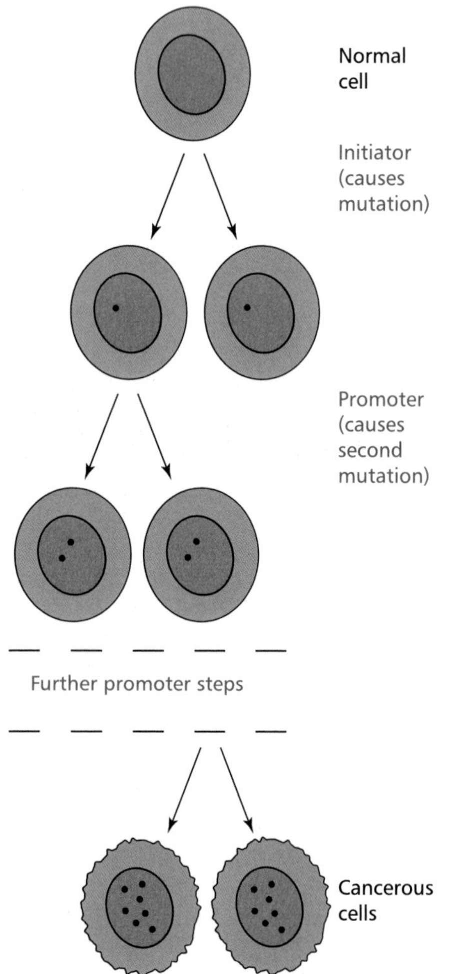

Figure 17.4 Schematic illustrating the multistep theory of cancer.

In 2004, a census of published scientific literature showed that 291 genes, representing approximately 1% of the total number of human genes, are associated with cancers. For 90% of these genes, somatic mutations were detected in the cancer cells but not in normal tissue. Approximately 20% of 'cancer genes' show mutations in the germline DNA, which predispose an individual to cancer, while 10% have been associated with mutations in both germline and somatic DNA.

Genes that are associated with cancer include those that control the cell cycle, including cell division and differentiation or molecules involved in signal transduction and associated growth factors, as well as those that control the process of programmed cell death that results in apoptosis (*Chapter 16*). Mutations in some of these genes may lead to increased proliferation and failures in apoptosis. Two major groups of genes linked to cancers are the **oncogenes** and the **tumor suppressor genes**.

ONCOGENES

Oncogenes are mutated forms of normal genes, called proto-oncogenes, which stimulate the increased proliferation of abnormal cells by encoding factors including growth factors and receptors, as well as proteins involved in signal transduction (*Chapter 7*). Examples include *MYC, FOS* and the *RAS* family of oncogenes. The mutation of a proto-oncogene to form an oncogene usually results in the production of a protein that has increased activity, or in the synthesis of greater than normal amounts of the protein, as, for example, when the gene is continually active. Such mutations are otherwise known as 'gain-of-function' mutations. Oncogenes were first discovered in certain viruses that cause cancer (*Section 17.5*). Some viruses which cause cancer have a gene that is the equivalent of a cellular proto-oncogene. These genes are thought to have been derived initially from host DNA into which the viral DNA was integrated and to have undergone mutation during viral passage. When DNA derived from the virus becomes inserted into the host genome, the viral oncogene may provide the additional signal for cellular proliferation, or may override the normal cellular controls, resulting in the unregulated division of infected cells. The viral and cellular forms of the oncogene are usually distinguished by the prefix '*v*' and '*c*' respectively as in, for example, *v-MYC* and *c-MYC*.

Some examples of oncogenes and associated tumors are shown in *Table 17.2*.

Proto-oncogene	Codes for	Disease
ERB-B1	epidermal growth factor receptor (EGFR)	squamous cell carcinoma of the lung
ERB-B2 (*HER*-2)	growth factor receptor	breast cancer ovarian cancer cancer of the salivary gland
H-*RAS*	GTPase	bladder cancer stomach cancer breast cancer thyroid cancer and others
K-*RAS*	GTPase	stomach cancer pancreatic cancer melanoma bladder cancer neuroblastoma thyroid cancer and others
BCR-ABL	tyrosine kinase	chronic myelogenous leukemia acute lymphoblastic leukemia
SRC	tyrosine kinase	colon cancer
MYC	transcription factor	breast, stomach and lung cancer leukemias
FOS	transcription factor	lung cancer breast cancer

Table 17.2 Some oncogenes and associated tumors

TUMOR SUPPRESSOR GENES

Tumor suppressor genes, in contrast to proto-oncogenes, encode proteins that inhibit the proliferation of cells that contain deleterious mutations. Mutations in the tumor suppressor genes themselves may then lead to a loss of this inhibition that is, they are 'loss of function' mutations. Some examples of tumor suppressor genes are the *TP53* gene, the retinoblastoma susceptibility gene, *RB*, and the Wilms' tumor gene, *WT1*. The *TP53* gene is

Margin Note 17.2 Gene *E2F3*

The gene *E2F3*, which is associated with the development of prostate cancer (*Section 17.8*), was discovered in 2004. The gene encodes a protein that controls cell division. It is produced in appropriate amounts in healthy prostate cells but is overexpressed in prostate cancer cells, leading to their excessive proliferation. The discovery is useful because this protein can be used as a marker for the more aggressive forms of prostate cancer, allowing treatment to be tailored and monitored appropriately.

Margin Note 17.3 *EMSY* and breast cancer

EMSY is a gene involved in regulating the expression of other genes involved in the repair of DNA. It appears to have a role in the development of breast cancer. In 2003, an analysis of more than 500 breast cancers and over 300 ovarian cancers showed that multiple copies of this gene were present in 13% of breast cancers and 17% of ovarian cancers. Women whose tumors carried multiple copies of the gene survived on average for six years after diagnosis whereas those with normal amounts of the gene had a mean survival time of 14 years. Thus, the presence of multiple copies is associated with an aggressive form of breast cancer and a poor prognosis.

found on the short arm of chromosome 17 and encodes a phosphoprotein, called tumor suppressor protein p53 (*Figure 17.5*) which has been called the 'guardian of the genome'. The gene becomes activated in cells where DNA has become damaged, leading to the production of p53 protein. The protein can bind to DNA, blocking division of damaged cells and inducing apoptosis, thus preventing replication of potential tumors. Mutations in *TP53* can lead to the production of a defective p53 which cannot recognize binding sites on DNA. Thus the replication of the cell is not inhibited, leading to a failure to remove damaged and potentially malignant cells.

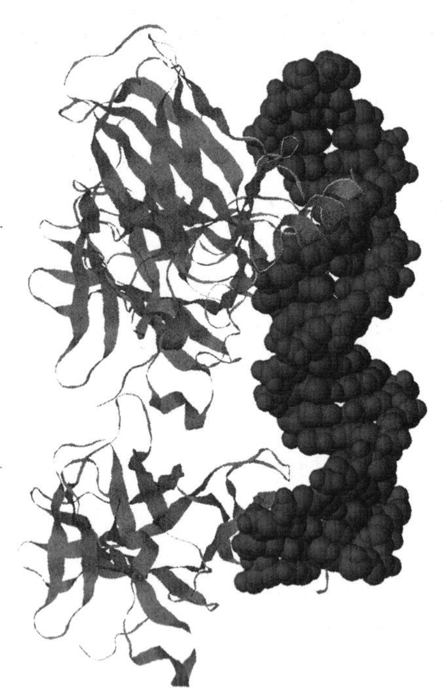

Figure 17.5 A molecular model showing the binding of the p53 protein to DNA (red), this inhibits the division of damaged cells that are potentially cancerous. PDB file 1TUP.

17.5 CAUSES OF CANCER

The mutations that lead to cancerous states are caused by or associated with a number of factors. These include mutations caused by errors in replicating DNA or failures in repairing damaged DNA, or they may be induced by a variety of environmental agents, including chemical carcinogens and ionizing radiation or by the action of some viruses. Mutations in cancer associated germ-line genes may also mean that offspring carrying the gene are more susceptible to developing cancer.

INHERITED CANCERS

The association between the inheritance of a mutated gene and the development of cancer genes may be direct or indirect. For a direct association, the offspring must inherit a mutated gene which confers increased susceptibility to a specific type of cancer. With an indirect association, the inherited gene is associated with defective mechanisms for the repair of DNA that, in turn, leads to a greater likelihood of cancer. An example of the latter occurs in people with the disease xeroderma pigmentosum in whom a failure to repair ultraviolet light-induced mutations in the DNA of skin cells leads to the development of multiple skin cancers.

Examples of cancer associated genes with a direct effect include the genes *BRCA1* and *BRCA2*. Mutations in these genes, which are associated with an increased susceptibility to cancer of the breast and ovaries, account for less than 2% of all breast cancers. However, patients with a defective *BRCA* gene have a much greater risk of developing breast cancer than those who do not. *BRCA1* was mapped to chromosome 17q21 in 1990 and at least 100 mutations have been identified in the gene. Women who inherit a mutated *BRCA1* gene have a 60–83% chance of contracting breast cancer at some stage and a 20–40% chance of developing ovarian cancer. This gene has now been sequenced and its product identified. It encodes a transcription factor (*Figure 17.6*) that regulates the expression of, amongst others, the tumor suppressor gene *TP53*. Thus the presence of a mutated gene encoding a defective transcription factor leads to a failure to eliminate damaged cells. Women carrying a mutated form of *BRCA1* may be offered a **prophylactic bilateral mastectomy**, that is the removal of the apparently healthy breasts to prevent the development of breast cancer.

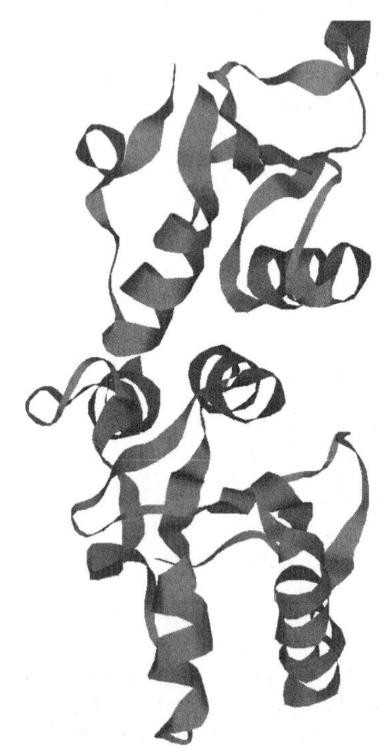

Figure 17.6 Molecular model of a *BRCA1* protein. PDB file 1L0B.

The *BRCA2* gene has been mapped to chromosome 13q12. Mutations in this gene confer a 40–60% chance of developing breast cancer and a 10–20% risk of ovarian cancer at some stage in their lives. A mutated *BRCA2* gene also increases the risk for male breast cancer.

Another gene associated with breast cancer is *CHK2*, which encodes a protein kinase, called the checkpoint kinase. This enzyme is involved in the control of the cell cycle (*Chapter 15*). Inheriting the abnormal variant of *CHK2* doubles the risk of a woman developing breast cancer, in that the variant form is present in 1.9% of women with breast cancer and 0.7% of healthy women.

BOX 17.1 Li-Fraumeni syndrome

Li-Fraumeni syndrome (LFS) is a rare autosomal dominant syndrome that is inherited in typical Mendelian fashion (*Figure 17.7* and *Chapter 15*) which predisposes the patient to cancer. Li-Fraumeni syndrome has been linked to mutations in *TP53*. More than 60% of members of LFS families have inherited mutations in one of the two copies of this gene. Mutations can occur spontaneously in germ cells of one parent or may occur early in the development of the embryo. Some LFS families have a mutation in the *CHK2* gene rather than that encoding *TP53*.

Li-Fraumeni syndrome was first described in 1968, and fewer than 400 families had been identified worldwide by 2006. In LFS families, affected individuals develop cancer at an early age and a wide range of cancers is seen amongst family members. These include cancers of the breast, brain soft tissue, bone and adrenal cortex as well as leukemias, and these are diagnostic. Patients also present with multiple primary tumors at these sites. However, patients within LFS families have also been found to develop tumors at other sites in addition to those used in diagnosis. Children who survive an initial cancer have a high risk of developing a second one.

To diagnose familial LFS, patients must have a number of family members diagnosed with childhood cancer, sarcoma, brain tumor or adrenal cortical carcinoma before the age of 45 years, a close relative (parents, siblings, first cousins) with any LFS associated cancer diagnosed at any age, and another close relative with any cancer diagnosed before the age of 60 years. The age at which patients present with an initial primary tumor is significant since more than half of patients with LFS present before the age of 45, compared with 10% of the general population.

The commonest cancer in females with LFS is breast cancer. Regular **mammography**, which is an X-ray examination of the breast, will detect a tumor as a shadow on the X-ray film or 'breast mammogram' and has been advocated for women in affected families. However, regular X-ray screening is controversial since ionizing radiation itself may increase the risk of cancer in these patients. Prophylactic bilateral mastectomy has also been proposed, though, again, this is controversial since the women remain at risk of developing a cancer at a different location. Genetic counseling of individuals in affected families is essential to inform patients of the potential risks of cancer and to help them develop strategies to avoid at-risk behaviors, such as exposure to radiation.

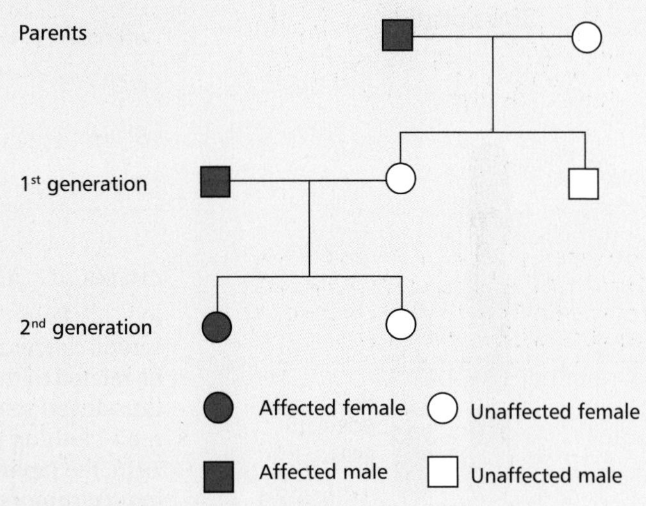

Figure 17.7 Scheme showing the autosomal dominant mode of inheritance of the Li-Fraumeni syndrome.

CHROMOSOMES AND CANCER

Chromosomal abnormalities are found in some cancers. For example, about 95% of patients with chronic myelogenous leukemia (CML) have a translocation (*Chapter 15*) involving chromosomes 9 and 22, that is, t(9:22). The translocation results in the production of a longer chromosome 9 and a shorter chromosome 22, commonly called the Philadelphia or Ph chromosome (*Figure 17.8*). The translocation results in the *BCR* gene on chromosome 22 becoming fused with part of the *ABL* gene on chromosome 9. The fused *BCR-ABL* encodes a tyrosine kinase that is continuously expressed, leading to continuous stimulation of proliferation and the development of CML (*Figure 17.9*). The Philadelphia chromosome is also found in 25–30% of adults with acute lymphoblastic leukemia (ALL; *Section 17.8*). In patients, the translocation originally occured in a single bone marrow cell. However, clonal expansion of the cell results in the blood becoming populated with cells bearing the Philadelphia chromosome. During active disease additional chromosomal abnormalities appear and these are indicative of a poor prognosis.

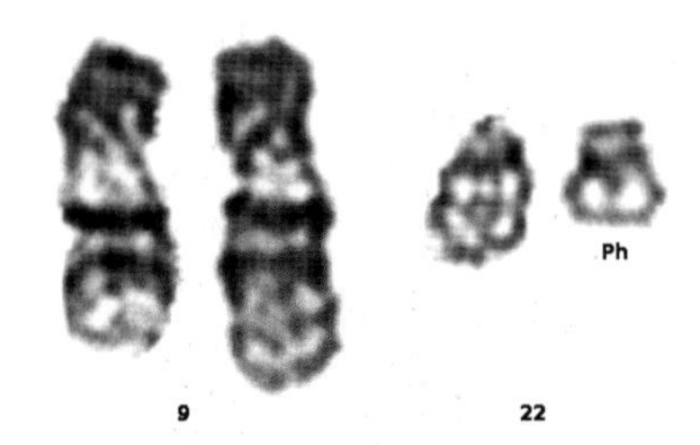

Figure 17.8 The abnormally long chromosome 9 and the Philadelphia chromosome (Ph, abnormal chromosome 22) from a patient with CML. Reprinted with permission of the Wisconsin State Laboratory of Hygiene, University of Wisconsin Board of Regents.

Other chromosome abnormalities have been associated with a wide range of cancers, including breast cancer, prostate cancer and neuroblastoma,

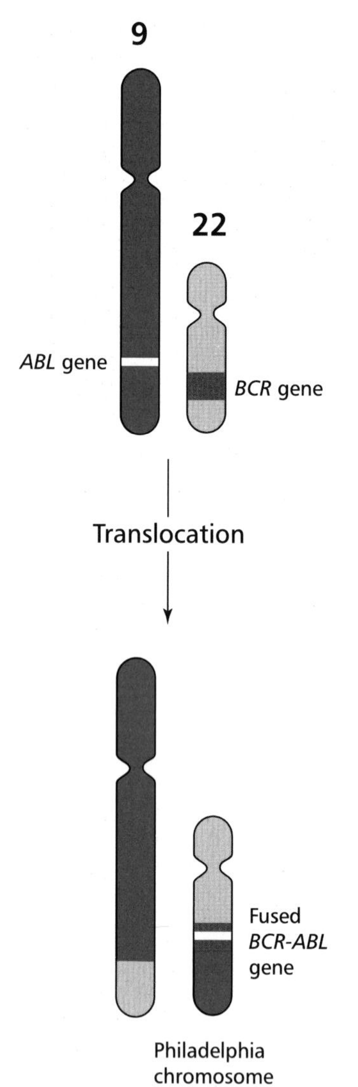

Figure 17.9 Schematic showing the translocation between chromosomes 9 and 22, which results in the Philadelphia chromosome.

but none to the same extent as that found in CML, where the presence of the Ph chromosome is diagnostic for the disease. Some other examples of chromosome abnormalities are shown in *Table 17.3*.

Abnormality	Associated with
Deletion in the short arm of chromosome 1 that is, del(1p)	colorectal adenoma
Deletion on the short arm of chromosome 3 between positions 12 and 14, that is, del(3)(p12;p14)	breast carcinoma
del(3p)	renal cell carcinoma; nonsmall cell carcinoma of lung
Translocation between chromosomes 1 and 17; between position 36 on the short arm of chromosome 1 and position 12 on the long arm of chromosome 17. That is, t(1;17)(p36;q12)	neuroblastoma
chromosome 1 rearrangements	breast carcinomas
del(11)(p13)	Wilm's tumors
t(15;17)	acute promyelocytic leukemia (APL)

Table 17.3 Some chromosomal abnormalities found in different cancers

CHEMICAL CARCINOGENS

In 1775 Potts (1714–1788), an English surgeon, noted the high incidence of scrotal cancer amongst chimney sweeps in London and suggested that this may be related to the accumulation of soot in their clothes. As a result of legislation introduced to ensure that chimney sweeps were able to bathe and to change their clothing regularly, scrotal cancer was eliminated in this profession. In 1918, the Japanese scientists, Yamagiwa and Ichikawa showed that they could induce tumors experimentally by painting coal tar on to the skin of rabbits. This ability of certain compounds to induce tumors experimentally (*Figure 17.10*) led to the identification of many carcinogenic chemicals, including those listed in *Table 17.4*.

The chemical induction of cancers is assumed to be a multistep process, probably involving mutations in several genes, possibly on different chromosomes. Carcinogenic compounds interact with DNA usually by one of a limited number of reactions. For example, alkylating agents such as dialkylnitrosamines and aflatoxin B1 lead to the addition of an alkyl group to electron-rich sites in DNA, as shown for aflatoxin B1 in *Figure 12.4 (B)*. Aromatic amines and amides form highly electrophilic aryl nitrenium ions which also interact with DNA. Polycyclic aromatic hydrocarbons transfer an alkyl group to DNA.

At its simplest, the process of chemical carcinogenesis can be thought to occur in three phases namely: tumor initiation, promotion and progression. During tumor initiation, the carcinogen, or a metabolite of the carcinogen, produces a mutation in the DNA. The cell may repair the damage but misrepair may lead to heritable changes. A cell that has undergone initiation, however, is not yet cancerous because the cell has to 'escape' from normal growth control and become autonomous. Tumor promoters stimulate clonal proliferation of an initiated cell. An example of a promoter is tetradecanoyl phorbol acetate, a constituent of croton oil. Croton oil will promote the development of carcinomas in the skin of mice that have been treated with a single dose of benzo[a]pyrene. Tumor progression involves the additional changes that lead to malignancy and the ability to form metastases.

Chemical	Source	Cancers caused
Aflatoxin	*Aspergillus flavus* and *Aspergillus parasiticum* growing on crops such as corn and peanuts	liver
Arsenic	in insecticides and herbicides	lung, skin
Asbestos	mineral fibers used in fire-insulation, brake linings	mesothelioma
Benzene	petrochemicals, dyes (industrial exposure)	bladder cancer
Benzo[a]pyrene	partial combustion of petroleum; tobacco smoke	lung
Polycyclic aromatic hydrocarbons	partial combustion of petroleum	lung
Polychlorinated biphenols	industrial processes, insecticides	liver, skin
Diethylstilbestrol	once used to prevent miscarriage in women	vaginal tumors in offspring of treated women
Vinyl chloride	industrial processes	liver

Table 17.4 Some carcinogenic chemicals

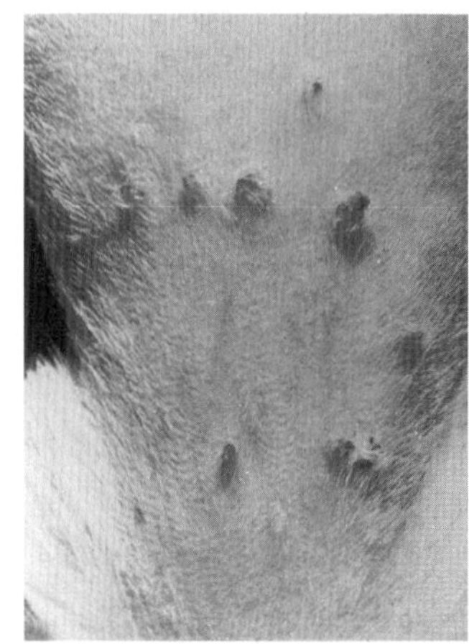

Figure 17.10 Carcinogen-induced tumors on the skin of a rat.

The targets for chemical carcinogens are the proto-oncogenes and the tumor suppressor genes described in *Section 17.4*. Mutations in *TP53* have been found in more than 500 types of human tumor. The mutations occur at vulnerable sites known as 'hotspots', though the position of the hotspot is not the same for all carcinogens. For example, the metabolite of benzo[a]pyrene preferentially forms adducts with guanine bases (*Figure 12.4 (A)*) at codons 157, 248 and 273 of *TP53*, which are the same mutational hotspots found in human lung cancers. This therefore supports the link between smoking and lung cancer that was established in the 1950s (*Box 17.2*). Some chemical carcinogens may work by promoting the generation of reactive oxygen intermediates which themselves attack DNA. Thus health food shops promote the sale of supplements which are known to remove or 'scavenge' these free radicals. Alternatively some carcinogens may interfere more directly with the regulation of cell proliferation or receptor-mediated cell signaling processes.

Testing for potential carcinogens

The commonest test for the ability of chemicals to cause mutations in DNA is the Ames test which uses a mutant of *Salmonella typhimurium* that is unable to grow on growth media in the absence of the amino acid histidine (*Figure 17.11*). The test involves exposing the bacterium to the chemical in question, usually in the presence of an extract of mammalian liver to provide enzymes that may activate any procarcinogens present. Mutations caused by the potential carcinogen may result in the reversion of the mutant bacterium into one that can synthesize histidine and thus grow on the histidine deficient medium.

Mutagenicity can also be tested by determining the ability of the chemical in question to cause cytogenetic changes in the bone marrow of rodents. Traditional tests for carcinogenicity have ultimately relied on the use of laboratory animals, though this has obvious ethical implications. Some chemicals result in the induction of tumors in the majority of experimental animals after a single dose. The required time for a tumor to develop is known as the latency period and this may be shortened by administering several doses of the carcinogen.

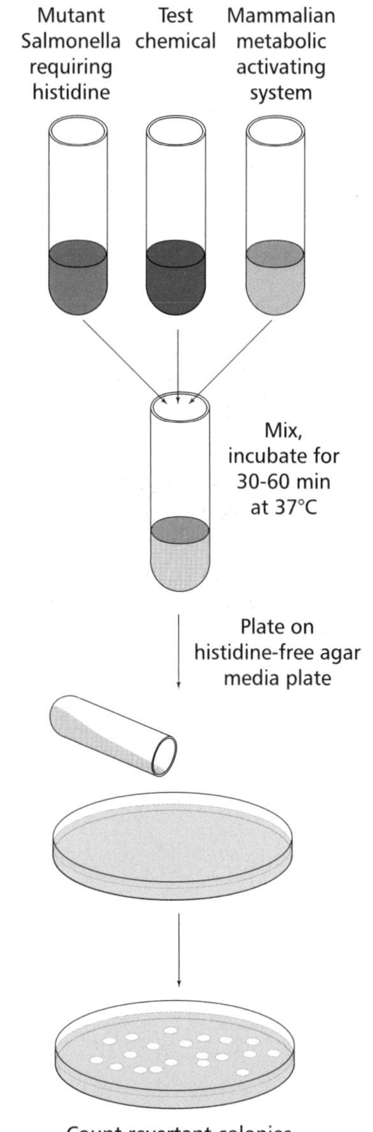

Figure 17.11 Schematic to show the Ames test. See main text for details.

BOX 17.2 Cigarettes and cancer

In the 21st century the link between smoking and cancer is well known. In the USA, smoking accounts for about 30% of all deaths from cancer. As well as being an established cause of lung cancer, smoking is also implicated in the development of cancers of the mouth, esophagus, bladder and pancreas. Other associations link smoking to increased risk of cancer of the stomach, liver and kidney and in the development of CML.

The link between smoking and lung cancer was first established by the renowned scientist and epidemiologist, Doll (1912–2005). Doll qualified in medicine in 1937 and, after World War II, began work on a project to determine the causes of a sharp increase in the number of deaths from lung cancer over the previous 30 years. He embarked on a study in which he gave questionnaires to lung cancer patients about their habits. Originally, he thought the increase might be due to exhaust fumes from cars. From a relatively small scale study to one which involved questionnaires sent to over 60 000 doctors, he and his colleague, Hill (1897–1991), were able to show that the risk of lung cancer was proportional to the number of cigarettes smoked by the patient (*Figure 17.12*) and that long-time smokers had three times the mortality rate of nonsmokers. In addition, he established the link between cigarette smoking and a number of other serious diseases such as coronary thrombosis (*Chapter 14*) and chronic bronchitis.

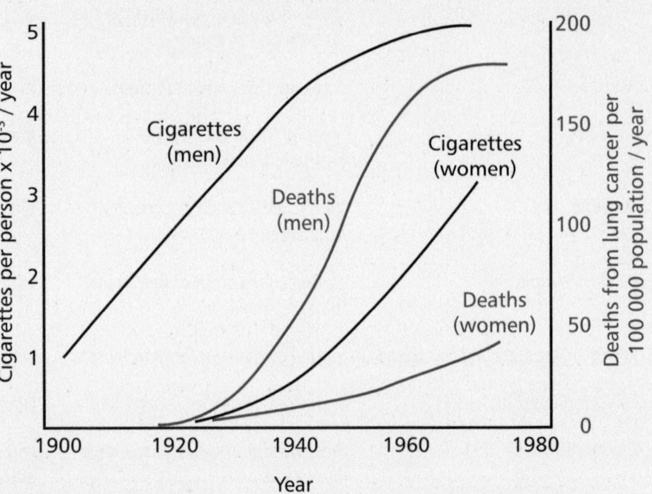

Figure 17.12 The relationship between deaths from lung cancer and the number of cigarettes smoked. Note the 20-year lag between the increases in cigarettes smoked and the increasing numbers of deaths.

Cigarette smoking accounts for nearly 90% of deaths from lung cancer and contributes to deaths from other forms of cancer including those of the larynx, mouth and esophagus; smokers are twice as likely to develop bladder cancer as nonsmokers. Fortunately, knowledge of the link between smoking and cancer has led to fewer people smoking in the developed world although the number of younger women who smoke is still increasing, as it is in parts of the developing world.

Doll used epidemiological studies to establish other links between social habits and clinical conditions. For example, he showed that imbibing alcohol during pregnancy can have undesirable effects on the unborn baby and that exposure to relatively small amounts of ionizing radiation increases the risk of leukemia.

DIET AND CANCER

It has been estimated by bodies such as the World Cancer Research Fund that between 30 and 40% of cancers could be prevented by eating a healthy diet, and by maintaining a healthy body weight, (*Chapter 10*) and participating in adequate physical activity. Conversely, prospective studies, in which researchers analyzed the diet and activity of a group of individuals, then monitored the frequency of cancer deaths in that group, have indicated that being overweight or obese contributes to 14% and 20% of deaths in men and women respectively. In addition, obesity has been strongly linked to a variety of cancers, including those of the GIT, liver, prostate, breast, uterus, cervix and ovary. There is also evidence to suggest a link between the consumption of foods with a high glycemic index (*Chapter 10*) and an increased incidence of cancer.

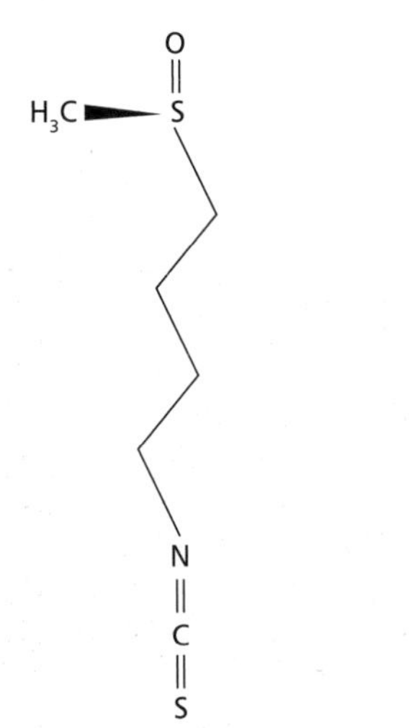

Figure 17.13 The structure of sulfurophane.

The consumption of low fiber, highly processed foods has a well-established association with the incidence of colorectal cancer (*Section 17.8*) though the link may be more complex than was originally thought. Indeed, an increased consumption of fiber-rich foods, such as fresh fruit and vegetables, has been correlated with a reduced incidence of several types of cancer, including those of the mouth, esophagus, lung and stomach in addition to those of the colon and rectum. Such associations have led to the recommendation by the UK Department of Health that individuals should consume at least five portions of fruit and vegetables a day. Cruciferous vegetables such as cabbage, cauliflower, broccoli and brussel sprouts contain sulfurophane (*Figure 17.13*), a chemical with anticancer properties. Increased consumption of such vegetables is inversely related to the incidence of breast and bladder cancers.

A number of nutritional interventions appear to offer some protection against cancer. These include decreasing the consumption of red meat and processed red meat, the intake of which is significantly linked to colorectal cancer. Increasing the consumption of ω-3 fats such as α-linolenic acid, (ALA), eicosapentaenoic acid (EPA) and docosahexaenoic acid (DHA) and decreasing that of ω-6 fats such as linoleic acid (*Chapter 10*) is also recommended. Flax seeds, which are a good source both of dietary fiber and ALA, have been shown to reduce the development of carcinogen-induced tumors and to reduce the rates of metastasis in animal models. Numerous studies have indicated that a sufficient vitamin D (*Chapter 10*) status offers protection against a variety of cancer types, and that improvements in its intake by, for example supplementation, could reduce the incidence of cancer and the associated mortality. Consumption of selenium, which is present in the active sites of several enzymes including the antioxidant enzyme glutathione peroxidase (*Chapter 10*), has been associated with a decreased risk of prostate cancer in men. Other studies have suggested that antioxidants, such as those found in green tea, are protective against cancer, although this is controversial. In addition a high consumption of lycopene (*Figure 17.14*), a carotenoid found in tomatoes, has been shown in several studies to be associated with a decreased incidence of prostate cancer.

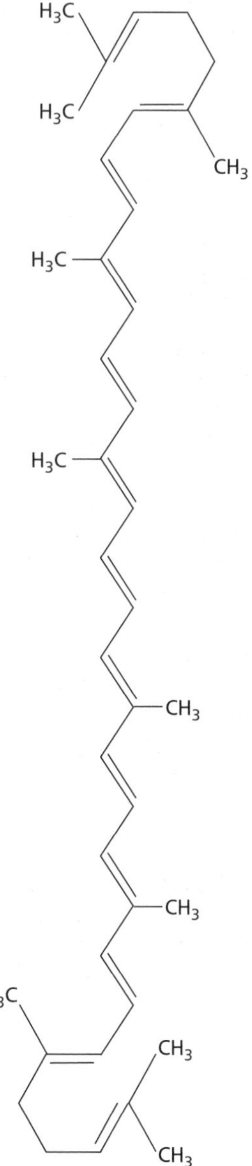

Figure 17.14 The structure of lycopene.

RADIATION AND CANCER

Ionizing radiation can promote the production of tumors *in vivo*, and the transformation of cultured cells *in vitro* from a normal to a malignant phenotype. Some of the evidence for radiation-induced carcinogenesis comes from studies of Japanese people irradiated during the atomic bomb explosions in Hiroshima (*Figure 17.15*) and Nagasaki, and from populations irradiated in nuclear accidents such as that which occurred in Chernobyl in 1986. Long-term studies of Japanese atom bomb survivors showed an increase in the incidence of leukemias in the first 5–10 years following exposure. The risk of solid tumors in these people was also increased significantly. Studies by Doll (*Box 17.2*) showed that infants previously subjected to X-irradiation *in utero* had an increased risk of developing leukemias and solid tumors.

Figure 17.15 Hiroshima: the Peace Memorial Park.

The condition xeroderma pigmentosum also provides evidence for radiation-induced carcinogenesis. These patients suffer multiple skin cancers caused by the failure of their cells to repair ultraviolet light-induced damage to DNA. In addition, normal cells in culture may be transformed by irradiation into cells with a cancerous phenotype. Such studies have been used to analyze the nature of radiation-induced carcinogenesis.

DNA is also the target of ionizing radiation in radiation-induced carcinogenesis and the damage caused includes deletions, inversions and translocations (*Chapter 15*). Irradiation is also known to induce gene amplification and to increase chromosomal instability and these, in turn, increase the likelihood of mutations occurring. Ultimately, mutation events involving proto-oncogenes and/or tumor suppressor genes are the most likely causes of radiation-induced carcinogenesis.

With some exceptions, cells are more susceptible to radiation-induced damage when dividing and this fact has been utilized in the use of radiation to treat cancer (*Section 17.7*).

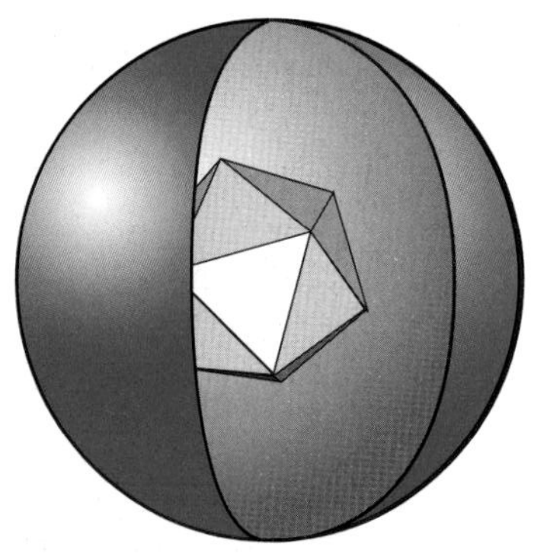

Figure 17.16 Schematic showing an Epstein-Barr virus. The particle is approximately 100 nm in diameter.

VIRUSES AND CANCER

Viruses were first implicated in the development of some types of cancer when it was shown by Rous (1879–1970) in 1911 that leukemia in chickens was caused by a 'filterable agent'. This virus, which causes sarcomas in chickens, is now called the Rous sarcoma virus and has been used extensively in research into oncogenic, or cancer-inducing, viruses.

Today, viruses are associated with between 10 and 20% of all human cancers. In the 1960s, Burkitt's lymphoma, which is a tumor found in the jaws of children in certain regions of Africa, was found to be induced by a virus later identified as the Epstein-Barr virus (EBV; *Figure 17.16*) or human herpesvirus 4 (*Table 3.1*) which also causes glandular fever, or infectious mononucleosis. The EBV infects epithelial cells and B lymphocytes, causing their transformation. This virus is associated with a number of cancers, including nasopharyngeal carcinoma and Hodgkin's lymphoma, in addition to Burkitt's lymphoma.

Cancer associated viruses belong to several groups, including the retroviruses (retroviridae), the papillomaviruses (papillomaviridae), the hepadnaviruses (hepadnaviridae), the flaviviridae and the herpesviridae (*Table 17.5*).

Class of virus	Examples	Associated cancer(s)
Hepadnaviridae	hepatitis B virus	hepatocellular cancer
Flaviviridae	hepatitis C virus	hepatocellular cancer
Herpesviridae	Epstein-Barr virus Kaposi's sarcoma associated herpes virus	Burkitt's lymphoma Kaposi's sarcoma
Papillomaviridae	human papilloma viruses	cervical cancer
Retroviridae	human T-cell leukemia virus (HTLV)	adult T-cell leukemia

Table 17.5 Some viruses associated with human cancer

Retroviruses are RNA containing viruses that have reverse transcriptase activity. When they infect a cell, the reverse transcriptase transcribes their RNA into DNA which may become incorporated into the host genome. Not all retroviruses cause cancer but those which do are called 'transforming retroviruses'. The human immunodeficiency virus (HIV) is a retrovirus that is also associated with cancer, although in this case the association is indirect. People with HIV who develop full-blown acquired immunodeficiency syndrome (AIDS; *Chapter 3*) have an increased incidence of tumors, such as lymphoma and Kaposi's sarcoma, which are associated with EBV and Kaposi's sarcoma virus (KSV). It is likely that the immunosuppression caused by HIV allows latent viruses such as EBV to escape immunological control. Kaposi's sarcoma virus is also associated with nonHodgkin's lymphoma and oral squamous cell carcinoma.

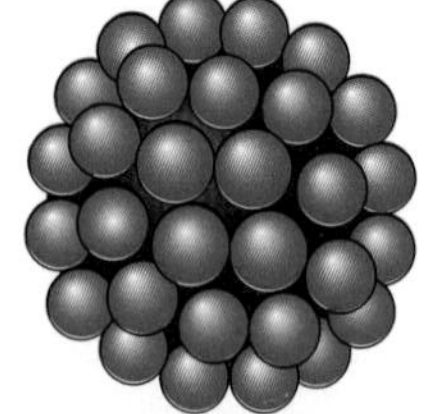

Figure 17.17 Schematic showing a human papilloma virus. The particle is approximately 55 nm in diameter.

The most widely recognized association between viruses and cancer occurs with certain strains of human papilloma virus (HPV; *Figure 17.17*) that are linked to the development of cervical cancer. More than 70 papilloma viruses have been found in humans. The genomes of these viruses all have a similar structure and contain at least seven 'early' genes (E1–E7) and two 'late' genes, L1 and L2 (*Figure 17.18*). Some HPV subtypes invade epithelial cells of the skin, causing warts while others infect the genital tract and cause benign warts, with low risk of cancer. Others, such as HPV 16, 18, 31, 33 and 45, are associated with the development of cervical cancer in women and are regarded as 'high risk' for inducing cancer. DNA from at least one of these 'high-risk' types is detected in approximately 90% of human cervical cancers. The HPV E6 and E7 genes are

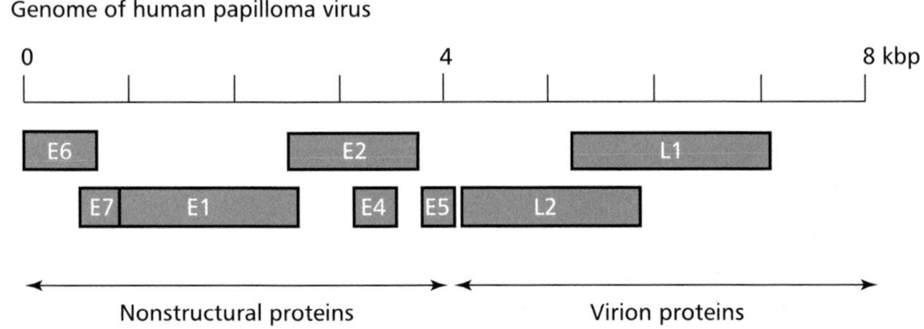

Figure 17.18 The genes of a human papilloma virus.

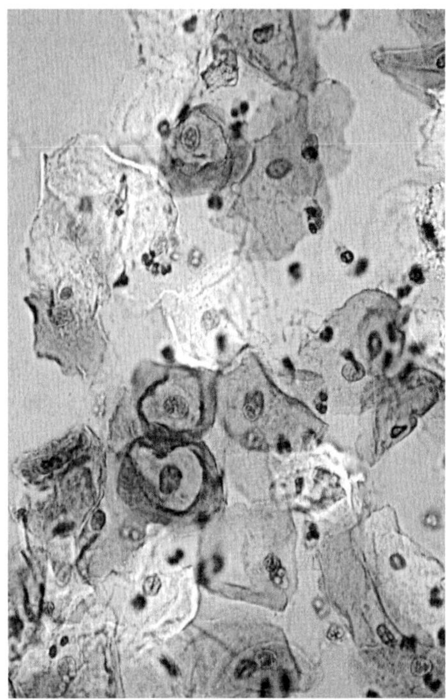

Figure 17.19 Light micrograph of a cervical smear showing lightly stained squamous cells from the superficial layer of the cervix and more darkly stained cells from the layer of the cervical wall immediately below that of the squamous cells. Note that these cells are abnormal in having comparatively large nuclei and show that the patient is at risk of developing cancer of the cervix. See also *Figure 1.15*. Courtesy of H. Glencross, Manchester Cytology Centre, Manchester Royal Infirmary, UK.

viral oncogenes and the proteins they encode inactivate certain regulators of cell division, such as the tumor suppressor protein p53 (*Section 17.4*). Cervical cancers develop from precursor 'lesions' that are graded high to low according to how much disruption of epithelial cell differentiation has occurred (*Figure 17.19*). In cervical carcinomas, viral DNA becomes integrated into the host genome. However, many benign lesions also contain these strains of HPV, so the presence of the virus alone is insufficient to cause cancer. It seems that long-term infection with HPV predisposes the individual to cervical cancer and that other agents, for example the carcinogens in cigarette smoke, are required to allow the tumor to progress to full malignancy.

17.6 GENERAL DIAGNOSIS OF CANCER

The symptoms of cancer at presentation depend on the location and extent of the tumor. Depending on the cancer, symptoms may develop when the cancer is relatively small as, for example, when even a small tumor of the brain causes pressure to develop. Other tumors, such as those developing in the ovary may not produce symptoms until it is relatively large.

A number of general signs may alert an individual to the presence of an undiagnosed tumor, for example, an unexplained weight loss of about 5 kg or more. Weight loss that occurs as a result of cancers is due to the release of cytokines from cells of the immune system. In end-stage cancer, this weight loss is known as **cachexia** (*Box 17.3*). Fever, which is also induced by cytokines, is often found in patients with advanced cancer, although this may also be an early warning of certain types of cancer. An example of this is Hodgkin's lymphoma, which is characterized by fever, often during sleep, and is accompanied by drenching night sweats. Fatigue may also be a symptom, particularly where the cancer causes a loss of blood with concomitant anemia, as may occur, for example, with stomach cancer. Other signs and symptoms include the presence of a lump, as for example in the breast or testicles, or unusual bleeding or discharges.

CLINICAL TESTS FOR CANCER

Clinical tests for cancer are used to screen for cancer in an 'at-risk' population, to detect cancer in a patient presenting indicative symptoms, to monitor the success of treatment or to detect recurrence in a patient who has been in remission. These tests fall into a number of categories. They may involve the detection and/or quantification of tumor associated molecules, the detection and localization of tumors within the body, and the histological examination of biopsies from suspect tissue to determine the nature of the tumor and/or to detect precancerous conditions.

BOX 17.3 Cachexia

Cachexia (derived from the Greek words *kakos* meaning bad and *hexis*, condition) is characterized by a severe loss of weight, largely of skeletal muscle mass, in someone who is not actively trying to lose weight. It is marked by fatigue, weakness, ill health and decreased appetite. It is usually a sign of various underlying disorders and is associated with chronic disease such as cancer, AIDS (*Chapter 3*), chronic infectious diseases, such as tuberculosis (*Chapter 4*) or severe malnutrition (*Chapter 10*). About half of all cancer patients lose weight although the extent varies with the type of cancer. At diagnosis, 80% of patients with upper GIT cancers and 60% with lung cancer are already suffering cachexia, although patients with hematological malignancies and breast cancer usually avoid such a substantial loss of weight. Most other solid tumors are associated with a higher frequency of cachexia.

Cachexia is commoner in children and elderly patients and becomes more pronounced as the disease progresses. Its prevalence increases from 50% to over 80% before death. In addition to compromizing a patient's quality of life, it is correlated with poor outcomes. Indeed, cachexia is the main cause of death in over 20% of patients, often due to degeneration of the respiratory muscles.

Cachexia also occurs secondarily because of an inability to ingest or use nutrients, for example from obstruction in the GIT or clinical malabsorption. In cancer, it is often associated with a disordered metabolism, and both tumor and host factors appear to play a major role in its development, although surgery or treatment-related disorders, as in the nausea and vomiting associated with chemotherapy or radiation therapy may also be involved. Patients tend to be insulin resistant and have high basal metabolic rates. Furthermore, the patient's response is analogous to that of a chronic infection (*Chapter 4*) and the immune system secretes Interleukins 1 and 6 (IL-1, 6) and tumor necrosis factor α (TNF-α) also called cachectin (*Figure 17.20*), which stimulate fever, protein and lipid breakdown and the production of acute phase proteins by the liver. Protein and lipid catabolism are also stimulated by the release of proteolysis-inducing factor and a lipid-mobilizing factor called zinc α2-glycoprotein (ZAG) from tumor cells which lead to the degeneration of skeletal muscle and adipose tissues respectively, resulting in cancer cachexia, while supplying the tumor with fuels such as glucose (*Figure 17.21*). This can be detected in a patient by evaluating his or her nutritional status, usually with a combination of clinical assessment and anthropometric tests as described in *Chapter 10*. Body weight, with reference to the normal adult weight, is the usual measure of nutritional status.

Obviously the best way to treat cancer cachexia would be to cure the cancer. Unfortunately, in adult patients with advanced solid tumors this is often not possible. Treatment should therefore be aimed at improving the quality of life and, for many patients, this means improving appetite and food intake and trying to inhibit muscle and fat wasting. Unfortunately, treatment is limited. Hypercaloric feeding, either enteral or parenteral, does not generally increase lean mass. Glucocorticoids are widely used as palliatives because they inhibit the synthesis and/or release of proinflammatory cytokines such as TNF-α and IL-1, which decrease food intake directly, or through other anorexigenic mediators, such as leptin (*Chapter 10*), corticotrophin releasing factor (CRF) and serotonin and have some limited effect in improving appetite and food intake. Corticosteroids have significant antinausea effects and improve asthenia (weakness) and pain control. However, studies have not shown any beneficial effect on body weight. Indeed, prolonged treatment can lead to weakness, delirium, osteoporosis and immunosuppression, all of which are significant problems initially in advanced cancer patients. The synthetic derivatives of progesterone, megestrol acetate (MA) and medroxyprogesterone acetate (MPA) taken orally have some effect in improving appetite, energy intake and nutritional status. One novel approach under investigation is to use supplements, such as ω-3 fatty acids (*Chapter 10*) that reduce IL-1 and TNF-α production and which may improve the efficacy of nutritional support.

Detection and estimation of tumor-associated molecules

Some types of tumor are associated with increases in tumor associated molecules in the blood. These 'tumor associated antigens' are detected and quantified by using monoclonal antibodies specific for the antigen in question. The antibody may be used in an enzyme-linked immunosorbent assay (ELISA) or in radioimmunoassay (RIA) as described in *Chapter 4*. Examples of tumor associated antigens include prostate-specific antigen (PSA; *Section 17.8*) which is elevated in the blood of patients with cancer of the prostate, and CA 125 antigen, which is found in the blood of women with ovarian cancer. Carcinoembryonic antigen (CEA) is a glycoprotein that is overproduced in most colon carcinomas and in carcinoma of the lung and breast. Serum concentrations are measured and used to monitor treatment and to predict prognosis.

Unfortunately, increases in tumor associated antigens are not exclusive to cancer so that the tests can only give an indication of cancer. Increased

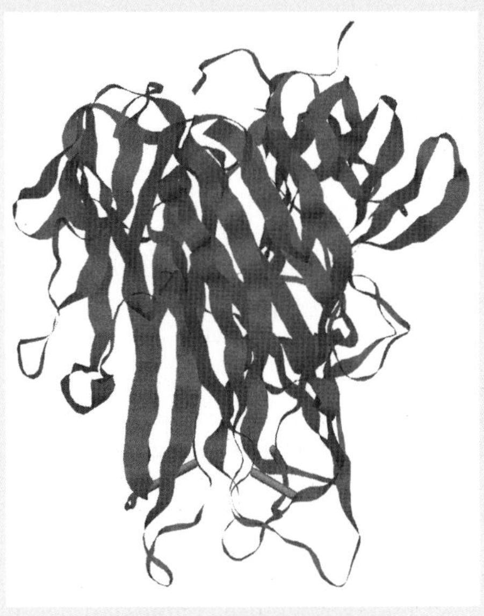

Figure 17.20 Molecular model of cachectin (Tumor Necrosis Factor α). PBD file 1TNF.

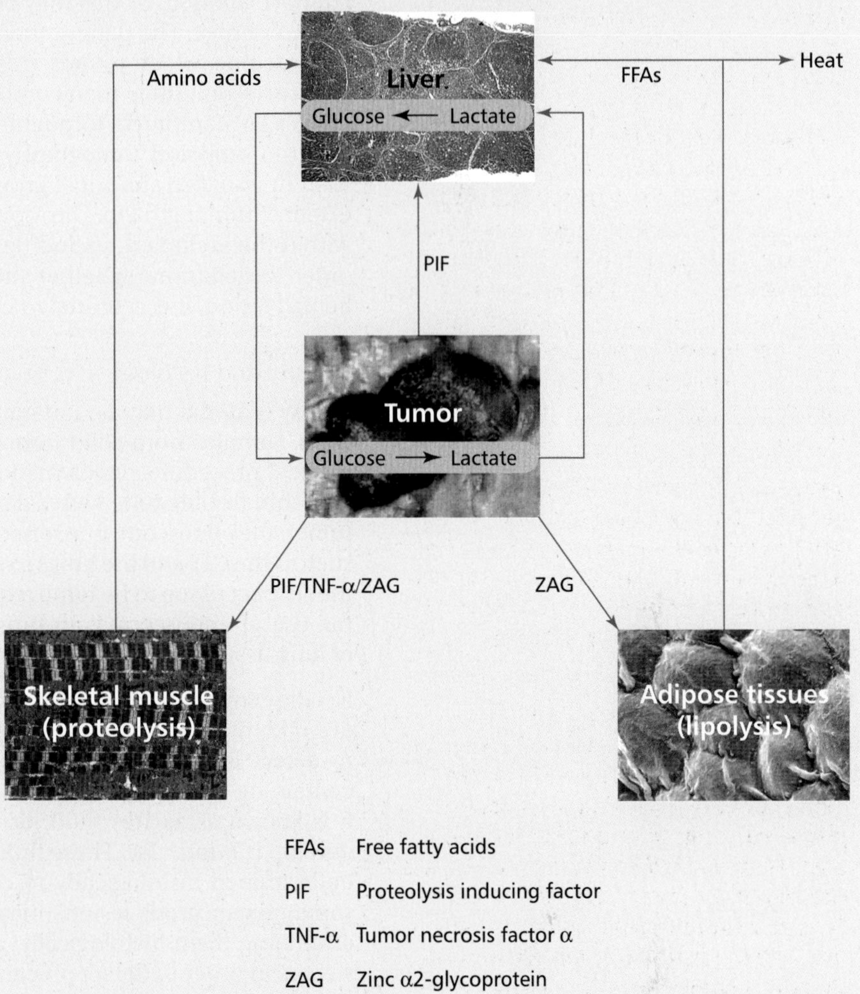

Figure 17.21 Some of the physiological activities involved in cancer cachexia.

concentrations should therefore always be followed by further tests, including physical examination and diagnostic imaging.

Detection and localization of tumors within the body

Patients who present with symptoms indicative of cancer may be examined using diagnostic imaging procedures in order to localize the potential tumor or to determine the extent of metastasis. Diagnostic imaging may involve the use of X-rays, computed tomography (CT) also known as computed axial tomography (CAT), magnetic resonance imaging (MRI), positron emission tomography (PET) and ultrasound, all of which are undertaken by clinical radiologists.

The routine screening for breast cancer by mammography is offered to women in the UK over the age of 50 years. A tumor will show up as a shadow on a breast mammogram. Patients who have lung cancer may have X-rays taken of their bones, in order to determine whether the tumor has spread to them.

Computed axial tomography scanners were introduced into hospitals in the UK in 1975. The technique uses X-rays to take sequential pictures of the body from different directions. In practice, the patient may be required to drink a contrast solution, or this may be administered intravenously, to enhance the tissue contrast. The patient then lies still on a table that passes through the X-ray machine, which rotates around the patient taking pictures of thin 'slices' of tissue. Computers then combine the images to produce three-dimensional images or computed tomograms. Magnetic resonance imaging (MRI) and positron emission tomography (PET), outlined in *Chapter 18*, may also be used to visualize abnormal growths within the body and for determining the extent of tumor growth. Ultrasound (*Chapter 14*) may be used to locate tumors within the abdomen, including tumors of the liver and ovaries. However, in order to determine whether the abnormal growth represents a cancer or a benign lesion, it is essential to examine biopsies of the growth.

Biopsies and histology

Biopsy material may be obtained from a variety of sources in one of several ways. Samples from solid tumors may be obtained by endoscopy, or during surgical procedures involving local or general anesthesia. An endoscope is a long thin flexible tube with a camera and light on the end. Depending on the tumor, the endoscope is inserted into a body cavity and allows internal tissues such as the GIT and the lungs to be viewed. Endoscopes also enable samples of the suspect tissue to be removed without the need for surgery. One example of the use of endoscopy is in bronchoscopy, which is used to obtain biopsies of lung tissue in suspected cases of lung cancer (*Figure 17.22*; *Section 17.8*).

Needle biopsy allows small amounts of tissue to be obtained from a variety of solid tissues. Samples of blood and bone marrow may be examined to detect leukemias. Some solid tumors may cause the build up of fluid containing cancer cells within the peritoneal cavity (*Chapter 10*), where it is known as **ascites** fluid, or in the thorax, where a pleural effusion may develop (*Chapter 14)*. These fluids contain tumor cells in suspension and may be examined histologically to confirm the presence of cancer cells. Finally, some precancerous lesions may be detected by taking smears of tissue and examining them histologically, as for example in the preparation of cervical smears in order to detect precancerous lesions as outlined below.

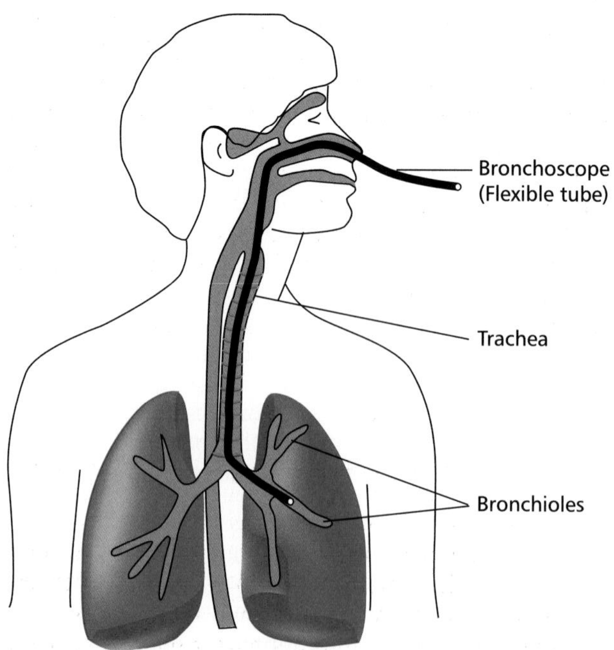

Figure 17.22 Schematic to show a bronchoscopy.

Solid tissue obtained for histological examination must be processed in order to obtain thin sections for microscopic examination. The material is first preserved by fixation in, for example, formalin, and then dehydrated and embedded in paraffin wax to support the tissue. This allows sections between 2 and 7 μm thick to be cut using a microtome. The sections are then mounted on slides, dewaxed with xylene, rehydrated and then stained in the manner appropriate to the investigation. The commonest stains for tissue sections are hematoxylin and eosin. Hematoxylin stains the nucleus purple/black depending on the formulation, while the eosin stains the cytoplasmic contents pink. Hematoxylin and eosin allow good differentiation within the tissue (*Figure 17.23*). Smears of blood and bone marrow are most frequently stained with May-Grunwald-Giemsa, which stains the nuclei blue, while the cytoplasmic contents stain differentially depending on the cell type. This stain therefore allows differentiation between different types of leukocytes. The stain commonly used to detect precancerous cells in cervical smears is the Papanicolaou stain, or Pap stain as it is more frequently known. This formulation contains five different stains: hematoxylin, which stains the nucleus, Orange G (OG-6) and EA, which contains light green, eosin Y and Bismarck brown Y. Orange G and EA are counterstains. The Pap stain is often used to stain buccal and sputum smears as well as those obtained from the cervix.

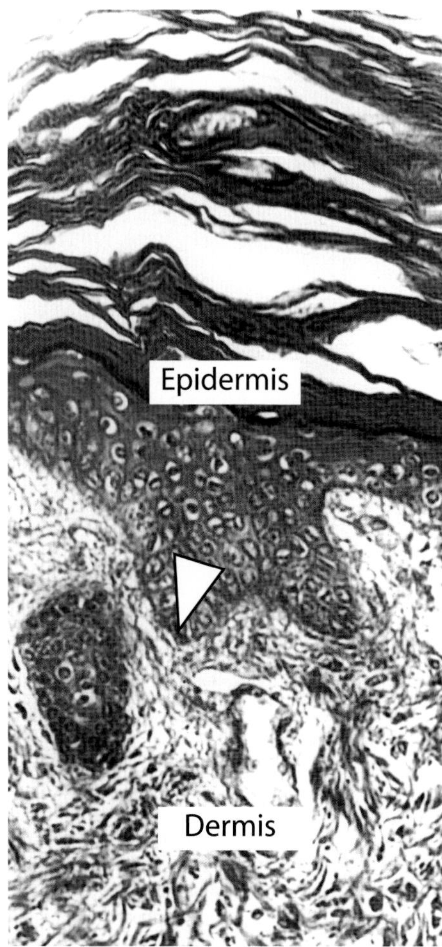

Figure 17.23 Section of skin stained with hematoxylin and eosin. Note the germinative layer of rapidly dividing cells, which is damaged by many anticancer drugs, is indicated by the arrow head.

The preparation of material from solid tissues for histological examination is a long process. Sometimes a more rapid examination may be required as, for example, when the result of a biopsy is needed during an operation in order to determine the extent of surgery required. In such cases, the biopsy is rapidly frozen to –176°C by immersion in liquid nitrogen. This process hardens the material and allows it to be cut into 5–10 μm sections using a freezing microtome or cryostat. Once the sections have been prepared they can be stained with hematoxylin and eosin without the long procedures required with paraffin sections, often within minutes of removal from the body. Examination will then, hopefully, ensure the appropriate surgical procedure.

It is possible to stain cancer cells more specifically if they bear a tumor-specific marker, such as CEA (*see above*). Cryostat sections are often used for this process, because fixation, wax embedding and clearing can destroy some antigenic sites on the tumor. The sections are stained by immunohistochemistry (*Chapter 4*). In this process, the sections are incubated with a monoclonal antibody to a tumor associated antigen. This specific binding is visualized by incubation with an enzyme-labeled anti-immunoglobulin, followed by a further incubation with the appropriate substrate. For enzyme immunohistochemistry, a colorless substrate is chosen that produces a colored insoluble product. It is now possible to use paraffin embedded sections for immunohistochemistry because the antigenic sites that were destroyed during the preparation process, can be 'retrieved' by microwaving the sections in water, subjecting them to heat at pressure using a pressure cooker or by allowing the sections to be partially digested with a proteolytic enzyme. The length of time required to retrieve the antigenic sites has to be determined by trial and error, using positive and negative control slides.

Molecular diagnosis

The histological diagnosis of potentially cancerous cells is increasingly being supported by molecular diagnostic techniques and it is probable that molecular methods will be used more frequently across a greater range of tumors once the genetic basis for the development of these tumors has been established. The determination of the genetic profile of a patient's primary tumor will no doubt become routine and this will inform treatment and be a predictor of prognosis. One way in which tumors have been investigated is to extract DNA using a fine needle biopsy, to amplify the DNA by using the polymerase chain reaction (PCR; *Chapter 3*), and to analyze the DNA obtained for mutations known to be implicated in cancer. Fluorescence *in situ* hybridization technique (FISH; *Chapter 4*) is also applied to diagnosis. Thin sections of the tumor are treated to separate the DNA strands which are then hybridized *in situ* with fluorochrome labeled probes for relevant mutated, cancer associated genes (*Figure 17.24*). The slides are then examined using a fluorescence microscope. The presence of the mutation is indicated by fluorescent spots in the nucleus.

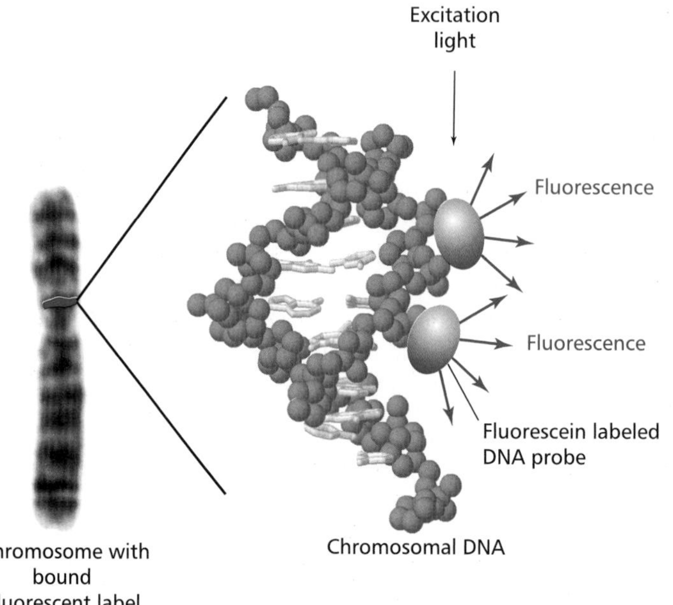

Figure 17.24 Schematic to outline the binding of a fluorescently-labeled DNA probe to target specific chromosomal DNA sequences (FISH). The use of FISH in the specific staining of telomeres is shown in *Figure 18.9*.

17.7 GENERAL TREATMENT OF CANCER

The traditional treatment for primary tumors that have not metastasized to other locations in the body has been surgical excision of the tumor, followed by chemotherapy or radiation therapy (radiotherapy). Chemotherapy involves the use of drugs to kill cancer cells while radiotherapy uses ionizing radiation. However, where a tumor has been diagnosed in an advanced state with metastases in other parts of the body, palliative care only may be given, to ease pain and discomfort.

'Staging' of the cancer is essential as it determines the treatment that the patient receives. Staging of a cancer involves looking at the extent of the cancer within the body. Though staging varies between different types of tumor, a simple system classifies the tumor into one of four stages as shown in *Table 17.6*. Other staging systems include the TNM classification, where T records how far the primary tumor has grown in its original location, N defines whether the tumor has spread to local lymph nodes and M describes whether the tumor has metastasized. The staging of cancers will be discussed further in *Section 17.8*.

The aims of chemotherapy and radiotherapy are to stop the cancer cells from dividing. Other treatments which have been attempted, with varying success include **immunotherapy**, where the objective is to stimulate the body's immune system to eliminate the cancer, as well as a number of 'alternative' therapies all of which are controversial.

Stage	Definition
1	the tumor is small, has not spread to other locations and cannot be felt. The patient is usually free of symptoms and the tumor has been detected by chance during routine medical examination
2	the tumor has not spread from its original location but may be felt during examination or shows up on scans
3	the tumor has spread to adjacent tissues
4	the tumor has metastasized to distant locations

Table 17.6 Staging of tumors

CHEMOTHERAPY

More than 100 different drugs are used to treat different cancers. They are usually given in combination (*Chapter 3*) and with radiation therapy and/or surgery. Most of the chemicals used in the treatment of cancer are, traditionally, those that kill cells, that is they are cytotoxic. Most are active on cells that are dividing, that is they are cycle-dependent. Drugs which kill cells that are not dividing are noncycle-dependent. There is also some evidence that at least some cytotoxic drugs act by inducing terminal differentiation or apoptosis in cancer cells. Many different types of chemicals target cell division in various ways. For example, several categories of drug interfere with DNA synthesis. These include folic acid antagonists, alkylating agents and purine and pyrimidine analogs and the topoisomerase inhibitors (*Chapter 3*). Other drugs inhibit cell division by disrupting the polymerization or depolymerization of microtubules, thus interfering with the separation of chromosomes during mitosis. Unfortunately, drugs that work by preventing cell division do not discriminate between dividing cancer cells and dividing healthy cells. Thus, these drugs have considerable toxicity, particularly towards bone marrow and the epithelial cells of the skin and GIT. Thus, chemotherapy is associated with anemia, nausea and damage to the actively dividing cells of the hair follicles, leading to considerable, but usually reversible, hair loss. A role of the hospital pharmacist is to advise on treatment to minimize the discomfort caused by this therapy. In the treatment of some cancers, aggressive chemotherapy may destroy bone marrow, such that patients require a bone marrow transplant (*Chapter 6*).

The route of administration of chemotherapeutic agents depends on the drug. Many drugs are administered intravenously by infusion. Others are administered intrathecally, that is, by injection into the innermost membrane surrounding the central nervous system. This is usually achieved by lumbar puncture. Administration of chemotherapy in hospitals requires the hospital pharmacist to work alongside the physician so that the most appropriate dose is administered in the most suitable manner. Some of the drugs used to treat cancer are discussed below, although this list is by no means exhaustive. It is worth noting that a natural selection process often takes place within tumors treated with chemotherapy, in that some of the tumor cells may develop resistance to the drug. In cases of drug resistance it is necessary to change the chemotherapeutic agent.

Dihydrofolate reductase (DHFR) is active in the synthesis of tetrahydrofolate, which is required for the synthesis of purines and pyrimidines, themselves required for the synthesis of nucleotides and DNA. Folic acid antagonists inhibit the DHFR and some, such as methotrexate, which is cytotoxic in concentrations between 10^{-7} and 10^{-8} mol dm^{-3} are used in the treatment of cancer. Methotrexate closely resembles the substrate for dihydrofolate reductase and can bind to it, inhibiting its action (*Figure 17.25*).

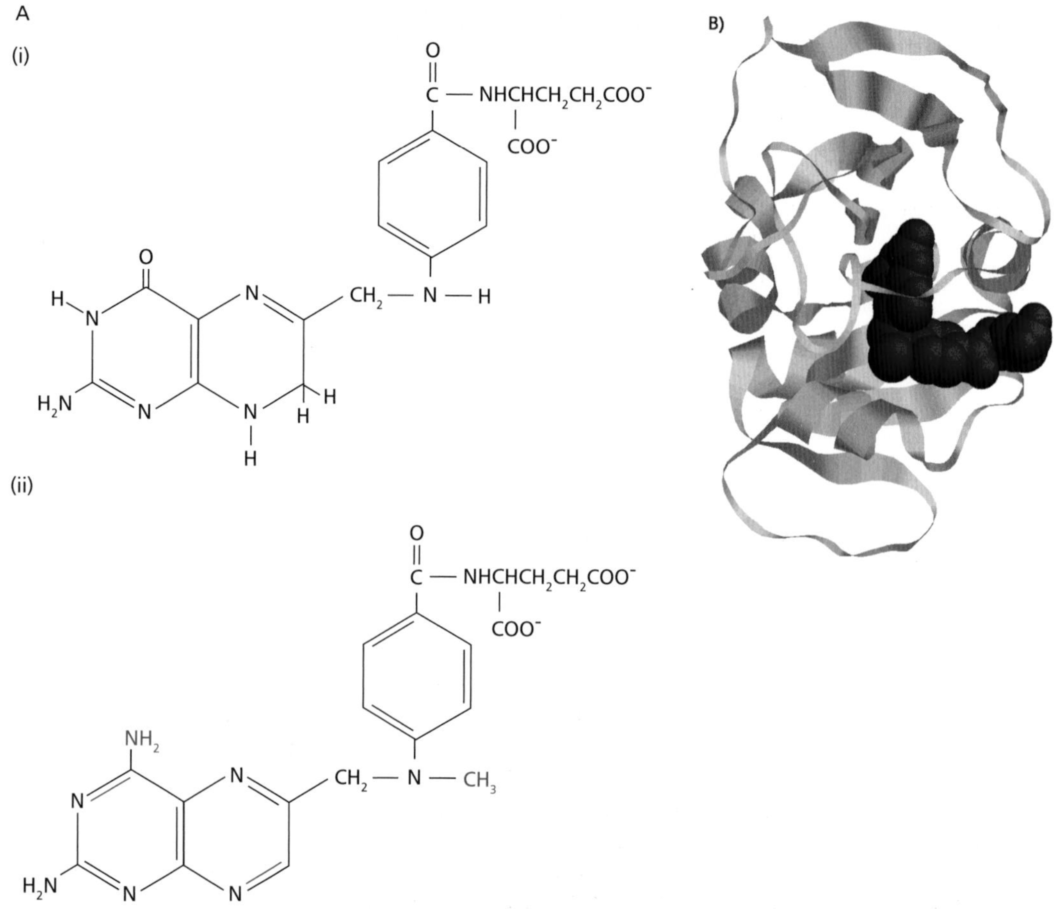

Figure 17.25 (A) Note the similar structures of i) dihydrofolate and ii) the anticancer drug, methotrexate. (B) Molecular model of dihydrofolate reductase with a bound methotrexate molecule. PDP file 1RG7.

Alkylating agents that have two reactive groups are bifunctional and can cross-link two biomolecules. Cross-linking the two strands of DNA is the major cause of toxicity of these drugs since this prevents the separation of the strands which is required for the synthesis of new DNA. Examples of alkylating agents include the nitrogen mustards, such as cyclophosphamide, melphalan, ifosfamide and chlorambucil, and the nitrosoureas, for example *bis*-chloroethyl nitrosourea (BCNU; carmustine) and cyclohexyl-chloroethyl nitrosourea (CCNU; lomustine). Some examples of their use is shown in *Table 17.7.*

Purine and pyrimidine analogs are drugs that resemble one of the bases found in DNA and/or RNA. When present during nucleic acid synthesis, they interfere with the synthesis of DNA, though the site at which they exert their effects depends on the drug itself. Pyrimidine analogs include 5-fluorouracil or 5-FU (*Figure 17.26*) and cytidine arabinoside (ara-C). Purine analogs include 6-mercaptopurine (6-MP) and 6-thioguanine. *Table 17.8* lists some of their uses in cancer treatment.

Microtubules form the mitotic spindle, which is essential for the process of chromosome separation during mitosis and meiosis. The formation of a spindle requires microtubules to polymerize from tubulin subunits, whereas separation of chromosomes requires depolymerization. Any drug which interferes with either of these processes will interfere with cell division. The first drug shown to prevent polymerization was colchicine, originally obtained from the autumn crocus, *Colchicum officinale*, though this is too toxic for

Drug	Examples of clinical use
Chlorambucil	chronic lymphocytic leukemia slow growing nonHodgkin's lymphomas;
Melphalan	multiple myeloma
Cyclophosphamide	breast cancer lymphatic cancer
Ifosfamide	testicular cancer lung cancer sarcoma
BCNU	Hodgkin's lymphoma nonHodgkin's lymphoma malignant melanoma multiple myeloma
CCNU	brain tumors Hodgkin's and nonHodgkin's lymphoma

Table 17.7 Alkylating agents in cancer treatment

therapeutic use. The vinca alkaloids vincristine, vinblastine and vinorelbine (*Figure 17.27*), all derived originally from the periwinkle plant *Vinca rosea*, work in a similar manner to colchicine. Vinblastine is used in combination with other drugs to treat testicular cancer, while vincristine is used in treating leukemias. Vinorelbine is used in lung and breast cancer therapy. Paclitaxel, otherwise known as taxol, and docetaxel or Taxotere (*Figure 17.28*) are derived from the bark and needles of the Pacific yew tree, *Taxus brevifolia* (*Figure 15.8 (B)*). These drugs also target microtubules but in this case they prevent their depolymerization. Both drugs are used in the treatment of ovarian, breast and lung cancer.

Topoisomerase inhibitors form a complex with DNA topoisomerase II and inhibit DNA replication. Examples of this class of drugs include VP-16, or etoposide, and VP-26, or teniposide, both of which are derivatives of podophyllotoxin, which is derived from the mandrake plant, *Madragora*. VP-16 is used in the treatment of small cell lung cancer (*Section 17.8*), testicular cancer and lymphomas and VM-26 is used to treat childhood leukemia.

Vinblastine R = CH_3

Vincristine R = CHO

Figure 17.27 Structures of the vinca alkaloids, vinblastine and vincristine.

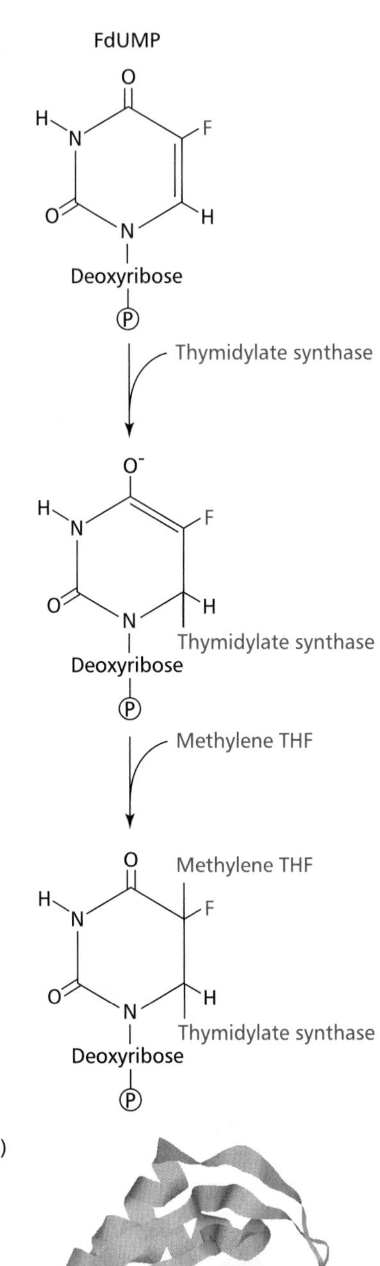

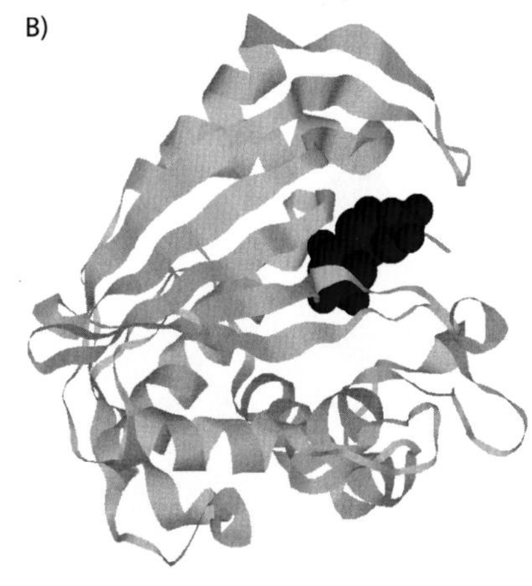

Figure 17.26 (A) 5-Fluorouracil is converted *in vivo* into the anticancer drug to fluorodeoxyuridylate (FdUMP). In the presence of methylene tetrahydrofolate (THF) this binds irreversibly to thymidylate synthase, an essential enzyme for DNA synthesis. (B) A molecular model of thymidylate synthase with a bound FdUMP. PDB file 1TSN.

Drug	Examples of clinical use
5-FU	breast and GIT cancers
Ara-C	acute leukemia
6-MP	acute myeloid leukemia acute lymphoblastic leukemia.
6-TG	acute myeloid leukemia acute lymphoblastic leukemia.

Table 17.8 Purine and pyrimidine analogs in cancer treatment

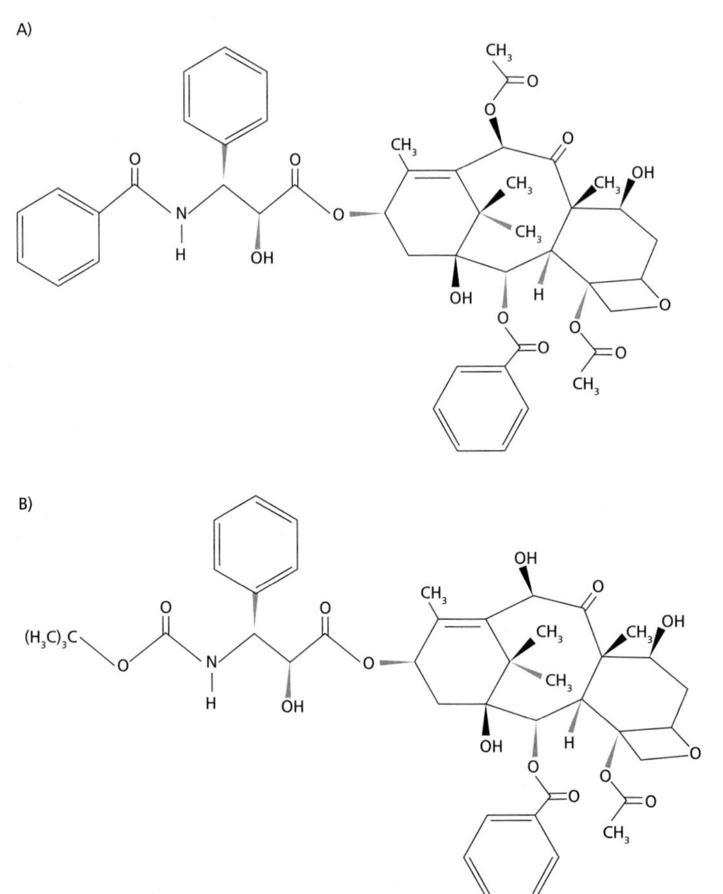

Figure 17.28 Structures of (A) paclitaxel (taxol) and (B) docetaxel (taxotere).

A)

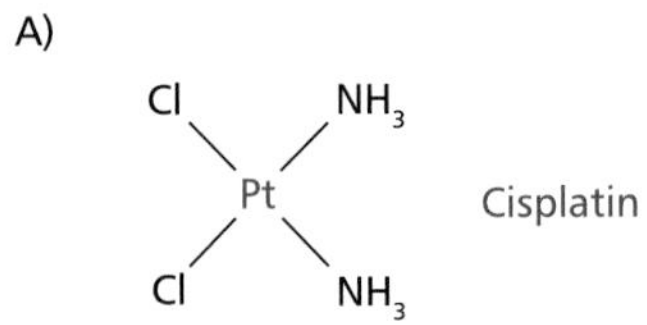

B)

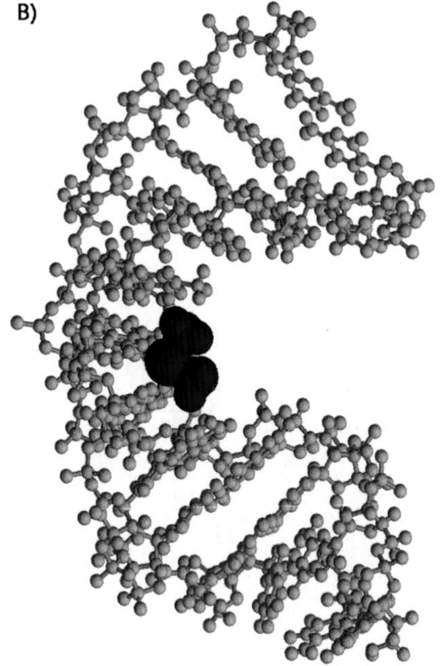

Figure 17.29 (A) Structures of the platinum-based anticancer drugs, cisplatin, carboplatin and oxaliplatin. (B) A molecular model of cisplatin bound to DNA. Note how the double helical structure has been bent and distorted by the binding of the drug. PDB file 1A84.

Cisplatin (*cis*-diamminedichloroplatinum) was first discovered in experiments that showed that the growth of bacteria was inhibited by an electric current delivered by platinum electrodes. This drug is now used in combination therapy to treat testicular cancer. It is also used in treating lung cancer. Other platinum-based drugs include carboplatin and oxaliplatin and all function by binding to DNA, cross-linking the strands and distorting its double helical shape (*Figure 17.29 (A) and (B)*). This facilitates the binding of other proteins to the DNA molecule that mediate toxicity of the drug.

Tumor-specific molecules, on the surfaces of cancer cells may be targeted with monoclonal antibodies (*Chapter 4*). In this way treatment should be more directed at the tumor and general toxicity reduced. Over the years, several approaches have been made to producing targeted therapy using these antibodies. For example, antibodies have been used to target cytotoxic drugs and toxins such as ricin directly at tumor cells, the so-called 'Magic bullet' therapy. However, since monoclonal antibodies are mouse immunoglobulin they stimulate an immune response in humans. Hence, antibodies have been engineered to contain mouse binding sites but which are carried on human constant regions (*Chapter 4*). Examples of monoclonal antibodies currently in use are trastuzumab (Herceptin), which is licensed for the treatment of secondary breast cancer and which is directed at growth factor receptors on the tumor cells (*Section 17.8*), and rituximab (Rituxan) which is used in the treatment of nonHodgkin's lymphoma.

RADIATION THERAPY

Radiation therapy, which involves the use of high energy electromagnetic waves, is used to treat between 50–60% of cancers. The treatment is based on the fact that most cells are susceptible to radiation when they are dividing. The common side effects of radiation therapy are fatigue, nausea and some external burns to the skin, therefore the dose of radiation needs to be carefully calculated in order to give the optimum antitumor dose with minimal side effects.

Radiation treatment is delivered in one of two ways. With external beam therapy, the area is irradiated with X-rays from an external source. The dose of X-rays is given in short, fractionated daily doses over a period of time, a regimen known as continuous hyperfractionated radiotherapy 'CHART', to allow normal cells to recover. Treatment regimens depend on the size and location of the tumor and the purpose of the treatment, that is, whether it is intended to cure, to shrink the tumor prior to surgery or chemotherapy or to palliate an incurable tumor. With internal therapy the radioisotope is placed near or inside the tumor for a short period of time, a process known as brachytherapy. For example, intracavitary radiotherapy involves the insertion of ^{137}Cesium into a body cavity in an applicator. This form of treatment is used for cancer of the vagina, cervix or uterus. Alternatively, thin radioactive wires may be inserted directly into the tumor, as, for example, in the treatment of prostate cancer. Internal therapy may also involve giving the patient a radioactive liquid either orally or intravenously. For example, a drink of radioactive iodine may be given to treat thyroid cancer. Since the thyroid preferentially takes up iodine (*Chapter 7*), the radioactivity becomes concentrated at the site where it is required. Intravenous radioactive liquids are used to treat metastatic bone cancer.

IMMUNOTHERAPY

The term *immunotherapy* in relation to cancer refers to processes that manipulate the immune system to improve the body's response against a tumor. Immunotherapy has a long history and began in the era before the advent of cytotoxic drugs when cancer patients were treated with Coley's toxin. This contained a mixture of killed *Streptococcus pyogenes* and *Serratia marcescens* bacteria that stimulated the immune system nonspecifically (*Chapter 4*). The BCG vaccine, which contains killed mycobacteria, was used in the 1960s and 1970s, to treat malignant melanoma and has since been used to treat bladder cancer. Mycobacteria are potent stimulators of the immune response and increase the production of several cytokines, including interferon γ (IFN-γ) and tumor necrosis factor α (TNF-α). Nowadays, recombinant cytokines may be given directly to enhance the immune response. For example, interferon α (IFN-α) has been used successfully to treat multiple myeloma, CML, hairy cell leukemia (*Section 17.8*) and malignant melanoma. Interleukin 2 (IL-2) appears to exert an anticancer effect through the prolonged stimulation of Natural Killer cells; it has been used to treat renal cancer and malignant melanoma (*Chapter 4*).

An exciting development in immunotherapy has been the production and clinical trials of a number of anticancer vaccines based on tumor associated antigens. Examples include vaccines based on peptides derived from carcinoembryonic antigens, and the BiovaxID™ vaccine for the treatment of follicular lymphoma, a type of nonHodgkin's lymphoma involving B lymphocytes.

17.8 SPECIFIC TYPES OF CANCERS

Cancer patients, naturally, may present with a variety of symptoms depending on the origin and extent of the tumor. The treatment administered for any cancer patient therefore depends to a large extent upon the type of cancer present, its location, the stage of the disease and the age and health of the patient. Major concerns in the developed world include lung, breast, prostate and colorectal cancers and these are discussed below. The leukemias will also be discussed as examples of tumors arising from bone marrow stem cells.

LUNG CANCER

Almost 29 000 deaths were attributed to lung cancer (*Figure 17.30*) in England and Wales in 2002 and around 37 000 new cases are diagnosed annually. In the USA lung cancer is, again, the second most common malignancy after prostate cancer in men and breast cancer in women, with over 160 000 new cases occurring each year. Worldwide, about one million new cases occur annually, with an incidence of 37.5 new cases per million population amongst men, and 10.8 cases per million amongst women.

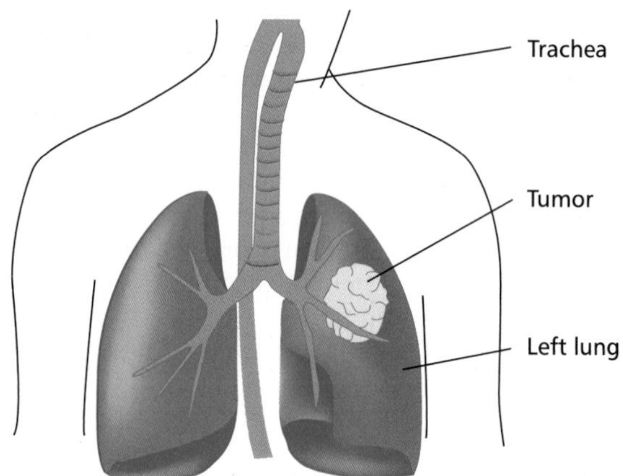

Figure 17.30 Schematic illustrating the presence of a cancer of the lung.

Lung cancer is a fatal disease and, in England and Wales, only approximately 20% and 6.0% of patients survive one year and five years respectively after diagnosis. These figures are higher in the USA, where the five-year survival figure stands at about 14%. Overall, lung cancer is the most frequent cause of death from cancer in men, who make up around 60% of all lung cancer cases. In women, lung cancer is the second commonest cause of cancer death after breast cancer. The risk of lung cancer increases with age with approximately 75% of deaths occurring after the age of 65.

A link between smoking and cancer is indisputable and smoking and passive smoking are the cause of 90% of lung cancers. The carcinogens in cigarette smoke include benzo[a]pyrene, a polycyclic hydrocarbon and *N*-nitrosamines, both of which are metabolized by the cytochrome P-450 enzymes in the liver to carcinogens that form adducts with DNA as described in *Chapter 12*. Radon, a naturally occurring radioactive gas produced from the decay of uranium and which is found at relatively high indoor levels in homes built of, and on, granite, is also responsible for a small but statistically significant number of cases each year. People who work in uranium mines are also at risk of radon-induced lung cancer. Exposure to the mineral asbestos is also strongly associated with the development of lung cancer and increases the risk of developing lung cancer fivefold.

Classification of lung cancer

Lung cancer is classified into two main groups namely small cell lung carcinoma (SCLC) and nonsmall cell lung carcinoma (NSCLC). Small cell lung carcinoma accounts for about 20% of all lung cancers. The cancer cells are small with a high nucleus to cytoplasm ratio. Nonsmall cell lung carcinomas are comprised of three types, namely squamous cell carcinoma, adenocarcinoma and large cell carcinoma. Squamous cell carcinomas develops from the epithelial cells lining the respiratory tract; they form 35% of all lung cancers. Adenocarcinomas develop from the mucus-secreting cells in the lining of the respiratory tract and account for 27% of lung cancers. Finally, large cell carcinoma, so called because the cells are relatively large and rounded compared with the other forms, accounts for 10% of total lung cancers.

Signs, symptoms, diagnosis and staging of lung cancer

Lung cancer does not generally cause symptoms in the early stages and by the time these occur the disease is generally in an advanced state. Some patients are asymptomatic and may only be diagnosed following a routine chest X-ray. Symptoms at presentation include a persistent and nagging cough, shortness of breath, recurrent chest infections such as pneumonia and bronchitis, coughing up blood-containing sputum or **hemoptysis**, chest pain when breathing or coughing, and an unexplained loss of weight. A patient showing these symptoms should be referred urgently for a chest X-ray and/or a CT scan. If these indicate cancer, the tumor should be staged by scanning patients using positron emission tomography (PET). Other tests include examination of the chest by inserting an endoscope through a small cut at the base of the neck. In addition, biopsy may be taken using a fine needle inserted into the lungs, guided by a CT scanner or X-ray machine.

Staging of the disease is required to determine treatment. The TNM staging system, mentioned in *Section 17.7*, for NSCLC classifies the primary tumor from T1 to T4, where T1 represents a tumor less than 3 cm diameter with no invasion of the main bronchus. Tumors greater than 3 cm that may also involve the main bronchus are classified as T2, while T3 represents a tumor of any size which has invaded the chest wall, diaphragm, mediastinal pleura, parietal pericardium or main bronchus. A T4 stage tumor is one of any size that has invaded any of a range of tissues, such as the heart, trachea or esophagus. The regional lymph nodes are staged as N0 to N3, where N0 represents no regional lymph node metastasis, N1 and N2 represents increasing metastasis to lymph nodes on the same side or opposite side to the tumor respectively. Where distant metastasis has occurred, this is classified as M1. The TNM is further classified into subsets as shown in *Figure 17.31*.

The staging for SCLC is somewhat different with patients being classified as having limited stage disease or extensive stage disease. Limited stage disease is used if the tumor is restricted to one hemithorax and may include patients with lymph node metastasis. Extensive disease is defined as disease at sites beyond the definition of limited disease.

Treatment of lung cancer

For NSCLC, surgery is used to remove the tumor, as directed by the staging. How much of the lung is removed depends on the stage of disease and the health

	T1	T2	T3	T4
N0	1A	1B	IIB	IIIB
N1	IIA	IIB	IIIA	IIIB
N2	IIIA	IIIA	IIIA	IIIB
N3	IIIB	IIIB	IIIB	IIIB

Key
Patient offered surgery if no medical contraindications
Surgery may be suitable for some patients, based on clinical judgment
Not suitable for surgery

Figure 17.31 The TNM classification of nonsmall cell lung cancer. Summarized from NICE guidelines www.nice.org.uk

of the patient. During surgery, lymph nodes are sampled to aid more accurate staging of disease. Radiotherapy is recommended for patients with stage I, II or III disease. Patients with Stage III or IV may be offered chemotherapy although NSCLC is only moderately sensitive to chemotherapy. Chemotherapy for advanced NSCLC uses a combination of drugs such as paclitaxel or docetaxel, together with a platinum drug such as cisplatin. Patients with SCLC are treated with a platinum drug and multidrug regimens. The drug regimen may be given in cycles to patients who respond to treatment. Patients with limited stage SCLC are given radiotherapy concurrently with chemotherapy. Those who have inoperable lung cancer should be given palliative treatments to ease their symptoms. This may include radiotherapy to relieve breathlessness and chest pain, opioids to ease the cough, and procedures to alleviate large airway obstruction.

BREAST CANCER

Breast cancer (*Figure 17.32*) is the commonest cancer in women in both the UK and the USA, and the second leading cause of cancer deaths in women. The incidence of new cases in the UK is about 41 000 per year and in the USA it is just over 210 000, while the annual incidence worldwide is approximately 1.2 million. It is estimated that about one in nine women will develop breast cancer during their lifetime. Breast cancer also occurs in men, though with a much lower incidence. In the UK, around 250 men are diagnosed with the disease each year and approximately 70 die annually of the disease.

Risk factors for the development of breast cancer include increased age, childlessness, early menarche or late menopause, hormone replacement therapy, being overweight or obese, use of the contraceptive pill and regular consumption of alcohol over a long period of time. Breastfeeding reduces the risk of contracting the disease. Familial breast cancer accounts for 5% of all breast cancers and is related to the inheritance of mutated forms of genes such as the *BRCA1* and *BRCA2* genes as described in *Section 17.4.*

Signs, symptoms, diagnosis and staging of breast cancer

Breast cancer develops either in the milk-producing glands or in the ducts that deliver milk to the nipple. Symptoms of breast cancer include a new painless lump occurring in a breast, changes in the size or shape of the breast or in the position of the nipple, a discharge from the nipple, an eczematous rash

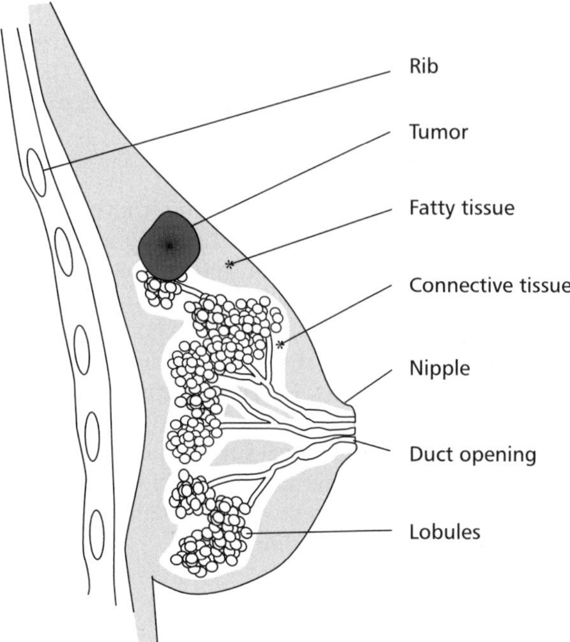

Figure 17.32 Schematic illustrating the presence of a cancer of the breast.

around the nipple or a thickening behind the nipple, puckering or dimpling of the skin of the breast or a swelling or lump in the armpit.

Mammography uses X-rays to locate the position of the potential tumor. Mammography is also used to screen for cancer and, in the UK, this is offered every three years to all women over the age of 50 years. The lump may also be examined by ultrasound or by Color Doppler ultrasound, which gives a picture of the blood supply to the lump. Microscopic examination of cells from a fine needle aspirate of the lump may also be helpful. Alternatively, the lump may be excised under a general anesthetic for pathological examination.

The staging system for breast cancer describes two stages of noninvasive and four stages of invasive breast cancer. Noninvasive stages include ductal carcinoma *in situ* (DCIS) in which cancer cells are contained within the ducts. If diagnosed at this stage, the disease is almost completely curable. Lobular carcinoma *in situ* occurs when cancer cells are restricted to the lining of the breast lobules. Stage 1 invasive cancer describes a tumor measuring less than 2 cm diameter, with no spread to the lymph nodes. In Stage 2 invasive cancer the tumor measures between 2 and 5 cm and/or there are affected lymph nodes. In Stage 3 invasive cancer the tumor is larger than 5 cm diameter and may be attached to muscle or skin. The lymph nodes at this stage are usually affected. Stage 4 invasive cancer describes a tumor of any size, the lymph nodes are usually affected and the cancer has metastasized.

A microscopic examination of cancer cells allows their appearance to be graded. Low grade or Grade 1 cancer cells have the appearance of differentiated normal cells, whereas high grade (Grade 3) tumor cells have an abnormal appearance and are characteristic of fast growing and aggressive cancers. The tumor cells may also be examined immunohistochemically or by FISH (*Section 17.6*) to detect expression of estrogen receptors or HER2 proteins on their surface. Knowledge of the presence of either of these molecules allows the treatment regimen to be determined more appropriately. Cells with estrogen receptors may be stimulated to divide by naturally occurring estrogen, and hormone therapy is indicated. The HER2 protein is a receptor for the human epidermal growth factor (hEGF). Breast cancers that are positive for HER2 proteins are stimulated to divide by naturally occurring hEGF and some treatments prevent this stimulation and reduce the growth of the cells. In the UK all women with early stage breast cancer are tested for HER2 status of the tumor.

Treatment of breast cancer

The first line of treatment for breast cancer is to remove the tumor. This may involve excision of the lump and some of the surrounding tissue or it may mean mastectomy. Chemotherapy or hormonal therapy may be given to reduce the size of the tumor prior to excision. During surgery, lymph nodes are also removed from the armpit and used for accurate staging of the tumor. Two to four weeks after excision or mastectomy, radiation therapy is used to destroy any remaining cancer cells. Chemotherapy, using a combination of drugs, may be used before and after surgery. If the tumor cells express estrogen receptors hormone therapy, using drugs such as tamoxifen, is given to block the estrogen receptors or to lower the amount of estrogen in the blood. Tumors that are HER2 positive may be treated with trastuzumab otherwise known as Herceptin which was described in *Section 17.7*.

PROSTATE CANCER

The prostate is a small gland, surrounding part of the urethra, which produces semen that mixes with the sperm produced by the testes as described in *Chapter 7* (*Figure 17.33 (A)*). Prostate cancer is a disease that principally affects men over the age of 50 years. Many prostate cancers go undiagnosed because the tumor remains latent for long periods of time. The annual incidence of prostate cancer in the UK is about 30 000 men and around one in 12 men will develop the condition. In the USA the incidence is half that of the UK. Its incidence has increased in recent years, although this may reflect the increasing age of the population. While most prostate cancers are slow growing, a small proportion of tumors grow and metastasize more quickly. The risk of a man contracting prostate cancer are increased if close relatives have had the disease, or if female family members have had breast cancer, especially if they were diagnosed at an early age. Men of Afro-Caribbean and African–American origin are at greater risk of getting the disease while Asian men have the lowest risk. A diet high in animal fat and dairy products and low in fresh fruit and vegetables increases the risk. Consumption of the carotenoid, lycopene (*Figure 17.14*), which is abundant in tomatoes and tomato products, reduces the risk of contracting prostate cancer. Lycopene has been shown to lower the amount insulin-like growth factor 1 (IGF-1; *Chapter 7*), which may otherwise stimulate the growth of cancer cells.

Signs, symptoms, diagnosis and staging of prostate cancer

The symptoms of prostate cancer are not usually present when the tumor is small. However, as the tumor grows there is difficulty and pain on passing urine, coupled with a more frequent need to urinate, particularly at night. There may also be blood in the urine. If cancer cells have spread to the bone there may be pain in the back and pelvis.

A patient presenting with symptoms of prostate cancer will be given a digital rectal examination (DRE; *Figure 17.33 (B)*). During this examination a gloved finger is inserted into the rectum, from where the prostate can be felt. An enlarged prostate that feels round and smooth is most likely a benign prostate hyperplasia. In contrast, a prostate tumor will make the gland feel hard and lumpy. Blood samples are also taken for a PSA test (*Section 17.6*). The PSA test measures the level of prostate specific antigen. The normal level of PSA is approximately 2.8 ng cm^{-3} in men of 50 and 5.3 ng cm^{-3} in men of 70 years of age. Men with levels of 10 ng cm^{-3} or above require referral for further

A)

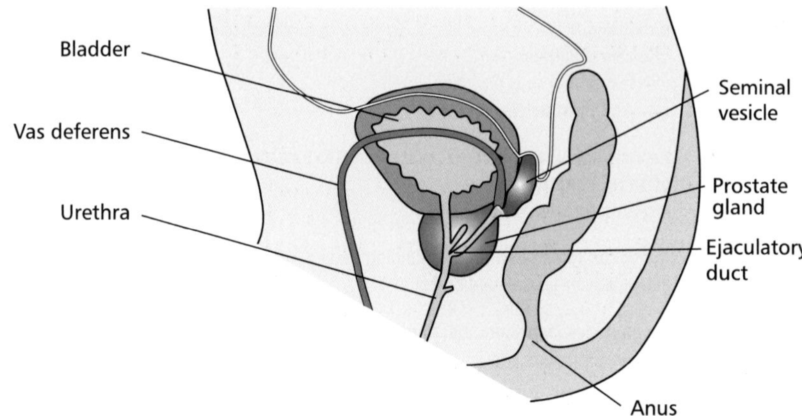

B)

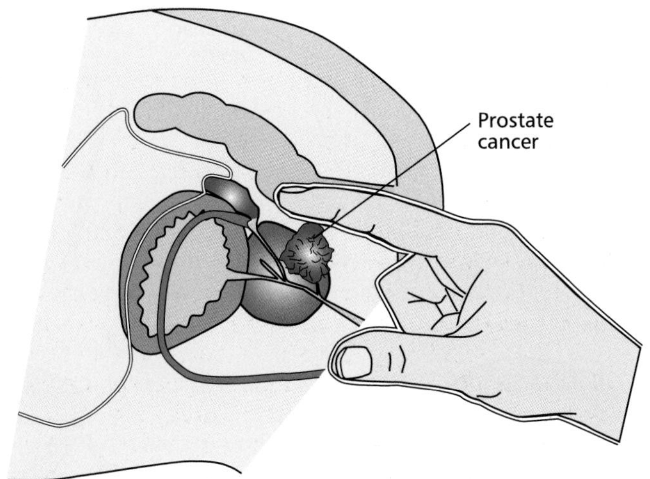

Figure 17.33 Diagrams illustrating (A) the normal appearance of the prostate gland and (B) a rectal examination of a cancerous prostate gland.

tests because above normal levels of PSA are found in conditions other than cancer, for example in urinary infections. Following referral, additional tests undertaken to determine the extent of the problem include an isotopic scan of bones and an MRI or CT scan. Biopsies of the prostate are obtained by passing a needle through the rectum into the prostate, aided by an ultrasound scan.

Prostate cancers are commonly graded using the Gleason system which scores the cancer according to the growth pattern and the arrangement of the cancer cells within the prostate. The higher the score, the more likely the tumor is to spread. The tumors are staged as shown in *Table 17.9* and *Figure 17.34.*

Treatment for prostate cancer

The treatment for prostate cancer is determined by the grade and stage of the tumor, including metastasis, the age and health of the patient and the concentration of PSA in the blood. The range of treatments for early prostate cancer includes surgical removal of the prostate gland or prostatectomy, radiation therapy, hormonal therapy, or combinations of any or all of these. Chemotherapy is rarely used to treat prostate cancer. Depending on the age of the patient, an early stage cancer may not require treatment, but may be actively monitored until treatment is needed. Prostate cancers often grow slowly and the treatments, which have considerable side effects, should

Stage	Description
T1	tumor within the prostate gland too small to be detected during a rectal examination often no symptoms PSA test or biopsy is positive localized prostate cancer
T2	tumor within prostate gland large enough to be detected at rectal examination often no symptoms localized prostate cancer
T3/4	tumor has spread into the surrounding tissues locally advanced prostate cancer
Metastatic cancer	lymph nodes, bones or other parts of the body affected advanced prostate cancer

Table 17.9 Staging of cancers of the prostate

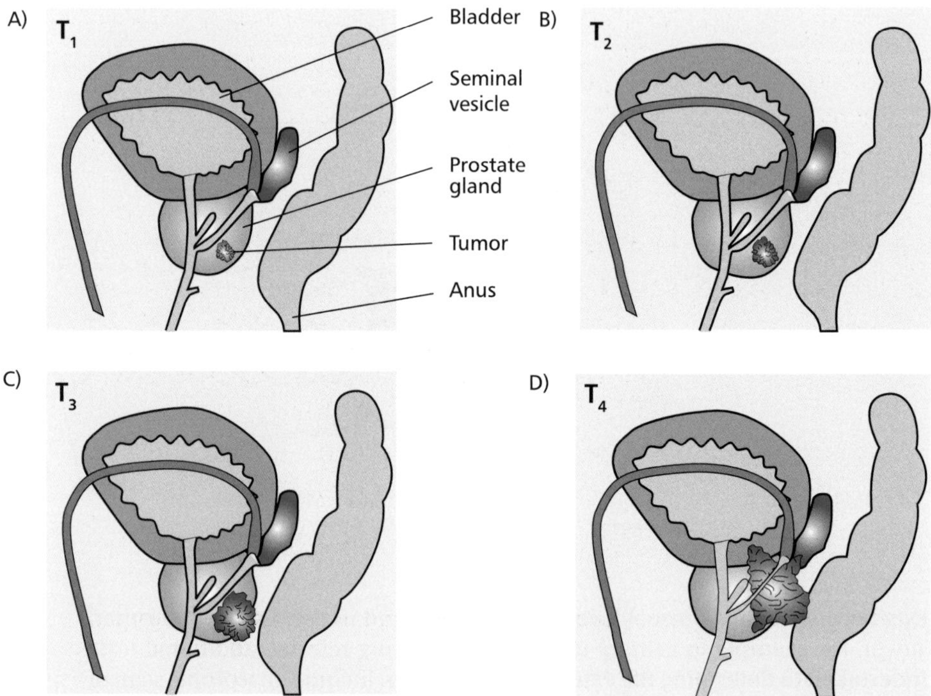

Figure 17.34 Diagrams (A–D) illustrating the four stages of prostate cancer.

be restricted to the more aggressive forms of the disease. Prostatectomy is the removal of the whole prostate gland and may result in impotence and incontinence.

External radiation therapy may be given following surgery to ensure that any remaining cancer cells are destroyed. Radiation therapy is also used as a first line of treatment and is an equally effective alternative to surgery. Therapeutic approaches include external beam and brachytherapy (*Section 17.6*). A form of external radiotherapy, known as conformational radiotherapy (CRT), allows the radiation beams to be shaped to match the shape of the prostate itself, which lowers the side effects caused by irradiation of the surrounding healthy tissue. Brachytherapy is delivered by the implantation of radioactive iodine 'seeds' or iridium wires in the prostate itself. If the cancer has spread to bone, radiation therapy using ^{89}Strontium, which is preferentially taken up by bone tissue, is given by intravenous injection.

Prostate cancer cells have receptors for the active form of testosterone and their growth requires a supply of testosterone from the testes. The aim of hormone therapy is to lower testosterone levels. Some of the drugs used are analogs of gonadotrophin releasing hormone (GnRH), examples being goserelin, leuprorelin and triptorelin. Goserelin is administered as a subcutaneous pellet while the others are injected subcutaneously or intramuscularly in liquid form. Other drugs used are antiandrogens, which block the interactions of hormone and receptor. The side effects of hormonal therapy include sexual impotence, flushes and sweating which may be reduced by intermittently stopping and starting the therapy.

The prognosis for men with prostate cancer is generally good, since this type of cancer usually occurs in older men and grows slowly.

COLORECTAL CANCER

Colorectal cancer, otherwise known as bowel cancer, is one of the three most common cancers in men and women, both in the UK and in the USA. In the UK there are about 35 000 new cases each year, with a slight majority occurring in men. Approximately 60% of new cases are cancers of the colon while the remainder have cancers of the rectum. In the USA there are roughly 135 000 cases each year, about 70% being colon cancers. The five-year survival rate for colorectal cancer is 50–60%. The vast majority of colorectal cancers are adenocarcinomas and this section will concentrate on these. The remainder fall into several groups as shown in *Table 17.10*.

Colorectal cancer	Description
Adenocarcinoma	95% of colorectal tumors, they arise from mucus-secreting cells in epithelium of GIT lining. Most produce mucin. Between 1 and 2% are **signet-ring** tumors where the intracellular mucus pushing nucleus to one side.
Squamous cell carcinoma	arise from epithelial cells of GIT lining.
Carcinoid tumor	rare, slow growing tumor of neuroendocrine origin.
Sarcoma	majority are leiomyosarcomas arising from smooth muscle in GIT wall.
Lymphomas	1% of colorectal cancers arising from lymphoid cells in GIT wall.

Table 17.10 Types of colorectal cancer

The risk factors for colorectal cancer include increasing age, a diet rich in fat and low in fiber, a history of inflammatory bowel disease and a hereditary predisposition. More than 80% are diagnosed in those aged over 60 years. A familial history of bowel cancer is also a strong risk factor as is the presence of **polyps**. A polyp is a benign tumor that arises from the epithelium of the GIT. Polyps range in size from a small bump to a lesion measuring 3 cm in diameter; most are asymptomatic.

Adenocarcinomas are known to originate from pre-existing polyps and the sequence of their development into a cancer is well documented. Individuals with familial adenomatous polyposis (FAP), a rare condition responsible for 1% of colorectal cancers, have a mutated form of the adenomatous polyposis coli (*APC*) gene. This gene encodes a protein that degrades β-catenin, which activates growth-promoting oncogenes such as *c-MYC*. Mutations in *APC* lead to the production of inactive β-catenin, hence oncogenes are continuously activated. Patients with FAP are almost certain to develop colorectal cancer in middle age and they are recommended to have their colon removed by the age of 25 years. The development of colorectal cancer is also associated with both hypomethylation and hypermethylation of DNA and this has been detected at the polyp stage. Methylation of DNA is involved in switching genes off, hence hypomethylation can result in activation of oncogenes, whereas hypermethylation could lead to inactivation of tumor suppressor genes. Deletions in the long arm of chromosome 18 and in the short arm of chromosome 17 are also seen in colorectal cancers. Mutations in the *RAS* gene have been detected in cells in the feces of patients with colorectal cancer and it is possible that this may be used as the basis for a noninvasive diagnostic procedure.

The condition, hereditary nonpolyposis colorectal cancer (HNPCC) is associated with inherited mutations in a number of genes involved in DNA repair that also predispose an individual to colon cancer. Hereditary nonpolyposis colorectal cancer is estimated to cause between 2 and 5% of colorectal cancers and is linked to 40% of cases of colorectal cancer occurring in people below the age of 30 years. Other risk factors for colorectal cancer include being diabetic (*Chapter 7*) and being of Ashkenazi Jewish origin. This group also have a higher incidence of mutated *BRCA1* and *BRCA2* genes (*Section 17.5*) and are therefore also more likely to develop breast cancer. The genetic link to colorectal cancer has not been fully elucidated.

Signs, symptoms, diagnosis and staging of colorectal cancers

Patients with colorectal cancer may present with symptoms that include loss of weight, bleeding from the rectum and/or blood in the feces, anemia resulting in fatigue and breathlessness, changes in bowel habits, including increased diarrhea, and pain in the abdomen. The tumor may cause a bowel obstruction and if this occurs the patient suffers abdominal pain, nausea and constipation. An initial detection of a lump in the abdomen or rectum means that the patient must be referred for further specialist examination. Colonoscopy is used to examine the whole colon and to obtain a biopsy for further investigation. An alternative to colonoscopy is flexible sigmoidoscopy, an endoscopic technique which examines only the lower part of the colon and the rectum. The concentration of CEA (*Section 17.6*) in the blood can be measured although increased values are not entirely specific for cancer and a poorly differentiated tumor may not produce CEA. This measurement may, however, be useful for monitoring disease progress and treatment. Abdominal CT scans are helpful in diagnosis of metastatic spread to the lymph nodes and liver, and chest X-rays are routinely used to detect metastases in the lungs. Prognosis is poor for patients with metastatic spread to these two organs.

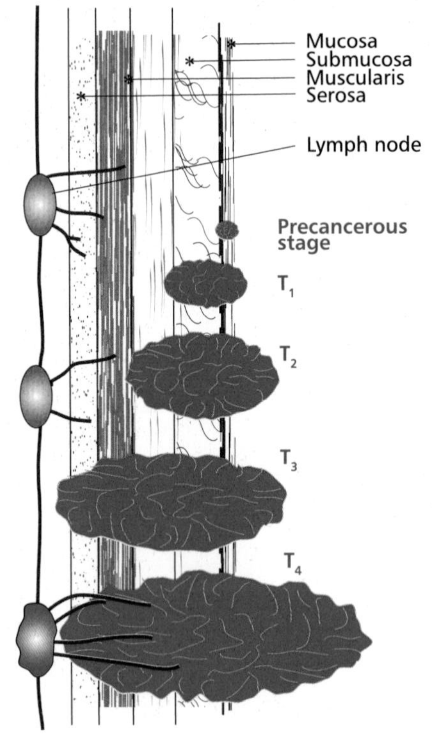

Figure 17.35 A diagram illustrating the four stages (TNM) of colorectal cancer.

Several staging systems have been used for colorectal cancer. The Dukes system is shown in *Table 17.11* which also documents the proportion of patients who present at this stage, together with the survival rates at five or two years.

The TNM staging system is also used increasingly for colorectal cancers (*Figure 17.35*). T1, T2 and T4 are equivalent to Dukes A, B and D respectively. T3 describes the situation if the tumor has broken the outermost membrane of the GIT. The lymph nodes are classified as N0 if there are none containing cancer cells, N1 where between one and three nearby lymph nodes are involved, and N3, where there are cancer cells in four or more lymph nodes that are more than 3 cm from the main tumor, or where there are cancer cells in lymph nodes connected to the main blood vessels around the GIT.

Treatment of colorectal cancer

The usual treatment for colorectal cancer is surgery. Various procedures may be used but, if possible, a resection, in which the length of GIT containing the tumor is excised and the two cut ends are stapled together to reinstate the integrity of the GIT, will be performed. The advantage of resection is that

Dukes system	Description	Proportion of patients diagnosed/%	Prognosis (5-year survival as %)
Dukes A	cancer confined to innermost lining of colon/rectum	10	80
Dukes B	cancer grown into muscle layer of wall of colon/rectum	35	60–70
Dukes C	cancer spread to lymph nodes surrounding the colon/rectum	25	30–60
Dukes D	cancer has metastasized to other parts of the body	30	15 (survival at 2 years after diagnosis)

Table 17.11 Classification of colorectal cancers

the patient does not require a colostomy bag. Patients with rectal cancer may be given radiation treatment after surgery to reduce the risk of the tumor recurring locally. Radiation therapy may also be used in the palliative care of late stage metastatic cancer. Chemotherapy, 5-FU and leucovorin is sometimes offered postsurgery to patients with a Dukes B, and usually to those with Dukes C cancer. Chemotherapy of metastatic colorectal cancer uses 5-FU, leucovorin and irinotecan (CPT11). Combined oxaliplatin and 5FU treatment may also be used. Patients successfully treated for colorectal cancer should be regularly checked for several years after surgery to detect any recurrence of the tumor.

LEUKEMIAS

Leukemias are tumors of bone marrow cells that give rise to the blood cells and platelets (*Chapter 13*). Leukemias were historically classified according to the speed of onset and progression, with those of sudden onset and rapid progression termed acute leukemias, while those that develop slowly over months and years are called chronic leukemias. The current classification is based on specific typing of the blood cells, although the terms acute and chronic are still used. The main types of leukemias encountered clinically are acute myeloblastic and chronic myelogenous leukemias and acute lymphoblastic and chronic lymphocytic leukemias. Another rarer leukemia is hairy cell leukemia named according to the microscopic 'hair-like' appearance of the plasma membrane.

Most leukemias are characterized by symptoms such as excessive bleeding of the gums and nose, bruising, fatigue, breathlessness and increased susceptibility to infections. These are due to inadequate production of erythrocytes, platelets and lymphocytes owing to the abnormal proliferation of leukemic cells in the bone marrow. Blood and bone marrow films may show immature blast cells (*Figures 17.36*). Treatments for leukemia may involve chemotherapy, radiation therapy and stem cell transplantation (*Chapter 6*).

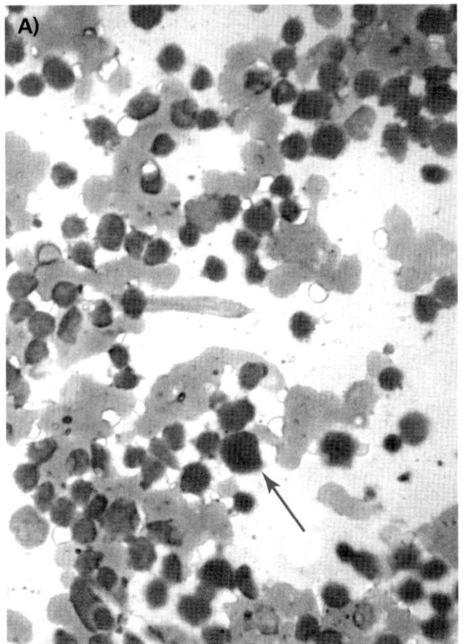

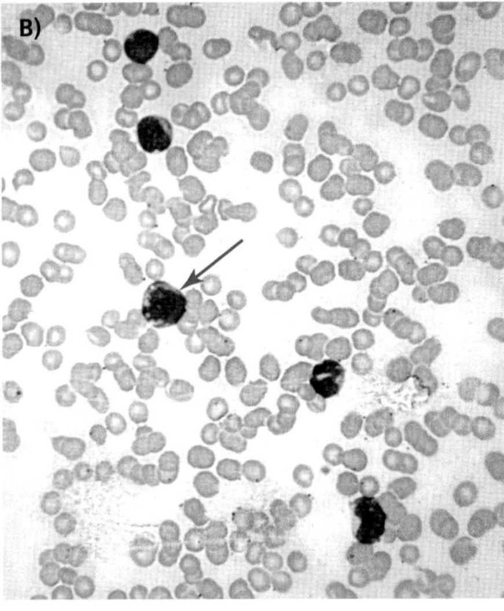

Figure 17.36 Blast cells, indicated by arrows, in (A) the bone marrow and (B) the blood of a patient with leukemia.

Chronic myelogenous leukemia

Chronic myelogenous leukemia (CML) is a rare disorder that affects approximately 500 people annually in the UK and accounts for between 7 and 15% of all leukemias. Patients are usually adults between the ages of 40 and 60 years. The disease is caused by malignant transformation of a myeloid stem cell which would normally give rise to the polymorphonuclear leukocytes, monocytes, basophils, erythrocytes and megakaryocytes. The clonal proliferation of a myeloid stem cell results in the accumulation of immature and mature myeloid cells within the bone marrow and spleen. Chronic myelogenous leukemia usually follows a clinical pattern in which there is a benign chronic phase, an accelerated phase and a blastic or blast crisis phase. During the chronic phase there are few signs and symptoms and many patients are asymptomatic. Patients may present with fatigue, anorexia and weight loss, and splenomegaly, with the condition being revealed by a routine blood count during the investigation of these symptoms. The blood count shows a leukocytosis, that is, an increased number of leukocytes, ranging from 10×10^9 to more than 500×10^9 cells dm^{-3}. A blood smear also shows large numbers of neutrophils and immature myelocytes. The Philadelphia chromosome is present in 95% of cases (*Section 17.4*). Splenomegaly may be present on examination. The presence of myeloblasts within a peripheral blood smear indicates a worsening of the disease and progression into the accelerated phase and the rapidly fatal blast crisis. This may take between three and five years from diagnosis.

Two major treatments are available for CML. The first is stem cell transplantation following radiation therapy to destroy the patient's own bone marrow (*Chapter 6*). The second is to use imanitib mesylate (Glivec) together with IFN-α. Imanitib mesylate is an inhibitor of tyrosine kinase that is administered orally. It is directed against the bcr-abl protein (*Section 17.5*) and is successful in inducing disease remission in approximately 90% of patients. However, the remainder are resistant, while up to 25% of patients also develop resistance during treatment. Chemotherapy with hydroxyurea may also be used. This is a temporary measure that reduces the leukocyte count by suppressing the division of the malignant cells. The effectiveness of treatment can be monitored by detecting the proportion of cells with the Philadelphia translocation.

Acute myeloid (myeloblastic) leukemia

Acute myeloid leukemia (AML) is a rare disease that affects approximately 2000 adults, mostly over the age of 60 years, and 100 children annually in the UK. The disease is characterized by the presence of immature myeloid cells, or myeloblasts, in the blood, due to clonal proliferation of an aberrant cell in the bone marrow. The cause of AML is unknown, though it has been linked to exposure to industrial solvents. There is no genetic link and cases are sporadic.

Acute myeloid leukemias are classified according to the type of cell which is involved, using the FAB (French–American–British) classification (*Table 17.12*).

Classification	Description
M0	AML with minimal myeloid differentiation
M1	AML without maturation
M2	AML with maturation
M3	acute promyelocytic leukemia
M4	acute myelomonocytic leukemia
M5	acute monocytic/monoblastic leukemia
M6	acute erythroleukemia
M7	acute megakaryoblastic leukemia

Table 17.12 FAB classification of acute myeloblastic leukemias

Patients present with anemia and bruising due to inadequate production of erythrocytes and platelets respectively. Poor platelet production is also associated with bleeding from the gums and nose. An inadequate production of appropriate leukocytes means that patients suffer repeated infections. Acute myeloid leukemia requires rapid diagnosis and immediate treatment. Chemotherapy is effective in bringing many patients into remission although high dose chemotherapy followed by a stem cell transplant may be required in some cases. Patients may require platelet transfusions before and during their treatment and erythropoietin may be given to treat the anemia. The use of the monoclonal antibody, gemtuzumab ozogamicin, directed at CD33 proteins present on the surfaces of the leukemic cells may be of some efficacy.

The prognosis for AML depends on the age at diagnosis. For adults under the age of 55 years, the five-year survival rate is between 40 and 60%, for those over 55 years it is unfortunately only 20%.

Chronic lymphocytic leukemia

Chronic lymphocytic leukemia (CLL) is the most common form of leukemia, affecting mostly those aged over 60 years. It is characterized by the clonal proliferation of lymphoid stem cells. However, the disease is slow growing and many patients are asymptomatic in the early stages. Signs and symptoms include frequent infections, anemia, bleeding and bruising. The spleen and the lymph nodes in the neck, axillae and groin may be swollen due to the accumulation of abnormal cells. The cause of CLL is unknown although in a minority of cases the disease may be familial. A diagnosis of CLL is by examination of blood and bone marrow. A lymph node biopsy may also be taken for histological examination. The disease is described as Stage A, B or C according to the degree of lymph node involvement. Stage C patients have enlarged lymph nodes in three or more areas and a low erythrocyte and/or platelet count. Patients with Stage A, with little or no lymph node enlargement may not need treatment. Patients with Stages B and C may be brought into remission with chemotherapy. Treatment is usually successful in maintaining remissions for several years.

Hairy cell leukemia is a rare form of CLL which occurs in people between 40 and 60 years of age. It constitutes 2–5% of leukemias. The leukemic cell is a B lymphocyte and is characterized by cells with outgrowths or projections which give the disorder its name (*Figure 17.37*) although the cause of the pathology is not known. Patients present with symptoms of breathlessness, weakness, weight loss and infections. There may also be splenomegaly. Diagnosis is by examination of blood smears and demonstration of splenomegaly on examination. Histological examination of a bone marrow specimen is also undertaken.

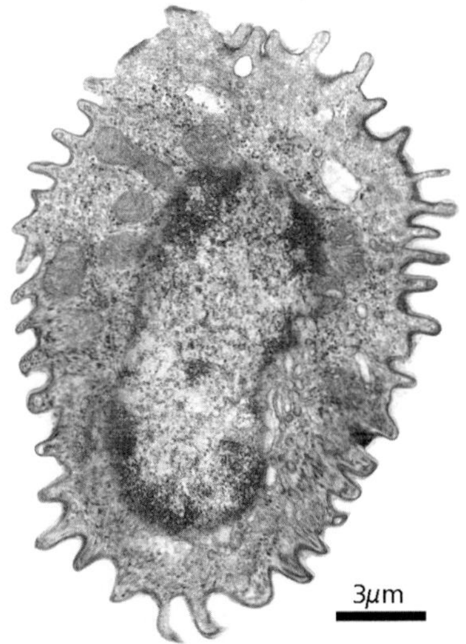

Figure 17.37 An electron micrograph of a single leukemic cell from a patient with hairy cell leukemia. Note the numerous extensions of the plasma membrane that give the condition its name.

Treatment of hairy cell leukemia is by chemotherapy, using cytotoxic drugs. Interferon α has also been used to treat this disorder. The prognosis for hairy cell leukemia is good. The leukemia is slow growing and chemotherapy may only be needed after regular monitoring of cell counts reveals the count of abnormal cells is rising and if the patient presents with symptoms.

Acute lymphoblastic leukemia

Acute lymphoblastic leukemia (ALL) is a rare disorder that occurs most frequently in children under the age of 15 years although it does affect around 600 adults annually in the UK, mostly between the ages of 15 and 25 and those over 75 years of age. The cause of ALL is unknown. The disease is characterized by a clonal proliferation of a lymphoid stem cell, leading to increased numbers of lymphoblasts in the blood. As with other leukemias, patients may present with unusual bleeding of the gums and nose, bruising, anemia, aching joints and bones and a general feeling of malaise. Diagnosis is achieved by microscopic examination of bone marrow taken from the pelvis or sternum. A lumbar puncture may be used to obtain a sample of cerebrospinal fluid, to detect the presence of leukemic cells.

Acute lymphoblastic leukemia is classified into three types according to the FAB classification. These are L1ALL, in which the lymphoblasts resemble mature lymphocytes, with L2ALL and L3ALL showing increasingly immature forms of lymphoblasts. The disease may also be classified according to whether the leukemic cells are T cells or B cells, or pre-B cells, which are immature forms. A diagnosis of these forms involves histological examination and immunofluorescence testing of cell surface characteristics, with analysis by flow cytometry (*Box 6.1* and *Chapter 4*). A proportion of ALL patients also have the Philadelphia chromosome, which can be shown by karyotyping.

The treatment for ALL involves chemotherapy and this achieves remission in 80% of patients. Chemotherapy may also require intrathecal injection of chemotherapeutic agents (*Section 17.7*) to destroy leukemic cells in the cerebrospinal fluid. Steroids, such as dexamethasone, may be combined with cytotoxic drugs and imanitib may also be used. Radiation therapy to the brain and the testes in men may also be required to prevent recurrence. When patients are at high risk of relapse, a stem cell transplant may be required.

The survival rates for ALL depend on the age at which it is diagnosed. Children with ALL have a five-year survival rate of between 65 and 75%, while for adults it is only 20–35%.

CASE STUDY 17.1

Chris, a 56-year-old man, recently noticed that he needed to urinate more frequently, especially during the night. Also, urination was painful. He visited his doctor where a rectal examination showed his prostate gland to be enlarged and smooth. A blood sample was taken for a PSA test. The results showed the level of PSA in his blood to be 5.3 ng cm^{-3}.

Questions

(a) What is the likely diagnosis for Chris?

(b) What would be the recommended treatment?

CASE STUDY 17.2

Rebecca is a 55-year-old woman. About four weeks ago she developed a bad cold and cough. While the cold cleared up, the cough has remained and has become more painful. During the last week she was alarmed to find that she was coughing blood into her handkerchief, especially first thing in the morning. Rebecca gave up smoking about two years ago, but was previously a smoker of 20 cigarettes a day, starting from the age of 23 years. Her husband also smokes. Rebecca visited her physician who referred her to a consultant. The consultant saw Rebecca two weeks later and she had a full examination, including a chest X-ray.

Questions

(a) What clinical findings might the physician be expecting to see?

(b) What is the risk of Rebecca having lung cancer if she is a nonsmoker?

CASE STUDY 17.3

Giles is a 14-year-old boy who, until recently, was a normal adolescent who loved sports, especially soccer. During the last few weeks, however, Giles has been feeling unwell. He has only played soccer twice in the last four weeks and, each time, he felt too tired during the first half to carry on for the whole match. Giles' mother noticed her son's tiredness and worried about the number of coughs and colds he suffered. However, when she noticed a number of large bruises on his legs and torso, she decided to take him to see his doctor who took a sample of blood for a full blood and platelet count. The blood count indicated a leukocytosis, with numerous blast cells present. Platelet numbers were reduced and there were indications of anemia.

Questions

(a) What disorders might be suspected?

(b) What further tests would confirm or refute your suspicions?

17.9 SUMMARY

Cancer is a group of diseases in which cells escape from the usual regulatory factors controlling cell division and proliferate to form tumors. Cells from tumors may break free and move through the blood and lymph to establish tumors at distant sites. Cancer is ultimately caused by successive mutations in the DNA that confer a selective advantage to the cancer cells. Examples of genes which, when mutated, may give rise to cancers include cellular proto-oncogenes and tumor suppressor genes. Mutations may be caused by chemical carcinogens or radiation, by oncogenic viruses or by intrinsic failures of DNA repair mechanisms. Some gene mutations that predispose to cancer are inherited. Examples include mutated forms of the *BRCA1* and *BRCA2* genes which predispose women to cancers of the breast and/or ovaries.

Diagnosis of cancer depends on the location and symptoms produced by the tumor. A whole battery of diagnostic procedures may be used, including tumor imaging, histological examination of tumor biopsies and immunoassays for tumor associated markers. Treatments for cancer include chemotherapy, radiation therapy and immunotherapy. In addition, hormone therapy may be useful for cancers of the prostate and breast.

QUESTIONS

1. Which of the following is the odd one out?
   a) Epstein-Barr virus;
   b) HIV;
   c) Kaposi's virus;
   d) human papilloma virus;
   e) human T-cell leukemia virus.

2. Which of the following statements is incorrect?
   a) A benign tumor does not metastasize.
   b) Oncogenes may be found in certain viruses.
   c) Tumor suppressor genes control cell division.
   d) Cancer may be treated with cytotoxic drugs.
   e) Leukemic cells originate in the blood.

3. Which of the following are the four most common tumor sites in adults in the UK?
   a) Lung, breast, colorectum and bone marrow;
   b) Breast, lung, colorectum and prostate;
   c) Lung, breast, stomach and prostate;
   d) Brain, lung, breast and bone marrow;
   e) Stomach, lung, skin and breast.

4. Arrange the two following lists into their most appropriate pairings.

Carcinoma	*BCR-ABL*
Metastasis	*TP53*
Oncogene	myeloblastic leukemias
Tumor suppressor gene	Hodgkin's lymphoma
Breast cancer	xeroderma pigmentosum
Li-Fraumeni syndrome	secondary tumors
Tyrosine kinase	*EMSY*
FAB	epithelia
Ultraviolet light	*E2F3*
Epstein-Barr virus	*CHK2*

5. Indicate whether the following statements are **TRUE** or are **FALSE**.
   a) Cachexia is defined as an unexplained loss of 2 kg of body weight.
   b) A viral equivalent of an abnormal human gene that is associated with cancer is denoted by the prefix *v-*.
   c) Cigarette smoke contains benzo[a]pyrene, which is a direct acting carcinogen.
   d) The use of a cryostat allows for a rapid histological examination of biopsy material from a suspected cancer.
   e) Cells undergoing mitosis are at their most resistant to radiation damage.
   f) Methotrexate is an antagonist to folic acid synthesis.
   g) Cyclophosphamide and 6-TA are used in the treatment of breast cancer.
   h) Vincristine, obtainable from *Colchicum officinale*, is used to treat leukemias.
   i) Coley's toxin contains killed Mycobacteria that stimulate the immune system.
   j) Hairy cell leukemia is the name of a rare form of ALL.

6. Write a 1500 word essay on the diagnosis and treatment of cancer.

FURTHER READING

Bhurani, D and Gottlieb, D (2004) The chronic leukaemias http://www.australiandoctor.com.au

Caldas, C (1998) Molecular assessment of cancer. *BMJ* **316:** 1360–1363.

Cooperberg, MR, Moul, JW and Carrol, PR (2005) The changing face of prostate cancer. *J. Clin. Oncol.* **10:** 8146–8151.

Doll, R (2000) Smoking and Lung Cancer. *Am. J. Crit. Care Med.* **162:** 4–6.

Donaldson, MS (2004) Nutrition and cancer: a review of the evidence for an anti-cancer diet. *Nutr. J.* **3:** 19–40.

El-Deiry, WS (2005) Colon cancer, adenocarcinoma. http://www.emedicine.com

Flaitz, CM and Hicks, MJ (1998) Molecular Piracy: the viral link to carcinogenesis. *Oral Oncol.* **34:** 448–453.

Futreal, PA, Coin, L, Marshall, M, Down, T, Hubbard, T, Wooster, R, Rahman, N and Stratton, MR (2004) A census of human cancer genes. *Nat. Rev. Cancer* **4**: 177–183.

Garland, CF, Garland, FC, Gorham, ED, Lipkin, M, Newmark, H, Sharif, BM and Holick, MF (2006) The role of vitamin D in cancer prevention. *Am. J. Public Health* **96:** 252–261.

Giordano, SH, Cohen, DS, Buzdar, AU, Perkins, G and Hortobagyi, GN. (2004) Breast cancer in men: A population-based study. *Cancer* **101:** 51–57.

Hoffbrand AV, Petit JE and Moss PAH (2001) *Essential Hematology.* 4th edn. Blackwell Science, Malden, MA.

Horig, H, Medina, FA, Conkright, WA and Kaufman, HL (2000) Strategies for cancer therapy using carcinoembryonic antigen vaccines. *Exp. Rev. Molec. Med.* **2000**: 1–24.

Jones, GR (2003) Li-Fraumeni syndrome. http://www.emedicine.com

King, JBR and Robins, M (2006) *Cancer Biology.* Prentice-Hall Ltd, Harlow, UK.

Latonen, L and Laiho, M (2005) Cellular UV damage responses-Functions of tumor suppressor p53. *Biochim. Biophys. Acta* **1755:** 71–89.

Liesner, RJ and Goldstone, AH (1997) ABC of clinical haematology: the acute leukaemias. *BMJ* **314:** 733–736.

Office for National Statistics (2005) Cancer statistics registration: Registrations of cancer diagnosed in 2002, England. Crown Copyright 2005. Also at http://www.statistics.gov.uk

Radtke, F and Clevers, H (2005) Self-renewal and cancer to the gut: Two sides of a coin. *Science* **307:** 1904–1909.

Ries, LAG, Eisner, MP, Kosary, CL, Hankey, BF, Miller, BA, Clegg, L, Mariotto, A, Feuer, EJ and Edwards, BK (eds). *SEER Cancer Statistics Review, 1975–2002,* National Cancer Institute. Bethesda, MD, http://seer.cancer.gov/csr/1975_2002/, based on November 2004 SEER data submission, posted to the SEER web site 2005.

Tisdale, MJ (2005) Molecular pathways leading to cancer cachexia. *Physiology* **20:** 340–348.

Turnbull, C and Hodgson, S (2005) Genetic predisposition to cancer. *Clin. Med.* **5:** 491–498.

Uauy, R and Solomons, N (2005) Diet, nutrition and the life-course approach to cancer prevention. *J. Nutr.* **135:** 2934S–2945S.

Various authors (2001) *Nature* **411:** 335–396. Contains a series of articles on cancer.

von Meyenfeldt, M (2005) Cancer-associated malnutrition: an introduction. *Eur. J. Oncol. Nurs.* **9:** S33–S38.

Wallis, G (1999) *The Genetic Basis of Human Disease.* The Biochemical Society, London. http://www.biochemistry.org/

Yuspa, SH (2000) Overview of carcinogenesis: past, present and future. *Carcinogenesis* **21:** 341–344.

Useful web sites:

The National Institute for Clinical Excellence (NICE) publishes its guidelines for treatment of patients at NICE guidelines www.nice.org.uk

http://www.cancerhelp.org.uk This is an excellent website with lots of useful information on many different types of cancer.

http://www.cancerbacup.org.uk Another excellent website with information on cancer.

chapter 18:

AGING AND DISEASE

OBJECTIVES

After studying this chapter you should be able to:

- describe some of the effects of aging on cells, tissues, organs and systems;
- list the possible causes of the effects of aging;
- describe a number of age-related clinical conditions;
- discuss how some of the clinical and social problems of aging may be managed.

18.1 INTRODUCTION

Aging is difficult to define but is perhaps most widely understood to be a decrease in the ability to survive. As people age (*Figure 18.1*), they are less able to perform strenuous physical activities which were relatively easy when they were younger. With aging comes a decline in the function of most organs in the body, making elderly people more susceptible to disease. Indeed, most major diseases of the developed world, such as coronary heart disease (*Chapter 14*), cancer (*Chapter 17*), and diabetes type 2 (*Chapter 7*) are age-related. With aging comes **senescence**, that is, a decline in functions of almost all parts of the body and at all levels of organization, from cells to organ systems. Senescence changes may be responsible for some disease of old age or may increase susceptibility to certain diseases. Thus elderly people make up a large proportion of patients in hospitals and in most countries a high proportion of the health budget is devoted to the care and treatment of the elderly.

Figure 18.1 Characteristic appearance of a 70-year-old man with his young grandchildren.

The average length of life of individuals in a population is known as the **life expectancy**. Life expectancy has increased over the past 100 years in the UK from 46 years in 1893 to 78 years in 2004. This increase has been due to developments in modern medicine, as well as to improvements in public health care, nutrition and housing. The increase in life expectancy has resulted in an enormous expansion of the elderly population in all industrialized countries, though this is less marked in developing countries. However, although the life expectancy has increased over the last 100 years, the human life span, that is, the maximum age that can be attained by members of a particular species, has not. Humans can live up to the age of about 120 years, but such longevity is exceptional. Humans live longer than other mammals: an elephant, for example has a life span of 70 years, while that of a mouse is a mere three years. While the life span of species is inherent, humans are able, to a certain extent, to increase their life expectancy by controlling their environment.

18.2 AGING OF CELLS, TISSUES, ORGANS AND SYSTEMS

The deleterious effects of the aging processes are numerous and diverse. They affect cells, tissues, organs and systems.

CELLS

Cellular functions decline in efficiency with advancing age. For example, the abilities of mitochondria to survive a hypoxic insult and perform oxidative phosphorylation, the synthesis of structural, enzyme and receptor proteins, the abilities of cells to take up nutrients and repair chromosomal damage all decline with age. Aged cells also have irregular and abnormally shaped organelles, particularly nuclei, Golgi apparatus and endoplasmic reticulum and accumulate waste products.

TISSUES

All tissues are affected with age. For example, muscle mass is subject to a condition known as muscle atrophy due to a reduction in size of muscle groups and to losses of individual muscle fibers. This results in a decreased capacity for work. Other factors, such as cardiovascular, respiratory and joint functions, also influence muscle strength. If the elderly are disabled by disease, for example, arthritis, mobility may also be restricted and muscles will atrophy unless specific exercises are undertaken.

ORGANS

Age-related changes to organs include a decrease in the size and activity of several major organs. There is a decrease, for example, both in the size and elasticity of the lungs, resulting in a reduced gas exchange capacity. In general, the function of the lungs is still sufficient for most activities although the capacity for strenuous activity will be reduced due to a decline in cardiovascular function (*Chapter 14*).

The weight and volume of kidneys may decrease between 20–30% with age as nephrons are lost and replaced by scar tissue. This results in a decrease in the rate of filtration and hence the excretory capacity (*Chapter 8*). Older people are thus more at risk of developing renal disease.

The liver also shrinks in size due to loss of cells. There is a concomitant decline in some liver functions especially in the metabolism and detoxification of drugs and xenobiotics (*Chapter 12*). This is clinically significant because it means that many medications are metabolized and cleared from the body more slowly, a fact that must be taken into account when prescribing drugs for the elderly.

Loss in the function of sense organs occurs with age. A decline in the ability of the lens to change shape makes focusing on near objects more difficult and wearing of spectacles becomes a necessity. Changes also occur in the lens proteins, the **crystallins**, causing them to become more cross-linked and browner in color. This results in more scattering and absorption of light, with less light reaching the retina. With age there is also a general physiological deterioration of the auditory system.

The brain loses weight with age, reducing from a typical mass of 1.4 kg at 20 years of age to about 1.3 kg at the age of 60. The loss is due to changes in composition that include an enlargement of the ventricles and a widening of the surface channels. Nerve cells are also lost and amyloid protein may be deposited. An accumulation of the pigment **lipofuscin** (*Section 18.3*) also occurs in certain neurons. These changes are believed to be responsible for a lengthening in reaction times, a decline in problem-solving and learning abilities and an impairment of memory.

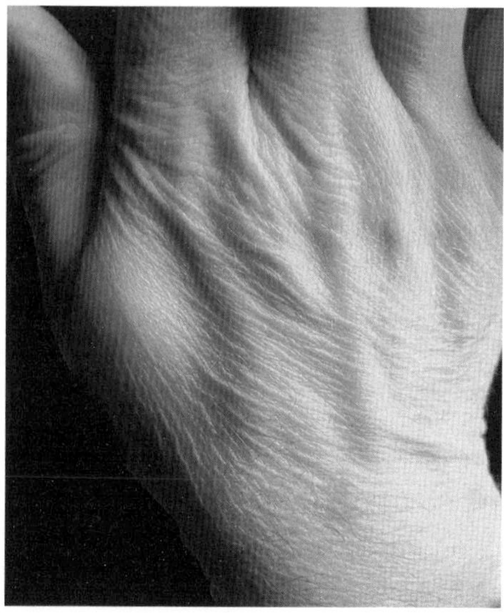

Figure 18.2 Wrinkled skin on the hand of an elderly person.

SYSTEMS

Changes to the skin are among the most easily recognized effects of aging (*Figure 18.2*). Indeed, many people use the appearance of skin and hair to assess the age of an individual. These changes include wrinkling, changes in skin pigmentation and graying and loss of hair. Skin wrinkling is caused by changes to collagen, with increased cross-linking and a reduction in elasticity. The follicles producing gray hair lack the pigment-forming melanocytes. There is a large variation in hair loss that is not surprising given the many genetic and hormonal influences involved.

Skin wounds heal more slowly in older individuals. Studies have compared healing of ischemic (reduced blood flow) and fully vascularized wounds in young and old rats. The fully vascularized wounds healed equally well in both populations whereas ischemic wounds took significantly longer to heal in older animals. It may be that impaired wound healing in older people may be related to diseases, such as **atherosclerosis**, or hardening of the arteries, which contributes to ischemia of the wounded tissue (*Chapter 14*).

There is evidence to suggest that endocrine function declines with age because of a reduction both of hormone production and of the numbers of hormone receptors on target cells. For example, there is an age-related decline in the functions of the reproductive organs, altered thyroid hormone status and an increase in the risk of developing diabetes (*Chapter 7*).

Immune function also declines with age. The thymus atrophies and there is a progressive decline in the function of T lymphocytes (*Chapter 4*). There is a concomitant decrease in the production of antibodies, possibly due to loss of regulatory T cell function together with an increase in the development of autoimmune reactions, leading to an increased susceptibility to infections in the elderly coupled with autoimmune injury to cells and tissues.

Numerous well-documented defects may occur in the cardiovascular system as it ages. Connective tissues, which are essential components of blood vessel walls, lose elasticity and this increases the rigidity of the vessels. Blood vessels are also prone to calcification and hardening of the arteries leading to atherosclerosis (*Chapter 14*). The narrowing of the lumen of blood vessels by arteriosclerosis leads to an increase in blood pressure in the elderly. With age, the heart muscle also becomes less efficient and the heart enlarges due to accumulation of fibrotic tissue, leading to a decline in cardiac output. A consequence of these changes is a reduced delivery of blood to peripheral tissues and to the heart itself.

BOX 18.1 Sunlight and skin

Type I collagen is the major collagen (*Figure 18.3A, B* and *C*) in the dermis of the skin. Its destruction, along with damage to other structural components of the skin occurs over decades and is thought to underlie the characteristic alterations in appearance of aged skin (*Figure 18.2*).

Ultraviolet light (UV) from the sun is a major factor contributing to the premature aging of skin. This effect is called **photoaging.** The clinical features of photoaging include fine and coarse wrinkling, blotchy pigmentation and rough skin texture, often described as 'leathery'. These sun-induced changes are additional to intrinsic aging processes. For this reason, people who spend long periods in the open air, either through occupation or choice, are strongly advised to use protective UV-blocking creams. The healthiest skin is found on areas usually well covered, such as that of the genitalia and buttocks.

The mechanisms underlying collagen degradation in photoaging are not fully understood but are at least in part due to the action of matrix metalloproteinases (MMPs) released from keratinocytes and fibroblasts. The levels of these collagen-degrading proteases in skin increase as a function of age and are also transiently increased in response to the action of UV light. In addition, the synthesis of new collagen decreases as aging takes place. All these effects mean there is a progressive damage and loss of function of the collagen in the dermis. The difference between young and old skin is, in part, a reflection of an intrinsic reduction in the capacity of old fibroblasts to synthesize collagen.

Overexposure of the skin to UV light, from the sun and from tanning beds, is also associated with an increase in skin cancer especially malignant melanoma, a particularly aggressive form of cancer (*Chapter 17*) which can also occur in the pigmented retina of the eye.

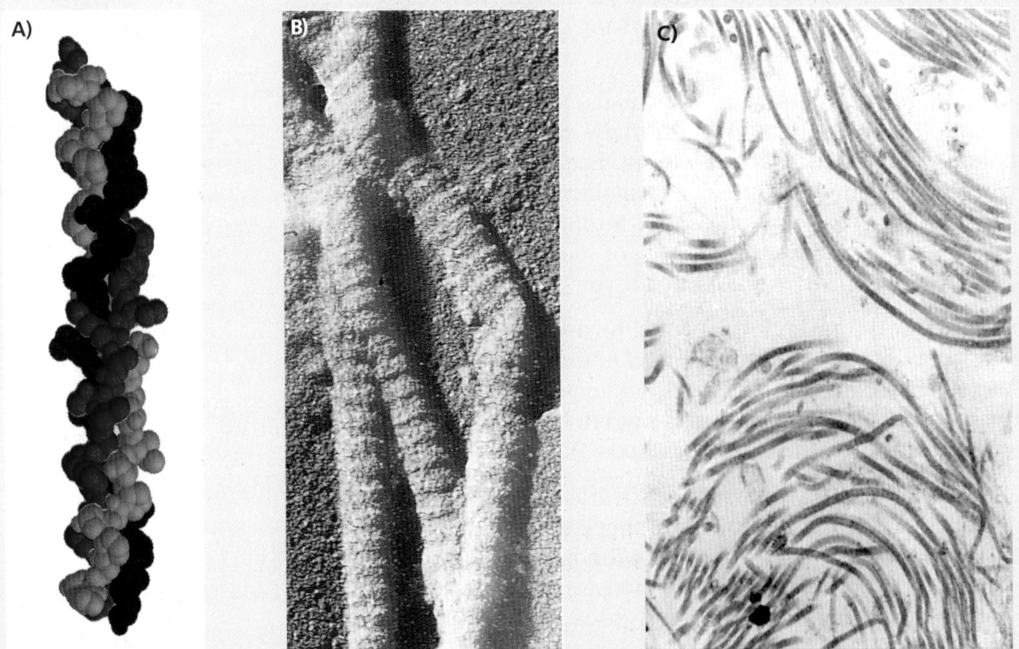

Figure 18.3 (A) Molecular model of portion of a collagen molecule. PDB file 1BKV. (B) Electron micrograph of collagen. (C) Collagen fibers in a sample of tissue.

18.3 CAUSES OF AGING

A number of theories have been proposed to explain cellular aging. These theories can be divided into two broad groups: those that are based on 'wear and tear' and those that propose a genetic basis.

'WEAR AND TEAR' THEORIES

'Wear and tear' theories suggest that aging processes in cells are due to a continual exposure to harmful agents from both inside and outside the cell throughout life. These agents include: free radicals, glycated proteins, waste products and products of erroneous biosynthesis, the error-catastrophe theory.

Free radicals

Free radicals are molecules that have an unpaired electron. This makes them highly reactive although they can be stabilized by the donation of electrons to, or removal from, other molecules. As a result of this process new radicals are produced and a chain reaction can be propagated. Free radicals are produced in phagocytic cells in processes aimed at destroying pathogens. They may also be produced during endogenous enzymatic reactions, especially oxidation–reduction reactions associated with hyperglycemia or following exposure to tobacco smoke or ionizing radiation.

The most studied free radical *in vivo* is the highly reactive hydroxyl radical ($OH^{\bullet}$) formed by the action of ionizing radiation and from some intermediates in biochemical processes. The superoxide radical ($O_2^{\bullet}$) is less toxic and is produced, for example, by metabolic reactions of the electron transport chain where oxygen is normally reduced to water by accepting electrons. During this process, a small proportion of this oxygen can be released as the superoxide radical after having accepted only one electron:

$$O_2 + e^- \longrightarrow O_2^{\bar{\bullet}}$$

While phagocytes routinely produce the superoxide radical as part of their antibacterial defence, it has been estimated that each cell in the body is exposed to attack by around 10 000 free radicals per day. This sustained exposure is thought to cause progressive damage to cells. The damaging chain reactions cease when two radicals meet and form a covalent bond, or when they react with a molecule that acts as a free radical trap. The latter includes vitamin E which acts as a free radical scavenger and, by virtue of its lipid solubility, may help to prevent damage to biological membranes. Glutathione (GSH), a tripeptide present in most cells, contains a thiol (–SH) group that is readily oxidized (*Figure 18.4*). Glutathione is usually maintained in a reduced state in the cytosol of cells and protects against free radical damage. The enzyme superoxide dismutase (*Figure 18.5*) removes superoxide radicals by converting them into hydrogen peroxide and dioxygen. The hydrogen peroxide is then oxidized to water by the catalase (*Figure 18.6*):

$$2H^+ + 2O_2^{\bar{\bullet}} \xrightarrow{\text{superoxide dismutase}} H_2O_2 + O_2$$

$$2H_2O_2 \xrightarrow{\text{catalase}} 2H_2O + O_2$$

There is evidence that dietary antioxidants, such as vitamins E and C (*Chapter 10*) may delay the aging process and increase life expectancy in rats, mice and some nonmammalian species but it is not known whether they act solely by reducing free radical damage.

A)

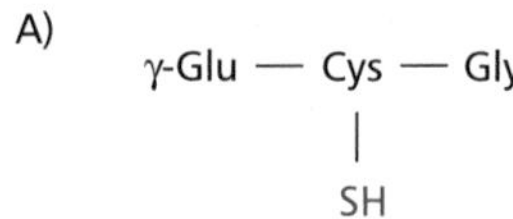

B)

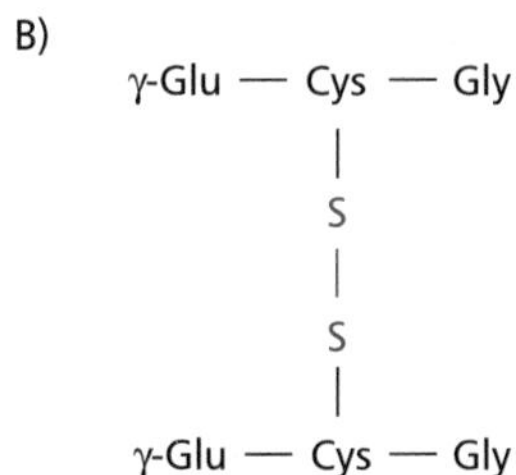

Figure 18.4 The (A) reduced and (B) oxidized forms of glutathione. See also *Figures 12.6* and *13.25*.

Figure 18.5 Molecular model of superoxide dismutase. Its Cu and Zn atoms are shown in red. PDB file 1PUO.

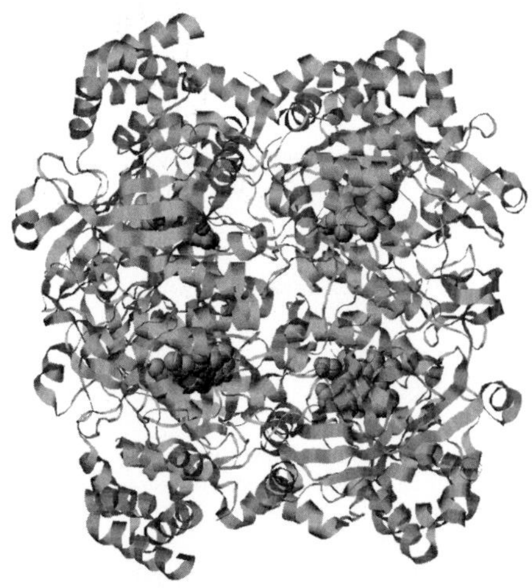

Figure 18.6 Molecular model of catalase. There are four heme groups and their associated Fe atoms are shown in red. PDB file 1DGF.

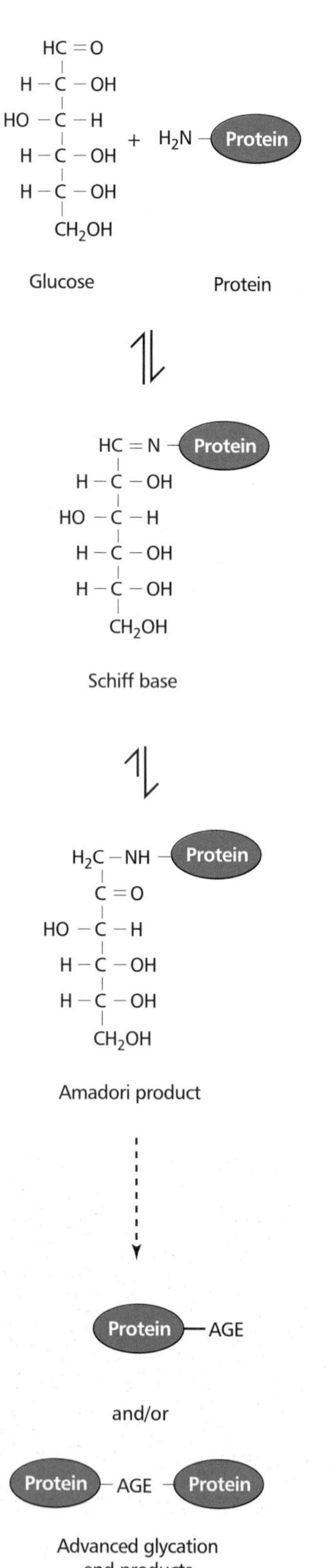

Figure 18.7 The reaction between glucose and a protein to form a glycated protein (Schiff base and Amadori product) and its subsequent conversion to an advanced glycation end product (AGE).

Glycated proteins

Both intra- and extracellular proteins are subject to posttranslational changes during aging. For example, proteins exposed to reducing sugars may undergo glycation. This occurs as a nonenzymatic reaction between free carbonyl groups of reducing sugars in their acyclic form and amino groups of the protein to form a Schiff base in a freely reversible reaction (*Figure 18.7*). The Schiff base is unstable and rearranges to a more stable Amadori product, the reaction being effectively irreversible.

The extent of glycation *in vivo* depends on the degree and duration of hyperglycemia. Glycated proteins may undergo further reactions to form cross-linked fluorescent structures called advanced glycation end products (AGEs). Advanced glycation end products accumulate with age, particularly on structural proteins, such as collagen which has a long half-life, and they can cause increased cross-linking of individual proteins. Such changes have deleterious effects since excessive cross-linking decreases the elasticity and permeability of the extracellular matrix and impairs the flow of nutrients into and waste products out of cells. Since a high proportion of the elderly population have diabetes or impaired glucose tolerance (*Chapter 7*), their proteins are more likely to be glycated.

Waste products

During aging, increasing amounts of waste material accumulate in the cytoplasm of cells. Many of these are waste products of normal cellular metabolism. For example, **lipofuscins** are yellow-brown pigments produced by degeneration of cell membranes and organelles, probably by the free radical peroxidation of membrane lipids. Lipofuscins accumulate with age in many types of cells, particularly nondividing cells such as those of muscle. Lipofuscins are chemically inert, strongly cross-linked molecules that are stored in lysosome-like structures (*Figure 18.8*). They are not susceptible to enzymatic digestion by the lysosomal enzymes (*Chapter 16*). It has been suggested that a gradual accumulation of substances like lipofuscins within cells interferes with their normal function, though there is no conclusive evidence for this. Furthermore, there is no correlation between the amount of lipofuscin accumulated and the reduction in cell function and survival.

Error-catastrophe theory

The error-catastrophe theory suggests that cellular dysfunction and, ultimately, cell death arises due to an accumulation of abnormal proteins. Protein synthesis involves transcription of DNA to give mRNA, which is transported to the cytoplasm and is translated to form polypeptides. Random errors in transcription and/or translation will lead to formation of abnormal proteins whose accumulation might impair cellular function. If the protein in question is an enzyme, such an error may lead to a malfunctioning enzyme and cellular dysfunction. Although enzyme activity is known to decline with aging, it has not always been possible to demonstrate any changes in enzyme structure with age. It seems that proteins are synthesized appropriately in older cells and most subsequent changes to their structures occur posttranslationally.

Some studies have indicated that certain enzymes have a changed conformation in an older cell. This suggests that enzyme molecules retained inside the cell for long periods are slowly denatured and consequently lose their biological activity. In younger cells, the original shapes of proteins can be restored by cycles of denaturation followed by renaturation. Weak interactions that confer shape to the denatured form of protein molecules are broken, allowing them to fold back to their original shape. This process therefore corrects the defective shape of denatured enzymes and produces molecules that are as efficient as a newly synthesized enzyme. Unfortunately, such repair mechanisms lose their efficacy as the cell ages.

GENOME-BASED THEORIES

There is evidence to suggest that aging is under genetic control. A number of genetic based theories have emerged, including those that suggest **programmed aging** and those that propose **gene mutations.**

Programmed aging

The theory of programmed aging suggests that each species has an in-built biological clock and that aging involves a genetically programmed series of events. In the 1960s, Hayflick demonstrated that cells are restricted in the number of times they can enter the cell cycle by an in-built genetic program of senescence. He showed that cultured fibroblast cells derived from human embryos could undergo 50 cell divisions, whereas those from adults were limited to about 20. In culture, the number of divisions is constant for each type of cell. This is referred to as the Hayflick limit. Furthermore, the factors that control the number of divisions are intrinsic to the cell and are not influenced by their environment. For example, if the nucleus of an old cell is transplanted into a young cell from which the original nucleus has been removed, the resulting cell has a lifespan that reflects that of the transplanted nucleus.

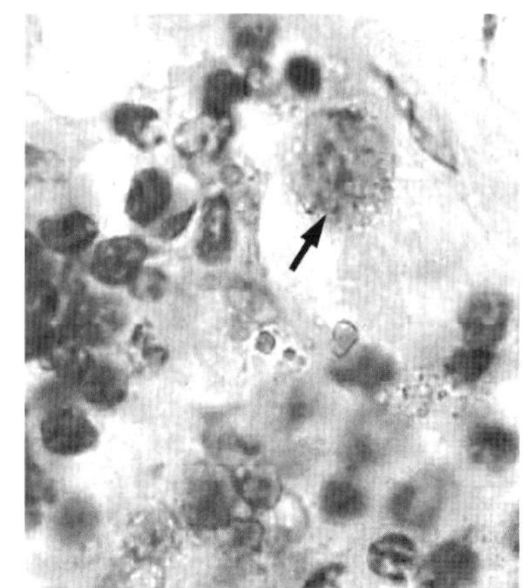

Figure 18.8 Light micrograph of a portion of a macrophage containing lipofuscin inclusions derived from lysosomes. A lipofuscin particle is indicated. Courtesy of Dr T. Caceci, Morphology Research Laboratory, Virginia-Maryland Regional College of Veterinary Medicine, USA.

When cells grown in culture are frozen and then recultured, they appear to retain the memory of the number of times they have already divided in the original culture. Hence they only complete the 'unused' number of cell divisions. It therefore appears that there is a biological clock within all cells. This biological clock, at least in part, resides in the telomeres, which are extensions of DNA found at the ends of chromosomes (*Figure 18.9*). Telomeric DNA protects the ends of the DNA molecule from damage. When DNA is replicated prior to cell division, telomeric DNA does not replicate. After each cell division the telomere becomes shorter in length. Once the telomeres shorten to a particular length, the cell can no longer divide and dies. The activity of telomerase can prevent the shortening of telomeres and enable the cell to divide continuously. Most somatic cells contain an inactive form of telomerase although a number of cell types, such as hemopoietic cells and cancer cells, have a permanent telomerase activity. These cells can divide indefinitely and are therefore potentially immortal.

The suggestion that aging is genetically programmed has received some criticism. For example, the number of divisions occurring *in vitro* may be different from those that occur *in vivo*. Furthermore, some cells, such as cardiac muscle cells and neurons, do not divide after birth, and so programmed aging may not apply to these cells.

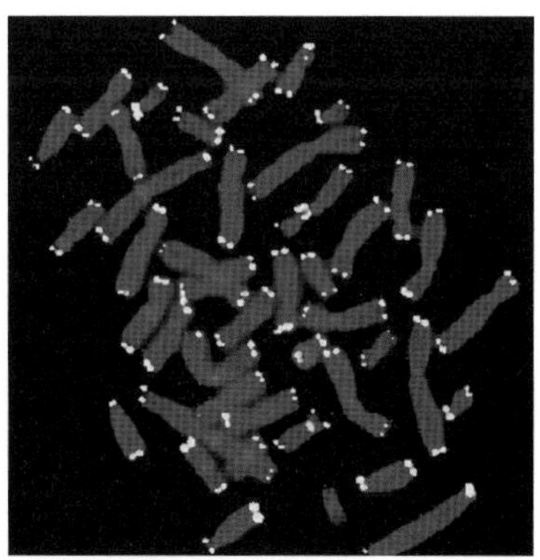

Figure 18.9 Telomeres of human chromosomes are stained to appear brighter than the rest of the chromosomes. Note these structural components of chromosomes are situated at the ends of the chromosomes. See also *Chapter 17* and *Figure 17.24*. Courtesy of Dr C. Counter, Department of Pharmacology and Cancer Biology, Duke University, USA.

Gene mutations

It is well known that mutations occur in genes during the lives of cells and that these mutations can alter the activities of the cells (*Chapter 15*). The gene mutation theory suggests that accumulation of mutations during the course of life leads ultimately to tissue and organ malfunctions and eventually death.

Genes are composed of DNA. The cell has several mechanisms to repair damaged, that is, mutated, DNA. Enzymes within the cell excise the damaged region of the gene and add back a new set of nucleotides using the undamaged DNA strand as a template. The gene mutation theory suggests that, with time, these DNA repair mechanisms become less efficient and some mutations are not repaired leading to functional changes.

In support of this theory, DNA obtained from liver cells of older mice has been found to have a greater number of mutations compared with similar cells from younger mice. In addition, liver cells obtained from strains of mice with a short lifespan show a higher incidence of mutations compared with similar cells from a strain of mice with longer lifespan. Radiation is known to cause mutations and shorten the lifespan of cells. It has been suggested that natural radiation might accelerate the aging process.

18.4 AGE-RELATED DISEASES

Factors that may contribute to causing disease in the elderly include the physiological and biochemical changes associated with normal aging, the cumulative exposure to harmful agents, and an increased sensitivity to agents or the environment. A number of diseases in particular show an increased incidence in older people. These include cancer, cardiovascular disease, type 2 diabetes, cataracts, arthritis, Parkinson's disease and Alzheimer disease. The latter, and Hutchinson-Gilford syndrome, which affects children, are described in *Box 18.2*.

CANCER

In general, the incidence of most cancers increases with age (*Figure 18.10*) with more than half of all cancers occurring in people over the age of 65 years. Two main hypotheses have been proposed to explain the link between cancer and age. First, an age-related accumulation of carcinogenic substances may increase the incidence of cancers in the elderly. This process is independent of the senescence changes described above that occur in the aging body. The second hypothesis proposes that age-related changes may make cells more vulnerable to becoming cancerous. Changes in immune, nutritional, metabolic and endocrine status occur with age and may create a more favorable environment for the induction of cancer. Such physiological changes may affect a number of cell processes such as the detoxification of mutagenic agents and the repair of damaged DNA (*Chapter 17*).

CARDIOVASCULAR DISEASE

Many of the changes in the cardiovascular system may be caused by disease rather than old age *per se*. The concentration of cholesterol in the plasma increases with age. Elevated levels over the years are thought to contribute to

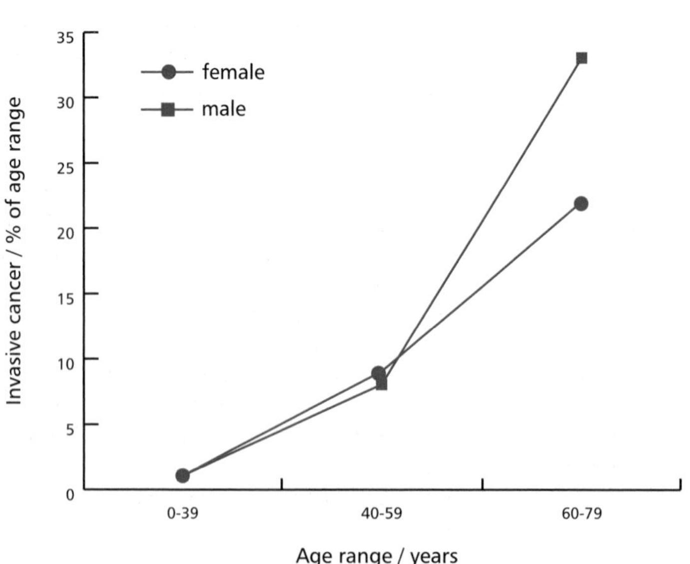

Figure 18.10 Graph showing the increasing incidence of cancers with age. Redrawn from DePinho, R.A. (2000) The age of cancer. *Nature* **408:** 248–254.

the high incidence of mortality from coronary heart disease especially if other risk factors are present. This risk may be decreased by changes to lifestyle, since eating an inappropriate diet, smoking and lack of exercise are known to be associated with atherosclerosis.

DIABETES MELLITUS TYPE 2

Many older people have some degree of impaired glucose tolerance that can be severe enough to be classified as type 2 diabetes mellitus. The main reason for high concentrations of blood sugar in the elderly is increased resistance to the effects of insulin in peripheral tissues that is associated with increased insulin levels after a meal. Diabetes in older people is strongly influenced by diet and exercise.

CATARACTS

A cataract is a partial or complete opacity of the lens of the eye that causes blurred vision. This affects the passage of light through the lens causing blindness. There are many different types of cataracts but one of the most common is senile cataract. Another risk factor for cataract is diabetes. Cataracts are treated by removal of the opaque lens and its replacement with a plastic lens.

ARTHRITIS

Arthritis is inflammation of the joints producing swelling, pain and restricted movement. Osteoarthritis affects the joint cartilage and underlying bone. It is particularly associated with increasing age, although it can occur in younger individuals who excessively use their joints in work or athletic activities. Osteoarthritis affects fingers, hip joints and knees (*Figure 18.11(A)*) but, unlike rheumatoid arthritis, does not always cause pain and inflammation. X-raying of joints usually shows some degree of osteoarthritis in nearly all elderly patients although few present with any symptoms. In severe cases, the joints of the fingers often show overgrowth referred to as Heberden's nodes (*Figure 18.11(B)*) although these tend not to be painful. Osteoarthritis cannot be cured although mild exercise can improve joint mobility.

Rheumatoid arthritis is characterized by a chronic inflammation of the joints that usually arises from an autoimmune reaction (*Chapter 5*). It is also more common in the elderly, although its onset can occur in any age group. The result is severe pain and disability. What initiates rheumatoid arthritis is not clear although a variety of bacteria, especially mycobacteria, have been implicated. The treatment of rheumatoid arthritis involves using nonsteroidal anti-inflammatory drugs. The surgical replacement of hip or knee joints may also be required in patients who become severely disabled.

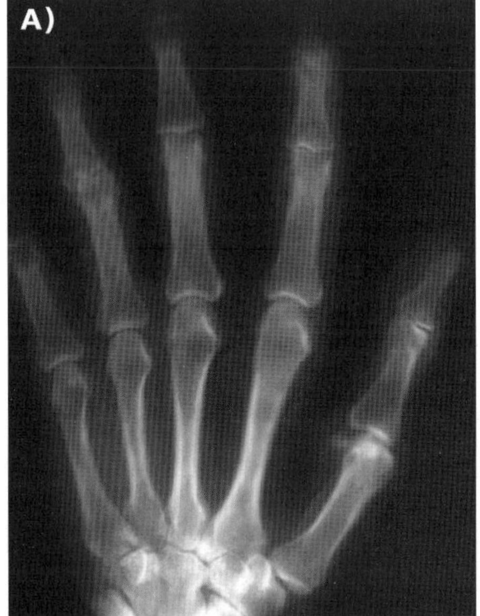

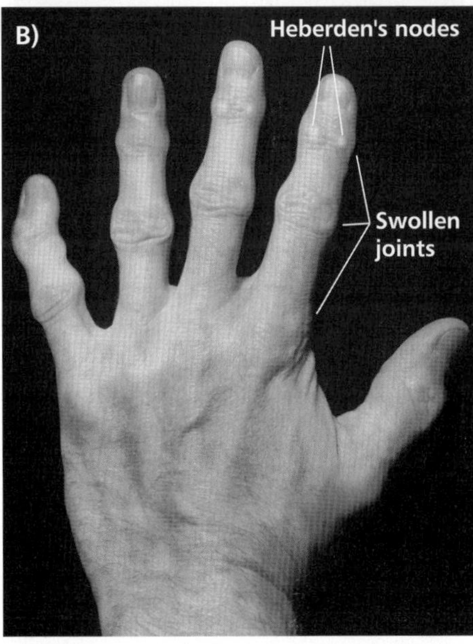

Figure 18.11 (A) X-ray image of a middle-aged female patient with osteoarthritis of the left hand who presented with pain and swelling in the finger joints. (B) The left hand of a sufferer of osteoarthritis. Courtesy of Dr. P. Young, Department of Radiology, University Hospitals of Cleveland, USA.

PARKINSON'S DISEASE

Parkinson's disease affects between 1 and 2% of individuals over the age of 70. The major defect in Parkinson's disease is degeneration of dopamine-secreting nerve cells although other neurons and neurotransmitters may also be affected. Patients have severe attacks of tremors that affect one hand and then spread to the leg on the same side and then to other limbs. The average survival time is eight to 10 years after diagnosis. Parkinson's is distinct from Alzheimer's disease (*Box 18.2*) in that different nerve cells are affected and there is loss of motor function, which is usually unaffected in Alzheimer's disease. A further feature of Parkinson's disease is the presence of cytoplasmic inclusions called Lewy bodies in some of the surviving neurons. Some researchers believe that an excess of free radicals causes the degeneration of these neurons.

BOX 18.2 The young and old of aging: Hutchinson-Gilford syndrome (progeria) and Alzheimer's disease

Hutchinson-Gilford syndrome or progeria is a disorder that causes premature aging. The name progeria comes from the Latin and Greek words *pro* and *geraios* that mean early and old age respectively. The syndrome was described by two British doctors in 1886 (Hutchinson) and 1904 (Gilford). Children with progeria age about 10 times faster than normal; thus a child of eight to ten will look like an 80-year-old (*Figure 18.12*). The development and appearance is seemingly normal in the first two years of life after which the characteristic aging changes take place with a rapidity that can be shocking. The appearance of children with progeria is remarkably similar. Clinical features include thinning and wrinkling of skin, prominent scalp veins, loss of subcutaneous fat, alopecia (loss of hair), beak-like nose, short stature, thin limbs with stiff swollen joints, severe arthritis, osteoporosis, high-pitched (squeaky) voice and normal or high intelligence. Some features of aging are absent and these children often present with delayed development of teeth, delayed sexual maturity but no increase in incidence of cancers, diabetes or cataracts. The life expectancy is approximately 13 years with a range of 7–27 years. Death usually occurs from a heart attack or cerebrovascular disease (*Chapter 14*).

The diagnosis of progeria is made on clinical grounds and can be difficult due to the rarity of the condition and its insidious onset in the early stages. Given its rarity, children with progeria often think they are the only ones with this disorder. At school these children perform very well and usually have a cheerful and open nature. However, because of their appearance, they are usually stared at by strangers and must learn to cope with such social problems from an early age. No treatment is available for progeria. Patients may be placed on low-dose aspirin therapy to delay symptoms of atherosclerosis.

Progeria is a rare condition affecting one in 10 million people and about 100 cases have been identified to date. In 2005, Europe had about 10 cases, with approximately 30 known cases worldwide. The disease is not restricted to any particular race or geographical area but males are affected one and a half times more frequently than females. Although originally classified as an autosomal recessive condition (*Chapter 15*), the precise mode of inheritance is still unclear. More recent studies have suggested a sporadic dominant mutation. This mutation results in the production of a truncated form of lamin A, a protein necessary to maintain the structure of the nucleus and control the movement of materials between the nucleus and cytoplasm (*Chapter 16*). Indeed, a single base change in the lamin A gene (*LMNA*) on chromosome 1 can cause the syndrome. The identification of the mutation has enabled a diagnostic genetic test to be developed that should allow an earlier identification or elimination of progeria in symptomatic children. In most cases, the mutation is probably 'fresh' and has occurred only by chance in the child and is not found in either parent.

Fibroblasts isolated from patients with progeria and grown in tissue culture have a shorter life span than fibroblasts from normal individuals of a similar age. This is due to the chromosomes of progeria sufferers having short telomeres (*Figure 18.9*), which results in cells having a lower Hayflick limit. Furthermore, some studies have shown a decrease in the ability of cells from patients with progeria to repair damaged DNA, although this finding has not been supported by other studies.

Alzheimer's disease (AD) is a degenerative condition of the brain in which some nerve cells lose function and die. Late-onset AD is the most common cause of dementia in the elderly and accounts for about half of such cases of dementia. In the UK, about 5 to 10% of the population over the age of 65 develop AD and this increases to over 20% of those over the age of 80 (*Figure 18.13*).

Alzheimer's disease is characterized by the presence of extracellular plaques in the brain (*Figure 18.14*), usually in the hippocampus, temporal and parietal regions. The patches are resistant to enzymatic or chemical digestion and remain in the brain

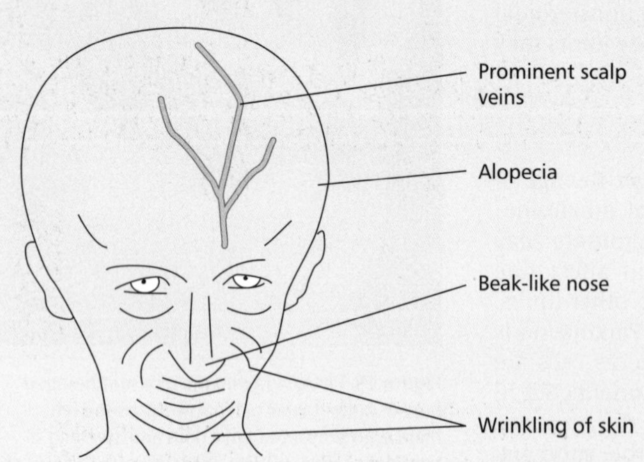

Figure 18.12 Schematic of the head of a progerian child.

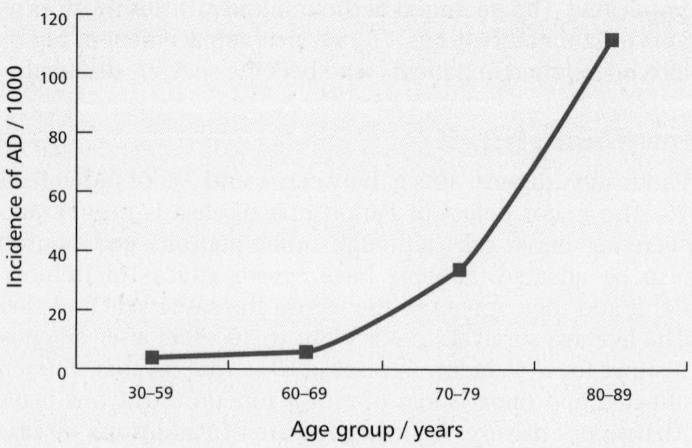

Figure 18.13 The increasing incidence of Alzheimer's disease with age. Redrawn from The Association of the British Pharmaceutical Industry website.

tissue even after neuron death. They consist mainly of a core of β-amyloid peptides (Aβ) consisting of 40–42 amino acid residues (*Figure 18.15*) entangled with tau protein and surround degenerating nerve terminals. The Aβ peptides are formed by two specific hydrolytic cleavages of a β-amyloid precursor protein (APP) and catalyzed by β-secretase and γ-secretase respectively (*Figure 18.16*). The function of APP is unclear, although it shows some resemblance to certain cell-surface membrane receptors.

Approximately 5 to 10% of AD cases are familial, that is inherited forms of the condition. However, most cases are sporadic or senile AD and the risk of developing the disease increases with age. Familial or early-onset AD is associated with mutations in two presenilin (PS) genes, PS1 and PS2. The pathological mechanisms by which these mutations cause AD is unclear. Mutations in PS1 are more common and appear to cause more aggressive forms of AD, in some cases with onset occurring before the age of 30 years, although 45 to 60 would be more likely for early-onset AD. Sporadic cases of AD are more likely to occur in people with the gene for the variant apolipoprotein E (*Chapter 14*), called apoEε4, especially if they are homozygous. A number of environmental risk factors, including exposure to aluminum, head injuries and viral infections are also associated with AD. Alzheimer's disease is also associated with a decline in choline acetyltransferase, an enzyme required for the synthesis of acetylcholine. Indeed, there is a correlation between a reduction of choline acetyltransferase activity, the number of plaques and severity of dementia.

The clinical features of AD can be divided into three stages. The first stage may last two to four years and is associated with memory loss, personality changes and disorientation to time and date. The memory loss in this first stage is difficult to distinguish from the normal forgetfulness that occurs in the elderly. The second stage may last for several years and includes confusion, depression, inappropriate social behavior, agitation and inability to carry out the activities of daily living. The memory lapses become more frequent and the patients often forget what they were doing only a few minutes previously. During this stage, personal hygiene is often neglected and vocal communication becomes impaired as the patient has difficulty in remembering words. The final stage lasts for one to two years although it can last for as long as 10 years. During this stage the affected individual fails to recognize their family, suffers from urinary and fecal incontinence and cannot communicate. The affected individuals are usually institutionalized at this stage.

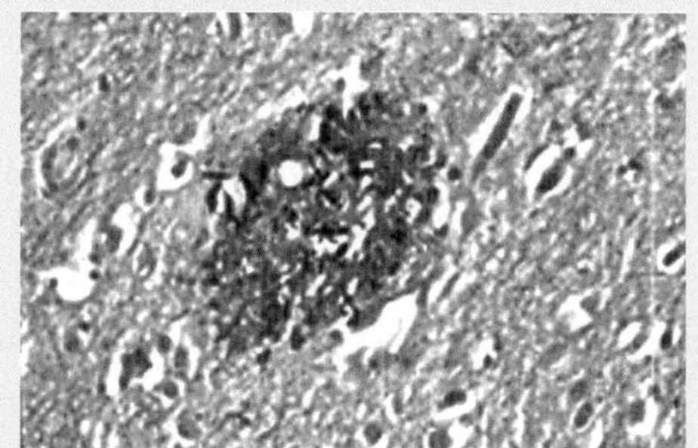

Figure 18.14 An amyloid plaque in the brain of a patient who died from Alzheimer's disease. Courtesy of Alzheimer's Disease Education & Referral Center, National Institute on Aging, USA.

Margin Note 18.1 Tau protein 

The protein tau is mainly expressed in the brain where it stabilizes and orientates microtubules (MTs) necessary for transport of materials in axons. The dephosphorylation of tau promotes the rapid and extensive polymerization of MTs. In contrast, the phosphorylation of tau decreases its ability to promote their assembly. In Alzheimer's brains, tau is hyperphosphorylated, and the ability to stabilize MTs is impaired.

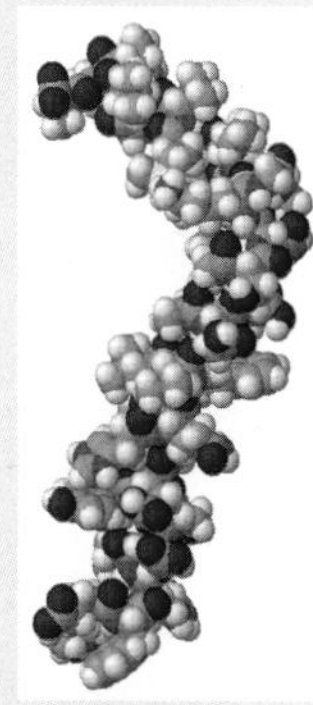

Figure 18.15 Molecular model of β-amyloid peptide. PDB file 1IYT.

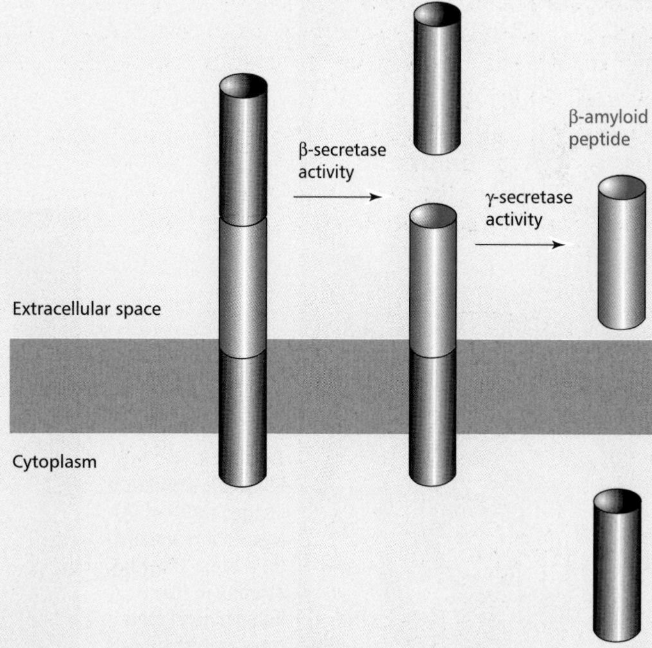

Figure 18.16 The release of β-amyloid peptide by the proteolytic activities of secretases. Redrawn from Haass, C. and Steiner, H. (2002) Alzheimer's disease – secretase: a complex story of GxGD-type presenilin proteases. *Trends in Cell Biol.* **12:** 556–562.

BOX 18.2 The young and old of aging: Hutchinson-Gilford syndrome (progeria) and Alzheimer's disease – *continued*

The onset of AD is insidious, starting with periods of forgetfulness, leading to a confused state and eventually frank dementia. Once severe dementia develops, life expectancy is about two or three years. The clinical course is about eight years and patients often die due to infections such as pneumonia, accidents and occasionally respiratory arrest.

There is no simple and completely accurate test to diagnose AD, although the need for one was identified in the USA (The Surgeon General's Report on Mental Health, 1999). A firm diagnosis of AD can only be given by a histological examination of brain tissue after death. However, brain imaging techniques are proving increasingly useful in diagnosis. Two particular imaging techniques have been developed that allow the structure and activities of the body, including the brains of AD patients, to be assessed. Magnetic resonance imaging (MRI) allows the structure of the brain to be studied in a noninvasive manner. The technique is based on the principle of nuclear magnetic resonance and uses powerful magnetic fields to obtain chemical and physical information about the molecules within the brain. A computer then uses this information to generate an image of the internal structure of the brain. Physical lesions to the brains of AD patients are clearly visible using MRI (*Figure 18.17(A)* and *(B)*). Positron emission tomography (PET) is also an imaging technique but one that allows the activities in different parts of the brain to be estimated. This is achieved by adding labeled glucose or water to the blood and then monitoring the flow of blood through the brain or the rate of glucose metabolism in the different parts of the brain. Again, a computer is able to analyze this information to produce digital images that highlight differences in the activities of the brains of normal and AD patients (*Figure 18.18(A)* and *(B)*).

Given the difficulty in diagnosis, cases of AD are under reported. Early diagnosis is of considerable benefit since it would allow all concerned to make informed, early social, legal and medical decisions about treatment and care for a patient. An early diagnosis would allow drug treatment and care that could delay institutionalization and substantially reduce costs. Conversely, an early test that indicated an absence of AD in suspected cases would alleviate the uncertainty and anxiety faced by the patients and their families. However, even if a quick diagnosis were possible, there is no effective treatment for AD. Sufferers may be placed on medication to alleviate symptoms such as depression and anxiety. Drugs that inhibit the degradation of acetylcholine within synapses, such as acetylcholinesterase inhibitors, are used in treatment. These drugs can delay the impairment of cognition, behavior and functional abilities. Vitamin E treatment has also been used, although some studies have suggested it is of little benefit.

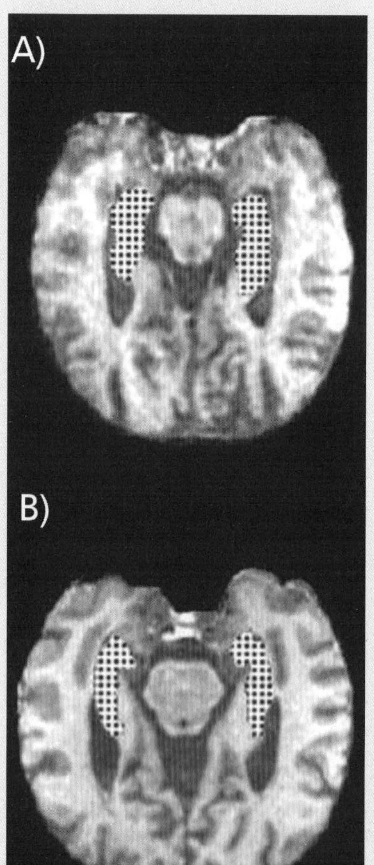

Figure 18.17 Magnetic resonance images (MRI) of (A) a normal brain and (B) a brain showing atrophy in the hippocampal area from an Alzheimer's disease patient. Courtesy of Dr M. de Leon, New York University School of Medicine, USA.

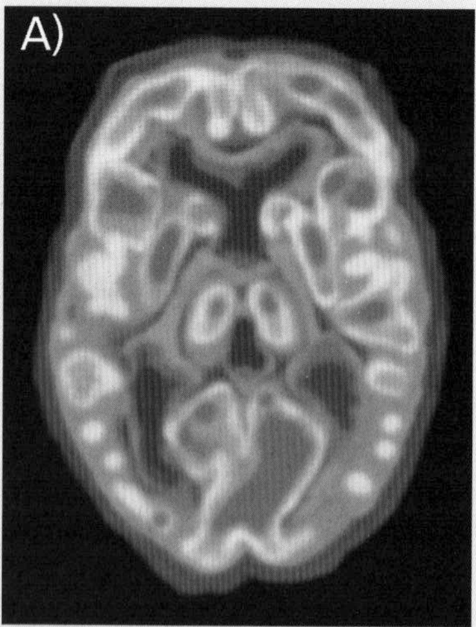

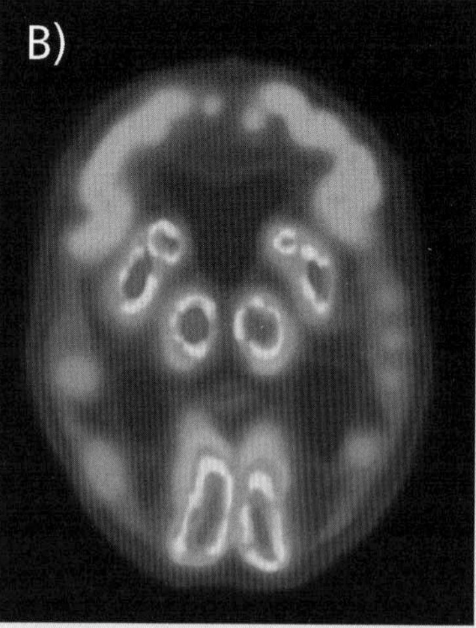

Figure 18.18 Positron emission tomography (PET) images showing activity of (A) a normal brain and (B) the brain of an Alzheimer's disease patient. Courtesy of Alzheimer's Disease Education & Referral Center, National Institute on Aging, USA.

18.5 CALORIE RESTRICTION AND AGING

A reduced energy intake ('calorie restriction') is known to slow down the rate of aging and onset of age-related disorders, such as cancer (breast, lymphomas, prostate), nephropathy, cataract, diabetes, hypertension, hyperlipidemia and autoimmune diseases. This has been demonstrated in a variety of species including chickens and rodents and is also believed to be true for humans. The effects of calorie restriction were demonstrated in the 1930s using laboratory rats. Rats were divided into two groups. One group was allowed to feed freely while the other was fed on a diet containing 30% of the calories of the first group, although they were provided with sufficient protein, fats, vitamins and minerals to maintain normal health. The calorie-restricted rats lived for four years compared with three years for those allowed to feed freely. In addition, the calorie-restricted rats developed fewer age-related diseases.

Studies on calorie restriction have been performed in primates with encouraging results. Long-term studies on rhesus monkeys showed that calorie restriction reduced the incidence of heart disease, diabetes and hypertension and was associated with a decreased concentration of blood cholesterol. Calorie restriction may, however, be difficult to apply to humans because many people may be unable to reduce their calorie intakes by an appreciable amount for the extended period of time required. However, it may be possible to motivate people to do this, especially those with family histories of age-related diseases such as cancer and neurodegenerative disorders.

The mechanism by which calorie restriction increases the life span is unclear but studies have shown that it is associated with a reduction in age-associated mutations when compared with normal diets. This was demonstrated by examining mutations in lymphocytes at four weeks, six months and one year of age.

A high calorie diet may increase free radical-mediated damage as the increased availability of nutrients to mitochondria increases the production of the superoxide radical. Thus, a calorie-restricted diet appears to reduce free radical damage to lipids, protein and DNA and improves the antioxidant status. Calorie restriction in animals has also been shown to reduce levels of tissue AGEs. The benefits of calorie restriction, however, depend on preventing malnutrition and reducing overall calorie intake rather than a particular nutrient.

18.6 INVESTIGATIONS AND MANAGEMENT OF THE ELDERLY

Investigation and management of illness in the elderly poses a number of problems. Many conditions are more common in the elderly and often the presentation of some of these diseases may differ from that in younger people. For example, diabetes mellitus in the elderly often presents as a complication of, for example, renal failure or impaired healing of wounds, instead of the classical signs of polydipsia or polyuria (*Chapters 7* and *8*) first seen in younger patients. Elderly people, particularly those with poor mobility, may also suffer from poor nutrition. Furthermore, they are often on multiple medications that may affect test results. The high incidence of many diseases in the elderly population justifies screening programs for such conditions to increase the chances of detecting disease at a more treatable stage. Clinicians, hospitals and geriatric clinics can all carry out screening. Some of the common investigations are listed in *Table 18.1*.

Investigation	Abnormality detected
Plasma creatinine	renal impairment
Plasma calcium	hyperparathyroidism / osteomalacia
Plasma glucose	diabetes mellitus
Thyroid hormones	hypo- and hyperthyroidism
Fecal occult blood	large bowel carcinoma
Blood pressure	hypertension
Cholesterol	coronary heart disease
Mammography	breast cancer

Table 18.1 Common clinical tests used in elderly patients

CASE STUDY 18.1

Joyce, a 70-year-old woman presented with an ulcer on the sole of her right foot. This was causing her considerable pain and on examination her foot was cold and appeared ischemic. The doctor suspected Joyce might be diabetic but she did not complain of thirst or polyuria. A random blood and urine specimen were taken and tested for glucose. The following results were obtained:

	Results	Reference range
Urine [glucose]	positive	negative
Plasma [glucose]	16 mmol dm^{-3}	(3.0–5.5 mmol dm^{-3})

Questions

(a) Does Joyce suffer from diabetes?

(b) If so, what type is it likely to be (*Chapter 7*)?

CASE STUDY 18.2

Harry, a 64-year-old priest, was noted for making mistakes while delivering his sermons. His wife noticed that these were becoming more common and that he was suffering from increasing lapses of memory. Harry was persuaded to visit his doctor and during his examination he appeared to be fully orientated and held a normal conversation. However, he was only able to recall one out of every three words after five minutes. He was referred for a more detailed neurological examination that was largely normal, although an MRI scan of his brain showed some possible hippocampal atrophy. No other significant changes were noted.

Questions

(a) What is a possible diagnosis?

(b) What advice should Harry and his family be given?

(c) How should Harry be treated?

18.7 SUMMARY

The deleterious effects of aging processes are numerous and diverse. They affect cells, tissues, organs and systems. Aging may be caused by a combination of 'wear and tear' with the accumulation of harmful metabolic products, damaged proteins, genetic mutations and the in-built aging process that can be demonstrated in cultured cells. Calorie restricted diets have been shown to increase life span in animals. A number of diseases are associated with increasing age. These include cardiovascular disease, type 2 diabetes, Alzheimer's and Parkinson's disease. There is some evidence that antioxidants in the diet may delay the onset of the aging process. Ideally, elderly patients should be routinely screened for diseases associated with old age.

QUESTIONS

1. Which of the following enzymes is an established intracellular anti-oxidant?

   a) lactate dehydrogenase;

   b) phenylalanine hydroxylase;

   c) superoxide dismutase;

   d) sucrase;

   e) γ secretase.

2. Which of the following statements is the correct definition of the term 'life span'?

   a) the mean age of individuals in a population;

   b) the minimum age of individuals in a population;

   c) the age when individuals in a population reach puberty;

   d) the maximum age that can be attained by members of a population;

   e) the maximum age that can be attained by most members of a population.

3. Free radicals can be defined as:

   a) molecules with a positive charge;

   b) enzymes;

   c) cross-linked proteins;

   d) atoms or molecules with an unpaired electron;

   e) ions with a negative charge.

4. In less than 100 words, outline the role of protein glycation in aging.

5. List FOUR metabolic or molecular changes you might reasonably expect to see in aging cells.

6. List FIVE disorders that show an increased incidence with age.

7. Which of the following is NOT a characteristic feature of progeria?

   a) accelerated aging;

   b) increased incidence of cancers;

   c) osteoporosis;

   d) alopecia;

   e) short stature.

8. Complete the following table with respect to familial (early onset) and senile (late onset) Alzheimer's disease.

Feature	Familial AD	Senile AD
Age of onset / years		
Percentage of cases		
Associated gene or risk factor		

FURTHER READING

Cummings, JL (2004) Alzheimer's disease. *N. Engl. J. Med.* **351:** 56–67.

Fisher, GJ, Kang, S, Varani, J, Bata-Csorgo, Z, Wan, Y, Dhatta, S and Voorhees, JJ (2002) Mechanisms of photoaging and chronological skin aging. *Arch. Dermatol.* **138:** 1463–1470.

Kipling, D, Davis, T, Ostler, EL and Faragher, RGA (2004) What can progeroid syndromes tell us about human aging? *Science* **305:** 1426–1431.

Kirkwood, TB (2005) Understanding the odd science of aging. *Cell* **120:** 437–447.

Kirkwood, TBL and Austad, SN (2000) Why do we age? *Nature* **408:** 233–238.

Koubova, J and Guarente, L (2003) How does calorie restriction work? *Genes Dev.*, **17:** 313–321.

Marks, R (1999) Spotlight on solar elastosis. *Retinoids* **15:** 6–10.

Pollex, RL and Hegele, RA (2004) Hutchinson-Gilford progeria syndrome, *Clin. Genet.* **66:** 375–381.

Robert, L and Robert, AM (2003) Aging, from basic research to pathological applications. *Pathol. Biol.* **51:** 543–549.

Sarkar, PK and Shinton, RA (2001) Hutchinson-Guildford progeria syndrome. *Postgrad. Med. J.* **77:** 312–317.

Selkoe, DJ (2001) Alzheimer's disease: genes, proteins and therapy. *Physiol. Rev.* **81:** 741–766.

Spence, AP (1999) *Biology of Human Aging*, Prentice Hall, New Jersey, USA.

Weindruch, R and Sohal, RS (1997) Calorie intake and aging, *N. Engl. J. Med.* **337:** 986–994.

Weinhert, BT and Timiras, PS (2003) Theories of aging. *J. Appl. Physiol.* **95:** 1706–1716.

CASE STUDY DISCUSSIONS

CHAPTER 1

DISCUSSION OF CASE STUDY 1.1

The result is most likely to be a transcription error (a 'blunder') since the result obtained is not compatible with life. Perhaps the true result was 3.5mmol dm^{-3} and not 35mmol dm^{-3}. In any case, a second K^+ determination is required.

DISCUSSION OF CASE STUDY 1.2

Sensitivity = (95/95 + 5) × 100 = 95%

Specificity = (95/95 + 5) × 100 = 95%

This means that the test will be positive in 95% of patients with prostate cancer and negative in 95% of people without the disease. That is, 5% of people with prostate cancer will be missed. An ideal diagnostic test will have 100% sensitivity and specificity but this is rarely achieved.

CHAPTER 2

DISCUSSION OF CASE STUDY 2.1

Brian is most likely in the early stages of a malarial infection. The disease is caused by several species of *Plasmodium*. This could be confirmed by a microscopic examination of a blood smear to show the presence of the parasite. Brian had traveled to an area where malaria is endemic and he should have taken appropriate prophylatic drugs.

CHAPTER 3

DISCUSSION OF CASE STUDY 3.1

The infection probably began as folliculitis that became inflamed and developed into an abscess. The most likely cause consistent with the appearances of the cultured microorganism is *Staphylococcus aureus*. The formation of an abscess restricts the flow of blood to the center of the infection, reducing the effectiveness of antibiotics, hence the need for excision and drainage.

DISCUSSION OF CASE STUDY 3.2

Chris had pharyngitis, most likely caused by *Streptococcus pyogenes* or one of a number of viruses. Taking throat swabs and culturing the bacterium can confirm this. Antibiotic treatment is desirable to effect a rapid cure and prevent the poststreptococcal sequelae, rheumatic fever and glomerulonephritis.

DISCUSSION OF CASE STUDY 3.3

All the symptoms are typical of whooping cough caused by *Bordetella pertussis.*

CHAPTER 4

DISCUSSION OF CASE STUDY 4.1

Immunoglobulin M is always first to be produced in an immune response and is the predominant antibody of the primary response. The fact that Maria had only IgM indicates that she has been exposed to this virus for the first time and is likely to be infected. Such an infection may harm the baby, particularly as Maria is in the first three months of gestation. Rubella infection of an early fetus may cause fetal death or the development of congenital rubella syndrome (CRS), which can cause mental retardation and impaired sight and hearing. If IgG antibodies had been detected instead of IgM, this would have indicated that Maria was immune to the virus and that the baby is unlikely to be exposed to this microorganism. Maria should be counseled as to the possible harm to her baby. The possibility of a therapeutic abortion should be discussed.

DISCUSSION OF CASE STUDY 4.2

Electrophoresis of the patient's serum, followed by staining of the electrophorogram, would show a thick band, referred to as a paraprotein, in the γ globulin region (*Figure. 1.9*). This would indicate myeloma, although other tests would be needed to confirm the class of antibody being produced. If this were an IgG myeloma, levels of IgG in the blood would be high, and a highly sensitive assay would not be required. Radial immunodiffusion using an antibody to IgG incorporated into the agar could be used to measure IgG levels. Nephelometry would be quicker and could confirm the IgG secreting nature of the myeloma.

CHAPTER 5

DISCUSSION OF CASE STUDY 5.1

Repeated bacterial infections suggest a defect in humoral immunity. This could be transient hypogammaglobulinemia, caused by a delay in the production of antibodies. However, his small tonsils suggest the more serious disorder, X-linked agammaglobulinemia (XLA). George needs to have a full blood count and an assessment of his B lymphocyte count. His immunoglobulin levels should also be measured to confirm a humoral deficit. If B cell numbers are greatly reduced, then it is likely that he has XLA. Molecular biological techniques would confirm mutations of the *BTK* gene. If XLA is confirmed, George will need immunological replacement therapy and antibiotics to control any bacterial infections.

DISCUSSION OF CASE STUDY 5.2

Peter has most likely had an anaphylactic shock reaction to the wasp sting. It is possible that he has been stung previously, but the speed of the collapse and the symptoms look likely to be due to a severe Type I hypersensitivity. Peter should be treated as a medical emergency. Signs of cardiac arrest need to be checked and Peter needs to be given intramuscular adrenaline. He would be advised to carry an adrenaline 'gun' in future.

DISCUSSION OF CASE STUDY 5.3

The tiredness, cold intolerance, weight gain and goiter all point towards an autoimmune disease of the thyroid. Tests to confirm this diagnosis would include an indirect immunofluorescence test on her serum, to demonstrate autoantibodies to thyroid antigens. ELISA tests would demonstrate antibodies to thyroglobulin, and a biopsy of her thyroid would show infiltration of the tissue with small lymphocytes. If confirmed, Jane should be treated with thyroxine to replace the deficiency caused by her autoimmune thyroiditis.

CHAPTER 6

DISCUSSION OF CASE STUDY 6.1

Given Maria is blood group O, she has anti-A antibodies in her plasma. These will destroy any fetal erythrocytes that enter her circulation. Thus she is unlikely to become sensitized and there is little chance of her baby developing hemolytic disease of the newborn. As Marie has never become sensitized despite several pregnancies, it is likely that each previous fetus was O negative, A negative or A positive. An O negative fetus would not sensitize Marie as the Rh group is identical to hers. If a fetus was A negative, Marie would not become sensitized, and any A cells would be destroyed immediately by the Anti-A in her plasma. Similarly, any A positive cells would be destroyed before Marie could be sensitized to the Rh antigen. Marie is only likely to become sensitized, following the birth of the present fetus, if that fetus is O positive.

DISCUSSION OF CASE STUDY 6.2

There are two approaches that could be made. First, John should be tissue typed and a search made on any available bone marrow registries. Alternatively, John could have an autologous transplant, with tumor cells eliminated from the bone marrow prior to reinfusion. John would avoid any likelihood of GVHD with this procedure.

DISCUSSION OF CASE STUDY 6.3

Michael has probably suffered a transfusion related acute lung injury (TRALI), possibly caused by antileukocyte antibodies in the transfused plasma. This should not be confused with an anaphylactic reaction (*Chapter 5*), which would also cause respiratory distress, but where pulmonary edema would not be expected.

CHAPTER 7

DISCUSSION OF CASE STUDY 7.1

Sarah could be suffering from some form of diabetes. However, her symptoms of frequent thirst and urination are not due to diabetes mellitus because of absence of glycosuria. The clinical features together with a finding of a raised plasma osmolality combined with a dilute urine, suggests diabetes insipidus.

She was investigated for diabetes insipidus with a fluid deprivation test. At the end of the test, her plasma osmolality was still high with a low urine osmolality suggesting diabetes insipidus as she has failed to conserve water during water deprivation. However, she responded to desmopressin by producing a more concentrated urine, suggesting a deficiency of ADH rather than renal insensitivity to ADH. It is most likely that Sarah is suffering from cranial diabetes insipidus.

DISCUSSION OF CASE STUDY 7.2

Ian has subclinical or compensated hypothyroidism, that is, he is at a higher risk of developing clinical hypothyroidism. Indeed, some patients with subclinical or compensated hypothyroidism show clinical features of hypothyroidism. An increased release of TSH is required to stimulate the release of thyroid hormones from a deteriorating thyroid gland. Ian's condition should be reviewed every few months and antibodies against his thyroid gland should be measured to detect whether there is an autoimmune cause.

DISCUSSION OF CASE STUDY 7.3

Amelia has hyponatremia with hyperkalemia. This indicates deficiency of mineralocorticoids that is typical of Addison's disease. The clinical features are also consistent with Addison's disease and Amelia has presented with an adrenal crisis. Amelia is dehydrated and this explains the high serum urea caused by a low glomerular filtration rate. She should be treated by correcting her fluid balance and then placing her on a steroid regimen that includes hydrocortisone and fludrocortisone.

DISCUSSION OF CASE STUDY 7.4

Jaclyn is a type 1 diabetic and presented with diabetic ketoacidosis (DKA), a medical emergency. This has probably occurred because Jaclyn has failed to take her insulin therapy although it may be that it has been precipitated by infections or even excessive stress. An overview of DKA is provided in *Figure 7.25*. The patient should be treated immediately. This will include infusions of isotonic saline to restore blood volume, insulin infusions to reduce hyperglycemia, K^+ supplements to replace lost K^+ and possibly hydrogen carbonate infusions to alleviate the severe acidosis present in this case.

DISCUSSION OF CASE STUDY 7.5

Rachel's symptoms indicate a possible PCOS. The moderate increase in testosterone combined with a raised LH and typical clinical findings are strongly supportive of this diagnosis. The pathogenesis of this condition is unknown. Normal ovaries synthesize androgens, such as testosterone and androstenedione. In PCOS their secretion increases and in the liver and adipose tissue they are converted to estrogens that inhibit release of FSH and prevent ovulation. Luteinizing hormone release is stimulated and in turn stimulates further release of androgens leading to hirsutism.

CHAPTER 8

DISCUSSION OF CASE STUDY 8.1

Ted has acute renal failure as a result of his accident. This acute renal failure may have been caused by release of myoglobin from crushed muscles together with loss of blood. This explains the high serum concentration of urea and its osmolality. The hyperkalemia is a result of renal failure but possibly also due to release of K^+ from damaged tissues. The low HCO_3^- reflects the metabolic acidosis that is a feature of acute renal failure.

DISCUSSION OF CASE STUDY 8.2

Arnie suffers from hypomagnesemia. The low serum K^+ may reflect its loss in diarrhea. Arnold presents with symptoms typical of hypocalcemia even though his total serum Ca^{2+} is normal. Patients with symptoms of hypocalcemia, but normal serum Ca^{2+}, should be investigated for hypomagnesemia. Arnold was treated with magnesium supplements given parenterally rather than orally because of his diarrhea and his symptoms resolved.

DISCUSSION OF CASE STUDY 8.3

John presents with nocturnal, acute, severe pain in the large toe together with signs of inflammation: all classical features of gout. The finding of hyperuricemia supports this diagnosis. However, some patients with gout have a normal serum urate concentration. The toe joint should be aspirated and examined for urate crystals. The aspirate should also be examined in a microbiology laboratory to exclude the possibility of an infection. John should be treated with indomethacin until his symptoms resolve. He should be advised to reduce his dietary intake of protein and alcohol.

CHAPTER 9

DISCUSSION OF CASE STUDY 9.1

John is suffering from a partially compensated metabolic alkalosis. The low blood H^+ concentration indicates an alkalosis while the high HCO_3^- is indicative of metabolic alkalosis. This alkalosis is associated with a loss of acid from the stomach by vomiting, hence the dyspepsia. John will have involuntarily reduced his ventilation rate to correct the alkalosis by increasing blood PCO_2 and so there has been a partial respiratory compensation.

DISCUSSION OF CASE STUDY 9.2

Tom's low $[H^+]$ indicates alkalosis. Given the low PCO_2, this is a respiratory alkalosis probably caused by his asthmatic attack. The HCO_3^- is still within the reference range so the condition is uncompensated. This is consistent with the slow nature of renal compensation.

DISCUSSION OF CASE STUDY 9.3

Terry has a high H^+ which is indicative of an acidosis. The patient has a low HCO_3^- which is indicative of a metabolic acidosis. His body has attempted to correct the acidosis by respiratory compensation causing hyperventilation leading to a reduced PCO_2. Thus Terry has a partially compensated metabolic acidosis.

CHAPTER 10

DISCUSSION OF CASE STUDY 10.1

Andrew is obese with a BMI of 31.72 kg m^{-2}. Health risks of obesity include heart disease, hypertension and type 2 diabetes (*Figure 10.25* and *Table 10.7*). His total cholesterol concentration is acceptable but not desirable. The triacylglycerol concentration is increased while the HDL concentration is low, increasing the risk of heart disease. His blood pressure is also relatively high. Since Andrew has a family history of heart disease, he would be strongly advised to eat a more appropriate diet including oily fish and vitamin rich food, exercise regularly and stop smoking.

DISCUSSION OF CASE STUDY 10.2

Emily's ideal weight for her height is 120 pounds, so she is only approximately 8% below this value. Her blood glucose is only slightly reduced but the ketone body content, indicative of a switch to fat metabolism (*Chapter 7*), is increased although not excessively. However, given her dieting, exercising and the evidence of a preoccupation with her weight, it is possible that Emily is starting to suffer from AN, although she should be examined to eliminate any other possible clinical causes, such as a neoplasm. She should receive specialized counseling, especially in nutrition and for any emotional problems perhaps related to academic issues that have led to her anorexic behavior. Her family should be involved whenever possible.

DISCUSSION OF CASE STUDY 10.3

Erythrocyte transketolase activity depends upon the thiamin derived coenzyme, TPP. The activation coefficient of transketolase activity in John's sample was about 23% which is evidence for a moderate to approaching severe deficiency. John should be placed on thiamin, and probably other vitamin, supplements. Appropriate nutritional advice should also be given to the nursing home.

CHAPTER 11

DISCUSSION OF CASE STUDY 11.1

Alice has celiac disease where the degree of malabsorption depends on the severity of the disease. Malabsorption of fat, protein and iron gave rise to her steatorrhea, hypoalbuminemia and anemia respectively. The xylose absorption test indicates defective carbohydrate absorption. The jejunal biopsy confirms diagnosis of celiac disease. Alice should be placed on a gluten-free diet and her symptoms should, hopefully, improve within three weeks.

DISCUSSION OF CASE STUDY 11.2

The urine bilirubin is negative and this means that the excess bilirubin in serum must be unconjugated otherwise it would be in a water-soluble form and lost in the urine. Normal AST and ALP activities suggest an absence of hepatocellular damage. Hemolysis cannot be the cause of the raised bilirubin as the hemoglobin concentration and reticulocyte count are both within their reference ranges. It is likely that Sadaf is suffering from Gilbert's syndrome where there is reduced activity of UDP-glucuronyltransferase and often decreased uptake of unconjugated bilirubin by liver cells. Gilbert's disease is characterized by mild jaundice, which is consistent with the clinical findings.

DISCUSSION OF CASE STUDY 11.3

The very high amylase activity suggests acute pancreatitis and this is consistent with the clinical findings. Hyperglycemia occurs quite often in acute pancreatitis but is transient in nature. Mark has a high serum urea concentration. This is due to shock where the reduced blood flow to the kidneys gives a reduced GFR and a resultant high serum urea concentration.

CHAPTER 12

DISCUSSION OF CASE STUDY 12.1

Helen has taken an overdose of aspirin. She has a metabolic acidosis being compensated by the hyperventilation. Initially, such patients present with a respiratory alkalosis and this is followed by a metabolic acidosis. The prothrombin time (*Chapter 13*) is prolonged due to decreased hepatic activation of clotting factors. Helen's treatment will be supportive and may include gastric lavage for up to 24 h after ingestion, possibly oral activated charcoal and infusions of sodium hydrogen carbonate to promote excretion of salicylate. Hemodialysis is not required, given the concentration of salicylate in Helen's plasma.

DISCUSSION OF CASE STUDY 12.2

Olga had taken a paracetamol overdose with vodka. Liver function tests were not abnormal on admission. The transient elevation of the plasma alanine aminotransferase suggests that some mild degree of liver damage had taken place. The time of the paracetamol overdose was not known accurately. Olga was treated with intravenous *N*-acetylcysteine. She should also receive counseling regarding the problems that led to the attempted suicide.

DISCUSSION OF CASE STUDY 12.3

Edward's abnormal clotting time and high bilirubin and alkaline phosphatase levels indicate a serious derangement of liver function. He probably has advanced cirrhosis of the liver. Extensive fibrosis of the liver has distorted the intraheptic bilary tree increasing the bilirubin and alkaline phosphatase levels (whose synthesis is induced by alcohol). γ-glutamyltransferase activity is frequently used as a marker for excessive alcohol consumption since its synthesis is also induced by alcohol. However, the test is not specific since γ-glutamyltransferase can also be induced by a number of drugs including some antiepileptics and oral contraceptives. Hyperbilirubinemia is not always a feature of chronic liver disease and can occur in other disease states.

Maureen would seem to have moderate alcoholic liver disease but she shows a somewhat different pattern in the liver function tests. Her GGT level is clearly increased and the transaminases are also abnormal but her bilirubin and alkaline phosphatase levels are only slightly above normal. Even relatively moderate levels of alcohol can produce structural changes in the liver cells including accumulation of fatty droplets, proliferation of the endoplasmic reticulum and enlargement of the mitochondria.

DISCUSSION OF CASE STUDY 12.4

The high $[H^+]$ and low HCO_3^- indicate a metabolic acidosis (*Chapter 9*) due to increased production of lactate from poisoning with carbon monoxide. Carbon monoxide binds more avidly to hemoglobin than oxygen and decreases the ability of hemoglobin to transport oxygen. John was unconscious because of reduced delivery of oxygen to his brain. John can be treated with hyperbaric oxygen. The high PO_2 causes displacement of carbon monoxide from hemoglobin.

CHAPTER 13

DISCUSSION OF CASE STUDY 13.1

Daniel clearly has sickle cell anemia and is homozygous for the condition. Both parents are heterozygotes and carry the sickle cell trait. Sickle cell individuals experience severe pain and need to be given strong painkillers.

Blood transfusions can help but these will lead to iron overload unless other means, such as iron chelating drugs, are used to reduce the iron load. Other treatments are available which aim to increase the amount of HbF in the blood. The chance of a further child having the homozygous condition is one in four, and the chance of their being a heterozygote one in two.

DISCUSSION OF CASE STUDY 13.2

The MCV is 100 fdm^3 (reference 76–96 fdm^3) therefore high; MCH is 22.2 pg $cell^{-1}$ (26–33 pg $cell^{-1}$) and low; the MCHC is 222 g dm^{-3} cells (310–350 g dm^{-3} cells) and also low. The high MCV and low primary indices indicate a macrocytic anemia. The presence of basophilic megaloblasts suggests pernicious anemia from lack of vitamin B_{12} or folate. This is probably due to atrophy of the gastric mucosa resulting in a lack of the glycoprotein, intrinsic factor, which enables vitamin B_{12} to be absorbed and also protects the vitamin from destruction in the gastrointestinal system. Hence Bill is vitamin B_{12} deficient and cannot therefore make DNA and consequently new erythrocytes (failure of erythropoiesis). The sore tongue, difficulties with swallowing, and epigastric pain are possibly related to an abnormality in the upper gastrointestinal system. A course of injections of vitamin B_{12} should lead a to complete remission of the condition and restoration of a normal blood picture.

DISCUSSION OF CASE STUDY 13.3

DNA analysis could be carried out (*Figure 13.23*). The chances of any of their children being homozygous for HbS would be 25%, and 50% for the heterozygous state.

CHAPTER 14

DISCUSSION OF CASE STUDY 14.1

Jim has hypercholesterolemia. This is due to hypothyroidism as indicated by the high TSH concentration (*Chapter 7*). The hypercholesterolemia is corrected by treatment of the underlying hypothyroidism.

DISCUSSION OF CASE STUDY 14.2

The modest rise in CK was probably due to the heavy fall, resulting in bruising and damage to muscle tissues, which then fell sharply. In myocardial infarctions, AST normally peaks 10 to 100-fold after 20–30 h and then declines to normal levels within two to six days, but this did not occur here. There is therefore no evidence for a myocardial infarction.

DISCUSSION OF CASE STUDY 14.3

Roger has high LDL cholesterol and together with clinical features of corneal arcus and xanthomata, which are diagnostic of familial hypercholesterolemia. Roger should be treated with lipid lowering drugs as well as a modified diet.

CHAPTER 15

DISCUSSION OF CASE STUDY 15.1

Christine is suffering from PKU hence the high serum concentration of phenylalanine. She should be placed on a diet low in phenylalanine. However, this diet must contain adequate amounts of tyrosine. Phenylalanine is broken down into several metabolites and these are lost in the urine and produce the characteristic odor. If Christine is not treated immediately, she may suffer irreversible brain damage resulting in a very low IQ. In addition, she will develop other clinical features characteristic of PKU, such as, a pale skin, blonde hair and blue eyes.

DISCUSSION OF CASE STUDY 15.2

The karyogram shows trisomy 21, the commonest cause of Down syndrome (*Box 15.4*). Its frequency increases substantially in the fetuses of older women. It is essential to obtain a karyogram of the child to determine whether the cause is a Robertsonian translocation to aid genetic counseling, because it tends to run in families, that is it is hereditable, while trisomy 21 is sporadic in occurrence. The advice to Jane would be that her son is not a familial Down syndrome child. However, if she wishes to have another child, she has a greater risk of carrying a Down syndrome child than younger mothers and so she should have cytogenetic testing of the baby during pregnancy.

CHAPTER 16

DISCUSSION OF CASE STUDY 16.1

A diagnosis of cystic fibrosis is usually made following a sweat test where there is a high concentration of chloride, since 99% of cystic fibrosis patients have concentrations of Cl^- in their sweat exceeding 70 mmol dm^{-3}. This test could be performed on the siblings but cannot detect heterozygous carriers. DNA diagnosis would detect children with cystic fibrosis and those who are heterozygous carriers.

DISCUSSION OF CASE STUDY 16.2

Janice is suffering from adult, type 1, Gaucher's disease. If untreated she will become anemic, her hepatosplenomegaly will increase and bouts of severe generalized pain in the body and limbs will occur with more intensity and frequency. Degeneration of vertebral bodies and hip joints are common leading to distortion of the spine and the need for hip replacement surgery. She can expect bouts of anemia and recurrent pulmonary infections. Treatment should consist of counseling to explain the nature of the disease, analgesics for the bone pains, blood transfusions to relieve the anemia and physiotherapy to keep the body mobile. Regular exercise, such as walking and swimming, should be encouraged. The major treatment should be intravenous enzyme therapy using β glucocerebrosidase to prevent the visceral complications. With appropriate treatment, Janice can expect a life of reasonable quality and length.

CHAPTER 17

DISCUSSION OF CASE STUDY 17.1

It is unlikely that Chris has cancer but rather benign prostate hyperplasia, as indicated by the rectal examination. The PSA is slightly elevated. It would be useful to monitor the level over a period of time as increasing concentrations would require further investigation to ensure that the condition was not becoming cancerous. If the benign hyperplasia begins to cause too much discomfort, it may be necessary to hospitalize Chris for treatment.

DISCUSSION OF CASE STUDY 17.2

From Rebecca's history, her consultant might be expecting to see a shadow on the lung indicative of carcinoma. The hemoptysis is a strong indicator, though some infectious diseases, such as tuberculosis, might also have this effect. Rebecca has smoked for 30 years and although she stopped smoking two years ago she still has a higher risk of lung cancer than somebody who has not smoked. Since she gave up smoking she would still have been passively inhaling smoke from her husband's cigarettes.

DISCUSSION OF CASE STUDY 17.3

Fatigue and lethargy in an adolescent boy is not in itself unusual but the leukocytosis might indicate an infectious disease such as glandular fever. The bruising and anemia are more worrying signs, especially as blast cells were seen, and may well indicate that Giles has developed a leukemia. Acute lymphoblastic leukemia is more common in children than adults and could be confirmed by microscopic examination of a sample of his bone marrow. Further tests include immunohistochemical staining of leukocytes in order to ascertain whether ALL is present and if it is a B or a T cell leukemia.

CHAPTER 18

DISCUSSION OF CASE STUDY 18.1

The high random plasma glucose is diagnostic of diabetes mellitus type 2. The classic symptoms of polyuria and polydipsia of diabetes are not always present in the elderly where the renal threshold for glucose is often elevated due to a decline in the glomerular filtration rate. This may be due to declining renal function with age, but can be made worse by the presence of renal failure, which is a chronic complication of diabetes.

DISCUSSION OF CASE STUDY 18.2

Harry is possibly (probably?) suffering from early stage Alzheimer's disease. He and his family should be advised of this and its consequences. He could be treated with acetylcholinesterase inhibitor therapy but this will only delay the progress of the symptoms.

ANSWERS TO QUESTIONS

CHAPTER 1

1. c
2. b
3. c
4. When using reference ranges, it is necessary to appreciate that only 95% of healthy individuals lie within them. This means there is a 5% chance of a healthy person having a value outside the range, so it is 95% probable that the individual is Mg^{2+} deficient.

CHAPTER 2

1.

HIV	virus
Treponema pallidum	bacterium
Trypanosoma brucei	protozoan
Candida albicans	yeast
Microsporum species	fungus

2.

Clostridium perfringens	gas gangrene
Epstein-Barr virus	glandular fever
HIV	CD4
Human rhinovirus 91	intracellular adhesion molecule
Microsporum spp, *Trichophyton* species	athlete's foot
Plasmodium	malaria
Sarcoptes scabies	scabies
Shigellae dysenteriae	shigellosis
Trypanosoma brucei	sleeping sickness
Varicella zoster	chicken pox
Yersinia pestis	bubonic plague

3.

Endotoxins	Exotoxins
Produced by Gram negative bacteria only.	exotoxins are proteins secreted by both Gram positive and negative bacteria.
Consist of portions of lipopolysaccharides embedded in the outer wall membrane.	three groups: 1) damage membranes, causing the cell to lose water and ions and disrupt ion gradients across the membrane; 2) those with specific host targets – wide variety of types; 3) superantigens synthesized by Gram-positive pathogens stimulate the immune system by binding MHC class II molecules and T_H cell receptors.
Damage caused varies with the susceptibility of the host.	
Endotoxins stimulate host cells to release pyrogens therefore fever is a common outcome.	
Endotoxic shock can result.	

CHAPTER 3

1. Viral — Herpes, HIV, papillomaviruses
   Bacterial — *Candida*, *Chlamydia*, *Neisseria gonorrhoeae* and *Treponema pallidum*
   Protozoal — *Trichomonas*
   Fungal — *Candida*

2. Culture of specimens; microscopy, serological detection and biochemical and molecular techniques.

3. The student is suffering from a form of food poisoning. Given the presence of blood, a possible cause is enterohemorrhagic *Escherichia coli* (EHEC). Treatment should consist of antibiotic therapy and intravenous fluids if dehydration is severe.

4. Colistin sulfate (CO), nalidixic acid (NA) and nitrofurantoin (NI)

5. Both griseofulvin and albendazole interfere with microtubule functions but griseofulvin is used to treat fungal infections and albendazole helminth parasites.

6. They may inhibit the synthesis of the bacterial nucleic acids.
   They may inhibit the synthesis of bacterial proteins.
   They may inhibit the synthesis the bacterial cell wall.
   They act by a variety of miscellaneous mechanisms.

7. Combination therapy

CHAPTER 4

1. c

2. a

3. b

4. All mature T lymphocytes have CD3;
   all B lymphocytes have surface immunoglobulin;
   macrophages have receptors for complement;
   helper T lymphocytes are CD4+;
   cytotoxic T lymphocytes are CD8+.

5. a) IgG; b) IgG; c) IgA, d) IgE; e) IgM
6. MHC encoded Class 1 proteins are concerned largely with defense against viruses. Erythrocytes are not susceptible to viral attack because they are anucleated and therefore do not have the genetic machinery to support viral replication.

CHAPTER 5

1. (c)
2. (a) False
   (b) True
   (c) True
   (d) False
   (e) True
3. (b)
4. By removing self-reactive cells during fetal development; by keeping immunogens away from the immune system; by the production of regulatory cells.
5.

Selective IgA deficiency	may be asymptomatic
Farmer's lung	is caused by immune complexes
Chediak Higashi disease	affects melanocytes
Anaphylactic shock	is due to IgE
Delayed hypersensitivity	is caused by T lymphocytes

CHAPTER 6

1. a, c, and e
2. b (because MHC I proteins are found on all cells, including antigen presenting cells) and e; c is untrue because complement is a set of proteins not a single one.
3. All except c. A donor who is pregnant may continue to donate blood after her pregnancy. All the others would be precluded from donating blood.
4. Previous transplant; previous transfusion with whole blood or blood/blood product with residual leukocytes; previous pregnancies if female.
5. It is technically much easier to obtain PBSCT by leukapheresis than it is to obtain bone marrow and the donor does not have to be anesthetized.

CHAPTER 7

1. d
2. a
3. b
4. d
5. d

6 c

7. (a) Sajida has hyperprolactinemia;
   (b) Probably treatment with the antidopamine;
   (c) Sajida should be advised to stop taking the antidopamine drug which should resolve both the hyperprolactinemia and the galactorrhea.

8. The glucose load has not suppressed the levels of GH. This is typical of, and confirms, the diagnosis of acromegaly.

9.

Type 1	Type 2
Deficiency of insulin	insulin resistance
Acute onset	chronic onset
Usually presents in childhood or adolescence	occurs in middle aged or elderly
Loss of weight	strong association with obesity
Prone to DKA	prone to HONK

10. Human GH promotes the linear growth of bones and stimulates an increased uptake of amino acids by cells and increased synthesis of proteins. It also stimulates increases in blood glucose and fat metabolism.

11. Congenital hypothyroidism is common with an incidence of one in 3500 newborns in the UK. Clinical evidence at birth that the child is abnormal is often absent. If the condition is not detected and treated, then the affected children suffer from cretinism which is characterized by irreversible brain damage, short stature, deafness and neurological signs.

CHAPTER 8

1. After a couple of days the kidneys adapt to decreased water intake and conserve water and electrolytes. However, insensible loss of some water does occur and both ECF and ICF values decline. As a consequence Robin will have hypernatremia. The decline in ECF volume will reach a critical level and will be unable to maintain blood circulation resulting in death. It is not a good idea to drink seawater as this contains high concentrations of salt (450 mmol dm^{-3}) and Mg^{2+} (50 mmol dm^{-3}) that aggravate the hypernatremia. The high salt can induce vomiting whereas the high Mg^{2+} may cause diarrhea and so further fluid loss.

2. The symptoms and high Na^+ and urea concentrations are all indicative of dehydration, presumably as a consequence of Jane not eating and drinking for several days.

3. e

4. 2.0 mmol dm^{-3}

5. d

6. Given that Alice's hypocalcemia was difficult to control and the doses of vitamin D were gradually increased, her hypercalcemia is probably iatrogenic.

7. Hypercalcemia commonly occurs in tumors that metastasize to the bones, such as those of the lung, breast and kidneys. Hypercalcemia

may be due to increased bone resorption or to local cytokine and prostaglandin synthesis. Also, tumors such as squamous cell carcinomas of the lungs, may produce PTHrp that has PTH like activity leading to hypercalcemia.

CHAPTER 9

1 d

2. (a) 10^3 or 1000
(b) pH = 4.94

3. (a) $[HCO_3^-]$ is 32 mmol dm^{-3}
(b) Alkalosis
(c) Metabolic

4. c

CHAPTER 10

1. c

2. d

3.

ATP7A	Menkes' disease
bulimia nervosa	dental erosion
copper	Wilson's disease
iodide	goiter
fluoride	dental caries
folic acid	green leafy vegetables
leptin	*OBESE/obese*
molybdenum	xanthine oxidase
edema	kwashiorkor
pantothenic acid	acetyl CoA
saturated fats	heart disease
thiamin	Wernicke-Korsakoff syndrome
vitamin B_1	thiaminase
vitamin B_6	tryptophan
vitamin B_{12}	Crohn's disease
vitamin C	scurvy
vitamin D	osteomalacia
vitamin E	hemolytic anemia
vitamin K	prothrombin time
zinc	phytanic acid

4. All the plant foods are deficient in some essential amino acids. It is therefore vital that your daughter eats a broad mix of cereals, nuts and vegetable, especially combining cereals with legumes and/or nuts, to ensure she receives adequate amounts of each.

5. There is some clinical evidence of PEMT and iron deficiency. The petechial hemorrhages and gum disease are typical of scurvy, which arises due to vitamin C deficiency. This is often seen in elderly patients who are on a poor diet.

6. Values for BMI for A and B are 23.13 and 37.04 kg m^{-2} respectively.
Theoretical ideal body weights for A and B of 178 and 166 lb respectively.

The value for skinfold thickness for A is within the standard range, while B's value is above the normal range.

Arm circumference for both A and B are within the standard range.

The waist:hip ratio for B is greater than 0.95.

In general, the results for A are associated with a low health risk; those for B show definite obesity with all its attendant health risks as indicated in *Figure 10.25*.

CHAPTER 11

1. b
2. d
3. (a) The most likely diagnosis is secondary acquired lactose intolerance following damage to the intestinal mucosa by the viral attack.
   (b) Tests include a lactose-free diet, hydrogen breath test, the stool acidity test and lactose tolerance test. A biopsy of the mucosa could be taken and tested for lactase activity if necessary. Since the symptoms are common to numerous other GIT conditions, care must be taken with the diagnosis.
   (c) Qiuyu's treatment should be a lactose-free diet that should quickly relieve her symptoms.
4. Some causes of jaundice are:
   (a) hemolysis, for example autoimmune hemolytic anemia, drug-induced hemolysis
   (b) drugs, for example rifampicin which interferes with bilirubin uptake
   (c) Gilbert's syndrome
   (d) Criggler-Najjar syndrome
   (e) gallstones
   (f) pancreatic carcinoma
5. Causes of malabsorption include:
   (a) deficiencies in pancreatic enzymes
   (b) deficiencies in bile salts
   (c) intestinal disease(s)
   (d) bacterial colonization of GIT

CHAPTER 12

1. (c)
2. (c)
3. The earlier the treatment the better the prognosis as early treatment reduces the extent of hepatotoxicity. A delay in treatment can be fatal.
4. Aldehyde will build up in hepatocytes and blood following the oxidation of ethanol. This will produce a range of unpleasant symptoms in the patient (headaches, nausea, vomiting etc.) that will eventually be associated with the use of alcoholic drinks and so produce a deterrent effect.
5. Many actions of ethanol may be attributed to its effects in disordering membranes, and changes in membrane fluidity can affect membrane-bound enzymes such as the Na^+/K^+-ATPase and adenylate cyclase. It

also affects several neurotransmitter systems in the brain including dopamine, GABA, glutamate, serotonin, adenosine, noradrenaline and opioid peptides. Many drugs are metabolized, in part, by the P-450 system and consequently there may be competition between ethanol and other drugs, such as barbiturates, whose hydroxylation at the smooth endoplasmic reticulum is inhibited. This increases and prolongs their action.

6. Many cosmetics obtained outside the UK contain lead and this is true for surma. Children using such cosmetics often develop lead poisoning. Investigations should include determining the levels of lead in whole blood and precursors of heme in erythrocytes.

7. Removal from the mercury source is important and will ensure the patient's recovery. A chelating agent such as dimercaprol may be used to rapidly reduce mercury in the body.

CHAPTER 13

1. (a) T
   (b) F
   (c) T
   (d) F
   (e) F

2. (A) Granulocyte; (B) basophil

3. A has both HbS and HbC but no HbA; B is homozygous for HbS; and D is an HbC heterozygote. Note the Glu to Val (HbS) means a change of charge from – to 0, and the Glu to Lys (HbC), a change from – to +.

4. Fetal Hb carries oxygen perfectly well but with a somewhat higher affinity than HbA.

5.

Index	iron deficiency anemia	pernicious anemia
PCV	D	NA or D
MCV	D	I
MCH	D	NA or D
MCHC	D	NA or D

6. The production of erythrocyte cell precursors has been stimulated by Elsa's severe anemia. As a result, the bone marrow spaces expanded in all areas of the skeleton leading to the deformities. It is these within the spine that caused compression of the spinal cord itself and the pain. Local irradiation may bring immediate relief. In addition, blood transfusions must be given to keep the Hb at near normal levels.

7. When the lining of a blood vessel is damaged, collagen fibers are exposed and circulating platelets stick to them. This causes them to degranulate releasing serotonin and ADP and other active biological molecules. Serotonin causes blood vessels to constrict restricting blood flow until clotting takes place. Clumping of platelets is triggered by ADP and the clump prevents further blood loss. The blood clotting cascade is then initiated.

8. Simon has hemophilia A. The manifestations encountered are caused by bleeding into soft tissues, such as muscle or into the peritoneal

cavity. Bleeding episodes can be managed by injections of purified Factor VIII, the antihemophilia factor.

CHAPTER 14

1. d
2. b
3. b
4.

Blood pressure	sphygmomanometer
Bundle of His	Purkinje fibers
Coronary arteries	myocardium infarction
Pericardial diseases	aspirin, ibuprofen
Palpitations	ventricular tachycardia
Right ventricle	tricuspid valve
SA node	pacemaker
Thrombolysis	tenecteplase
Vasodilation	lowered blood pressure
Xanthoma	familial hypercholesterolemia

5. The changes in the normal production of adrenaline lead to arrhythmias.
6. It is likely that Barry has had a myocardial infarction. However, his chest pain has only lasted for 30 min so he will not show any abnormal enzyme or protein levels, therefore there is no need to investigate serum enzyme levels at this stage.
7. Lowering the amount of carbohydrate in the diet may decrease triacylglycerol and cholesterol levels, because the triacylglycerols in the VLDL particles are synthesized by the liver mainly from dietary carbohydrate. The VLDL particle secreted into the blood will, of course, carry cholesterol.
8. LDL receptors recognize ApoB-100, whereas chylomicrons contain ApoB-48, apoC and apoE that are not recognized by LDL receptors.

CHAPTER 15

1. b
2. e
3. 25%
4. The children have equal chances of receiving the stable or mutant gene. Fifty percent of female children will be affected. Most of the males who receive the mutation will have the full mutation. However, there have been cases which the number of repeats was reduced.
5. In autosomal dominant syndromes an affected father would pass it on to an average of 50% of his sons and daughters. X-linked dominant syndromes do not show male-to-male transmission but all the daughters of an affected father would show the syndrome. If many families were studied, the ratios of affected males to females for an autosomal dominant would be 1:1, but 1:2 for X-linked dominant syndromes since males have only half the chance of inheriting a defective X chromosome.
6. (A) Turner (X0), (B) XYY and (C) Klinefelter (XXY) syndromes.

CHAPTER 16

1. b
2. c
3.

Adrenoleukodystrophy	VLCFAs
α-spectrin	hereditary spherocytosis
CFTR	chloride channel
Dystrophin	BMD, DMD
EPB41	hereditary elliptocytosis
Hexoseaminidase A	Tay-Sachs disease
Mitochondrial gene for $tRNA^{leu}$	MELAS
Niemann-Pick disease	sphingomyelin
Refsum's disease	phytanic acid
Sarcoglycans	LGMD

4. Myoglobin is an intracellular protein. However, the muscle myopathy associated with a number of mitochondrial diseases leads to its release from the damaged tissue and its subsequent loss in the urine.
5. (a) B
   (b) A
6. The excessive destruction of misshapen erythrocytes releases bilirubin in quantities that exceeds the livers conjugating capacity (*Chapter 11*). The bilirubin, a bile pigment, enters the gall bladder in high concentrations and precipitates to form gallstones.

CHAPTER 17

1) b
2) e
3) b
4)

Carcinoma	epithelia
Metastasis	secondary tumors
Oncogene	*E2F3*
Tumor suppressor gene	*TP53*
Breast cancer	*EMSY*
Li-Fraumeni syndrome	*CHK2*
Tyrosine kinase	*ABL-ACR*
FAB	myeloblastic leukemias
Ultraviolet light	xeroderma pigmentosum
Epstein-Barr virus	Hodgkin's lymphoma

5)
   (a) F
   (b) T
   (c) F
   (d) T
   (e) F
   (f) T
   (g) F
   (h) F
   (i) F
   (j) F

CHAPTER 18

1. c
2. d
3. d
4. Proteins become glycated on exposure to reducing sugars. These glycated proteins undergo further reaction forming fluorescent, cross-linked structures called advanced glycation end products (AGEs). These are formed irreversibly and accumulate on long-lived proteins, for example, collagen and as the body ages such proteins in particular become cross-linked. This is believed to contribute towards the aging process.
5. Any four from:
   decreased oxidative phosphorylation;
   decreased protein synthesis;
   decreased absorption of nutrients;
   increased damage (mutations) to DNA;
   accumulation of oxidative stress damage;
   accumulation of glycated proteins and AGEs;
   changes to the morphology of organelles.
6. Any five from:
   cataract;
   diabetes mellitus type 2;
   osteoarthritis;
   Parkinson's disease;
   Alzheimer's disease;
   cardiovascular disease;
   cancers.
7. b
8.

Feature	Familial AD	Senile AD
Age of onset / years	45 to 60 (sometimes as low as 30)	60 to 65
Percentage of cases	5–10	90–95
Associated gene or risk factor	PS-1 and PS-2 genes	apoEε4

INDEX

Note: Entries which are simply page numbers refer to the main text. Other entries have the following abbreviations immediately after the page number: B, Box; C, Case Study; F, Figure; M, Margin Note; T, Table.

F

H

M